P9-DEX-796

The Enzymes

VOLUME V

HYDROLYSIS

Sulfate Esters
Carboxyl Esters
Glycosides

HYDRATION

Third Edition

CONTRIBUTORS

ROBERT H. ABELES
A. S. BHARGAVA
KONRAD BLOCH
STEPHEN DYGERT
HARRY C. FROEDE
L. GLASER
JENNY PICKWORTH GLUSKER
ALFRED GOTTSCHALK
DONALD J. HANAHAN
L. E. HENDERSON
ROBERT L. HILL
TOSHIZO ISEMURA
K. K. KANNAN
KLAUS KRISCH
J. OLIVER LAMPEN
E. Y. C. LEE
A. LILJAS
S. LINDSKOG
KARL MEYER
R. G. NICHOLLS
P. O. NYMAN
A. B. ROY
JOSEPH E. SPRADLIN
B. STRANDBERG
TOSHIO TAKAGI
JOHN W. TEIPEL
JOHN A. THOMA
HIROKO TODA
AKIRA TSUGITA
W. J. WHELAN
D. R. WHITAKER
IRWIN B. WILSON
FINN WOLD
W. A. WOOD
H. ZARKOWSKY

ADVISORY BOARD

THE ENZYMES

Edited by PAUL D. BOYER

Molecular Biology Institute and
Department of Chemistry
University of California
Los Angeles, California

Volume V

HYDROLYSIS

Sulfate Esters
Carboxyl Esters
Glycosides

HYDRATION

THIRD EDITION

ACADEMIC PRESS New York and London 1971

ACADEMIC PRESS, INC.
111 Fifth Avenue, New York, New York 10003

United Kingdom Edition published by
ACADEMIC PRESS, INC. (LONDON) LTD.
24/28 Oval Road, London NW1 7DD

Library of Congress Catalog Card Number: 75-117107

PRINTED IN THE UNITED STATES OF AMERICA

Contents

List of Contributors

Numbers in parentheses indicate the pages on which the authors' contributions begin.

ROBERT H. ABELES (481), Graduate Department of Biochemistry, Brandeis University, Waltham, Massachusetts

A. S. BHARGAVA (321), Experimental Research Pharma, Schering A.G., Berlin, Germany

KONRAD BLOCH (441), Department of Chemistry, James Bryant Conant Laboratory, Harvard University, Cambridge, Massachusetts

STEPHEN DYGERT (115), Department of Chemistry, Cornell University, Ithaca, New York

HARRY C. FROEDE (87), School of Pharmacy, University of Colorado, Boulder, Colorado

L. GLASER (465), Department of Biochemistry, Washington University School of Medicine, St. Louis, Missouri

JENNY PICKWORTH GLUSKER (413), The Institute for Cancer Research, Philadelphia, Pennsylvania

ALFRED GOTTSCHALK (321), Max-Planck-Institute for Virus Research, Tübingen, West Germany

DONALD J. HANAHAN (71), Department of Biochemistry, College of Medicine, University of Arizona, Tucson, Arizona

L. E. HENDERSON (587), Department of Biochemistry, University of Göteborg, and Chalmers Institute of Technology, Göteborg, Sweden

ROBERT L. HILL (539), Department of Biochemistry, Duke University Medical Center, Durham, North Carolina

TOSHIZO ISEMURA (235), Institute for Protein Research, University of Osaka, Osaka, Japan

K. K. KANNAN (587), The Wallenberg Laboratory, University of Uppsala, Uppsala, Sweden

KLAUS KRISCH (43), Institute of Physiological Chemistry, University of Kiel, Kiel, West Germany

J. OLIVER LAMPEN (291), Institute of Microbiology, Rutgers University, The State University of New Jersey, New Brunswick, New Jersey

E. Y. C. LEE (191), Department of Biochemistry, University of Miami, Miami, Florida

A. LILJAS* (587), The Wallenberg Laboratory, University of Uppsala, Uppsala, Sweden

S. LINDSKOG (587), Department of Biochemistry, University of Göteborg, and Chalmers Institute of Technology, Göteborg, Sweden

KARL MEYER (307), Belfer Graduate School of Science, Yeshiva University, New York, New York

R. G. NICHOLLS† (21), Department of Physical Biochemistry, John Curtin School of Medical Research, Australian National University, Canberra, Australia

P. O. NYMAN (587), Department of Biochemistry, University of Göteborg, and Chalmers Institute of Technology, Göteborg, Sweden

A. B. ROY (1, 21), Department of Physical Biochemistry, John Curtin School of Medical Research, Australian National University, Canberra, Australia

JOSEPH E. SPRADLIN (115), Miles Laboratory, Elkhart, Indiana

B. STRANDBERG (587), The Wallenberg Laboratory, University of Uppsala, Uppsala, Sweden

TOSHIO TAKAGI (235), Institute for Protein Research, University of Osaka, Osaka, Japan

* Present address: Department of Biological Sciences, Purdue University, Lafayette, Indiana.

† Present address: Department of Agricultural Biochemistry, Waite Agricultural Research Institute, Glen Osmond, Australia.

JOHN W. TEIPEL* (539), Department of Biochemistry, Duke University Medical Center, Durham, North Carolina

JOHN A. THOMA (115), Department of Chemistry, University of Arkansas, Fayetteville, Arkansas

HIROKO TODA (235), Institute for Protein Research, University of Osaka, Osaka, Japan

AKIRA TSUGITA (343), Laboratory of Molecular Genetics, University of Osaka, Osaka, Japan

W. J. WHELAN (191), Department of Biochemistry, University of Miami, Miami, Florida

D. R. WHITAKER (273), Department of Biochemistry, University of Ottawa, Ottawa, Canada

IRWIN B. WILSON (87), Department of Chemistry, University of Colorado, Boulder, Colorado

FINN WOLD (499), Department of Biochemistry, University of Minnesota Medical School, Minneapolis, Minnesota

W. A. WOOD (573), Department of Biochemistry, Michigan State University, East Lansing, Michigan

H. ZARKOWSKY (465), Department of Biochemistry, Washington University School of Medicine, St. Louis, Missouri

* Present address: Department of Chemistry, School of Chemical Sciences, University of Illinois, Urbana, Illinois.

Preface

This volume completes the presentation of the largest and best understood group of enzymes—those catalyzing hydrolyses. It covers hydrolysis of sulfate and carboxyl esters and of glycosides. Also included is a complete coverage of the group of enzymes covering hydration. The objective of this volume, as of the others, is to present enzymes about which considerable information at the molecular level is known. Authors are also encouraged to summarize pertinent information about related enzymes for which no separate chapters are included. The volume presents considerably more information about these important enzymes than is available from any other single source.

The completion of each volume is a reminder of the richness of available information for each enzyme. General structural and catalytic features emerge and elegant intricacies of structure and reaction become visible. These are well illustrated in the structure–catalysis relationships for glycosides and in the striking correlations of genetic probes of structure and function exemplified by phage lysozyme discussed in this volume. And we see the power of the X-ray technique combined with other approaches as demonstrated by the fine chapter on carbonic anhydrase.

Because of unavoidable circumstances, Professor Desnuelle's chapter on lipases was not completed in time to be included in this volume. It will, however, appear in Volume VII.

Again, it is a pleasure to acknowledge the invaluable assistance of Advisory Boards. For the sections on hydrolysis, which are an extension of the coverages in Volumes III and IV, we are indebted to C. B. Anfinsen, I. R. Lehman, Alton Meister, and Stanford Moore. For the section on hydration, R. H. Abeles, B. L. Horecker, E. G. Krebs, and P. R. Vagelos served as Advisory Board Members. The continued capable assistance of the staff of Academic Press and of Lyda Boyer as editorial assistant for the treatise made this volume possible. We are especially grateful to Dr. Klaus Krisch for preparing his chapter on such short notice.

Paul D. Boyer

Contents of Other Volumes

Volume I: Structure and Control

Volume IV: Hydrolysis: Other C–N Bonds, Phosphate Esters

Volume VI: Carboxylation and Decarboxylation (Nonoxidative), Isomerization

The Enzymes

VOLUME V

HYDROLYSIS

Sulfate Esters
Carboxyl Esters
Glycosides

HYDRATION

Third Edition

1

The Hydrolysis of Sulfate Esters

A. B. ROY

I. Introduction

Within the narrowest sense of the title this chapter should only deal with the sulfatases which catalyze the hydrolysis of the anion of a sulfate ester [reaction (1)].

$$R{\cdot}OSO_3^- + H_2O \rightarrow R{\cdot}OH + H^+ + SO_4^{2-} \quad (1)$$

However, it can conveniently be extended to consider enzymes hydrolyzing sulfatophosphates, the sulfatophosphate sulfohydrolases, which catalyze reaction (2),

$$R{\cdot}O{-}\underset{\underset{O}{|}}{\overset{\overset{O^-}{|}}{P}}{-}O{-}SO_3^- + H_2O \rightarrow R{\cdot}OPO_3^{2-} + 2\,H^+ + SO_4^{2-} \quad (2)$$

and enzymes hydrolyzing sulfamate anions, the sulfamatases, which catalyze reaction (3).

$$R{\cdot}NHSO_3^- + H_2O \rightarrow R{\cdot}NH_3^+ + SO_4^{2-} \quad (3)$$

It might be expected that enzymes hydrolyzing thiosulfate esters [reaction (4)] would exist but none is yet known although potential substrates

$$R{\cdot}SSO_3^- + H_2O \rightarrow R{\cdot}SH + H^+ + SO_4^{2-} \quad (4)$$

certainly occur naturally (e.g., *S*-sulfocysteine and *S*-sulfoglutathione).

The sulfatases form by far the largest group of the enzymes to be considered here, and many different types have been recognized. Nevertheless, only two examples of one group, the arylsulfatases, have been obtained as homogeneous preparations; thus, considerable doubt exists about the specificity, or indeed about the separate existence, of some of these enzymes.

Methods for the assay of many of the sulfatases were reviewed some years ago (*1*), and a general discussion of them has recently appeared (*2*).

II. Sulfatases

A. Arylsulfatases

Arylsulfatases have been quite extensively studied, but only two have been obtained in a homogeneous state. The properties of these and of some related enzymes are considered in Chapter 2 by Nicholls and Roy, this volume, and the present discussion will be restricted to general topics.

The reaction catalyzed by the arylsulfatases is Eq. (1) where R·OH is a phenol. The specificity of the various arylsulfatases varies somewhat, but most aryl sulfates are hydrolyzed although at very different rates. Typical substrates are 2-hydroxy-5-nitrophenyl sulfate [(I)

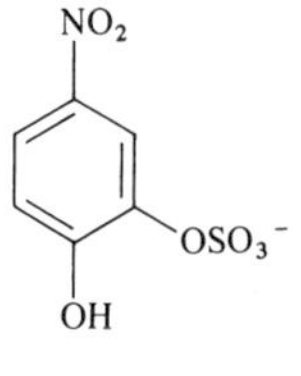

NO_2 / OSO_3^- / OH (I) NO_2 / OSO_3^- (II)

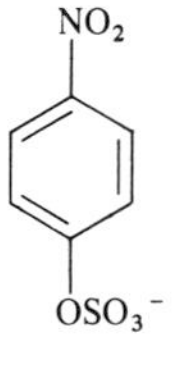

1. K. S. Dodgson and B. Spencer, *Methods Biochem. Anal.* **4**, 211 (1957).

2. A. B. Roy and P. A. Trudinger, "The Biochemistry of Inorganic Compounds of Sulphur." Cambridge Univ. Press, London and New York, 1970; K. S. Dodgson and F. A. Rose, *in* "Metabolic Conjugation and Metabolic Hydrolysis" (W. H. Fishman, ed.), Vol. I, p. 239. Academic Press, New York, 1970.

nitrocatechol sulfate] and 4-nitrophenyl sulfate (II). Since the arylsulfatases catalyze the fission of the O–S bond (*3*) they might be expected to use sulfate acceptors other than water and so act as sulfotransferases. There is no well-documented example of such a transfer, but it has been suggested that certain carbohydrates (*4*) may under some conditions serve as acceptors.

Two different types of arylsulfatase can be distinguished: those which are not inhibited by sulfate (type I) and those which are (type II). It is convenient to consider these separately.

1. *Type I Arylsulfatases*

Most of the microbial arylsulfatases appear to be type I enzymes, but only one example has been adequately characterized: this is the arylsulfatase of *Aerobacter aerogenes* (*5*). The function of the arylsulfatases in microorganisms, where they are of sporadic occurrence (*6*), is not known but they appear to be related in some way to the general sulfur metabolism of the cell.

When *A. aerogenes* is grown with limiting amounts of sulfate, or with methionine as the sole sulfur source, a greatly increased synthesis of arylsulfatase occurs: under conditions of limiting sulfate about 700 times as much arylsulfatase is synthesized as under sulfur-sufficient conditions (*7*). The synthesis of arylsulfatase is repressed by sulfate, sulfite, thiosulfate, *S*-sulfocysteine and cysteine, and less effectively by cystine and lanthionine (*7, 8*). The nature of the repressor has not been established, and it could be either sulfate or cysteine. An interesting feature is the derepression caused by tyramine (*8*); this phenomenon explains the previous suggestion (*9*) that tyramine was an inducer of arylsulfatase in *A. aerogenes.*

A similar type of control mechanism probably occurs in fungi although here there is no derepression by tyramine (*10*).

Type I arylsulfatases also occur in at least some vertebrates (*11*) where they (the sulfatases C) are localized in the microsomes, apparently

3. B. Spencer, *BJ* **69,** 155 (1958).
4. S. Suzuki, N. Takahashi, and F. Egami, *J. Biochem.* (*Tokyo*) **46,** 1 (1959).
5. L. R. Fowler and D. H. Rammler, *Biochemistry* **3,** 230 (1964).
6. M. Barber, B. W. L. Brooksbank, and S. W. A. Kuper, *J. Pathol. Bacteriol.* **63,** 57 (1951); J. E. M. Whitehead, A. R. Morrison, and L. Young, *BJ* **51,** 585 (1952).
7. D. H. Rammler, C. Grado, and L. R. Fowler, *Biochemistry* **3,** 224 (1964).
8. T. Harada and B. Spencer, *BJ* **93,** 373 (1964).
9. T. Harada, *Bull. Agr. Chem. Soc. Japan* **21,** 267 (1957).
10. T. Harada and B. Spencer, *BJ* **82,** 148 (1962).
11. A. B. Roy, *BJ* **68,** 519 (1958); *Australian J. Exptl. Biol. Med. Sci.* **41,** 331 (1963).

on the membrane system (*12*). They have not been obtained in solution, and they are always accompanied by steroid sulfatases: indirect evidence (see below) suggests that the latter are distinct from the arylsulfatases, but this has not been proved. The function of the arylsulfatases in animals is unknown.

2. *Type II Arylsulfatases*

Type II arylsulfatases are found principally in animal tissues, where they are very widespread (*2, 11*), and only one example is known from microorganisms, that from *Proteus vulgaris* (*13*).

In the mammals, and probably in vertebrates generally, the type II arylsulfatases, the sulfatases A and B, are found in the lysosomes (*14*) and so have an intracellular distribution quite different from that of the type I arylsulfatases. In the invertebrates the situation is not so clear: there is some evidence that here also they may be particle-bound (*15*), but at least in mollusks and insects they are found in the digestive juices.

The function of the arylsulfatases in animals is quite unknown. Lysosomal enzymes are generally considered to be involved in degradative processes and part of the difficulty in assigning a physiological—or pathological—role to the arylsulfatases is the lack of information about those aryl sulfates which occur *in vivo*, and about the role which these play in the living organism. As will be considered later (Section II,C,4), under certain conditions sulfatase A can show cerebroside sulfatase activity, and it may be that the true function of this enzyme is not to act as an arylsulfatase but as a cerebroside sulfatase. If this be the case then the specificity of the other arylsulfatases will require careful examination.

B. Steroid Sulfatases

Steroid sulfatases catalyze the hydrolysis of the sulfate esters of several types of steroid. The hydroxyl group esterified may be that of a phenol or of a primary or secondary alcohol. None of this group of enzymes has been purified despite considerable interest in the role of steroid sulfates, and hence steroid sulfatases, in metabolism (*16*).

12. D. W. Milsom, F. A. Rose, and K. S. Dodgson, *BJ* **109,** 40P (1968).
13. K. S. Dodgson, *Enzymologia* **20,** 301 (1959).
14. R. Viala and R. Gianetto, *Can. J. Biochem. Physiol.* **33,** 839 (1955); A. B. Roy, *BJ* **77,** 380 (1960).
15. C. Jackson and R. E. Black, *Biol. Bull.* **132,** 1 (1967).
16. E-E. Baulieu, C. Corpéchot, F. Dray, R. Emiliozzi, M-C. Lebeau, P. Mauvais-Jarvis, and P. Robel, *Recent Progr. Hormone Res.* **21,** 411 (1965); S. Bernstein and S. Solomon, "The Chemistry and Biochemistry of Steroid Conjugates." Springer-Verlag, Berlin and New York, 1970.

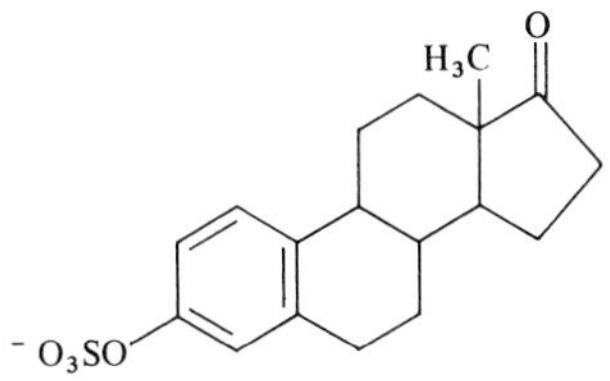

(III) 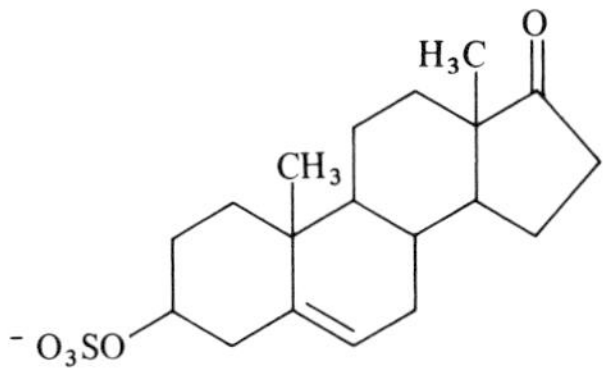 (IV)

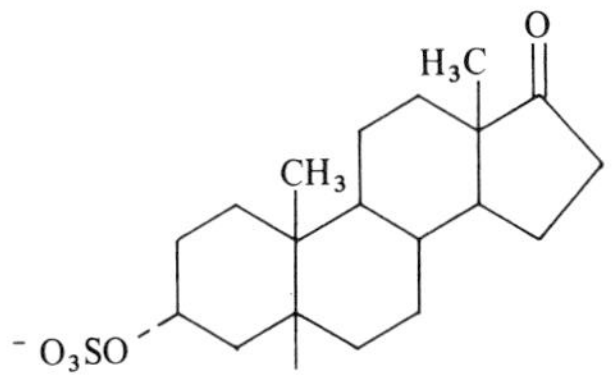

(V) 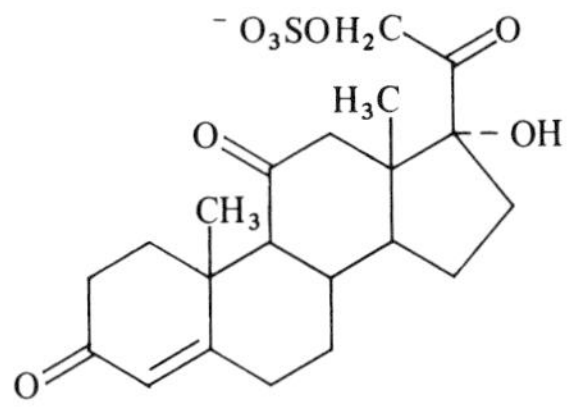 (VI)

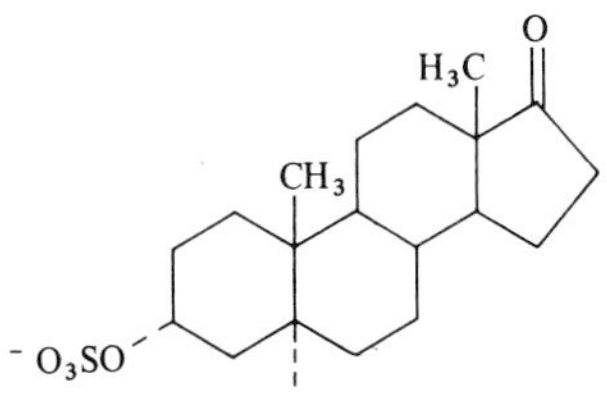

(VII) 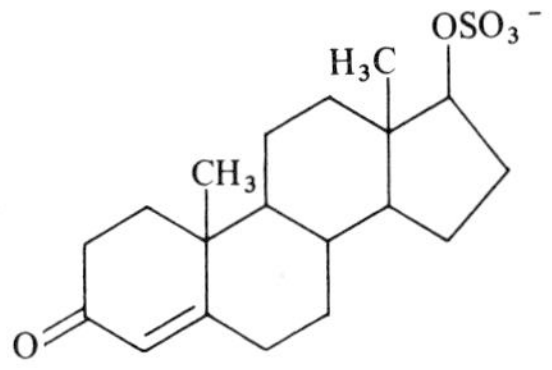 (VIII)

Indirect evidence suggests the existence of four types of steroid sulfatase which hydrolyze estrone sulfate (III), androstenolone sulfate [dehydroepiandrosterone sulfate (IV)], etiocholanolone sulfate (V), and cortisone 21-sulfate (VI), respectively. It must be stressed that the separate identity of these activities has by no means been proved and there are quite opposing views on this subject. For example, it has been claimed on the one hand that in the human placenta estrone sulfate, androstenolone sulfate and 4-nitrophenyl sulfate are hydrolyzed by separate enzymes (*17*) and on the other that all steroid sulfates are hydrolyzed by a single enzyme in rat testis (*18*). The evidence for the latter view is not convincing, and bearing in mind the very different struc-

17. A. P. French and J. C. Warren, *BJ* **105,** 233 (1967).
18. A. D. Notation and F. Ungar, *Steroids* **14,** 151 (1969).

tures of the substrates it is difficult not to conclude that each of them (III)–(VI) must be hydrolyzed by a different enzyme.

Whatever the number of enzymes involved, of the commonly occurring steroid sulfates only androsterone sulfate (VII) and testosterone sulfate (VIII) are resistant to hydrolysis by presently known enzymes.

The function of the steroid sulfatases in mammals needs no discussion. It can be presumed that they play a part in the general pathways of steroid metabolism, and in one case a disordered estrogen metabolism has been shown to be associated with a deficiency of placental androstenolone sulfatase (*19*). In mollusks, many of which are rich sources of steroid sulfatases, the situation is obscure. The function of these molluskan enzymes is quite unknown, and the fact that they are secreted in the digestive juices makes it difficult to believe that they can be concerned only with steroid metabolism.

1. *Estrone Sulfatase*

Estrone sulfatase catalyzes the hydrolysis of estrone sulfate (III) and is therefore an arylsulfatase. The possibility of its being one of the known arylsulfatases must be stressed and certainly the arylsulfatase of *Aspergillus oryzae*, which is devoid of androstenolone sulfatase activity (*20*), can hydrolyze estrone sulfate (*20, 21*). The likelihood of there being a separate estrone sulfatase in animal tissues was first noted by Pulkkinen (*22*), and more recent studies (*17*) seem to have established that the estrone sulfatase activity of placenta does not result from the associated arylsulfatase, sulfatase C, which hydrolyzes 4-nitrophenyl sulfate. In rat kidney also (*23*) estrone sulfatase seems to be distinct from sulfatase C and androstenolone sulfatase. Both these estrone sulfatases occur in the microsomes, and all attempts to obtain them in solution have failed.

The specificity of estrone sulfatase has not been studied in detail, but all estrogen 3-sulfates appear to be hydrolyzed at an optimum pH between 7 and 8, and with a K_m value of about 0.02 mM. The 2-sulfate of 2-hydroxyestrone is also hydrolyzed, probably more rapidly than the corresponding 3-sulfate, and this may be of importance in estrogen metabolism (*24*).

19. J. T. France and G. C. Liggins, *J. Clin. Endocrinol. Metab.* **29**, 138 (1969).
20. K. H. Ney and R. Ammon, *Z. Physiol. Chem.* **315**, 145 (1959).
21. A. Butenandt and H. Hofstetter, *Z. Physiol. Chem.* **259**, 222 (1939).
22. M. O. Pulkkinen, *Acta Physiol. Scand.* **52**, Suppl. 180 (1961).
23. N. G. Zuckerman and D. D. Hagerman, *ABB* **135**, 410 (1969).
24. M. Miyazaki, I. Yoshizawa, and J. Fishman, *Biochemistry* **8**, 1669 (1969).

Extracts from molluskan tissues hydrolyze estrogen sulfates, but there is no indication of what particular type of sulfatase is involved.

2. *Androstenolone Sulfatase*

The usual substrate for androstenolone sulfatase (often simply known as steroid sulfatase) is androstenolone sulfate (IV), and its hydrolysis occurs by fission of the O–S bond (*25*) so that the configuration around C-3 is retained during the reaction.

There are two main sources of this type of sulfatase. The most useful is the digestive gland (hepatopancreas) and digestive juices of many mollusks of which the limpet, *Patella vulgata*, and the snail, *Helix pomatia*, have been the most studied. This enzyme is readily obtained in solution and should be amenable to purification although little has been achieved (*26*). The other source is mammalian tissues: liver (*27*), testis (*28*), and especially placenta (*17, 29*). These enzymes are localized in the microsomes (*27*) and are difficult to obtain in solution. The androstenolone sulfatase of rat testis has been solubilized by digestion with heat-treated snake venom, but such treatment gives an unstable enzyme which tends to aggregate so that little purification has been possible (*30*).

The pH optima are about 4.5 for the molluskan enzyme and about 8 for the mammalian enzyme, and while the former is noncompetitively inhibited by sulfate ($K_i = 0.4$ mM) or phosphate ($K_i = 0.8$ mM) (*31*) the latter is not inhibited by sulfate and only weakly by phosphate. It is generally accepted that both types of androstenolone sulfatase hydrolyze only the 3-sulfates of 3β-hydroxy-5α- and 3β-hydroxy-Δ^5-steroids (*27, 31, 32*); that is, the sulfate esters of steroids having essentially planar molecules with the hydroxyl group at position 3 in the equatorial position. It must be pointed out, however, that no androstenolone sulfatase has been studied under conditions where low rates of hydrolysis would have been readily detected, and detailed specificity studies must await the availability of a pure enzyme. This would allow, for example, confirmation of the statements (*18, 32*) that cyclohexyl sulfate is not hydrolyzed by androstenolone sulfatases. The apparent inability of 3α-sulfates of 5α-steroids [e.g., androsterone sulfate (VII)], of steroid 17-sulfates

25. G. G. Logan and J. C. Warren, *BJ* **114,** 707 (1969).
26. P. Jarrige, *Bull. Soc. Chim. Biol.* **45,** 761 (1963).
27. A. B. Roy, *BJ* **66,** 700 (1957).
28. S. Burstein and R. I. Dorfman, *JBC* **238,** 1656 (1963).
29. J. C. Warren and A. P. French, *J. Clin. Endocrinol. Metab.* **25,** 278 (1965); A. P. French and J. C. Warren, *Steroids* **8,** 79 (1966).
30. S. Burstein, *BBA* **146,** 529 (1967); G. Bleau, A. Chapdelaine, and K. D. Roberts, *Can. J. Biochem.* **49,** 234 (1971).
31. P. Jarrige, J. Yon, and M. F. Jayle, *Bull. Soc. Chim. Biol.* **45,** 783 (1963).
32. A. B. Roy, *BJ* **62,** 41 (1956).

TABLE I

KINETIC CONSTANTS FOR THE HYDROLYSIS OF SOME STEROID SULFATES BY UNFRACTIONATED PREPARATIONS OF STEROID SULFATASES FROM *Helix pomatia*[a] AND FROM RAT TESTIS[b] AT pH 4.5 AND 7.4, RESPECTIVELY

	Helix		Rat	
	K_m (mM)	V^c	K_m (mM)	V^c
17-Oxoestra-1,3,5(10)-trien-3-yl sulfate (estrone sulfate)	0.11	—	0.020	—
17-Oxo-5α-androstan-3β-yl sulfate	0.032	1.3	—	—
17-Oxoandrost-5-en-3β-yl sulfate (androstenolone sulfate)	0.028	1.0	0.010	1.0
17β-Hydroxyandrost-5-en-3β-yl sulfate	—	—	0.005	3.3
20-Oxo-5α-pregnan-3β-yl sulfate	0.023	1.8	—	—
20-Oxopregn-5-en-3β-yl sulfate (pregnenolone sulfate)	0.02	1.6	0.003	15
Cholest-5-en-3β-yl sulfate	—	—	0.050	0.92
17-Oxo-5β-androstan-3α-yl sulfate (etiocholanolone sulfate)	3.07	0.80	—	—
17-Oxo-5β-androstan-3β-yl sulfate	6.35	0.78	—	—
17α-Hydroxy-3,11,20-trioxopregn-4-en-21-yl sulfate (cortisone 21-sulfate)	0.2	182	—	—

[a] From Jarrige, Yon, and Jayle (*31*).

[b] From Notation and Ungar (*18*).

[c] The values of V are referred to that of androstenolone sulfate, but it is not implied that a single enzyme is responsible for the hydrolysis of all these substrates.

[e.g., testosterone sulfate (VIII)], and of steroid 20-sulfates to act as substrates for androstenolone sulfatase must also be reinvestigated with pure enzymes.

Some kinetic constants for the androstenolone sulfatases from *H. pomatia* (*31*) and from rat testis (*18*) are given in Table I. The values for the latter enzyme must be accepted with caution because they were determined in a reaction mixture containing sulfate, ATP, and NAD, all of which are possible modifiers of the enzyme, and certainly determinations (*33*) under different conditions have given lower values of K_m. The androstenolone sulfatase of rat testis is powerfully inhibited by many unconjugated steroids (*33*, *33a*): derivatives of estrane, androstane, and pregnane inhibit to varying degrees but the most potent inhibitor is 5α-androstan-3β,17β-diol with a K_i of 1.7 μ*M* (*33*). Other values of K_i mostly fall in the range 1–10 μ*M*. The inhibition has been reported as competitive (*33a*) or partially competitive (*33*), and possible control functions have been envisaged. Much of this kinetic work is complicated by the particulate nature of mammalian androstenolone sulfatase, and this makes it even more unfortunate that statistical methods have not been used in the interpretation of the data.

It is interesting that the inhibition of molluskan androstenolone sulfatase by sulfate is noncompetitive (*31*) and of mammalian androstenolone sulfatase by androstenolone is competitive (*33a*). These findings would, independently or together, be consistent with androstenolone sulfatase catalyzing an ordered uni–bi reaction with the products being released from the enzyme in the order sulfate and then steroid. More detailed kinetic studies will be required to show whether this is indeed the case.

3. *Etiocholanolone Sulfatase*

The separate existence of etiocholanolone sulfatase is more doubtful, but the activity attributed to it has been found in only a few mollusks (*34*) and comparative studies suggest the participation of an enzyme distinct from androstenolone sulfatase. A typical substrate is etiocholanolone sulfate (V) and in view of the nonplanar structure of 5β steroids it seems unlikely that a single enzyme could react both with it and with planar steroids such as androstenolone sulfate (IV). Evidence for the

33. A. H. Payne, M. Mason, and R. B. Jaffe, *Steroids* **14,** 685 (1969).
33a. A. D. Notation and F. Ungar, *Biochemistry* **8,** 501 (1969).
34. Y. A. Leon, R. D. Bulbrook, and E. D. S. Corner, *BJ* **75,** 612 (1960); E. D. S. Corner, Y. A. Leon, and R. D. Bulbrook, *J. Marine Biol. Assoc. U. K.* **39,** 51 (1960).

participation of two separate enzymes also comes from kinetic studies (*31*) although this interpretation was not put on them by their author. Enzyme preparations from *H. pomatia* hydrolyze etiocholanolone sulfate with a K_m of 3.1 mM (*31*). Etiocholanolone sulfate also competitively inhibits the hydrolysis of androstenolone sulfate by this preparation with a K_i of 0.09 mM (*31*). The difference between these values of K_m and K_i seems to be too great for them to refer to a single enzyme.

The most useful source of etiocholanolone sulfatase is the digestive juice of *Helix pomatia*, preparations from which will hydrolyze both the 3α- and 3β-etiocholanolone sulfates, but only at about 10% of the rate of androstenolone sulfate (*31*). Values of K_m and V are given in Table I. It is obvious from these that the low rate of hydrolysis of etiocholanolone sulfate which is found in practice is a result of the very high K_m. The value of V differs little from that of androstenolone sulfate, but the limited solubility of steroid sulfates precludes this rate being attained in practice.

It is interesting that the presumed etiocholanolone sulfatase will, unless there be two such enzymes, hydrolyze both the 3α- (equatorial) and 3β- (axial) sulfates of 5β-androstan-17-one whereas androstenolone sulfatase will hydrolyze only the 3β- (equatorial) sulfate of 5α-androstan-17-one. The geometry of the active sites of two types of sulfatase must, it would seem, be very different.

4. *Cortisone Sulfatase*

Again the separate identity of this enzyme has not been proven but the structure of its substrate, the primary alkyl sulfate cortisone 21-sulfate (VI), suggests that this will be the case. The activity is widely distributed in mollusks but again *H. pomatia* and *P. vulgata* are the usual sources, the former being a particularly rich one (*31*). It has not been detected in mammalian tissues.

Preparations from *H. pomatia* will hydrolyze several corticosteroid 21-sulfates so that the presumed enzyme cannot be absolutely specific. The pH optimum is about 4.5, and the K_m is 0.2 mM cortisone 21-sulfate (*31*).

C. Glycosulfatases

Glycosulfatases hydrolyze carbohydrate sulfates of various types, and again several different such enzymes must exist although none of them has been purified to any significant extent so that their precise specificities must remain in doubt.

1. *Glucosulfatase*

Enzymes of the type glucosulfatase hydrolyze simple monosaccharide and disaccharide sulfates as well as compounds such as adenosine 5′-sulfate (*35*). They have been found in microorganisms and in mollusks, but their occurrence in higher animals is doubtful (*2*). The glucosulfatases from mollusks appear to be rather nonspecific; for example, crude preparations from the periwinkle *Littorina littorea* will hydrolyze not only glucose and galactose 6-sulfates but also glucose 3-sulfate (*36*). Further glucose and galactose 6-sulfates appear to be hydrolyzed by a single enzyme in *Patella vulgata* (*37*). The pH optimum is between 5 and 6 and the K_m values for the above three substrates are 17, 72, and 30 mM, respectively (*36*).

The enzymes from microorganisms may be more specific. In particular, the glucosulfatase produced by the mold *Trichoderma viride* grown on a medium containing glucose or galactose 6-sulfates will hydrolyze these esters but not glucose 3-sulfate (*38*). A glucosulfatase from *Pseudomonas carrageenovora* also shows a high specificity (*39*).

The function of the glucosulfatases is unknown although it is possible that in microorganisms they play a part in the general sulfur metabolism of the cell. In mollusks they may be involved in the degradation of the polysaccharide sulfates present in many species (*40*), and it has been suggested that they may even be involved in the synthesis of these esters (*4*).

2. *Cellulose Polysulfatase*

Cellulose polysulfatase has been found only in the triton *Charonia lampas*. It hydrolyzes the polysaccharide charonin sulfate, a complex sulfated glucan from this species, and the structurally related cellulose polysulfate (*41*). Dextran polysulfate is hydrolyzed at a much slower

35. T. Soda, *J. Fac. Sci., Univ. Tokyo, Sect. I* **3,** 149 (1936); I. Yamashina and F. Egami, *J. Japan. Biochem. Soc.* **25,** 281 (1953); P. F. Lloyd and C. H. Stuart, *BJ* **99,** 37P (1966).
36. K. S. Dodgson, *BJ* **78,** 324 (1961).
37. R. J. Fielder and P. F. Lloyd, *BJ* **109,** 14P (1968).
38. A. G. Lloyd, P. J. Large, M. Davies, A. H. Olavesen, and K. S. Dodgson, *BJ* **108,** 393 (1968).
39. J. Weigl and W. Yaphe, *Can. J. Microbiol.* **12,** 874 (1966).
40. F. Egami and N. Takahashi, *in* "Biochemistry and Medicine of Mucopolysaccharides" (F. Egami and Y. Oshima, eds.), p. 53. Maruzen Co., Ltd., Tokyo, 1962.
41. N. Takahashi and F. Egami, *BBA* **38,** 375 (1960); *BJ* **80,** 384 (1961).

rate. This enzyme may also play a part in the carbohydrate metabolism of the mollusk (*40*).

3. *Chondrosulfatase*

Probably several types of enzyme must be included under chondrosulfatase. The name would imply that the substrate is a chondroitin sulfate, but at least one of the so-called chondrosulfatases will not attack intact chondroitin sulfate but only the disaccharides produced by the degradation of the polysaccharide.

The best known source of chondrosulfatase is *Proteus vulgaris*. This enzyme will not hydrolyze intact chondroitin sulfates but only the disaccharides (or oligosaccharides) produced therefrom by the action of chondroitinase or hyaluronidase (*42*). It will also hydrolyze *N*-acetylchondrosine 6-sulfate which can be prepared chemically (*43*) and provides a well-defined substrate. Extracts of *P. vulgaris* contain two separable chondrosulfatases, chondroitin-4-sulfatase and chondroitin-6-sulfatase, which specifically hydrolyze the unsaturated disaccharides produced through the action of bacterial hyaluronidase on chondroitin 4-sulfate and chondroitin 6-sulfate, respectively (*44*). They also hydrolyze the corresponding *N*-acetylchondrosine 4- and 6-sulfates, but they do not hydrolyze higher oligosaccharides. Although there two enzymes have not been purified they have already proved most useful tools for investigations of the chemistry of mucopolysaccharides (*45*).

Chondrosulfatases of a rather different specificity occur in animal tissues. For example, the mollusk *Patella vulgata* contains a chondrosulfatase which will attack intact chondroitin 4-sulfate. The corresponding oligosaccharides, especially the tetrasaccharide, are also hydrolyzed (*46*). The lysosomes of rat liver contain a chondrosulfatase (*47*) which will attack oligosaccharides, at least as large as the octasaccharide, produced by the action of testicular hyaluronidase on chondroitin 4-sulfate. Neither intact chondroitin 4-sulfate nor *N*-acetylchondrosine 6-sulfate is hydrolyzed. Calf aorta, on the other hand, contains a chondrosulfatase which apparently hydrolyzes intact chondroitin 4-sulfate but not chondroitin 6-sulfate, dermatan sulfate, or keratan sulfate (*48*).

42. K. S. Dodgson and A. G. Lloyd, *BJ* **66**, 532 (1957).
43. A. G. Lloyd, A. H. Olavesen, P. A. Woolley, and G. Embery, *BJ* **102**, 37P (1967).
44. T. Yamagata, H. Saito, O. Habuchi, and S. Suzuki, *JBC* **243**, 1523 (1968).
45. H. Saito, T. Yamagata, and S. Suzuki, *JBC* **243**, 1536 (1968).
46. P. F. Lloyd and R. J. Fielder, *BJ* **109**, 14P (1968).
47. N. Tudball and E. A. Davidson, *BBA* **171**, 113 (1969).
48. E. Held and E. Buddecke, *Z. Physiol. Chem.* **348**, 1047 (1967).

4. *Cerebroside Sulfatase*

Cerebroside sulfatase was first detected (*49*) in the abalone, *Haliotis* sp., but it has since been found in mammalian tissues. The cerebroside sulfatase from the lysosomes of pig kidney has been considerably purified (*50*) and shown to hydrolyze natural and synthetic cerebroside 3-sulfates ($K_m = 0.1$ mM). Synthetic cerebroside 6-sulfates are not hydrolyzed (*51*). Galactose 3-sulfate is also slowly hydrolyzed but galactose 6-sulfate is not.

The most interesting feature of the cerebroside sulfatase from kidney is its association with the arylsulfatase, sulfatase A. Electrophoresis of the preparation separated two fractions one of which was an arylsulfatase, probably sulfatase A, having little activity on cerebroside sulfates. The other fraction, the "complementary fraction," was itself devoid of sulfatase activity but increased tenfold the cerebroside sulfatase activity of the arylsulfatase without altering its arylsulfatase activity (*50*). Because of this, and of other indirect evidence, it has been suggested that sulfatase A and cerebroside sulfatase are one and the same enzyme (*51*). This is supported by the weak cerebroside sulfatase activity of pure sulfatase A from ox liver being greatly increased by complementary fraction from pig kidney (*51*). Crude preparations of other arylsulfatases do not show this dual specificity (*51*). Further, H. Jatzkewitz and K. Stinshoff (personal communication) have now shown that, under suitable conditions, sulfatase A can show its maximum cerebroside sulfatase activity in the absence of any complementary fraction. The latter is certainly not required for the detection of cerebroside sulfatase in crude tissue extracts (*50, 51a*).

That cerebroside sulfatase is involved in the normal catabolism of cerebroside sulfates seem fairly clear. The congenital disease metachromatic leukodystrophy is characterized by the deposition of large amounts of cerebroside sulfates in the tissues, and cerebroside sulfatase is lacking from the tissues of patients suffering from this condition (*52*). It is most interesting that metachromatic leukodystrophy is also characterized by a complete lack of sulfatase A, and sometimes of sulfatases B and C (*53*); this seems further evidence for the view that sulfatase

49. Y. Fujino and T. Negishi, *Bull. Agr. Chem. Soc. Japan* **21,** 225 (1957).
50. E. Mehl and H. Jatzkewitz, *Z. Physiol. Chem.* **339,** 260 (1964).
51. E. Mehl and H. Jatzkewitz, *BBA* **151,** 619 (1968).
51a. M. T. Porter, A. L. Fluharty, J. Trammell, and H. Kihara, *BBRC* **44,** 660 (1971).
52. E. Mehl and H. Jatzkewitz, *BBRC* **19,** 407 (1965).
53. J. Austin, D. McAfee, D. Armstrong, M. O'Rourke, L. Shearer, and B. Bachhawat, *BJ* **93,** 15C (1964); J. Austin, D. Armstrong, and L. Shearer, *Arch. Neurol.* **13,** 593 (1965).

A and cerebroside sulfatase are one and the same, but this cannot yet be regarded as proved.

D. Choline Sulfatase

As far as is known choline sulfatase, which hydrolyzes choline sulfate ester (IX), is found only in certain fungi and bacteria where it seems to play an important role in general sulfur metabolism (*54*).

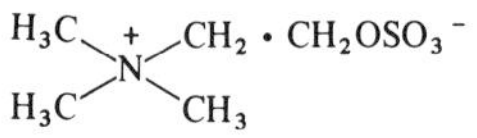

$$(H_3C)_2\overset{+}{N}(CH_3)CH_2 \cdot CH_2OSO_3^-$$

(IX)

Choline sulfatase is formed in large amounts in the mycelium of *Aspergillus nidulans* grown under sulfur-deficient conditions; subsequent growth in the presence of sulfur compounds causes the enzymic activity to fall (*55*). The co-repressor of the synthesis of choline sulfatase is probably cysteine although this has not been unambiguously shown. The enzyme is rather unstable and has not been significantly purified. Its pH optimum is about 7.5, and the K_m is 35 m*M* choline sulfate. Sulfite, phosphate, and cyanide are quite powerful inhibitors and sulfate a less effective one. Cysteine inhibits the enzyme by 75% at 1 m*M* and *in vivo* could be a feedback inhibitor of choline sulfatase (*55*) as well as the co-repressor of its synthesis.

Neurospora crassa contains a similar repressible choline sulfatase but here the co-repressor may be sulfide (*56*).

Pseudomonas nitroreducens (*57*) and *Pseudomonas aeruginosa* (*58*) form an intracellular choline sulfatase when grown on a medium containing choline sulfate; this enzyme is therefore inducible by its substrate and so differs from the corresponding fungal enzyme which is repressible by sulfur-containing compounds. The bacterial enzyme has a pH optimum of 8.3, a K_m of 40 m*M* choline sulfate, and is inhibited by sulfite but not by sulfate, phosphate, or cyanide. The preparation from *P. nitroreducens* showed no gluco-, alkyl- or myrosulfatase activities.

54. B. Spencer and T. Harada, *BJ* **77,** 305 (1960); T. Harada and B. Spencer, *J. Gen. Microbiol.* **22,** 520 (1960).
55. J. M. Scott and B. Spencer, *BJ* **106,** 471 (1968).
56. R. L. Metzenberg and J. W. Parson, *Proc. Natl. Acad. Sci. U. S.* **55,** 629 (1966).
57. I. Takebe, *J. Biochem.* (*Tokyo*) **50,** 245 (1961).
58. T. Harada, *BBA* **81,** 193 (1964).

E. ALKYLSULFATASES

The first suggestion of the occurrence of an alkylsulfatase came from the demonstration (*59*) that *Bacillus cereus mycoides* could hydrolyze 3′,5′-dichlorophenoxyethyl sulfate. More recently extracts containing an alkylsulfatase hydrolyzing dodecyl sulfate and related primary alkyl sulfates have been prepared from *Pseudomonas* sp. grown in the presence of the former ester (*60*). The enzyme has a pH optimum of about 7.5, is inhibited by phosphate, and has the remarkably high temperature optimum of 70° for a reaction time of 10 min (*61*). It showed no arylsulfatase activity and could not hydrolyze 3′,5′-dichlorophenoxyethyl sulfate.

A further type of alkylsulfatase hydrolyzing the sulfate esters of secondary alcohols also occurs in microorganisms. Extracts hydrolyzing pentan-3-yl sulfate (as well as dodecyl sulfate) have been prepared from *Aerobacter cloacae* cultured in the presence of C_{10}–C_{20} secondary alkyl sulfates (*62*).

Although these alkylsulfatases have been obtained from microorganisms grown in the presence of foreign alkyl sulfates, the enzyme may have a physiological role in view of the occurrence of the disulfate of docosan-1,14-diol in *Ochromonas danica* (*63*), and probably in other microorganisms. This ester would presumably be a substrate for both the primary and the secondary alkylsulfatases mentioned above.

There is a suggestion that mammalian tissues may show alkylsulfatase activity (*64*), but it must be recalled that the C–O–S system can be split by mechanisms not involving sulfatases. Serine *O*-sulfate is desulfated *in vivo* (*65*), but studies of the enzyme system *in vitro* have shown that the reaction products are pyruvate, ammonia, and sulfate (*66*) which probably arise through a β elimination and certainly not through a simple hydrolysis.

F. MYROSULFATASE

Myrosulfatase differs from the other sulfatases considered above in that it attacks not the C–O–S system but the N–O–S system found in

59. A. J. Vlitos, *Contrib. Boyce Thompson Inst.* **17,** 127 (1953).
60. Y-C. Hsu, *Nature* **200,** 1091 (1963); **207,** 385 (1965).
61. W. J. Payne, J. P. Williams, and W. R. Mayberry, *Appl. Microbiol.* **13,** 698 (1965).
62. W. J. Payne, J. P. Williams, and W. R. Mayberry, *Nature* **214,** 623 (1967).
63. G. L. Mayers and T. H. Haines, *Biochemistry* **6,** 1665 (1967).
64. J. B. Knaak, S. J. Kozbelt, and L. J. Sullivan, *Toxicol. Appl. Pharmacol.* **8,** 369 (1966).

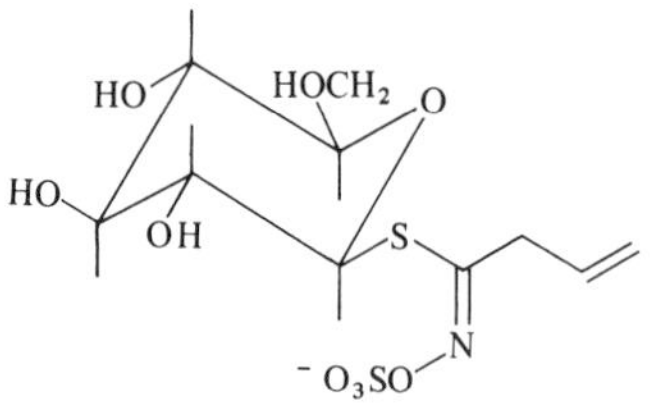

(X) 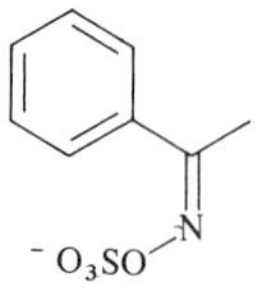 (XI)

mustard oil glycosides such as sinigrin (X). The formation of sulfate from such a glycoside can occur through the action of a thioglycosidase followed by a spontaneous Losen rearrangement or through the action of the enzyme system myrosinase (myrosin) of which myrosulfatase is a component. Comprehensive reviews of the myrosinase reaction are available (*67*).

Myrosinase, and therefore myrosulfatase, seems to be of common occurrence in the seeds and vegetative tissues of the Cruciferae: it occurs in a few related families but is not of general occurrence in plants (*68*).

Myrosinase from *Brassica juncea* has been separated into a glycosidase and a sulfatase which together catalyze the classic degradation of sinigrin (*69*). The sulfatase has a pH optimum of 6, is not inhibited by phosphate, and can hydrolyze not only sinigrin but also several oxime *O*-sulfonates, such as acetophenone oxime *O*-sulfonate (XI), which contain the N–O–S system. It shows no arylsulfatase or glucosulfatase activity.

There have been reports of the occurrence of myrosulfatase in molluskan tissues (*70*), but these must be viewed with reserve because of the ability of glycosidases to catalyze the myrosinase reaction. The need for caution is stressed by the fact that both tetracetyl sinigrin and tetramethyl sinigrin are hydrolyzed by molluskan preparations (*71*) al-

---

65. N. Tudball, *BJ* **85,** 456 (1962).

66. J. H. Thomas and N. Tudball, *BJ* **105,** 467 (1967); N. Tudball, J. H. Thomas, and J. A. Fowler, *ibid.* **114,** 299 (1969).

67. F. Challenger, "Aspects of the Organic Chemistry of Sulphur." Butterworth, London and Washington, D. C., 1959; A. Kjaer, *Fortschr. Chem. Org. Naturstoffe* **18,** 122 (1960).

68. Z. Nagashima and M. Uchiyama, *Nippon Nogei Kagaku Kaishi* **33,** 881 (1959); *Chem. Abstr.* **57,** 3787h (1962).

69. R. D. Gaines and K. J. Goering, *ABB* **96,** 13 (1962).

70. N. Takahashi, *J. Biochem.* (*Tokyo*) **47,** 230 (1960).

71. Z. Nagashima and M. Uchiyama, *Nippon Nogei Kagaku Kaishi* **33,** 1068 (1959); *Chem. Abstr.* **58,** 4769a (1963).

though these are not substrates for plant myrosinase. Obviously this problem is ideally suited for study with oxime *O*-sulfonates as substrates.

III. Sulfatophosphate Sulfohydrolases

Two different sulfohydrolases are responsible for the hydrolysis of the sulfatophosphate bond [reaction (2)] in adenylyl sulfate and 3′-phosphoadenylyl sulfate, respectively. These reactions are of considerable practical importance because they destroy the sulfate-containing nucleotides required in many metabolic studies.

Adenylyl sulfate sulfohydrolase occurs in both the lysosomal and soluble fractions of rat liver (*72*), and it has been partially purified from the latter. It has a pH optimum of about 5.2 in acetate, a K_m of 0.5 m*M* adenylyl sulfate, and is inhibited by phosphate, pyrophosphate, and thiol-binding reagents. Adenosine triphosphate and related nucleoside triphosphates are powerful inhibitors, 0.1 m*M* ATP causing a 70% inhibition. The partially purified enzyme does not hydrolyze 3′-phosphoadenylyl sulfate although crude preparations from rat liver do. There has been a recent report (*72a*) that several tissues contain two adenylyl sulfate sulfohydrolases which have rather interesting properties, but many aspects of the work require more detailed investigation.

3′-Phosphoadenylyl sulfate sulfohydrolase is of widespread distribution (*2*), but there is some disagreement about its intracellular localization. It has been reported to occur in the soluble (*73*) and lysosomal (*72*) fractions of rat liver and in virtually all the subcellular fractions of pig kidney (*74*). The pH optimum is about 6 and the K_m is 0.045 m*M* 3′-phosphoadenylyl sulfate. Alone among the sulfatases, the 3′-phosphoadenylyl sulfate sulfohydrolase of sheep brain is activated, some sevenfold, by 5 m*M* Co^{2+} or Mn^{2+} (*75*); the activation of the enzyme from rat liver is not statistically significant (*73*) and although the enzyme from pig kidney has been reported to be activated by Co^{2+} (*74*) figures have not been given. The enzyme is not inhibited by EDTA but it is by fluoride (70% at 0.01 *M*), an effect of some practical importance in minimizing the destruction of 3′-phosphoadenylyl sulfate in crude tissue

72. R. Bailey-Wood, K. S. Dodgson, and F. A. Rose, *BBA* **220,** 284 (1970).

72a. D. Armstrong, J. Austin, T. Luttenegger, B. Bachhawat, and D. Stumpf, *BBA* **198,** 523 (1970).

73. T. Koizumi, T. Suematsu, A. Kawasaki, K. Hiramatsu, and N. Iwabori, *BBA* **184,** 106 (1969).

74. J. Austin, D. Armstrong, D. Stumpf, T. Luttenegger, and M. Dragoo, *BBA* **192,** 29 (1969).

75. A. S. Balasubramanian and B. K. Bachhawat, *BBA* **59,** 389 (1962).

preparations (*76*). Unlike adenylyl sulfate sulfohydrolase, 3′-phosphoadenylyl sulfate sulfohydrolase is inhibited by reagents containing sulfhydryl groups: 2,3-dimercaptopropanol is a particularly powerful inhibitor, causing virtually complete inhibition at 0.01 *M* (*75*).

Although these enzymes are regarded as sulfohydrolases this has not been proven. Certainly their action leads to the production of sulfate from a sulfatophosphate, but this could occur by fission of either the O–S or the P–O bond. If the former be the case then the enzymes are certainly sulfohydrolases, but if the latter be true then they should be regarded as phosphatases. That they might indeed be phosphatases is perhaps suggested by the apparent dependence of 3′-phosphoadenylyl sulfohydrolase on Co^{2+}: no other sulfatase is known to be metal dependent whereas phosphatases in general are. This problem could be solved by studying these hydrolyses in $H_2{}^{18}O$ (*76a*).

IV. Sulfamatases

Sulfamatases, often called sulfamidases, catalyze reaction (3) above, a reaction which must involve fission of the N–S bond. As far as is known at present they are not general sulfamatases but hydrolyze only 2-deoxy-2-sulfoamino-D-glucose or polysaccharides such as heparin which contain this sugar.

The only known microbial source is *Flavobacterium heparinum* which adaptively produces sulfamatase when grown on a medium containing heparin or 2-deoxy-2-sulfoamino-D-glucose (*77*). The enzyme is particle bound and has not been obtained in a soluble form (*78*). It preferentially hydrolyzes 2-deoxy-2-sulfoamino-D-glucose; some simple derivatives of this, and intact heparin, are hydrolyzed much more slowly. The pH optimum is 7.2, the K_m is 5.3 m*M* 2-deoxy-2-sulfoamino-D-glucose. Sulfate and phosphate are inhibitors, but some simple sulfamates are not (*79*).

In mammals the only known sources are lymphoid tissues such as spleen, ileum, and lung (*79*). This sulfamatase, in striking contrast to the microbial enzyme, hydrolyzes heparin and the oligosaccharides de-

76. S. Suzuki and J. L. Strominger, *JBC* **235**, 257 (1960).
76a. S. J. Benkovic and R. C. Hevey, *JACS* **92**, 4971 (1970).
77. A. G. Lloyd, B. A. Law, L. J. Fowler, and G. Embery, *BJ* **110**, 54P (1968).
78. C. P. Dietrich, *BJ* **111**, 91 (1969); B. A. Law, A. G. Lloyd, G. Embery, and G. B. Wisdom, *ibid.* **115**, 10P (1969).
79. A. G. Lloyd, L. J. Fowler, G. Embery, and B. A. Law, *BJ* **110**, 54P (1968).

rived therefrom, but not 2-deoxy-2-sulfoamino-D-glucose. The pH optimum is 5.0; sulfate and phosphate are inhibitors as are also a number of simple sulfamates such as serine *N*-sulfate or phenyl sulfamate. Particularly interesting is the suggestion by Dietrich (*80*) that only one sulfate group is removed by the enzyme, no matter how many may be present in the macromolecular substrate.

80. C. P. Dietrich, *Can. J. Biochem.* **48**, 725 (1970).

2

Arylsulfatases

R. G. NICHOLLS • A. B. ROY

I. Introduction

The arylsulfatases (arylsulfate sulfohydrolasés, EC 3.1.6.1) are the best known of the sulfatases. Two representatives have been obtained as homogeneous proteins and others have been highly purified but, despite a considerable amount of information on the physical, chemical, and enzymological properties of these enzymes, a full understanding of the action of any one of them is remote. The present discussion will be restricted to those enzymes which have been significantly purified. More general topics have already been considered in Chapter 1 by Roy, this volume.

The arylsulfatases catalyze the hydrolysis of an aryl sulfate anion by fission of the O–S bond (*1*). The reaction (1) is apparently irreversible and there is no evidence that any sulfate acceptor other than water can

$$R{\cdot}OSO_3^- + H_2O \rightarrow R{\cdot}OH + H^+ + SO_4^{2-} \quad (1)$$

be utilized nor that any metal ion is involved. The thermodynamics of the arylsulfatase reaction have not been studied, but the activation energy is probably about 12–14 kcal/mole.

1. B. Spencer, *BJ* **69**, 155 (1958).

In the following discussion it will be assumed that the enzymic reaction takes the form of an ordered uni–bi reaction as represented in scheme (2).

$$E + S \rightleftharpoons ES \rightleftharpoons EP_1P_2 \rightleftharpoons EP_2 + P_1 \rightleftharpoons E + P_2 \quad (2)$$

The arylsulfatases form a homogeneous group of enzymes but two rather different types, the type I and type II arylsulfatases, can be distinguished by their behavior toward sulfate ions, one of the products of the arylsulfatase reaction. The type I arylsulfatases are hardly inhibited by sulfate, whereas the type II enzymes are strongly inhibited. In the original distinction between the type I and type II arylsulfatases (*2*) it was stated that the former were not inhibited by sulfate or phosphate but were by cyanide or fluoride, whereas the latter were inhibited by sulfate and phosphate but not by cyanide or fluoride. This multiple definition has led to difficulties, and it seems that a simple distinction based on the action of sulfate may be theoretically sounder as well as practically simpler.

The assay of the arylsulfatases is, in principle, simple and several types of methods are available (*2*). First, spectrophotometric methods for the determination of the liberated phenol. The most commonly used substrate is nitrocatechol sulfate (potassium or dipotassium 2-hydroxy-5-nitrophenyl sulfate) (*3*) although potassium 4-nitrophenyl sulfate is also useful (*4*), especially in continuous assays (*5*). Other chromogenic substrates are phenyl sulfate, 4-acetylphenyl sulfate, and phenolphthalein disulfate; but the latter is not hydrolyzed by all arylsulfatases (*2*). Fluorimetric methods have also been described (*6*), but they have not yet been widely used and their only advantage lies in their great sensitivity. Second, methods following the liberation of H^+ ions in the pH stat (*7, 8*). These are particularly valuable because they allow the determination of initial velocities and they can, in principle, be used with any substrate. Third, methods based on the determination of sulfate. These should be as useful as those measuring H^+ ions, but at present they are limited by the tedious determination of small quantities of sulfate (*9*).

2. K. S. Dodgson and B. Spencer, *Methods Biochem. Anal.* **4,** 211 (1957).
3. A. B. Roy, *BJ* **77,** 380 (1960).
4. D. H. Rammler, C. Grado, and L. R. Fowler, *Biochemistry* **3,** 224 (1964).
5. E. C. Webb and P. F. W. Morrow, *BJ* **73,** 7 (1959).
6. W. R. Sherman and E. F. Stanfield, *BJ* **102,** 905 (1967); G. G. Guilbault and J. Hieserman, *Anal. Chem.* **41,** 2006 (1969).
7. S. O. Andersen, *Acta Chem. Scand.* **13,** 120 (1959).
8. L. W. Nichol and A. B. Roy, *J. Biochem.* (*Tokyo*) **55,** 643 (1964).
9. K. S. Dodgson and B. Spencer, *BJ* **55,** 436 (1953); K. S. Dodgson, *ibid.* **78,** 312 (1961).

The availability of a sulfate-specific electrode could, however, allow the development of useful methods. Finally, in certain circumstances it may be advantageous to determine unhydrolyzed substrate. Such a method has been described (*10*), but it cannot be recommended for general use.

It should be stressed that although the methods are simple in principle, and can give reliable assays with pure or partially purified enzyme preparations, the assay of arylsulfatases in crude tissue extracts is difficult because of the presence of potential inhibitors such as phosphate or sulfate. Also, most tissues contain several arylsulfatases which do not differ sufficiently to allow their individual determination by assays using a single substrate at different concentrations and pH. In some cases differential assays are possible by suitable choice of substrate (*11*), and in others methods based on the differential inhibition of the different enzymes have been developed (*12, 13*), but only one (*12*) of these has been widely used. It would seem that the chromatographic separation of the different enzymes might be a preferable approach (*14*) although the insolubility of some arylsulfatases must be kept in mind.

II. Type I Arylsulfatases

Type I arylsulfatases, which are not significantly inhibited by sulfate, are found in both microorganisms and in animals. The typical example of the latter is sulfatase C, present in the liver microsomes of many mammalian species. This enzyme will not be considered further because it has not been obtained in a soluble form and it is always accompanied by other types of sulfatase (*15*). The best known of the microbial type I arylsulfatases is that in commercial preparations of *Aspergillus oryzae* (e.g., Mylase P, Clarase, and Takadiastase), but it has not been purified and there has been no systematic study of its properties. The type I arylsulfatase of *Aerobacter aerogenes* has been obtained in an apparently

10. A. B. Roy, *BJ* **62,** 41 (1956).
11. D. W. Milsom, F. A. Rose, and K. S. Dodgson, *BJ* **109,** 40P (1968).
12. H. Baum, K. S. Dodgson, and B. Spencer, *Clin. Chim. Acta* **4,** 453 (1959).
13. L. M. Dzialoszynski, W. Bleszynski, and J. Lewosz, *Zeszyty Nauk. Uniw. Mikolaja Kopernika Toruniu* **9,** 15 (1966).
14. A. B. Roy and P. Trudinger, "The Biochemistry of Inorganic Compounds of Sulphur." Cambridge Univ. Press, London and New York, 1970.
15. A. B. Roy, *BJ* **66,** 700 (1957); A. P. French and J. C. Warren, *ibid.* **105,** 233 (1967).

homogeneous form (*16*), but little is known of its enzymology. In contrast, the enzyme from *Alcaligenes metalcaligenes* has not been significantly purified although quite detailed studies have been made of its specificity (*17*).

What little information is available on the physical properties of the enzymes from *A. aerogenes* and *A. oryzae* is summarized in Table I. Both are rather heat stable. That from *A. oryzae* has a temperature optimum of about 50° (with a reaction time of 3 hr) (*18*) and that from *A. aerogenes* showed little decrease in specific activity below 60° at which temperature its half-life was about 10 min (*16*). The enzyme from *A. metalcaligenes* does not have this heat stability.

The specificity of the arylsulfatase of *A. aerogenes* is apparently rather great. Although it will hydrolyze 4-nitrophenyl sulfate ($K_m = 3.3$ mM, pH optimum 7.1) and phenyl sulfate it will not hydrolyze 1-naphthyl sulfate or phenolphthalein disulfate (*16*). The latter is certainly not a general substrate for arylsulfatases but, as far as known, 1-naphthyl sulfate is. In contrast, the arylsulfatases from *A. oryzae* and *A. metalcaligenes* are rather nonspecific enzymes which can hydrolyze a wide range of aryl sulfates (*17, 19*) and the former at least can hydrolyze the heterocyclic "aryl" sulfate, indoxyl sulfate (*20, 21*). The pH optima are in the region of 6 and 8, respectively. Some pertinent data on their

TABLE I
SOME PHYSICAL PROPERTIES OF THE TYPE I ARYLSULFATASES FROM *A. aerogenes*[a] AND *A. oryzae*

Physical property	*A. aerogenes*	*A. oryzae*
$s_{20,w}$ (S)	3.5	—
Molecular weight	40,700[b]	90,000[c]
f/f_0	1.24	—
Isoelectric point	—	4.7–5.0[d]

[a] From Fowler and Rammler (*16*).
[b] From approach-to-equilibrium sedimentation.
[c] From chromatography on Sephadex.
[d] Probably three isoenzymes have isoelectric points in this range.

16. L. R. Fowler and D. H. Rammler, *Biochemistry* **3**, 230 (1964).
17. K. S. Dodgson, B. Spencer, and K. Williams, *BJ* **64**, 216 (1956).
18. L. D. Abbott, *ABB* **15**, 205 (1947).
19. D. Robinson, J. N. Smith, B. Spencer, and R. T. Williams, *BJ* **51**, 202 (1952).
20. C. Neuberg and J. Wagner, *Biochem. Z.* **161**, 492 (1925).
21. The statement (*19*) that Takadiastase can hydrolyze quinolinyl 2-sulfate arises from a mistranslation. The compound hydrolyzed is the sulfate ester of "*o*-oxychinolin" (*20*) or quinolinyl 8-sulfate. This is essentially a substituted phenyl sulfate.

specificity are summarized in Table II. With a series of substituted phenyl sulfates there are rectilinear relationships between the Hammet substitution constants σ and both V and K_m for the hydrolysis of these compounds by the arylsulfatase of *A. metalcaligenes* (*17*). With this enzyme, therefore, the maximum rate may be governed by the efficiency of the phenol, R·OH, as a leaving group. As is clear from Table II, however, the same relationship does not hold for the enzyme from *A. oryzae*.

The type I arylsulfatases are scarcely inhibited by sulfate; for example, the arylsulfatase of *A. oryzae* is inhibited to the extent of only about 5% by 50 mM sulfate with 5 mM nitrocatechol sulfate as substrate. Phosphate is likewise generally a poor inhibitor although the enzyme

TABLE II
KINETIC CONSTANTS FOR SOME TYPE I ARYLSULFATASES

	A. metalcaligenes[a]		*A. oryzae*[b]	
Substrate	K_m (mM)	V[c]	K_m (mM)	V[c]
4-Nitrophenyl sulfate	0.48	1.0	0.17, 0.2*	1.0
4-Acetylphenyl sulfate	0.90	1.0	—	—
4-Aldehydophenyl sulfate	—	—	0.21	0.73
3-Nitrophenyl sulfate	0.98	0.48	0.84	1.4
2-Nitrophenyl sulfate	0.68	0.94	0.34	0.87
2-Chlorophenyl sulfate	1.3	0.38	—	—
4-Chlorophenyl sulfate	0.94	0.34	0.97	1.3
Phenyl sulfate	5.9	0.059	1.7*	0.83*
3-Methylphenyl sulfate	3.1	0.016	0.8*	0.81*
2-Methylphenyl sulfate	6.3	0.008	0.95*	0.70*
3-Aminophenyl sulfate	5.7	0.006	—	—
4-Methylphenyl sulfate	3.1	0.048	0.75*	0.74*
4-Methoxyphenyl sulfate	6.2	0.007	—	—
4-Aminophenyl sulfate	11	0.016	—	—
2-Aminophenyl sulfate	11	0.006	—	—
2-Hydroxy-5-nitrophenyl sulfate	1.7	0.14	0.35	0.67
4-Hydroxy-5-nitrophenyl sulfate	—	—	0.13	1.1
2-Naphthyl sulfate	—	—	0.3*	0.81*

[a] For the enzyme from *Alcaligenes metalcaligenes* the conditions were pH 8.75 in 0.1 M phosphate buffer at 37°. The values must be accepted with caution because incubation times of up to 18 hr were used (*17*).

[b] For the enzyme from *Aspergillus oryzae* the conditions were pH 6 in 0.5 M acetate (*19*) or pH 6.2 in 0.1 M citrate (values marked *), both at 37°. Incubation times were from 1 to 3 hr.

[c] Values for V are given relative to that for 4-nitrophenyl sulfate.

from *A. aerogenes* does seem to be somewhat more sensitive than the others (*16*). Sulfite is an inhibitor, but not a particularly powerful one, concentrations of about 5 m*M* being required to give a significant inhibition. These type I arylsulfatases, and arylsulfatases in general, are powerfully inhibited by "carbonyl" reagents such as cyanide, hydroxylamine, hydrazine, and phenylhydrazine (*16, 19, 22, 23*). It was claimed (*22*) that the inhibition of the enzyme from *A. metalcaligenes* by these reagents was uncompetitive because reciprocal plots at different concentrations of inhibitor gave a family of parallel lines. The original interpretation of this behavior was that the inhibitor combines not with the free enzyme but only with the enzyme–substrate complex. More recently it has been confirmed (*24*) that the presence of substrate (4-nitrophenyl sulfate) is necessary for the inhibition of the sulfatase of *A. metalcaligenes* by cyanide and further shown that the inhibitor is bound to the enzyme to give a complex which does not dissociate on dialysis or on chromatography on Sephadex. The inhibition is therefore essentially irreversible and so cannot be strictly uncompetitive. The inhibition of the arylsulfatase of *A. aerogenes* by cyanide is likewise dependent upon the presence of substrate (*16*) and is irreversible although it has not been shown that binding occurs. Further investigation of the cyanide binding is certainly required, especially in view of the finding (*25*) that metal ions are involved in the inhibition of sulfatase A (a type II arylsulfatase) by "carbonyl" reagents.

It is obvious that much further work remains to be done with these enzymes and it is unfortunate that most enzymological studies continue to be carried out with quite impure preparations when a pure enzyme, that from *A. aerogenes*, is available.

III. Type II Arylsulfatases

Type II arylsulfatases are, as has already been stated, characterized by their being inhibited by sulfate ions. They seem to be principally of animal origin but one such enzyme has been found in a microorganism, *Proteus vulgaris* (*26*). Only one type II arylsulfatase has been obtained as a homogeneous protein although several others have been considerably purified.

22. K. S. Dodgson, B. Spencer, and K. Williams, *Nature* **177**, 432 (1956).
23. L. M. Dzialoszynski, *Bull. Soc. Amis Sci. Lettres Poznan* **B11**, 87 (1951).
24. I. S. Shaw, W. D. Stein, and C. H. Wynn, *BJ* **112**, 14P (1969).
25. A. B. Roy, *BBA* **198**, 76 (1970).
26. K. S. Dodgson, *Enzymologia* **20**, 301 (1959).

The type II arylsulfatases of mammals are typical lysosomal enzymes. Those of invertebrates may also occur in cell organelles but at least some are present in the digestive fluids in the gut.

A. Sulfatase A

1. *From Ox Liver*

Sulfatase A has been obtained from ox liver in a yield of about 0.5–1.0 mg/kg liver, amounting to about 20% of the activity originally present (*8*). The preparation is a glycoprotein and is homogeneous with respect to sedimentation coefficient, as shown by boundary analyses (*8*), and to mobility in both zone and boundary electrophoresis (*27*). It is a rather stable protein and can be heated at 65° for 5 min without loss of activity (*7*).

The approximate amino acid composition of the enzyme is shown in Table III (*28*). Only two points require mention. First, the presence of

TABLE III
THE APPROXIMATE COMPOSITION OF SULFATASE A FROM OX LIVER (MW 107,000)

Amino acid	No. of residues
Lysine	14
Histidine	39
Arginine	34
Aspartic acid	72
Threonine	56
Serine	59
Glutamic acid	88
Proline	90
Glycine	109
Alanine	94
½-Cystine	21
Valine	49
Methionine	16
Isoleucine	16
Leucine	133
Tyrosine	28
Phenylalanine	50
Tryptophan	11
Glucosamine	9
Ammonia	89

27. L. W. Nichol and A. B. Roy, *Biochemistry* **5**, 1379 (1966).
28. L. W. Nichol and A. B. Roy, *Biochemistry* **4**, 386 (1965).

large amounts of proline, and, second, the presence of glucosamine which is presumably accompanied by other carbohydrates although scarcity of material has prevented investigation of this point. The rather large amounts of proline may account for the anomalous ORD shown by the enzyme (*28*).

a. Molecular Weight. The physical properties of sulfatase A are summarized in Table IV. At ionic strength 0.1 and at pH values between 6.5 and at least 9.3 it exists as a monomer of molecular weight 107,000 ($s^0_{20,w}$, 6.4_6 S) (*8, 28*). Below pH 5.5 at ionic strength 0.1 it exists as a tetramer of molecular weight 411,000 ($s^0_{20,w}$, 14.2 S) which is stable until the pH approaches the isoelectric point of 3.4 when heterogeneous slowly sedimenting material (s about 3 S) is formed (*8, 28*). Between pH 5.5 and 6.5, and at an ionic strength of 0.1, the enzyme exists as an equilibrium mixture of polymeric species of which the dimer is probably quantitatively the most important (*28*). The dimer is also formed at pH 7.5 when the ionic strength is raised to 2 (*28*). The mechanism of the polymerization is not fully understood, but it has been suggested (*27*) that hydrophobic interactions are responsible for maintaining the structure of the tetramer. This view is based on the slight increase of $s_{20,w}$

TABLE IV
THE PHYSICAL PROPERTIES OF SULFATASES A AND B FROM OX LIVER

Physical property[a]	Sulfatase A[b] pH 7.5	Sulfatase A[b] pH 5.0	Sulfatase B[c] pH 7.5
$s^0_{20,w}$ (S)	6.5	14.2	—
$D^*_{20} \times 10^7$ (cm² sec⁻¹)[d]	4.9	3.2	—
Molecular weight	107,000[e]	411,000[e]	45,000[f]
Mobility $\times 10^5$ ($cm^2\ sec^{-1}\ V^{-1}$)	−9.9	−5.3	—
Net charge	−30	−9	—
$E^{1\%}_{280\ m\mu}$	7.0	6.6	14
$\bar{V}$	0.71_4	0.71_6	—
Isoelectric point	3.4[g]		8.3[h]

[a] The parameters were measured at ionic strength 0.1 and 20° except for the electrophoreses which were carried out at 4°.
[b] From Nichol and Roy (*8, 27, 28*).
[c] The molecular weight of sulfatase B is the same at pH 5 and 7.5, and the two isoenzymes Bα and Bβ do not differ in the properties listed (*41*).
[d] An apparent diffusion coefficient calculated from boundary spreading during sedimentation.
[e] Equilibrium and approach-to-equilibrium sedimentation.
[f] Chromatography on Sephadex.
[g] Moving boundary electrophoresis.
[h] Isoelectric focusing.

with increase in temperature (suggesting a strengthening with increasing temperature of the bonds maintaining the tetramer), on the increased tendency to form polymers at high ionic strengths, and on the disruption of the tetramer by sodium dodecyl sulfate. The model which is proposed (*27*) suggests that the monomeric units are held apart by their large net negative charge at pH values greater than about 6.5 (—30 per monomer unit at pH 7.5) while at pH values below about 5.5 the net charge is so decreased (—9 per monomer unit at pH 5.0) that hydrophobic attraction can overcome electrostatic repulsion and allow the formation of the tetramer. The evidence in favor of this hypothesis is purely circumstantial and the importance of other interactions, especially hydrogen bonds, cannot be excluded. In particular, it would seem that hydrogen bonding between un-ionized carboxyl groups could well be important. Sulfatase A contains large numbers of dibasic amino acid residues (Table III) and the ionization of their carboxyl groups would certainly be decreasing rapidly in the pH region where polymerization becomes important; this is reflected in the decreasing electrophoretic mobility below pH 5.5 (Fig. 1).

The subunit structure of the monomer (molecular weight 107,000) is not completely understood. From an investigation (*29*) of the action of urea, with and without dithiothreitol, and of sodium dodecyl sulfate on the enzyme it was suggested that subunits of molecular weight about 25,000 were present. These were apparently linked in pairs by disulfide

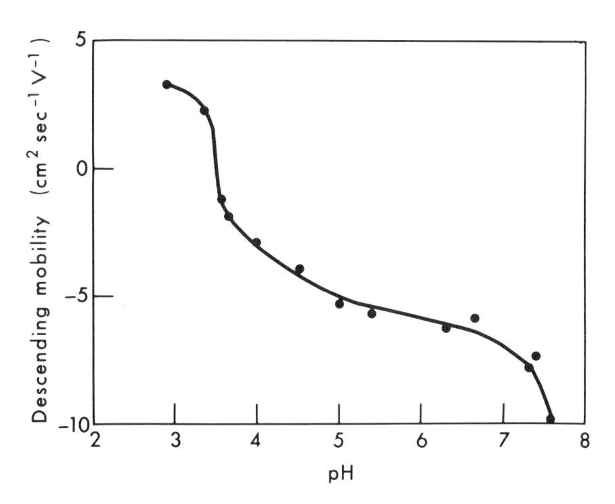

FIG. 1. The electrophoretic mobility of sulfatase A as a function of pH. Uni-univalent buffers, $I = 0.1$, temperature 4°.

29. A. B. Roy and A. Jerfy, *BBA* **207,** 156 (1970).

bonds and then again in pairs by hydrogen bonds and/or hydrophobic interactions. The subunits showed no enzymic activity and on removal of the denaturant they formed enzymically inactive nonspecific aggregates.

The above discussion refers to the polymerization of sulfatase A at protein concentrations of between 0.005 and 1% (and at ionic strength 0.1 and at 20°). At "catalytic" concentrations (less than 0.00005% or 0.5 μg/ml) the tetramer is unstable, even at pH 5 and ionic strength 0.1, and dissociates to form lower polymers and/or monomer. This has clearly been shown by boundary analysis on Sephadex (*28*). Figure 2 shows the relationship between elution volume (or molecular weight) and protein concentration. Because of difficulties in assaying the enzyme (see below) the data are not sufficiently accurate to define the products of the dissociation of the tetramer, but it is clear that in most assays the enzyme must be present as a mixture of polymers of which the monomer could well be dominant.

b. Kinetics. The kinetic properties of sulfatase A are extremely complex. The original work (*30, 31*) was carried out with quite impure enzyme preparations so that its significance is doubtful, but homogeneous preparations of sulfatase A show the same anomalous kinetics although here

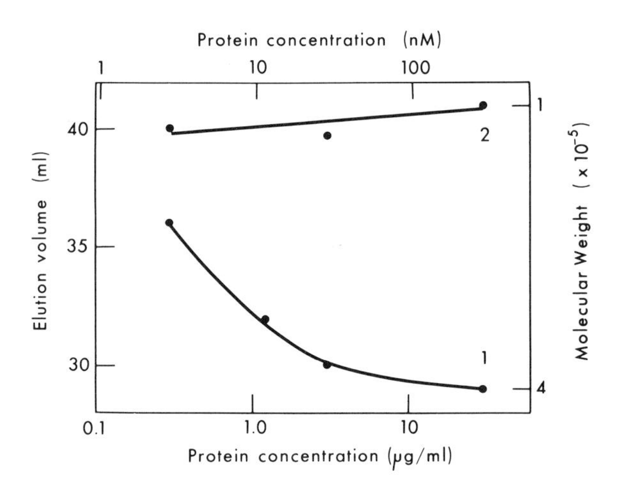

FIG. 2. The dissociation of the tetramer of sulfatase A (curve 1, pH 5) at low concentrations. Curve 2, monomer at pH 7.5. Temperature 25°, $I = 0.1$. Molar concentrations are in terms of the monomer.

30. A. B. Roy, *BJ* **55**, 653 (1953); *Experientia* **13**, 32 (1957).
31. S. O. Andersen, *Acta Chem. Scand.* **13**, 884 and 1671 (1959).

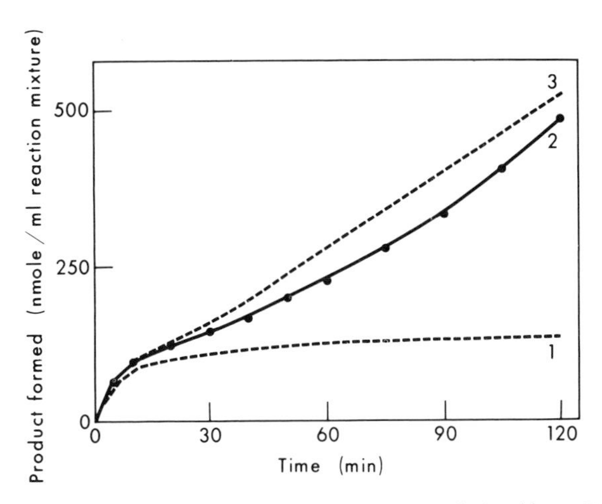

FIG. 3. Progress curves for the hydrolysis of nitrocatechol sulfate (5 mM) by sulfatase A (0.14 μg/ml) at pH 5 and 37°. Curve 1, pH stat assay (stirred); curve 2, spectrophotometric assay (not stirred); curve 3, pH stat assay in the presence of 0.01% bovine serum albumin (stirred).

the situation is further complicated by the surface denaturation which occurs when dilute solutions (< 5 μg/ml) of the enzyme are stirred, as during assays in the pH stat. This denaturation can be prevented by having 0.01% bovine serum albumin present in all assays. Examples of typical progress curves are shown in Fig. 3.

It was early suggested (*30*) that these anomalous kinetics could be caused by the polymerization of the enzyme to a more active form. This is incorrect, and it is likely that the monomer is the more active species. The anomalous kinetics do not, however, depend upon the occurrence of simple polymerization–depolymerization reactions of the type described above. In a reaction mixture of pH 5.0, ionic strength 0.1, containing enzyme at such a concentration that the tetramer is the stable species (30 μg/ml) the specific activity of enzyme added as monomer initially falls more rapidly than does that of enzyme added as tetramer (Fig. 4), but the subsequent course of the reaction is the same in both cases. This suggests that the change from monomer to tetramer is accompanied by a decrease in catalytic activity. Further, as is clear from Fig. 4, the change must be complete within about 4 min of mixing, confirming evidence from sedimentation studies (*28*) that equilibration between the polymeric forms of sulfatase A is rapidly attained.

A complete interpretation of the anomalous kinetics is not yet possible, but it is clear that during the enzymic reaction a catalytically inactive form of the enzyme is produced and that this can be subsequently reac-

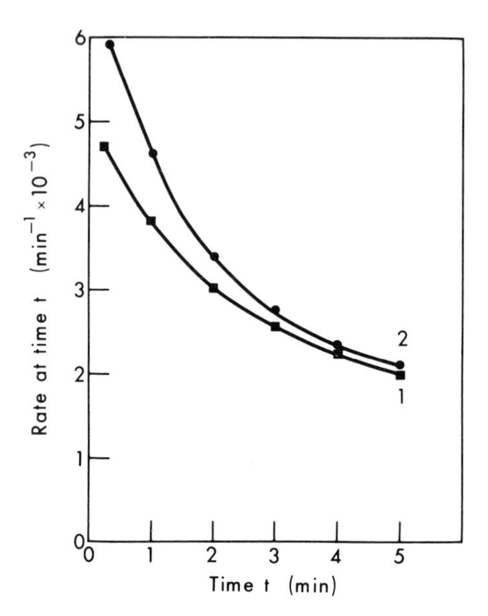

FIG. 4. Changes in rate during the early stages of sulfatase A reactions started by adding the tetramer (curve 1) or the monomer (curve 2) to give final concentrations of 30 μg/ml. Substrate, 4-nitrophenyl sulfate (30 mM); pH 5; $I = 0.1$; temperature 20°. The tetramer is the stable species under these conditions.

tivated by sulfate ions, one of the reaction products. Some evidence for this is shown in Fig. 5. It should be noted that the inactivation is dependent upon the formation of an enzyme–substrate complex because it is not prevented by the addition of sulfate at the beginning of the reaction although this does alter the shape of the progress curve (*30*). This is, of course, exactly the explanation given by Dodgson for the anomalous kinetics shown by a crude sulfatase A from human liver (*32*). It differs from that of Andersen (*31*) in that it ascribes no role to the second product of the sulfatase reaction, a phenol (4-nitrocatechol in most cases). Finally it should be pointed out that the substrate- and product-induced changes could be changes in the molecular weight of the enzyme but if they are, they are of a quite different type to those discussed in the previous section. A detailed kinetic study has now been made (*32a*).

Much further work is required to elucidate this situation, but it should be clear that in any studies of the kinetics of the reaction catalyzed by sulfatase A it is essential that the absolute concentration of the

32. H. Baum, K. S. Dodgson, and B. Spencer, *BJ* **69**, 567 (1958); H. Baum and K. S. Dodgson, *ibid.* p. 573.

32a. R. G. Nicholls and A. B. Roy, *BBA* **242**, 141 (1971).

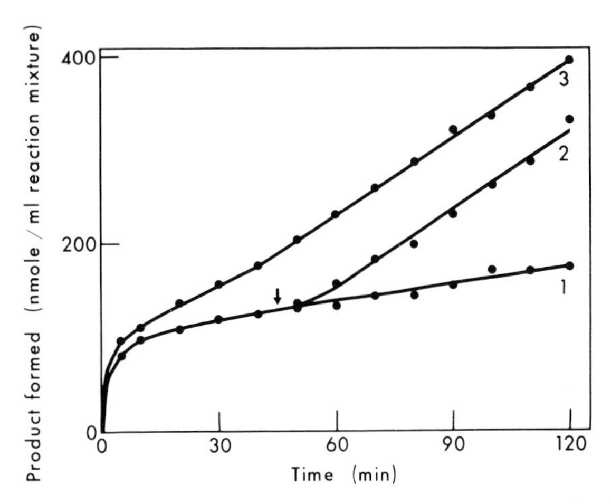

FIG. 5. The effect of sulfate ions on the hydrolysis of nitrocatechol sulfate (3 mM) by sulfatase A (0.1 μg/ml) in 0.25 M acetate, pH 6.0, temperature 37°. Curve 1, control; curve 2, sulfate added at arrow to give a concentration of 2 mM; and curve 3, sulfate (2 mM) present at beginning of reaction.

enzyme be known and, if possible, selected so that only one molecular species of the enzyme is initially present. It is also essential that initial velocities be measured. If such is not the case then the results will apply not to native sulfatase A but to a modified form thereof. In the following sections the data refer only to true initial velocities and therefore may differ somewhat from earlier estimates (*30*).

c. Catalytic Properties. Some data on the enzymic properties of sulfatase A are summarized in Table V and some information on its specificity in Table VI. Again the presence of a carbocyclic aromatic ring is not essential because heterocyclic "aryl" sulfates, the pyridyl sulfates, are hydrolyzed by sulfatase A (*33*). A number of substrates give similar values for V and this suggests that in such cases the rate-limiting step of reaction (2) may be the liberation of sulfate. With substrates which give lower values for V the rate limiting step must be one of the earlier reactions, perhaps the interconversion of the enzyme-substrate–enzyme-product complex because there is no obvious relationship between V and the structure of the phenol liberated. There are insufficient data to draw any conclusion regarding the relationship between structure and K_m except that the latter definitely decreases as the number of rings in the substrate increases. The enzyme, however, shows no steroid sulfatase activity (*33a*). The possibility of sulfatase A acting as a cerebroside sul-

33. A Jerfy and A. B. Roy, *Australian J. Chem.* **23,** 847 (1970).
33a. A. B. Roy, *Z. Physiol. Chem.* **333,** 166 (1963).

TABLE V
SOME KINETIC CONSTANTS FOR SULFATASES A AND B FROM OX LIVER

Kinetic constant	Sulfatase A[a] (1)	Sulfatase A[a] (2)	Sulfatase A[a] (3)	Sulfatase B[b]
K_m (mM), 2-Hydroxy-5-nitrophenyl sulfate	0.30 ± 0.01	0.41 ± 0.03	0.49	1.86 ± 00.9
K_m (mM), 4-Nitrophenyl sulfate	—	—	77	10.8 ± 1.0
K_i (mM), SO_4^{2-}	0.14 ± 0.004[c]	0.26 ± 0.02[c]	—	1.2, 4.6[d]
V[e]	11,000	22,600	16,400	—

[a] The three sets of values for sulfatase A were obtained in the pH stat under the following conditions: (1) pH 5.0, I = 0.1, 20°; (2) pH 5.0, I = 0.1, 37°; and (3) pH 5.6, 37°, ionic strength uncontrolled and set by the concentration of the substrate.

[b] With sulfatase B the values for 2-hydroxy-5-nitrophenyl sulfate were determined spectrophotometrically in 0.5 M acetate buffer, pH 5.4, and for 4-nitrophenyl sulfate in the pH stat at pH 6.6, I = 0.15 in KCl.

[c] Simple competitive inhibition.

[d] Noncompetitive inhibition: the values of K_i were calculated from the slopes and intercepts, respectively, of reciprocal plots.

[e] Expressed as moles of nitrocatechol sulfate hydrolyzed per minute per mole of enzyme.

TABLE VI
KINETIC CONSTANTS FOR THE SULFATASE A OF OX LIVER[a]

Substrate	K_m (mM)	V[b]
Phenyl sulfate	310	0.33
2-Hydroxyphenyl sulfate	38	2.7
2-Nitrophenyl sulfate	6.5	2.7
3-Nitrophenyl sulfate	77	3.0
4-Nitrophenyl sulfate	53	1.0
2-Hydroxy-5-nitrophenyl sulfate	0.47	3.0
2-Naphthyl sulfate	53	0.45
5,6,7,8-Tetrahydro-2-naphthyl sulfate	63	0.18
2-Phenanthryl sulfate	9.5	0.18
Pyridyl 3-sulfate	31	0.36
2-Nitropyridyl 3-sulfate	4.4	3.3
Pyridyl 4-sulfate	0.76	0.63

[a] Reaction velocities were measured in the pH stat at 37° with the ionic strength uncontrolled and governed by the substrate concentration.

[b] Values for V are given relative to 4-nitrophenyl sulfate.

fatase has already been discussed (see Chapter 1) and will not be considered here. The pH optimum is in the region of 5–5.5 but varies somewhat with changes in conditions (*8*, *30*).

Sulfate and sulfite ions are competitive inhibitors of sulfatase A with K_i values of approximately 0.15 mM and 2 μM, respectively. The fact that sulfate gives linear competitive inhibition of the sulfatase A reaction is consistent with the reaction sequence shown in reaction (2) if P_2 is sulfate (*34*). The predicted product-inhibition pattern of such a mechanism is linear competitive inhibition by the last-released product (sulfate) and a noncompetitive inhibition by the first-released product (a phenol). Unfortunately, the latter cannot be confirmed because 4-nitrocatechol does not significantly inhibit sulfatase A, even at a concentration of 10 mM.

The structural analogs of nitrocatechol sulfate, 2-hydroxy-5-nitrobenzoic acid and 2-hydroxy-5-nitrobenzene sulfonic acid, have no significant inhibitory effect. Likewise when 2-naphthyl sulfate is used as substrate, neither 2-naphthyl thiosulfate nor 2-naphthyl sulfamate significantly inhibits sulfatase A.

Certain "carbonyl" reagents are powerful inhibitors of sulfatase A although there is no evidence for the occurrence of carbonyl groups in the enzyme. The inhibition is dependent upon the presence of metal ions, especially Cu^{2+}, in the system and is prevented by, or subsequently reversed by, the addition of EDTA (*25*). It appears that these "carbonyl" reagents—hydroxylamine, hydrazine, *N*-phenylhydroxylamine and phenylhydrazine—act by reducing traces of Cu^{2+} in the reagents (or enzyme) to Cu^{+} which, like Ag^{+} (which inhibits by 85% at 10 μM), could be a powerful inhibitor of sulfatase A. Ascorbic acid, another powerful reducing agent, also inhibits sulfatase A in the presence of traces of metal ions but not if EDTA is present. Although the above mechanism seems the most likely one, it cannot be excluded that the inhibition is caused by the formation of inactive ternary complexes of enzyme-metal-reagent. This type of inhibition would also be prevented, or reversed, by EDTA.

d. Reactive Groups. It was suggested (*35*) that sulfatase A is an SH enzyme but this is unlikely unless the SH groups are very unreactive. Although sulfatase A is inactivated by treatment with *p*-chloromercuribenzoate at pH 5 it is not so inactivated at pH 7.5, nor is it inactivated by *N*-ethylmaleimide at either pH (*36*). Also, SH groups cannot be

34. W. W. Cleland, *BBA* **67,** 104 (1963).
35. A. B. Roy, *BJ* **59,** 8 (1955).
36. A. Jerfy and A. B. Roy, *BBA* **175,** 355 (1969).

detected in the enzyme using Ellman's reagent. The mercurial is therefore almost certainly reacting with some group other than an SH group, perhaps a histidyl residue which might also react with the inhibitory Ag^+ or Cu^+ ions.

The participation of a histidyl residue, directly or indirectly, in the reaction catalyzed by sulfatase A is also suggested by the fact that the enzyme is inactivated by treatment with acetic anhydride and that this modified enzyme reactivates spontaneously on standing at pH 6 (*36*). The reactivation has a half-time of 90 min, similar to that for the hydrolysis of *N*-acetylimidazole. The inactivation of sulfatase A by photooxidation in the presence of Rose Bengal is not inconsistent with the participation of a histidyl residue in the enzyme reaction, but this treatment causes quite drastic changes in the structure of the protein (*36*). Recent studies (*36a*) of the nonenzymic hydrolysis of 2,4(5)-imidazolylphenyl sulfate have increased the interest in a possible role for histidyl residues in sulfatase A. It has been shown that the hydrolysis of this ester proceeds by an intramolecular catalysis, probably through the imidazole moiety acting as a general acid–base catalyst; thus, the reaction could serve as a useful model for an arylsulfatase-catalyzed hydrolysis (*36a*).

Tyrosyl residues may also be involved, directly or indirectly, in the reaction because sulfatase A is irreversibly inhibited by treatment with *N*-acetylimidazole or with tetranitromethane (*36*). The latter has been shown to nitrate about eight tyrosyl residues in the protein without drastically altering its structure, as judged by its sedimentation in the ultracentrifuge. Although tyrosyl residues were undoubtedly nitrated, a simultaneous oxidation of histidyl residues cannot be excluded.

There is no evidence for the participation of amino (*36*) or tryptophanyl residues in the reaction catalyzed by sulfatase A. The position of the pH optimum (about 5) might suggest the participation of carboxyl groups, but this has not been shown although the treatment of sulfatase A with diethylamine in the presence of a water-soluble carbodiimide under forcing conditions does cause some inactivation of the enzyme. Sulfatase A is not inactivated by diisopropylphosphorofluoridate which suggests that the enzyme does not contain any reactive seryl residues.

2. *From Other Sources*

Enzymes similar to the sulfatase A of ox liver appear to be quite widely distributed in mammals (*37*), but few such enzymes have been

36a. S. J. Benkovic and L. K. Dunikoski, *Biochemistry* **9**, 1390 (1970).
37. A. B. Roy, *BJ* **68**, 519 (1958).

studied in detail. One has been obtained from ox brain in a form homogeneous to electrophoresis on acrylamide gel (*38*), and a considerable amount of information is available on its enzymic properties (*38, 39*) which are obviously very similar to those of the corresponding liver enzyme. Human liver also contains a sulfatase A quite comparable to that of ox liver (*32*). To judge by the chromatographic behavior of the crude enzyme, it has a molecular weight of about 100,000 at pH 7.5 and at pH 5 it forms a polymer which may be a tetramer or perhaps a trimer. Crude preparations of the sulfatase A from human liver show anomalous kinetics quite analogous to those of the ox enzyme. These have been investigated in considerable detail (*32*) and the conclusions thereby drawn seem to be directly applicable to the ox enzyme.

The livers of lower vertebrates may contain arylsulfatases of a rather different type (*40*). A sulfatase showing some resemblance to the sulfatase A of ox liver has been partially purified from the liver of the red kangaroo, *Megaleia rufa*. Although there are obvious analogies between this enzyme and the sulfatase A of ox or human liver, there are important differences. In particular, there is no suggestion that the enzyme (molecular weight 100,000; isoelectric point 5) polymerizes. Its kinetic behavior resembles, but is not identical with, that of the enzymes from eutherian mammals (*40a*).

B. Sulfatase B

1. *From Ox Liver*

Although sulfatase B has been obtained with a specific activity similar to that of sulfatase A, it has not been shown that the preparation consists of a single protein. The enzyme occurs in much smaller amounts than does sulfatase A and the situation is further complicated by the existence of at least two isoenzymes, sulfatases Bα and Bβ, which are separable by chromatography on CM-Sephadex (*41*). No other difference in the physical or enzymic properties of these isoenzymes has been found.

The physical properties of sulfatase B have been summarized in Table IV. It is clearly distinguished from sulfatase A both by its higher

38. W. Bleszynski and L. M. Dzialoszynski, *BJ* **97,** 360 (1965); W. Bleszynski, *Enzymologia* **32,** 169 (1967); W. Bleszynski, A. Leznicki, and J. Lewosz, *ibid.* **37,** 314 (1969).
39. W. Bleszynski and A. Leznicki, *Enzymologia* **33,** 373 (1967).
40. A. B. Roy, *Australian J. Exptl. Biol. Med. Sci.* **41,** 331 (1963).
40a. A. B. Roy, *BBA* **227,** 129 (1971).
41. E. Allen and A. B. Roy, *BBA* **168,** 243 (1968).

isoelectric point and its lower molecular weight although it tends to form aggregates at low ionic strengths near its isoelectric point (*41*). Because samples of proven homogeneity have not been available the amino acid composition of sulfatase B is not known, but it is likely to be rather different from that of sulfatase A because the latter has a specific absorbance only half that of sulfatase B (*28, 41*).

The enzyme has been little investigated, and the scanty information available on its kinetic properties is summarized in Table V. It must be stressed that sulfatase B shows none of the anomalous kinetics characteristic of sulfatase A. The most interesting feature of the kinetics is the strong activation by chloride, and other monovalent anions, of the hydrolysis of 4-nitrophenyl sulfate (*5, 41*). In 0.15 M NaCl the activity is increased some sixfold over that in the absence of chloride. On the other hand, the hydrolysis of nitrocatechol sulfate is inhibited by chloride (*5*). Too few substrates have been investigated to allow any interpretation of this activation, which is caused by a change in V and not in K_m (*5*). However, it is found not only with 4-nitrophenyl sulfate but also with 1-naphthyl sulfate, 4-nitro-1-naphthyl sulfate and 6-benzoyl-2-naphthyl sulfate, whereas only the hydrolysis of nitrocatechol sulfate is known to be inhibited by chloride.

The inhibition of sulfatase B by sulfate is noncompetitive (*42*), in sharp contrast to the competitive inhibition of sulfatase A by sulfate (*30*). More detailed studies using purified preparations of sulfatase B showed that the inhibition was linear noncompetitive with different values of K_i calculated from the slopes and intercepts of reciprocal plots (*42a*). This behavior would be consistent with sulfate being P_1, but not P_2, in reaction (2). Like sulfatase A, sulfatase B is not significantly inhibited by 4-nitrocatechol so that further information on the pattern of product inhibition is not available. Nevertheless, the fact that the inhibition of sulfatase B by sulfate is noncompetitive shows that the mechanism of this reaction must be very different from that of sulfatase A.

2. *From Other Sources*

Again sulfatases comparable to the sulfatase B of ox liver seem to be widespread in mammals (*37*) but few have been studied. Three isoenzymes of sulfatase B have been separated from human brain. Two have been obtained homogeneous on acrylamide gel and their properties investigated (*38, 39*). The kinetics of sulfatase B of human liver have been

42. A. B. Roy, *BJ* **57**, 465 (1954).
42a. A. B. Roy, *BBA* **198**, 365 (1970).

studied (*43*), but the enzyme has not been significantly purified. An interesting feature it shares with the sulfatase B of ox liver is the considerable influence of acetate on the substrate concentration–activity curves with nitrocatechol sulfate.

The liver of the red kangaroo contains small amounts of a sulfatase B which has many properties in common with the ox enzyme. In particular it shows the same type of complex noncompetitive inhibition by sulfate (*40a*).

C. Other Type II Arylsulfatases

Type II arylsulfatases occur in invertebrate tissues and secretions, but none of these enzymes has been purified so that there is no information on their physical properties. The best known of these enzymes is that from the digestive juices of *Helix pomatia* (*44*). This, to judge by its behavior on paper electrophoresis, must have an isoelectric point close to pH 6 so that it is more akin to the sulfatase of kangaroo liver than to those of the higher mammals. It also differs from sulfatase A in showing normal kinetic behavior. The rate of hydrolysis of 4-nitrophenyl sulfate is not increased by chloride so that this enzyme shows no affinity with sulfatase B.

On the other hand, the sole example of a type II arylsulfatase from microorganisms, that from *Proteus vulgaris*, does resemble sulfatase B because the hydrolysis of 4-nitrophenyl sulfate by this enzyme is activated by chloride (*26*).

IV. General Considerations

The validity of the subdivision of the arylsulfatases into the type I and type II enzymes requires consideration. If the definition be that type I arylsulfatases are not inhibited by sulfate, whereas the type II enzymes are inhibited by sulfate, then there would be a clear-cut distinction between the two groups. It must be stressed, however, that this may simply be a reflection of the paucity of data. If, on the other hand, the original (*2*) multiple definition be adopted (see Section I) then difficulties might arise. For example, the arylsulfatase of *Aerobacter aerogenes* is clearly a type I enzyme on the basis of its response to sulfate yet it is inhibited

43. K. S. Dodgson and C. H. Wynn, *BJ* **68**, 387 (1958).
44. K. S. Dodgson and G. M. Powell, *BJ* **73**, 666 and 672 (1959); P. Jarrige, *Bull. Soc. Chim. Biol.* **45**, 761 (1963).

by phosphate. In this case, therefore, the multiple criteria lead to an inconsistency.

The significance of this differing response to sulfate is not known, but, taken at its face value, it would indicate different reaction mechanisms for the two types of enzyme.

The subdivision of the type II arylsulfatases into sulfatases A, B, and others may be less justifiable. The distinction between sulfatases A and B is quite clear when only the enzymes from the higher mammals are considered. Sulfatase A has an acid isoelectric point, shows anomalous kinetics, and has a K_m for nitrocatechol sulfate considerably smaller than that for any other substrate so far studied; sulfatase B has an alkaline isoelectric point, shows normal kinetics, has similar K_m values for nitrocatechol sulfate and 4-nitrophenyl sulfate, and hydrolyzes the latter substrate at a significant rate only in the presence of chloride. Difficulties arise when type II arylsulfatases from other sources are considered, and it could be that sulfatases A and B represent only the extremes of a range of such enzymes. For example, the isoelectric points range from 3.4 for ox liver sulfatase A through 5 and 6 for the enzymes from the kangaroo and *Helix pomatia,* respectively, to 8.3 for ox liver sulfatase B. On the other hand, if the different pattern of sulfate inhibition shown by sulfatases A and B of ox and kangaroo liver is a general phenomenon then there may indeed be a fundamental difference between these types of enzyme, despite claims that they are interconvertible (*44a*).

The use of nitrocatechol sulfate as a substrate can introduce difficulties, which are usually overlooked, in attempting to compare the activities and K_m values of different enzymes. These arise through the presence in nitrocatechol sulfate of a hydroxyl group with a pK of 6.4. When sulfatase A is assayed at pH 5 this hydroxyl group is almost completely un-ionized; when the sulfatase of *Aspergillus oryzae* is assayed at pH 6.2 it is about 50% ionized; and when the sulfatase of *Alcaligenes metalcaligenes* is assayed at pH 8.8 it is fully ionized. For practical purposes, therefore, different substrates are used with sulfatase A and with the sulfatase of *A. metalcaligenes*, that of the former being singly charged, that of the latter being doubly charged. If the particular ionic species attacked by the different enzymes were known, the appropriate corrections could be made to the apparent values of K_m. For example, the values of 0.41 and 0.49 mM nitrocatechol sulfate given in Table V for the K_m of sulfatase A at pH 5.0 and 5.6, respectively, suggest that it is the singly ionized species which combines with the enzyme. On this assumption the values of K_m become 0.39 and 0.42 mM, values certainly identical within experimental error.

44a. A. Goldstone, P. Konecny, and H. Koenig, *FEBS Letters* **13**, 68 (1971).

An analogous situation must arise with the reaction product, 4-nitrocatechol [pK 6.6 and 11.3 (*45*)], and with the inhibitors sulfite and phosphate (pK_2 sulfurous acid and phosphoric acid, 7.2). Over the above pH range the dominant ionic species of the latter inhibitors change from singly to doubly charged anions, a change which must be reflected in their inhibitory effects. This could be taken as a further argument against using the response to phosphate to distinguish between type I and type II arylsulfatases. The sulfate ion is doubly charged over the whole pH range of interest in enzymic studies so that no such problem exists with its use.

The insensitivity of sulfatases A and B to product inhibition by phenols in general and 4-nitrocatechol in particular may also be a consequence of changes in ionization. If these enzymes catalyze a nucleophilic displacement on sulfur [reaction (3)] then the phenol (which may re-

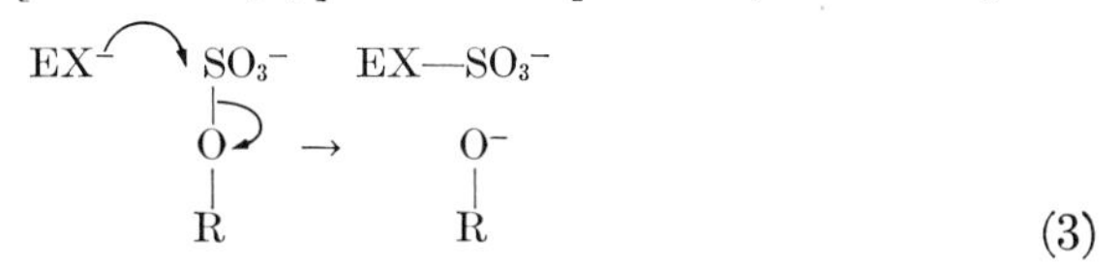

main enzyme-bound) must be produced in the anionic form which, at pH 5–5.5, will be protonated to give the un-ionized phenol. If this be the case then these sulfatases would not be expected to be inhibited by 4-nitrocatechol because this cannot be added, at least at the pH normally used, in the ionic form of the reaction product.

45. S. R. Cooper and V. J. Tulane, *Ind. Eng. Chem., Anal. Ed.* **8**, 210 (1936).

3

Carboxylic Ester Hydrolases

KLAUS KRISCH

I. Introduction

Esterases, in the stricter sense of the word, catalyze the hydrolysis of a large number of uncharged carboxylic esters. In contrast to the

$$R_1{-}C({=}O){-}O{-}R_2 + H_2O \longrightarrow R_1{-}C({=}O){-}OH + R_2{-}OH$$

lipases, their action is generally restricted to short chain fatty acid esters (for further differentiation see the chapter by Desnuelle, Volume VII). In 1953, Aldridge (*1*) proposed a frequently used classification of esterases based on the behavior toward organophosphorus compounds, such as diethyl *p*-nitrophenyl phosphate (=paraoxon; E 600). Accordingly, A-esterases (EC 3.1.1.2) are not inhibited by organophosphorus compounds but hydrolyze them as substrates. They also split aromatic esters such as phenyl acetate and, therefore, have been designated as arylesterases. In contrast, B-esterases (EC 3.1.1.1) are inhibited stoichiometrically by organophosphates without hydrolyzing them. These enzymes have been formerly known as "ali-esterases" or, because of their wide specificity, as unspecific esterases. It is this group of esterases with which this review will be mainly concerned. In pig kidney a third type of esterase has been described. This enzyme is neither inhibited by organophosphorus compounds nor does it hydrolyze them and, therefore, has been named C-esterase (*2*). It should be emphasized and will become evident from this chapter, however, that none of the present classifications of esterases is completely satisfactory and unambiguous.

Because of the participation of a serine residue at the active site the B-esterases belong to the group of serine hydrolases. These also comprise cholinesterase, acetylcholinesterase, and several important endopeptidases, e.g., chymotrypsin, trypsin, elastase, thrombin, and subtilisin, which also show esterase activity toward certain substrates and will be covered elsewhere. For a survey of earlier literature on esterases the reader is referred to previous reviews (*3–7*).

Esterases are widely distributed in vertebrate tissues, blood serum (*1, 8–14,* and others), insects (*15–18*), plants (*19*), citrus fruits (*20, 21*), mycobacteria (*22*), and fungi (*23*). In mammals, as in the pig, the highest activities are found in liver, kidney, duodenum, and brain (*24*). A fluoride-sensitive, tributyrin hydrolyzing esterase has been isolated from rat adipose tissue (*25*). In male animals testis and epididymides are also rich in esterase (*26, 27*).

1. W. N. Aldridge, *BJ* **53,** 110 and 117 (1953).
2. F. Bergmann, R. Segal, and S. Rimon, *BJ* **67,** 481 (1957).
3. R. Ammon, *in* "Handbuch der Enzymologie" (F. F. Nord and R. Weidenhagen, eds.), Vol. 1, p. 350. Akad. Verlagsges., Leipzig, 1940.
4. W. N. Aldridge, *in* "Biochemists' Handbook" (C. Long, ed.), p. 273. Spon, London, 1961.
5. R. Ammon and M. Jaarma, "The Enzymes," 1st ed., Vol. 1, p. 390, 1950.
6. D. K. Myers, "The Enzymes," 2nd ed., Vol. 4, Part A, p. 475, 1960.
7. B. H. J. Hofstee, "The Enzymes," 2nd ed., Vol. 4, Part A, p. 485, 1960.
8. K. B. Augustinsson, *Acta Chem. Scand.* **13,** 571, 1081, and 1091 (1959).

As is generally agreed, the esterase activity in liver (*28–33*) and kidney (*24, 32*) is mainly associated with the microsomal fraction. In addition a lysosomal arylesterase with an acid pH optimum has been described. This enzyme, however, contributes only 1.2% to the total esterase activity of a rat liver homogenate (*34*). In rat brain, nonspecific esterases are preferentially localized in the 100,000 $\times$ *g* sediment. Twelve percent of the activity, however, have been found in the crude mitochondrial fraction, where most of it is bound to myelin (*35*).

II. Molecular Properties of Carboxylesterases (EC 3.1.1.1)

A. Multiple Molecular Forms

Following separation of aqueous extracts from various mammalian organs (especially from liver, kidney, and brain) by electrophoresis in starch

9. S. H. Lawrence, P. J. Melnick, and H. E. Weimer, *Proc. Soc. Exptl. Biol. Med.* **105**, 572 (1960).
10. K. B. Augustinsson, *Ann. N. Y. Acad. Sci.* **94**, 844 (1961).
11. R. L. Hunter and D. S. Strachan, *Ann. N. Y. Acad. Sci.* **94**, 861 (1961).
12. A. R. Hess, R. W. Angel, K. D. Barron, and J. Bernsohn, *Clin. Chim. Acta* **8**, 656 (1963).
13. F. Margolis and P. Feigelson, *JBC* **238**, 2620 (1963).
14. W. Pilz, E. Stelzell, and I. Johann, *Enzymol. Biol. Clin.* **9**, 97 (1968).
15. K. A. Lord and C. Potter, *Nature* **172**, 679 (1953).
16. J. E. Casida, *BJ* **60**, 487 (1955).
17. G. W. Beckendorf and W. P. Stephen, *BBA* **201**, 101 (1970).
18. W. P. Stephen and I. H. Cheldelin, *BBA* **201**, 109 (1970).
19. T. P. Singer, *JBC* **174**, 11 (1948).
20. E. F. Jansen, R. Jang, and L. R. MacDonald, *ABB* **15**, 415 (1947).
21. R. L. Blakeley, J. de Jersey, E. C. Webb, and B. Zerner, *BBA* **139**, 208 (1967).
22. D. K. Myers, J. W. Tol, and M. H. T. de Jonge, *BJ* **65**, 223 (1957).
23. R. J. W. Byrde and A. H. Fielding, *BJ* **61**, 337 (1955).
24. H. C. Benöhr, W. Franz, and K. Krisch, *Arch. Pharmakol. Exptl. Pathol.* **255**, 163 (1966).
25. A. N. Glazer, *JBC* **243**, 3693 (1968).
26. D. K. Myers and S. P. Simons, *BJ* **53**, xvii (1953).
27. R. S. Holmes and C. J. Masters, *BBA* **132**, 379 (1967).
28. A. B. Novikoff, E. Podber, J. Ryan, and E. Noe, *J. Histochem. Cytochem.* **1**, 27 (1953).
29. E. Underhay, S. J. Holt, H. Beaufay, and C. de Duve, *J. Biophys. Biochem. Cytol.* **2**, 635 (1956).
30. K. Krisch, *Biochem. Z.* **337**, 531 and 546 (1963).
31. K. Hayase and A. L. Tappel, *JBC* **244**, 2269 (1969).
32. W. S. Schwark and D. J. Ecobichon, *Can. J. Physiol. Pharmacol.* **46**, 207 (1968).

gel, a multiplicity of esterase bands has been detected (*27*, *32*, *36–46*, and others). Thus in rat liver as many as 13 and in rat kidney as many as 11 electrophoretically different esterases have been described (*37*). There are also numerous reports about multiple forms of nonspecific esterases in plasma or serum of many species (*8–14*, and others). Further information on esterase isozymes are given in a recent review (*47*). Esterase activities in starch or polyacrylamide gels may be easily detected by substrate staining techniques as adopted from histochemistry (*48*, *49*). In most cases unspecific substrates, such as α-naphthylacetate, are used. Subsequently, the liberated α-naphthol is coupled with a diazonium salt yielding an insoluble azo dye. Interestingly, immunoprecipitates formed by reaction of esterases with homologous antisera are still enzymically active and therefore can be sensitively stained in the same way (*50*).

In most cases the molecular basis of the multiplicity of tissue esterase bands is still unknown. Highly purified preparations of bovine and pig liver esterase at lower concentrations may show one or more weak additional bands which probably result from dissociation into subunits (*33*, *51–53*). Ecobichon (*54*) has recently shown by gel filtration that all esterase isozymes of a bovine liver extract possess the same molecular weight of 53,000–55,000.

---

33. H. C. Benöhr and K. Krisch, *Z. Physiol. Chem.* **348,** 1102 and 1115 (1967).
34. S. Shibko and A. L. Tappel, *ABB* **106,** 259 (1964).
35. A. H. Koeppen, K. D. Barron, and J. Bernsohn, *BBA* **183,** 253 (1969).
36. S. I. Read, E. J. Middleton, and W. P. McKinley, *Can. J. Biochem.* **44,** 809 (1966).
37. W. S. Schwark and D. J. Ecobichon, *Can. J. Biochem.* **45,** 451 (1967).
38. C. L. Markert and R. L. Hunter, *J. Histochem. Cytochem.* **7,** 42 (1959).
39. D. J. Ecobichon and W. Kalow, *Can. J. Biochem. Physiol.* **39,** 1329 (1961).
40. D. J. Ecobichon and W. Kalow, *Biochem. Pharmacol.* **11,** 573 (1962).
41. D. J. Ecobichon and W. Kalow, *Can. J. Biochem. Physiol.* **42,** 277 (1964).
42. D. J. Ecobichon, *Can. J. Biochem. Physiol.* **43,** 595 (1965).
43. D. J. Ecobichon, *Can. J. Biochem. Physiol.* **44,** 225 (1966).
44. R. S. Holmes and C. J. Masters, *BBA* **159,** 81 (1969).
45. S. F. Estrugo, J. Bernsohn, K. D. Barron, and A. Hers, *BBA* **171,** 265 (1969).
46. N. Kingsbury and C. J. Masters, *BBA* **200,** 58 (1970).
47. A. L. Latner and A. W. Skillen, "Isoenzymes in Biology and Medicine," p. 66. Academic Press, New York, 1968.
48. R. L. Hunter and C. L. Markert, *Science* **125,** 1294 (1957).
49. J. Uriel, *Ann. Inst. Pasteur* **101,** 104 (1961).
50. P. Hain and K. Krisch, *Z. Klin. Chem. Klin. Biochem.* **6,** 313 (1968).
51. D. L. Barker and W. P. Jencks, *Biochemistry* **8,** 3879 and 3890 (1969).
52. D. J. Horgan, J. K. Stoops, E. C. Webb, and B. Zerner, *Biochemistry* **8,** 2000 (1969).
53. M. T. C. Runnegar, E. C. Webb, and B. Zerner, *Biochemistry* **8,** 2018 (1969).
54. D. J. Ecobichon, *Can. J. Biochem.* **47,** 799 (1969).

B. Enzyme Preparations and Criteria of Purity

Since the last edition of "The Enzymes" considerable progress has been made in the isolation and characterization of carboxylesterases (EC 3.1.1.1), especially of those from mammalian liver. Although the ubiquitous occurrence of unspecific esterases has been known for a long time and several pioneers of enzymology have early studied them (*3*), well-defined enzyme preparations of high purity have not been obtained until the beginning of the last decade. This has prevented for a long time the elucidation of the molecular and physical properties of this enzyme group. Among other reasons it is the introduction of column chromatographic techniques which has led to the successful separation of contaminating hemoproteins.

In the years from 1950 to 1960 the best, though still not homogeneous, preparations have been those of Connors *et al.* (*55*) and of Burch (*56*) from horse liver. Today pig liver esterase can be considered the best-defined preparation of this enzyme. Starting from an acetone powder and using a DEAE-cellulose column, the procedure of Adler and Kistiakowsky (*57*) in 1961 first led to a preparation of high purity with an estimated molecular weight of 150,000–200,000.

In 1963, the author (*30*) isolated an acetanilide-hydrolyzing amidase from pig liver microsomes which then turned out to be identical with pig liver esterase. After extraction of the microsomal fraction with glycerol the enzyme was purified by fractionation with ammonium sulfate, an acid precipitation step at pH 4.2, and column chromatography on DEAE-Sephadex A-50. If 2–3 preparative ultracentrifuges for the large-scale isolation of microsomes are available, the procedure can be easily scaled up. As a typical example, 100–200 mg highly purified enzyme are obtained from about 1500 g pig liver within 2–3 weeks. The final preparation is homogeneous according to the following criteria: electrophoresis on paper (*30*), starch gel (*58*), and polyacrylamide (*59*), ultracentrifugation (*30*), immunodiffusion and immunoelectrophoresis (*50*). During the final ammonium sulfate precipitation the preparation shows a typical silky luster. Microscopic examination of the sediment, however, proved to be somewhat disappointing insofar as the platelet- and needle-like crystals to be seen are extremely thin and fragile and, in addition,

55. W. M. Connors, A. Pihl, A. L. Dounce, and E. Stotz, *JBC* **184,** 29 (1950).
56. J. Burch, *BJ* **58,** 415 (1954).
57. A. J. Adler and G. B. Kistiakowsky, *JBC* **236,** 3240 (1961).
58. W. Franz and K. Krisch, *Z. Physiol. Chem.* **349,** 575 (1968).
59. E. Pahlich, K. Krisch, and K. Borner, *Z. Physiol. Chem.* **350,** 173 (1969).

amorphous material has always been present (*30*). The same difficulties were also reported by Barker and Jencks (*51*) when they attempted to crystallize the Adler-Kistiakowsky enzyme (*57*). So far, the preparation has also resisted all attempts to recrystallize it. It is, therefore, a matter of definition whether, at present, pig liver esterase is considered as a crystalline enzyme or not. This purification procedure (with some minor modifications such as omission of the pH 4.2 step and an additional gel filtration on Sephadex G-200) was later also successfully employed in the isolation of similar esterases from pig kidney (*58*) and beef liver (*33*).

In 1966, Webb and his associates (*52, 60*) reported another large-scale purification procedure of pig liver esterase starting from a chloroform–acetone powder. Their final preparation looked microscopically quite similar to that of the author's (*30*) and was designated as crystalline (*52*). Starch gel electrophoresis showed only one single protein band, whereas in disc electrophoresis up to four faster moving minor components were detected by substrate staining. In ultracentrifugation experiments the purity of this preparation was estimated to be 88% (*61*). In addition, esterases have been purified by similar procedures from the liver of rats (*31*), beef (*33, 62*), chicken (*63, 64*), and sheep (*64*).

C. Molecular Weight

During the past decade several papers on the molecular weight of liver carboxylesterases have appeared. The available data, including sedimentation and diffusion coefficients and partial specific volumina, are presented in Table I. They are based on ultracentrifugation studies, either by sedimentation by diffusion, and approach to sedimentation equilibrium (*65*), or by equilibrium sedimentation. As can be seen from Table I the values of pig liver esterase reported from various laboratories are in good agreement, all being in the range of 162,000–168,000. It should be mentioned, however, that Kibardin (*66*) has found molecular weights of 180,000 at pH 7.2 and 45,000 at pH 8.0. The partial specific volumes,

60. D. J. Horgan, E. C. Webb, and B. Zerner, *BBRC* **23,** 18 and 23 (1966).
61. D. J. Horgan, J. R. Dunstone, J. K. Stoops, E. C. Webb, and B. Zerner, *Biochemistry* **8,** 2006 (1969).
62. M. T. C. Runnegar, K. Scott, E. C. Webb, and B. Zerner, *Biochemistry* **8,** 2013 (1969).
63. G. I. Drummond and J. R. Stern, *JBC* **236,** 2886 (1961).
64. R. C. Augusteyn, J. de Jersey, E. C. Webb, and B. Zerner, *BBA* **171,** 128 (1969).
65. W. J. Archibald, *J. Phys. & Colloid Chem.* **51,** 1204 (1947).
66. S. A. Kibardin, *Biokhimiya* **27,** 82 (1962).

TABLE I
MOLECULAR WEIGHTS AND EQUIVALENT WEIGHTS OF CARBOXYLESTERASES

Source	$s^{\circ}_{20,w}$ (Svedberg units)	$D^{\circ}_{20,w} \times 10^7$ (cm^2 sec^{-1})	$\bar{V}$ (ml g^{-1})	Molecular wt	Equivalent wt	Inhibitor	References
Horse liver	—	—	—	—	96,000	[^{32}P]-DFP	(*68*)
Pig liver	8.2	4.6	0.738	166,000	86,000	Diethyl *p*-nitrophenyl phosphate	(*67, 69, 70*)
Pig liver	—[a]	4.3[b]	0.733	163,000	78,000, 79,000	*p*-Nitrophenyl dimethyl carbamate, [^{32}P]-DFP, diethyl *p*-nitrophenyl phosphate	(*60, 61, 64*)
Pig liver	7.36[b]	4.12[b] 4.3[c]	0.740	168,000	78,200		(*51*)
Ox liver	—[a]	—	0.742	168,000	78,000–83,000	Diethyl *p*-nitrophenyl phosphate	(*33*)
Ox liver	—	—		~150,000	~68,000, ~70,000	*p*-Nitrophenyl dimethyl carbamate, diethyl *p*-nitrophenyl phosphate, [^{32}P]-DFP	(*53, 62, 64*)
Pig kidney	8.25	—	0.737	162,000	80,100	Diethyl *p*-nitrophenyl phosphate	(*58*)

[a] When the Archibald (*65*) method has been used, no values for s and D are given, since thereby the ratio s/D is directly obtained.
[b] Not extrapolated to zero protein concentration.
[c] Determined by immunodiffusion (*70a*).

calculated from the amino acid composition (*51, 67*) or by direct determination, (*60*) are also similar. Thus, with regard to the molecular properties, the pig liver esterases isolated by different procedures appear to be closely related. On the other hand, some differences in the immunological and kinetic behavior between the "Australian" enzyme (*52, 60*) and the Adler-Kistiakowsky preparation (*57*) have been observed (*51*).

Within the limits of experimental error, esterases of different origin, such as those from bovine liver (*33*) and pig kidney (*58*), seem to have the same molecular weight. The Webb group, however, found somewhat lower molecular and equivalent weights for their ox liver preparation (*53, 62*). Kingsbury and Masters (*46*) have recently determined the molecular weights of electrophoretic variants of numerous vertebrate carboxylesterases by disc electrophoresis in polyacrylamide gels of different composition.

D. Equivalent Weight

The equivalent weight of B-type esterases can be elegantly determined on the basis of their stoichiometric reaction with organophosphorus inhibitors or carbamates. The corresponding values and the inhibitors used for the titration of the enzymes are also listed in Table I. In 1949, Boursnell and Webb (*68*) gave the first estimate of the equivalent weight of horse liver esterase using [^{32}P]-DFP. They found that 1 g-atom [^{32}P] combined with 96,000 g enzyme. This figure is obviously too high owing to the inhomogeneity of their preparation. Later values of the order of 80,000 have been reported. Occasionally a tendency toward somewhat lower values has been observed (*71*). It is evident from Table I that this figure corresponds closely to one-half the molecular weight. This, of course, means that pig liver esterase contains two active sites as was found simultaneously and independently in 1966 by the author (*70*) and by Horgan *et al.* (*60, 61*). The occurrence of two active sites on the pig enzyme was confirmed later also by using bis(*p*-nitrophenyl) phosphate, a new and more specific esterase inhibitor (*72*). Analogously, the equivalent weights of pig kidney (*58*) and ox liver esterase (*33*) were found to be about one-half the corresponding molecular weights.

67. B. Klapp, K. Krisch, and K. Borner, *Z. Physiol. Chem.* **351,** 81 (1970).
68. J. H. Boursnell and E. C. Webb, *Nature* **164,** 875 (1949).
69. W. Boguth, K. Krisch, and H. Niemann, *Biochem. Z.* **341,** 149 (1965).
70. K. Krisch, *BBA* **122,** 265 (1966).
70a. A. C. Allison and J. H. Humphry, *Nature* **183,** 1590 (1959).
71. K. Krisch, E. Heymann, and W. Junge, unpublished observations (1970).
72. E. Heymann and K. Krisch, *Z. Physiol. Chem.* **348,** 609 (1967).

E. Subunit Structure

The assumption of two active sites is generally consistent with the reported subunit composition. Thus, dissociation of pig and ox liver esterases into active half-molecules at high dilutions has been observed independently in several laboratories by gel chromatography (*33, 51, 53, 73*). In a detailed study, Barker and Jencks (*51*) showed that pig liver esterase [preparation according to Adler and Kistiakowsky (*57*)] at pH 4.5 reversibly dissociates to active half-molecules with a molecular weight of 75,000–85,000. As shown by gel filtration on Sephadex G 100 dissociation to half-molecules also occurs at neutral pH in extremely dilute solutions or in the presence of salts (0.5 *M* NaCl; 0.5 *M* LiBr). Below pH 4 the enzyme undergoes irreversible denaturation to inactive half-molecules of altered shape. In 6 *M* guanidine–HCl the uncorrected molecular weight was found to be 53,500. This was interpreted as an indication of a further dissociation into quarter molecules. It should be mentioned, however, that the assumption of four polypeptide chains with a molecular weight of 42,000 (*51, 69*) still needs further experimental confirmation. Horgan *et al.* (*61*) obtained no evidence for a dissociation into species of a molecular weight of about 40,000 in the presence of 8 *M* urea.

The subunit composition of the beef liver enzyme is still the subject of some controversy. Benöhr and Krisch (*33*) have reported the separation of two enzymically active fractions of ox liver esterase by column chromatography on DEAE-Sephadex A-50, which have been assumed, on the basis of gel filtration experiments, to be the monomeric and dimeric form of the enzyme. In contrast, Runnegar *et al.* (*53*), while confirming the reversible dissociation into an active monomeric form, suggested that the two peaks obtained by DEAE-Sephadex chromatography (*33*) represent mixtures of three electrophoretic variants with the same molecular weights.

There is also some uncertainty with regard to the number of active sites on ox liver carboxylesterase. Thus, in addition to the reported equivalent weight of highly purified preparations of the order of 70,000 (*53, 62, 64*), titration of three electrophoretic variants of a less pure sample with *o*-nitrophenyldimethyl carbamate gave equivalent weights of 54,000–59,000 only (*62*). This observation is difficult to understand because with increasing purity of the enzyme the equivalent weights should be expected to decrease, whereas Runnegar *et al.* (*62*) pointed out "that the titration is overestimating the active enzyme in the system." These lower equivalent weights, however, would be consistent with the

73. D. Barker and W. P. Jencks, *Federation Proc.* **26**, 452 (1967).

observation of Ecobichon (*54*) that all esterases of a crude extract of bovine liver have an apparent molecular weight of 53,000–55,000 as determined by gel filtration. This author, therefore, suggested that during the purification of ox liver esterase an association occurs and that the associated form with a molecular weight of 168,000 (*33, 67*) consists possibly of three subunits.

Further support for the assumption of a trimer came from a recent reinvestigation of the subunit weights of carboxylesterases from pig liver, pig kidney, and ox liver (*73a*). By disc electrophoresis in the presence of sodium dodecylsulfate, by gel filtration in 6 *M* guanidine, and by quantitative determination of the N-terminal groups values of about 60,000 were obtained for all three preparations. In the esterases from pig liver and pig kidney glycine was found as N-terminal group, whereas leucine was N-terminal in ox liver esterase. No other N-terminal dinitrophenylated amino acids were detected. Thus, the subunits of each of the three esterases seem to be identical with regard to their N-terminal groups. They also possess the same chain weights. This, of course, does not exclude the possibility that there might exist differences in the amino acid sequence within the peptide chains. If one assumes the associated form of the mentioned carboxylesterases to be a trimer, this is obviously difficult to reconcile with the reported number of two active sites (*60, 61, 70*). A possible explanation would be that either one of the three subunits does not bear an active site or that the reported equivalent weights obtained in two laboratories by titration with organophosphorus compounds for some unknown reason may be considerably too high (80,000 instead of 55,000–60,000) (*71*). Obviously, this unsatisfactory present situation requires further experimental work for clarification.

F. Amino Acid Composition

The amino acid compositions of highly purified carboxylesterase preparations (*51, 67*) are given in Table II. The analysis of pig liver esterase by Barker and Jencks (*51*) does not take into account the carbohydrate content of the enzyme. This introduces, however, only a slight error of 2–3%. The agreement of their data with the results from the author's laboratory is generally satisfactory. Larger deviations, however, have been found in the proline (14%), methionine (31%), and tryptophan (100%) content. The amino acid analyses are further evidence for a structural relationship of esterases from various sources. Klapp *et al.* (*67*) have made fingerprint studies of tryptic digests from

73a. E. Heymann, W. Junge, and K. Krisch, *FEBS Letters* **12,** 189 (1971).

TABLE II
AMINO ACID COMPOSITION OF SOME ESTERASES[a]

Amino acid	Pig liver esterase [Ref. (*51*)]	Pig liver esterase [Ref. (*67*)]	Pig kidney esterase[b] [Ref. (*67*)]	Ox liver esterase [Ref. (*67*)]
Asp + Asn	129	128	126	129
Thr	76	74	72	67
Ser	80	78	85	99
Glu + Gln	150	155	144	127
Pro	93	107	98	103
Gly	129	116	126	101
Ala	117	110	107	115
Val	110	117	111	109
½-Cys–Cys + Cys	14	17	10	11
Met	36	25	31	26
Ile	52	49	53	58
Leu	147	143	139	152
Tyr	40	35	37	34
Phe	78	71	69	71
Lys	98	92	91	89
His	31	35	34	40
Arg	49	45	49	47
Trp	60	30	30	53

[a] Residues per molecule (nearest integer) based on a molecular weight of 168,000 (*51*) or 166,000 (*67*), respectively.

[b] By addition of the amino acid residues of pig kidney esterase the recovery has been 81.3% only (*67*). The values given here are corrected for a recovery of 99%.

carboxymethylated carboxylesterases from pig liver, pig kidney, and ox liver. According to the peptide maps there are significant structural differences between the porcine and bovine liver esterases. The esterases from pig liver and pig kidney appear to be closely related but not completely identical. These results are consistent with earlier enzymological and immunological findings (*50*). From optical rotary dispersion (ORD) measurements the α-helix content of pig liver esterase was calculated as 19.3% (*59*). The isoelectric point is 5.0 (*51, 52, 57*).

III. Catalytic Properties

A. SUBSTRATE SPECIFICITY

1. *Hydrolysis of Carboxyl Esters*

Nonspecific carboxylesterases hydrolyze a wide variety of carboxyl esters. It is therefore almost impossible to give a complete list of their

TABLE III

Substrate Specificity of Carboxylesterases

	References
1. Carboxyl Esters	
(a) Esters of unsubstituted fatty acids	
Phenyl formate (acetate, propionate, butyrate, valerate)	(*84*)
Ethyl formate (acetate, propionate, valerate)	(*89*)
m-Carboxyphenyl esters of a homologous series of *n*-fatty acids (chain length from C_2 to C_{12})	(*90*)
Ethyl acetate	(*33, 58, 91*)
Glyceryl triacetate (triacetin)	(*30, 58*)
p-Nitrophenyl acetate	(*33, 51, 58, 70, 89*)
o-Nitrophenyl acetate	(*33, 70*)
2,6-Dichlorobenzenone-indophenyl acetate	(*92*)
Vitamin A acetate	(*87*)
Methyl butyrate	(*33, 58, 91*)
Methyl butyrate (3-methylbutyrate, pentanoate, 3- and 4-methylpentanoate, hexanoate, heptanoate)	(*86*)
Glyceryl tributyrate (tributyrin)	(*30, 58*)
Ethyl butyrate	(*30, 33, 62, 85, 93*)
o-, *m*-, and *p*-Nitrophenyl butyrate	(*91*)
2,4-Dinitrophenyl butyrate	(*91*)
m-(*n*-Pentanoyloxy)benzoic acid	(*51*)
m-(*n*-Heptanoyloxy)benzoic acid	(*51*)
(b) Esters of other acids	
Ethyl benzoate (benzenesulfonate, lactate, acetoacetate; diethyl succinate, fumarate, aspartate; *p*-hydroxybenzoate, bromomalonate, terephthalate, and other ethyl esters)	(*89*)
Procaine (2-diethylaminoethyl *p*-aminobenzoate)	(*30, 33, 58, 70*)
1-Tyrosine ethyl ester (and many other amino acid esters)	(*30, 33, 58, 85–87, 89*)
2. Thioesters	
6-*S*- and 8-*S*-Acetoacetyl monothioloctanoate	(*63*)
8-*S*-Acetoacetyl, 6-ethyl monothioloctanoate	(*63*)
8-*S*-Acetoacetyl dihydrolipoic acid	(*63*)
6-*S*- and 8-*S*-Acetyl dihydrolipoic acid	(*63*)
6-*S*-Acetoacetyldecanoate	(*63*)
S-Acetyl- and *S*-acetoacetyl-BAL[a]	(*63*)
p-Nitrothiophenyl hippurate	(*85*)
Thiophenyl acetate	(*94*)
3. Aromatic Amides	
Acetanilide[b]	(*30, 33, 58, 70, 91*)
Phenacetin	(*58, 70, 95*)
Monoethylglycine 2,6-xylidide	(*30, 33, 58*)
Diethylglycine 2,6-xylidide (Xylocaine; lidocaine)	(*30*)

TABLE III (*Continued*)

	References
N-(*n*-Butylamino)acetyl 2-chloro, 6-methylanilide·HCl (Hostacaine; butacetoloid)[c]	(*93, 96*)
L-Leucyl *p*-nitroanilide	(*58, 97*)
L-Leucyl β-naphthylamide	(*30, 58, 97*)
2-(N^4-*n*-Propylaminoacetyl)-sulfanilamido 4,6-dimethylpyridine HCl	(*93, 96*)

[a] BAL stands for 2,3-dimercaptopropanol.

[b] And some other acylanilides (*N*-propionyl-, *N*-butyryl-, *N*-valeroyl-, and *N*-benzoylanilide).

[c] And some other derivatives with different chain length of the *N*-*n*-alkyl residue.

substrate specificity. Several more important compounds with carboxyl ester bonds, which have been employed during the past decade as substrates of partially or highly purified esterase preparations in various laboratories, are compiled in Table III. These include not only some long-known "ali-esterase" (e.g., aliphatic) substrates such as tributyrin and methyl butyrate but also aromatic esters, many of which are hydrolyzed with high velocities. Therefore, the widely used distinction between carboxylesterases (formerly ali-esterases, EC 3.1.1.1) and arylesterases (EC 3.1.1.2) appears to be somewhat misleading.

Although some labile ester substrates (such as *o*- and *p*-nitrophenyl acetate) are also split by several endopeptidases (*74–76*), by other enzymes [e.g., 3-phosphoglyceraldehyde dehydrogenase (*77, 78*) carbonic anhydrase (*79–82*)], and even by nonenzymic proteins (*83, 84*), the corresponding molecular activities are generally low as compared to the action of "true" esterases (*60, 70, 85*). There is another overlap in the substrate specificity of esterases and α-chymotrypsin since liver and kidney esterases also hydrolyze L-tyrosine ethyl ester (*30, 33, 58, 85*) as well as several other amino acid esters (*30, 86, 87*).

74. B. S. Hartley and B. A. Kilby, *BJ* **50,** 672 (1952).
75. B. A. Kilby and G. Youatt, *BJ* **57,** 303 (1954).
76. M. L. Bender and F. J. Kézdy, *Ann. Rev. Biochem.* **34,** 49 (1965).
77. E. L. Taylor, B. P. Meriwether, and J. H. Park, *JBC* **238,** 734 (1963).
78. M. T. A. Behme and E. H. Cordes, *JBC* **242,** 5500 (1967).
79. Y. Pocker and J. T. Stone, *JACS* **87,** 5497 (1965).
80. Y. Pocker and J. T. Stone, *Biochemistry* **6,** 668 (1967); **7,** 2936 (1968).
81. J. A. Verpoorte, S. Mehta, and J. T. Edsall, *JBC* **242,** 4221 (1967).
82. M. Liefländer and R. Zech, *Z. Physiol. Chem.* **349,** 1466 (1968).
83. B. S. Hartley and B. A. Kilby, *BJ* **56,** 288 (1954).
84. W. K. Downy and P. Andrews, *BJ* **96,** 21c (1965).

The reported catalytic rates vary considerably. High molecular activities have been observed, for instance, for phenyl acetate, *p*-nitrophenyl butyrate, *o*- and *p*-nitrophenyl acetate, and methyl butyrate, most of them being in the order of 20,000–60,000 ($t = 25°C$). Examples of poor substrates include procaine (*30, 33, 58*) and vitamin A acetate (*88*). Proteins (casein and hemoglobin), dipeptides, D,L-acetylmethionine, olive oil, β-naphthylpalmitate and stearate, and acetylcholine are not attacked by pig liver carboxylesterase (*30*).

Several studies have been concerned with the influence of chain length, as well as of the acyl as of the alcohol moiety, on the reactivity of esterase substrates (*85, 90, 98, 99*). Hofstee (*90*) has shown for several *n*-fatty acid esters of *m*-hydroxybenzoic acid that an elongation of the acyl carbon chain up to C_{12} results in an increase in reactivity. Another investigation on the specificity of horse liver esterase has been undertaken by Webb (*98*). When varying the chain length of the alcohol residue a threefold change in reactivity of acetate and butyrate esters has been found with a maximum at a C_4 alkyl chain. The activity toward the ethyl esters of several *n*-alkyl fatty acids, however, was increased 14-fold from the acetate to the valerate. Recently, the rates of hydrolysis of butyric acid esters of several phenols by various liver carboxylesterases have been reported. In addition, the reactivity of phenyl esters of formic, acetic, trimethylacetic, propionic, butyric, and valeric acid has been compared in this study (*85*). The substrate specificity of a goat intestinal esterase will be dealt with later (see Section IV). Levy and Ocken (*89*) distinguished three groups of carboxyl ester

85. J. K. Stoops, D. J. Horgan, M. T. C. Runnegar, J. de Jersey, E. C. Webb, and B. Zerner, *Biochemistry* **8**, 2026 (1969).
86. T. A. Krenitsky and J. S. Fruton, *JBC* **241**, 3347 (1966).
87. M. I. Goldberg and J. S. Fruton, *Biochemistry* **8**, 86 (1969)
88. J. Bertram and K. Krisch, *Eur. J. Biochem.* **11**, 122 (1969).
89. M. Levy and P. R. Ocken, *ABB* **135**, 259 (1969).
90. B. H. J. Hofstee, *JBC* **207**, 219 (1954).
91. W. Franz and K. Krisch, *Z. Physiol. Chem.* **349**, 1413 (1968).
92. A. J. J. Ooms and J. C. A. E. Breebaart-Hansen, *Biochem. Pharmacol.* **14**, 1727 (1965).
93. I. Reimann, Ph.D. Thesis, University of Münster/Germany (1969).
94. P. Greenzaid and W. P. Jencks, *Biochemistry* **10**, 1210 (1971).
95. E. Bernhammer and K. Krisch, *Biochem. Pharmacol.* **14**, 863 (1965).
96. T. Eckert, I. Reimann, and K. Krisch, *Arzneimittel-Forsch.* **20**, 487 (1970).
97. E. Bernhammer and K. Krisch, *Z. Klin. Chem.* **4**, 49 (1966).
98. W. Dixon and E. C. Webb, "Enzymes," 2nd ed., p. 220. Longmans, Green, New York, 1964.
99. P. Malhotra and G. Philip, *Indian J. Biochem.* **3**, 7 (1966); *Biochem. Z.* **346**, 386 (1966).

substrates, namely, (1) unsubstituted monocarboxylate esters, which have the highest relative velocity; (2) substituted monocarboxylate esters; and (3) dicarboxylate diesters and substituted diesters, of which only one ester group is hydrolyzed. This, among many other indications, is evidence that positively charged polar compounds, (e.g., choline esters) as well as negatively charged compounds are either not attacked at all or are poor substrates for carboxylesterases (*30, 89, 95, 98*). It can be assumed, therefore, that hydrophobic bonds play a major role in enzyme–substrate binding of carboxylesterases (*85, 98*).

Dating back to the work of Dakin the stereospecificity of esterases has been the subject of many earlier investigations with impure preparations, in most cases D- and L-mandelic acid esters being used (*3*). Liver esterases from pig (*89*) and monkey (*100*) show no optical specificity toward D,L-ethyl mandelate. Lately, Stoops *et al.* (*85*) have shown that L- and D-tyrosine ethyl ester are hydrolyzed at the same rate by pig liver esterase, whereas the carbobenzoxy-L-tyrosine *p*-nitrophenyl ester is attacked 27 times faster than the corresponding D-derivative. Further substrates include L-and D-β-butyrolactone (*56*) and oxazolinones (*85, 101, 102*). A γ-lactonase found in rat liver microsomes (*103*) is obviously not related to liver esterase. In the older literature there have been many reports that esterases under appropriate conditions also catalyze the synthesis of esters from acid and alcohol (*3*).

2. *Hydrolysis of Thioesters*

Drummond and Stern (*63*) partially purified an enzyme from chicken liver which hydrolyzed *S*-acetoacetyl dihydrolipoic acid, *S*-acetyl dihydrolipoic acid, and some other related thioesters. With regard to the pH dependence, the degree of sensitivity to DFP and physostigmine, and the constant ratio of activities (*S*-esterase/*O*-esterase) during purification, this enzyme is obviously very similar to the well-defined esterases from other sources. Highly purified esterases from pig liver (*71, 85, 94*) and goat intestinal mucosa (*99*) are also capable of splitting thioester linkages.

3. *Hydrolysis of Aromatic Amides*

The first indication that carboxylesterases from liver and some mycobacteria may not only hydrolyze ester bonds but also amides came

100. E. Bamann and H. Gelber, *Z. Physiol. Chem.* **318,** 82 (1960).
101. J. de Jersey, M. T. C. Runnegar, and B. Zerner, *BBRC* **25,** 383 (1966).
102. J. de Jersey and B. Zerner, *Biochemistry* **8,** 1967 (1969).
103. W. N. Fishbein and S. P. Bessman, *JBC* **241,** 4835 and 4842 (1966).

from Myers *et al.* (*22*). As shown in **1963** by the author, highly purified pig liver esterase catalyzes the hydrolysis of certain aromatic amides such as acetanilide, phenacetine, and several other anilide derivatives (*30*; Table III). This reaction is inhibited by organophosphorus compounds as well (*30, 70, 95*). The molecular activities towards amide substrates, however, are generally very low. The possibility of an amidase contaminating the esterase preparation has been ruled out. Thus the hydrolysis of *N*-(*n*-butylamino)acetyl 2-chloro-6-methylanilide (Hostacaine), a well-known local anesthetic, is competitively inhibited by ethyl acetate and vice versa (*93*). Furthermore, pig liver esterase was titrated with stoichiometric amounts of diethyl *p*-nitrophenyl phosphate (Fig. 1). Simultaneous activity determinations toward an ester (procaine) and some amides (acetanilide, phenacetin, and L-leucyl β-naphthylamide) gave identical inhibition curves. In all cases the activities reached zero after reaction of one mole of enzyme with two moles of the inhibitor (*70*). This suggests that the above mentioned substrates are not only hydrolyzed by the same enzyme but at the same active site as well.

Reimann (*93, 96*) has recently made an extensive study on the relationship between the chemical structure of aminoacyl arylamides and their corresponding rates of hydrolysis by pig liver esterase as isolated according to (*30*). By varying the length of the *N*-*n*-alkyl chain of the glycine residue from C_1 to C_6 the highest maximal velocities have been found for the *n*-propyl- and *n*-butylaminoacetyl 2-chloro-6-methylanilides the molecular activities being about 17,000. At the pH optimum of

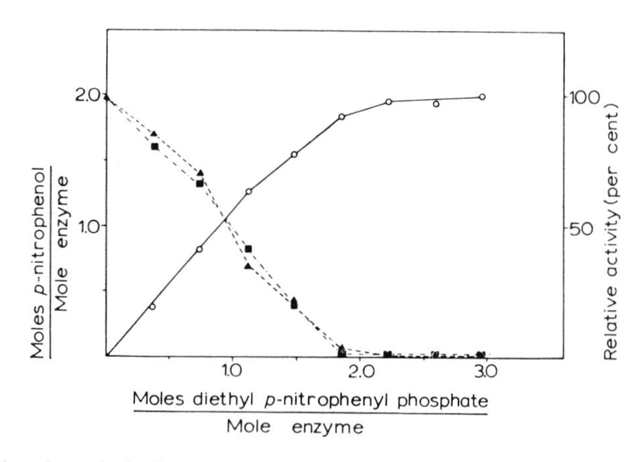

FIG. 1. Titration of pig liver carboxylesterase with diethyl *p*-nitrophenyl phosphate. Comparison of *p*-nitrophenol *liberation* (○) with the relative enzymic activities toward acetanilide (▲) and procaine (■). From Krisch (*70*). Reproduced from *BBA* **122,** 265 (1966).

8.6 these amides are mainly nonionized and are probably bound to the enzyme in this form. Aromatic amides, while being significantly split by pig liver and pig kidney esterases, are much less efficiently hydrolyzed by ox liver esterase (*33*).

4. *Acyl Group Transfer*

Many hydrolases are known to be able to transfer acyl groups not only to water but also to several other nucleophilic acceptors. This is true for unspecific carboxylesterases as well. In 1953, Bergmann and Wurzel (*104*) demonstrated a direct acyl group transfer by a partially purified dog liver esterase from glycine ethyl ester to hydroxylamine yielding the corresponding hydroxamic acid. In 1966, Krenitsky and Fruton (*86*) observed that a partially purified enzyme preparation from beef liver, which was tentatively named "aminoacyl transferase," did not only hydrolyze esters such as methyl butyrate and L-leucine methyl ester but was also an efficient catalyst in the synthesis of dipeptides from suitable amino acid esters. Both reactions were strongly inhibited by DFP. As intermediates in the reaction the corresponding dipeptide esters were probably involved. In 1967, Benöhr and Krisch (*33*) found that a highly purified preparation of beef liver esterase also catalyzes the formation of dipeptides from certain amino acid esters (L-phenylalanine methyl ester and L-tyrosine ethyl ester). These observations were extended in a detailed study by Goldberg and Fruton (*87*), which provided additional evidence that the transfer of amino acyl groups is indeed an intrinsic property of beef liver esterase.

Analogously, carboxylesterases from pig liver, pig kidney, and beef liver catalyze the enzymic acylation of aniline and *p*-phenetidine to the corresponding acylamides. Ethyl acetate and methyl butyrate, but not acetyl-CoA, have been shown to be suitable acyl donors (*91*). The reaction rates of this acyl group transfer are, however, slow. Recently, a pronounced stimulation of the turnover of phenyl acetate in the presence of methanol as acyl acceptor has been observed (*94*). These observations have been interpreted by assuming a competition between water and another acceptor for the acyl–enzyme intermediate [see Eq. (1)].

B. pH Optimum

The pH optima of most carboxylesterases from mammalian sources have been consistently reported to be in the range of pH 7.5–9.0 (*4*, *30*,

104. F. Bergmann and M. Wurzel, *BBA* **12**, 412 (1953).

33, 55, 58, 70, 99). Acidic pH optima have been found for a lysosomal esterase from rat liver (*34*) and for citrus fruit acetylesterase (*20*).

C. Kinetics

It might be expected, by analogy with other serine hydrolases, that the enzymic hydrolysis of esters and amides by carboxylesterases proceeds via an acyl–enzyme intermediate as shown in the following equation (*70, 85, 91, 94*):

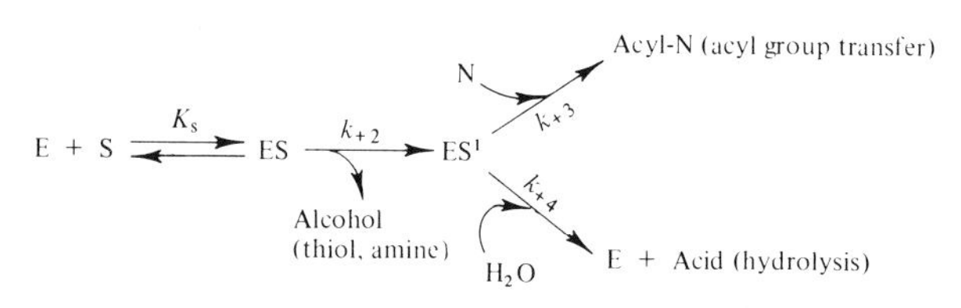

where ES is the enzyme–substrate complex, ES′ the acyl–enzyme, and N a nucleophilic acceptor, which may be acylated instead of water (see Section III,A,4). The best direct evidence would be the isolation of the acyl–enzyme intermediate of a carboxylesterase as it has been reported at acid pH for α-chymotrypsin (*105*). This, however, encounters difficulties owing to the instability of most carboxylesterases at low pH and to the high velocity of the hydrolytic deacylation of ES′. Lately, a kinetic study provided strong support for the acyl–enzyme mechanism in reactions catalyzed by beef liver esterase with amino acids as acyl acceptors (*105a*). Another detailed investigation on the mechanism of the reaction of pig liver esterase with alcohols gave further kinetic evidence for an acyl–enzyme intermediate which partitions between water and methanol (*94*). These authors assume, however, that their enzyme preparation (*52*) contains at least two different types of active sites. Quite analogously to Eq. (1) the reaction of esterases with organophosphorus inhibitors may be formulated. In contrast to the acyl–enzyme ES′, however, no significant hydrolysis of the phosphorylated enzyme generally occurs, although a very slow real turnover of diethyl *p*-nitrophenyl phosphate by pig liver esterase has been reported (*60, 61, 70*). Thus it is obviously a matter of the relative magnitude of the rate of dephosphorylation (k_4), whether an enzyme can be considered an A-type or a B-type esterase.

In studies on the influence of substrate concentration on the initial reaction velocity, deviations from the normal Michaelis-Menten kinetics

105. A. K. Balls and N. H. Woods, *JBC* **219**, 245 (1956).
105a. M. I. Goldberg and J. S. Fruton, *Biochemistry* **9**, 3371 (1970).

have been described (*51, 57, 73, 89, 106*). Thus Adler and Kistiakowsky obtained a nonlinear Lineweaver-Burk plot (substrate methyl butyrate) for their pig liver esterase, which was independent of the degree of purity of the preparation (*57, 89*). Interestingly, a normal kinetics has been found after partial heat denaturation of the enzyme (*57*). For an explanation these authors discuss five possibilities of which substrate activation and assumption of two interacting catalytic sites are considered the most probable ones.

Substrate activation has also been observed with the same preparation by other authors (*51, 73, 89*) and with the "Australian" enzyme as well (*85*). As found by Barker and Jencks (*51*) the maximal velocity is increased up to 2.4-fold in the presence of acetone and dioxane. It is concluded that each half-molecule of pig liver esterase contains an activator site, which can be occupied either by substrate or by other activators, as well as a catalytic site (*51*). A similar substrate activation, as indicated by a nonlinear Eadie plot, has been reported by Stoops *et al.* (*85*) using phenyl butyrate as a substrate. There was no decrease in substrate activation under conditions where the enzyme can be assumed to exist preferentially in the monomeric form (*51*). With increasing concentrations of benzene, maximal velocity increases until finally Michaelis-Menten kinetics are observed over the entire substrate concentration range. When studying the hydrolysis of ethyl butyrate, however, a linear Eadie plot was obtained (*85*).

It is interesting to note that significant differences in the kinetic behavior of the Adler-Kistiakowsky (*57*) and the "Australian" (*52*) pig liver esterase preparation have been observed such as different activity ratios toward various substrates, a lag period in the hydrolysis of *p*-nitrophenyl acetate, and the degree of activation by organic solvents (*51*). Other authors working with different esterase preparations (*30, 55*) and other substrates have found no indication of deviations from the normal Michaelis-Menten kinetics (*30, 33, 52, 55, 85, 95*). The reasons for these discrepancies are not clear at present, although conformational changes or partial denaturation during the enzyme isolation, for instance, by organic solvents, may be discussed in this context.

D. Studies on the Active Site by Means of Organophosphorus Compounds

All serine hydrolases are strongly inhibited by organophosphorus compounds, such as diethyl *p*-nitrophenyl phosphate and DFP which have

106. A. J. Adler and G. B. Kistiakowsky, *JACS* **84,** 695 (1962).

proved to be valuable tools in the elucidation of the active site and the mechanism of action of serine hydrolases. Recently, a survey on their application in the study of carboxylesterases has been given (*107*). As mentioned earlier the number of active sites and the normality of esterase solutions may be determined by titration with organophosphorus compounds (Fig. 1; *60, 61, 70*). Furthermore, radioactive inhibitors such as [^{32}P]-DFP have been used for affinity-labeling of the active site (*21, 64, 68, 108, 109*). After peptic digestion of the phosphorylated enzyme derivative the isolation of octapeptides from the active site has been reported. In all cases the β-hydroxyl group of a serine residue has been identified as the site to which the diisopropylphosphoryl residue is bound. The amino acid sequence in the vicinity of this "active serine" has been established by conventional methods in peptide chemistry. The results of several laboratories are presented in Table IV. In 1959, the sequence of a peptide from the active site of horse liver esterase was reported by Jansz *et al.* (*108*). Later, Webb *et al.* (*21, 64*) made a comparative study on the active site peptides of liver esterases from various animal species. It was found that the pig and sheep octapeptides were identical with those of horse liver esterase, whereas in both ox and chicken one of the eight amino acids was replaced (Table IV).

Recently, the sequences of radioactive octapeptides from pig liver and kidney esterases prepared according to Krisch (*30*) and Franz and Krisch (*58*) have been reported (*109*). They differ from the results of

TABLE IV
AMINO ACID SEQUENCE AROUND THE REACTIVE SERINE RESIDUE OF CARBOXYLESTERASES[a]

Carboxylesterase from	Sequence of active site peptide	Ref.
Horse liver	Gly–Glu–Ser(P)–Ala–Gly–Gly–(Glu, Ser)	(*108*)
Ox liver	Gly–Glu–Ser(P)–Ala–Gly–Ala–Glu–Ser	(*64*)
Chicken liver	Gly–Glu–Ser(P)–Ala–Gly–Gly–Ile–Ser	(*64*)
Sheep liver	Gly–Glu–Ser(P)–(Ala, Gly, Gly, Glu)–Ser	(*64*)
Pig liver	Gly–Glu–Ser(P)–(Ala, Gly, Gly, Glu)–Ser	(*64*)
Pig liver	Gly–Glu–Ser(P)–Ala–Gly–$\frac{\text{Asp}}{\text{Glu}}$–Gly–Ser	(*109*)
Pig kidney	Gly–Glu–Ser(P)–Ala–Gly–$\frac{\text{Asp}}{\text{Glu}}$–Gly–Ser	(*109*)

[a] Data from Heymann *et al.* (*109*). Reproduced from *Z. Physiol. Chem.* **351**, 931 (1970).

107. K. Krisch, *Z. Klin. Chem. Klin. Biochem.* **8**, 545 (1970).
108. H. S. Jansz, C. H. Posthumus, and J. A Cohen, *BBA* **33**, 396 (1959).
109. E. Heymann, K. Krisch, and E. Pahlich, *Z. Physiol. Chem.* **351**, 931 (1970).

other laboratories in one position near the C-terminal end and in the finding of an hitherto unreported aspartic acid residue. The isolated product is probably still a mixture of two similar peptides [Gly–Glu–Ser(P)–Ala–Gly–Glu–Gly–Ser and Gly–Glu–Ser(P)–Ala–Gly–Asp–Gly–Ser]. The sequence of the two peptides from pig liver and pig kidney esterase is identical. It is assumed that the "Asp peptide" and the "Glu peptide" are probably derived from a mixture of two closely related isoenzymes. In contrast to the endopeptidases, in all "true" esterases studied so far the acidic amino acid adjacent to the active serine has been found to be glutamic acid instead of aspartic acid.

In analogy to the mechanism of ester hydrolysis by endopeptidases as studied in detail in many laboratories [for a review, see Bender and Kézdy (*76*)], the participation of a histidine residue at the active site of carboxylesterases has been frequently discussed (*59, 110*). It should be mentioned, however, that direct experimental evidence for this is still lacking. As shown by ORD measurements, photooxidation of pig liver esterase in the presence of methylene blue not only destroys histidine residues but also leads to conformational changes and partial denaturation (*59*). For this reason the positive identification of a histidine residue as being involved in the catalytic reaction has not yet been possible.

Whereas the typical anticholinesterase agents are highly toxic and inhibit all serine hydrolases, there exist some other phosphate esters which are less toxic and act more selectively on carboxylesterases. Thus tris(*o*-cresyl) phosphate, while a potent inhibitor of "ali-esterases" (*111*), has only a slight effect on cholinesterases in the rat [cf. Myers and Mendel (*112*)]. On prolonged application, tris(*o*-cresyl) phosphate and some related phosphoric acid triesters produce characteristic neurotoxic symptoms such as ataxia and polyneuritic paralysis, especially in chicken (*113, 114*). As the receptor site a brain esterase has been assumed. On the other hand, not all inhibitors of brain esterases are also neurotoxic (*113, 114*). Recently, bis(*p*-nitrophenyl) phosphate has been found to inhibit stoichiometrically and irreversibly microsomal carboxylesterases from liver and kidney. It may be used, therefore, for the determination of the number of active sites. This anionic compound does not block other

110. G. H. Dixon, H. Neurath, and J. F. Pechère, *Ann. Rev. Biochem.* **27**, 489 (1958).
111. B. Mendel and D. K. Myers, *BJ* **53**, xvi (1953).
112. D. K. Myers and B. Mendel, *BJ* **53**, 16 (1953).
113. W. N. Aldridge and I. M. Barnes, *Biochem. Pharmacol.* **15**, 541, 549 (1966).
114. M. K. Johnson, *BJ* **114**, 711 (1969).

serine hydrolases, e.g., the cholinesterases, chymotrypsin, and trypsin. Its toxicity in mice (LD_{50} 410 mg/kg) is about 100 times less than that of diethyl *p*-nitrophenyl phosphate (*72*).

E. Other Inhibitors

Compared to the significance of the organophosphorus compounds other esterase inhibitors are of minor importance. It has been known for a long time that unspecific esterases are inhibited by quinine, atoxyl (*p*-arsanilate) and fluoride ions in a concentration range of 0.1–10 m*M* (*3, 30, 33, 58*). Recently, the mechanism of inhibition of some serine esterases by phenylarsonic acids has been investigated (*25*). According to inhibition experiments with *p*-chloromercuribenzoate and other SH blocking agents there is no evidence that SH groups are essential for esterase activity (*4, 30, 58*). Metal ions and other cofactors of low molecular weight are obviously not involved in the catalytic reaction (*30, 58*). Cholinesterases have been frequently distinguished from other esterases since they are completely inhibited by 10^{-5} *M* physostigmine, whereas for the inhibition of other esterases higher concentrations are required (*8, 33, 47, 58, 95*).

F. Assay Procedures

Today there are many reliable methods available for measuring carboxylesterase activities. Perhaps the most widely used assay consists in the titration of liberated acid equivalents by the pH-stat technique employing mostly aliphatic esters as substrates (*24, 30, 33, 52, 56–58, 85–87*). Acid formation by esterases may also be followed by manometric determination of CO_2 liberated from a bicarbonate buffer (*1, 55, 115, 116*). Other convenient substrates are the nitrophenyl esters of several acids. The yellow (*p*- or *o*-)nitrophenolate ion liberated may be measured in a continuous photometric test (*33, 58, 70, 85, 117*). In addition, spectrophotometric assays in the UV range have been reported for ester (*30, 33, 51, 58, 90*) and amide substrates (*91, 93, 96*). After enzymic hydrolysis of naphthyl esters the (α- or β-) naphthol formed may be coupled with diazonium salts to azo dyes (*27, 48, 49, 118*). This method is suitable in particular for substrate staining of esterase bands after electrophoretic

115. W. N. Aldridge and A. N. Davison, *BJ* **51**, 62 (1952).
116. W. N. Aldridge, *BJ* **57**, 692 (1954).
117. C. Huggins and J. Lapides, *JBC* **170**, 467 (1947).
118. H. A. Ravin and A. M. Seligman, *ABB* **42**, 337 (1953).

separation. Analogously the formation of aniline from acetanilide is measured by subsequent diazotization (*30*). Whereas all methods mentioned so far measure product formation, in the hydroxamic acid method of Hestrin (*119*) the disappearance of unhydrolyzed substrate is determined. The stalagmometric assay (*120*), which has been frequently used earlier, is obsolete today.

G. Possible Physiological and Pharmacological Significance

At present the metabolic function of most carboxylesterases is still obscure. Almost all substrates listed in Table III (with the exception of *S*-acetyl dihydrolipoic acid) are foreign compounds which do not occur normally in the intermediary metabolism. Some of them (procaine, lidocaine, butacetoloid, and phenacetin) are of pharmacological significance. Thus, unspecific esterases obviously play a role in the so-called detoxification system of the body by their ability to metabolize numerous foreign compounds with ester and amide bonds, being the hydrolytic counterpart of the microsomal mixed function oxidase system (*121*). The activity of esterases may be of importance for the duration of action and for the toxicity of drugs with susceptible bonds. *In vivo* and *in vitro* studies in the rat have shown that blocking of liver esterases by bis(*p*-nitrophenyl) phosphate markedly inhibits the formation of methemoglobin as induced by high doses of acetanilide and phenacetin. After a single dose of bis(*p*-nitrophenyl) phosphate (100 mg/kg ip) the half-time of the reappearance of the inhibited esterase–amidase activity in rat liver has been found to be 50–60 hr (*122*).

The problem of the unknown metabolic function of carboxylesterases has been approached by using organophosphorus compounds as tools. In 1953, Myers and Mendel reported that oxygen consumption and acetoacetate formation in rat liver homogenates and slices are not influenced by diethyl *p*-nitrophenyl phosphate, DFP, tetraethyl pyrophosphate and tris(*o*-cresyl) phosphate. These results fail, as the authors concluded, to give any indication of an essential function of ali-esterases in the normal lipid metabolism of rat liver (*112*). In another preliminary study the

119. S. Hestrin, *JBC* **180,** 249 (1949).

120. R. Willstätter and F. Memmen, *Z. Physiol. Chem.* **129,** 1 (1923).

121. J. R. Gillette, A. H. Conney, G. J. Cosmides, R. W. Estabrook, J. R. Fouts, and G. J. Mannering, eds., "Microsomes and Drug Oxidations." Academic Press, New York, 1969.

122. E. Heymann, K. Krisch, H. Büch, and W. Buzello, *Biochem. Pharmacol.* **18,** 801 (1969).

influence of organophosphorus compounds on the metabolism of the isolated perfused rat liver was investigated. In liver samples taken by freeze stop the concentrations of numerous substrates from essential pathways of the intermediary metabolism were determined (glycogen, glucose, glucose 6-phosphate, fructose 1,6-diphosphate, dihydroxyacetone phosphate, 3-phosphoglycerate, phosphoenol pyruvate, pyruvate, lactate; citrate, α-ketoglutarate, malate; α-glycerophosphate, total lipids; as well as ATP, ADP, AMP, and inorganic phosphate). After 60-min perfusion with diethyl *p*-nitrophenyl phosphate or DFP, however, significant differences were not found in any of the above-mentioned metabolites as compared to control livers whereas, according to expectation, esterase activity was completely abolished (*123*). Because of its higher specificity and low toxicity bis(*p*-nitrophenyl) phosphate should be expected to be a useful tool for *in vivo* studies on the unknown function of carboxylesterases. So far the results after application of this compound (100 mg/kg ip daily, up to 4 weeks) in rats, however, were mostly negative as well, the animals showing normal growth and no gross histological alterations in several tissues (*72, 124*). Thus all hopes for a clue to the metabolic function of esterases by means of organophosphorus compounds have not yet been realized.

An interesting hypothesis has been put forward recently by Okuda and Fujii (*125*). These authors postulate that the protein moieties of rat liver esterase and lipase are immunologically identical and that liver lipase is easily converted into an esterase by treatment with acetone or pancreatic lipase. Conversely, a partially purified esterase preparation, when sonicated with a lipid extract from rat liver, is transformed to some extent (25%) into a lipase as evidenced by the ability to hydrolyze a coconut oil emulsion (Ediol) and by a different behavior in gel filtration. It is concluded that rat liver lipase is in fact a complex of liver esterase and lipid and that a single enzyme is responsible for the hydrolysis of both short and long chain fatty acids. If this is true the physiological function of native liver esterase obviously would be the hydrolysis of long chain fatty acid triglycerides. Because of the ability of some liver esterases to effect the synthesis of dipeptides from amino acid esters, it has been speculated that they might possibly play a role in the biosynthesis of peptide bonds (*86, 87*).

Pilz *et al.* have extensively studied the arylesterases of human and animal sera and separated them into several fractions by preparative starch gel electrophoresis. It was reported that arylesterases catalyze a

123. H. Schimassek and K. Krisch, unpublished observations (1966).
124. W. Junge and K. Krisch, unpublished observations (1970).
125. H. Okuda and S. Fujii, *J. Biochem.* (*Tokyo*) **64**, 377 (1968).

transesterification, e.g., the formation of β-naphthylstearate or oleate from β-naphthylpropionate and free fatty acids. This would mean an alkyl group transfer of the alcohol moiety. These authors assumed tyrosine esters to be the physiological substrates and postulated a participation of serum arylesterases in the transport of free fatty acids, serving as an auxiliary "coenzyme system" for lipoproteid lipase (*14*, *126–128*).

IV. Some Other Individual Esterases

The following survey on some other individual esterases can by no means be considered to be complete but is restricted to some well-defined enzymes of this group which have been highly or partially purified.

An apparently homogeneous esterase from goat intestinal mucosa has been extensively characterized with regard to its substrate specificity using, besides others, the *p*-nitrophenyl esters of acids of different chain length from C_2 to C_{18} (*99*). It is relatively specific for esters of short chain fatty acids, *n*-butyrates being the best substrates. Esters with a chain length of the acyl moiety of 12 and more carbon atoms were not further hydrolyzed. It is suggested that the alkyl chain of the acyl residue is attached by hydrophobic bonds to a narrow hydrophobic region at the active site.

As mentioned in Section I it has been assumed for a long time that the A-esterase in mammalian serum hydrolyzing phenyl esters (arylesterase, EC 3.1.1.2) and the activity towards organophosphorus compounds (*129–131*) are the result of the same enzyme (*1*). In the light of recent findings, however, this has become doubtful. Thus, Main (*132*) purified 385-fold a diethyl *p*-nitrophenyl phosphate hydrolyzing enzyme from sheep serum and showed that his purified preparation did not hydrolyze phenyl acetate any more. *p*-Nitrophenyl acetate and DFP were probably hydrolyzed by still other enzymes. The molecular weight of the "paraoxonase" was estimated to be 35,000–50,000. Pilz *et al.* (*14*) separated 10 arylesterases from rabbit serum, none of which was able

126. W. Pilz and H. Hörlein, *Z. Physiol. Chem.* **335**, 221 (1964).
127. W. Pilz, *Z. Physiol. Chem.* **338**, 238 (1964).
128. W. Pilz, H. Hörlein, and E. Stelzl, *Z. Physiol. Chem.* **345**, 65 (1966).
129. E. G. Erdös and L. E. Boggs, *Nature* **190**, 716 (1961).
130. K. Krisch, *Z. Klin. Chem. Klin. Biochem.* **6**, 41 (1968).
131. M. Skrinjaric-Spoljar and E. Reiner, *BBA* **165**, 289 (1968).
132. A. R. Main, *BJ* **74**, 10 (1960); **75**, 188 (1960).

to hydrolyze organophosphorus compounds. It would be beyond the scope of this review to deal further with the complex problem of differentiation of the numerous esterase activities in serum and the enzymic hydrolysis of organophosphorus compounds.

Acetylesterases (EC 3.1.1.6) occurring mainly in higher plants, citrus fruits, and fungi (*20, 23*) but also in animal tissues preferentially attack acetyl esters. They are relatively resistant to inhibition by diethyl *p*-nitrophenyl phosphate and seem to be related to the A-esterases of serum (*23*). These criteria do not seem to be very satisfactory, and it might be questioned whether they suffice for listing the acetylesterases as a separate group of esterases. Further information on this enzyme, as well as on some other carboxylesterases, such as chlorophyllase, tannase, pectinesterase, and lactonases, should be obtained from previous reviews (*6, 7*).

The existence of an atropine esterase (EC 3.1.1.10) in serum and tissues of some but not all rabbits has been known for a long time since it is a favorite subject for pharmacogenetical studies. A genetically determined atropine hydrolyzing enzyme was purified 122-fold from rabbit serum by Margolis and Feigelson (*13*). Its molecular weight was estimated to be approximately 65,000. With regard to its substrate specificity it proved to be a rather nonspecific B-type esterase hydrolyzing a variety of esters including butyrylcholine, benzoylcholine, glyceryl tributyrate, phenyl butyrate, and *p*-nitrophenyl acetate. The activities toward several substrates showed the same elution pattern on DEAE-cellulose chromatography and were inhibited to the same extent by diethyl *p*-nitrophenyl phosphate.

A review on vitamin A esterase (EC 3.1.1.12) has recently been given by Olson (*133*). Most studies on vitamin A esterase have been carried out with homogenates, subcellular fractions, and partially purified preparations. Up to now there exists no highly purified and well-characterized esterase specific for short chain vitamin A esters. On the other hand, it was shown in a recent communication that vitamin A acetate is also a substrate of highly purified carboxylesterases from liver and kidney (*88*). It is questionable, therefore, whether the distinction between vitamin A esterase (EC 3.1.1.12) and unspecific carboxylesterases (EC 3.1.1.1) is still justified.

Enzymes catalyzing the hydrolysis or synthesis of sterol esters with no apparent requirement for a high energy source, such as ATP and CoA, are referred to as cholesterol esterases (EC 3.1.1.13). Earlier reports on this enzyme have been concerned with its partial purification from a

133. J. A. Olson, *Pharmacol. Rev.* **19,** 559 (1967).

porcine pancreas acetone powder (*134, 135*). Recently, a cholesterol esterase has been purified 600-fold from rat pancreatic juice. The purified enzyme was homogeneous on disc electrophoresis and did not hydrolyze methyl butyrate, triolein, or secondary alcohol esters. The pH optima are 6.2 for synthesis and 6.6 for hydrolysis of cholesterol esters (*136*). Cholesterol esterase attacks preferentially long chain fatty esters of cholesterol and therefore resembles more a lipase than a true esterase.

A pethidine (= dolatin)-hydrolyzing esterase was purified 11-fold from rabbit liver by Ammon and Kamphues (*137*). The isolated enzyme behaved homogeneously as judged from electrophoresis on cellulose acetate, from rechromatography on DEAE-Sephadex, and from a velocity run in the analytical ultracentrifuge. It was inhibited by E-600, by DFP, and by higher concentrations of sodium fluoride, atoxyl, and physostigmine sulfate. Besides pethidine also norpethidine, tributyrin, propanidid, glycyl-L-proline, L-leucylglycine, and L-tyrosine ethyl ester were hydrolyzed.

ACKNOWLEDGMENT

The author wishes to thank Drs. E. Heymann and W. Franz for stimulating discussions during the preparation of this manuscript.

Addendum

Bauminger and Levine (*138*) recently reported on immunological studies with beef liver carboxylesterase isolated by a new procedure. The enzyme was homogeneous on disc electrophoresis and had a molecular weight of 54,000 as determined by gel filtration and electrophoresis on sodium dodecyl sulfate containing acrylamide. Further evidence for the existence of two different types of active site in pig liver esterase was obtained by kinetic studies with the irreversible inhibitor phenylmethansulfonyl fluoride. In addition, the enzyme isolated according to Krisch (*30*) could be separated into 4 closely related electrophoretic variants with the same molecular weight by disc electrophoresis in 4–6% acrylamide gel (*139*).

134. H. H. Hernandez and I. L. Chaikoff, *JBC* **228**, 447 (1957).
135. M. Korzenovsky, E. R. Diller, A. C. Marshall, and B. M. Auda, *BJ* **76**, 238 (1960).
136. J. Hyun, H. Kothari, E. Herm, J. Mortenson, C. R. Treadwell, and G. V. Vahouny, *JBC* **244**, 1937 (1969).
137. R. Ammon and V. Kamphues, *Enzymologia* **37**, 343 and 356 (1969).
138. S. Bauminger and L. Levine, *BBA* **236**, 639 (1971).
139. E. Heymann, W. Junge, and K. Krisch, in preparation.

4

Phospholipases

DONALD J. HANAHAN

I. Introduction

The phospholipases, which are classified as hydrolases, are of quite widespread occurrence in nature. There are four well-documented established types which can attack a typical phosphoglyceride, phosphatidylcholine, illustrated below at the indicated positions:

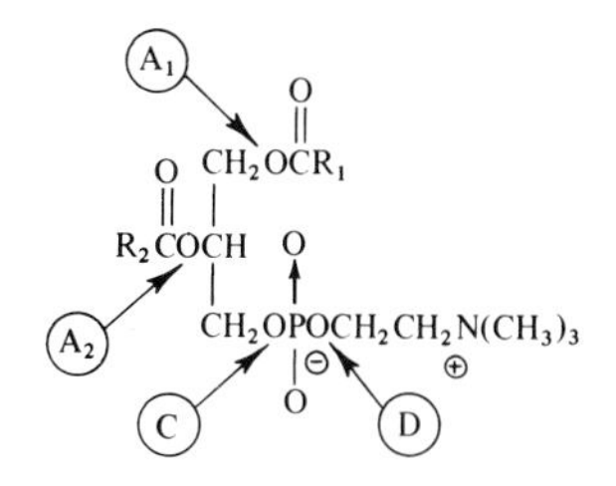

Thus, using the above molecule as a model substrate, phospholipase A_1 would catalyze release of a fatty acid from the 1 position, phospholipase

A_2 the fatty acid from the 2 position, phospholipase C the phosphorylated base, and phospholipase D the nitrogenous base only. Other compounds may serve as substrates for these enzymes, as shall be discussed below. In addition to the above-mentioned there is another recognized activity, phospholipase B, which has as its primary substrate a monoacylphosphoglyceride (a "lyso" compound). The products of this reaction are a fatty acid and the corresponding glycerylphosphoryl base. There appears to be some slight activity of this enzyme toward the diacylphosphoglycerides. Evidence has also been presented for the occurrence of a phosphatidate phosphatase (phosphohydrolase) which has as its substrate, diacylglycerylphosphoric acid (phosphatidic acid). The products of the reaction are diglyceride and inorganic phosphate.

Certain historical facets related to isolation and identification of the phospholipases may be obtained by reference to reviews by Hanahan (*1*) and by Kates (*2*). A description of progress in this field as of 1965 was provided by van Deenen and deHaas (*3*). Recently, a detailed consideration of the approaches used for purification and assay of the phospholipases was published and essentially covered the field through 1967 (*4*). On the basis of information presented in this latter review, it was decided to focus attention here primarily on those phospholipases which have been purified to a significant degree and/or on which detailed physical and chemical constants are available, together with any new information regarding their mode of action. Thus, this presentation essentially represents a status report on two enzymes, namely, phospholipase A_2 and phospholipase C.

Phospholipases A_2 and C have been the most frequently studied and widely used enzymes in this class of hydrolases. Phospholipase A_2, with its specific attack on the 2 position of phosphoglycerides, such as phosphatidylcholine and phosphatidylethanolamine, has provided a wealth of information regarding not only the specific positioning of fatty acids but also the relative metabolic behavior of these acids. In many instances phospholipase C, which releases diglyceride and phosphorylcholine from phosphatidylcholine, has allowed further examination of the structure of the phosphoglycerides and a convenient route to preparation of diglycerides. Currently, considerable interest has been

1. D. J. Hanahan, *Prog. Chem. Fats Lipids* **4,** 141 (1957).
2. M. Kates, *in* "Lipid Metabolism" (K. Bloch, ed.), p. 165. Wiley, New York, 1960.
3. L. L. M. van Deenen and G. H. deHaas, *Ann. Rev. Biochem.* **35,** 157 (1966).
4. J. M. Lowenstein, ed., "Methods in Enzymology," pp. 131–213, 1969.

manifested in the use of these phospholipases to establish a role of phosphoglycerides in the structure and behavior of membranes. All too often, though, investigators have employed quite impure enzyme preparations in such studies and any resulting alterations to a membrane may provide little or no information of import as to the role of phosphoglycerides in membrane structure. Furthermore, the action of a phospholipase on a phosphoglyceride can cause a considerable change in the physical state of the lipids, which then undoubtedly influences the physical state of contiguous proteins and other components. Artifacts, which are not unknown to the lipid or membrane field, can arise and conclusions regarding the role of lipids would be of little value. There is little doubt that these enzymes could prove useful in certain types of membrane studies, yet the problems alluded to above should cause some pause for reflection on the meaning of experimental results using these enzymes.

With the availability of highly purified and crystalline phospholipases, the opportunity to investigate the three-dimensional structure of these enzymes by X-ray crystallographic analysis (coupled with sequencing of their amino acids) provides an excellent opportunity to explore and explain certain of their unusual properties (heat stability, activity in solvents such as diethyl ether, etc.). Of equal importance these enzymes provide a fine opportunity for study of lipid–protein interactions. The availability of a wide range of stereochemically pure phosphoglycerides provides the necessary components for a highly varied experimental attack.

II. Phospholipase A_2

Phospholipase A_2 (phosphatide acyl-hydrolase, EC 3.1.1.4) catalyzes the specific removal of the acyl group from the 2 position of an *sn*-3 phosphoglyceride (*5*) which represents the predominant stereochemical form of these phospholipids in nature. It has been the most widely studied of all the phospholipases and, as might be expected, has been obtained in the highest state of purity. Two sources, snake venom and pancreas, have provided the best preparations of this enzyme, and a description of the results derived from these studies is given below.

5. IUPAC-IUB Commission on Biochemical Nomenclature, *Biochemistry* **6**, 3287 (1967).

A. Sources

Phospholipase A_2 activity has been reported in rat liver mitochondria (*6*), in many different tissues of the rat (*7*), and in the venoms of seven different species of snakes (*8*).

B. Substrates and Mode of Attack

Under proper conditions, this enzyme can attack a wide variety of phosphoglycerides, provided their stereochemical configuration is that of an *sn*-glycero-3-phosphate. Phosphatidylcholine, phosphatidylethanolamine, phosphatidylglycerol, and diphosphatidylglycerol are examples of compounds attacked by phospholipase A_2. The attack in each instance is the same in that it catalyzes release of one fatty acid per glycerol chain and in the *sn*-3 series, only from the 2 position (*9–11*). It is of interest to note that a "β-lecithin," an *sn*-2 phosphoglyceride, could be attacked stereochemically with release of only one fatty acid (*12*).

It is also of interest to note that the phospholipase A_2 isolated from the venom of *Crotalus atrox* (Western diamondback rattlesnake) attacks the acyl ester (in the 2 position) of the vinyl ether (OCH=CHR in the 1 position) containing phosphoglycerides (plasmalogen) much more slowly than the diacyl acyl ester derivative (*13*). Similarly, the acyl ester of the saturated ether ($O–CH_2CH_2R$ in 1 position) phosphoglycerides appears to be more slowly attacked by phospholipase A_2 from *C. adamanteus* than the diacyl ester compound (*14*). Preliminary evidence suggests that the phosphono(–C–O–$\underline{\text{P–C}}$)glyceride may be quite resistant to attack by this enzyme (*14*).

C. Isolation and Purification of Enzyme

Recently, three different laboratories have isolated and purified phospholipase A_2, using snake venom and pig pancreas as the source of en-

6. G. L. Scherphof and L. L. M. van Deenen, *BBA* **98,** 204 (1965).
7. J. J. Gallai-Hatchard and R. H. S. Thompson, *BBA* **98,** 128 (1965).
8. L. J. Nutter and O. S. Privett, *Lipids* **1,** 258 (1966).
9. N. H. Tattrie and J. R. Bennett, *Can. J. Biochem. Physiol.* **41,** 1983 (1963).
10. D. J. Hanahan, H. Brockerhof, and E. J. Barron, *JBC* **235,** 1917 (1960).
11. L. L. M. van Deenen and G. H. deHass, *BBA* **70,** 211 (1963).
12. G. H. deHaas and L. L. M. van Deenen, *BBA* **84,** 569 (1964).
13. E. L. Gottfried and M. M. Rapport, *JBC* **237,** 329 (1962).
14. D. J. Hanahan, unpublished observations.

zyme. The salient features of each of these isolation procedures is outlined below.

1. *Crotalus adamanteus Venom*

Wells and Hanahan (*15*) employed commercially available lyophilized venom of *Crotalus adamanteus* (Eastern diamondback rattlesnake) as starting material. The venom was suspended in tris-ethylenediaminetetra-acetate (EDTA) buffer and then subjected to gel filtration on Sephadex G-100. The activity was eluted as a single peak and was further chromatographed on BioRex 70 and subsequently on DEAE-cellulose (Whatman DE52) from which two well-separated peaks of activity were eluted (*16*). Final purification could be achieved on SE-Sephadex. These two activities represented 2–4% of the starting protein and approximately 30% of the original activity in each peak. The two enzymic activities could be crystallized in the cold by slow addition of ammonium sulfate.

The assay procedure for phospholipase A_2 activity was based on an earlier observation (*17*) that this enzyme was active, in diethyl ether, on highly purified phosphatidylcholine. The assay method involves titration of the liberated fatty acid. The increase in specific activity of the enzyme over the starting material was found to be 14-fold. The final specific activity was 3150 and 3250 units (microequivalents of fatty acid liberated in 1 min) per milligram protein, respectively, for each fraction. The turnover number of both enzymes is 1600 moles of substrate hydrolyzed per mole of enzyme per second.

The presence of two forms of this enzyme appear unique to this venom. The same two forms were obtained whether fresh or lyophilized venom was used as the starting material, and they did not appear to be equilibrium forms since one could not be converted to the other. No phosphodiesterase, protease, or amino acid oxidase activity could be detected. These two forms of phospholipase A_2 are very stable and can be maintained for several months at 4° without loss in activity. Each enzyme showed a single, but of different mobility, band on gel electrophoresis. A combination of the two forms could be resolved on gel electrophoresis. Sedimentation velocity patterns of the two enzymes indicated no detectable asymmetry.

An interesting facet of the assay for this enzymic activity is the influence of water. In the usual procedure, diethyl ether, which contains a small amount of added methanol, is used as solvent for the substrate,

15. M. A. Wells and D. J. Hanahan, *Biochemistry* **8,** 414 (1969).
16. K. Saito and D. J. Hanahan, *Biochemistry* **1,** 521 (1962).
17. D. J. Hanahan, *JBC* **195,** 199 (1952).

and the enzyme in an aqueous solution is added to it. The amount of water required for maximal activity of the enzyme is 100 times that consumed in the reaction. Thus, far more water than necessary for hydrolysis was needed before the reaction was initiated. The relationship of the water to the physical structure of the substrate is complex and is under current investigation (*18*). In addition to the water requirement, both sodium chloride and a divalent cation such as Ca^{2+} are required for maximal activity (*18*). In this latter study, other cations such as Mg^{2+}, Mn^{2+}, and Cd^{2+} also activate to varying degrees, whereas Zn^{2+}, Co^{2+}, Fe^{2+}, Al^{3+}, and Ba^{2+} are potent inhibitors. The mode of action of calcium, for example, in activation is not simply as a cofactor per se or in the removal of liberated fatty acids but appears to be involved possibly in a conformational change in the enzyme. Furthermore, zinc, in inhibiting the enzymic action, may affect regions or groups on this protein quite removed from the calcium active site(s).

2. *Pig Pancreas*

DeHaas *et al.* (*19*) used freshly prepared homogenates of pig pancreas for the source of enzyme. The homogenate is allowed to autolyze for 12–24 hr at room temperature and then subjected (after adjustment of pH and removal of precipitate) to ammonium sulfate fractionation. The material insoluble at 0.6 saturation was dissolved in 0.75 *M* NaCl and chromatographed on Sephadex G-50. The eluate containing enzymic activity was then passed through a DEAE-cellulose column and the single peak of activity then chromatographed on CM-cellulose. The activity from the latter chromatographic procedure moved as a single band on disc gel electrophoresis. The enzyme was obtained in a 31% yield. In this fractionation procedure, an egg yolk assay system was employed and phospholipase A_2 activity was reported as alkali uptake in microequivalents per minute per milligram protein. Apparently calcium was necessary for optimal activity in this system, but zinc, cadmium, and lead, as well as EDTA, strongly inhibited. The purified enzyme was not inhibited by diisopropylphosphofluoridate (DFP). It showed no change in activity after 5 min at 98° at pH 4.0, and 45% of its activity was recovered after heating for 1 hr at 98° at pH 4.0. Furthermore, storage of the enzyme in 8 *M* urea at 20° for 22 hr caused no decrease in activity.

The mode of action of the purified enzyme on several synthetic mixed

18. M. A. Wells, unpublished observations.
19. G. H. deHaas, N. M. Postema, W. Nieuwenhuizen, and L. L. M. van Deenen, *BBA* **159**, 103 (1968).

acid phosphoglycerides (in a deoxycholate-diethyl ether medium) was reported in this same study. This enzyme preparation had no activity toward an *sn*-glycerol-1-phosphatidylcholine and, as expected then, only 50% activity towards a racemic phosphatidylcholine. However, the reactivity toward *sn*-glycero-3-phosphatidylcholine(s) and ethanolamine(s) was maximal with no preference shown for chain length or unsaturation of the fatty acyl group in the 2 position. No specific activity value for this enzyme on these synthetic substrates was provided. Of considerable interest, this enzyme exhibited very high activity toward anionic substrates such as phosphatidic acid, phosphatidylglycerol, and diphosphatidylglycerol (cardiolipin). This latter action is in contradistinction to that found for the enzyme obtained from snake venom.

DeHaas *et al.* (*19*) further noted that freshly prepared homogenates of pig pancreatic tissue contained a low level of phospholipase A_2 activity, but during autolysis a considerable rise in enzymic activity was noted. It was reasoned that the increase in activity in pancreas homogenates was related to proteolytic activation. This activation process was increased by trypsin and decreased by DFP. Subsequent investigation (*20*) led to the isolation of a pre-phospholipase A_2, a zymogen of phospholipase A_2, through use of the same procedure as noted above for (pancreas) phospholipase A_2 preparation. Upon treatment with trypsin, this material yielded phospholipase A_2. Diisopropylphosphofluoridate was found to inhibit activation during isolation. Further information on this zymogen is provided later.

3. *Crotalus atrox Venom*

Wu and Tinker (*21*) used this venom for their starting material. The venom was subjected to ammonium sulfate fractionation and then Sephadex G-75 column chromatography. Active protein appeared as a single band on polyacrylamide gel disc chromatography (at pH values from 4.5 to 8.4) and was eluted from Sephadex G-75 with a constant specific activity some 35 times that of the crude venom. The final yield of material was 0.48% of the starting protein and 16.8% of the initial activity. It exhibited exclusive action at the 2 position of (liver) lecithin. No detectable protease, phosphodiesterase, or monoesterase activity was found. The assay system was a titrimetric one employing purified lecithin as substrate (in diethyl ether, chloroform, or as a sonicated lecithin dispersion in water) and cresol red as the indicator.

20. G. H. deHaas, N. W. Postema, W. Nieuwenhuizen, and L. L. M. van Deenen, *BBA* **159**, 118 (1968).
21. T. W. Wu and D. O. Tinker, *Biochemistry* **8**, 1558 (1969).

Contrary to the results with phospholipases from pancreas and *Crotalus adamanteus* venom as described above, Wu and Tinker reported that their preparation was inactivated by DFP. It did not show any isozymic forms. However, there was a drastic loss of enzymic activity in the initial ammonium sulfate fractionation and though it is doubtful that any isozyme would be lost specifically at this point, it is worthy of note. The purified phospholipase A_2 withstands heating to 80° for 30 min and pH 3.0, but similar treatment at pH 7 immediately inactivated it. Lyophilization caused inactivation of the enzyme, but activity was restored to a considerable extent in an aqueous medium over a 24 hr period.

Michaelis-Menten kinetics apparently are applicable to this enzyme in ether and in chloroform, K_m 8.3 and 8.5 mM, respectively. The enzyme was activated to varying degrees by Ca^{2+} and other divalent cations. However, certain divalent cations such as Zn^{2+} inhibited its action. Trace amounts of inhibitory cations required that EDTA be present in the assay mixture.

D. Physical and Chemical Characteristics

A considerable amount of data have now become available on the physical and chemical characteristics of the phospholipase A_2 from *Crotalus adamanteus* venom and pig pancreas. Inasmuch as there are many aspects of these data that may prove of importance in explaining certain of their unusual properties, e.g., reactivity in solvents such as diethyl ether and chloroform, it was considered of import to collate these results at this point.

1. *Molecular Weight*

Molecular weights have been obtained on phospholipase A_2 preparations from *C. adamanteus* venom, *C. atrox* venom, and pig pancreas. In addition, data on the molecular weight of pre-phospholipase A_2 (zymogen) isolated from pig pancreas are available. The figures obtained from sedimentation equilibrium, Sephadex G-75 chromatography, and amino acid assays are recorded in Table I.

It is of considerable interest to note that the molecular weight of the isozyme forms of this enzyme from *C. adamanteus* venom is approximately twice that of the enzyme isolated from *C. atrox* venom and pig pancreas. No explanation for this difference can be provided at this time.

The lower molecular weight of the phospholipase A_2 of pig pancreas, as compared to its zymogen form, can be attributed to the loss of a hepta-

TABLE I
MOLECULAR WEIGHT VALUES

Technique	Source of enzyme				
	Crotalus adamanteus[a] venom		*Crotalus atrox*[b] venom	Pig pancreas[c]	
	α	β		Zymogen (Pre)	Active form
Sephadex G-75 chromatography	Not run	Not run	14,500 ± 500	14,800 ± 500	13,900 ± 450
Amino acid content (calculated)	29,864	29,864	Not run	14,749	14,150
Sedimentation equilibrium (M_s, D)	29,800	29,800	Not run	Not run	13,500 ± 5%

[a] From Wells and Hanahan (*15*).
[b] From Wu and Tinker (*21*).
[c] From deHaas *et al.* (*19, 20*).

peptide during the enzymic conversion of the pre-phospholipase A_2 (nonactive) to the active form.

2. *Amino Acid Content*

The amino acid content of the phospholipase A_2 obtained from *C. adamanteus* venom and of the zymogen and active form of this enzyme isolated from pig pancreas is presented in Table II. Certain aspects of these results and comments on other applicable observations are presented below.

a. Phospholipase A_2 of Crotalus adamanteus Venom. Both forms of this enzyme showed identical amino acid content. One is immediately impressed by the high content of cystine residues, and since no free sulfhydryl groups could be detected in the intact protein, this indicates the presence of nearly 15 disulfide bonds per mole. This value is confirmed by results of treatment of the protein with mercaptoethanol. This

TABLE II
AMINO ACID COMPOSITION

Amino acid	Pig pancreas[a] Pre-phospholipase A_2 (nearest integer/14,000 g)	Pig pancreas[a] Phospholipase A_2 (nearest integer/14,000 g)	*C. adamanteus* venom[b] Phospholipase A_2 (nearest integer/ 29,864 g)
Asp	23	23	30
Thr	7	7	13
Ser	12	10	13
Glu	9	7	24
Pro	6	6	16
Gly	7	6	24
Ala	8	8	15
Val	2	2	11
Cys	14	14	30
Met	2	2	2
Ile	6	5	11
Leu	7	7	11
Tyr	8	8	16
Phe	5	5	10
Lys	9	9	16
His	3	3	5
Arg	5	4	12
Trp	2	2	7
NH_3	—	—	16

[a] From Maroux *et al.* (*22*).
[b] From Wells and Hanahan (*15*).

latter procedure, which eliminated all the enzymic activity, produced 29 residues of free sulfhydryl groups per mole of protein. These data agreed well with results on cysteic acid content after performic acid oxidation. Of considerable interest, no N-terminal groups on either native or modified protein could be detected.

b. Pig Pancreas. The amino acid content of the pre and active forms of phospholipase A_2, as might be expected, are quite comparable. However, close inspection of the data in Table II shows that the active form contains approximately 6–7 less amino acids. This difference is attributable to the cleavage (by trypsin) of a heptapeptide from the prephospholipase A_2. It is now believed (*22*) that pre-phospholipase A_2 contains only one tryptophan residue instead of two, and 12 half-cystine residues instead of the 14 listed in Table II.

Pre-phospholipase A_2 contains a single polypeptide chain, with 7 disulfide bonds and no N-terminal amino acid, but it does have a cystine carboxyl terminal group. On treatment with trypsin, cleavage of the seventh bond (Arg–Ala) in the chain occurs. The released heptapeptide was shown by mass spectrometry to have the following structure: PyroGlu–Glu–Gly–Ile–Ser–Ser–Arg. The origin of the N-terminal pyroglutamic acid is unknown, i.e., whether it was present in the original molecule or whether it derived from a cyclization of a glutamate or glutamine residue during the isolation period. The resulting major peptide fragment, phospholipase A_2, had full enzymic activity and was indistinguishable from that found in autolyzates of pancreas. It contained an N-terminal alanine and a C-terminal cystine. On disc gel electrophoresis it migrated as a single band which was distinct from phospholipase A_2 but could be converted to the phospholipase A_2 band by treatment with trypsin.

3. *Amino Acid Sequence*

The sequence of amino acids in the active phospholipase A_2 obtained by tryptic attack on the pre-phospholipase A_2 from pig pancreas has been reported by Maroux *et al.* (*22*). These results are presented in Fig. 1 and illustrate the unique structural characteristics of this protein. The peptide released from the pre-phospholipase A_2 was attached through its arginine residue to the alanine (N-terminal) present in the active form of this enzyme. Four of the six disulfide bridges have been identified and are connected between residues 18 and 83, 34 and 129, 68 and 98, and 91

22. S. Maroux, A. Puigserver, V. Dlouha, P. Desnuelle, G. D. deHaas, A. J. Slotboom, P. P. M. Bonsen, N. Nieuwenhuizen, and L. L. M. van Deenen, *BBA* **188**, 351 (1969).

8 9 10 11 12 13 14 15 16 17 18 19
Ala-Leu-Trp-Gln-Phe-Arg-Ser-Met- Ile-Lys-Cys-Ala-
20 21 22 23 24 25 26 27 28 29 30 31
Ile-Pro-Gly-Ser-His-Pro-Leu-Met-Asp-Phe-Asn-Asn-
32 33 34 35 36 37 38 39 40 41 42 43
Tyr-Gly-Cys-Tyr-Cys-Gly-Leu-Gly-Gly-Ser-Gly-Thr-
44 45 46 47 48 49 50 51 52 53 54 55
Pro-Val-Asn-Glu-Leu-Asn-Arg-Cys-Glu-His-Thr-Asp-
56 57 58 59 60 61 62 63 64 65 66 67
Asn-Cys-Tyr-Arg-Asp-Ala-Lys-Asn-Leu-Asn-Asp-Ser-
68 69 70 71 72 73 74 75 76 77 78 79
Cys-Lys-Phe-Leu-Val-Asp-Asn-Pro-Tyr-Thr-Glu-Ser-
80 81 82 83 84 85 86 87 88 89 90 91
Tyr-Ser-Tyr-Cys-Ser-Ser-Asn-Thr-Glx-Ile-Thr-Cys-
92 93 94 95 96 97 98 99 100 101 102 103
Asn-Ser-Lys-Asn-Asn-Ala-Cys-Glu-Ala-Phe-Ile-Cys-
104 105 106 107 108 109 110 111 112 113 114 115
Asn-Arg//Asn-Ala-Ala-Ile-Cys-Phe-Ser-Lys-Ala-Pro-
116 117 118 119 120 121 122 123 124 125 126 127
Tyr-Asn-Lys-Glu-His-Lys-Asn-Leu-Asn-Thr-Lys-Lys-
128 129
Tyr-Cys

FIG. 1. Amino acid sequence of reduced phospholipase A_2. Ala_8 represents the N-terminal amino acid in the active enzyme, which was derived by tryptic cleavage of a heptapeptide from the zymogen form. The source of this material was pig pancreas. Data from the work of Maroux *et al.* (*22*). Reproduced from *BBA* **188,** 351 (1969).

and 103. Undoubtedly, the high cross-linking via the disulfide bonds contributes to the unusual stability of this enzyme.

III. Phospholipase C

The enzymic activity of phospholipase C (phosphatidylcholine:Choline-phosphohydrolase, EC 3.1.4.3) catalyzes the hydrolytic cleavage of diacyl-*sn*-glycero-3-phosphorylcholine, as an example, to a 1,2-diacyl-glycerol (diglyceride) and phosphorylcholine.

A. SOURCES

This enzyme is found mainly in bacteria, with *Clostridium welchii* and *Bacillus cereus* being the most frequently used sources. It has been reported to be present in the marine planktonic chrysomonad *Monochrysis lutheri* (*23*). No evidence for its presence in mammalian tissues has been noted to date.

B. Isolation and Purification

Two procedures have been reported for the isolation and partial purification of this enzyme from bacterial sources and are outlined below.

1. *Bacillus cereus*

Kleiman and Lands (*24*) used a late-log phase (20–24 hr) static culture of *Bacillus cereus*, strain 7004, as starting material. This was centrifuged to remove cells and the supernatant fluid brought to 70% saturation with ammonium sulfate. The precipitated material contained all the phospholipase C activity. Subsequent dialysis and lyophilization yielded material for further purification by two different approaches. The first procedure involved chromatography of the crude enzyme preparation, in tris-buffer pH 7.4 at 0°, directly on a polyethyleneimine cellulose column with elution of enzyme activity in the solvent front. This resulted in a 22-fold purification. The second technique involved careful addition of protamine to the crude enzyme preparation with formation of a precipitate. The enzyme was recovered in the soluble fraction and a 2–3-fold increase in specific activity was obtained. The latter fraction was then chromatographed on DEAE-cellulose columns using a gradient of salt and buffer. This latter approach allowed a good recovery of activity and a 23-fold purification. Two different systems were used for assay of this enzymic activity, one of which utilized an ether-containing mixture of the substrate (lecithin) to which the enzyme was added. At the end of the incubation time, extraction with chloroform–methanol and washing with water yielded one product of the reaction, namely, water-soluble organic phosphorus or phosphorylcholine, which was then assayed for total phosphorus. An alternative method involved use of sonically dispersed phospholipid substrate, to which the enzyme was added. Subsequent to incubation, the extraction and remainder of the procedure was the same as described above. The activity of this purified phospholipase C preparation toward various phosphoglyceride substrates was undertaken and activities were evaluated. It was evident that this enzyme was different in its specificity as compared to the phospholipase C from *Clostridium perfringens* toxin in that it was active toward a

23. E. Bilinski, N. J. Antia, and Y. C. Lan, *BBA* **159**, 496 (1968).
24. J. H. Kleiman and W. E. M. Lands, *BBA* **187**, 477 (1969).

diacylglycerylphosphorylcholine (phosphatidylcholine), diacylglycerylphosphorylethanolamine (phosphatidylethanolamine) and diacylglycerylphosphoryl monomethylethanolamine. It was inactive toward diacylglycerylphosphate (phosphatidic acid). No evidence on the presence of other enzymic activity, e.g., proteases, in this phospholipase C preparation was presented by these investigators. This enzyme preparation had relatively little or no activity toward sphingomyelin.

Activators and Inhibitors. Dithiothreitol showed an inhibitory effect at a range of 1 to 10 mM. p-Chloromercuribenzoate was not inhibitory at concentrations near 0.1 mM; similarly, 5,5′-dithiobis(2-nitrobenzoic acid) was not inhibitory in the range of 5000–4 μM. Zinc ions showed little effect on this reaction. There was some question as to the importance of calcium ion in this reaction system. No chemical or physical parameters such as amino acid content, sedimentation equilibrium values, or molecular weights were presented.

2. *Clostridium perfringens*

Pastan *et al.* (*25*) have outlined a procedure wherein a phospholipase C specific for sphingomyelin and one specific for phosphoglycerides (for phosphatidylcholine) can be isolated in reasonable purity from cultures of *C. perfringens.*

Clostridium perfringens (ATCC 10543) was grown in liquid medium and at the proper time the cell culture was centrifuged and the cells discarded. The supernatant was then subjected to ammonium sulfate fractionation and later to chromatography on Sephadex G-100. In the latter instance, interestingly, two peaks of activity were obtained, one of which was active primarily toward sphingomyelin and the other of which was active primarily toward lecithin. Subsequent chromatography of these two fractions on DEAE-cellulose columns (0.01 M tris-HCl, pH 6) showed that the lecithin hydrolyzing activity was not adsorbed, whereas the sphingomyelin hydrolyzing activity was adsorbed and could be eluted later with sodium chloride. However, as noted by these authors, the sphingomyelinase activity was not present in all preparations.

Phospholipase C activity was assayed by the release of water-soluble organic phosphorus (phosphorylcholine) from a sonically dispersed substrate. The unreacted substrate and the other product ceramide (acyl sphingosine) were removed by solvent extraction. Specific comments on the individual hydrolyases are as follows.

25. J. Pastan, V. Macchia, and R. Katzen, *JBC* **243,** 3750 (1968).

a. Sphingomyelin Hydrolyase Activity. The optimal pH for this hydrolysis was 7.8–8.8 and the apparent K_m was 8×10^{-4} M. The enzyme was completely inhibited by preincubation with EDTA and was partially inhibited by preincubation with mercaptoethanol. Iodoacetic acid, iodoacetamide, DFP, and *N*-ethylmaleimide did not decrease this enzymic activity. There was apparently no product inhibition in this reaction.

Diethyl ether added to the reaction mixture increased the enzymic activity twofold. This latter observation would be in accord with an earlier report that a crude preparation of phospholipase C from *C. perfringens* toxin would attack lecithin in a diethyl ether solution (*26*). Interestingly, Mg^{2+} activated the sphingomyelin hydrolyzing activity whereas Ca^{2+} inhibited it. Zinc, copper, and iron were also inhibitory to varying degrees. This preparation did exhibit some activity toward lysolecithin and lecithin, but the major activity (on a rate basis, a tenfold difference) was toward sphingomyelin. It showed no activity toward phosphatidylserine or (dipalmitoyl) phosphatidylethanolamine. No activity toward an impure phosphatidylinositol was observed.

b. Lecithin Hydrolyase Activity. This enzyme fraction showed high activity toward lecithin with some reactivity toward sphingomyelin (approximately 20% of the rate): Ca^{2+} activated this enzyme system which was only 10% as active in its absence. Interestingly, most previous assays for phospholipase C on sphingomyelin included calcium in the medium, which would inhibit sphingomyelinase. Thus, this latter activity would not have been noted. Addition of Mg^{2+} to a calcium-inhibited preparation of sphingomyelinase restored its activity toward sphingomyelin. Phosphatidylethanolamine and phosphatidylserine were not hydrolyzed by this lecithin hydrolyase.

26. D. J. Hanahan and R. Vercamer, *JACS* **76,** 1804 (1954).

5

Acetylcholinesterase

HARRY C. FROEDE • IRWIN B. WILSON

I. Introduction

The enzymic hydrolysis of acetylcholine in nervous tissue [Eq. (1)]

$$(CH_3)_3\overset{+}{N}C_2H_4O\overset{O}{\overset{\|}{C}}CH_3 + H_2O \rightleftharpoons (CH_3)\overset{+}{N}C_2H_4OH + CH_3COOH \qquad (1)$$

was suggested 57 years ago by Dale (*1*) as a method for the rapid removal of this substance after it had served as a neurohumoral transmitter in nervous function. Later, cholinesterase activity was found in blood (*2*) and it was shown that there are at least two distinct enzymes, one in red cells and one in serum (*3*). The kinetic properties of these two en-

1. H. H. Dale, *J. Pharmacol. Exptl. Therap.* **6,** 147 (1914).
2. E. Stedman and L. H. Easson, *BJ* **26,** 2056 (1932).
3. G. A. Alles and R. C. Hawes, *JBC* **133,** 375 (1940).

zymes have become the basis for classifying enzymes in other tissues. The red cell type, called acetylcholinesterase (acetylcholine hydrolase, EC 3.1.1.7) hydrolyzes acetylcholine far more rapidly than butyrylcholine whereas the serum type, called butyrylcholinesterase, hydrolyzes butyrylcholine about four times more rapidly than acetylcholine (*4*).

In contrast to butyrylcholinesterase, acetylcholinesterase is subject to marked substrate inhibition and hydrolyzes D-β-methyl acetylcholine. Neither enzyme is completely specific for choline esters. For example, acetylcholinesterase can hydrolyze ethyl acetate and phenyl acetate. The former is a poor substrate while the latter is a good substrate.

At present there is no reason to suppose that all enzymes that readily hydrolyze acetylcholine must closely resemble one of these two enzymes. Possibly one day we may find an enzyme that hydrolyzes acetylcholine far more readily than butyrylcholine yet does not show marked substrate inhibition. There are also marked differences between different acetylcholinesterases. The eel enzyme, for example, is not inhibited by phenylmethanesulfonyl fluoride (*5*) but the bovine red cell enzyme is readily inhibited (*6*).

Acetylcholinesterase is found in nervous tissue of all species of animals (*7*) and besides its hydrolytic activity there has been some suggestion that it may function as a physiological receptor (*8*). Inhibitors of the enzyme are toxic, and some anticholinesterases are used as war gases and as insecticides. Others find use in the treatment of glaucoma and myasthenia gravis.

In this chapter we shall try to outline the main course of research in acetylcholinesterase catalysis. This includes the kinetics of catalysis and the possibility of regulation of catalysis via regulatory sites and subunit interaction. We shall not discuss the hydrolytic role of the enzyme in physiological functions nor report experimental work probing the question of the relationship of the enzyme to the acetylcholine receptor in the nervous system. An excellent review of earlier work on these subjects has appeared (*7*). There are a large number of interesting observations in this field.

4. K. B. Augustinnson and D. Nachmansohn, *Science* **100,** 454 (1944).
5. D. E. Fahrney and A. M. Gold, *JACS* **85,** 997 (1963).
6. P. Turini, S. Kurooka, M. Steer, A. N. Carbascio, and T. P. Singer, *J. Pharmacol. Exptl. Therap.* **167,** 98 (1969).
7. G. B. Koelle, sub ed., "Handbuch der experimentellen Pharmakologie," Vol. 15. Springer, Berlin, 1963.
8. J-P. Changeux, T. Podleski, and J.-C. Meunier, *J. Gen. Physiol.* **54,** *225S* (1969).

II. Molecular Properties

A. Purification

Only the enzyme from the electric organ of electric eel, *Electrophorus electricus*, has been highly purified and crystallized. Nachmansohn first recognized the electric eel as an excellent source of acetylcholinesterase, and this enzyme was brought to over 50% purity by ammonium sulfate fractionation and high-speed centrifugation (*9*). The enzyme had a specific activity of 7 mmoles of acetylcholine hydrolyzed per milligram of protein per minute at 25°C and pH 7.0.

Kremzner and Wilson obtained the enzyme at 90% purity, S.A. = 11, by column chromatography using, in part, an especially prepared cellulose derivative that was made as an affinity support (*10*). This cellulose derivative, B-DEAE cellulose, was prepared by treating diethylaminoethyl (DEAE) cellulose with benzyl bromide. It was hoped that a quaternary benzyldiethylaminoethyl cellulose derivative would be formed. Although the affinity of acetylcholinesterase for the cellulose was greatly enhanced, few quaternary groups were actually formed, and the cellulose did not function as a specific affinity column as judged by the observation that the affinity of acetylcholinesterase for the cellulose was only slightly decreased in the presence of quaternary ammonium ions. Nonetheless, the cellulose did afford a good purification of the enzyme, presumably because of hydrophobic properties introduced into the cellulose.

The enzyme, purified by the Kremzner-Wilson method, was crystallized by Leuzinger and Baker (*11*). The crystalline enzyme had a specific activity of 12 and was judged homogeneous by disc electrophoresis and high-speed centrifugation. Purification steps yielding the greatest increase in specific activity were chromatography on B-DEAE and on cellulose phosphate.

Other affinity columns that are effective in purification have been prepared (*12*, *12a*).

9. M. A. Rothenberg and D. Nachmansohn, *JBC* **168,** 223 (1947).
10. L. T. Kremzner and I. B. Wilson, *JBC* **238,** 1714 (1963).
11. W. Leuzinger and A. L. Baker, *Proc. Natl. Acad. Sci. U. S.* **57,** 446 (1967).
12. N. Kalderon, I. Silman, S. Blumberg, and Y. Dudai, *BBA* **207,** 560 (1970).
12a. J. D. Berman and M. Young, *Proc. Natl. Acad. Sci. U. S.* **68,** 395 (1971).

B. Physical Properties

The Rothenberg and Nachmansohn preparation had a very high sedimentation coefficient, in excess of 40 S, suggesting a molecular weight in excess of 2 million. By using the same method, Lawler obtained even higher molecular weights, sometimes in excess of 30 million (*13*).

The Kremzner and Wilson preparation and the crystalline enzyme have low molecular weights of about 250,000. Kremzner and Wilson (*14*) found an $s_{20,w}$ of 10.8 S and a diffusion coefficient of 4.3×10^{-7} cm^2/sec. These values yielded a molecular weight of 230,000 and a friction coefficient of 1.25. This friction coefficient indicates a compact or globular structure. Leuzinger *et al.* (*15*) obtained 260,000 molecular weight by equilibrium sedimentation. Since this value is approximately the same as found by Kremzner and Wilson, it confirms the value for the diffusion coefficient, the calculated friction coefficient, and the conclusion that the enzyme is a globular protein.

Since most of the enzyme is membrane bound in tissues, the appearance of high molecular weight species is of physiological interest. The molecular weight of the enzyme *in situ* is not known. In some preparations (but not those purified by the method of Kremzner and Wilson) the molecular size, as determined by sucrose gradient velocity sedimentation, depends upon the salt concentration and is very high (60 S) in low salt ($<0.1\ M$) and readily dissociates in higher salt concentrations (*16–18*). However, the smaller weight forms do not usually associate unless the enzyme is extensively dialyzed and then only if the preparation is fresh. Whether these aggregates are active per se or whether they become dissociated under the conditions of assay has not been investigated.

There have been several reports of a number of different sedimenting forms of acetylcholinesterase in rather impure preparations (*19–21*). In some cases these "peaks" were determined by protein measurements and therefore do not necessarily indicate enzyme. In other cases the enzymic activity was measured, and there can be little doubt that multiple en-

13. H. C. Lawler, *JBC* **238,** 132 (1963).
14. L. T. Kremzner and I. B. Wilson, *Biochemistry* **3,** 1902 (1964).
15. W. Leuzinger, M. Goldberg, and E. Cauvin, *JMB* **40,** 217 (1969).
16. M. A. Grafius and D. B. Millar, *BBA* **110,** 540 (1965).
17. J-P. Changeux, *Compt. Rend.* **D262,** 937 (1966).
18. M. A. Grafius and D. B. Millar, *Biochemistry* **6,** 1034 (1967).
19. A. B. Hargreaves, A. G. Wanderley, F. Hargreaves, and H. S. Goncalves, *BBA* **67,** 641 (1963).
20. J. Massouli and F. Rieger, *European J. Biochem.* **11,** 441 (1969).
21. J. Kates, personal communication (1970).

zyme forms are obtained in solution. The antibiotic, tyrocidine, aggregates acetylcholinesterase, but in this case the enzyme is inhibited (*22*). An aggregating protein factor has been reported (*22a*).

Although the lowest molecular weight for the active form of highly purified eel enzyme is 250,000 dalton ($s_{20,w} = 10.8$), forms from various sources with a lower sedimentation coefficient ($s_{20,w} = 7\text{–}9$) have been observed in a number of cases (*20, 21*). However, the 10.8 S form has not been broken down to smaller yet active particles.

Information about the number of active sites in a particular molecular form depends on means of evaluating the concentration of active sites in solution. This can be done by labeling the enzyme with ^{32}P-DFP (diisopropyl fluorophosphate) even with impure preparations, if suitable precautions are observed. When the concentration of active sites (normality of the solution) is known, the enzymic activity per site or turnover number can be calculated. If now the specific activity of pure enzyme is known, the weight per active site or the equivalent weight can be evaluated. The first such studies gave a turnover number of 4.5×10^5 per minute per site (*23*). Even though the specific activity of pure enzyme was not known at that time, results indicated that there were about 50 active sites in the 2×10^6 molecular weight form, the only form known at that time. Kremzner and Wilson, using titration of active sites with the irreversible inhibitor *o*-ethyl-*S*-dimethylaminoethylphosphothiolate, found a turnover number of 6×10^5 per minute (*14*).

Wilson and Harrison (*24*), using a method based upon the hydrolysis of the very poor substrate, dimethylcarbamyl fluoride, found a turnover number of 7×10^5 per minute and Bender and Stoops (*25*), using *o*-nitrophenyl dimethyl carbamate to obtain a burst of *o*-nitrophenol by reaction with acetylcholinesterase, found a turnover number of 1×10^6 min^{-1} for phenyl acetate hydrolysis V_{max}, which corresponds to about 8.5×10^5 min^{-1} for acetylcholine hydrolysis at assay conditions. Froede and Wilson (*26*), using ^{14}C-DFP, found 8×10^5 min^{-1}. These values indicate an equivalent weight of roughly 60,000 and point to the existence of four or more active sites in the 250,000 molecular weight enzyme. However, Leuzinger *et al.*, working with the crystalline enzyme, reported only two active sites in the 250,000-molecular weight enzyme. Their method of

22. J-P. Changeux, A. Ryter, W. Leuzinger, P. Barrand, and T. Podleski, *Proc. Natl. Acad. Sci. U. S.* **62,** 986 (1969).
22a. L. T. Kremzner, personal communication (1971).
23. H. O. Michel and S. Krop, *JBC* **190,** 119 (1951).
24. I. B. Wilson and M. A. Harrison, *JBC* **236,** 2292 (1961).
25. M. L. Bender and J. K. Stoops, *JACS* **87,** 1622 (1965).
26. H. C. Froede and I. B. Wilson, *Israel J. Med. Sci.* **6,** 179 (1970).

determination was not reported (*15*). Evidently there remains some question concerning the number of active sites, although the weight of evidence suggests four.

The possible presence of four active sites per molecule and the existence of isozymes from various sources (*27–29*) suggest that the 250,000-dalton enzyme might consist of at least four subunits. The enzyme has been dissociated into inactive subunits (*15, 26*). With guanidine hydrochloride, two equal subunits are obtained and with guanidine hydrochloride plus mercaptoethanol, four equal subunits are obtained. The role of mercaptoethanol in the dissociation does not appear to involve the reduction of intersubunit disulfide bonds since the dissociation readily occurs with quite dilute sodium dodecylsulfate (*26*), and even smaller subunits are obtained. Subunits of 20,000 and 40,000 have been observed recently (*29a*).

Other evidence shows that the subunits are not identical, however. Leuzinger *et al.* (*15*) found that the C-terminal amino acids are serine and glycine and therefore there must be at least two types of subunits. They have designated the enzyme as an $\alpha_2\beta_2$ form. It is not known if the 120,000 molecular weight form is α_2 or β_2.

The amino acid composition as reported by Leuzinger and Baker (*11*) is given in Table I. The enzyme also contains some carbohydrate;

TABLE I
AMINO ACID ANALYSIS OF ACETYLCHOLINESTERASE[a]

	I	II		I	II
Lysine	8	8	Alanine	10	10
Histidine	4	4	Half-cystine	2	2
Arginine	10	10	Valine	12	12
Tryptophan	4	4	Methionine	5	5
Aspartic acid	20	21	Isoleucine	6	6
Threonine	8	8	Leucine	16	16
Serine	12	12	Tyrosine	7	7
Glutamic acid	16	16	Phenylalanine	10	10
Proline	14	14	Hexosamine	3	5
Glycine	14	13	Ammonia	20	18

[a] Data were obtained from the analysis of two different preparations (I and II) of the enzyme. The values shown are based on the assumption of four histidine residues per minimum molecular weight (*11*).

27. J. Bernsohn, K. D. Barron, and M. T. Hedrick, *Biochem. Pharmacol.* **12,** 761 (1963).
28. D. J. Ecobichon and Y. Israel, *Can. J. Biochem.* **45,** 1099 (1967).
29. G. A. Davis and B. W. Agranoff, *Nature* **220,** 277 (1968).
29a. D. B. Millar and M. A. Grafius, *FEBS Letters* **12,** 61 (1970).

glucosamine, galactosamine, and possibly talosamine were identified (*11*).

The molecular weight of red cell enzyme has been estimated as 180,000 by electrophoresis of the sodium dodecyl sulfate denatured protein using ^{3}H-DFP as a label. Further treatment with mercaptoethanol yielded a 90,000-molecular weight subunit. The turnover number was 6×10^5 per active site (*30*).

III. The Catalytic Mechanism

A. Substrate Binding

The active site of acetylcholinesterase consists of two subsites, an anionic site and an esteratic site (*31, 32*) (Fig. 1). The anionic site determines specificity with respect to the alcohol moiety, and the esteratic site is involved in the actual catalytic process. Since an acetyl enzyme derivative of the esteratic site is involved in the catalytic process (*32, 33*) it is apparent that the structure about this subsite determines specificity with respect to the acid function of the substrate.

1. *Anionic Site*

As its name implies, the anionic site is the locus of an electrically negative potential which attracts the quaternary ammonium head of acetylcholine. This interaction makes a binding contribution of about

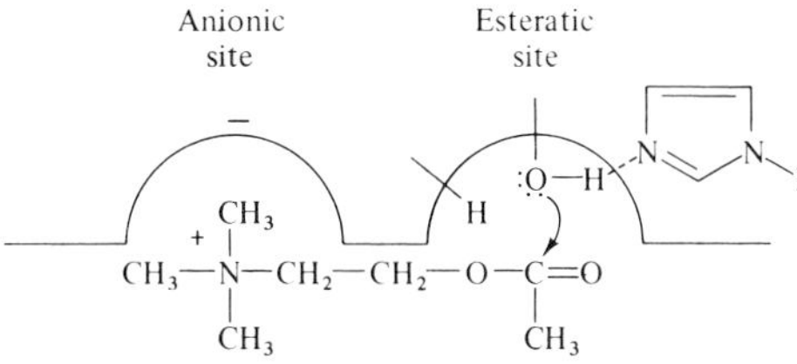

Fig. 1. The enzyme–substrate complex and one of the more commonly suggested mechanisms whereby the hydroxyl group of serine functions as a nucleophile to displace choline and form an acetyl enzyme. The imidazole of histidine serves as a general base, and a proton is shown at the esteratic site to envisage the possibility that a proton may be involved in the catalysis, perhaps by transfer to make choline (rather than the choline dipolar ion) the leaving group.

30. M. B. Bellhorn, O. O. Blumenfeld, and P. M. Gallop, *BBRC* **39,** 267 (1970).
31. D. H. Adams and V. P. Whittaker, *BBA* **4,** 543 (1950).
32. I. B. Wilson and F. Bergmann, *JBC* **185,** 479 (1950).
33. I. B. Wilson, F. Bergmann, and D. Nachmansohn, *JBC* **186,** 781 (1950).

2 kcal/mole of free energy corresponding to the measured difference of a factor of 30 in the binding of a charged molecule as compared to its uncharged isosteric analog (*31, 32*). Hydroxyethyldimethylammonium ion (I) and isoamylalcohol (II) are such a pair of inhibitors. A factor

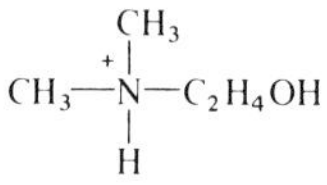

$$CH_3-\overset{+}{N}(CH_3)(H)-C_2H_4OH \quad \text{(I)}$$

$$CH_3-C(CH_3)(H)-C_2H_4OH \quad \text{(II)}$$

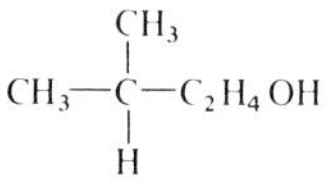

of 30 corresponds to a single negative charge at a distance of 5 Å from the positive center of charge. This distance corresponds to close contact between tetramethylammonium ion, $r = 3.5$ Å and a negatively charged atom, say, oxygen, $r = 1.5$ Å. These data will of course fit other arrangements, as, for example, two negative charges at a distance of 7 Å. From pH effect, Krupka has concluded that the anionic site has a pK_a of 6.3 (*34*).

The binding contributions of the methyl groups in acetylcholine were evaluated from a number of inhibitors including a series obtained by progressively replacing the hydrogen atoms of ammonium ion by methyl groups (*35*). Five compounds were involved: ammonium ion, methylammonium ion, dimethylammonium ion, trimethylammonium ion, and, finally, tetramethylammonium ion. A similar series was obtained starting with ethanolamine. The result was that on the average each replacement of H by CH_3 increased the binding by a factor of seven, or 1.2 kcal of free energy except for the last replacement. The last replacement contributed only a factor of three.

Since the N–H group is surely involved in hydrogen bonding with water, the origins of the contribution arising from the replacement of H by CH_3 are evidently complicated, but the tendency has been to refer to this effect as hydrophobic bonding. This assignment appears to be reasonable since the effect of the fourth replacement would correspond close to a statistical effect if the tetrahedral structure of tetramethylammonium ion requires one CH_3 group to project into the solution when the ion is bound. Similarly, the replacement of a methyl group by an ethyl group or by a larger hydrocarbon chain increases binding.

Most of the binding of acetylcholine is accounted for by Coulombic and hydrophobic interactions at the anionic site, and of these the sum of the hydrophobic interactions are the larger.

34. R. M. Krupka, *Biochemistry* **5,** 1988 (1966).
35. I. B. Wilson, *JBC* **197,** 215 (1952).

The role of the anionic site may not be limited to merely binding and orienting the substrate although that in itself has a large catalytic effect.

B. Esteratic Site

Although pH changes may affect a protein in many ways, it was suggested that the bell-shaped pH curve for enzymic hydrolyses might reflect the ionization of two essential groups in the esteratic site [Eq. (2)]

$$\underset{\text{inactive}}{EH_2^+} \underset{+H^+}{\rightleftharpoons} \underset{\text{active}}{EH} \underset{-H^+}{\rightleftharpoons} \underset{\text{inactive}}{E^-} \qquad (2)$$

with pK_a values of roughly 6.5 and 10.5 (*34, 36, 37*). The first ionization which corresponds to the imidazole group of histidine suggested a catalytic role for unprotonated imidazole in the hydrolytic process. This view is now widely held, but the original assumption that the basic group with a pK_a of 6.5 is the enzymic nucleophile has been discarded in favor of the hydroxyl group of serine. Imidazole is envisaged as playing an auxiliary role in promoting the nucleophilicity of the serine side chain; a number of similar proposals have been made (*36–43*), particularly for chymotrypsin. One of these is shown in Fig. 1. As already indicated, in its reaction with acetylcholine a group on the enzyme functions as a nucleophile, choline is displaced, and an acetyl enzyme derivative is formed. In the present view it is the hydroxyl group of serine that is acetylated although the transient existence of *N*-acetyl imidazole has not been completely ruled out. The mechanistic scheme is given by Eq. (3a),

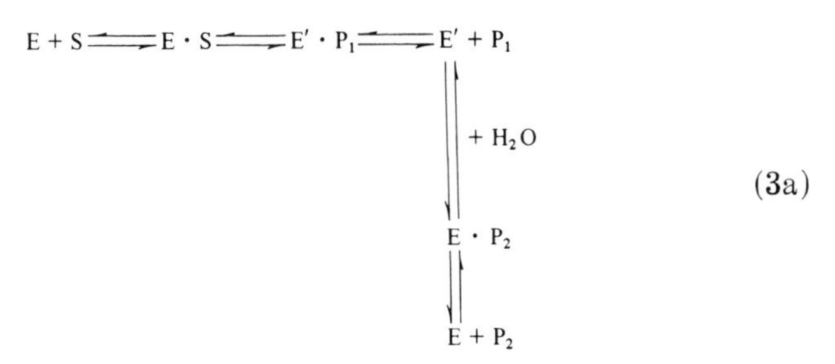

(3a)

36. R. M. Krupka, *Biochemistry* **5**, 1983 (1966).
37. I. B. Wilson and F. Bergmann, *JBC* **186**, 683 (1950).
38. L. W. Cunningham, *Science* **125**, 1145 (1957).
39. A. P. Brestkin and E. V. Rozengart, *Nature* **205**, 388 (1965).
40. L. Polgar and M. L. Bender, *Proc. Natl. Acad. Sci. U. S.* **64**, 1335 (1969).
41. R. M. Krupka, *Biochemistry* **6**, 1183 (1967).
42. M. Caplow, *JACS* **91**, 3639 (1969).
43. G. R. Hillman and H. G. Mautner, *Biochemistry* **9**, 2633 (1970).

where S is the substrate, E·S is the Michaelis complex between enzyme and substrate, $E' \cdot P_1$ is the Michaelis complex between acetyl enzyme and choline, E′ is the acetyl enzyme, $E \cdot P_2$ is the Michaelis complex between acetic acid and enzyme, P_1 is choline, and P_2 is acetic acid. In the usual case where the concentrations (P_1) and (P_2) are not sufficiently high to affect the kinetics and the dissociations of complexes are rapid, the pattern of Eq. (3b) holds.

$$\mathrm{E + S} \underset{k_{-1}}{\overset{k_1}{\rightleftharpoons}} \mathrm{E \cdot S} \xrightarrow{k_2} \mathrm{E' + P_1} \xrightarrow[k_3]{+\ \mathrm{H_2O}} \mathrm{E + P_2} \tag{3b}$$

The velocity of this reaction has the Michaelis form [Eq. (4)]

$$v = \frac{k_{cat}E^\circ}{1} + \frac{K_m}{(S)} \tag{4}$$

where

$$k_{cat} = \frac{k_2}{1} + \frac{k_2}{k_3} \qquad K_m = \frac{(k_{-1} + k_2)/k_1}{1 + k_2/k_3}$$

For acetylcholine k_{cat} (turnover number per active site) $= 7 \times 10^5$ min^{-1} and $K_m = 9 \times 10^{-5}$ *M* at 25°C, pH 7.0, 0.1 *M* NaCl. With this value of k_{cat}, E° is the enzyme normality rather than the enzyme concentration. At low (S), E° becomes E, the rate controlling process is the acetylation of the enzyme, and the velocity is given by Eq. (5):

$$v = \frac{k_{cat}}{K_m}(S)(E) \tag{5}$$

Thus, the second-order rate constant for the acetylation of the enzyme is $k_{cat}/K_m = 8 \times 10^9$ 1 mole^{-1} min^{-1}. This is an extremely high value. For ethyl acetate the corresponding quantities are k_{cat} 10^5, $K_m \sim 0.5$, and $k_{cat}/K_m \approx 2 \times 10^5$. The second-order rate constant for the hydrolysis of acetylcholine is 4×10^4 times greater than the second-order rate constant for the hydrolysis of ethyl acetate.

The difference in rate can be explained in a formal way. In absolute rate theory all transition states (with some exceptions) react at the same rate, k_bT/h. A difference in rate is then simply a reflection of a difference in the concentration of transition states. We can explain this greater concentration of transition states with acetylcholine by greater binding in the transition state arising from the quaternary ammonium function. The binding factors are known for interaction with the free enzyme. Let us assume that the same factors apply in the transition state. The binding contribution of a quaternary ammonium function in acetylcholine as compared to ethyl acetate involves 7×7 for two CH_3 groups, $3\times$ for the statistical effect of the third CH_3 group, and $30\times$ for the Coulombic attraction, making a total binding contribution of 4500. In this picture the existence of appropriate binding features contributes to the catalytic process by stabilizing the transition state, and this is the purpose of the anionic site. These interactions may be greater ($10\times$) in the transition state.

Acetylcholine, acetylthiocholine, and phenyl acetate are about equally good substrates, and there is no substrate known that is substantially better. Thus for these substrates it appears that k_2 is substantially larger than k_3; and the existence of a better substrate (acetate ester) is not likely.

The only estimate of the relative values of k_2 and k_3 have been made from an Arrhenius plot of log v (at high S) vs. T^{-1} (*44*). A straight line was obtained with poor substrates, but a curved line was obtained with acetylcholine. The curve was interpreted as representing the two steps of the hydrolytic process, k_2 and k_3, with different energies of activation. The ratio of k_2/k_3 was 6 at 25°C. Thus it appears that deacetylation is the rate limiting step for good substrates. For acetylcholine then, the dissociation constant of E·S is ~ 7 times the value of K_m ($k_{-1} > k_2$).

The conformation of acetylcholine in solution is evidently pertinent to its role as a substrate of the enzyme as well as its role as a receptor effector. A number of papers involving calculations of the conformation have appeared and X-ray diffraction has revealed the conformation of acetylcholine, acetylthiocholine, and acetylselenocholine in the crystalline state (*45–47*). In the crystal the C–N and O–C bonds have a gauche relationship, but C–N and S–C, C–N, and Se–C have a trans relationship. These relationships are believed to exist also in solution.

Hillman and Mautner (*43*) studied the enzymic hydrolysis of the three compounds and concluded that the rate of acylation of the enzyme decreased when S and Se replace O. They discussed the role of conformational, electronic, and other effects on the kinetics of hydrolysis. The gauche conformation seems to be the preferred orientation for acylation.

44. I. B. Wilson and E. Cabib, *JACS* **78,** 202 (1956).
45. F. P. Canepa, P. Pauling, and H. Sorum, *Nature* **210,** 907 (1966).
46. C. C. J. Culvenor and N. S. Ham, *Chem. Commun.* p. 537 (1966).
47. E. Shefter and H. G. Mautner, *Proc. Natl. Acad. Sci. U. S.* **63,** 1253 (1969).

C. Inhibitors

Only those inhibitors that react with the active site or appear to do so are discussed.

1. *Anionic Site Inhibitors*

From what has already been said it is apparent that any substituted ammonium ion, especially a tertiary or quaternary ammonium ion, is a potential inhibitor because it may be capable of binding at the anionic site. Tris (tris-trihydroxyethyl amino methane), which is a primary amine and a common buffer, is not an inhibitor. A sizable area comprising the anionic site seems to be hydrophobic because the introduction of larger hydrocarbon chains or rings almost always increases binding (*48–51*). Thus phenyltrimethylammonium, *N*-methylpyridinium, *N*-methylquinolinium and isoquinolinium, and *N*-methylacridinium ions are increasingly potent inhibitors.

3-Hydroxyphenyltrimethylammonium ion, $K_i \sim 2 \times 10^{-7}\ M$, is a rather interesting inhibitor in that the properly positioned OH group makes a sizable contribution of a factor of 120 to binding (*52*). This binding contribution has been ascribed to hydrogen bonding, perhaps with the imidazole group of the active site.

One might hazard the guess that except for special circumstances as described above the introduction of polar groups in an anionic site inhibitor would lead to decreased potency.

A new type of inhibitor of acetylcholinesterase was introduced when Belleau and Tani (*53*) reported the irreversible inhibition of erythrocyte enzyme by *N,N*-dimethyl-2-chloro-2-phenethylamine. The active form of inhibitor is the aziridinium (ethyleniminium) ion which apparently alkylates the enzyme at or near the anionic site [Eqs. (6) and (7)].

$$\phi\text{—}\underset{\text{Cl}}{\underset{|}{\text{CH}}}\text{—CH}_2\text{—N}\begin{matrix}\diagup R_1\\ \diagdown R_2\end{matrix} \longrightarrow \begin{matrix}\phi\text{—CH}_2\text{—CH}_2\\ \diagdown\ \diagup\\ \text{N}^+\\ \diagup\ \diagdown\\ R_1\quad R_2\end{matrix} + \text{Cl}^- \tag{6}$$

48. B. Belleau and G. Lacasse, *J. Med. Chem.* **7,** 768 (1964).
49. B. Belleau, *Ann. N. Y. Acad. Sci.* **144,** 705 (1967).
50. J. V. A. Auditore, *Intern. J. Neuropharmacol.* **3,** 599 (1964).
51. B. V. Rama-Sastry and E. C. White, *BBA* **151,** 597 (1968).
52. I. B. Wilson and C. Quan, *ABB* **73,** 131 (1958).
53. B. Belleau and H. Tani, *Mol. Pharmacol.* **2,** 411 (1966).

$$\phi-\underset{\substack{|\\ \mathrm{N}^+}}{\mathrm{CH}}\!\!-\!\!\mathrm{CH_2}\ (\text{aziridinium ring};\ \mathrm{N}^+ \text{ bearing } \mathrm{R_1},\ \mathrm{R_2}) + \mathrm{E} \longrightarrow \phi-\underset{\substack{|\\ \mathrm{E}}}{\mathrm{CH}}-\mathrm{CH_2}-\mathrm{N}\begin{matrix}\mathrm{R_1}\\ \mathrm{R_2}\end{matrix} \ \text{ or } \ \phi-\underset{\substack{|\\ \mathrm{N(R_1)(R_2)}}}{\mathrm{CH}}-\mathrm{CH_2}-\mathrm{E} \quad (7)$$

Other "alkylating" inhibitors include *N,N*-dibenzyl-β-chloroethylamine (dibenamine) (*54*) and *N*-(2-chloroethyl)-*N*-methyl-2-chloro-2-phenethylamine (*55*). The aziridinium ions form reversible complexes prior to alkylation, and the alkylation is blocked by quaternary ammonium ions such as tetramethylammonium ion (*53*) and by substrates (*56*).

Although the alkylated enzyme has little hydrolytic activity for acetylcholine, it has an enhanced activity for hydrolyzing the very poor substrate indophenyl acetate (*56, 57*).

In addition to enhanced hydrolysis of indophenyl acetate the alkylated enzyme is still capable of reacting with organophosphates such as sarin, soman, and tetraethylpyrophosphate, but not with amiton (*55*). The latter, which contains a cationic amine function, is thought to interact with the anionic site prior to phosphorylation of the enzyme. This again would indicate that alkylation of the enzyme by aziridinium ions prevents the interaction of cationic groups with the anionic site. Added evidence for alkylation near this site comes from the observation that tetramethylammonium ion and (3-hydroxyphenyl)trimethylammonium ion do not inhibit the hydrolysis of indophenyl acetate by alkylated enzyme, whereas these compounds are inhibitors of the hydrolysis of indophenyl acetate by untreated enzyme (*56*).

Another irreversible inhibitor that reacts with acetylcholinesterase to produce an enzyme with altered hydrolytic activity is *p*-(trimethylammonium)benzenediazonium fluoroborate (*58–60*). This compound was first reported by Wofsy and Michaeli to react rapidly with the enzyme to

54. F. Beddoe and H. J. Smith, *Nature* **216,** 706 (1967).
55. R. D. O'Brien, *BJ* **113,** 713 (1969).
56. J. E. Purdie, *BBA* **185,** 122 (1969).
57. J. E. Purdie and R. A. McIvor, *BBA* **128,** 590 (1966).
58. L. Wofsy and D. Michaeli, *Proc. Natl. Acad. Sci. U. S.* **58,** 2296 (1967).
59. J.-C. Meunier and J-P. Changeux, *FEBS Letters* **2,** 224 (1969).
60. T. Podleski, J.-C. Meunier, and J-P. Changeux, *Proc. Natl. Acad. Sci. U. S.* **63,** 1239 (1969).

form a coupled azo compound. Presumably the quaternary nitrogen is directed to the anionic site to form a complex prior to coupling. The coupling reaction is much faster than with other diazonium compounds. The enzyme is protected by quaternary ammonium ions such as phenyltrimethylammonium ion, tetraethylammonium ion, decamethonium, and D-tubocurarine against inactivation by *p*-(trimethylammonium)benzenediazonium fluoroborate (*8*, *58*). Reacting the enzyme with this reagent causes a marked loss in hydrolytic activity for acetylcholine, but an enhanced activity for hydrolyzing indophenyl acetate. Thus, there are similarities between aziridinium- and diazonium-inhibited enzyme.

2. *Esteratic Site Inhibitors*

In this section we include organophosphates, methane sulfonates, and carbamates. These compounds are called irreversible inhibitors or acid transferring inhibitors.

a. Organophosphates. A large number of compounds with structure (III) are inhibitors where R is an aryl or alkyl group; R_1 is an aryloxy,

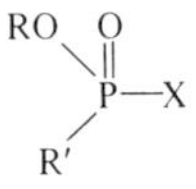

RO O
\ ||
P—X
/
R′

(III)

alkoxy, aryl, alkyl, or substituted amino function; and X is a leaving group. It has generally been thought that X must be a good leaving group and in the more potent inhibitors X is a group such as *p*-nitrophenol, CN, F, and phosphate diester but recently it has been shown that X may be a moderately long alkoxy function (*61*).

During inhibition the enzyme acts as a nucleophile and the hydroxyl group of serine is phosphorylated (*62–70*). In a limited series of organo-

61. P. Bracha and R. D. O'Brien, *Biochemistry* **9**, 741 (1970).
62. A. S. V. Burgen, *Brit. J. Pharmacol.* **4**, 219 (1949).
63. W. N. Aldridge, *Chem. & Ind. (London)* p. 473 (1954).
64. W. N. Aldridge, *BJ* **46**, 451 (1950).
65. B. J. Jandorf, H. O. Michel, N. K. Schaffer, P. Egan, and W. H. Summerson, *Discussions Faraday Soc.* **20**, 134 (1955).
66. C. Hansch, *J. Org. Chem.* **35**, 620 (1970).
67. W. N. Aldridge and A. Davison, *BJ* **51**, 62 (1952).
68. W. N. Aldridge, *BJ* **54**, 442 (1953).
69. K. B. Augustinsson, *Arkiv. Kemi* **6**, 331 (1953).
70. I. B. Wilson, *JBC* **190**, 111 (1951).

phosphates there is a positive and strong relationship between the rate of reaction with hydroxide ion and with the enzyme (*64–68*).

Using diethyl fluorophosphate as an example, the reaction scheme with the enzyme can be represented by Eq. (8):

$$\begin{array}{c} (EtO)_2POF + E \underset{K_1}{\rightleftharpoons} (EtO)_2POF \cdot E \\ k_{-2} \downarrow\uparrow k_2 \\ (EtO)_2POE \cdot F \\ \downarrow\uparrow \\ (EtO)_2POE + F^- \xrightarrow[H_2O]{k_3} (EtO)_2POOH + E \end{array} \tag{8}$$

The associations and dissociations of the Michaelis complexes are taken to be rapid with respect to the covalent reaction steps.

In the usual case the reaction is carried out in the absence of a kinetically significant concentration of the leaving group and the scheme simplifies to Eq. (9):

$$\begin{array}{c} (EtO)_2POF + E \rightleftharpoons (EtO)_2POF \cdot E \\ \downarrow k_2 \\ (EtO)_2POE + F^- \end{array} \tag{9}$$

Usually the inhibitor is present in large excess over the enzyme and then the inhibition is pseudo first order with (*71*) [Eq. (10)]

$$k_i = \frac{k_2}{1} + \frac{K_I}{(I)} \tag{10}$$

If $(I) \ll K_I$, which is very often the case, $k = k_2(I)/K_I$ and the second-order rate constant is $k_i = k_2/K_I$.

Organophosphates can be very potent inhibitors and there are many compounds with k_i values in the range 10^4–10^8 liters mole^{-1} min^{-1}. It is difficult to demonstrate the reversible complex and measure K_I because even for concentrations of inhibitor that are very much smaller than K_I the inhibition is very fast. However, the measurements have been made in some cases (*72–76*).

The organophosphates are hemisubstrates in that they react with the

71. R. J. Kitz, S. Ginsburg, and I. B. Wilson, *Mol. Pharmacol.* **3,** 225 (1967).
72. A. R. Main, *Science* **144,** 992 (1964).
73. E. Reiner and W. N. Aldridge, *BJ* **105,** 171 (1967).
74. Y. C. Chiu and W. C. Dauterman, *Biochem. Pharmacol.* **18,** 1665 (1969).
75. P. E. Braid and M. Nix, *Can. J. Biochem.* **47,** 1 (1969).
76. A. R. Main and F. Iverson, *BJ* **100,** 525 (1966).

enzyme to form a phosphoryl enzyme derivative analogous to the acetyl enzyme formed by the reaction of acetylcholine with the enzyme, but unlike the acetyl enzyme which reacts with water in 0.1 msec the phosphoryl enzyme reacts slowly with water. If the reaction with water were rapid, these substances would be substrates. The dimethylphosphoryl enzyme reacts with water in an hour or more, the diethylphosphoryl enzyme in several hours, and the diisopropylphosphoryl enzyme reacts extremely slowly (*65, 77–84*). Good nucleophiles such as NH_2OH react with the phosphoryl enzyme much more rapidly than water to dephosphorylate the enzyme and restore enzymic activity (*85*). This process is called reactivation.

The reaction of organophosphate inhibitors with the enzyme involves the catalytic machinery of the enzyme. The evidence for this is that although these substances hydrolyze only slowly and the hydroxyl group of serine with which they react is intrinsically a poor nucleophile, although better than other alcohols, the inhibitory reaction occurs with enzyme speed. The reaction is blocked by reversible anionic site inhibitors such as tetramethylammonium ion and the more specific (3-hydroxyphenyl)trimethylammonium ion (*86, 87*). The pH dependence of the inhibitory reaction approximates the pH dependence of the hydrolysis of acetylcholine and is precisely the same as the pH dependence of the hydrolysis of β-bromoethyl acetate (*88*). With the latter substrate, acetylation is rate controlling and its hydrolysis is therefore a better reference for the phosphorylation reaction than is the hydrolysis of acetylcholine. The phosphorylation reaction is inhibited by fluoride and fluoride also inhibits the hydrolysis of substrates (*89–92*). Again, compounds containing a quaternary nitrogen function are much better in-

77. W. N. Aldridge, *BJ* **55,** 763 (1953).
78. I. B. Wilson, S. Ginsburg, and C. Quan, *ABB* **77,** 286 (1958).
79. F. Hobbiger, *Brit. J. Pharmacol.* **6,** 21 (1951).
80. H. S. Jansz, D. Brons, and M. G. P. J. Warringa, *BBA* **34,** 573 (1959).
81. W. N. Aldridge and A. Davison, *BJ* **55,** 763 (1953).
82. D. R. Davies and A. L. Green, *BJ* **63,** 529 (1956).
83. I. B. Wilson, S. Ginsburg, and E. K. Meislich, *JACS* **77,** 4286 (1955).
84. F. W. Hobbiger, *Brit. J. Pharmacol.* **11,** 295 (1956).
85. I. B. Wilson, *JBC* **190,** 111 (1951).
86. G. B. Koelle, *J. Pharmacol. Exptl. Therap.* **88,** 232 (1946).
87. F. Bergmann and A. Shimoni, *BBA* **8,** 520 (1952).
88. D. F. Heath, *Intern. Ser. Monographs Pure Appl. Biol., Mod. Trends Physiol. Sci. Div.* **13,** 156 (1961).
89. E. Heilbronn, *Acta Chem. Scand.* **19,** 1333 (1965).
90. G. Cimasoni, *BJ* **99,** 133 (1966).
91. R. M. Krupka, *Mol. Pharmacol.* **2,** 558 (1966).
92. I. B. Wilson and R. A. Rio, *Mol. Pharmacol.* **1,** 60 (1965).

hibitors than would be expected from the pK_a of the leaving group (*71*, *93–96*). A number of these compounds such as (IV), (V), and (VI)

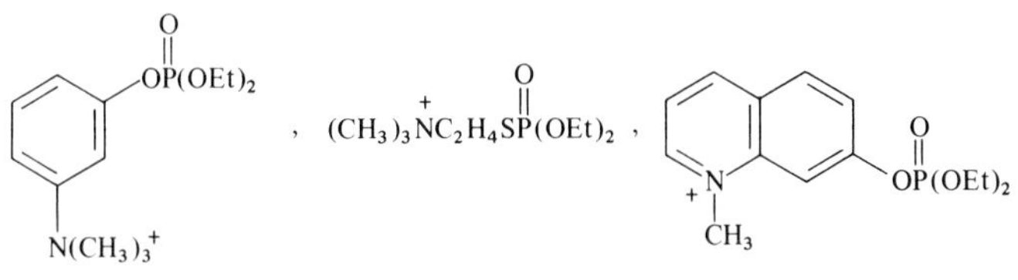

$$(CH_3)_3\overset{+}{N}C_2H_4S\overset{O}{\overset{\|}{P}}(OEt)_2$$

(IV) (V) (VI)

are very potent inhibitors. In today's terminology, they would be called active-site-directed inhibitors. Also, the reaction of the enzyme with sarin and tabun is stereospecific (*97*, *98*, *98a*). Finally, the intrinsically poor nucleophile, choline, readily reactivates the inhibited enzyme (*85*). In its interaction with the enzyme-active site, choline is of course promoted to a good nucleophile, for the enzyme must catalyze the synthesis of acetylcholine as well as its hydrolysis.

There has been considerable interest in designing good reactivators for the inhibited enzyme. Examination of Eq. (8) shows that the phosphorylation reaction which often proceeds to completion is nonetheless reversible, and for every inhibitor the leaving group is a conjugate reactivator. Thus during inhibition, the enzyme functions as a nucleophile but during reactivation it serves as a leaving group.

During a reactivation experiment the concentration of reactivator, (R), will almost always be much larger than the enzyme concentration and the pseudo-first-order rate constant for reactivation is given by Eq. (11)

$$k = \frac{k_{-2}}{1} + \frac{K_R}{(R)} \tag{11}$$

and the second-order rate constant is $k_r = k_{-2}/K_R$. The equilibrium constant for the phosphorylation reaction is $K_p = k_i/k_r$. In many cases both k_i and k_r are large enough to measure. Thus for the reaction of diethylfluorophosphate with the enzyme $k_i = 2.3 \times 10^5$ liters mole^{-1} min^{-1} and $k_r = 10$ liters mole^{-1} min^{-1} so that $K_p = 2.3 \times 10^4$ at 25°C pH 7.0

93. B. E. Hackley, Jr., G. M. Steinberg, and J. C. Lamb, *ABB* **80,** 211 (1959).
94. B. E. Hackley, Jr. and O. O. Owens, *J. Org. Chem.* **24,** 1120 (1959).
95. A. S. V. Burgen and F. Hobbiger, *Brit. J. Pharmacol.* **6,** 593 (1951).
96. L. E. Tammelin, *Acta Chem. Scand.* **11,** 1340 (1957).
97. P. J. Christen and J. A. C. M. van den Muysenberg, *BBA* **110,** 217 (1965).
98. J. H. Keijer and G. F. Wolring, *BBA* **185,** 465 (1969).
98a. F. C. G. Hoskin and G. S. Trick, *Can. J. Biochem.* **33,** 963 (1955).

(*92*). Evidently if the free energy of hydrolysis of diethylfluorophosphate were known, the free energy of hydrolysis of the phosphoryl enzyme could be calculated. In the absence of this knowledge, Wilson and Rio (*92*) assumed that the free energy of hydrolysis of the "P–F bond" was the same in the inhibitor as in fluorophosphoric acid, the value of which was known (*99, 100*). This assumption has since proved to be erroneous, and the value given by them for the free energy of hydrolysis of the diethylphosphoryl enzyme from eel is wrong. The correct value is —18 (*101*). This value is more negative for eel enzyme than for bovine erythrocyte enzyme which in turn is more negative than for chymotrypsin. The ratio of k_r to k_i is set by the free energy of hydrolysis of the inhibitor. Thus, if two inhibitors have the same free energy of hydrolysis, the leaving group of the better inhibitor will be the better reactivator. We can make a more general statement with rather less certainty. Since equilibrium constants vary over a much smaller range than rate constants, more often than not the leaving group of a good inhibitor will be a good reactivator.

In those cases that have been studied, the reaction of a closely related series of nucleophiles with an organophosphate follows a linear free energy relationship (*102–104*), $\log k = \log G + \beta pK_a$. Thus, for the reaction of sarin with oximes, $\log G = -3.25$ and $\beta = 0.642$ and with hydroxamic acids $\log G = -3.87$ and $\beta = 0.80$, where the value of k is calculated for the anion, which is the presumed nucleophile. The nucleophilicity is determined by the pK_a of the nucleophile. In the reactivation of the inhibited enzyme we are restricted to a pH close to 7.0 if our goal is an antidote for these compounds. For reaction at pH 7.0 there are two opposing effects. If the pK_a is high the anion will be a good nucleophile, but there will be little anion present. The actual rate constant is given by the above k times the fraction of anion at the pH selected [Eq. (12)]. Thus,

$$k_{obs} = GK_a^{-\beta} \Big/ \left(1 + \frac{(H^+)}{K_a}\right) \qquad \underset{\text{(optimum)}}{pK_a} = pH - \log\left(\frac{1}{\beta} - 1\right) \tag{12}$$

and there is a value of K_a at which the rate is optimal for a given pH.

99. W. Lange, *Chem. Ber.* **62B**, 1084 (1929).
100. D. P. Ames, S. Ohashi, C. F. Callis, and J. R. von Wazer, *JACS* **81**, 6350 (1959).
101. H. C. Froede and I. B. Wilson, unpublished observations (1970).
102. J. Epstein, P. L. Cannon, Jr., H. O. Michel, B. E. Hackley, Jr., and W. A. Mosher, *JACS* **89**, 2937 (1967)
103. G. M. Steinberg and R. Swidler, *J. Org. Chem.* **30**, 2362 (1965).
104. Y. Ashoni, Ph.D. Thesis, Hebrew University (1970).

For the usual values of β, the optimal pK_a is about the same as the pH $\pm$ 1 and the rate is not overly sensitive to the value of K_a. Thus, for the reaction of sarin at pH 7.0, we can do little better than to select a nucleophile with a pK_a of about 7.5, and if the inhibited enzyme behaved in the same way we should be similarly restricted.

This is not at all the case and the search for effective reactivators of the inhibited enzyme has been guided by the properties of the enzyme. The ability of choline to act as a nucleophile in reactivating inhibited enzyme was taken as an indication that the enzymic machinery in all or in part could be exploited in the reactivation process. Choline was assumed to bind to the anionic site in the diethylphosphoryl enzyme, positioned so that the oxygen atom fell one bond length from and directed toward the phosphorus atom. The binding of choline was easily proved by showing that choline reactivation has an enzyme saturation effect and that anionic site inhibitors such as tetramethylammonium ion block reactivation by choline. It was hoped that by combining an intrinsically good nucleophilic group with a suitably placed quaternary structure, a molecule could be devised that would very readily reactivate the enzyme. A large number of good reactivators were found culminating in pyridine-2-aldoxime methiodide, 2-PAM, PAM-2 (VII) (*105, 106*). [The value of k_r for this compound is 1.4×10^4 liters mole^{-1} min^{-1} (*78*).] This compound

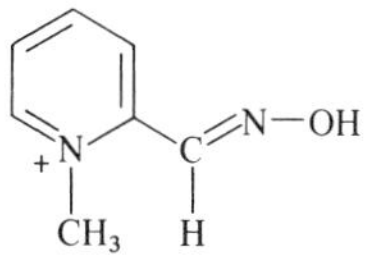

(VII)

was originally thought to have the *anti* configuration but is now known to be *syn* (*107*).

The two 4-PAM compounds, *syn* and *anti*, are respectively 20 and 70 times less active than 2-PAM (*108*). However, bisquaternary *syn* compounds (VIII) with an *n*-methylene bridge (or with other kinds of bridges) are not only 100 times more active than 4-PAM but are also considerably more active than 2-PAM (*109–113*). The length of the bridge

105. I. B. Wilson and S. Ginsburg, *BBA* **18**, 168 (1955).

106. A. F. Childs, D. R. Davies, A. L. Green, and J. P. Rutland, *Brit. J. Pharmacol.* **10**, 462 (1955).

107. D. Carlstrom, *Acta Chem. Scand.* **20**, 1240 (1966).

108. E. J. Poziomek, D. N. Kramer, W. A. Mosher, and H. O. Michel, *JACS* **83**, 3916 (1961).

109. E. J. Poziomek, B. E. Hackley, Jr., and G. M. Steinberg, *J. Org. Chem.* **23**, 714 (1958).

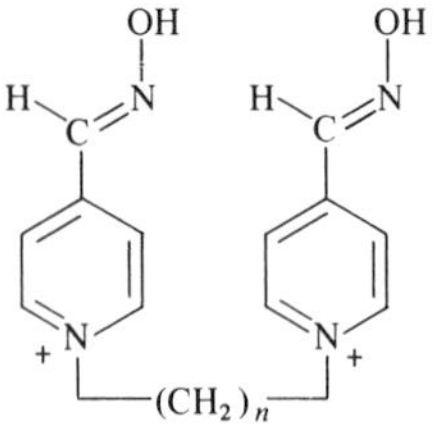

(VIII)

is not critical but the compound where $n = 3$, TMB_4, is the most active. The second oxime function is not necessary, a pyridine ring or trimethyl amine function serves as well.

Similar bridged diquaternary compounds based upon 2-PAM are also more active than 2-PAM, but in this case the length of the bridge is more critical and $n = 5$ is the best compound. The improvement over 2-PAM is only a factor of four, which is little more than the statistical factor of two.

The enhanced reactivating power of the diquaternary reactivators is even more dramatic in the phenylketoxime derivatives of the 4-PAMs (replace C–H by C–ϕ) and occurs with both *syn* and *anti* compounds (*114*). The possibility of a second anionic site arises in thinking about the greater activity of TMB_4 as compared to 4-PAM.

These compounds, 2-PAM and TMB_4, have been used successfully in animals and man as antidotes for a number of organophosphate poisons.

In the reaction of nucleophiles with compounds such as sarin, it is the anion that is the actual reactant and it is tempting to think that in the reactivation of the enzyme the anion is again the actual reactant. However, this may not always be the case. Consider a possible mechanism for the reaction of sarin with the enzyme, in which a proton from the enzyme is transferred to fluoride. In this case the reverse reaction, reactivation by fluoride, would require HF to be the actual reactant.

There is some evidence that there may be such an electrophilic component in the enzyme reaction. *N*-Phenylcarbamyl fluorides react much more readily with the enzyme than do the corresponding chlorides (*115*). Similarly methanesulfonyl fluoride reacts with the enzyme almost as

---

110. F. Hobbiger and P. W. Sadler, *Nature* **182,** 1672 (1958).
111. I. B. Wilson and S. Ginsburg, *Biochem. Pharmacol.* **1,** 200 (1959).
112. A. Luettringhaus and I. Hagedorn, *Arzneimittel-Forsch.* **14,** 1 (1964).
113. F. Hobbiger and V. Vojodic, *Biochem. Pharmacol.* **15,** 1677 (1966).
114. R. J. Kitz, S. Ginsburg, and I. B. Wilson, *Biochem. Pharmacol.* **14,** 1471 (1965).
115. H. P. Metzger and I. B. Wilson, *Biochemistry* **3,** 936 (1964).

rapidly as the chloride and bromide even though the last two would be expected to react very much more readily (*116*). These results can be explained by the much greater tendency of fluoride to accept a proton. However, these observations also can be explained by an addition–elimination mechanism in which the central atom, carbon or sulfur, will have greater electrophilic properties in the fluorine compounds.

When the enzyme inhibited by diisopropylfluorophosphate is allowed to stand, it becomes increasingly difficult to reactivate the enzyme (*7*). This phenomenon, called aging, results from the loss of an isopropyl group to leave monoisopropylphosphoryl enzyme (*117–119*). The loss of an alkyl group is more rapid with branched groups and is peculiar to cholinesterase—aging does not occur with chymotrypsin. Aging occurs less rapidly at higher pH. It has been proposed that aging involves the loss of a carbonium ion in a process involving an electrophilic assist by an acidic group in the enzyme [Eq. (13)]:

$$\text{E}(\text{H}\cdots\text{O}{=})\text{–O–P(O–R)(O–R)} \longrightarrow \text{E–O–P(OH)(OR)(=O)} + \text{R}^+ \qquad (13)$$

Enzyme that has been alkylated by an aziridinium ion and then phosphorylated does not show aging (*120*).

Since phosphate diesters are far more stable to nucleophilic attack than phosphate triesters it is understandable that the aged enzyme is refractory toward reactivation. For the same reason only one phosphate diester is known to inhibit the enzyme desmethylamiton, *O*-ethyl-*S*-(2-diethylaminoethyl)phosphorothiolate (*121*). This compound forms an immediately aged enzyme.

b. Carbamates. A number of compounds with the general structure (IX) are inhibitors of acetylcholinesterase, where R and R_1 may be hy-

$$\begin{matrix}\text{R} \\ \text{R}_1\end{matrix}\!\!>\text{N–C(=O)–X}$$

(IX)

116. R. J. Kitz and I. B. Wilson, unpublished observations (1965).
117. T. E. Smith and E. Usdin, *Biochemistry* **5,** 2914 (1966).
118. F. Berends, C. H. Posthumus, I. van der Sluys, and F. A. Deierkauf, *BBA* **34,** 576 (1959).
119. J. A. Cohen, R. A. Oosterbaan, H. S. Janz, and F. Berends, *J. Cellular Comp. Physiol.* **54,** Suppl. 1, 231 (1959).
120. R. A. McIvor, *BBA* **198,** 143 (1970).
121. A. H. Aharoni and R. D. O'Brien, *Biochemistry* **7,** 1538 (1968).

drogens, alkyl or aryl groups, and X is a leaving group. These compounds react with the enzyme to form a carbamyl enzyme. The reaction scheme is the same as for a substrate except that the hydrolysis of the carbamyl enzyme is very slow compared to the acetyl enzyme. Thus, in analogy to Eq. (8), the reactions may be depicted by Eq. (14):

$$R_1RN\overset{O}{\overset{\|}{C}}{-}X + E \underset{K_I}{\rightleftharpoons} R_1RN\overset{O}{\overset{\|}{C}}X \cdot E \xrightarrow{k_2} R_1RN\overset{O}{\overset{\|}{C}}{-}E + X \xrightarrow{k_3} R_1RN\overset{O}{\overset{\|}{C}}OH + E \qquad (14)$$

The value of k_3 depends upon R_1 and R is fairly rapid when these functions are H and/or CH_3, as in the case of many of the pharmacologically important carbamates. When one or both of these groups are aryl, $k_3 \rightarrow 0$ (*122*).

A number of people have suggested that carbamates might form a carbamyl enzyme derivative, and Myers and Kemp (*123*) showed that dimethylcarbamyl fluoride is an inhibitor.

When a carbamate inhibitor such as dimethylcarbamyl fluoride is added to an enzyme solution, a steady state is approached in which the rate of carbamylation equals the rate of decarbamylation. This steady state may be depicted by Eq. (15):

$$\left(\frac{E}{E'}\right)_{ss} - 1 = \frac{k_3}{k_2}\left(1 + \frac{K_I}{(I)}\right) \qquad (15)$$

where E is the active enzyme and E′ is the carbamyl enzyme (*122*). If this solution is extensively diluted, carbamylation virtually stops but decarbamylation continues unabated and the result is that enzymic activity returns slowly in accordance with the value of k_3. The half-time for the decarbamylation of dimethylcarbamyl enzyme is 30 min; for the monomethylcarbamyl enzyme, 40 min; and for the carbamyl enzyme, 2 min (*124*). Thus dilution or dialysis will restore enzymic activity, and it was at one time thought that carbamates were reversible inhibitors in contrast to organophosphates that were thought to be irreversible inhibitors (*32, 64*). Actually, the mechanism of inhibition is the same for both groups of compounds, but the pharmacologically interesting carbamates have a value of k_3 that is kinetically significant whereas the pharmacologically interesting organophosphates have a value of k_3 that is much smaller. However, one should note that enzyme inhibited by

122. I. B. Wilson, M. A. Hatch, and S. Ginsburg, *JBC* **235,** 2312 (1960).
123. D. K. Myers and A. Kemp, *Nature* **173,** 33 (1954).
124. I. B. Wilson, M. A. Harrison, and S. Ginsburg, *JBC* **236,** 1498 (1961).

dimethylfluorophosphate is readily restored to activity by dilution but enzyme inhibited by diphenylcarbamyl fluoride and certain bisquaternary carbamates is not.

Evidence that a carbamyl enzyme is indeed formed was obtained by showing that the half-time for recovery of enzymic activity is independent of the group X and that inhibited enzyme can be reactivated by hydroxylamine and by choline (*123*). Further evidence was obtained in which the rapid mixing of the dimethylcarbamate of *o*-nitrophenol and enzyme yielded a burst of *o*-nitrophenol (*25*).

The formation of a carbamyl enzyme could have been deduced from the observation that butyrylcholinesterase slowly hydrolyzes physotigmine (*125*), since the enzyme theory requires the formation of an acyl enzyme intermediate in the hydrolysis of any substrate. There are many structures that make suitable leaving groups. Among them are Cl, F, *p*-nitrophenol, choline, 3-hydroxypyridinium, (3-hydroxyphenyl)trimethylammonium, and other quaternary phenols based upon quinoline, isoquinoline, etc. (*126*). However, in comparison with the organophosphates there is less dependence upon X being an intrinsically good leaving group (low pK_a) and more upon the actual structure. Thus it makes a big difference whether the leaving group is a 2-, 3-, or 4-hydroxyphenyltrimethylammonium. Although 3-hydroxypyridinium is a better leaving group than (3-hydroxyphenyl)trimethylammonium, the dimethyl carbamate of the latter (neostigmine) is a better inhibitor than the dimethyl carbamate of the former (pyridostigmine). Although choline is a poor leaving group, and diethylphosphoryl choline is not an inhibitor, dimethylcarbamyl choline is an effective inhibitor.

c. Methane Sulfonate. A number of compounds with the structure CH_3SO_2X are inhibitors of acetylcholinesterase (*122, 127*). As in the case of the organophosphates and carbamates, the methane sulfonates react with the hydroxy group of serine to form in this case, a methane sulfonyl enzyme derivative. The reaction [Eq. (16)], however, is much slower and k_3 is barely demonstrable. The reversible complex is easily demonstrated with *N,N*-pentamethylene-bis-3-hydroxypyridinium iodide methane sulfonate because K_1 and k_2 are relatively small.

Insofar as it has been investigated, the structure of X is rather limited. Only X = Cl, Br, F, 3-hydroxyphenyltrimethylammonium, 3-hydroxypyridinium and diquaternary compounds based on 3-hydroxy-

125. A. Goldstein and R. E. Hamlisch, *ABB* **35,** 12 (1952).
126. R. J. Kitz, S. Ginsburg, and I. B. Wilson, *Biochem. Pharmacol.* **16,** 2201 (1967).
127. R. J. Kitz and I. B. Wilson, *JBC* **237,** 3245 (1962).

$$\begin{array}{c} CH_3SO_2\text{—}X \; + \; E \underset{K_I}{\rightleftharpoons} CH_3SO_2X \cdot E \\ \downarrow k_2 \\ CH_3SO_2E \; + \; X \xrightarrow{k_3 \approx 0} \end{array} \tag{16}$$

pyridinium have been shown to be effective inhibitors. Compounds with X = *p*-nitrophenol, choline, and thiocholine are not inhibitors.

The formation of a methane sulfonyl enzyme derivative was demonstrated by the reactivation of the enzyme with pyridine oximes and thiocholine at rates that were independent of the inhibitor, i.e., independent of the leaving group (*128*).

3. *Fluoride Inhibition*

In 1965, Heilbronn (*89*) reported the first detailed analysis on the inhibition of acetylcholinesterase by sodium fluoride. The inhibition is readily reversible by dilution or dialysis (*89–91*). Although inhibition increases with decreasing pH, the pK_a for inhibition was between 6.8 and 5.7. It was suggested that inhibition was by F^- and not by HF ($pK_a = 3.17$) or HF^{-2} ($pK_a = 0.59$) because the concentration of these species would be very low. The inhibition of enzymic activity has been described as mixed, noncompetitive, competitive, and uncompetitive, depending on the substrate used, fluoride concentration, and conditions for assay (*89–91*). According to Krupka (*91*), fluoride can bind to free enzyme (E), acetyl-enzyme (E′) or to the enzyme substrate complex (E·S) to give the different types of inhibition. For example, inhibition of acetylcholine hydrolysis is of the mixed type and the noncompetitive component is attributed to inhibition of deacetylation of the enzyme. Binding of fluoride to the E·S complex is indicated by the observation that inhibition of *N*-methylaminoethyl acetate hydrolysis for which acetylation is rate controlling is noncompetitive (*91*). On the other hand, from the effect of temperature on the rate of hydrolysis of acetylcholine in the absence and presence of fluorides, Heilbronn (*89*) concluded that only acetylation was affected by fluoride.

Fluoride inhibits phosphorylation (*92*), carbamylation (*129*), and methanesulfonylation (*130*) of acetylcholinesterase. The phosphoryl enzyme is reactivated by fluoride serving as a nucleophile (*92, 130a*), but decarbamylation whether by water or added nucleophiles is unaffected. Similarly, reactivation of the methane sulfonyl enzyme is not influenced by fluoride. Thus it appears that the effect of fluoride on deacetylation and on

128. J. Alexander, I. B. Wilson, and R. J. Kitz, *JBC* **238,** 741 (1963).
129. C. M. Greenspan and I. B. Wilson, *Mol. Pharmacol.* **6,** 266 (1970).
130. C. M. Greenspan and I. B. Wilson, *Mol. Pharmacol.* **6,** 460 (1970).
130a. E. Heilbronn, *Acta Chem. Scand.* **18,** 2410 (1964).

decarbamylation may be different. Possibly there is some facet of catalysis that is exercised in deacetylation and blocked by fluoride that is not involved in decarbamylation.

Studies of the effect of fluoride on methanesulfonylation indicated that fluoride and quaternary ammonium ions could simultaneously bind to the enzyme. Some ammonium ions decreased the binding of fluoride, some were without effect, but some increased the binding of fluoride. The simultaneous binding of fluoride and ammonium ion in these studies supports the contention that fluoride and substrate may simultaneously bind to the enzyme and may even be cooperative. This latter possibility tends to weaken the evidence that deacetylation can be inhibited by fluoride because a large (7×) cooperative effect if unlikely, nonetheless could account for the fluoride inhibition of the hydrolysis of acetylcholine.

It would be of interest to know how the fluoride is bound and why this binding interferes with the hydrolytic process. Since it is known that fluoride can be involved in hydrogen bonding, it is reasonable to assume that it binds to the active enzyme by this mechanism. Since it has been proposed by Metzger and Wilson (*131*), and can be inferred from the results of Bergmann *et al.* (*132*), that an electrophilic mechanism may be involved in the catalytic process, fluoride may interfere with the electrophilic mechanism that involves a hydrogen transfer to the O–ester linkage of the substrate.

D. Acceleration

The carbamylation and methanesulfonylation reactions of the enzyme are inhibited by substituted ammonium ions as would be expected. There are some interesting exceptions however. The reactions of dimethylcarbamyl fluoride and methanesulfonyl fluoride (chloride and bromide) with the enzyme are not inhibited by ammonium ions, but on the contrary these reactions are increased in rate (*133–135*). This acceleration can be quite substantial, and accelerations as high as 50-fold or greater have been observed. Tetraethylammonium ion is one of the better accelerators, and there are also ions such as *N*-methylpyridinium that bind to the enzyme but neither accelerate nor inhibit the reaction. This compound can prevent acceleration by other ions as can also 3-hydroxy-

131. H. P. Metzger and I. B. Wilson, *Biochemistry* **3,** 926 (1964).
132. F. Bergmann, S. Rimon, and R. Segal, *BJ* **68,** 493 (1958).
133. H. P. Metzger and I. B. Wilson, *JBC* **238,** 3432 (1963).
134. B. Belleau, V. Ditullio, and Y. H. Tsai, *Mol. Pharmacol.* **6,** 41 (1970).
135. R. Kitz and I. B. Wilson, *JBC* **238,** 745 (1963).

phenyltrimethylammonium which inhibits carbamylation and methanesulfonylation.

The cause of this phenomenon is not known, but it would appear to involve a conformational change in the enzyme. Acetylhomocholine, a substrate of the enzyme, also accelerates. It is of interest that a substrate such as acetylcholine may carry its own accelerating structure. Conformational changes of the enzyme have been observed when acetylhomocholine, 3-hydroxyphenyldimethylethylammonium chloride or tetraethyl pyrophosphate were added to the protein (*136*).

The kinetics for acceleration with methanesulfonyl fluoride follows the formulation given by Eq. (17):

$$\begin{aligned} E + A &\underset{K_A}{\rightleftharpoons} E \cdot A \\ E + I &\xrightarrow{k_i} E' \\ E \cdot A + I &\xrightarrow{\alpha k_i} E' \end{aligned} \tag{17}$$

with the first-order rate constant, k, for the formation of E at constant (A) and (I) as given by Eq. (18):

$$k = \frac{k_i[1 + \alpha(A)/K_A]}{1 + (A)/K_A} \tag{18}$$

If $\alpha > 1$ the ion A is an accelerator, if $\alpha < 1$ the ion inhibits sulfonylation, and if $\alpha = 1$ the ion although bound does not affect the rate of sulfonylation. All three circumstances have been observed.

The locus of the accelerator site is not known. The simplest model would assume that the accelerator site is the anionic subsite of the active site, but various investigators have been increasingly emphasizing the possibility of a second anionic regulatory site in acetylcholinesterase reactions (*8*).

How do ammonium ions affect decarbamylation? Under the usual conditions of measurement, ammonium ions inhibit decarbamylation (*133*). Thus the potency of dimethylcarbamyl fluoride as an inhibitor can be enormously increased by the presence of an ammonium ion because the ion on the one hand accelerates carbamylation and at the same time decreases the rate of decarbamylation.

Under conditions of low salt decarbamylation is slower. However, tetraethylammonium ion accelerates decarbamylation fourfold at *low* salt, but tetramethylammonium ion has little or no effect (*137*). This accelerating effect of TEA^+ on decarbamylation is readily demonstrated

136. R. J. Kitz and L. T. Kremzner, *Mol. Pharmacol.* **4**, 104 (1968).
137. B. D. Roufogalis and J. Thomas, *Mol. Pharmacol.* **5**, 28 (1969).

with enzyme from red cells but is very slight at best with the eel enzyme.

At this point we might discuss the effect of ammonium ions on the kinetics of the hydrolysis of acetylcholine (and other substrates). The inhibitor 3-hydroxyphenyltrimethylammonium is a competitive inhibitor as judged by the observation that in a v^{-1}, s^{-1} plot the slope is greatly increased by the inhibitor, but the intercept in the presence and absence of inhibitor is quite precisely the same, i.e., the apparent $V_{\max}$ is not affected. Other inhibitors decrease the apparent $V_{\max}$ moderately, but some inhibitors such as trimethylammonium ion have a large effect. The decrease in the apparent $V_{\max}$ is readily accounted for by assuming that the ion binds at the anionic site in the acetyl enzyme, interferes with the directed approach of water, and thereby prevents deacetylation. This conclusion is supported by the observation that decarbamylation is prevented by ammonium ions. Also, there is a lesser decrease of the apparent $V_{\max}$ with poorer substrates, where deacetylation is not slower than acetylation; therefore, inhibition of deacetylation would not have such a great effect on the overall rate (*138*).

The binding constants of trimethylammonium and dimethylammonium to the anionic site of the acetyl enzyme are about the same, but in most cases the binding to the free enzyme is greater than to the acetyl enzyme. In the case of the eel enzyme the acetyl enzyme·inhibitor complex can deacetylate at only a very slow rate at best, but in the case of the red cell enzyme the rate with some inhibitors may be slowed very little (*34*).

Substrate inhibition can be explained by assuming that a molecule of acetylcholine can bind at the anionic site of the acetyl enzyme; in support of this idea, it has been shown that decarbamylation is also inhibited by acetylcholine (*44*, *139*). Thus Eqs. (19) and (20) hold

$$\mathrm{E}' + \mathrm{S} \underset{K_s'}{\rightleftharpoons} \mathrm{E}' \cdot \mathrm{S} \tag{19}$$

$$v = k_{\mathrm{cat}}\mathrm{E}^\circ \Big/ \left[1 + \frac{K_m}{(\mathrm{S})} + \frac{(\mathrm{S})}{K_s'(1 + k_3/k_2)}\right] \tag{20}$$

The substrate inhibition constant is made up of two parts, one K_S' depends upon the ability of the substrate to bind to the acetyl enzyme and the other upon the extent to which deacetylation is rate controlling.

In this formulation E′·S was considered completely inactive, but it appears that E′·S may hydrolyze at a slow rate, perhaps 10% of the rate of hydrolysis of E′.

The more traditional method of explaining substrate inhibition assumes

138. R. M. Krupka, *Biochemistry* 3, 1749 (1964).
139. I. B. Wilson and J. Alexander, *JBC* **237**, 1323 (1962).

an inactive E·S·S complex and leads to the same form of velocity dependence on substrate concentration.

We have noted that in low salt the hydrolysis of the carbamyl enzyme (red cell) is accelerated by tetraethylammonium ion. It appears that this ion has a similar effect on the rate of hydrolysis of the red cell acetyl enzyme in low salt because the apparent V_{max} is found to increase (*140, 141*). Whether tetraethylammonium ion can increase the rate of acetylation in low salt with acetylcholine as a substrate is not known. If this were to occur it would imply that tetraethylammonium ion can bind to a second site besides the anionic site.

The existence of such an allosteric site has been invoked to explain the inhibition by compounds like gallamine and *d*-tubocurarine. These compounds readily inhibit acetylcholine hydrolysis but only up to about 50%. This indicates that a ternary complex of inhibitor, substrate, and enzyme is possible and suggests that the inhibitor is binding at a peripheral anionic site (regulatory site) and not at the active site since the substrate must occupy the active site. Binding studies with *p*-trimethylammonium benzene diazonium fluoroborate suggested ". . . two active sites and at least two regulatory sites per 2.6×10^5 molecular weight" (*8*). Gallamine accelerates the decarbamylation reaction in low salt (*142*). If gallamine does not bind at the anionic site as suggested above, it would appear that the accelerator site is the same as the proposed regulatory site.

140. B. D. Roufogalis and J. Thomas, *J. Pharm. Pharmacol.* **20,** 135 (1968).
141. B. D. Roufogalis and J. Thomas, *Mol. Pharmacol.* **4,** 181 (1968).
142. R. J. Kitz, L. M. Braswell, and S. Ginsburg, *Mol. Pharmacol.* **6,** 108 (1970).

6

Plant and Animal Amylases

JOHN A. THOMA • JOSEPH E. SPRADLIN • STEPHEN DYGERT

I. Introduction

The amylase-catalyzed hydrolysis of an $\alpha, 1 \rightarrow 4$-linked glucose polymer is achieved by the transfer of a glucosyl residue to water. Amylases are either referred to as α- or β-amylases or as endo- or exoamylases. α-Amylases ($\alpha, 1 \rightarrow 4$-glucan 4-glucanohydrolase, EC 3.2.1.1) were named by Kuhn (*1*) because the hydrolytic products possess the α config-

1. R. Kuhn, *Ann. Chem.* **443**, 1 (1925).

uration. α-Amylases attack large linear substrates at most internal bonds (*2*) and are properly classified as endoenzymes. Later, Ohlsson (*3*) discovered another amylase which hydrolyzed starch to yield a single product, β-maltose. He named it β-amylase (α, 1 → 4-glucan maltohydrolase, EC 3.2.1.2). β-Amylases are exoenzymes and attack alternate linkages from the nonreducing end of a substrate (*4*). The prefix α or β conveys information about product stereochemistry not substrate specificity. Some typical characteristics of these enzymes are presented in Table I.

TABLE I
SIMILARITIES AND DIFFERENCES OF AMYLASES[a]

Characteristic	α-Amylase	β-Amylase
Cleavage point	α, 1 → 4 Glucosidic bond Cleave C_1—O_4' bond	α, 1 → 4 Glucosidic bond Cleave C_1—O_4' bond
Configuration of new reducing unit	α	β
Mechanism	endo-Attack	exo-Attack
End products	Oligosaccharides mixture	Maltose
Decrease in viscosity and iodine staining	Rapid	Slow
Action at branch point	Can bypass	Cannot bypass
Transferase activity	Insignificant	Insignificant
Origin	Plant and animal	Plant

[a] Modified after Robyt and Whelan (*12*).

Recently, a third class of amylases has been discovered which are poorly characterized and commonly called glucoamylases or γ-amylases (α, 1 → 4-glucan glucohydrolase, EC 3.2.1.3). These amylases like β-amylase are exoenzymes but remove a single glucose unit from the nonreducing end of its substrates with inversion of configuration. In animals glucoamylase occurs mainly in the liver and to a smaller extent in other organs (*5, 6*). It appears to be very similar to the fungal glucoamylase (*7*). These enzymes cleave both α, 1 → 4 and α, 1 → 6 linkages and convert starch-like substrates completely to D-glucose. Very little

2. C. T. Greenwood and E. A. Milne, *Advan. Carbohydrate Chem.* **23,** 281 (1968), and references therein.
3. E. Ohlsson, *Z. Physiol. Chem.* **189,** 17 (1930).
4. D. French, "The Enzymes," 2nd ed., Vol. 4, p. 345, 1960.
5. E. L. Rozenfeld, I. S. Lukomskaya, N. K. Rudakova, and A. N. Shubina, *Biokhimiya* **24,** 1047 (1959).
6. E. L. Rozenfeld and I. A. Popova, *Bull. Soc. Clin. Biol.* **44,** 129 (1962).
7. J. H. Pazur, *in* "Starch: Chemistry and Technology" (R. Whistler and E. F. Paschall, eds.), Vol. 1, p. 133. Academic Press, New York, 1965.

information is available for animal glucoamylases, and they will not be discussed further in this chapter.

Some confusion has arisen in classifying the amylases since they have common substrates and products. Several authors (*8, 9*) have offered guidelines. The conception that all amylases act the same is grossly erroneous. Such a view has undoubtedly contributed to disinterest in carbohydrate enzymology.

In the past, impure enzymes, poorly characterized substrates, and inadequate analytical tools were frequently employed to investigate amylase action. In the last decade and a half a whole new arsenal of sophisticated techniques have been placed at the disposal of the carbohydrate enzymologist. Significant advancements have resulted. This chapter will emphasize recent developments in the amylase field, with discussion of the emerging fields of biosynthesis and the control of amylase action. For other recent reviews, see references *2, 4, 10–12*.

II. Assay Methods

A large number of modifications of amylase assay procedures are reported in the literature. The following section gives brief comments on the merits and utility of different assays.

Amylases fragment large substrates decreasing the average chain length with a concomitant increase in chain ends. Amylase measurements are designed to follow either substrate loss or product formation. The International Commission on Enzymes (*13*) has recommended a standard method for expressing enzymic activity as the number of micromoles of substrate transformed under defined conditions by an enzyme. Since polymeric substrates possess many potential points of cleavage, Robyt and Whelan (*12*) have proposed that "one micro-equivalent of group concerned" should be substituted for "one micromole of substrate." If one accepts this definition, then assay methods based

8. E. T. Reese, A. H. Maguire, and F. W. Parrish, *Can. J. Biochem.* **46,** 25 (1968).
9. K. K. Tung and J. H. Nordin, *Anal. Biochem.* **29,** 84 (1969).
10. E. H. Fischer and E. A. Stein, "The Enzymes," 2nd ed., Vol. 4, p. 333, 1960.
11. J. A. Thoma and J. E. Spradlin, *Brewers Dig.* **45,** No. 2, 58 and 66 (1970), and references therein.
12. J. F. Robyt and W. Whelan, *in* "Starch and its Derivatives" (J. A. Radley, ed.), 4th ed., Chapters 13, 14, and 15. Chapman & Hall, London, 1968.
13. International Union of Biochemistry, Commission on Enzymes. Rept., p. 8. Pergamon Press, New York, 1961. (*I.U.B. Symp. Ser.* **20.**)

upon the appearance of product (new reducing groups) are preferred for treatment of amylase kinetics (*14*). A rigorous and unambiguous treatment of amylase kinetics, however, is very complicated because bond susceptibility to hydrolysis depends upon its position in the substrate (*2, 11, 15*) as well as environmental conditions (*2, 12*) and other factors mentioned below.

The production of new reducing ends is most frequently followed by the 3,5-dinitrosalicylate (*16–18*), alkaline ferricyanide (*19–23*), or alkaline copper (*24–26*) methods, or by oxidation of glucose with glucose oxidase (*26a*). The simplest but by far the least desirable alkaline oxidation method involves reduction of 3,5-dinitrosalicylate. With this method, reducing equivalents for equimolar amounts of substrates increases significantly with increasing chain length (*18*). Measurements are subject to interference by Ca^{2+} used to stabilize amylases (*18*) and are relatively insensitive. This latter problem can create severe experimental problems with substrates that have very low dissociation constants. The other alkaline procedures display much less dependence of reducing equivalents on chain length (*18, 25, 26*). The best reagent for detailed enzyme kinetics is the copper-neocuproine reagent (*25*). It has the advantage of highest sensitivity and precision, low and stable background, and is relatively insensitive to a wide range of contaminating substances (*25*). Ironically, its greatest disadvantage is its high sensitivity leading to high substrate backgrounds. This problem can be circumvented by borohydride reduction of the substrate (*27*). Also, the use of poorly characterized substrate such as whole or acid modified

14. M. Dixon and E. C. Webb, "Enzymes," 2nd ed., Chapters 1–4. Academic Press, New York, 1964.
15. J. A. Thoma, C. Brothers, and J. E. Spradlin, *Biochemistry* **9,** 1768 (1970).
16. E. H. Fischer and E. Stein, *Biochem. Prep.* **8,** 27 (1961).
17. P. Bernfeld, "Methods in Enzymology," Vol. 1, p. 149, 1955.
18. J. R. Robyt and W. J. Whelan, *BJ* **95,** 10P (1965).
19. W. S. Hoffman, *JBC* **120,** 51 (1937).
20. R. M. McCready and W. Z. Hassid, *JACS* **65,** 1154 (1943).
21. J. F. Robyt and S. Bemis, *Anal. Biochem.* **19,** 56 (1967).
22. G. K. Adkins, W. Banks, C. T. Greenwood, and A. W. MacGregor, *Staerke* **21,** 57 (1969).
23. J. F. Robyt and D. French, *ABB* **122,** 8 (1967).
24. N. Nelson, *JBC* **153,** 375 (1944).
25. S. Dygert, L. H. Li, D. Florida, and J. A. Thoma, *Anal. Biochem.* **13,** 367 (1965).
26. W. Hoover, Ph.D. Thesis, University of Illinois, Champaign, Illinois, 1954.
26a. A. S. Keston and R. Brandt, *Anal. Biochem.* **6,** 461 (1963).
27. D. H. Strumeyer, *Anal. Biochem.* **19,** 61 (1967).

starch should be avoided for careful amylase kinetics because of the difficulty of interpreting the significance of the kinetic parameters (*28, 29*).

For routine analysis of biological fluids and bacteriological cultures the starch iodine reaction (*30–34*), paper disc (*35*), or chromogenic substrates (*36–38*) are recommended because of their speed and convenience. The decrease in iodine staining capacity during polysaccharide breakdown primarily reflects cleavage of large chains and is rather insensitive to rupture of polymers with less than 25 glucose units (*39*). This procedure, like the more cumbersome viscometric procedure (*40–43*), has the advantage of being sensitive to initial random amylase attack but insensitive to repetitive or multiple attack (*23, 44*). In contrast, reducing equivalents yielded by a good alkaline oxidation procedure are relatively insensitive to fragmentation patterns. Thus coupled viscosity-reducing value or blue value (iodine stain)-reducing value plots reveal important differences about the behavior of amylases (*23*) (see Fig. 1). If changes in viscosity and loss of iodine staining are quantitatively correlated to increase in reducing ends, these former two methods are suitable for evaluation of kinetic constants. However, calibration curves will be required for each set of experimental conditions and for each amylase.

A recent innovation for amylase assays involves the use of insoluble chromogenic substrates (*36–38*). Dyes coupled to a cross-linked in-

28. K. Hanson, *Biochemistry* **1,** 723 (1962).
29. J. A. Thoma, *Biochemistry* **5,** 1365 (1966).
30. G. B. Manning and L. L. Campbell, *JBC* **236,** 2952 (1961).
31. J. W. Van Dyk, J. T. Kung, and M. L. Caldwell, *JACS* **78,** 3345 (1956).
32. J. Robyt and D. French, *ABB* **100,** 451 (1963).
33. D. E. Briggs, *J. Inst. Brewing* **73,** 361 (1967).
34. J. Mestecky, F. W. Kraus, D. C. Hurst, and S. A. Voight, *Anal. Biochem.* **30,** 190 (1969).
35. E. Stark, R. Wellerson, P. A. Tetrault, and C. F. Kossack, *Anal. Biochem.* **30,** 190 (1969).
36. M. Ceska, E. Haltman, and B. G. A. Ingelman, *Experientia* **25,** 255 (1969).
37. H. Rinderknecht, P. Wilding, and B. J. Haverback, *Experientia* **23,** 805 (1967).
38. F. F. Hall, T. W. Culp, T. Hayakawa, C. R. Ratliff, and N. C. Hightower, *Am. J. Clin. Pathol.* **53,** 627 (1970).
39. J. Hollo and J. Szeitli, *in* "Starch and its Derivatives" (J. A. Radley, ed.), 4th ed., Chapter 7. Chapman & Hall, London, 1968.
40. C. T. Greenwood and A. W. MacGregor, *J. Inst. Brewing* **71,** 405 (1965).
41. C. T. Greenwood and E. A. Milne, *Staerke* **20,** 139 (1968).
42. C. T. Greenwood, A. W. MacGregor, and E. A. Milne, *Carbohydrate Res.* **1,** 303 (1965).
43. C. T. Greenwood, A. W. MacGregor, and E. A. Milne, *ABB* **112,** 466 (1965).
44. R. Bird and R. H. Hopkins, *BJ* **56,** 86 (1954).

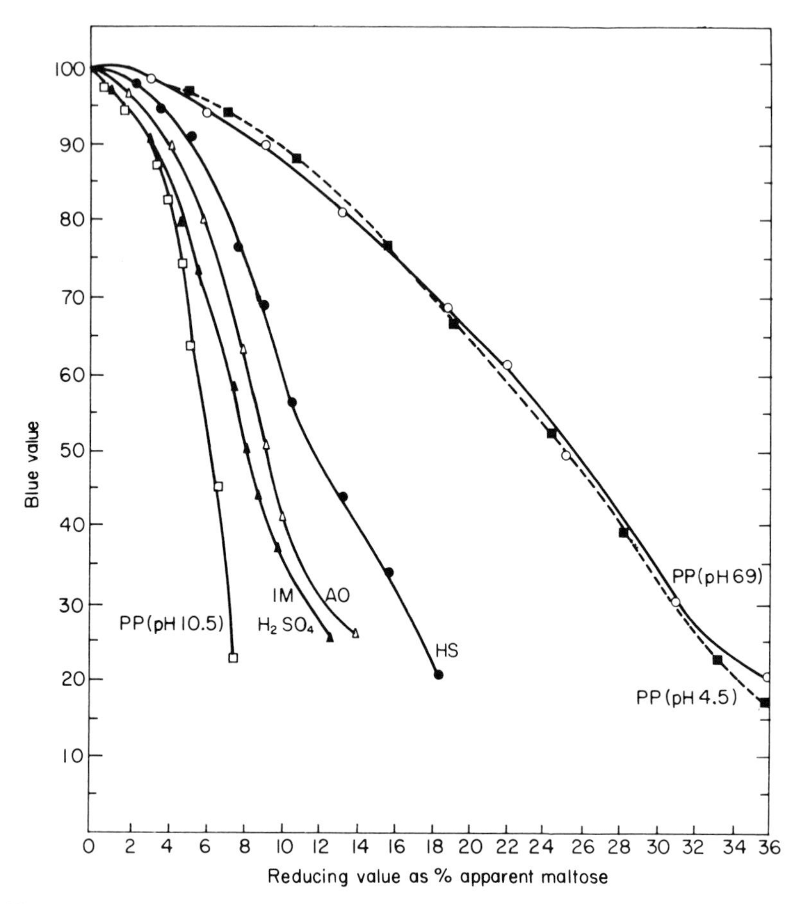

FIG. 1. Comparison of the drop in iodine color (blue value) with the increase in reducing value for the hydrolysis of amylose by various catalysts. Blue value is defined as $(A_t/A_0) \times 100$, where A_0 and A_t are the absorbancies (620 nm) of the iodine complex of the digest at zero time and at t minutes of hydrolysis. (○) Porcine pancreatic α-amylase at pH 6.9, 10 mM Cl^-, and 40°; (●) human salivary α-amylase at pH 6.9, 10 mM Cl^-, and 40°; (□) porcine pancreatic α-amylase at pH 4.5, 10 mM Cl^-, and 40°; (△) *Aspergillus oryzae* α-amylase at pH 5.5, and 40°; (▲) 1 M sulfuric acid at 60°; (■) porcine pancreatic α-amylase at pH 10.5, 10 mM Cl^-, and 40°. From Robyt and French (*23*).

soluble polysaccharide are released into solution attached to oligosaccharide fragments when the substrate is digested by amylases. The chief merits of this technique are its extreme simplicity and stability of the substrate. This method as well as drop in viscosity and iodine staining methods can conveniently assay α-amylase in the presence of β-amylase.

III. Origins, Purification, and Molecular Variants

Starch and related polymers are universal sources of dietary carbon throughout the plant and animal kingdoms. This fact probably accounts for the ubiquitous presence of α-amylases in living organisms. α-Amylases are located in almost every fluid and tissue to some extent. In mammals the salivary gland and pancreas secrete the main supplies of amylases which are utilized for the initial degradation steps in starch metabolism.

In contrast, β-amylase appears to be confined to the seeds of higher plants and sweet potatoes and is present only in limited amounts in oats, corn, rice, and sorghum. In the mature seed, β-amylase is stored adjacent to the aleurone layer (*45, 46*) chemically attached to glutenin (*47*).

The amylases are typical protein enzymes and are purified by conventional protein fractionation techniques (*10, 40, 48–71*). Freeing β-

45. C. Engel, *Rec. Trav. Chim.* **64,** 318 (1945).
46. K. Linderstrøm-Lang and C. Engel, *Enzymologia* **3,** 138 (1937).
47. E. V. Rowsell and L. J. Goad, *BJ* **84,** 73P (1962).
48. H. A. Soper, R. W. Hartley, Jr., W. R. Carroll, and E. A. Peterson, *Proteins, Composition, Struct. Function* **3,** 1–97 (1965).
49. For specific details, see A. A. Green, and W. L. Hughes, S. P. Colowick, R. R. Porter, C. H. W. Hirs, L. Heppel, and G. Gomori, *in* "Methods in Enzymology," Vol. 1, pp. 67–138, 1955; E. A. Peterson, and H. A. Sober, Ö. Levin, M. Bier, and A. F. Brodie, Vol. 5, pp. 3–51, 1962.
50. W. Banks, C. T. Greenwood, and J. Thomson, *Makromol. Chem.* **31,** 197 (1959).
51. J. E. Kruger and R. Tkachuk, *Cereal Chem.* **46,** 219 (1969).
52. R. Shainkin and Y. Birk, *BBA* **122,** 153 (1966).
53. D. P. Botes, F. J. Joubert, and L. Novellie, *J. Sci. Food Agr.* **18,** 409 and 415 (1967).
54. A. Gertler and Y. Birk, *BJ* **95,** 621 (1965).
55. K. H. Tipples and R. Tkachuk, *Cereal Chem.* **42,** 111 (1965).
56. R. R. Barton, French Patent, 1,497,880 (1967).
57. A. H. Blandamer and R. B. Beechey, *Nature* **197,** 591 (1963).
58. B. Gelotte, *Acta Chem. Scand.* **18,** 1283 (1964).
59. P. Wilding, *Clin. Chim. Acta* **8,** 918 (1963).
60. S. Schwimmer and A. K. Balls, *JBC* **179,** 1063 (1949).
61. A. Markovitz, H. P. Klein, and E. H. Fischer, *BBA* **19,** 267 (1956).
62. S. K. Dube and P. Nordin, *ABB* **94,** 121 (1961).
63. S. Schwimmer and A. K. Balls, *JBC* **180,** 883 (1949).
64. S. Schwimmer and A. K. Balls, *JBC* **179,** 1036 (1949).
65. W. Muller, H. Gralfs, and G. Manilz, German Patent, 1,284,391 (1968).

amylase from traces of α-amylase contamination has been a vexing purification problem. Several techniques have proved valuable for this chore, which take advantage of the sensitivity of α-amylases to selective denaturation. These methods involve controlled denaturation of α-amylase at acid pH (*50*) or by heat (*51*) or by proteolytic digestion (*10*) coupled with the removal of structure stabilizing Ca^{2+} from α-amylase with ethylenediaminetetraacetate (EDTA) (*40, 41*). Affinity chromatography on starch (*60–64*) and glycogen precipitation (*40, 66–69*) are powerful tools for amylase purification. Adsorptivity changes inversely with granule size and is directly proportional to surface area (*63, 64*). Amylopectin is a more effective adsorbant than either amylose or whole starch granules (*63, 64*). β-Amylase and glucoamylase exhibit a much lower affinity than α-amylase for starch, and this offers a simple basis to separate these related enzymes (*62*). Glycogen precipitation, a technique pioneered by Schramm and co-workers (*66*), is a valuable technique for purification of α-amylase. The utility of this method will probably be limited to enzymes capable of forming insoluble multimolecular aggregates. α-Amylases relatively free from β-amylase can be selectively extracted from ungerminated grain seeds by salt extraction. β-Amylase, stored as a zymogen, can be extracted after 2-mercaptoethanol and papain treatment (*72*).

A large number of amylases have been extensively purified or purified to the point of crystallinity. The following amylases are in this category: α-amylases—hog pancreas (*73, 74*), human pancreas (*75*), and saliva (*76–78*), rat pancreas (*79*), malted barley (*60*), malted sorghum

66. A. Loyter and M. Schramm, *BBA* **65,** 200 (1962).
67. D. French and M. Abdullah, *Cereal Chem.* **43,** 555 (1966).
68. C. T. Greenwood, A. W. MacGregor, and E. A. Milne, *Carbohydrate Res.* **1,** 229 (1965).
69. C. T. Greenwood, A. W. MacGregor, and E. A. Milne, *ABB* **112,** 459 (1965).
70. M. Nummi, R. Vilhunen, and T. M. Enari, *Acta Chem. Scand.* **19,** 1793 (1965).
71. P. Cuatrecasas, M. Wilchek, and C. B. Anfinsen, *Proc. Natl. Acad. Sci. U. S.* **61,** 636 (1968).
72. N. Mugibayashi and R. Shinke, *Nippon Nogei Kagaku Kaishi* **41,** 295 (1967); CA **67,** 113925g (1967).
73. E. H. Fischer and P. Bernfeld, *Helv. Chim. Acta* **31,** 1831 (1948).
74. M. L. Caldwell, M. Adams, J. T. Kung, and G. C. Toralballa, *JACS* **74,** 4033 (1952).
75. E. H. Fischer, F. Duckert, and P. Bernfeld, *Helv. Chim. Acta* **33,** 1060 (1950).
76. K. H. Meyer, E. H. Fischer, A. Staub, and P. Bernfeld, *Helv. Chim. Acta* **31,** 2158 (1948).
77. J. Muus, *Compt. Rend. Trav. Lab. Carlsberg, Ser. Chim.* **28,** 317 (1953); CA **47,** 8133 (1953).
78. H. Mutzbauer and G. V. Schulz, *BBA* **102,** 526 (1965).
79. N. G. Heatley, *Nature* **181,** 1069 (1958).

(*53, 80*), soy beans (*68*), broad beans (*69*), malted wheat (*41*), pigeon pancreas (*81*), and shore crab (*57*); β-amylases—sweet potato (*82*), malted barley (*83*), barley (*70*), wheat (*84–86*), soybeans (*54, 87, 88*), and malted sorghum (*53*).

The amino acid composition of amylases as given in Tables II (*53, 54, 86, 89–91*) and III (*92*) depends upon their source. Compositional differences are reflected in their physical, chemical, and enzymological properties. A fascinating difference between amylases is the hydrolytic activity they display toward starches of various sources; for example, the amylases from serum, urine, and saliva digested potato starch more rapidly than cornstarch while cornstarch is more susceptible to the amylases from pancreatic serum, secretin-stimulated duodenal fluid, and pancreatic extracts (*93*). Hall *et al.* (*94*) have extended these observations and have characterized amylases by their ability to digest different starch preparations. They found the relative rate of initial attack for salivary amylase on Lintner soluble starch compared to amylose Azure was at least twice as great as the relative rates obtained for pancreatic amylase. Similar results (*94*) were obtained for the hog saliva and hog pancreatic α-amylases.

The literature is replete with similar observations, but no satisfying rationalization of the results is offered. The interaction of many factors, the history of the starch sample, degree of branching, extent of swelling, granule size, permeability to the amylase, enzyme affinity for the substrate, degree and polarity of multiple attack, and fragmentation patterns of the enzymes are probably influencing factors.

80. H. G. Griffin and Y. V. Wu, *Abstr. Papers, 154th ACS Meeting, Chicago,* p. 45D (1967).

81. F. B. Straub, *Acta Physiol. Acad. Sci. Hung.* **12,** 295 (1957).

82. A. K. Balls, M. K. Walden, and R. R. Thompson, *JBC* **173,** 9 (1953).

83. A. Piguet and E. H. Fischer, *Helv. Chim. Acta* **35,** 257 (1952).

84. K. H. Meyer, P. F. Spahr, and E. H. Fischer, *Helv. Chim. Acta* **36,** 1924 (1953).

85. C. C. Walden, Ph.D. Dissertation, University of Minnesota, Minneapolis, Minnesota, 1954.

86. T. Tkachuk and K. H. Tipples, *Cereal Chem.* **43,** 62 (1966).

87. J. Fukumoto and Y. Tsujisaka, *Kagaku To Kogyo (Osaka)* **28,** 282 (1954).

88. J. Fukumoto and Y. Tsujisaka, *Kagaku To Kogyo (Osaka)* **29,** 124 (1955).

89. J. Muus, *JACS* **76,** 5163 (1954).

90. M. L. Caldwell, E. S. Dickey, V. M. Hanrahan, H. C. Kung, J. T. Kung, and M. Misko, *JACS* **76,** 143 (1954).

91. J. A. Thoma, D. E. Koshland, R. Shinke, and J. Ruscica, *Biochemistry* **4,** 714 (1965).

92. G. M. Malacinski and W. J. Rutter, *Biochemistry* **8,** 4382 (1969).

93. S. Meiter and S. Rogols, *Clin. Chem.* **14,** 1176 (1968).

94. F. F. Hall, C. R. Ratcliff, T. Hayakawa, T. W. Culp, and N. C. Hightower, *Am. J. Dig. Dis.* **15,** 1031 (1970).

TABLE II
AMINO ACID COMPOSITION OF AMYLASES[a]

Amino acid	Human saliva α [Ref. (*89*)]	Hog pancreas α [Ref. (*90*)]	Malted sorghum α [Ref. (*53*)]	Sweet potato β [Ref. (*91*)]	Soybean β [Ref. (*54*)]	Malted sorghum β [Ref. (*53*)]	Wheat β [Ref. (*86*)]
Glycine	6.8	6.7	7.0	4.7	6.5	4.9	5.6
Alanine	4.4	6.9	6.6	5.4	6.1	7.1	6.0
Valine	6.9	7.8	5.4	5.9	7.2	6.6	7.3
Leucine	5.8		7.9	7.5	13.2	9.1	9.7
Isoleucine	5.8	11.5	6.7	4.6	6.6	4.9	4.3
Serine	7.8	4.1	3.4	3.0	5.6	2.6	3.5
Threonine	4.5	3.9	4.2	2.8	4.4	5.1	3.1
½-Cystine	4.4	2.3	0.4	1.3	1.0	*b*	1.5
Methionine	2.4	2.1	1.4	4.1	2.2	2.5	2.9
Proline	3.6	3.6	4.5	5.4	7.5	5.4	6.0
Phenylalanine	7.2	10.1	6.2	5.7	5.9	5.7	6.1
Tyrosine	5.5	5.3	6.0	6.2	8.8	8.1	6.9
Tryptophan	7.2	6.7	*b*	4.0	4.0	2.2	4.4
Histidine	3.2	3.9	5.0	2.0	3.0	2.7	4.8
Lysine	6.3	4.9	5.2	7.4	7.6	6.8	4.7
Arginine	8.7	5.8	5.8	5.4	5.7	6.3	8.0
Aspartic acid	19.3	14.5	15.7	12.7	17.9	14.6	12.4
Glutamic acid	9.6	10.5	9.9	9.9	16.7	13.9	13.8
Amide	1.3	1.6	1.2	*b*	1.7	1.8	1.3

[a] Composition in grams per 100 g of protein.
[b] Not reported.

Amylases from a given organ in different species often display greater similarity in their enzymic behavior than amylases from different organs within the same species. According to limited investigations, the human and monkey amylase antienzymes register more organ than species specificity (*95*, *96*) (Table IV). The salivary, pancreatic, and serum amylases also differ electrophoretically (*97*, *98*).

The following section on pancreatic amylase isozymes presents this area in terms of emerging patterns. The generalizations here may require some modification as additional data are reported.

Molecular variants of amylases from pancreas have been investigated in several laboratories where workers are beginning to clarify the origins and properties of these isozymes. Fractionation of the multiple forms

95. T. Hayakawa, F. F. Hall, and N. C. Hightower, *Clin. Res.* **17**, 527 (1969).
96. R. McGeachin, *Ann. N. Y. Acad. Sci.* **151**, 208 (1968).
97. J. T. Nielsen, *Hereditas* **61**, 400 (1969).
98. F. F. Hall, M. J. Gulig, and N. C. Hightower, *Clin. Res.* **17**, 303 (1969).

TABLE III
AMINO ACID COMPOSITION OF RABBIT PANCREATIC AMYLASE ISOZYMES

Amino acid	Amylase[a] one	Amylase[a] two	Amylase[a] three
Aspartic acid and asparagine	14.9	15.5	16.0
Threonine	4.26	4.44	4.86
Serine	6.10	5.34	5.67
Glutamic acid and glutamine	8.37	8.48	8.94
Proline	3.56	3.67	3.78
Glycine	4.52	4.58	4.40
Alanine	3.55	3.55	3.51
½-Cystine	1.85	2.02	2.05
Valine	6.98	7.33	7.49
Methionine	1.67	1.56	1.90
Isoleucine	5.63	5.83	5.24
Leucine	5.46	6.18	5.51
Tyrosine	6.07	6.42	6.06
Phenylalanine	6.34	6.45	6.86
Lysine	5.34	4.86	4.57
Histidine	2.73	2.53	2.35
Arginine	8.47	7.92	7.58
Tryptophan	3.63	2.77	2.84

[a] Grams per 100 g of protein calculated from data in Malacinski and Rutter (*92*).

is invariably achieved either by electrophoresis or ion exchange chromatography (*92, 98–108*) pointing to a charge difference between the isozymes. The level of acidic amino acids appears solely responsible for the differences in migration rates (*92, 107*) since no significant differ-

TABLE IV
INHIBITION OF AMYLASES BY RABBIT ANTISERA TO HUMAN SALIVARY AMYLASE[a]

Amylase from	Inhibition by 1:20 antisera (%)	
	Man	Monkey
Pancreas	75	75
Saliva	78	77
Serum	45	60
Urine	50	—
Liver	7	0

[a] From McGeachin (*96*).

99. M. T. Szabo and F. B. Straub, *Acta Biochim. Biophys. Acad. Sci. Hung.* **1,** 397 (1966).
100. P. Juhasz and M. T. Szabo, *Acta Biochim. Biophys. Acad. Sci. Hung.* **2,** 217 (1967).

ences are found for any other amino acids. Many physical, chemical and enzymic properties of the variants are either very similar or identical (*92, 101*). Attempts to interconvert the multiple forms or implicate preparation artifacts invariably failed, inferring that the difference was of genetic origin (*92*). This supposition is strongly supported by demonstrated differences in amino acid composition and in tryptic peptide maps (*92*). The most convincing data are those of Heller and Kulka (*104*) who reported that pooled pancreatic homogenates of chicks revealed a fast and a slow amylase band upon electrophoresis but electrophoretograms of individual chick isozymes revealed three phenotypes. Either a single fast, a single slow, or a combination of bands was observed, which implicates two allelic genes. A similar situation is found for the mouse (*109*). In many higher animal species three electrophoretic bands are often observed (*92, 101–108*). Since the amylases from diverse sources have the same approximate molecular weight and (as noted later) are dimeric, hybridization of two genetically controlled monomers is hypothesized to explain the three bands.

The isozymes from human parotid gland present quite a different story (*109a*). The amylases are fractionated electrophoretically into five major and two minor bands. Chromatographically, the isozymes separate into two fractions which differ in carbohydrate content. Similarity in amino acid composition and peptide maps has led to the speculation that the variants are of a common genetic origin.

Amylase variants in saliva and human colostrum, although known, have not been extensively investigated and do not merit further review here (*103*). Multiple forms of amylases may occur in insects and have been shown to be under genetic control in *Drosophila* where six forms are observed (*105, 106*). Six isozymes are also reported for germinating barley and appear in timed sequence (*102*). Their production was greatly stimulated by gibberellic acid, but only five could be stimulated in the aleurone layers.

---

101. J. J. M. Rowe, J. Wakim, and J. A. Thoma, *Anal. Biochem.* **25,** 206 (1968).
102. H. A. Van Onckelen and R. Verbeek, *Planta* **88,** 255 (1969).
103. R. Got, C. Bertagnolio, M. B. Pradal, and J. Frot-Coutaz, *Clin. Chim. Acta* **22,** 545 (1968).
104. H. Heller and R. G. Kulka, *BBA* **165,** 393 (1968).
105. W. W. Doane, *J. Exptl. Zool.* **164,** 363 (1967).
106. W. W. Doane, *Proc. 12th Intern. Congr. Entomol., London, 1964* p. G2 (1965).
107. P. Cozzone, L. Pasiro, and G. Marchis-Mouren, *BBA* **200,** 590 (1970).
108. G. Marchis-Mouren and L. Pasiro, *BBA* **140,** 366 (1967).
109. K. Sick and J. T. Nielsen, *Hereditas* **51,** 291 (1964).
109a. D. L. Kauffman, N. I. Zager, E. Cohen, and P. J. Keller, *ABB* **137,** 325 (1970).

The multiple forms of β-amylase reported for cereal seeds (*47, 72, 110*) are probably artifactual. Efficient extraction requires treatment with papain and a reducing agent. Partial proteolysis and disulfide reduction is likely to result in variations of molecular size and activity (*70, 111*).

IV. Composition, Structure, and Modification

A. Amino Acid Content

In contrast to proteases and nucleases, amylases have received little attention. For example, the amino acid sequence of a single amylase has not yet been determined, compelling evidence for covalent intermediates has not been recorded, and only scant information about the nature of the active sites is available.

The amino acid composition of mammalian and plant amylases are recorded in Table II and may be compared to the amino acid composition of rabbit pancreas isozymes in Table III. The amylases appear to be typical proteins, and no outstanding compositional differences between them are evident.

B. Subunits and Multiple Binding Sites

Equilibrium dialysis (*112*) against a solution of a competitive inhibitor or a substrate under nonhydrolytic conditions and ultracentrifugation are the common methods used to establish the number of binding sites on the enzyme. Loyter and Schramm showed that hog pancreatic α-amylase possesses two maltotriose binding sites per mole (*113*). However, it has not been ascertained whether this enzyme is composed of one or two polypeptide chains. The existence of two independent binding sites accounts for the ability of this enzyme to form multimolecular complexes with limit dextrins (*113*).

The molecular weight of native sweet potato β-amylase is 197,000 which drops to 47,500 (*111*) in 8 *M* guanidinium chloride, suggesting that sweet potato β-amylase is tetrameric. Equilibrium dialysis against

110. M. Nummi, P. Seppala, R. Vilhumen, and T. M. Enari, *Acta Chem. Scand.* **19,** 2003 (1965).
111. J. E. Spradlin and J. A. Thoma, *JBC* **245,** 117 (1970), and references therein.
112. I. M. Klotz, F. M. Walker, and R. B. Pivan, *JACS* **68,** 1486 (1946).
113. A. Loyter and M. Schramm, *JBC* **241,** 2611 (1966).

cyclohexaamylose, a competitive inhibitor (*114*), suggests that the four sites bind independently. It has been inferred but not proved that the inhibitors bind at catalytic sites.

β-Amylases are formed as insoluble zymogens during ripening (*72*) in grains and the molecular weight of the active enzyme depends upon the technique of extraction from the cereal grains (*86, 70, 115*). Nummi *et al.* (*70, 110*) found that the activity of the enzyme extracted from ungerminated barley was inversely proportional to molecular size. The larger less active enzymes probably have portions of their insolubilizing matrix still appended. It appears that the variable molecular weight is artifactual and not the result of monomer association.

C. Size and Shape

The molecular weights of several α- and β-amylases are listed in Table V. With the exception of native sweet potato β-amylase, molecular weights are in the range of 50,000 and the enzymes have sedimentation

TABLE V
Molecular Weights of Amylases

Source of α-amylase	Ref.	Molecular wt	Source of β-amylase	Ref.	Molecular wt
Hog pancreas	(*116*)	45,000[a]	Germinated sorghum	(*53*)	55,900[a]
Human saliva	(*52*)	69,000[a]			54,200[d]
Human saliva	(*78*)	55,200[a]	Soya bean	(*54*)	61,700[a]
Ungerminated oats, rye, wheat, soybeans, and broad beans	(*41*)	45,000[b]	Sweet potato	(*111*)	197,000[d]
Germinated barley and wheat	(*41*)	45,000[b]	Sweet potato monomer	(*111*)	47,500[e]
Germinated sorghum	(*53*)	48,000[c]	Wheat	(*86*)	64,200[a]

[a] Calculated from velocity sedimentation.
[b] Determined by gel permeation.
[c] Extrapolated from published data.
[d] Calculated from sedimentation equilibrium.
[e] Determined in 8 *M* guanidinium chloride by sedimentation equilibrium.

114. J. A. Thoma, D. E. Koshland, J. Ruscica, and R. Baldwin, *BBRC* **12**, 184 (1963).
115. A. H. Cook and J. R. A. Pollock, *J. Inst. Brewing* **63**, 24 (1957).
116. C. E. Danielsson, *Nature* **160**, 899 (1947).

coefficients of approximately 4.5. The much smaller sedimentation coefficient of 1.3 reported for shore crab (*57*) α-amylase is indicative of a smaller protein. The action patterns of the enzymes (see below) suggest that binding sites span from 4 to 9 glucose units. French (*117*) has noted that in contrast to most enzymes, amylases will be dwarfed by their natural substrate, starch.

No detailed knowledge about the three-dimensional structure of amylases is available. Optical rotatory dispersion studies indicate that the right-hand α-helix content of pancreatic α-amylase does not exceed 15% (*118*). Perturbation difference spectra show that 80% of the tryptophan in pancreatic amylase and 30% of the tyrosine residues are accessible to solvent (*118*). In 90% D_2O all of the tryptophan residues but only 30% of the tyrosine residues were exposed (*118*). It was concluded that all the tryptophan residues lie on or near the surface while 70% of the tyrosine residues are buried within the molecule. The physical properties of the enzyme are typical of globular proteins (*119, 120*).

D. Chemical Modification

A large volume of literature describing the modification of side chains of amylases has accumulated. Frequently the time dependence of activity was the only parameter monitored when the enzymes were modified. Few quantitative correlations between the number and type of side chains destroyed and changes in V_{max}, K_m, or structure appear to exist (*121*). Consequently, most of the enzyme inactivation kinetics are interpretable only at a phenomenological level and not at a molecular level. Reaction of enzymes with selective reagents (*122, 123*) is of course intended to specify the role of side chain residues in enzymic action.

Practically every reactive side chain in proteins has been implicated in amylase activity after submission to such harsh and unspecific reagents as nitrous acid, fluorodinitrobenzene, ketene, acetic anhydride, iodine,

117. D. French, *Baker's Dig.* **31,** 24 (1957).
118. M. Krsteva and P. Elodi, *Acta Biochim. Biophys. Acad. Sci. Hung.* 3, 275 (1968).
119. B. Jirgensons, *JBC* **240,** 1064 (1965).
120. B. Jirgensons, *ABB* **74,** 70 (1958).
121. W. J. Ray and D. E. Koshland, Jr., *JBC* **236,** 1973 (1961).
122. A. N. Glazer, *Annu. Rev. Biochem.* **39,** 101 (1970).
123. C. H. W. Hirs, "Methods in Enzymology," Vol. 11, Sects. VIII and IX, Academic Press, New York, 1967.

and diazobenzenesulfonic acid (*41, 68, 69, 124–129*). These experiments will not be discussed in detail because the experimental design does not pinpoint the cause for inactivation, for example, steric blockage of substrate binding, or destruction of catalytic amino acids.

In many cases, the more definitive experiments are the ones which yielded negative results; for example, iodoacetamide, *p*-mercuribenzoate, and phenylmercuric chloride do not inactivate hog pancreatic α-amylase (*130*) and only partial activity loss in cereal and human salivary amylase occurs upon exposure to mercurials (*41, 131, 132*). These experiments tend to rule out critical roles for sulfhydryl groups. In the case of sweet potato β-amylase, however, specific alkylation of the thiol groups has been pinpointed as the cause of partial inactivation and thermal instability (*111*). Activity loss appears to be influenced by the size of the alkylating agent. Modification induces small but significant changes in the physical and chemical properties of the protein. A minor conformation change concomitant with modification is the explanation offered for these observations. It was speculated that the thiol groups play a regulatory and not a catalytic role (*111*).

The influence of pH on activity is also frequently employed in an attempt to assign pK values and implicate catalytic amino acids (*68, 69, 117, 133–137*). Almost invariably carboxyl and imidazole groups have been assigned catalytic functions from pH data (Table VI). The hazards of interpreting these profiles are discussed later. Interpretation of pK data is on much firmer ground if the heats of ionization are measured. Even though dissociation constants may be perturbed by a hundredfold on a protein, the heat of ionization remains fairly constant and characteristic of the ionizing side chain (*138, 139*). One such careful

124. J. E. Little and M. L. Caldwell, *JBC* **142,** 58 (1942).
125. J. E. Little and M. L. Caldwell, *JBC* **147,** 229 (1943).
126. F. J. Di Carlo and S. Redfern, *Arch. Biochem.* **15,** 333 (1947).
127. K. Benner and K. Myrback, *Arkiv Kemi* **4,** 7 (1952).
128. I. Radichevich, M. M. Becker, M. Eitingon, V. H. Gettler, G. C. Toralballa, and M. L. Caldwell, *JACS* **81,** 2845 (1959).
129. C. E. Weill and M. L. Caldwell, *JACS* **67,** 212 and 214 (1945).
130. M. L. Caldwell, C. E. Weil, and R. S. Weil, *JACS* **67,** 1079 (1945).
131. E. H. Fischer and C. H. Haselbach, *Helv. Chim. Acta* **34,** 325 (1951).
132. J. Muus, F. P. Brockett, and C. C. Connelley, *ABB* **65,** 268 (1956).
133. J. A. Thoma, J. Wakim, and L. Stewart, *BBRC* **12,** 350 (1963).
134. J. Wakim, M. Robinson, and J. A. Thoma, *Carbohydrate Res.* **10,** 487 (1969).
135. N. A. Zherebtsov, *Biokhimiya* **33,** 435 (1968).
136. S. Ono, K. Hiromi, and Y. Yoshikawa, *Bull. Chem. Soc. Japan* **31,** 957 (1958).
137. J. A. Thoma and D. E. Koshland, *JMB* **2,** 169 (1960).
138. J. T. Edsall and J. Wyman, "Biophysical Chemistry," Vol. 1, Chapter 6, p. 494. Academic Press, New York, 1958.

TABLE VI
APPARENT IONIZATION CONSTANTS DETERMINED FROM VARIATIONS OF REACTION VELOCITY WITH pH

Source of amylase	pK_a	pK_b	Ref.
	α-Amylases		
Human saliva	4.6	6.7[a]	(*143*)
	5.3	7.8[b]	
Hog pancreas	4.6	6.2[a]	(*133*)
	5.7	8.9[b]	
Soybean	4.3	8.1	(*68*)
Broad bean	4.1	7.1	(*69*)
Cereals (oats, rye, and wheat)	3.2–3.3	6.1–6.8	(*41*)
	β-Amylases		
Sweet potato	3.7	8.0	(*91, 137*)
Barley	3.4	8.1	(*135*)

[a] Chloride-free enzyme.
[b] Chloride-activated enzyme.

study has been carried out with barley β-amylase (*135*) where the heats of ionization of the acidic and basic group were found to be +2565 and +6970 cal/mole, respectively. These heats of ionization are characteristic of carboxyl and imidazole side chains (*138, 139*).

The only amylases for which the assignments of catalytic amino acids can be made with some measure of confidence are barley β-amylase and porcine pancreatic α-amylase. Both enzymes are purported to employ carboxyl and imidazole side chains as catalytic agents. Assignments for the former enzyme rest primarily on photooxidation experiments and upon heats of ionization mentioned above (*135*). Because of its availability, porcine pancreatic amylase has probably received more attention than any other amylase. The following line of evidence (*118, 124, 133, 134*) points to carboxyl and imidazole participation in catalysis: the enzyme is relatively unaffected by thiol agents, phosphate is absent from the enzyme, the α-amino group is acetylated (*108*), the pH dependence of photooxidation is characteristic of histidine (*140*), tyrosine ionization is anomalous and irreversible with pK values of 11.5 and 10.95 in the forward and reverse direction, respectively, and reversible in the absence of calcium with pK values of 10.9, V_{max}-pH profile gives pK_a in the range of 4.6 and pK_b of 6.2 in the absence of chloride. Although

139. E. J. Cohn and J. T. Edsall, "Proteins, Amino Acids and Peptides," p. 445. Reinhold, New York, 1943.
140. E. W. Westhead, *Biochemistry* **4**, 2139 (1965).

these two groups are probably required for activity the exact nature of their participation remains to be unraveled.

V. pH, Temperature, and Salt Effects

Enzyme stability and catalytic efficiency are pH-dependent parameters of amylases that attract the greatest attention. A brief review of the effects of hydrogen ion concentration on amylases is followed by a discussion of the problems of interpreting pH-activity data.

The α-amylases from higher plants (*40, 68, 69*) and mammals (*141, 142*) are generally stable from pH 5.5 to 8.0. Cereal α-amylases exhibit optimal activities between pH 5 and 6 (*2*) and undergo rapid irreversible inactivation below pH 5 (*2*). Animal α-amylases usually exhibit a maximum activity around neutral pH (*2*) but shift down scale when the chloride is removed (*134, 143*). Sather (*144*) reported that an amylase from the crab, *Metapograpsue messor*, has a basic pH optimum of 8.0. The optimum pH of animal glucoamylases is fairly low for animal enzymes, pH 4.8–5.0 (*5, 6*).

The pH optimum for β-amylases lies between pH 4 and 6 (*2*), and they are relatively stable at pH 3.6. The stability of an amylase is a function of protein concentration (*145, 146*), salt composition (*11*), and salt concentration (*11*). The profound effect of calcium salts on stabilization of amylases, discovered by Fischer and co-workers (*10*), is shown in Fig. 2. Stabilizing and destabilizing effects of adjuncts and protein concentration are almost always overlooked in studies of amylase stability.

The change in activity of enzymes as a function of hydrogen ion concentration is often employed to measure the pK_a values of the catalytic side chains. The activity of an enzyme is sensitive to the degree of substrate saturation (*147*), duration of the assay if denaturation occurs (*148*), the mechanism of the reaction (*149*), the conformation of the

141. P. Bernfeld, A. Staub, and E. H. Fischer, *Helv. Chim. Acta* **31,** 2165 (1948).
142. P. Bernfeld, F. Duckert, and E. H. Fischer, *Helv. Chim. Acta* **33,** 1064 (1950).
143. K. Myrback, *Z. Physiol. Chem.* **159,** 1 (1926).
144. B. T. Sather, *Comp. Biochem. Physiol.* **28,** 371 (1969).
145. S. Nakayama and Y. Kono, *J. Biochem.* (*Tokyo*) **44,** 25 (1957); **45,** 243 (1958).
146. S. Nakayama, *J. Biochem.* (*Tokyo*) **49,** 328 (1961).
147. H. R. Mahler and E. Cordes, "Biological Chemistry," Chapter 6. Harper, New York, 1966.
148. M. J. Selwyn, *BBA* **105,** 193 (1965).
149. W. P. Jencks, *Ann. Rev. Biochem.* **32,** 639 (1963).

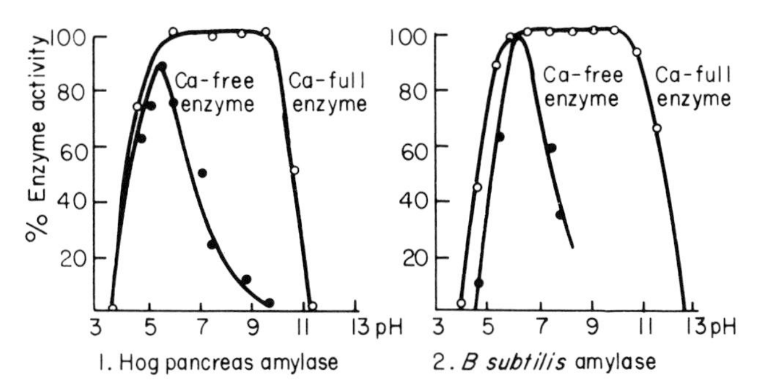

FIG. 2. Stability of Ca-containing and Ca-free amylases. The enzymes were incubated for 20 hr at 25° at various pH values, then tested for activity in the presence of Ca^{2+}. From Fischer and Stein (*10*).

protein, ionic strength (*150*), and other factors beside the intrinsic ionization constants of the catalytic groups. Unfortunately, in most investigations of pH-dependent amylase activity, these effects have not been sorted out and make identification of catalytic groups a most hazardous venture. The slope and intercepts of double reciprocal plots rather than enzymic activity are theoretically the proper object of a pH study (*147*). However, even when $\tilde{K}_m$, $\tilde{V}_{\max}$ and $\tilde{K}_m/\tilde{V}_{\max}$ are plotted vs. pH, interpre-

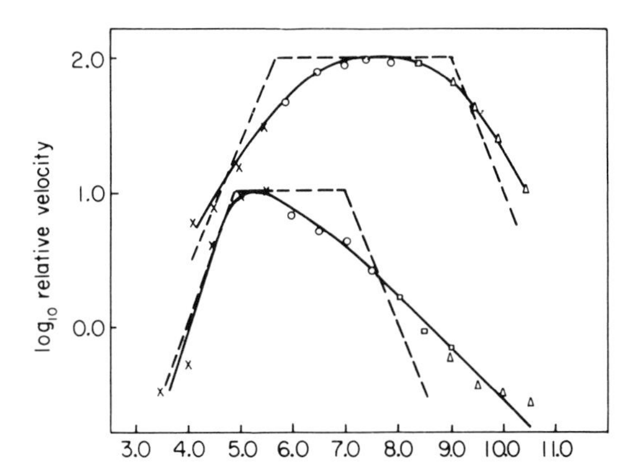

FIG. 3. pH dependence of velocity of α-amylase at saturating starch concentration in the presence and absence of chloride ion. Upper curve, 0.025 M KCl at 25°; lower curve, absence of chloride at 25°. (×) acetate, (○) phosphate, (□) pyrophosphate, and (△) carbonate buffers. The velocities were measured sequentially at high substrate concentration and the maximum velocities computed by Eq. (1). Straight dashed lines represent limiting theoretical slopes of zero or ± unity. From Wakim *et al.* (*134*).

150. R. A. Alberty and V. Bloomfield, *JBC* **238**, 2804 (1963).

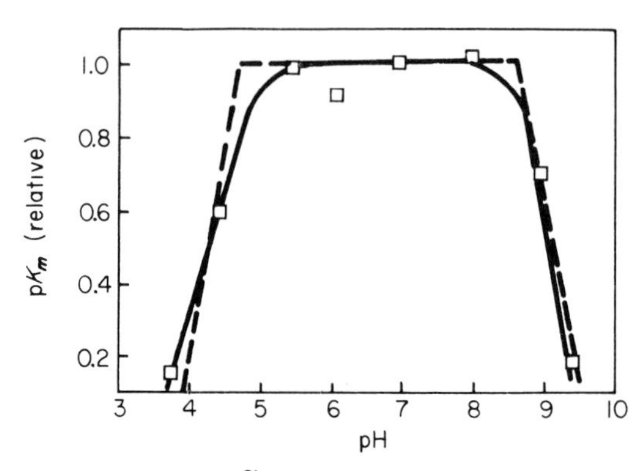

FIG. 4. pH dependence of —log $\tilde{K}_m$ of starch–α-amylase system. Velocity measurements were made under standard assay conditions (*134*) except for adjustment of pH with pyrophosphate-acetate buffer and presence of 0.01% bovine-serum albumin in 0.1 *M* chloride. Straight dashed lines represent theoretical slopes of zero or ± unity. From Wakin *et al.* (*134*).

tation of the curves is not straightforward (*149*) for a two proton model (Scheme 1). The work of Wakim *et al.* (*134*) illustrates further complications of the problem. The influence of pH on $\tilde{V}_{\max}$ and $\tilde{K}_m$ for porcine pancreatic α-amylase is depicted in Figs. 3 and 4. The slopes of the acidic and basic limbs of the curves in Fig. 3 are experimentally less than unity, indicating that the simple model of the pH dependence (*147*) of the enzymic action does not adequately describe the system. The divergence of $\tilde{V}_{\max}$ from unit slopes may arise from electrostatic perturbation of ionization constants, multiple intermediates, or conformational effects (*147, 150, 151*). Interestingly, the slopes of the log $\tilde{V}_{\max}$-pH profiles for β-amylase (*91*) and lysozyme (*152*) are also less than unity and approximate 0.6 ± 0.1. This deviation from theory may be fortuitous, but if it is not

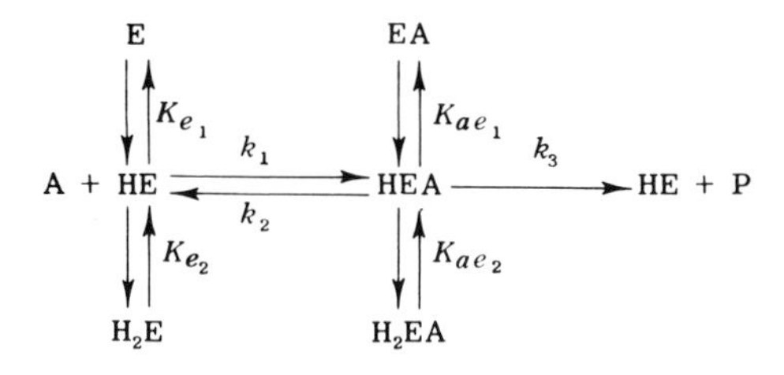

SCHEME 1. Two proton model to explain the pH dependence of enzymic action.

151. R. A. Alberty, *J. Cell. Comp. Physiol.* **47,** Suppl. 1, 245 (1956).
152. T. Rand-Meir, F. W. Dahlquist, and M. A. Raftery, *Biochemistry* **8,** 4206 (1969).

the significance of this observation has escaped explanation. The simple, two proton model (Scheme 1) is also incapable of accounting for the bell-shaped dependence of both $\tilde{V}_{max}$ and $\tilde{K}_m$ on pH.

According to the analysis of Scheme 1:

$$\tilde{v}_0 = \frac{\tilde{V}[A]}{\tilde{K}_m + [A]} \tag{1}$$

$$\tilde{V}_{max} = \frac{V}{1 + K_{ea1}/H^+ + H^+/K_{ea2}} \tag{2}$$

$$\tilde{K}_m = \frac{K_m(1 + K_{e1}/H^+ + H^+/K_{e2})}{1 + K_{ea1}/H^+ + H^+/K_{ae2}} \tag{3}$$

$$\frac{\tilde{K}_m}{\tilde{V}_{max}} = \frac{K_m}{V}\left(1 + \frac{K_{e1}}{H^+} + \frac{H^+}{K_{e2}}\right) \tag{4}$$

where the tilde implies apparent constants; V and K_m are the pH-independent Michaelis constants; the subscripts e and ea denote properties of the enzyme and intermediates (*152a*), respectively; and the subscripts 1 and 2 correspond to acid and basic dissociation constants, respectively. It is quite obvious after assigning reasonable values to the dissociation constants that the curves corresponding to Eqs. (3) and (4) cannot have the same shape. Theory predicts that $\tilde{V}_{max}$ and $K_m/\tilde{V}_{max}$ are bell-shaped functions of pH. In order to account for the observations with α-amylase it is necessary to invoke a four-proton model in which one set of protons is associated with a pair of catalytic residues and the other set is associated with a pair of dissociable amino acid side chains required for attachment of substrate. This requires the addition of Eqs. (5)–(7) to Scheme 1.

$$E \underset{}{\overset{K_3}{\rightleftharpoons}} EH \overset{K_4}{\rightleftharpoons} EH_2 \tag{5}$$

$$HE \overset{K_3}{\rightleftharpoons} HEH \overset{K_4}{\rightleftharpoons} HEH_2 \tag{6}$$

$$H_2E \overset{K_3}{\rightleftharpoons} H_2EH \overset{K_4}{\rightleftharpoons} H_2EH_2 \tag{7}$$

In this set of equations, the protons on the right are associated exclusively with binding and those protons on the left are associated exclusively with catalysis. Since the catalytic groups are not involved in binding, it is reasonable to expect that $K_a \approx K_{ae}$. If substrate only combines with the enzyme forms given in Eq. (6), then the familiar Michaelis–Menten relationships are found to be:

152a. A and S are alternatively used to designate substrates in conformity with the original references.

$$\tilde{v}_0 = \frac{V/\gamma[\mathrm{A}]}{K_m\alpha + [\mathrm{A}]} \tag{8}$$

$$\tilde{V}_{\max} = \frac{V}{\gamma} = \frac{V}{1 + K_{\mathrm{ae1}}/\mathrm{H}^+ + \mathrm{H}^+/K_{\mathrm{ae2}}} \tag{9}$$

$$\tilde{K}_m = K_m\alpha = K_m(1 + K_3/\mathrm{H}^+ + \mathrm{H}^+/K_4) \tag{10}$$

$$\frac{\tilde{K}_m}{\tilde{V}_{\max}} = \frac{K_m}{V}(1 + K_3/\mathrm{H}^+ + \mathrm{H}^+/K_4)(1 + K_{\mathrm{ae1}}/\mathrm{H}^+ + \mathrm{H}^+/K_{\mathrm{ae2}}) \tag{11}$$

For this oversimplified situation, theory predicts $\tilde{V}_{\max}$ and $\tilde{K}_m$ but not $\tilde{K}_m/V_{\max}$ give bell-shaped curves. However, when multiple intermediates are included in the scheme, the inflection points in the $\tilde{V}_{\max}$ and $\tilde{K}_m$ curves will be weighted averages of dissociation constants, and their significance is difficult to specify. It has been emphasized that proton dissociation constants evaluated by kinetic techniques may be seriously perturbed from their intrinsic values and not amenable to rigorous interpretation (*149*). The difficulties inherent in the interpretation of experimental data in terms of the two-proton model are also associated with the more elaborate model. Assuming formation of a substrate complex does not greatly perturb the ionization constants of the enzyme, apparent equilibrium constants for the catalytic groups can be approximated from the log $\tilde{V}_{\max}$ vs. pH profiles, and apparent ionization constants for the protons involved in substrate binding can be estimated from $\tilde{K}_m$ dependence on pH. The pK_a values of the catalytic side chains in the presence and absence of chloride are approximated to be 5.6–5.9 and 4.7–4.9, respectively. The pK_b values in the presence and absence of chloride are estimated to be 8.8–9.9 and 6.0–6.4, respectively. Some of the shift in the pH-rate profile upon addition of chloride may arise from a change in the rate-limiting step.

Table VI lists the apparent ionization constants which have been reported for plant and animal amylases. The ionization constants have been interpreted by most authors to implicate carboxyl and imidazole residues in catalysis, but this conclusion requires confirmation by independent techniques.

Since rate constants universally exhibit temperature dependence, and enzyme mechanisms involve a sequence of several steps, including possible side reactions, one can anticipate some difficulty in sorting out and rigorously interpreting temperature effects on amylase action. Temperature-induced changes of amylase activity have often been considered to operate exclusively on k_{cat} in spite of the fact that it is widely recognized that protein stability, substrate affinity, and ionization constants are also affected. More advanced treatments of temperature effects on enzymic action are available (*14*, *28*). In an attempt to evaluate the

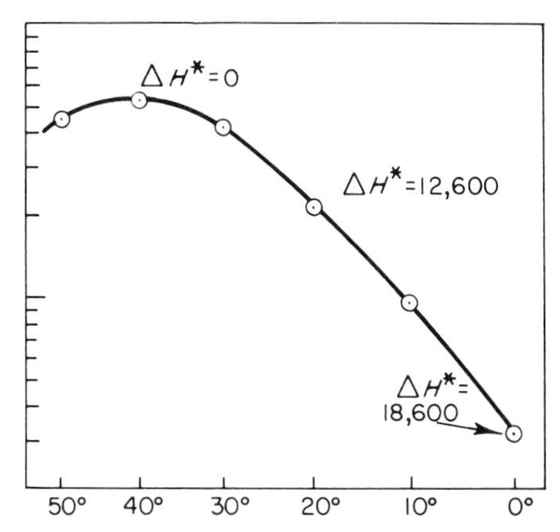

FIG. 5. Arrhenius plot for salivary amylase. The activity in arbitrary units is plotted on a logarithmic scale against the reciprocal of the absolute temperature (temperatures are indicated in °C). The apparent heats of activation are calculated from the slopes of the curve at different temperatures. From French (*117*).

activation energy for amylase-catalyzed hydrolysis many investigators (*40, 68, 80, 83, 84, 153*) employ Arrhenius plots of activity. Curvature of the plots has frequently been observed and has been interpreted as a decrease in activation energy with increasing temperature (see Fig. 5). Activation energies of 14 ± 2 kcal are usually found about 10°C decreasing to a few kilocalories around 40° to 50°C (*2*). Repeated observation of curvature of Arrhenius plots under widely different conditions probably means that part of the decrease in activation energy with rising temperature is real but part is probably the result of an increase in the dissociation constant of the enzyme substrate and part the result of enhanced denaturation rates (*117*).

Temperature optima for amylases are consistently reported (*2*) in the literature but have little significance because they represent compensating effects upon denaturation and k_{cat}. Furthermore, the temperature optimum is partly regulated by the duration of the assay (*14*) and thermal protection afforded by the substrate. The work of Kalnitsky and Resnick on ribonuclease is recommended as a model for future experimentation in this area for sorting out thermal denaturation effects (*154*).

The apparent heats of formation of sorghum and (*53*) hog pancreas α-amylase (*31*) and sorghum β-amylase (*53*) are —3.4, —7, and —2.6 kcal/

153. H. M. Levy, N. Sharon, and D. E. Koshland, Jr., *Proc. Natl. Acad. Sci. U. S.* **45,** 758 (1959), and references therein.
154. G. Kalnitsky and H. Resnick, *JBC* **234,** 1714 (1959).

mole, respectively. The heat of formation of the sweet potato β-amylase–cyclohexaamylose complex is essentially zero (*115*). These apparent thermodynamic quantities were measured from the dependency of K_m or K_i upon temperature. Since K_m is a complex function of productive and nonproductive complexes and k_{cat}, the significance of the heats of formation in many instances are not directly related to an affinity constant (*28, 29*). The thermodynamic quantities associated with amylase substrate and inhibitor interactions can be expected to be influenced by chain length, degree of branching, long range ordering of the substrate chain, and other variables (*28*).

Nakayama *et al.* (*145, 146*) published the most extensive studies on thermal denaturation of amylases. According to their work

$$k_{obs} = \alpha C^{-\beta} \tag{12}$$

where k_{obs} is the apparent first-order inactivation constant with respect to time, C is the enzyme concentration, and α and β are environmentally sensitive constants. As might be anticipated, α is a minimum in the region of the isoelectric pH. Alpha tends to increase with salt concentration in the region of 0.01–1 M. The exponential factor β increases as temperature is lowered. Although Nakayama reported β-amylase denaturation is first order with respect to time, Spradlin (*111*) has found that the alkylated enzyme denatures biphasically, and Wakim (*134*) has reported similar behavior for porcine pancreatic α-amylase. These observations are typical of the denaturation kinetics reported for other enzymes (*155*).

All plant and animal α-amylases appear to contain calcium (*2, 10, 156*) and its removal results in both reversible (*2, 10*) and irreversible inactivation or in great loss of thermal stability. Fischer and Stein (*10, 156*) noted that this inactivation could probably be ascribed to proteolytic degradation or to exposure to unfavorable conditions used for the release of the metal rather than a spontaneous unfolding of the apoenzyme. The loosely bound Ca^{2+} is easily removed by dialysis against EDTA. The one gram atom of metal per mole which remains is then only released by extensive dialysis against EDTA or more efficiently by electrodialysis or by denaturation (*156*). The association constant for Ca^{2+} at this high affinity site has been estimated to be 10^{12} to 10^{15} M^{-1} (*10*). β-Amylases, in contrast, do not appear to require any metal for activation (*4*). Generally, mammalian α-amylases bind calcium more tightly than those of higher plants (*2*). Calcium-deficient amylases are fre-

155. K. Laidler, "The Chemical Kinetics of Enzyme Action," Chapt. XIII, p. 336. Oxford Univ. Press, London and New York, 1958.

156. B. L. Vallee, E. A. Stein, W. N. Summerwell, and E. H. Fischer, *JBC* **234, 2901 (1959)**.

quently found to be quite susceptible to protease digestion while the metalloenzymes are resistant to proteolysis (*157, 158*). High concentrations of calcium also retard thermal inactivation (*41*) at temperatures above 50°C, particularly for cereal α-amylases which bind calcium weakly.

Removal of Ca^{2+} from α-amylase does not result in any significant changes in hydrodynamic or rotatory properties (*10*); thus, gross structural alterations are unlikely but minor conformational changes cannot be excluded. Based upon many observations two roles are relegated to calcium; first, it dramatically stabilizes the compact architecture of the molecule and, second, it helps to maintain an enzymically active conformation (*10*). Further evidence that Ca^{2+} is important in controlling the structure of the enzyme comes from the work of Krsteva and Elodi (*118*). They have shown that tyrosine ionization in calcium-rich hog pancreatic α-amylase is anomalous and irreversible with pK values of ionization of 11.5 and 10.95 in the forward and reverse direction, respectively. Upon the removal of calcium from the enzyme, tyrosine ionizes normally with a pK of 10.9.

The problem in establishing the role of calcium participation in α-amylase action can be traced to the extreme difficulty of removing the ubiquitous Ca^{2+} from glassware, reagents and substrates. Hsui *et al.* (*159*) were able to reversibly remove one tightly bound Ca^{2+} ion from porcine and human α-amylases. Replacement of Ca^{2+} by Sr^{2+}, Mg^{2+}, Ba^{2+}, and Na^{+} resulted in some activation. However, it should not be assumed that metals are an essential cofactor since it has not been definitely established that the metal deficient enzyme is devoid of activity. Because of the ability of amylases to selectively concentrate Ca^{2+} in the presence of other metal ions it cannot be rigorously accepted that the other cations mentioned above are the real activating agents. An involvement of Ca^{2+} either in substrate binding or in the catalytic event must await further experimentation.

It has long been recognized that mammalian α-amylases require chloride for maximum activity (*2*). Muus (*132*) observed that 1–10 mM NaCl gave maximum activity with salivary amylase and that the chloride ion depressed the solubility of the enzyme but enhanced its stability against inactivation at elevated temperatures with heavy metals. The optimum NaCl concentration for porcine pancreatic α-amylase is 10 mM with higher concentrations producing an inhibitory effect (*160*).

157. E. A. Stein and E. H. Fischer, *JBC* **232,** 867 (1958).
158. E. A. Stein, J. Hsui, and E. H. Fischer, *Biochemistry* **3,** 56 (1964).
159. J. Hsui, E. H. Fischer, and E. A. Stein, *Biochemistry* **3,** 61 (1964).
160. K. H. Meyer, E. H. Fischer, and P. Bernfeld, *Helv. Chim. Acta* **30,** 64 (1947).

The activating power of halides for some mammalian amylases has never been fully delineated. Removal of halides from the enzyme is reported not to alter the K_m for substrate binding and probably does not change the substrate affinity (*133*). Hence, activation must result from an increase in k_{cat}. The anion may enhance catalytic efficiency by stabilizing a positive oxycarbonium intermediate, inducing a realignment of the catalytic groups or perhaps by direct participation via a glucosyl chloride. A partial activation mechanism has been definitely established: a separation of the acid and basic ionization constants of the enzyme. This effect is illustrated in Fig. 3. The difference in the pK values of the catalytic groups of hog pancreatic α-amylase are estimated to be 3.4 in the presence of chloride and 2.2 in the absence of chloride. Alberty (*150*, *151*) and Dixon and Webb (*14*) have discussed in some detail how broadening of pH-activity profiles leads to enhancement of enzymic activity at the optimum pH. The separation of the apparent ionization constants of porcine pancreatic α-amylase although contributing to enhancement of activity can account for only a small fraction of it. It should be emphasized that halides activate but are not mandatory for amylase activity (*134*).

Several heavy metal ions (*2*) have been shown to inhibit amylases, e.g., mercury, silver, copper, and lead. In general, mercury is a more effective inhibitor than copper or lead. Ammonium molybdate and ascorbic acid (*2*, *161*) also inhibit amylases. As might be expected, inhibition has generally been blamed on sulfhydryl modification. Almost invariably this explanation is unconfirmed speculation.

VI. Mechanism of Action

Discussions about the mechanism of amylase catalysis have primarily centered on the stereochemical relationship between substrate and products, ring distortion upon binding of the substrate to the enzyme, and the involvement of amino acid side chains in polyfunctional catalysis. These three facets of amylase chemistry will be discussed sequentially.

Carbohydrates give either quantitative retention or inversion of configuration at the anomeric center undergoing hydrolysis (*8*, *134*, *162*–

161. C. T. Greenwood and E. A. Milne, *Staerke* **20**, 101 (1968).
162. J. A. Thoma and D. E. Koshland, *JBC* **235**, 2511 (1960).
163. S. Ono, K. Hiromi, and Z. Hamauzu, *J. Biochem.* (*Tokyo*) **57**, 35 (1965).
164. Z. Hamauzu, K. Hiromi, and S. Ono, *J. Biochem.* (*Tokyo*) **57**, 39 (1965).
165. T. E. Nelson, *JBC* **245**, 869 (1970).
166. F. W. Parrish and E. T. Reese, *Carbohydrate Res.* **3**, 424 (1967).

167); no exceptions to this rule have been reported. The absence of racemization is firm evidence that solvolysis must precede release of product from the enzyme surface but does not imply anything about the timing of the water attack along the reaction coordinate. This observation is consistent with participation of solvent with either a pre- or posttransition state enzyme–substrate complex. For the reasons discussed below we believe it is hazardous to use the stereochemistry of products as a criterion for the mechanism of amylase hydrolysis.

For many years enzymes that invert configuration were classified as single displacement catalysts and enzymes that retain configuration were classified as double displacement catalysts (*168*). This speculation focused interest upon the stereochemistry of amylase catalysis which presumably furnished a guide to the mechanism of bond transformation. A double displacement reaction giving rise to an even number of Walden inversions would normally be expected to proceed via a covalent glycosyl intermediate. In spite of the search for such an intermediate, even by rapid kinetic methods, none has ever been detected (*169*) for glycosidases. However, covalent catalysis has been conclusively demonstrated for a related enzyme, sucrose phosphorylase (*170, 171*). Affinity labeling of the active site of β-glucosidase has been achieved with conduritol epoxide, a substrate analog of glucose, but this experiment cannot be construed as strong support for a covalent substrate intermediate because of the high reactivity of the analog (*172*).

An enzyme-directed approach of the hydrolytic water molecule to the reaction center has been suggested as an alternative explanation of the stereochemistry of amylases (*134, 173*). For inverting enzymes, backside solvation of an oxycarbonium ion was postulated; for retaining enzymes, direct front-side approach of the attacking water molecule on the oxycarbonium ion was postulated. Sterically, the feasibility of front-side attack depends critically upon the transition state geometry of the reacting intermediate. If the glucopyranoside undergoing attack achieves a half-chair configuration along the reaction coordinate, solvent approach from above or from below the ring is clearly possible. Some idea of the steric restraints placed upon the system is given in Fig. 6 which portrays the hydrolysis of an α-glucoside. Front-side attack by

167. B. Capon, *Chem. Rev.* **69,** 407 (1969), and references therein.

168. D. E. Koshland, Jr., *Biol. Rev.* **28,** 416 (1953).

169. H. Gutfreund, "An Introduction to the Study of Enzymes," p. 239. Wiley, New York, 1965.

170. R. Silverstein, J. Voet, D. Reed, and R. H. Abeles, *JBC* **242,** 1338 (1967).

171. F. De Toma and R. H. Abeles, *Federation Proc.* **29,** No. 2, p. 461. Abstr. No. 1214 (1970).

172. G. Legler, *BBA* **151,** 728 (1968).

173. J. Thoma, *J. Theoret. Biol.* **19,** 297 (1968).

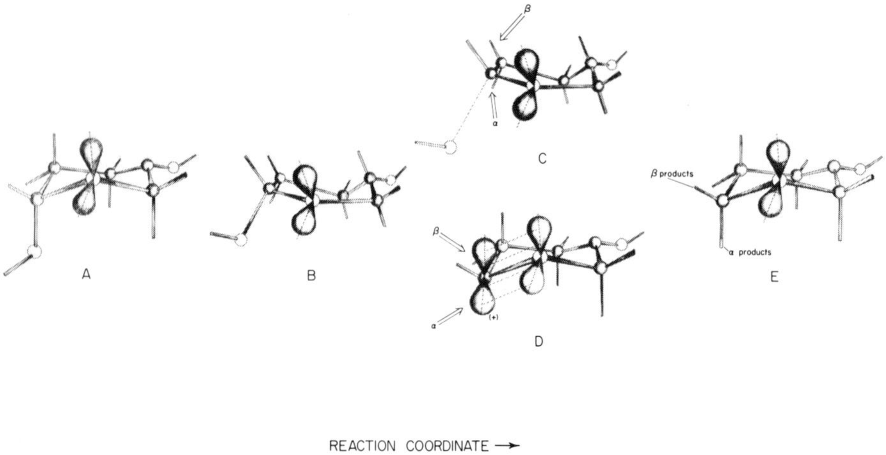

REACTION COORDINATE →

FIG. 6. A schematic representation of possible conformations and intermediates along the reaction pathway for amylase-catalyzed hydrolysis of glucopyranosides. The view is along a line passing approximately through the ring oxygen and a point approximately midway between C-3 and C-4 of the ring to undergo hydrolysis. The view is toward the nonreducing end of the molecule. For the sake of clarity the enzyme and functional groups are not pictured. Drawing (A) portrays the residue to undergo hydrolysis. Drawing (B) portrays a distorted pyranoside approximating a half-chair conformation. Note that the distortion increases the accessibility of the anomeric carbon to water on the "underside" (i.e., a front-side attack could be promoted by α-amylase). Drawings (C) and (D) portray structures approaching transition states. The former would occur if bond formation were synchronous to bond rupture and the latter if bond formation follows bond rupture. Drawing (E) portrays the structures of the final products (retained and inverted stereochemistry for α- and β-amylase, respectively). Shaded spheres, carbon; unshaded spheres, oxygen. The double-shafted arrows labeled α and β show possible directions of approach of the attacking water molecules allowed by α- and β-amylase, respectively. From Thoma (*173*).

water on a glucoside in the C-1 conformation is probably impossible because of steric crowding (Fig. 6D), but Fig. 6C shows how steric congestion is greatly relieved when the glucoside is distorted to a half-chair configuration and the aglycone has departed. Thus, retention of configuration probably involves capture of a solvent molecule subsequent to release of the aglycone and essentially precludes nucleophilic assistance by the solvent. Enzymes that retain configuration must permit some translocation of the aglycone from the binding site to permit the approach of solvent. This steric restraint is not imposed on enzymes that invert configuration (Fig. 6). Since a common intermediate can be envisioned for both α- and β-amylase, it is unlikely that the stereochemistry of the products is a useful guide to the timing of water attack along the reaction coordinate. We conclude that a front-side attack of solvent on an oxycarbonium ion must remain an important candidate for an α-amylase reaction mechanism. The mechanism of a related carbohydrase, lysozyme, has been interpreted in this fashion from X-ray crystallographic examination of inhibitor enzyme complexes (*174, 175*).

It might be mentioned that an analogous situation is found with electrophilic substitution on carbon catalyzed by different enzymes. Rose (*176*) has proposed that the steric course of the reaction is governed by enzyme guided protonation of the intermediate rather than from a difference in the stereochemistry of electrophilic substitution at saturated carbon.

After some years of dormancy, there is renewed belief in enzyme-induced substrate distortion as an enzyme activation mechanism. The importance of this factor in enzyme-assisted reactions has been the subject of a number of recent articles (*177–179*). Undoubtedly the distortion mechanism that has gained the greatest popularity and stimulated the most excitement is that proposed by Phillips (*180*) and his co-workers for hen egg-white lysozyme hydrolysis of bacterial cell wall.

The importance of the factors assisting acetal hydrolysis can quantitatively be assessed by conceptually dissecting the enzyme mechanism into its component parts and attempting to evaluate the contribution of each component to the overall catalytic event. At least two factors, distortion and side chain functional catalysis, are thought to be important contributing agents to catalysis. Many workers tacitly assume that acid catalysis of acetals is a reasonable model for the enzyme-catalyzed reaction since the transition states of both of these reactions supposedly possess substantial carbonium ion character. They also assume that there are no interaction mechanisms between ring distortion and functional group catalysis. Both of these assumptions are questionable.

Relative to acyclic analogs, cyclic acetals are decidedly more difficult

174. J. A. Rupley, V. Gates, and R. Bilbrey, *JACS* **90,** 5633 (1968).
175. F. W. Dahlquist, T. Rand-Meir, and M. A. Raftery, *Biochemistry* **7,** 3269 and 3281 (1969).
176. I. A. Rose, *Brookhaven Symp. Biol.* **15,** 293 (1962), and references therein.
177. W. P. Jencks, *in* "Current Aspects of Biochemical Energetics" (N. O. Kaplan and E. Kennedy, eds.), p. 273. Academic Press, New York, 1966.
178. R. Lumry and R. Beltonen, *in* "Structure and Stability of Biological Macromolecules" (S. Timashiff and G. Fasman, eds.), p. 65. Marcel Dekker, New York, 1969.
179. R. Lovrien and T. Linn, *Biochemistry* **6,** 2281 (1967).
180. D. C. Phillips, *Sci. Am.* **215,** 78 (1966).

to hydrolyze (*181*) because of ring strain. As the aglycone departs, the anomeric carbon shifts from sp^3 toward sp^2 hybridization requiring expansion of the C-2–C-1–O-5 bond angle (see Fig. 6). This conformation change introduces both torsional and bond angle strain (*181*). The bond angle strain and torsional strain are essentially eliminated in the acyclic analog because of the greater freedom of motion about the sigma bonds. The depressing effect of strain on hydrolysis of a six-membered ring can be assessed (*173, 181*) by comparing the relative rates of cleavage of α-methyl-D-glucoside (I) and a noncyclic analog (II). At 25°C (II) is cleaved approximately 10^6 times more rapidly than (I).

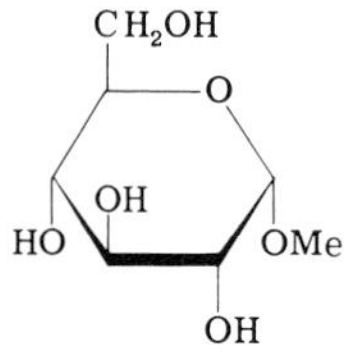

(I) 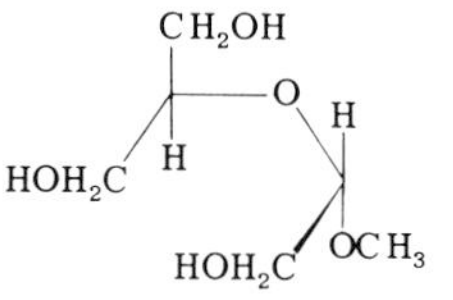 (II) 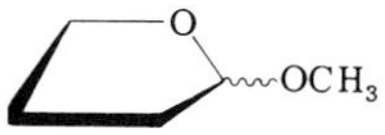 (III)

According to transition state theory, this rate enhancement is achieved by lowering the activation energy by approximately 8.7 kcal/mole. This energy difference approximates the activation barrier for interconversion of the cyclohexane chair forms estimated at approximately +10 kcal/mole (*182*). By way of comparison, the unitary free energy of binding *N*-acetylglucosamine into the distortion subsite on lysozyme is estimated to be in the vicinity of +6 kcal/mole (*183*). This energy is the composite of the actual distortion energy and the energy released upon binding the distorted monomer and represents a minimum strain energy. It is noteworthy that this strain energy is still 2.5 kcal/mole less than that required to compensate for the increase in the activation barrier introduced by cyclizing an acetal. One may speculate then that distortion of the pyranoside ring toward the half-chair conformation may simply be the enzymic device used to compensate for the increase in the activation barrier caused by cyclizing the acetal. The energy expended for substrate distortion must be acquired at the expense of favorable enzyme–substrate contacts. For a polymeric substrate the difficulty is easily solved since endothermic distortion of one monomer residue can be compensated for by exothermic interaction of adjacent monomers.

181. J. N. BeMiller, *Advan. Carbohydrate Chem.* **22,** 25 (1967).
182. E. L. Eliel, N. C. Allinger, S. J. Angyal, and G. A. Morrison, "Conformational Analysis," pp. 41–42. Wiley (Interscience), New York, 1954.
183. D. Chipman and N. Sharon, *Science* **165,** 454 (1969), and references therein.

However, for a glucosidase the distortion problem is quite severe because the substrate lacks adjacent groups for compensating exothermic interactions. Based on the lysozyme model with a distortion energy of 6 kcal/mole and a K_m of 1 mM, glucosidases must release at least 10 kcal/mole upon association with a distorted substrate, a permissible but very large energy change. If the speculation is correct that distortion compensates for angle and torsional strain in the ring, the basic problem of C–O bond heterolysis still remains unsolved.

Side chain catalysis will now be evaluated. As noted above, carboxyl anion and imidazolium cation have been implicated as the catalytic agents in a variety of amylases (*10, 173*). Reasons for preferring an ionized rather than the kinetically equivalent unionized residues have been offered (*173*). Suggestions for the probable modes of participation of these side chains rest strongly on the behavior of model systems. Extensive data on acetal hydrolysis have been reviewed and discussed elsewhere (*167, 181*). However, the following observations are pertinent to speculation about an enzymic mechanism. General acid catalysis of both aliphatic and aromatic acetals has been convincingly demonstrated (*167, 184*). Intramolecular nucleophilic assistance of acetal solvolysis by both sulfur and oxygen, anchimeric assistance by acetamido groups (*126, 167*), and electrostatic assistance by an ionized carboxyl group (*185*) are postulated. Specific acid- and base-catalyzed hydrolyses of aromatic glycosides, respectively, exhibit a negative and positive Hammett coefficient while the Hammett coefficient for enzyme-catalyzed hydrolysis is negative but of intermediate value (*167, 173*). Thus the enzymes appear to employ nucleophilic and/or general acid assistance (*173*) for departure of the aryloxy group. The solvent isotope effect at pH or pD 7.0, $V_{\max(H_2O)}/V_{\max(D_2O)} = 1.25$, for both α- and β-amylase while that for the general acid-catalyzed hydrolysis of 2-carboxyphenyl β-D-glucoside $k_{(D_2O)}/k_{(H_2O)} = 1.75$ (*167, 173*). Capon (*167*) uses these data to argue against general base catalysis with a carboxyl group with β-amylase and nucleophilic catalysis for α-amylase. Great care must be exercised in interpreting solvent isotope effects for enzyme-catalyzed reactions (*186*). In summary, general acid catalysis acting in concert with some sort of nucleophilic participation (electrostatic solvation or covalent bond formation) seems a reasonable process to invoke for amylase catalysis.

One goal of physical organic chemistry is to give a quantitative ac-

184. A. Kankaanpera and M. Lahti, *Acta Chem. Scand.* **23**, 3266 (1969).

185. J. Long, Ph.D. Dissertation, University of California, Berkeley, California, 1969.

186. W. P. Jencks, "Catalysis in Chemistry and Enzymology," p. 243. McGraw-Hill, New York, 1969.

counting of enzyme catalysis by extrapolation of catalytic factors estimated from model systems. Following the approach of Koshland (*187*) for estimating the proximity and orientation contribution of carboxyl and imidazole side chains to amylase catalysis, Thoma (*173*) found that the maximum computed enzyme rate coefficient at 25°C was approximately 10^5 times smaller than the observed coefficient. This factor is an estimation of enhancement brought about by aligning the substrate with the catalytic groups on the enzyme and decreasing the activation entropy. The calculation was predicated on the assumption that the general acid-catalyzed reaction would have a smaller pseudo-first-order rate coefficient than the specific hydronium ion-catalyzed acetal hydrolysis. The computed rate probably overestimated the rate enhancement that can be achieved by organization of the substrate and catalytic groups on the enzyme. An acceleration factor of 55 was assigned for loss of translational freedom, a factor of 10 for orientation and a factor of 10 for loss of rotation about the point of substrate-catalyst contact. Loss of rotational freedom will certainly be negligible for such groups of OH^- but may be significant for conjugated catalysts such as imidazole and carboxylate groups. In short, a factor of 100 for loss of two degrees of rotational freedom is most generous and constitutes a maximum estimate of entropy effects on the reaction.

Comparison of the hydrolysis rate of α-methyl D-glucoside (I) and its analog (II) and tetrahydro-2-methoxypyran (III), may help assign more reasonable estimates of computed enzyme catalytic coefficients. The latter compound (III) is cleaved 3×10^6 times more rapidly than the glucoside (*181*) at 30°C. The rate acceleration probably reflects lack of inhibiting inductive effects and diminished torsional strain in the transition state (*181*). Since tetrahydro-2-methoxypyran (III) has a lower activation barrier for hydronium ion-catalyzed hydrolysis (*181*) it should be the more susceptible than (I) to general acid catalysis. Since general acid catalysis has been observed for (III) (*181*, *188*), we propose that the rate coefficient computed for the enzyme is grossly overestimated. Shifting from specific acid catalysis to general acid catalysis will lower the catalytic rate by at least six orders of magnitude leaving an enzyme enhancement factor of 10^5 (proximity-orientation effect) $\times 3 \times 10^6$ (inductive and strain effects) $= 3 \times 10^{11}$ unexplained. Model compounds (I) and (II) suggest ring distortion can account only for an enhancement factor of 10^6 leaving an acceleration factor at least of 10^5 unaccounted for. Such computations do not give any insight into the

187. D. E. Koshland, Jr., *J. Theoret. Biol.* **2**, 75 (1962).
188. T. H. Fife and L. K. Jao, *JACS* **90**, 4081 (1968).

actual catalytic mechanism but only exclude certain simple extrapolations based on model systems and leave the impression that cooperative interplay of various phenomena occur at the enzyme-water "interface" that are difficult to reproduce in aqueous solution.

The rationale for these types of computations tacitly assumes additivity of thermodynamic activation factors or the absence of interaction mechanisms (*189, 190*). It is entirely possible, even likely, that synergistic effects cooperate on the enzyme to create entirely new reaction surfaces not normally used in aqueous solution. For example, forced twisting of the ring imparts angle strain into the enlarged C-2–C-1–O-5 bonds. The results of Schleyer (*191*) imply that a concomitant elongation and weakening of the C-1–O-1 bond occurs as the *p* character (*192*) of the bond increases. Compression of the reactive center and the catalytic groups can lead to changes in the degree of solvation and overcome the energy barrier caused by van der Waals repulsion which may lower the activation enthalpy. Such mechanisms could sensitize a normally inert bond to unusual types of catalysis. Solvent effects have been offered as an alternate explanation for the exceptional efficiency of amylases. It is widely recognized that the local environment at an active site may be substantially different from that of the bulk solvent and change the reaction pathways. Thus, ion pair stabilization of the oxycarbonium transition state in a local hydrophobic medium has been proposed to lead to a large rate acceleration for β-amylase (*193*) and lysozyme (*180*). However, the insensitivity of acetal hydrolysis over a wide range of dielectric constant (*194–196*) does not support this contention. The rate accelerations afforded by ion pairing on hydrophobic enzyme surface are only likely to compensate for the decelerating effect accompanying translocation of the substrate from water, a good ionizing solvent, to the lower dielectric medium on the enzyme (*197*).

In a single displacement reaction at enzyme saturation the velocity of the reaction takes the form k_{cat} [E_0] [H_2O] if the solvent participates

189. P. Wells, *Chem. Rev.* **63,** 171 (1963).
190. J. E. Leflar and E. Grunwald, "Rates and Equilibria of Organic Reactions as Treated by Statistical, Thermodynamic, and Extra Thermodynamic Methods." Wiley, New York, 1963.
191. P. von R. Schleyer, *JACS* **86,** 1854 (1964).
192. L. N. Ferguson, *J. Chem. Educ.* **47,** 46 (1970).
193. D. E. Koshland, Jr., J. A. Yankeelov, and J. A. Thoma, *Federation Proc.* **21,** 1031 (1962).
194. J. Koskikallio and A. Ervasti, *Acta Chem. Scand.* **16,** 701 (1962).
195. J. Koskikallio and I. Tarvainen, *Acta Chem. Scand.* **16,** 263 (1962).
196. R. Wolford, *J. Phys. Chem.* **68,** 3392 (1964).
197. D. E. Koshland, Jr. and K. E. Met, *Ann. Rev. Biochem.* **37,** 359 (1968).

as a nucleophile in the rate-limiting step. Recently, Nelson (*165*) showed that an exo-β-(1 → 3)D-glucanase obeys this rate law in water–glycerol and water–ethanol mixtures when laminarin is cleaved with inversion of configuration. Since the rate deceleration at high alcohol concentration was not caused by denaturation or transfer to the alcohol, he proposed that water participates as a back-side displacing group in the rate-limiting step. A related paper on the moderation of α-amylase activity by various alcohols was the subject of another investigation (*197a*). The activity of this retaining enzyme was also inversely proportional to alcohol concentration. Both ethanol and glycerol inhibited catalysis to approximately the same degree. At 1 *M* alcohol the logarithm of the enzyme rate constant was linearly related to the reciprocal of the dielectric constant for a series of alcohols. Since identical effects of glycerol (Fig. 7) manifest themselves for an enzyme that inverts and an enzyme that retains configuration upon hydrolysis, no convincing

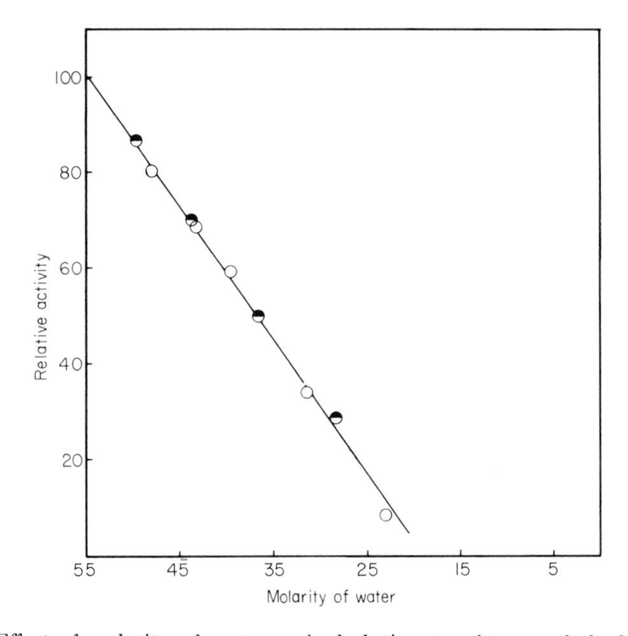

FIG. 7. Effect of molarity of water on hydrolytic rate of two carbohydrases: (○) data replotted from Nelson (*165*), enzyme used was exo-β-(1 → 3)-D-glucanase, which inverts configuration on hydrolysis; (◐) data replotted from Whitaker *et al.* (*197a*), enzyme used was porcine pancreatic α-amylase, which retains configuration upon hydrolysis. The molarity of water is expressed in terms of its concentration in solution.

197a. J. R. Whitaker, A. L. Tappel, and E. Wormser, *BBA* **62**, 300 (1962).

inferences about the mechanism of catalysis can be drawn from the dependence of the rate equation on solvent concentration.

At the present time, it must be concluded that no satisfying quantitative account of carbohydrase catalysis is at hand.

VII. Action Pattern

The term action pattern refers to the reaction pathway of an enzyme-catalyzed reaction and involves studies of structural and chemical relationships between ground states of substrates, intermediates, and products. For example, the relative susceptibility to attack of various bonds in a given molecule or of a given size molecule in a population of molecules or the stereochemical relationships between substrate and products fall within the scope of action pattern investigations. In contrast, mechanism of enzymic action refers to the process of bond transformation which is related to transition state chemistry.

A. Hydrolysis and Condensation Specificity

The anomeric configuration of oligosaccharide fragments resulting from amylase hydrolysis of α, $1 \rightarrow 4$-glucosyl bonds is normally measured polarimetrically. This is the common technique for distinguishing between α- and β-amylases (*198*). Gas–liquid chromatography (*199*), NMR (*200*), and oxidation of β-glucose by glucose oxidase (*201*) are alternately used to establish anomeric configuration. Interestingly, endoamylases commonly retain and exoamylases commonly invert configuration. It has been proposed without supporting evidence that these two classes of enzyme work by different reaction mechanisms (*200*). At the present moment the significance of this observation is obscure. As expected, amylases catalyze the incorporation of ^{18}O from $H_2{}^{18}O$ into C-1 and not the C-4 position of the amyolytic products (*202–204*). The iso-

198. K. Hiromi, T. Shibaoka, H. Fukube, and S. Ono, *J. Biochem.* (*Tokyo*) **66,** 63 (1969), and references therein.

199. G. Semenza, H.-Cu. Curtis, O. Raunhardt, P. Hori, and M. Müller, *Carbohydrate Res.* **10,** 417 (1969).

200. D. E. Eveleigh and A. S. Perlin, *Carbohydrate Res.* **10,** 87 (1969).

201. J. Wakim, Ph.D. Dissertation, Indiana University, Bloomington, Indiana, 1965.

202. F. C. Mayer and J. Larner, *BBA* **29,** 465 (1958).

203. F. C. Mayer and J. Larner, *JACS* **81,** 188 (1959).

204. M. Halpern and J. Leibowitz, *BBA* **36,** 29 (1959).

topic experiments identify the position of heterolysis as the C–1–O bond. However, fission of the C–1–O-5 bond and ring opening prior to hydrolysis is not excluded. For nonenzymic catalysis, rate-limiting heterolysis of the acyclic C–1–O bond is favored (*167*).

During the past few years (*205*) the importance of enzyme-catalyzed reversion during amylolysis (carbohydrate condensation) reactions has been emphasized. According to the principle of microscopic reversibility, all amylases must catalyze stereospecific substrate synthesis since, thermodynamically, this reaction is the reverse of hydrolysis. Hehre *et al.* (*205*) have definitely proved this point. They demonstrated that amylases possess stereochemical or anomeric specificity for glucosyl donors but not for acceptors. Anomalous product distribution ratios are sometimes linked to reversion reaction; for example, Robyt and French (*206*), in a paper destined to become a classic in carbohydrate enzymology, discovered that the specific radioactivity of ^{14}C maltose released from reducing end labeled maltotriose (Glc_3) by the action of pancreatic α-amylase increases with the concentration of substrate (see Fig. 8). At high triose concentration two substrate molecules condense to maltohexaose (Glc_6), a reaction that becomes kinetically competitive with triose hydrolysis. The hexaose is then rapidly cleaved to maltose and maltotriose

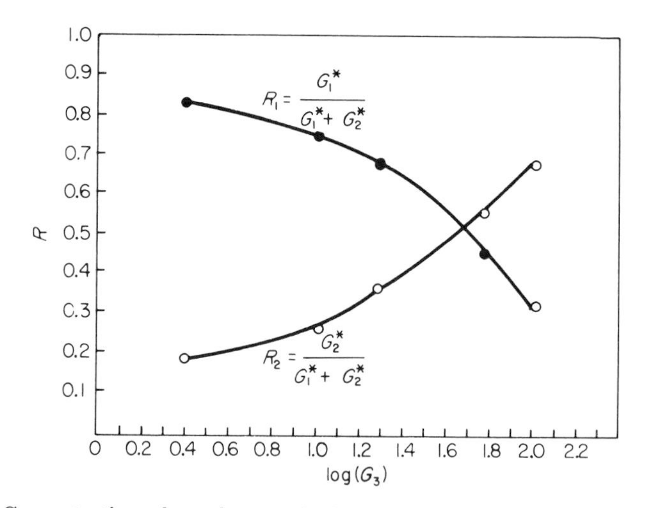

FIG. 8. Concentration dependence of the action pattern of maltotriose. The values G_1^* and G_2^* for each substrate concentration come from initial velocity measurements. The substrate concentration is expressed as m*M* (*206*).

205. E. J. Hehre, G. Okada, and D. S. Genghof, *ABB* **135,** 75 (1969), and references therein.

206. J. Robyt and D. French, *JBC* **245,** 3917 (1970).

(see Fig. 9). Since no glucose is produced from Glc_6 breakdown, maltose is the predominant product of Glc_3 fragmentation at very high substrate concentrations. Shifts in the fragmentation patterns for Glc_4 but not for higher oligosaccharides were also observed as substrate concentration was altered. This work illuminates some unexpected but intriguing complexities of amylase action. Convincing evidence is presented that the maltotriosyl enzyme complex derived from Glc_4 breakdown can partition between transfer to substrate to form Glc_7 or transfer to water. However, the maltotriosyl enzyme complex derived from Glc_5 breakdown is apparently transferred exclusively to water (transfer to Glc_2, the product, regenerates starting material and is not observable). A host of explanations can account for this phenomenon and it would appear that the Glc_3–amylase complex has a "memory" of the original substrate size which directs further reactions.

Based upon product analysis of maltodextrin digests, Robyt and French (*206*) claimed that the specificity site of porcine pancreatic α-amylase is composed of five subsites with the catalytic site located between

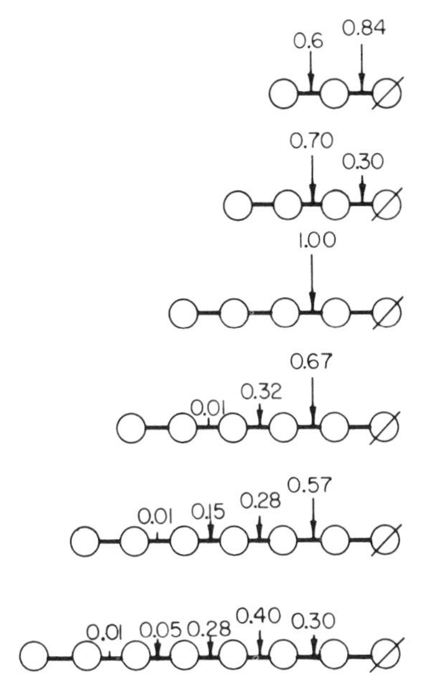

FIG. 9. The frequency distribution of bond cleavage during initial action of porcine pancreatic α-amylase on reducing end labeled oligosaccharides. The values shown for G_3* and G_4* are for infinitely dilute substrate concentrations; (○) nonreducing glucose unit; (∅) reducing glucose unit; and (——) α, 1 → 4-glycosidic bond. From Robyt and French (*206*).

subsites II and III. Although this proposal is consistent with the bulk of the cleavage patterns, some anomalies exist; for example, without modification, their subsite model cannot explain the failure of the enzyme to break the terminal bonds on substrates larger than Glc_5 nor the shift of maximum frequency of cleavage from bond 2 in Glc_7 to bond 3 in Glc_8 (Fig. 9). It is quite clear that monomer residues located at a considerable distance from the cleavage point exert a profound influence on the positional specificity of the enzyme. Fruton and his co-workers have called attention to similar phenomena for proteolytic enzymes (*207*). Jencks (*177*) has reviewed the theories which purport to explain such observations, but at the present moment one cannot distinguish between them. Careful, comprehensive, and quantitative work of this type is unfortunately a rarity in carbohydrate enzymology. At the present stage of development, the continuing publication of incomplete and qualitative work probably will not add significantly to our understanding of amylase chemistry.

B. Substrate Structural Requirements

Substrate analogs are frequently employed to discern the requirements for binding and reaction. Binding specificity and reaction specificity are frequently determined by different parts of the molecule (*207, 208*). Amylases generally act very slowly on maltotriose but cleave the larger oligosaccharides more rapidly (*206*, *209*). Because of the difficulty of modifying oligosaccharides, essentially no information is available about effects of substrate modification on amylase activity. However, a few reports for salivary, pancreatic α-amylase and sweet potato β-amylase and D enzyme have appeared (*206*, *209–212*). α-Amylase preferentially releases D-glucose from the nonreducing end of maltotriose (*206*) and the α- and β-methyl trioside (*210, 211*). Opening the reducing pyranoside of maltotetraose either by oxidation or reduction depresses the rate of cleavage with α-amylase (*209*) but has little effect on larger oligosaccharides. Maltotetratol and maltotetraonic acid are cleaved much more rapidly than α-methyl maltotrioside (*209*). It was concluded

207. K. Medzihradszky, I. M. Voynick, H. Medzihradsky-Schweiger, and J. S. Fruton, *Biochemistry* **9,** 1154 (1970).
208. G. Hein and C. Neimann, *Proc. Natl. Acad. Sci. U. S.* **47,** 1341 (1961).
209. F. W. Parrish, E. E. Smith, and W. J. Whelan, *ABB* **137,** 185 (1970).
210. J. H. Pazur and T. Budovich, *Science* **121,** 702 (1955).
211. J. H. Pazur, J. M. Marsh, and T. Ando, *JACS* **81,** 2170 (1959).
212. S. Peat, W. J. Whelan, and G. Jones, *JCS* p. 2490 (1957).

that three intact pyranoside residues and a remnant of a fourth are the minimum requirements of a moderately good substrate.

Methylation of 40% of the primary OH groups of amylose almost abolishes pancreatic α-amylase activity but the enzyme readily acts on 6-deoxyamylose (90% substitution) (*213*) and yields some 6-deoxy-D-glucose when treated with salivary amylase. Only D-glucose is found in a digest of "3,6-anhydro-amylose" (67% substitution). Limited information about the requirements for substrate structure or the specificity of the enzyme can be decided from studies with partially substituted substrates.

Measurement of the Michaelis constants for each substrate analog should also be reported. Systematic trends in K_m and V_{max} and k_{cat}/K_m with structural modification of substrates are helpful in delineating the requirements for binding and catalytic specificity. The extensive work on proteases with synthetic substrates illustrates the value of this approach (*207*, *214*, *215*). Carbohydrate enzymologists have frequently ignored correlations between these different types of specificity. Simpler syntheses of modified oligosaccharides will certainly augment more intensive mapping of enzyme specificity sites.

The catalytic specificity of β-amylase has generated substantial interest since it recognizes the difference between internal segments and the nonreducing terminus of a starch chain. The effects of modification at the nonreducing and reducing portions of maltodextrins (Glc_n) have been studied. Glc_4 is the smallest maltodextrin rapidly cleaved by β-amylase to maltose. β-Amylase is either retarded or inhibited by alteration at either terminus of Glc_4 (*216–218*). Thus, interactions with the substrate 5–10 Å on both sides of the catalytic site influence catalytic specificity. Methylation of the terminal C_4OH group converts a substrate into a good competitive inhibitor (*218*). Similar observations are frequently made for other enzymes and several hypotheses, binding site span, substrate alignment, induced fit, distortion, and nonproductive complex formation (*219*), have been purported to explain the data. At our present state of knowledge it is impossible to decide which hypothesis or combination of them

213. B. J. Bines and W. J. Whelan, *Chem. & Ind.* (*London*) p. 997 (1960).

214. M. L. Bender and J. Kézdy, *Ann. Rev. Biochem.* **34,** 49 (1965).

215. T. R. Hollands and J. Fruton, *Biochemistry* **7,** 2045 (1968), and references therein.

216. C. E. Weill and R. Rebhahn, *Carbohydrate Res.* **3,** 242 (1966).

217. R. E. Wing and J. N. BeMiller, *Carbohydrate Res.* **10,** 371 (1969).

218. A. Neeley, M. S. Thesis, University of Arkansas, Fayetteville, Arkansas, 1970.

219. W. P. Jencks, "Catalysis in Chemistry and Enzymology." McGraw-Hill, New York, 1969.

is in greatest accord with the experimental data. Any modification of internal portions of an amylose chain hampers hydrolysis by β-amylase, even though the degree of substitution is as low as 0.1–0.20 (*213, 220*) and reflects the inability of β-amylase to bypass a branch point or substituted monomer.

The presence of an $\alpha, 1 \rightarrow 6$ branch point in starch renders neighboring $\alpha, 1 \rightarrow 4$ linkages resistant to attack by α-amylase. The smallest rapidly produced branched dextrin (*221*) formed in the first stages of hydrolysis is a pentasaccharide, 6^3-α-maltosylmaltotriose [see Whelan (*222*) for nomenclature]. Extended action of salivary or pancreatic amylase yields a tetrasaccharide, 6^3-α-glucosylmaltotriose, as the smallest branched limit dextrin (*223, 224*). Whelan (*224a*) has reported that panose, 6^2-α-glucosylmaltose, is formed as the smallest limit dextrin by malted barley α-amylase. Mammalian amylase action around a branch point is summarized in Fig. 10 (*225*).

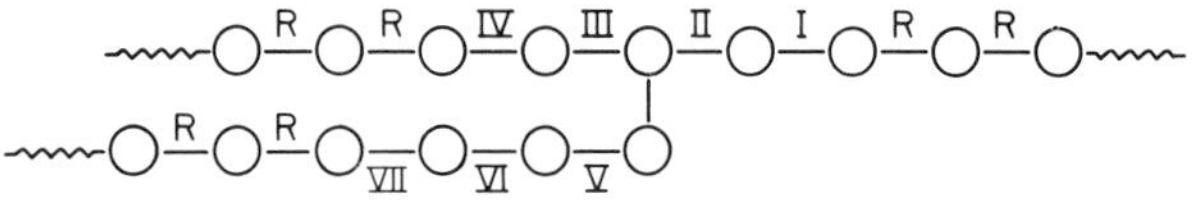

FIG. 10. Illustration of the effect of branch point on ability of salivary amylase to rupture neighboring bonds. O—O and $\overset{\text{O}}{\underset{\text{O}}{|}}$ represent two α-glucose residues bonded by $\alpha, 1 \rightarrow 4$ and $\alpha, 1 \rightarrow 6$ linkages, respectively. The bonds labeled R are readily cleaved while the links designated by the Roman numerals exhibit varying degrees of resistance. From French (*225*).

C. SUBSITE MODEL

The chain length dependence of the Michaelis parameters of polymer hydrolyzing enzymes originally led various authors to speculate that the binding region of these enzymes is composed of a set of tandem subsites geometrically complimentary to the monomer units of the polymer substrate (*15*). This idea had been solidly reinforced by the interactions of these enzymes with substrates and substrate analogs of various sizes and optical purity.

We now wish to examine some of the consequences of a subsite

220. E. Husemann and E. Lindemann, *Staerke* **6**, 141 (1954).
221. B. J. Bines and W. J. Whelan, *BJ* **76**, 246 (1960).
222. W. J. Whelan, *Ann. Rev. Biochem.* **29**, 105 (1960).
223. P. Nordin and D. French, *JACS* **80**, 1445 (1958).
224. R. C. Hughes, Ph.D. Dissertation, University of London, 1959.
224a. W. J. Whelan, *Staerke* **12**, 358 (1960).
225. D. French, *Bull. Soc. Chim. Biol.* **42**, 1677 (1960).

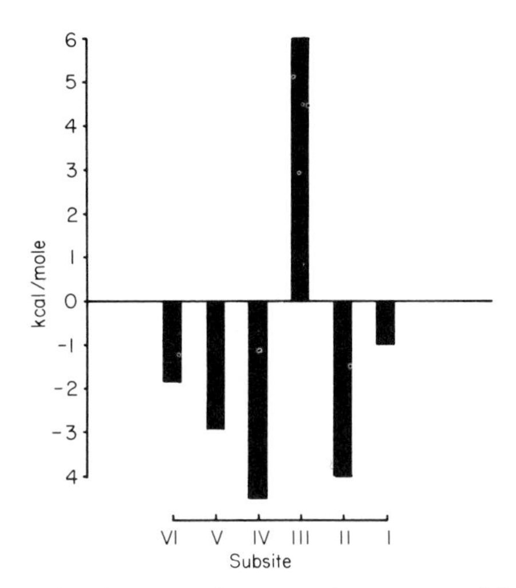

FIG. 11. Energy contributions of subsites on lysozyme to binding of N-acetyl-D-glucosamine polymers $(GlcNAc)_n$. Six monomers are required to fill the active site with the reducing end residing in site I. The catalytic groups lie between II and III. Preferential mode of binding of oligosaccharides is with reducing groups occupying site IV. Drawn from data in Chipman and Sharon (*183*).

energy contour on the action of the enzyme. For illustrative purposes we will use a contour similar to that proposed for lysozyme (see Fig. 11). Because each of the monomer residues of a homopolymer is similar and can occupy any site, each saccharide can form a variety of complexes with the enzyme. These are illustrated in Fig. 12. The relative concen-

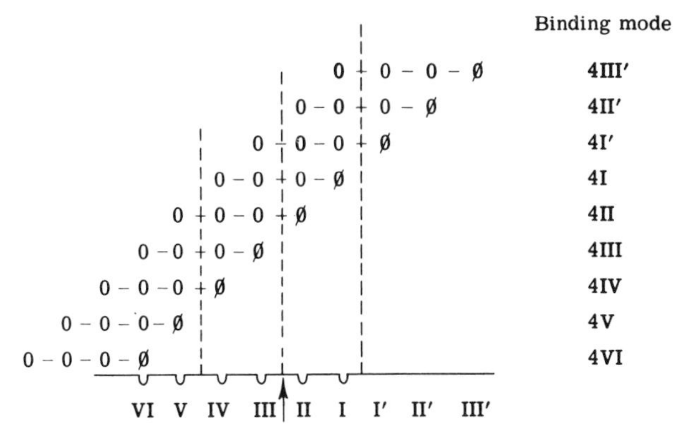

FIG. 12. Examples of productive and nonproductive complexes of $(GlcNAc)_4$ with lysozyme: U represents a subsite on the enzyme, O represents GlcNAc, and Ø is the hemiacetal reducing end of the molecule; ↑ is the position of the catalytic amino acids. Sites I′, II′, and III′ are imaginary. Dotted lines are visual aids.

TABLE VII
EFFECT OF BARRIER HEIGHT AT SUBSITE III ON THE MOLE FRACTION OF VARIOUS $(GlcNAc)_4$ COMPLEXES WITH LYSOZYME[a]

Binding mode	Height in calories/mole			
	0.0	2000	4000	6000
	Mole Fraction of More Highly Populated Complexes × 100			
4IV	11.1	40.1	94.9	99.7
4III	2.1	1.4	0.12	0.004
4II	83.3	56.1	4.63	0.17
4I	3.4	2.3	0.19	0.007

[a] The principal effect of distortion is to cause a dramatic shift from the productive mode 4II to the nonproductive mode 4IV. This effect, though prevalent for larger oligosaccharides, is less pronounced. Data computed according to the following equation:

$$\text{Mole fraction} = K_{a,i}/\Sigma_i K_{a,i}$$

where $K_{a,i}$ is the microscopic association constant for $(GlcNAc)_4$ with lysozyme in the ith mode.

trations of each of these complexes will be proportional to the sum of the energies of the sites occupied by the saccharide. Thus, the population of the complexes for a tetrasaccharide is dominated by 4III (*226*), which is, in fact an inhibited or nonreactive complex (see Table VII). Substrate self-competitive inhibition of this type was noted in 1960 by Thoma and Koshland (*227*). Later, Hein and Niemann (*208*) came to similar conclusions from their studies on chymotrypsin and called the self-inhibited complexes unproductive complexes. Complexes capable of reaction, e.g., 4II and 4I, were labeled productive complexes. This notation has been widely accepted.

The very unfavorable interaction at site III dramatically lowers the fraction of productive complexes for small oligosaccharides and has a very profound effect on $\tilde{K}_m$ and $\tilde{V}_{\max}$ as illustrated by Tables VIII and IX. Thus, the subsite model is most helpful in rationalizing effects of chain length on the Michaelis parameters. For a more complete discussion, see reference *11*. The variation of Michaelis constants with chain length for some enzymes can be nicely accommodated by the subsite

226. Carbohydrate enzymologists conventionally number oligosaccharides from the reducing end. The numbering of polysaccharides and enzyme subsites conforms to that convention in this review. Thus subsites I, II, III, IV, VI, and VII, respectively, conform to sites F, E, D, C, B, and A in Phillips' (*226a*) nomenclature. The numbers increase from right to left.

226a. C. C. F. Blake, L. N. Johnson, G. A. Mair, A. C. T. North, D. C. Phillips, and V. R. Sarma, *Proc. Roy. Soc.* **B167**, 378 (1967).

227. J. A. Thoma and D. E. Koshland, Jr., *JACS* **82**, 3329 (1960).

TABLE VIII

DEPENDENCE OF APPARENT ASSOCIATION CONSTANT FOR ENZYME–SUBSTRATE COMPLEX AS A FUNCTION OF CHAIN LENGTH[a]

Chain length	Barrier height of subsite III in calories/mole			
	0	3000	6000	Exptl.[b]
	Computed or Experimental K_a[c]			
1	51	51	51	20 — 50
2	4.6×10^2	4.5×10^3	4.5×10^3	5×10^3
3	1.2×10^5	0.9×10^5	0.9×10^5	1.0×10^5
4	3.9×10^6	1.1×10^5	0.9×10^5	1.0×10^5
5	9.3×10^7	7.0×10^5	0.9×10^5	1.0×10^5
6	4.9×10^9	3.3×10^6	1.1×10^5	1.0×10^5
7	8.9×10^9	5.9×10^6	1.3×10^5	—

[a] Energy contour of subsites identical to lysozyme with subsite III varied.

[b] Experimental values from lysozyme $(GlcNAc)_n$ interactions *(183)* and Fig. 11.

[c] Computed according to the following equation:

$$K_{a,n} = \sum_{i=1}^{n+5} K_i$$

where n is the chain length, K_i the microscopic association constant for ith mode, and $K_{a,n}$ the apparent macroscopic association constant.

TABLE IX

DEPENDENCE OF APPARENT V_{max} ON CHAIN LENGTH AND ENERGY BARRIER AT SUBSITE III FOR HYDROLYTIC ENZYME WITH SUBSITE ENERGY CONTOUR IDENTICAL TO LYSOZYME

Chain length	Barrier height in calories/mole			
	1000	3000	6000	Exptl.
	Relative Computed or Experimental V_{max}[a]			
2	3.2×10^{-3}	2.1×10^{-5}	1.4×10^{-7}	1.0×10^{-7} [b]
3	0.23	2.0×10^{-3}	1.3×10^{-5}	3.3×10^{-5} [b]
4	0.95	0.21	1.8×10^{-3}	2.7×10^{-4} [b]
5	1.0	0.87	4.2×10^{-2}	1.3×10^{-2} [c]
6	1.0	0.97	0.19	0.10 [c]
7	1.0	0.98	0.29	—
8	1.0	1.0	1.0	—

[a] Interpretation of these values is complicated by the transferase activity of lysozyme [see Chipman and Sharon *(183)*]. Computed according to the following equation:

$$V_{max} = \frac{\text{(productive complexes)}}{\text{(productive + nonproductive complexes)}}$$

[b] Determined under 1:1 complex saturation under first-order conditions for lysozyme.

[c] Same as for footnote *b* except zero-order conditions for lysozyme.

model. Elongating the substrate increases the ratio of internal elements to terminal elements and increases the probability that a bond lies across the catalytic site. Since it is proportional to the fraction of the enzyme associated productively with substrates, $\tilde{V}_{max}$ measures this effect. Long polymers contain a larger number of internal elements than shorter polymers and statistically have a greater possibility of complex formation. Experimentally, this effect manifests itself as a greater enzyme affinity for longer chains.

Because $\tilde{K}_m$ and $\tilde{V}_{max}$ are complex constants dependent upon both productive and nonproductive modes of binding, great caution must be exercised in interpreting the significance of $\tilde{K}_m$ and $\tilde{V}_{max}$ in terms of the size of the specificity site.

Substrate distortion or the energy barrier at site III and the affinity at site II, the two sites adjacent to the catalytic groups, play a most important role in determining the relationship of $\tilde{K}_m$ and $\tilde{V}_{max}$ to chain length (see Tables VIII and IX). As a rule of thumb, it can be stated that when the barrier to the left of the catalytic site is large compared to the potential well to the right, the hydrolysis rate is extremely sensitive to n, but when distortion is small, chain length effects are minimal. Thus, the sum of the free energies of association with these two sites closely controls the fraction of the enzyme associated productively with substrate. Consequently, $\tilde{V}_{max}$ dependence on chain length may be a useful guide to evaluating substrate distortion during hydrolysis when used in conjunction with binding data. In this regard, it has been noted that mammalian α-amylases cleave both large and small substrates rapidly, whereas cereal α-amylases cleave the larger sugars much more rapidly than the smaller ones (*2*). There are two obvious explanations for this observation: first, the distortion energy may be compensated for by an adjacent site in mammalian amylases; or, second, their substrate span may be much smaller, reducing the probability of nonproductive binding.

The subsite model can also explain the preference of enzymes to cleave certain bonds in small substrates. According to this model, the relative concentration of the products of digestion are partially dependent on positional specificity. With a tetrasaccharide (for example, Table VII) the bond penultimate to the nonreducing terminus is apparently very resistant to attack because of the low concentrations of mode 4II. Thus, the apparent preference for enzyme attack of the susceptible bonds is partially thermodynamic and not completely kinetic in origin.

The induced fit hypothesis offers an alternative thermodynamic explanation of these effects. Here, presumably, the size and structure of the substrate influences the positioning of the catalytic amino acids near the reactive substrate bond. While there is undoubtedly some merit to this hypothesis, it is our feeling that the subsite model offers the best and most comprehensive framework for the interpretation of the action of depolymerizing enzymes at the present time. Additional postulates should be invoked only in cases where this model cannot account for the data.

If the character of the products results from a population distribution of binding modes, then it follows that product patterns convey information about the nature and number of subsites. Recently, based on some simplifying assumptions, a mathematical development of this concept was presented (*15*) which shows how product patterns are helpful in the mapping of subsites.

A straightforward steady state analysis reveals that the product distribution ratio for a pair of adjacent modes gives apparent subsite energies,

$$RT \ln \frac{\dot{P}_i}{\dot{P}_{i+1}} = \Delta\tilde{G}_{u,i+1} - \Delta\tilde{G}_{u,i+1-n} \qquad (13)$$

where $\dot{P}$ is equal to $d(P)/dt$ and n represents substrate chain length. The substrate is bound with its reducing end at subsite i and the subscript u implies a unitary function.

The significance of Eq. (13) can be appreciated by reference to Fig. 12. This equation predicts that when $n = 4$, the apparent subsite association affinities for sites I and V govern the substrate distribution between binding modes 4II and 4I since sites II, III, and IV are occupied in common. In the case where $\Delta\tilde{G}_{\mathrm{V}}$ is more negative than $\Delta\tilde{G}_{\mathrm{I}}$, mode 4II will dominate 4I, and more monosaccharide than disaccharide (produced from the reducing end) should appear in the reaction digest. The product ratio, Glc_2/Glc_1, substituted into Eq. (13) gives the difference in the apparent unitary free energy for subsite association of positions I and V. By analogy the ratio, Glc_2/Glc_1, for the trisaccharide substrate gives site IV with respect to I. Thus, by varying the chain length, information about the energetics of all sites except the two sites adjacent to the catalytic site can be gleaned. The reason that no information is forthcoming for these two sites is that they are common to all productive complexes and therefore have no influence on the products released.

An investigation of the products produced by *Bacillus subtilis* (*227a*) liquefying α-amylase from homopolymers of different sizes reveals the power of this technique for subsite mapping (*15*). In such experiments great care must be taken to insure that the results are not complicated by transfer activity or reversion.

227a. According to Campbell the manufacturers have incorrectly identified the source of this enzyme which is in fact derived from *Bacillus amyloliquefaciens*.

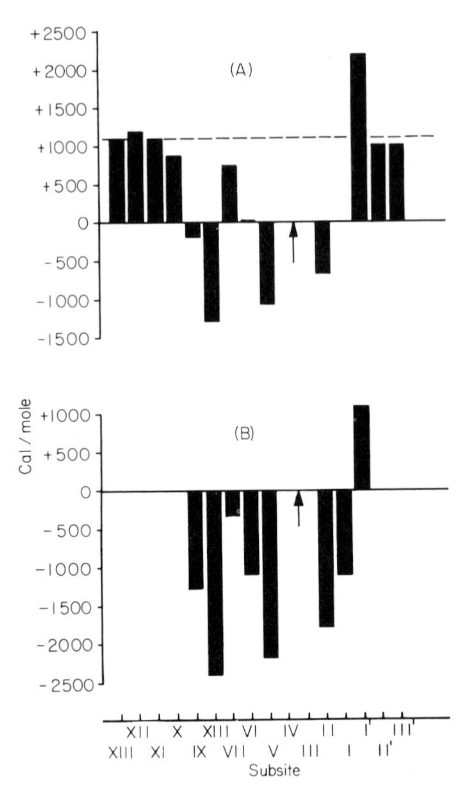

FIG. 13. Histogram of subsite interaction energy of glucose residues with *B. subtilis* liquefying amylase: (A) Site I used as thermodynamic reference; (B) imaginary sites used as reference. After Thoma *et al.* (*15*).

Analysis of the product distribution data according to Eq. (13) for *B. subtilis* (*227a*) amylase produced the upper histogram in Fig. 13A. For convenience, site I was selected as the reference and $\Delta\tilde{G}_{u,\text{I}}$ arbitrarily set equal to zero. An interesting feature revealed by the histogram is that subsites X to XIII and II′ and III′ are all approximately 1.1 kcal more positive than the reference site (Fig. 13). The obvious interpretation of these data is that residues occupying these "sites" do not make contacts with the enzyme. Reassignment of the other sites (lowering them 1.1 kcal) produces a thermodynamic outline of the active site seen in the lower histogram of Fig. 13B. Topologically, the site appears to span nine slots with the catalytic groups located between slots III and IV. If the enzyme interacts with an extended oligosaccharide, it must span about 45 Å. The product analysis also reveals marked heterogeneity of the subsites analogous to the observations made on lysozyme. Sweet potato β-amylase was depicted as containing four to six subsites (*227, 228*), plant α-amylases up to ten subsites (*2*), and porcine pancreatic α-amylase five subsites (*206*). In every case, it has been proposed that the cleavage site is nearer one end of the binding site than the other. It has been speculated that the polarity of multiple attack may be related to this asymmetry (*206*).

On the basis of the limited data available to date, it seems likely that product distributions are a more reliable index of the size of a binding site than $\tilde{K}_m$ and $\tilde{V}_{max}$ dependence on chain length, since product patterns are related only to productive modes of binding and are more easily interpreted in terms of the subsite model.

The subsite histogram, Fig. 13, gives a clear accounting of the preferential attack of *B. subtilis* α-amylase. The unfavorable energetics associated with occupancy of site I′ causes the preferred attack near the reducing end of the molecule. However, as monomer units are added to lengthen the substrate, internal rupture occurs with increasing frequency because of the increasing probability of forming productive complexes with longer chains. Eventually, for very long polymers the initial cleavage will appear to be random.

D. Single Chain or Multichain Attack

Some of the early quantitative investigations on α-amylases led various authors to propose that large polymeric substrates were randomly degraded [for lead references, see Thoma and Spradlin (*11*)]. The pathway in which the enzyme randomly encountered a substrate chain, cleaved it, dissociated and transferred its action to another chain became known as the multichain action pattern.

However, considerable evidence has accumulated to show that the probability of amylase attack of a bond in the vicinity of one which has just been hydrolyzed is frequently larger than expected if cleavage occurs at random (*11*). Two different action patterns, single chain and multiple attack, have been offered to account for this phenomenon. In the single chain action pattern, the enzyme degrades the polymer molecule completely to products after complexing with it. Multiple attack describes the situation where a variable number of hydrolytic cleavages occur during the lifetime of a given enzyme–substrate complex. Mathematically, single chain and multichain action patterns are limiting cases of the multiple attack scheme. Thus the probability that an enzyme–

228. R. W. Youngquist, Ph.D. Dissertation, Iowa State University, Ames, Iowa, 1962.

substrate complex will dissociate before reacting is very large for a multichain enzyme, very small for a single chain enzyme, and intermediate in value for a multiple attack enzyme.

Support for the single chain action pattern was offered in 1948 for wheat β-amylase by Swanson and Cori (*229, 230*) and for barley β-amylase by Kerr and co-workers (*231, 232*). Kerr and co-workers digested a "highly purified, corn crystalline amylose of chain length 235" to approximately 50% hydrolysis, recrystallized the undigested substrate with butanol, and compared the physical characteristics of the reprecipitated material to the properties of the original substrate. The data of this experiment are recorded in Table X.

TABLE X
PROPERTIES OF UNDIGESTED AND BARLEY β-AMYLASE-TREATED AMYLOSE

Measurement	Original amylose	Recovered amylose
DP[a] (osmotic pressure of acetate in chloroform)	235	235
Intrinsic viscosity	0.46	0.48
Ferricyanide number	3.43	3.53
Iodine affinity (% I_2 bound)	20.1	20.4

[a] Where DP stands for degree of polymerization or chain length.

Cowie *et al.* (*233*) have extended this experimental approach to a larger substrate, an amylose sample of DP 3200 (where DP stands for degree of polymerization). Sedimentation constants plotted vs. amylose concentration determined at 35, 45, 55, and 77% conversion into maltose by soybean β-amylase were independent of the degree of hydrolysis. Furthermore, the viscosity and iodine affinity of the original amylose and limit dextrin were identical. Other amylose samples gave similar results.

Swanson and Cori (*229–230*) claimed evidence for a single chain attack when digestion of amylose by wheat β-amylase failed to shift the wavelength of maximum absorption (λ_{max}) of the amylose–iodine complex. In a more carefully contrived experiment, Bourne and Whelan (*234*) reported that between 69 and 86% hydrolysis the λ_{max} of the amylose-I_2 complex shifted from 6500 to 5800 Å, and cited their results as support for a multichain action pattern.

229. M. A. Swanson, *JBC* **172,** 805 and 825 (1948).
230. M. A. Swanson and C. Cori, *JBC* **172,** 815 (1948).
231. F. C. Cleveland and R. W. Kerr, *Cereal Chem.* **25,** 133 (1948).
232. R. W. Kerr and H. Gehman, *Staerke* **3,** 271 (1951).
233. J. M. G. Cowie, I. D. Fleming, C. T. Greenwood, and D. J. Manners, *JCS* p. 697 (1958).
234. E. J. Bourne and W. J. Whelan, *Nature* **166,** 258 (1950).

In 1961, French (*235*) made the critical observation that the outcome of experiments described above will reflect the chain length distribution of the original amylose sample. Since natural amylose, the source of the materials used in the above experiments, is thought (*236*, *237*) to closely conform to a most probable distribution, it is important to examine the effects of this population on the product distribution.

French stated (*235*) that "if the molecular weight distribution of the original amylose is the most probable distribution, then each of the three suggested action patterns (single chain, multichain, or multiple attack) will lead to average molecular weight and size distribution of the residual amylose precisely the same as those of the original amylose." This is illustrated in Fig. 14 and may be simply verified.

French's statement says, in essence, that a constant fraction of each DP is removed in unit time for the most probable distribution even for the multichain action pattern. Thus one needs to show that (*238*)

$$\frac{dN_{i-1}/dt}{N_{i-1}} = \frac{dN_i/dt}{N_i} = \frac{dN_{i+1}/dt}{N_{i+1}} \tag{14}$$

where N_i represents the mole fraction of polymer of DP_i. For the multichain action pattern, assuming the microscopic constants to be independent of chain length and amylose to be a most probably distributed polymer and the sole product of β-amylase digestion to be maltose:

$$\frac{dN_i}{dt} = kN_{i+1} - kN_i \tag{15}$$

For the most probable distribution, $N_i = p_i(1 - p)$, where p is the proportion of possible monomer bonds that can form (*239*); thus,

$$\frac{dN_i/dt}{N_i} = \frac{kp^{i+1}(1-p) - kp^i(1-p)}{p^i(1-p)} = k(p-1) \tag{16}$$

which is, as we wanted to show, independent of the degree of polymerization.

For the Poisson distribution, on the other hand, where $N_i = (e^{-\nu}\ \nu^{i-1})/[(i-1)!]$ (*239*), $\nu + 1$ being the average DP, it can be shown that

$$\frac{dN_i/dt}{N_i} = \frac{k(\nu - i)}{i} \tag{17}$$

The value $(dN_i/dt)/N_i$ ranges from positive to negative as i passes through

235. D. French, *Nature* **190,** 445 (1961).
236. E. Husemann and B. Pfannemüller, *Makromol. Chem.* **87,** 139 (1965).
237. E. Husemann and B. Pfannemüller, *Makromol. Chem.* **43,** 156 (1961).
238. S. Dygert, Ph.D. Dissertation, Indiana University, Bloomington, Indiana, 1971.
239. C. Tanford, "Physical Chemistry of Macromolecules," Chapter 3. Wiley, New York, 1963.

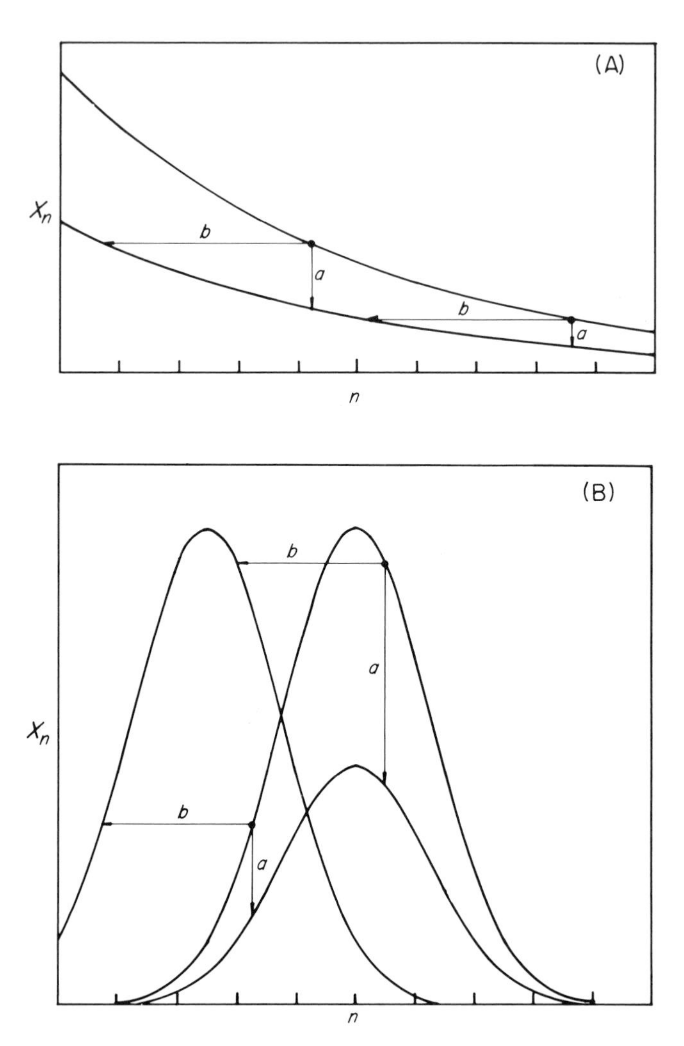

FIG. 14. Effect of substrate distribution and action pattern on product distribution. (A) Attack of most probable distribution of amylose by (*a*) single chain and (*b*) multichain patterns. In (*a*) the number X_n of chains of a given length n is reduced by a given proportion. In (*b*) each chain is shortened by a given amount. The two action patterns give rise to identical distributions at intermediate stages of digestion. (B) Attack of normal distribution of amylose by (*a*) single chain and (*b*) multichain patterns. In (*a*) the distribution is unchanged, but the number of moles of substrate is decreased. In (*b*) the average molecular weight falls in proportion to the extent of degradation. Thus, narrowly distributed substrates are best suited to test the enzyme's mode of action. After French (*235*).

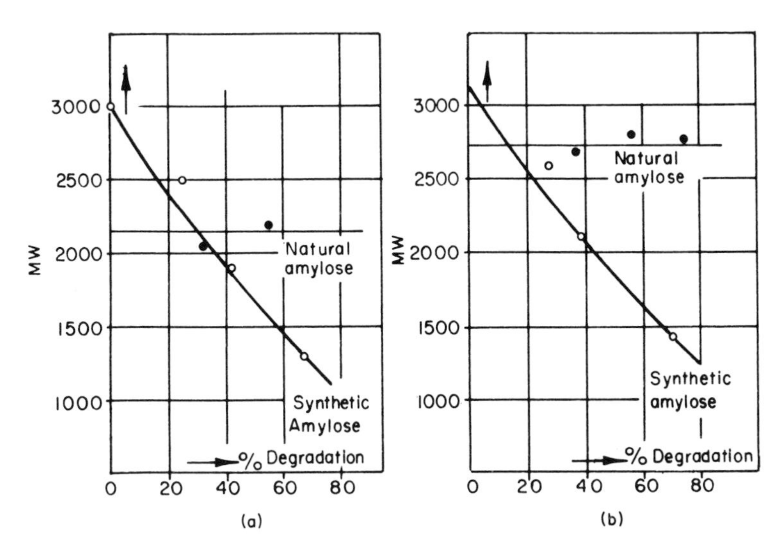

FIG. 15. Degradation of natural (most probable distribution) and synthetic (Poisson distribution) amylose with (a) phosphorylase and (b) β-amylase. Weight-average molecular weight plotted vs. degree of hydrolysis. After Husemann and Pfannemüller (*237*).

ν. Phrased alternately, the relative concentration of small molecules increases at the expense of the large molecules so that the average chain length drops as digestion proceeds. Banks and Greenwood have further discussed this problem (*240*). The elegant experimental work of Husemann and Pfannemüller (*237*) decisively bolsters French's logic. Phosphorylase-synthesized amylose preparations, which approximate a Poisson distribution, do show a very marked decrease in weight average DP with β-amylolysis, while natural amylose does not show any change in $\overline{DP}_w$ with β-amylolysis out to 70% hydrolysis (Fig. 15). (The source of the amylase is not clear.) A multichain (or multiple attack) action pattern is thus indicated.

E. MULTIPLE ATTACK

French *et al.* (*241*) were the first to show that the build up of the intermediate product (maltopentaose) during β-amylolysis of maltoheptaose varied with conditions; at high temperature (70°C) or high pH (pH 10) significant amounts of maltopentaose accumulated. However, at

240. W. Banks and C. T. Greenwood, *Staerke* **20,** 315 (1968).
241. D. French, D. W. Knapp, and J. H. Pazur, *JACS* **72,** 1866 (1950).

pH 4.7 and 26°C, very little intermediate was detected. These initial observations, correctly diagnosed by the authors, indicated that the action pattern is not completely single chain or completely multichain and that it can be varied by changing the experimental conditions.

In a more quantitative experiment, Bailey and Whelan (*242*) isolated the hydrolysis products on a charcoal-Celite column and analyzed their data according to the following kinetic scheme:

$$\text{Glc}_7 \xrightarrow{k_1'} \text{Glc}_2 + \text{Glc}_5 \xrightarrow{k_2'} 2\ \text{Glc}_2 + \text{Glc}_3$$

Under first-order conditions, it is possible to calculate the level of intermediate for multichain action at any stage of digestion. The ratio of the recovered intermediate to the calculated intermediate is called the "proportion of multichain action." The results varied between 26 and 98% multichain action as shown in Table XI.

TABLE XI

EXPERIMENTAL AND CALCULATED LEVELS OF INTERMEDIATES DURING β-AMYLASE DIGESTION OF MALTODEXTRINS[a]

Substrate	*T* (°C)	pH	Conversion to maltose (%)	Exptl. recovery of intermediate	Calc. recovery of intermediate	Multichain attack (%)
7	35	4.8	52	3.3	12.6	26
7	35	8.0	38	4.1	11.7	35
6	35	4.8	64.5	25.3	45.7	56
6	70	4.8	48	27.1	45.8	56
6	0.5	4.8	69.5	35.6	36.4	98

[a] After Bailey and Whelan (*242*).

The amylose-iodine absorption spectrum also changed during β-amylolysis (DP of substrate = 49) in a manner consistent with a multiple attack mechanism (*242*). No change in action pattern occurred in the presence of a large molar excess of maltotetraose, and it was reasoned that the multiple attack was not a result of slow diffusion of the large amylose molecules since the tetraose should be able to compete in a diffusion process and thus shift the action to multichain.

The first representational model for multiple attack was presented by Bailey and French (*243*) and is shown in Fig. 16.

This scheme (Fig. 16) implies a rapid isomerization of the enzyme–substrate complex after release of maltose from a nonproductive to a

242. J. M. Bailey and W. J. Whelan, *BJ* **67**, 540 (1957).
243. J. M. Bailey and D. French, *JBC* **226**, 1 (1957).

$$
\begin{array}{ccccccc}
 & & & \mathrm{E + Glc_7} & & \mathrm{E + Glc_5} & & \mathrm{E + Glc_3} \\
 & & & \uparrow k_{-1,7} & & \uparrow k_{-1,5} & & \uparrow k_{-1,3} \\
\mathrm{E + Glc_9} & \underset{k_{-1,9}}{\overset{k_{+1,9}}{\rightleftharpoons}} & \mathrm{E \cdot Glc_9} \xrightarrow{k_{r,9}} & \mathrm{E \cdot Glc_7} & \xrightarrow{k_{r,7}} & \mathrm{E \cdot Glc_5} & \xrightarrow{k_{r,5}} & \mathrm{E \cdot Glc_3}
\end{array}
$$

FIG. 16. Multiple attack on Glc_9 under conditions that reattack on released product is negligible according to simplified scheme of Bailey and French (*243*). E represents β-amylase. Under more extensive digestion conditions bimolecular association constants $k_{+1,7}$ and $k_{+1,5}$ must be added to the scheme to account for attack of Glc_7 and Glc_5 after an initial dissociation.

productive mode which can then undergo dissociation or further reaction. Hanson (*28*) and Thoma (*29*) have introduced a more general scheme, Fig. 17, that allows substrate dissociation from the nonproductive complex formed after hydrolysis of a bond, but before rearrangement of the substrate. Steady state kinetics cannot distinguish between the two schemes since they give identical rate equations (*29*). A more thorough theoretical analysis of this scheme (*29*) casts doubt on the significance of the calculated absolute rate constants (*117*) for amylases.

By a clever application of probability theory, Bailey and French deduced the relationship between the number of molecules suffering n ruptures given the average number of encounters per molecule by the enzyme (*243*). For long chains, assuming that k_1 is independent of chain length and that the encounters between enzyme and substrate are completely at random, the distribution of encounters per molecule is given by

$$N_x = \frac{e^{-\nu} \nu^x}{x!} \tag{18}$$

where N_x is the proportion of substrate molecules which have been encountered a total of x times and ν is the average number of encounters per molecule. The enzyme–substrate complex may either react via a k_r

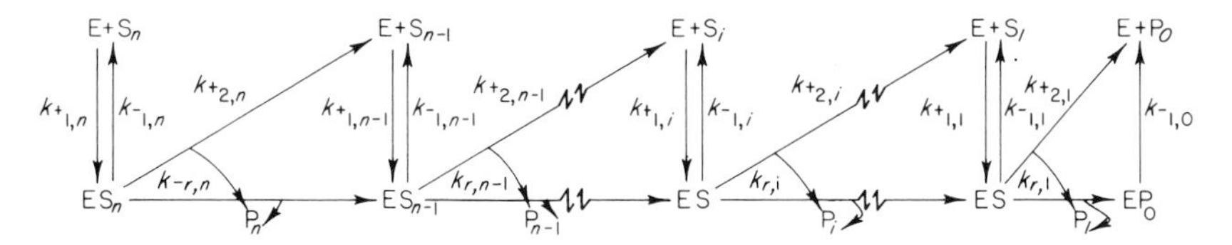

FIG. 17. Endwise cleavage of polymer mixture (S_n, S_{n-1}, . . . S_1 . . . S_0) composed of polymers of fixed sequence from end not attached to the enzyme. For β-amylase, polymer defined from reducing end and P_i is maltose. A k_{+2} process involves cleavage of a substrate and dissociation of both fragments from the enzyme. A k_r process involves cleavage, release of maltose, and rearrangement of the substrate fragment to form a new active complex. From Thoma (*29*).

process or dissociate via the k_{-1} process. The probability that it will decay before it reacts is $(1 - f)$ where $f = k_r/(k_{-1} + k_r)$. The probability (P_n) that it will react just n times is

$$P_n = (1 - f)f^n \tag{19}$$

The distribution of reactions per molecule can now be obtained by considering the total processes that can lead to 0, 1, 2, 3, 4, etc., reactions and is found to be

$$S_n = e^{-\nu f} f^n \sum_{L=1}^{n} \frac{(n-1)![(1-f)\nu]}{(n-L)!(L-1)!(L)!} \tag{20}$$

S_n is the probability that a given substrate molecule has reacted, with

TABLE XII
EFFECT OF CHAIN LENGTH AND pH ON THE DEGREE OF REPETITIVE ATTACK BY β-AMYLASE ON MALTODEXTRINS AT 25°C[a]

Chain length	pH	k_r/k_{-1}	$k_{r,5}/k_{-1,5}$	$k_{r,7}/k_{-1,7}$	$k_{r,9}/k_{-1,9}$
6	4.8	0.16			
7	3.5		1.3		
	3.6	0.82			
	4.8	0.85			
	4.8	0.81			
	4.8		1.4		
	8.0	0.61			
	9.0		0		
	9.2	0.10			
	10.0	−0.04[b]			
8	3.6	1.94			
	4.8	3.00			
	4.8	2.6			
	7.0	2.03			
	9.2	1.00			
9	3.5		1.7	2.7	
	4.8	1.5			
	4.8		1.5	2.4	
	9.0		0	0	
11	4.8	1.7			
	4.8		1.6	2.4	1.9
	9.0		0	0	0
13	3.5		1.7	3.1	3.0
	4.8		1.8	2.9	4.3
44	4.8	3.3[c]			

[a] From Thoma and Spradlin (*11*).
[b] Probably indistinguishable from 0.0.
[c] At 35°C.

product formation, just n times, and ν is the average number of encounters per molecule.

Bailey and French (*243*) determined the value of f for sweet potato β-amylase acting on a Poisson distributed amylose, ($\overline{DP}_n = 44$) labeled at the nonreducing end (pH 4.8 and $T = 35°C$). They found $f = 0.76$ which means that an average number of 3.3 maltose units are removed per enzyme–substrate encounter. A further experiment using an internally labeled substrate showed that the results could not be explained by four sequential active sites, even though the enzyme is known to be a tetramer (*111*). The f parameter has been measured for chromographically pure oligosaccharides as a function of pH (*228, 244, 245*) (see Table XII). End effects are apparent to approximately a chain length of 10 where the f value approximates the value of 0.75 found by Bailey and French for a $\overline{DP}_n$ 44 amylose preparation. At high pH, the enzyme definitely shifts toward multichain action. The data at low pH are not extensive enough to permit a firm conclusion to be drawn about a shift in action.

Recently, the action of β-amylase on the maltodextrins has been reanalyzed by computer simulation (*11*). The technique requires numerical integration of the differential equations of Fig. 16 using various values of k_1, k_r, and k_{-1} and searching for the best least squares fit of the data. It was found that the levels of intermediates during hydrolysis are extremely sensitive to k_r/k_{-1} (Fig. 18). Hence the levels of intermediates are a good quantitative index of the extent of repetitive attack. Within experimental error the kinetic and statistical approaches are in accord. Further experimental substantiation of the French model comes from initial steady state studies of the degradation of ^{14}C-labeled oligosaccharides. For the hydrolysis of Glc_9 when the extent of hydrolysis is held low enough to make reattack of a previously encountered substrate very improbable, it is easily shown at steady state that

$$\frac{Glc_3}{Glc_5} = \frac{k_{r,5}}{k_{-1,5}} \tag{21}$$

and

$$\frac{Glc_5 + Glc_3}{Glc_7} = \frac{k_{r,7}}{k_{-1,7}} \tag{22}$$

Thus, product ratios furnish a direct route to the ratio, $k_{r,n}/k_{-1,n}$.

The experimental values shown in Table XII are in fairly good agree-

244. D. French and R. W. Youngquist, *Staerke* **15**, 425 (1963).

245. J. E. Spradlin, Ph.D. Dissertation, Indiana University, Bloomington, Indiana, 1970.

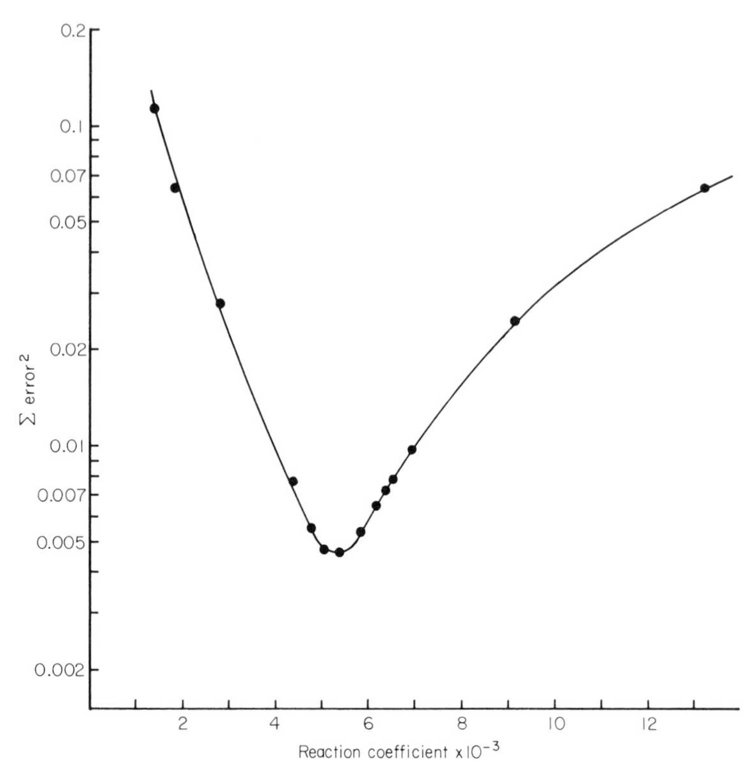

FIG. 18. Sensitivity of the fit of a multiple attack scheme to experimental data for β-amylolysis of Glc_7. The index of fit is the sum of the errors squared (Σe^2) when simulated profiles were compared to French and Youngquist data (*244*). Σe^2 represents composite of errors from Glc_7, Glc_5, and Glc_3 curves. Raw data for input into program were read from smoothed curves and are not actual experimental points. If the smoothed data are accurate to $\pm 2\%$ then Σe^2 less than 0.01 is within experimental error for 27 points. Constants as in Fig. 3 for repetitive attack except $k_{r,7}$ and $k_{r,5}$ varied. The apparent minimum occurs at $k_r/k_{-1} = 1.3$. Since Σe^2 is plotted on a log scale, it can be seen that the fit is very sensitive to k_r/k_{-1}. From Thoma and Spradlin (*11*).

ment with the statistically determined values, and the fact that β-amylase operates by a multiple attack pattern on small polysaccharides (DP 50 or less) appears inescapable.

Work by Pazur *et al.* (*246*) published in 1950 originally led to the concept of multiple attack by α-amylases. These authors followed the progress of a porcine pancreatic amylase digest at pH 7 and 10.3 by

246. J. H. Pazur, D. French, and D. W. Knapp, *Proc. Iowa Acad. Sci.* **57,** 203 (1950).

iodine staining and reducing power. At pH 10.3 the blue value (iodine staining capacity of long saccharides) dropped much more rapidly as a function of the extent of hydrolysis (reducing value) than at pH 7.0. Furthermore, at the iodine achroic point, the distribution of products at pH 7.0 was distinctly different from that at pH 10.3. The results were explained by repeated cleavage of a given substrate before the substrate had time to diffuse away from the enzyme. Figure 19 illustrates the multiple attack process envisioned for α-amylases and Fig. 1 the experimental consequences.

The smaller the slope of the lines in Fig. 1, the greater the degree of multiple attack. The small molecules released after the initial cleavage do not appreciably decrease the blue value but add substantially to the

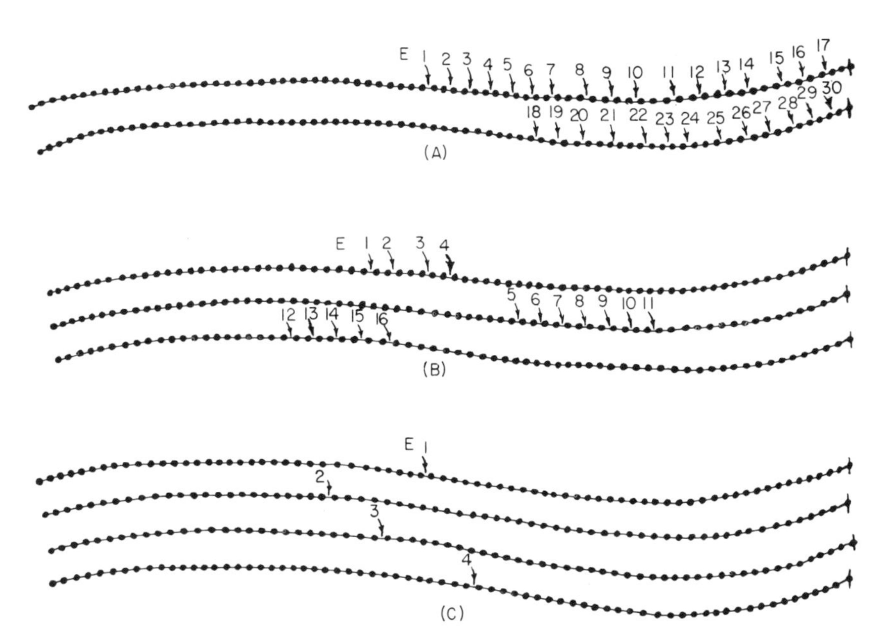

FIG. 19. Types of attack patterns for endoamylases: (A) single chain, (B) multiple attack, and (C) multichain. Each case illustrates the action of a single enzyme molecule. The arrows represent the catalytic hydrolysis of a glycosidic bond; (●) amylose molecule made up of glucosyl units linked $\alpha, 1 \rightarrow 4$; (●) reducing hemiacetal end group. The numbers refer to the sequence of hydrolytic events by the enzyme. The oligosaccharide product specificity in cases (A) and (B) is arbitrarily assumed to be maltose and maltotriose. The authors have assumed a definite polarity of the enzyme action toward the reducing end. The actual direction of the action is now known to be toward the nonreducing end for pancreatic α-amylase. For illustrative purposes the amylose molecule is pictured as a long "string"; however, it undoubtedly possesses a certain amount of secondary helical structure that is not illustrated. From Robyt and French (*23*).

reducing value. Figure 1 shows that porcine pancreatic α-amylase exhibits the greatest degree of multiple attack. The fact that 1 *M* H_2SO_4 shows a greater degree of multiple attack than porcine pancreatic α-amylase (at pH 10.5) is rationalized by assuming porcine pancreatic α-amylase shows an avoidance of chain ends and/or that the acid shows a preference for chain ends.

Robyt and French (*23*) have quantitated the degree of multiple attack for several α-amylases. After various degrees of hydrolysis, an amylose digest was fractionated by precipitating the high molecular weight polymers with ethanol. Reducing groups in the precipitated fraction represent the number of effective encounters, and the oligosaccharide reducing groups in the supernatant fraction represent the subsequent reactions and end effects. The degree of multiple attack is related to the ratio of these two numbers, i.e., the number of subsequent attacks per effective encounter. This value is 6 for porcine pancreatic α-amylase (pH 6.9), 2 for human salivary α-amylase (pH 6.9), and 1.9 for *Aspergillus oryzae* α-amylase (pH 5.5). At pH 10.5, porcine pancreatic α-amylase shows a degree of multiple attack of 0.7, whereas 1 *M* H_2SO_4 gives a value of 0.9.

Greenwood and Milne (*2*) have investigated the point of initial encounter of an enzyme along a long chain. They found a linear relationship between the reciprocal of the viscosity average molecular weight and time when amylose is digested by cereal amylases. This experiment has now been accepted (*23*, *247*) as evidence that the initial cleavage occurs randomly along the amylose molecule without providing any evidence for or against multiple attack.

Banks *et al.* (*247*) estimated the degree of multiple attack by following the ratio, weight average degree of polymerization/number average degree of polymerization ($\overline{DP}_w/\overline{DP}_n$). Oligosaccharides released by multiple attack profoundly influence $\overline{DP}_n$ but have minor effect on $\overline{DP}_w$. Thus the ratio, $\overline{DP}_w/\overline{DP}_n$, should increase in proportion to the degree of multiple attack. According to their data, only porcine pancreatic α-amylase (pH 4.8, 37°C) showed multiple attack. In addition, no multiple attack was observed for α-amylases from *Bacillus subtilis*, human saliva, malted rye, or porcine pancreatic amylase when the latter digest was carried out in a medium of 40% glycerol. The blue value vs. reducing value curves were in accord with these findings, again affirming its usefulness as a measure of multiple attack. The fact that glycerol decreases multiple attack probably means it is a noncompetitive inhibitor as

247. W. Banks, C. T. Greenwood, and K. M. Khan, *Carbohydrate Res.* **12**, 79 (1970).

it is for sweet potato β-amylase (*238*) and forces dissociation after attack.

Inspection of the data of Banks *et al.* (*247*) suggests that their method is far less sensitive than the technique of Robyt and French (*23*) for estimating the degree of repetitive attack. The former method is based upon cleavage of the odd chain length maltodextrins to glucose and maltose and the detection of glucose by glucose oxidase. Thus even-membered oligosaccharide products are not registered, and an enzyme which released large amounts of maltose would appear to give a low degree of multiple attack.

Robyt and French (*248*) have recently published an outstanding paper aimed at elucidating the polarity or direction of multiple attack by porcine pancreatic α-amylase. In principle, after the initial cleavage, subsequent attacks might proceed either toward the nonreducing end, the reducing end, or alternate in direction. By comparing the rate of release of products of end-labeled Glc_8 at pH 10.5 (no multiple attack) and at pH 6.9, it was possible to deduce that the polarity of multiple attack is exclusively toward the nonreducing end. The logic and simplicity in this paper are masterful and illustrate the kinds of elegant and decisive experiments that can be preformed with end-labeled substrates. A careful perusal of this paper is recommended for all enzymologists interested in the action pattern of polymer degrading enzymes.

F. Effects of Chain Length

Thoma and Koshland (*227*) formulated a mathematical model for the action of sweet potato β-amylase that predicts the dependence of $\tilde{V}_{max}$ and $\tilde{K}_m$ on the degree of polymerization of substrates. Their model, based on Fig. 20, reflects the idea that the enzyme can form productive complexes with the nonreducing terminus of a starch chain and nonproductive complexes with the cycloamyloses or the topologically equivalent internal segments of starch chains. The model predicts linear dependence of $1/\tilde{K}_m$ and $1/\tilde{V}_{max}$ on DP:

$$1/\tilde{K}_m = 1/K_m + (\mathrm{DP} - m)/K_\mathrm{I} \tag{23}$$

$$1/\tilde{V}_{max} = 1/V_{max}\left[1 + (\mathrm{DP} - m) \times \frac{K_m}{K_\mathrm{I}}\right] \tag{24}$$

where K_I is the dissociation constant for an internal segment, the tilde implies an apparent constant, and K_m and V_{max} are the Michaelis con-

248. J. Robyt and D. French, *ABB* **138**, 662 (1971).

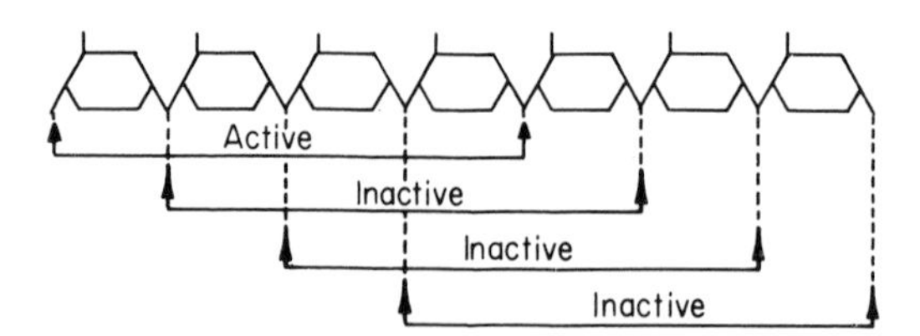

FIG. 20. Schematic representation of active and inactive complexes of a heptaose with β-amylase in which four consecutive groups are necessary for binding and in which a free terminal C_4 group is necessary for enzymic action after binding has occurred. From Thoma and Koshland (*227*).

stants characteristic of the terminal segments. $\tilde{V}_{max}/\tilde{K}_m$ should be constant.

This simplistic development has been criticized because it fails to consider multiple attack by β-amylase and the effect of polymer distribution on kinetic parameters (*238*). Recently, two rigorous steady state analyses (*28*, *29*) have treated this problem in more detail. The treatments are based upon the interactions visualized in Fig. 17. Steady state analysis gives rise to the following rather formidable dependence of the apparent initial velocity $\tilde{v}_0$ on substrate, S, and enzyme, E.

$$\frac{\tilde{v}_0}{[E_0]} = \frac{\sum_{i=1}^{n} \left\{ (k_{+2,i} + k_{r,i}) \sum_{p=i}^{n} [S_p] K_p \theta_p \right\}}{1 + \sum_{i=1}^{n} \sum_{p=i}^{n} [S_p] K_p \theta_p} \tag{25}$$

where K_p and θ_p are complex collections of rate constants. Substitution for the individual polymer substrates in terms of the number average concentration $[S_a]$ and collection of terms gives rises to a greatly simplified relationship homeomorphic with the Michaelis-Menten equation:

$$\tilde{v}_0 = \frac{[E_0][S_a \tilde{k}]}{1 + [S_a]/\tilde{K}_m} \tag{26}$$

Intuitively, it is obvious from such theoretical developments that in steady state endoenzymes as well as exoenzyme-mixed polymer systems will obey Michaelis-Menten kinetics regardless of the complexity of the system. However, the interpretation of the Michaelis parameters will be related to the enzyme's action pattern.

Specialization of this general model by including only one active complex and assuming $(n + 1 - m)$ inactive complexes, where m is the number of units bound to the active site and n is the number of cleavable bonds and that k_r is independent of chain length, gives the equation by Thoma and Koshland (*227*):

$$\tilde{v}_0 = \frac{[\mathrm{E}][\mathrm{S}]k_r/\tilde{K}_m}{\dfrac{[\mathrm{S}]}{\tilde{K}_m}\left[1 + \dfrac{(n+1-m)\tilde{K}_I}{[\mathrm{S}]}\right] + 1} \tag{27}$$

These equations apply to molecularly pure polymers. However, above DP 10–15, it is necessary to follow enzymic action on polymer mixtures. The ratio of $\tilde{V}_{\max}/\tilde{V}_{\max,\ \mathrm{distribution}}$ and $\tilde{K}_m/\tilde{K}_{m,\ \mathrm{distribution}}$ should

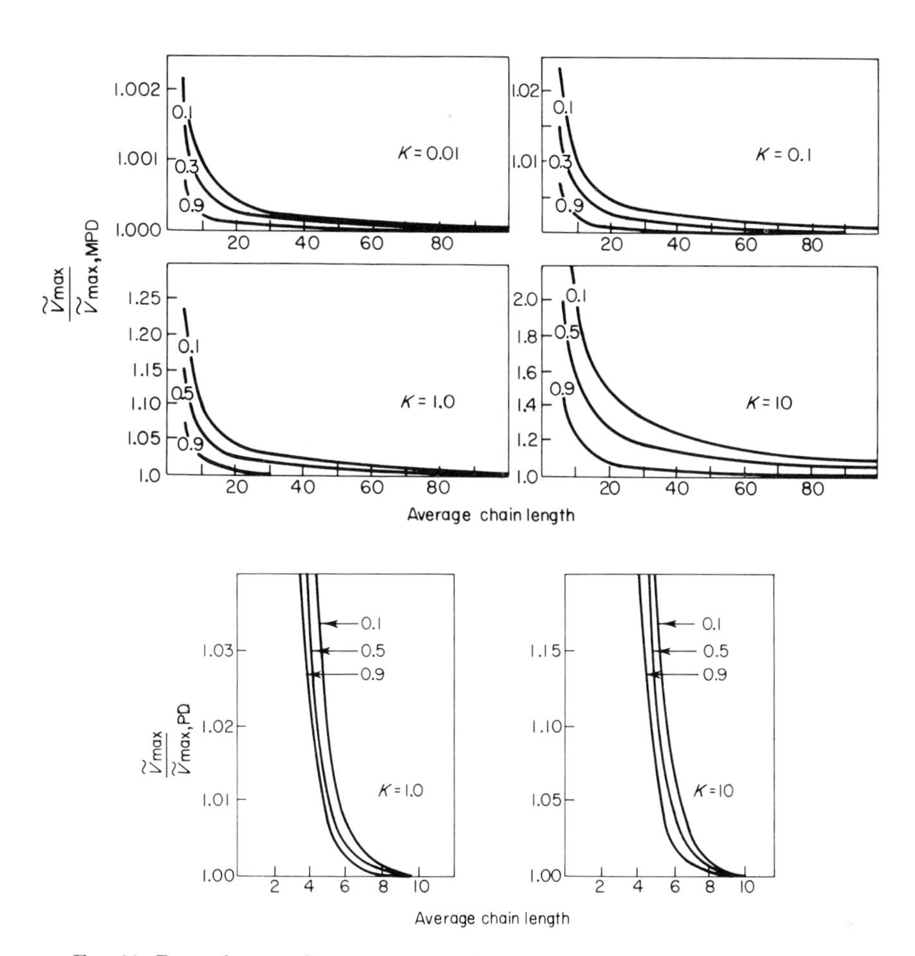

FIG. 21. Dependence of apparent maximum velocity upon chain length for molecularly pure homopolymer and homopolymers of most probable distribution (MPD) and Poisson distribution (PD). Small number inserts in the curves correspond to the degree of repetitive attack which is assumed independent of chain length. K is a measure of the effectiveness of the product inhibition which increases in proportion to K. It is obvious that a broad substrate distribution (MPD) can give rise to very serious end effects even at rather long chain lengths. End effects are more pronounced when the degree of multiple attack is high. From Thoma (*29*).

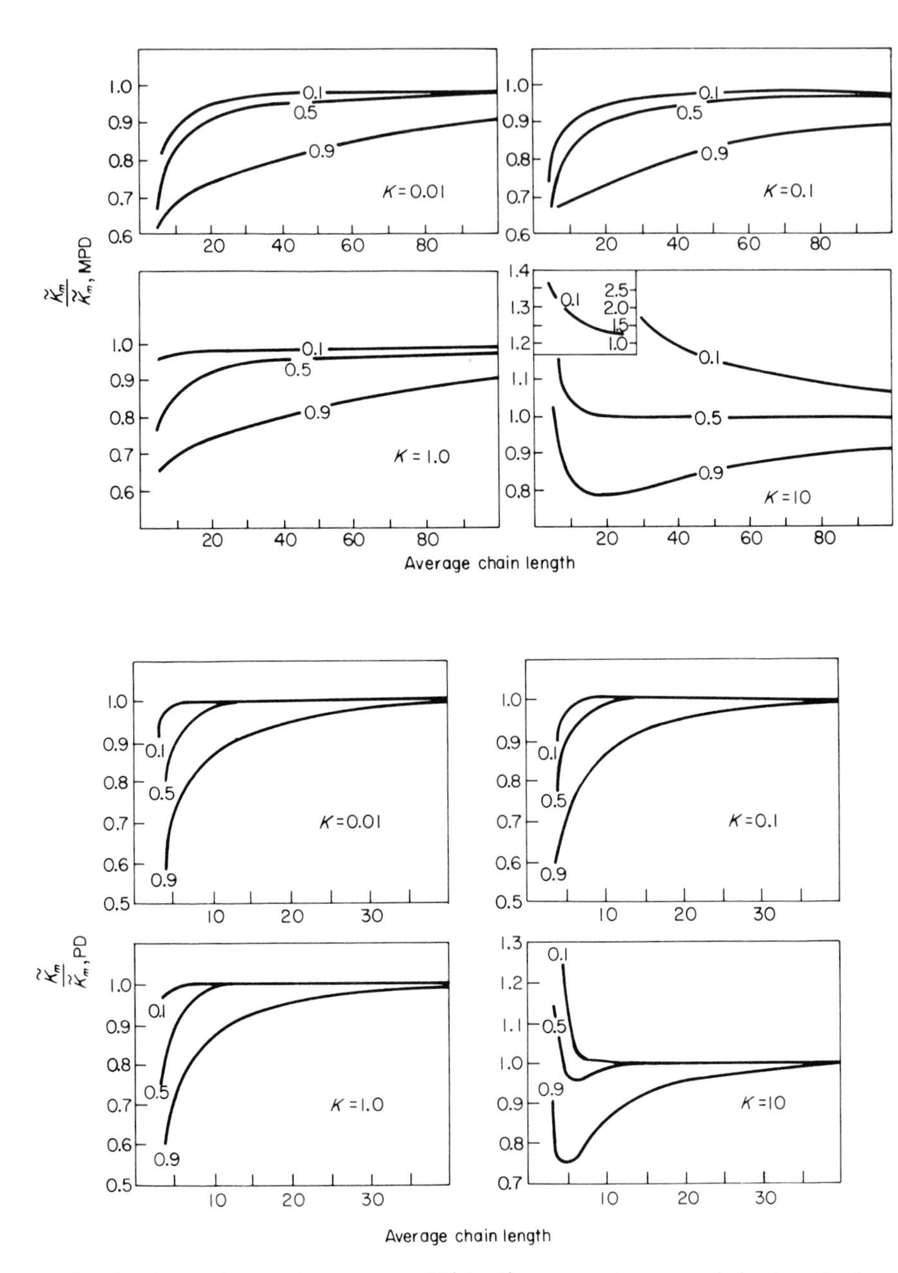

FIG. 22. Dependence of apparent Michaelis constant upon chain length for molecularly pure homopolymer and homopolymers of most probable distribution (MPD) and Poisson distribution (PD). Small number inserts in the curves correspond to the degree of repetitive attack which is assumed independent of chain

be very close to unity if the polymer mixture of average DP_n can be used to mimic a molecularly pure polymer of DP_n. The dependence of these ratios on chain length are shown in Figs. 21 and 22. From these graphs it was concluded that polymers with narrow distributions and average chain lengths of 20 or more are suitable for studies of the chain length dependence of Michaelis parameters when the degree of repetitive attack was 4 or less bonds. For α-amylases which cleave internal bonds, the $\tilde{K}_m$ and $\tilde{V}_{max}$ ratios are almost certain to require much longer substrates before the ratios approach unity. Therefore, caution must be exercised in the interpretation of Michaelis parameters for α-amylases.

With sweet potato β-amylase, the predicted dependence of $1/\tilde{K}_m$ on DP for three linear substrates of DP 4, 24, and 44 was found and interpreted in terms of competitive self-inhibition resulting from formation of nonreactive complexes with internal starch segments (*227*). Dygert (*238*), in a more careful and extensive study, also found that $\tilde{K}_m$ decreased with substrate size using pure oligosaccharides (DP 4–13) and Poisson distributed amylose between DP 24 to 480, as shown in Fig. 23. However, Fig. 24 shows, contrary to theory, that $\tilde{V}_{max}$ is essentially independent of chain length.

Ono *et al.* (*249*), studying glucoamylase, found that $\tilde{K}_m$ decreases by a factor of 10 as substrate is lengthened from DP 15 to 800, whereas the $\tilde{V}_{max}$ decreases by less than a factor of 2. This, in a general way, is consistent with the data of sweet potato β-amylase, i.e., $\tilde{V}_{max}/\tilde{K}_m$ is not constant and $\tilde{K}_m$ decreases much more rapidly than $\tilde{V}_{max}$ with increasing DP.

On the other hand, Husemann and Pfannemüller (*236*, *237*) found the predicted linear chain length dependence of $1/\tilde{V}_{max}$ and $1/\tilde{K}_m$ using narrow polymer distributions synthesized by potato phosphorylase; their results are recorded in Table XIII. As predicted, $1/\tilde{V}_{max}$ and $1/\tilde{K}_m$ are both linear functions of DP, and $\tilde{V}_{max}/\tilde{K}_m$ remains approximately constant. The internal inhibition constant calculated from these data is approximately 0.1 *M* while the binding for the terminal segment is 1.3×10^{-4} *M*. In terms of the model these data are interpreted to mean that the nonreducing terminus is bound several orders of magnitude more tightly than an internal segment and competitive self-inhibition is

249. S. Ono, K. Hiromi, and M. Jinbo, *J. Biochem.* (*Tokyo*) **55**, 315 (1964).

length. *K* is a measure of the effectiveness of the product inhibition which increases in proportion to *K*. It is obvious that a broad substrate distribution (MPD) can give rise to very serious end effects even at rather long chain lengths. End effects are more pronounced when the degree of multiple attack is high. From Thoma (*29*).

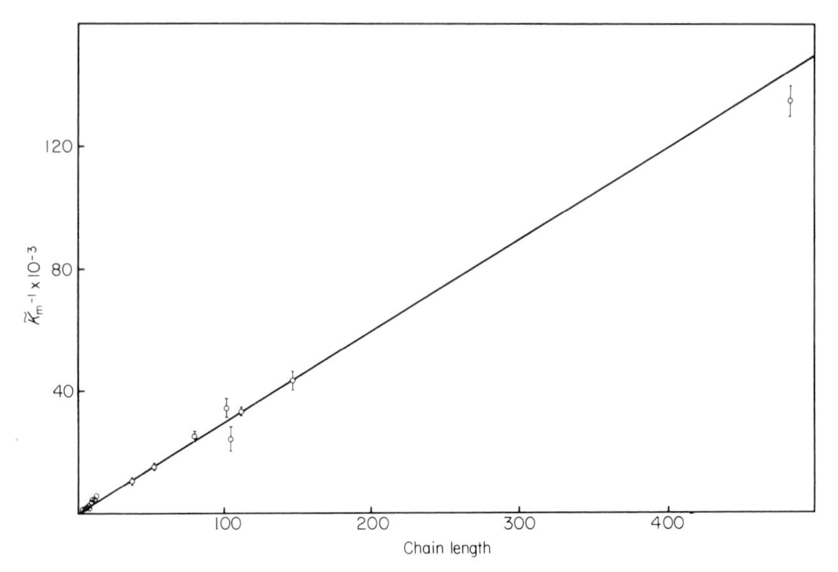

FIG. 23. Dependence of $\tilde{K}_m$ on chain length for sweet potato β-amylase. From Dygert (*238*).

very weak. In contrast sweet potato β-amylase has approximately equal affinity for a chain end and the chain center as judged by the dependence of $\tilde{K}_m$ on chain length (*227, 238*). $\tilde{V}_{max}$ vs. chain length plots give no indication of enzymic interaction with internal chain units (*238*). These

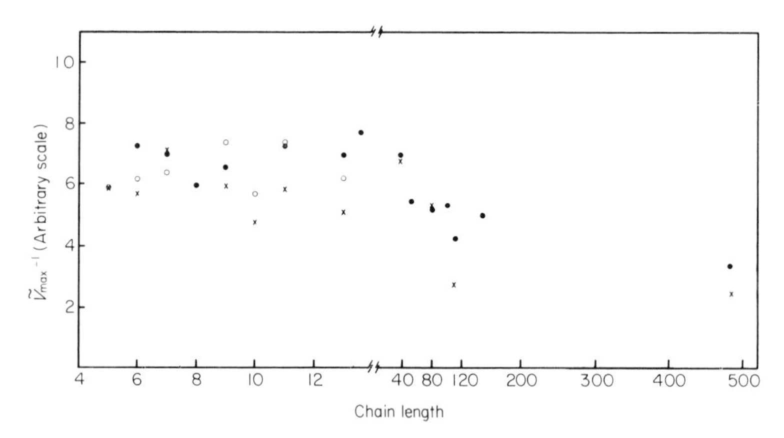

FIG. 24. Dependence of $\tilde{V}_{max}$ on chain length for sweet potato β-amylase. (●) from Lineweaver-Burk plot; (×, ○) separate velocity measurements at saturating substrate concentrations. From Dygert (*238*).

TABLE XIII
DEPENDENCE OF APPARENT MICHAELIS PARAMETERS FOR BARLEY β-AMYLASE ON CHAIN LENGTH[a]

$\overline{DP}_n$	$\tilde{V}_{max}$	$\tilde{K}_m$	$\tilde{V}_{max}/\tilde{K}_m$
750	1.03	6.66×10^{-5}	0.155
1775	0.625	3.94×10^{-5}	0.159
2875	0.40	2.96×10^{-5}	0.135
3815	0.345	2.33×10^{-5}	0.148

[a] After Husemann and Pfannemüller (*236, 237*).

experiments leave our understanding of β-amylase action in a very unsatisfactory state. There is at the present time no single model that can accommodate all of the experimental data. Development of a unifying scheme is complicated by the use of different enzymes and the possibility that the degree of repetitive attack is a function of chain length (*250*), and that the ability of enzyme to complex with various portions of the substrate may be influenced by the degree of aggregation, polydispersity, and molecular weight (*251*).

Although the importance of repetitive attack in depolymerase action appears beyond dispute, the question of inhibition by internal segments is not yet conclusively settled. More careful and detailed investigations will be required before a complete understanding of the action pattern of β-amylase is at hand.

The finding that $\tilde{K}_m$ decreases and $\tilde{V}_{max}$ remains constant for sweet potato β-amylase with increasing DP may mean that the tighter binding implied by the decrease in $\tilde{K}_m$ represents productive binding. This productive binding could result from the random binding along the amylose chain at one site on the enzyme accompanied by degradation of the nonreducing end at another site. A more appealing explanation is that the binding of internal segments of the enzyme is an integral step in the production of reactive complexes with the nonreducing end of the amylose molecule even though a satisfying pictorial representation of this scheme is difficult to depict.

Inhibition studies with substrate analogs is an accepted way to map the binding area and deduce an estimate of the minimum number of subsites and the relative affinity of the different subsites for monomer units. In addition, inhibition studies give information about the action pattern of the enzyme (*252*). In the case of α-amylase the effect of

250. E. Husemann and B. Pfannemüller, *Makromol. Chem.* **69**, 74 (1963).
251. E. Husemann and B. Pfannemüller, *Makromol. Chem.* **49**, 214 (1961).
252. W. W. Cleland, *BBA* **67**, 173 (1963).

inhibitors on the degree of multiple attack should provide information relating to the mechanism of multiple attack.

French and co-workers (*244, 253*) claimed that maltose competitively inhibits sweet potato β-amylase with a K_I of 6 mM and does not change the degree of multiple attack even in the presence of 5% maltose. The cyclohexaamyloses and α-methyl glucoside are all reported to be competitive inhibitors of sweet potato β-amylase (*162*).

A more extensive inhibition study of sweet potato β-amylase was undertaken by Dygert (*238*) with the intention of learning something about the subsites. The results are shown in Table XIV.

TABLE XIV
EFFECTS OF VARIOUS INHIBITORS ON SWEET POTATO β-AMYLASE[a]

Inhibitor	Conc. range (M)	Type of inhibition[b]	Comments
Glucose	0–0.5	I linear; S linear; noncompetitive	Slopes and intercepts vs. [I] curve upward at higher concentration
Maltose	0–0.1	I linear; S parabolic; noncompetitive	Intercepts vs. [I] curve upward at higher concentration
α-Methyl glucoside	0–0.25	I linear; S linear; noncompetitive	Slopes and intercepts vs. [I] curve upward at higher concentration
Cyclohexaamylose	$0–5 \times 10^{-4}$	S linear; competitive	
Cyclohexaamylose	0.0088–0.006	I linear; S linear; noncompetitive	
Cycloheptaamylose	$0–4.24 \times 10^{-4}$	S linear; competitive	
Cyclooctaamylose	$0–1.79 \times 10^{-3}$	S linear; competitive	

[a] From Dygert (*238*).
[b] For nomenclature, see Cleland (*252*).

α-Methyl glucoside is a noncompetitive inhibitor in conflict with an earlier report (*162*). This discrepancy is explained by the fact that these authors used the Dixon plot ($1/\tilde{V}_{max}$ vs. [I]) to determine the type of inhibition which assumes that $K_{I,slope}/K_{I,intercept}$ is constant. In the present case this is a false assumption. A simple reaction scheme that can account for these effects and a linear noncompetitive inhibition of β-amylase by maltose is shown in Fig. 25 where the product maltose is involved in normal product and dead-end inhibition (*252*). The cyclodextrins show decreasing inhibitory power with size, even though the

253. U. K. Misra and D. French, *BJ* **77**, 1p (1960).

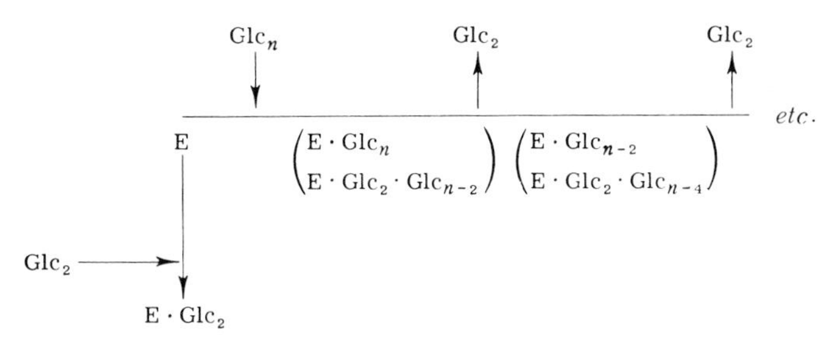

FIG. 25. Dead-end and normal product inhibition of sweet potato β-amylase by maltose; E represents β-amylase and n, the original substrate size.

number of glucose units available to inhibit becomes greater. This result suggests that the active site is most complementary to cyclohexaamylose. Note that the cyclohexaamylose also shows an intercept effect at high concentration.

Maltose was found (*238*) to be a noncompetitive inhibitor as opposed to the results of Misra and French (*253*). According to the multiple attack theory of Bailey and French (*243*), one expects anything that decreases V_{max} (intercept effect) to decrease the degree of multiple attack if substrate dissociation is not slowed proportionately. Thus it is surprising that Youngquist (*254*) did not find any effect on multiple attack upon addition of 5% maltose. In agreement with Table XIV that α-methyl glucoside is a noncompetitive inhibitor, Spradlin (*245*) has shown a decrease in multiple attack of sweet potato β-amylase acting on α-methyl Glc$_7$ in the presence of 0.25 M α-methyl glucoside.

This intercept effect of double reciprocal plots caused by maltose means that the product combines with an enzyme form other than that which complexes the substrate. A logical interpretation of this effect is that the product combines with a nonproductive enzyme–substrate complex which is formed when maltrose leaves the active site after bond rupture. Hence subsites must exist on both sides of the catalytic area, a conclusion corroborated by substrate analog interactions with β-amylase (see above).

Assuming that the glucoside can bind at the maltose site, we infer that saturation of the maltose site forces dissociation of the nonproductive enzyme–substrate complex before it rearranges to form a productive complex and thus interrupts the normal sequence of repetitive attack. This dissociation constitutes the k_2 process in Fig. 17.

Neeley (*218*) studied the specificity of sweet potato β-amylase toward

254. R. W. Youngquist, Ph.D. Dissertation, Iowa State University, Ames, Iowa, 1962.

maltooligosaccharides modified at the C4 hydroxyl position. The 4,*O*-methyl-α-methyl maltooligosaccharides were found to be strong competitive inhibitors.

VIII. Biosynthesis, Genetics, and Control

A rigorous demonstration that relates an observed increase in enzyme activity to *de novo* protein synthesis is a very difficult experiment task. The accepted experimental criteria to demonstrate this point have been ably reviewed by Ashmore and Weber (*255*). Enzymic activity increases may arise from *de novo* synthesis, stabilization of existing enzyme via various mechanisms, activation of existing enzyme, removal of inhibitors, or by combination of these or other phenomena. These authors wisely warn against accepting any single experimental criterion as unequivocal evidence for *de novo* protein synthesis. The remainder of this chapter must be viewed in light of the reservations outlined by Ashmore and Weber (*255*) and Schimke (*255a*).

The incorporation of ^{14}C into amylases may be demonstrated in whole animals (*256*), tissue cultures (*257*), or individual organs by perfusion techniques (*258*). However, the appearance of radioactive enzyme and buildup of enzymic activity may not be related to the increased rate of amylase synthesis. For example, Chignell (*256*) found that the amylase isolated from pilocarpine-treated rats fed leucine ^{14}C was less radioactive than that of control rats. He concluded that pilocarpine depresses the rate of amylase catabolism but does not affect the rate of synthesis enhancing the enzyme level. This experiment illustrates the importance of quantitatively comparing radioactivity incorporated to the activity level of the enzyme. Such experiments suggest that the biosynthesis of amylases occurs in the pancreas (*259*), the salivary gland (*256*), the parotid gland (*260*), and the liver (*258*).

The production of an enzyme by a tissue culture over and above that stored and extractable by the tissue is good evidence that *de novo* syn-

255. J. Ashmore and G. Weber, *in* "Carbohydrate Metabolism and its Disorders" (F. Dickens, P. J. Randle, and W. J. Whelan, eds.), Vol. 1, p. 336. Academic Press, New York, 1968.

255a. R. T. Schimke, *in* "Current Topics in Cellular Regulation" (B. L. Horecker and E. R. Stadtman, eds.), Vol. 1, p. 77. Academic Press, New York, 1969.

256. C. F. Chignell, *Biochem. Pharmacol.* **17**, 2225 (1968).

257. R. J. Grand and P. R. Gross, *JBC* **244**, 5608 (1969).

258. M. Arnold and W. J. Rutter, *JBC* **238**, 2760 (1963).

259. P. D. Webster and L. Zieve, *New Engl. J. Med.* **267**, 604 (1962).

260. N. Jacobsen, *Arch. Oral Biol.* **14**, 679 (1969).

thesis in the tissue occurs in the whole organism. Jacobsen (*260*) used this technique to show *de novo* synthesis of amylase by monkey parotid gland. The conclusion that parotid gland is a source of α-amylase is bolstered by the demonstration (*257*) that radioactive amino acids are incorporated into amylase by rat parotid gland slices and that the rate of incorporation is maintained at a constant rate for at least 3 hr in unstimulated slices. Epinephrine (30 μM) stimulates a 2-fold increase in incorporation of radioactive amino acids into amylase after 1 hr incubation. Epinephrine stimulation of protein synthesis is independent of its effects in secretory activity.

Amylase synthesized by liver tissue is apparently released into the circulation media under normal physiological conditions. Arnold and Rutter (*258*) perfused liver with a L-leucine-^{14}C medium and isolated labeled amylase with antiserum and found that amylase accumulated in the perfusate in a reproducible and linear manner. Amylase production was depressed under conditions where protein synthesis was repressed. Thus liver is clearly implicated as one source of serum amylase.

The production of amylases in digestive fluids occurs in two steps, first synthesis and then translocation. The enzymes after synthesis are stored in zymogen granules within the gland tissue and upon call by the organism are secreted into the digestive system. The functional unit of the exocrine portion of the pancreas is composed of large spherical or pyramidal cells arranged in grapelike clusters around terminal pancreatic ductules. The cells may be divided into two zones, an apical zone containing numerous enzymic or zymogen granules and a basal zone relatively free of granules. The number of zymogen granules in the cell decreases with secretory activity or prolonged fasting. The granules themselves are not secreted into the pancreatic juice, but the contents of the zymogen granules are transferred across the membrane of the pancreatic cell (*259*). From electron microscope studies (*259*) granules appear to arise as small spheres pinched off from the tubular ergastoplasmic sacs in the basal portion of the cell. The new ergastoplasmic sacs appear within cytoplasmic centers in relation to the nuclear membrane and then migrate to the apex of the cell from which their contents can be secreted.

By using pulse-chase experiments, Jamieson and Palade (*261*) determined the time required for unstimulated guinea pig pancreas slices to transport radioactive label from an amino acid pool to its zymogen granule. They found that it took approximately 50 min for the granules to become maximally labeled after a 3 min exposure to leucine-^{14}C. The

261. J. D. Jamieson and G. E. Palade, *J. Cell Biol.* **34**, 577 (1967).

production and storage of digestive enzymes in exocrine glands is becoming better defined, but much remains to be learned about the effects of hormones and drugs upon the production, storage, and secretion of the enzymes.

While the origin of the amylases in the digestive juices seems to be well established, the sources of the amylases in the other body fluids such as the serum, sweat, and urine are only partly elucidated. Information in this area is rapidly developing.

The comparison of the electrophoretic mobility of amylases of unknown origin with those of known origin on cellulose acetate strips has been the major method employed to implicate a site of synthesis (*262, 263*). This type of study yields only suggestive evidence of the synthesis site of amylase under investigation and should be corroborated by independent techniques. Amylases from the fingerprints of 187 persons of various ages (*262*) have mobilities electrophoretically identical to those of salivary amylases. A similar study of urine (*263*) reveals two electrophoretic variants that migrate at rates comparable to salivary and pancreatic amylases. Neither the salivary nor pancreatic isozymes are resolved on cellulose acetate. The salivary enzymes migrate together as a fast band and the pancreatic enzymes as a slower band. In normal adults the two bands contain equivalent amounts of enzymes, but in patients with pancreatitis the slow (pancreatic) band predominates. Mumps enhanced the fraction of amylase in the rapidly moving (salivary) band. Similar observations have been made on the serum of bank vole (mouse) (*97*).

Pinpointing the source of serum amylases is a complicated and difficult problem to unravel because the blood contacts most of the body tissue and exchanges chemicals with them. The level and specificity of amylase activity in the serum, however, is also altered by such varied disorders as pancreatitis, perforated peptic ulcer, liver and bilary disturbances, peritonitis, intestinal obstruction, and ruptured ectopic pregnancy (*259*). The work of Hall and co-workers (*94*) is of particular interest in this connection since they have shown that the substrate specificity of serum amylase depends upon the general health condition of the donor at the time of sampling. This suggests that the amylase present in serum arises from diverse sources in the body and that the fraction arising from a single location is partially controlled by general health, diet, environment, emotional state, and drugs.

As testing and comparison methods become more refined and capable

262. J. Kamaryt, *Cesk. Dermatol.* **43**, 302 (1968); *CA* **70**, 658722 (1969).
263. S. E. Aw, *Nature* **209**, 298 (1966).

of detecting minor variations in amylase levels and specificity these tests may evolve as a useful biochemical biopsy technique.

Genetics studies involving amylases are just now beginning to be reported (*264–266*). Electrophoretic profiles of amylases are now known to vary with genetic pedigree. Human saliva specimens from related individuals display (*264*) several isozyme patterns and contain up to eight enzymic bands.

Pedigree amylase correlations indicate that multiple gene loci are probably involved in the production of human salivary amylase isozymes. Developments in this area are almost exclusively dependent upon the superb resolving power of disc electrophoresis. Some exciting developments in the genetic relationships of amylases can be anticipated within the next few years.

Amylase variants may also be useful to the biochemical taxonomist. The amylases of *Drosophila melanogaster* (*265, 266*) have served as a criterion for identifying 11 new strains. Eight strains are characterized by electrophoretic banding patterns while the other three are characterized by differences in enzymic activity. Genetic crosses locate the *Amy* region in *Drosophila melanogaster* at 77.3 on the genetic map of the second chromosome and to the right, but near, section 52F on the salivary chromosome map. Ogita (*266*) found that amylases in the house fly, *Musca domestica*, fall into two groups: Group A consists of four isozymes which migrate toward the anode at pH 6.8, and group B consists of three isozymes which migrate toward the cathode. Genetic crosses between wild type and multimutant strains show that each group of isozymes is controlled by a separate gene on the third chromosome. The crossover data suggest that the amylase-A isozymes are under the control of four codominant alleles at the *Amy-A* locus 17.5 map units to the right of the loop wing locus. The amylase-B isozymes are controlled by three codominant alleles at the *Amy-B* locus 16 map units to the left of the loop wing locus.

β-Amylase, restricted to the plant kingdom, is synthesized only in the seed or tuber. It has long been recognized that β-amylase is laid down in ungerminated cereals in an insoluble and inactive form. As early as 1908, Ford and Gutherie (*267*) studied its extraction with cysteine-activated papain. Since that time other investigators have found that

264. B. Boettcher and F. A. De la Lande, *Australian J. Exptl. Biol. Med. Sci.* **47,** 97 (1969).
265. W. W. Doane, *J. Exptl. Zool.* **171,** 321 (1969).
266. Z. Ogita, *Ann. N. Y. Acad. Sci.* **151,** 243 (1968).
267. J. S. Ford and J. M. Gutherie, *J. Inst. Brewing* **14,** 61 (1908).

β-amylase from many sources is present as a zymogen (*47, 70, 72, 111*). The region between the aleurone and the endosperm in barley seeds contains most of the enzyme while the aleurone is almost devoid of amylase (*268*). Mugibayashi *et al.* (*269*) followed the appearance of soluble and insoluble β-amylase in ripening barley seeds and observed that the soluble form, detectable early in the ripening period, increased until the harvesting season and then decreased to a low constant level. The insoluble zymogen form gradually increased to its maximum levels during the latter stages of ripening. The accumulation of zymogen β-amylase parallels the synthesis of the insoluble proteins, hordein and glutenin. Rowsell and Goad (*47*) earlier had shown that the zymogen β-amylase is associated with the glutenin fraction and not with the starch granules. The zymogen can be converted to the fully active form by combined treatment with proteases and thiol reagents (*47, 70, 72*) or it can be partially activated by either treatment.

In accord with this observation recent data suggest that the sulfhydryl groups in β-amylase play a role in regulating enzymic activity but are not directly involved in substrate catalysis or binding (*111*). Physical and chemical studies revealed that a small conformation change accompanies thiol modification which profoundly depresses activity without substantially altering substrate affinity. The effects are in proportion to the size of the modifying group. Oxidation of the thiol groups to a mixed disulfide inhibits the enzyme but activity can be regenerated by reduction with cysteine. Thus it was speculated that in nature the role of the thiol groups is control of *in vivo* activity. Oxidation of the thiol groups is the mechanism which allows the plant to store the enzyme as an inactive zymogen (*47, 70, 72, 111, 270*). The enzyme is probably activated when demands are made on the energy reserves during germination. Activation may be achieved by a combination of proteolytic enzymes (*271*) and a disulfide reductase reportedly present in seeds (*272*).

Plant α-amylase, in contrast to the β-amylase, is not present in ungerminated seed (*273*) (see Fig. 26) but is synthesized *de novo* upon

268. C. Engel, *BBA* **1,** 42 (1947).

269. N. Mugibayashi, R. Shinke, and Y. Nichitai, *Proc. Symp. Amylase, Osaka, Japan* p. 80 (1965).

270. A. M. Allschul, L. Y. Yalsu, R. L. Ory, and E. M. Engleman, *Ann. Rev. Plant Physiol.* **17,** 113 (1966).

271. E. V. Rowsell and L. J. Goad, *BJ* **90,** 12p (1964).

272. M. D. Hatch and J. F. Turner, *BJ* **76,** 556 (1960).

273. R. Grabar and J. Daussant, *Cereal Chem.* **41,** 523 (1964).

273a. F. B. Salisbury and C. Ross, "Plant Hormones and Growth Regulators." Wadsworth Publ., Belmont, California, 1969.

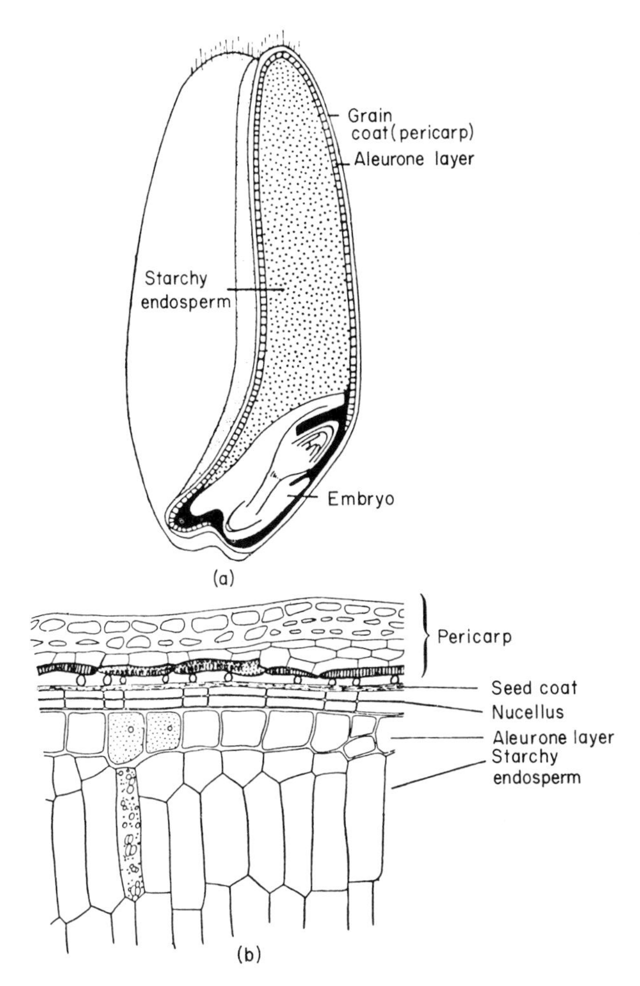

FIG. 26. The aleurone layer in wheat seeds. (a) Longitudinal section showing position of aleurone layer in the seed relative to the embryo and endosperm. (b) Enlarged cross section of a small portion of a mature wheat grain. From Salisbury and Ross (*273a*).

germination. Synthesis is initiated by the plant hormone, gibberellic acid, in the aleurone layer (outermost layer of the endosperm) of seeds. In fact this observation is the basis for (*274*) a bioassay of gibberellic acid using endosperm. Gibberellic acid stimulates the release of α-amylase into the starchy endosperm causing production of reducing sugar in the ambient medium. In proportion to the level of applied gibberellin, the

274. P. B. N. Nicholls and L. G. Paleg, *Nature* **199**, 823 (1963).

assay has a sensitivity of $3 \times 10^{-11} M$ and is simplier than the earlier method.

Varner and co-workers (*275–277*) sectioned barley seeds transversely and treated the endosperm portion with gibberellic acid in the presence of different ^{14}C-amino acids. DEAE chromatography gives one labeled protein peak exactly coincident with the α-amylase activity. The gibberellin-dependent stimulation of α-amylase is prevented by anaerobiosis, dinitrophenol, and parafluorophenylalanine (*275*) implicating oxidative phosphorylation as the energy source for *de novo* protein synthesis.

Filner and Varner (*276*) have presented additional evidence that all of the gibberellin-evoked α-amylase represents *de novo* synthesis. When isolated aleurone layers were incubated in $H_2{}^{18}O$ in the presence of gibberellic acid, the hydrolysis of reserve protein of the aleurone layers furnishes amino acids labeled with ^{18}O for new protein synthesis. The newly synthesized protein being more dense can be fractionated from preexisting proteins in a cesium chloride density gradient. No ^{16}O enzyme appeared when amylase synthesis was stimulated in the presence of $H_2{}^{18}O$, and the authors concluded that α-amylase is not produced from an inactive precursor but results from *de novo* synthesis.

Groat and Briggs (*278*) have detected a time lag between the appearance of gibberellin in the embryo, its translocation to the aleurone, and the onset of α-amylase synthesis in the endosperm. Plant growth regulators such as indoleacetic acid, benzylaminopurine, and kinetin (*279*) do not stimulate protein synthesis in the aleurone layer. At the onset of germination it appears that the embryo via hormone action stimulates the production of hydrolytic enzymes which can mobilize reserve materials in the endosperm for translocation to the embryo to act as nutrients (*280–282*).

Katsumi and Fukuhara (*283*) treated the emerging leaf sheath from dwarf maize seeds with gibberellin and observed an increase in α-amylase activity in the leaf which paralleled shoot elongation. The same results were obtained with seedlings devoid of endosperm indicative of some α-amylase synthesis in the sheath. Very little is known about

275. J. E. Varner, *Plant Physiol.* **39,** 413 (1964).
276. P. Filner and J. E. Varner, *Proc. Natl. Acad. Sci. U. S.* **58,** 1520 (1967).
277. J. E. Varner and G. R. Chandra, *Proc. Natl. Acad. Sci. U. S.* **52,** 100 (1964).
278. J. I. Groat and D. E. Briggs, *Phytochemistry* **8,** 1615 (1969).
279. D. E. Briggs, *Nature* **210,** 419 (1966).
280. D. E. Briggs, *Phytochemistry* **7,** 513 (1968).
281. D. E. Briggs, *Phytochemistry* **7,** 531 (1968).
282. D. E. Briggs, *Phytochemistry* **7,** 539 (1968).
283. M. Katsumi and M. Fukuhara, *Physiol. Plantarum* **22,** 68 (1969).

the synthesis of α-amylase in the growing tissues of plants although it probably does occur (*283, 284*).

The introduction of an alien genome from diploid rye into a tetraploid durum wheat to give a hexaploid hybrid has a most unusual influence on the amylase protein (*285*). The enzymes (parents and hybrid) were purified to homogeneity by ammonium sulfate and acetone fractionation, gel filtration, and preparative electrophoresis. All three amylases had identical sedimentation constants and elution volumes by gel filtration. Polyacrylamide gel electrophoresis incompletely resolved the mixture into three bands. The "hybrid" enzyme exhibited intermediate electrophoretic mobility, Michaelis parameters, and sensitivity to temperature and pH. However, the hybrid enzyme contained significantly lower amounts of lysine, histidine, aspartic acid, threonine, glycine, isoleucine, and phenylalanine than either parental amylase and significantly higher concentration of arginine, serine, and cysteine. The mechanism of hybrid enzyme generation is still unknown but should prove to be an exciting story as it develops.

284. R. R. Swain and E. E. Dekker, *Plant Physiol.* **44,** 319 (1969).
285. W. Y. Lee and A. M. Unrau, *J. Agr. Food Chem.* **17,** 1306 (1969).

7

Glycogen and Starch Debranching Enzymes

E. Y. C. LEE • W. J. WHELAN

I. Introduction

The enzymes that hydrolyze the $1 \rightarrow 4$ and $1 \rightarrow 6$ bonds of glycogen and starch can be divided into three groups as follows:

Hydrolyzing $1 \rightarrow 4$ bonds only: α-amylase (EC 3.2.1.1) and β-amylase (EC 3.2.1.2)
Hydrolyzing $1 \rightarrow 6$ bonds only: debranching enzymes
Hydrolyzing $1 \rightarrow 4$ and $1 \rightarrow 6$ bonds: glucoamylase (EC 3.2.1.3, *syn* glucamylase, amyloglucosidase, and γ-amylase)

This chapter is concerned with the second group of enzymes. We have not given any EC designations, though some exist (e.g., EC 3.2.1.9, EC 3.2.1.10, and EC 3.2.1.33). The nomenclature of this group is exceedingly confused but is under active revision by the IUB/IUPAC Commission on Biochemical Nomenclature. We are not concerned with enzymes that hydrolyze $1 \rightarrow 6$-α-glucosidic bonds in a "nonamylaceous" glucan environment, for example, isomaltases and dextranases. These may well have the capacity to split the glycogen/amylopectin branch linkages, but any such action is probably without physiological significance. What we now know about the debranching enzymes is that they are adapted to glycogen, amylopectin, pullulan, and derived oligosaccharides. It is an absolute requirement that the substrate should contain $\alpha, 1 \rightarrow 4$ bonds, as well as the $\alpha, 1 \rightarrow 6$ bond. These enzymes do not act on isomaltose or dextran to any detectable degree.

The debranching enzymes can be divided into two classes, and one of these is capable of further subdivision as follows:

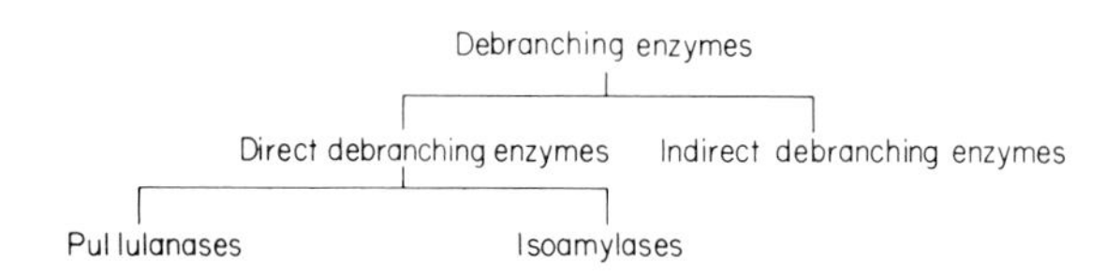

A "direct" debranching enzyme will attack unmodified glycogen and/or amylopectin, with hydrolysis of the $1 \rightarrow 6$ bond. For "indirect" debranching the polymer must first be modified by another enzyme or enzymes. The two types of debranching action are depicted in Fig. 1. This shows a segment of a glycogen or amylopectin molecule consisting of one A and one B chain. These are chains of $1 \rightarrow 4$-bonded α-glucose units joined by the (arrowhead) $1 \rightarrow 6$ bond. It is convenient to distinguish these two types of chain since both glycogen (*1*) and amy-

1. J. Larner, B. Illingworth, G. T. Cori, and C. F. Cori, *JBC* **199**, 641 (1952).

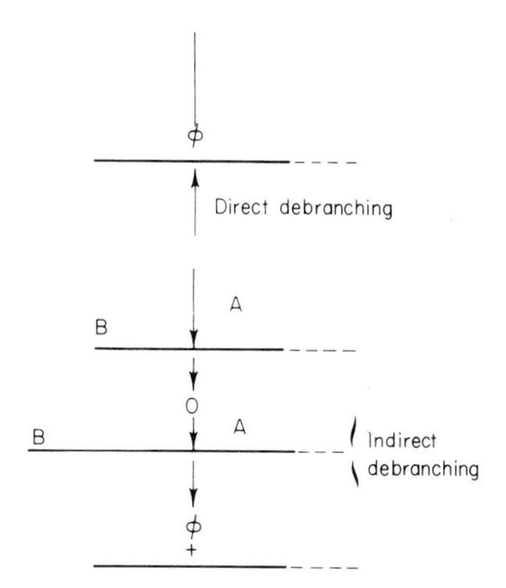

FIG. 1. The methods of direct and indirect debranching of glycogen and amylopectin. (O) α-Glucose unit and (Ø) glucose or reducing end glucose unit; (↓) 1 → 6 bond; A and B denote the two main types of chain (see text).

lopectin (*1, 2*) can be defined as consisting of about 50% of each. There is also in each polymer molecule one unique chain, the C chain, which is substituted one or more times at a primary hydroxyl group by a branch point but is terminated at one end by a free reducing glucose unit.

The direct debranching enzyme simply hydrolyzes the 1 → 6 bond in the substrate. If it is an A chain that is being split off, the minimum number of glucose units in that chain must be two (*3–5*), and for the best characterized debranching enzyme, *Aerobacter aerogenes* pullulanase, the (C) chain carrying the A chain must also have a minimum of two 1 → 4-bonded glucose units. Thus, the smallest substrate for pullulanase is the linear tetrasaccharide 6^2-α-maltosylmaltose, depicted thus (symbols as in Fig. 2):

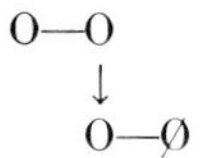

The smallest substrate for isoamylase is not known.

The indirect debranching process requires the concerted actions of

2. S. Peat, W. J. Whelan, and G. J. Thomas, *JCS* p. 3025 (1956).

3. W. J. Whelan, *Biochem. Soc. Symp.* **11,** 17 (1953).

4. M. Abdullah, B. J. Catley, E. Y. C. Lee, J. Robyt, K. Wallenfels, and W. J. Whelan, *Cereal Chem.* **43,** 111 (1966).

5. D. French and M. Abdullah, *BJ* **100,** 6P (1966); M. Abdullah and D. French, *Nature* **210,** 200 (1966); *ABB* **137,** 483 (1970).

two enzymes, as will be explained later in detail (Section III). It is sufficient here to state that the actions consist first in the operation of a transglycosylase that shortens the A chain to a single glucose unit, which is then removed by the hydrolytic action of amylo-1,6-glucosidase. Clearly, as the A chain decreases in length, another chain, shown in Fig. 1 as the B chain, must increase in length.

This system contrasts in an interesting manner with the direct debranching enzyme. As already noted, the latter can only remove an A chain which is at least two glucose units (α-maltose) in length. The indirect debranching enzyme (amylo-1,6-glucosidase) will only remove α-glucose. Therefore there is no common substrate for the direct debranching enzyme and amylo-1,6-glucosidase, though the presence of the transferase associated with the latter will convert a substrate for one into a substrate for the other. The smallest known substrate for amylo-1,6-glucosidase is the branched pentasaccharide, 6^3-α-glucosylmaltotetraose (*6*) (symbols as in Fig. 2):

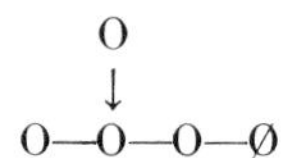

If we consider only animals and plants we can generalize that the indirect debranching system is associated with the occurrence of glycogen in mammals and the direct debranching enzyme with starch in plants. Direct debranching enzymes have not been reported in mammals or the indirect variety in plants. Microorganisms collectively contain all types, and yeast seems to have both the indirect system and isoamylase.

Only one direct debranching enzyme (a pullulanase) has been crystallized. There are two examples of pure, indirect debranching enzyme systems. A bacterial isoamylase has been purified to homogeneity (see Section II,C).

II. Direct Debranching Enzymes

Direct debranching enzymes have been detected in many sources. Each has its own special characteristics of specificity. Nevertheless, two classes of direct debranching enzyme can be distinguished, and most, if not all, of the individual direct debranching enzymes fall clearly into one class or the other. Section IV gives a detailed description of how the two

6. B. Illingworth and D. H. Brown, *Proc. Natl. Acad. Sci. U. S.* **48**, 1619 (1962).

classes may be distinguished. Suffice it here to say that one (pullulanase) attacks the α-glucan pullulan; the other (isoamylase) does not.

Of the two names, and others used for direct debranching enzymes, pullulanase is the only one that has not been debased by improper usage. Isoamylase is a name that has been used to describe a variety of activities, some of them definitely due to quite different enzymes (*6a*). It has, however, been used most recently by Yokobayashi *et al.* (*7*) to describe a new pseudomonad enzyme that seems to be the archetype of most earlier isoamylases, and we shall retain the name, describing the pseudomonad enzyme as the standard of reference for isoamylases. The reference standard for pullulanases is that elaborated by *Aerobacter aerogenes*. We shall present the information on pullulanase and isoamylase out of historical context, choosing instead to describe the best characterized enzymes, then comparing them with less well-understood enzymes.

A. Pullulanase from *Aerobacter aerogenes*

1. *Preparation and Physical Properties*

Pullulanese was discovered by Bender and Wallenfels (*8*). They had earlier described an α-glucan, pullulan, from *Pullularia pullulans* (*9, 10*). Though this is like amylopectin and glycogen, in containing $1 \rightarrow 4$ and $1 \rightarrow 6$ bonds, it could not be hydrolyzed by α-amylase. Pullulanase, however, was able to convert pullulan almost quantitatively into maltotriose, and this formed part of the proof that pullulan is a linear polymer of α-maltotriose joined endwise through $1 \rightarrow 6$ bonds. There are many variations on this structure, produced by different *Pullularia* strains, and one variety of pullulan contains 5% maltotetraose in addition to maltotriose (*11*). Wallenfels *et al.* (*10*) have described anomalous tetrasaccharide end groups. Catley (*12*) has found that α-amylase attacks the intramolecular maltotetraose residues and this could give rise to such end groups. Pullulanase is elaborated by *A. aerogenes* when grown on pullulan, starch, or starch-derived oligosaccharides. It is extracellular, but if glucose is in-

6a. This term has been used, improperly, to describe isoenzyme forms of α-amylase.

7. K. Yokobayashi, A. Misaki, and T. Harada, *Agr. Biol. Chem.* **33**, 625 (1969).

8. H. Bender and K. Wallenfels, *Biochem. Z.* **334**, 79 (1961).

9. H. Bender, J. Lehmann, and K. Wallenfels, *BBA* **36**, 309 (1959).

10. K. Wallenfels, G. Keilich, G. Bechtler, and D. Freudenberger, *Biochem. Z.* **341**, 433 (1965).

11. B. J. Catley and W. J. Whelan, *ABB* **143**, 138 (1971).

12. B. J. Catley, *FEBS Letters* **10**, 190 (1971).

cluded in the medium the enzyme is cell bound and may be extracted with a detergent and subsequently crystallized (*13–15*). Pullulanase has also been found in *Escherichia intermedia* by Ueda and Nanri (*16*), who chose to term it *isoamylase,* despite their observation that the already known yeast isoamylase does not attack pullulan. Isoamylase should not be used to describe an enzyme attacking pullulan.

Walker (*17*) has described an intracellular pullulanase from *Streptococcus mitis* that is seemingly similar to that from *A. aerogenes* with the exception that maltotriosyl A chains in dextrins are hydrolyzed 16 times faster than are maltosyl A chains (cf. Table II). The same author has shown that *S. mitis* also contains an α-1,6-glucosidase (see Section III).

Wallenfels and Rached (*13*) measured the molecular weight of *Aerobacter* pullulanase by sedimentation, reporting a value of 145,000. Mercier *et al.* (*18*) found values of 100,000–140,000 (sedimentation equilibrium), 105,000 (density gradient), and 55,000 (Sephadex G-200 gel filtration). The first value gives evidence of association, though the molecular weight was higher as the solution was diluted. The low value obtained by gel filtration probably reflects an association of the enzyme with the α-glucan (dextran) from which Sephadex is made. At present one can only say that the molecular weight probably lies between 10 and 15×10^4. Dissociation into subunits occurs on gel electrophoresis in sodium dodecyl sulfate (*18*).

Frantz *et al.* (*19*) described the occurrence of two forms of pullulanase, separable by gel filtration. The apparent molecular weights were 5×10^4 and 15×10^4, determined by passage through Sephadex. The heavier of these spontaneously changes to the lighter, and this process is accelerated on addition of trypsin. The crystalline enzyme is the lighter form (*18*).

A remarkable property of the enzyme is its ability to recover activity after being heated (*20*). On standing at room temperature the heat-inactivated enzyme (3 min at 100°) will, in 20 min, recover as much as

13. K. Wallenfels and J. R. Rached, *Biochem. Z.* **344,** 524 (1966).
14. K. Wallenfels, H. Bender, and J. R. Rached, *BBRC* **22,** 254 (1966).
15. H. Bender and K. Wallenfels, "Methods in Enzymology," Vol. 8, p. 555, 1966.
16. S. Ueda and N. Nanri, *Appl. Microbiol.* **15,** 492 (1967).
17. G. J. Walker, *BJ* **108,** 33 (1968).
18. C. Mercier, B. M. Frantz, and W. J. Whelan, *European J. Biochem.* (in press).
19. B. M. Frantz, E. Y. C. Lee, and W. J. Whelan, *BJ* **100,** 7P (1966).
20. M. Abdullah, B. J. Catley, W. F. J. Cuthbertson, and W. J. Whelan, *BJ* **100,** 8P (1966).

80% of the original activity (*21*). The optimum pH for this phenomenon is at 7; the optimum pH for activity is 5, but if heated at this pH no activity returns.

2. *Substrate Specificity and Action Pattern*

Recognizing that pullulanase was like the already known plant R-enzyme, the laboratories of Wallenfels and Whelan combined to determine the specificity of the enzyme. They found that the bacterial pullulanase acted on a range of oligo- and polysaccharide substrates in much the same way that bean R-enzyme had been reported to do (*3*, *22*, *23*), and showed that the smallest substrate was 6^2-α-maltosylmaltose (see above). The isomeric tetrasaccharides, 6-α-maltotriosylglucose and 6^3-α-glucosylmaltotriose, are not substrates. Both saccharide components joined by the α-$1 \rightarrow 6$ linkage must therefore contain at least two $1 \rightarrow 4$-bonded α-glucose units (maltose). Though the pullulanase preparation used was probably no more than 5% pure, it was subsequently found that all the poly- and oligosaccharase activities that the impure *Aerobacter* enzyme shared in common with the bean R-enzyme were retained in a preparation of bacterial pullulanase purified to homogeneity (*24*, *24a*). This point is of particular importance in relation to what will later be written about plant pullulanase (R-enzyme, Section II,B).

Table I records the relative abilities of *Aerobacter* pullulanase and bean R-enzyme to act on amylopectin, glycogen, and their β-dextrins, as judged by the increased ability of β-amylase to convert the polysaccharides into maltose, following debranching. There are also recorded the results of the simultaneous actions of β-amylase and the debranching enzymes.

Aerobacter pullulanase, acting alone, essentially debranches amylopectin and its β-dextrin almost completely. It has little or no action on rabbit liver glycogen and only partly debranches the β-dextrin. Abdullah *et al.* (*4*) had earlier recorded a slight degree of debranching of glycogen by amorphous pullulanase. The lack of action of pullulanase now recorded may be the result of using crystalline pullulanase or because the glycogen

21. B. J. Catley, B. M. Frantz, and W. J. Whelan, *4th FEBS Meeting Abstr.*, p. 77. Universitatsforlaget, Oslo, 1967.
22. P. N. Hobson, W. J. Whelan, and S. Peat, *JCS* p. 1451 (1951).
23. S. Peat, W. J. Whelan, P. N. Hobson, and G. J. Thomas, *JCS* p. 4440 (1954).
24. G. S. Drummond, Ph.D. Thesis, University of Miami (1969); *Dissertation Abstr. Intern.* **31**, 1669B (1970).
24a. G. S. Drummond, E. E. Smith, and W. J. Whelan, *FEBS Letters* **9**, 136 (1970).

TABLE I
A COMPARISON OF THE ACTIONS OF *Aerobacter* PULLULANASE AND BEAN PULLULANASE (R-ENZYME) ON AMYLOPECTIN, GLYCOGEN, AND THEIR β-LIMIT DEXTRINS AS JUDGED BY CHANGES IN DEGREE OF β-AMYLOLYSIS[a]

	Conversion into maltose (%)				
		Successive actions of pullulanase and β-amylase		Simultaneous actions of pullulanase and β-amylase	
Substrate	β-Amylase	Bean	*Aerobacter*	Bean	*Aerobacter*
Waxy maize amylopectin	52	64	92	101	99
Amylopectin β-dextrin	0	73	99	—	103
Rabbit liver glycogen	48	47	—	—	—
	47[b]	—	47[b]	—	97
	40	—	—	40	—
Glycogen β-dextrin	0	0	39	—	—

[a] Data from Abdullah *et al.* (*4*) and Peat *et al.* (*23*).
[b] C. Mercier, unpublished results.

was prepared under conditions that prevented degradation by endogenous α-amylase, or both. [Bean R-enzyme does not attack glycogen but will attack its products of α-amylolysis (*23*).]

When pullulanase and β-amylase acted in concert on glycogen, there was total hydrolysis (Table I). It seems therefore that pullulanase can remove the short (2- and 3-unit) A chains that β-amylase creates (cf. glycogen β-dextrin, Table I), and provided that these are continually regenerated by the presence of β-amylase, total hydrolysis ensues. It cannot be, however, that these α-maltosyl and α-maltotriosyl units are the only substrates for pullulanase since the enzyme almost completely debranches amylopectin (Table I), which has little or none of these short chains. The ability or otherwise of pullulanase to attack a polysaccharide must be determined by the overall tertiary structure of the polymer. Three features that contribute to the tertiary structure may be distinguished: (1) average outer chain length (the distance between the nonreducing chain end and the outermost branch point), (2) the average inner chain length (distance between branch points), and (3) whether the $1 \rightarrow 6$ bond constitutes a point of branching.

To some extent the outer chain length is important. Thus *Aerobacter* pullulanase attacks glycogen β-dextrin but not the parent glycogen (Table I). On the other hand, amylopectin, with a longer outer chain length than glycogen is attacked. The relative spacing of the branch points is presumably important since glycogen and amylopectin β-dextrins with equal outer chain lengths, but differing inner chain lengths, are attacked to unequal degrees. Yet pullulan, in which every third linkage is a $1 \rightarrow 6$ bond is totally hydrolyzed. This closeness of $1 \rightarrow 6$ bonds probably only occurs in the most highly branched regions of glycogen β-dextrin. The important difference here is that glycogen β-dextrin is branched, while pullulan is linear.

We therefore conjure up a picture of glycogen as presenting a hedgehog-like appearance to the enzyme, with the outer chains, the "spines," so closely packed as to prevent access by pullulanase. When the chains are trimmed by β-amylase the $1 \rightarrow 6$ bonds rise sufficiently near to the surface to become accessible to *Aerobacter* pullulanase, and continual trimming by β-amylase permits the eventual degradation of the molecule. It might be noted that bean R-enzyme cannot even attack glycogen β-dextrin (Table I), an important difference from its *Aerobacter* counterpart.

Both types of pullulanase attack amylopectin which, it is thought, has a structure similar to glycogen. The lower density of branch points, and hence of chain ends, means that the gaps between chains are wider than in glycogen and the debranching enzyme is therefore able more easily to approach the branch linkage. *Aerobacter* pullulanase is eventually able to reach most of these, but the majority elude the action of bean pullulanase (Table I).

Table IIA compares the rates at which *Aerobacter* pullulanase hydrolyzes various oligosaccharides. It will be noted that there is no major difference between the rates at which linear and branched oligosaccharides are attacked. Table IIB compares the relative initial rates of hydrolysis of polysaccharides. The greater ease of hydrolysis of amylopectin β-dextrin than of the parent polysaccharide is seen. It should be stressed that these are initial rates and that for the β-dextrins these probably represent the rate of removal of the 2- and 3-unit A chains. The structure of the polysaccharide remaining is constantly changing and the rate of hydrolysis will change in parallel, reaching zero in the case of the glycogen β-dextrins long before debranching is complete (see Table I).

Table IIA shows similar data for another bacterial pullulanase, that from *S. mitis* (*17*). It will be seen that although *Aerobacter* and *S. mitis* pullulanases both attack all seven substrates listed, the relative rates of attack vary widely. This is particularly so in the case of the action of the *S. mitis* enzyme on oligosaccharides carrying a maltose A chain, where the rate of attack relative to the analog carrying a maltotriose A chain is very low. This same feature will arise again when we describe pseudomonad isoamylase (Section II,C).

The relative rates of debranching in Table IIB were obtained by measuring the reducing powers of the chain ends set free on hydrolytic

TABLE IIA
RELATIVE RATES OF HYDROLYSIS OF OLIGOSACCHARIDES BY BACTERIAL PULLULANASES[a]

Substrate		Relative initial rates of hydrolysis	
Name	Structure	*A. aerogenes*	*S. mitis*
Pullulan	(O—O—O)$_n$ ↓	100	100
6^2-α-Maltosylmaltose	O—O ↓ O—Ø	23	3.8
6^3-α-Maltosylmaltotriose	O—O ↓ O—O—Ø	55	7.7
6^3-α-Maltotriosylmaltotriose	O—O—O ↓ O—O—Ø	91	128
6^3-α-Maltosylmaltotetraose	O—O ↓ O—O—O—Ø	171	23
6^3-α-Maltotriosylmaltotetraose	O—O—O ↓ O—O—O—Ø	112	282
6^3-α-Maltotetraosylmaltotriose	O—O—O—O ↓ O—O—Ø	57	208

[a] Data from Abdullah *et al.* (*4*) and Walker (*17*). Symbols as in Fig. 2.

TABLE IIB
RELATIVE RATES OF HYDROLYSIS OF POLYSACCHARIDES BY *Aerobacter* AND SWEET CORN PULLULANASE[a]

	Relative initial rates of hydrolysis		
		Sweet corn[b]	
Substrate	*A. aerogenes*	F1	F2
Pullulan	100	100	100
Amylopectin β-dextrin	54	213	164
Amylopectin	17	13	12
Glycogen β-dextrin (shellfish)	34	7	7
Glycogen (shellfish)	—	—	1.6
Glycogen α-limit dextrins	54	213	152

[a] Data from Mercier *et al.* (*18*) and Lee *et al.* (*36*).
[b] F1 and F2 refer to the two fractions of pullulanase isolated from sweet corn (*36*).

debranching. This is the most direct measure of pullulanase action. Another measure of debranching action on an amylaceous polysaccharide, and one of historical importance since it was this that led to the discovery of several direct and indirect debranching enzymes, is the increase in intensity of iodine stain and a change in λ_{max} of the iodine complex to higher wavelengths. In the case of direct debranching enzymes the change has been attributed to a greater ease of iodine complex formation by the liberated unit chains (*22*). In the case of indirect debranching enzymes, the change results from an increase in chain length (Section III). There is no direct correlation between iodine stain change and degree of debranching.

The action pattern of pullulanase on pullulan has been a subject of controversy (*24, 25*). It was reported to be endo (*26*), but Wallenfels *et al.* (*27*) claimed that it was exo. Drummond *et al.* (*28*) have subsequently confirmed that the pattern is indeed endo. There is found at intermediate stages of pullulanolysis a range of $1 \rightarrow 6$-linked oligomers of maltotriose, and quantitative assay of the maltotriose formed during hydrolysis suggests that the depolymerization is occurring in an entirely random manner. It must be noted that the pullulan molecule is linear. Endo action on this polymer does not necessarily mean that action on a branched polymer (amylopectin) will also be endo.

3. *Reversion Reactions of Pullulanase*

That hydrolytic carbohydrases can repolymerize the products of their action on poly- and oligosaccharides has long been known (*29*). Recently, Hehre (*30*) has shown that α-amylase, which forms α-maltose from starch, polymerizes α-maltose and not β-maltose. The reverse is true for β-amylase. Pullulanase liberates α-maltotriose from pullulan (*13*) and may be expected to repolymerize the α-anomer, though this has not been determined. The enzyme certainly polymerizes maltose, forming a tetrasaccharide, and joins maltose to amylose (*5*). It also joins maltose to cyclohexa-amylose [Schardinger α-dextrin (*5, 31*)].

25. W. J. Whelan, *BJ* **122,** 609 (1971).
26. B. J. Catley, J. F. Robyt, and W. J. Whelan, *BJ* **100,** 5P (1966).
27. K. Wallenfels, J. R. Rached, and F. Hucho, *European J. Biochem.* **7,** 231 (1969).
28. G. S. Drummond, E. E. Smith, W. J. Whelan, and H. Tai, *FEBS Letters* **5,** 85 (1969).
29. K. Nisizawa and Y. Hashimoto, *in* "The Carbohydrates" (W. Pigman and D. Horton, eds.), Vol. 2A, p. 242. Academic Press, New York, 1970.
30. E. J. Hehre, G. Okado, and D. S. Genghof, *ABB* **135,** 75 (1969).
31. P. M. Taylor and W. J. Whelan, *in* "Control of Glycogen Metabolism" (W. J. Whelan, ed.) p. 101. Academic Press, London, 1968.

B. Plant Pullulanase (R-Enzyme)

The first direct debranching enzyme of the pullulanase type was reported in 1951, in broad beans and potatoes, and termed R-enzyme (*22*). Its classification as a pullulanase came later, when pullulan and bacterial pullulanase were discovered. The plant preparation has, however, most of the properties subsequently ascribed to *Aerobacter* pullulanase. It debranches amylopectin, amylopectin β-dextrin and the same type of oligosaccharides as does the latter (*3*, *22*). It also hydrolyzes pullulan (*24a*). It does not, however, seem to attack glycogen or glycogen β-dextrin (*23*). Table I compares the degrees to which *Aerobacter* pullulanase and bean R-enzyme attack polysaccharides in terms of the ensuing increase in β-amylolysis.

Some confusion in the field has resulted from a report by MacWilliam and Harris (*32*) who in 1959 described the fractionation of R-enzyme from the original source of broad beans and from malted barley. By passage through alumina they obtained fractions that on the one hand debranched amylopectin and its macromolecular β-amylase limit dextrin, and on the other the oligosaccharide α-amylase limit dextrins. Thus R-enzyme seemed to be separable into a polysaccharase and an oligosaccharase. MacWilliam and Harris (*32*) took the unfortunate step of naming the polysaccharase "R-enzyme" and the oligosaccharase "limit dextrinase." This was doubly confusing since it had the effect of redefining what was meant by R-enzyme, while the second name was one that had earlier been used to describe a totally different activity, that of an α-glucosidase (*33*). Limit dextrinase is a name of ancient currency in the brewing industry that describes an enzyme(s) preparation yielding yeast-fermentable sugars from nonfermentable oligosaccharides. More than one type of enzyme could do this, and the assignment of the name to a particular type has not been agreed on.

In this narrative, the subsequent use within quotation marks, of "R-enzyme" and "limit dextrinase" refers to activities as defined by MacWilliam and Harris (*32*), whose findings were initially supported by Manners and co-workers. Thus, Manners and Sparra (*34*) reported similar findings for malted barley.

It seemed for a time that bacterial pullulanase was a single entity, acting on poly- and oligosaccharides, but the analogous system from plants was a mixture of separable activities. An apparent complexity

32. I. C. MacWilliam and G. Harris, *ABB* **84**, 442 (1959).
33. L. A. Underkofler and D. K. Roy, *Cereal Chem.* **28**, 18 (1951).
34. D. J. Manners and K. L. Sparra, *J. Inst. Brewing* **72**, 360 (1966).

came to light in our laboratory (*24a*) when the amylopectin-synthesizing system of the potato (Q-enzyme) was being purified, and the preparation was tested for its freedom from other starch-metabolizing enzymes. It was noted that the partly purified Q-enzyme hydrolyzed amylopectin β-limit dextrin and pullulan but not amylopectin. This hydrolytic activity was separated from the Q-enzyme and studied further. However, the ability to attack amylopectin was to be seen, if the enzyme preparation was present in sufficient concentration. Experiments with *Aerobacter* pullulanase revealed the same phenomenon, namely, that the ability to attack amylopectin could be "diluted out" while preserving the activities toward β-dextrin, pullulan, and α-limit dextrins.

Shortly afterwards, Manners *et al.* (*35*) reported that oat and malted barley "limit dextrinase" had additional properties, namely, the ability to hydrolyze amylopectin β-limit dextrin and pullulan. Thus "limit dextrinase," originally described as an oligosaccharase, was now a polysaccharase also. That the property to hydrolyze amylopectin β-limit dextrin was in conflict with the original MacWilliam and Harris report (*32*) was seemingly overlooked. In its new guise, "limit dextrinase" was a preparation that performed all the functions of the original R-enzyme, save one, the ability to attack amylopectin.

In reporting our results (*24a*), and comparing them with those of Manners *et al.* (*35*), we have suggested that "limit dextrinase" is simply the original R-enzyme, used in a concentration too low to notice attack on amylopectin, and that "limit dextrinase," as an entity separate from R-enzyme, does not exist. It is relevant to note that in the first description of R-enzyme (*22*), the much more rapid action on amylopectin β-dextrin than on amylopectin was seen. With our colleague, J. J. Marshall, we have subsequently obtained similar results with two pullulanases separated from sweet corn (*36*). The pullulanase activity was purified, and at the last purification stage, namely, fractionation on hydroxylapatite, was separated into two fractions, both with apparently identical activities and having specific activities of 5.5 and 6.5 (μmoles of maltotriose from pullulan)/min/mg of protein. It is a reflection on the state of the art that this is the first report of the specific activity of a plant pullulanase. Though still not homogeneous, the two fractions had been purified 3200-fold and 3800-fold, respectively, indicating the relatively small amounts present in sweet corn. Both fractions, even though purified to such an extent, have all the oligo- and polysaccharase activities of *Aerobacter* pullulanase (Table IIB) and the second fraction exhibited the

35. D. J. Manners, J. J. Marshall, and D. Yellowlees, *BJ* **116**, 539 (1970).
36. E. Y. C. Lee, J. J. Marshall, and W. J. Whelan, *ABB* **143**, 365 (1971).

same effect on dilution as did the potato enzyme; i.e., the action toward amylopectin selectively disappeared. This is added proof of the nonexistence of a separate "limit dextrinase."

We therefore conclude that the balance of evidence is that R-enzyme is a single entity and is the plant equivalent of bacterial pullulanase. The name pullulanase is to be preferred, but R-enzyme is useful in defining the original plant preparation. The fact that pullulanases from different sources may have somewhat different specificities, or attack substrates at different relative rates (Table I, IIA, IIB), can be accommodated by qualifying the name with the source of the enzyme. To refer to bean pullulanase will therefore convey a sense similar to that implied by "potato phosphorylase" when it is wished to distinguish the enzyme from, say, muscle phosphorylase.

There is, of course, the question of what is the polysaccharase that MacWilliam and Harris (*32*) and Manners and Sparra (*34*) refer to as R-enzyme. We suggest in a later section that this may be an isoamylase.

C. Pseudomonad and *Cytophaga* Isoamylases

The term *isoamylase* has been in the literature for many years, being used to describe a debranching enzyme from yeast. In Section III,D we discuss the controversy as to the precise nature of the debranching systems in yeast. The word isoamylase, as descriptive of a new type of debranching enzyme, has been used very recently as the name for an enzyme from a soil pseudomonad, *Pseudomonas* SB-15 (*37*). The properties of this enzyme have been very clearly delineated (*7, 37a*) and are reinforced by the even more recent discovery of a similar enzyme in a species of *Cytophaga* (*38*). The *Pseudomonas* enzyme has been purified to homogeneity (*37a*), and has a molecular weight of 94,000. We recommend that the designation isoamylase should henceforth be used for this type of enzyme.

Table IX in Section IV gives a detailed comparison of the specificities of *Aerobacter* pullulanase and the two bacterial isoamylases. The distinctive feature of these isoamylases is that they are the only known debranching enzymes which will totally debranch glycogen into its unit chains. They are distinguished from pullulanase by the fact that they have no action on pullulan.

37. T. Harada, K. Yokobayashi, and A. Misaki, *Appl. Microbiol.* **16**, 1439 (1968).
37a. K. Yokobayashi, A. Misaki, and T. Harada, *BBA* **212**, 458 (1970).
38. Z. Gunja-Smith, J. J. Marshall, E. E. Smith, and W. J. Whelan, *FEBS Letters* **12**, 96 (1970).

TABLE III

A COMPARISON OF THE ACTIONS OF *Pseudomonas* ISOAMYLASE AND *Aerobacter* PULLULANASE ON AMYLOPECTINS, GLYCOGENS, AND THEIR β-DEXTRINS AS JUDGED BY CHANGES IN DEGREE OF β-AMYLOLYSIS[a]

		Conversion into maltose (%)			
		Successive actions of debranching enzyme and β-amylase		Simultaneous actions of debranching enzyme and β-amylase	
Substrate	β-Amylase	Isoamylase	Pullulanase	Isoamylase	Pullulanase
Waxy maize amylopectin	50	99	95	95	103
Amylopectin β-dextrin	0	80	97	72	97
Potato amylopectin	47	96	98	97	103
Oyster glycogen	38	102	46	100	99
Glycogen β-dextrin	0	79	31	76	99
Rabbit liver glycogen	42	100	51[b]	99	98

[a] From Yokobayashi *et al.* (*7*).

[b] C. Mercier (unpublished result) finds no increase in β-amylolysis of rabbit liver glycogen after the action of *Aerobacter* pullulanase (see Table IIA).

Table III records a comparison of the actions of *Pseudomonas* isoamylase and *Aerobacter* pullulanase on amylopectin, glycogen, and their β-limit dextrins (*7*). The incomplete action on β-dextrin arises from the difficulty the enzyme has in cleaving the α-maltosyl stubs of the β-dextrin (*7, 37a*). This is also true of *Cytophaga* isoamylase (*38*). [The A chains of β-dextrin are two or three glucose units in length, depending on whether they originate from chains containing an even or odd number of glucose units, respectively (*2*).] This behavior is reminiscent of that of *S. mitis* pullulanase (Table IIA). The *Pseudomonas* and *Cytophaga* isoamylases also act on oligosaccharide α-limit dextrins (*37a, 38*; Table IX). Two differences between the bacterial and yeast isoamylases are (i) the far from complete debranching of glycogen by the latter (Section III,D), and (ii) the fact that the yeast enzyme does not seem to share the difficulty that the bacterial enzyme has in removing maltose stubs from β-limit dextrin (*39*).

Because the bacterial isoamylases will not attack pullulan, we may conclude that they will only hydrolyze $\alpha, 1 \rightarrow 6$ bonds that constitute points of branching in amylaceous molecules. This is also a feature of the specificity of the indirect debranching enzyme systems (Section III,D). The inability, or difficulty, that the bacterial isoamylases have in re-

39. G. N. Bathgate and D. J. Manners, *BJ* **107,** 443 (1968).

moving α-maltosyl stubs has been utilized in an exploration of glycogen and amylopectin structure, leading to a revision of the Meyer-Bernfeld formulation (Section VI,B). We may note that Yokobayashi *et al.* (*7*, *37a*) observed a slight liberation of maltose from amylopectin and glycogen β-limit dextrins, while Gunja-Smith *et al.* (*39a*) observed none from glycogen and amylopectin "φ-β"-dextrins, these being the products of successive actions of muscle phosphorylase and β-amylase. There are subtle differences between the nonreducing terminal structures in these two types of dextrin that may explain the differing results.

D. Yeast Isoamylase

The term yeast isoamylase was coined in 1951 by Maruo and Kobayashi (*40*) to describe an enzyme preparation from yeast that had earlier been termed amylosynthease. The latter name arose from the observation that when incubated with waxy (glutinous) rice starch (almost wholly amylopectin) the intensity of the iodine stain increased (*41*). We now know that this phenomenon is characteristic both of direct and indirect debranching enzymes, but Nishimura (*41*), the original discoverer of the enzyme thought, not unnaturally, that he was observing some kind of synthesis.

Maruo and Kobayashi (*40*) correctly concluded that the action was one of debranching and likened it to that of the (then) recently discovered R-enzyme (*22*). Looking back, it seems that this second definition of the enzyme was also incorrect. This judgment is based on the magnitude of the changes taking place in the rice starch on incubation with yeast isoamylase, as summarized in a recent review by Kobayashi (*42*). The amylopectin changed from one of average chain length 18 to 100, the iodine stain from red to blue, and the product was "highly susceptible" to retrogradation, and precipitation by butanol. The last three changes are consistent with the first, namely, there was an *increase* in chain length and a consequent decrease in the proportion of branches. To this extent Nishimura (*41*) was correct in thinking that synthesis was occurring. Chain elongation of this magnitude could not, however, be the result of action by a direct debranching enzyme. Also, the changed iodine stain caused by, say, R-enzyme (*22*), would not be

39a. Z. Gunja-Smith, J. J. Marshall, C. Mercier, E. E. Smith, and W. J. Whelan, *FEBS Letters* **12**, 101 (1970).

40. B. Maruo and T. Kobayashi, *Nature* **167**, 606 (1951).

41. S. Nishimura, *Biochem. Z.* **223**, 161 (1930); **225**, 264 (1930); **237**, 133 (1931).

42. T. Kobayashi, *J. Japan Soc. Starch Sci.* **17**, 61 (1969).

described as "blue." What these changes are strongly indicative of is the action of an indirect debranching enzyme, as described later in this chapter. That this is highly probable is indicated by the fact that Lee *et al.* (*43, 44*) have discovered an indirect debranching system in yeast. It seems, however, that despite the knowledge of this discovery, Kobayashi and his colleagues (*45*) still find it difficult to interpret their results. They have most recently described a yeast preparation that liberates both glucose and maltose from oligosaccharide limit dextrins of amylopectin prepared by the combined actions of α- and β-amylase. Glucose is not known to be liberated by any direct debranching enzyme.

Manners *et al.* (*46, 47*) had also described what they believed to be a direct debranching enzyme (isoamylase) from yeast. When Lee *et al.* (*43*) reported the discovery of the indirect debranching system, Bathgate and Manners (*39*) reexamined their yeast isoamylase preparation and agreed that amylo-1,6-glucosidase activity was present, this being a component of the indirect debranching system. They were, however, able to separate this activity from the isoamylase and to show the reality of the latter activity, which, from this report, and one by Kjølberg and Manners (*47*), seems to be capable of the partial debranching of glycogen, amylopectin, and their β-limit dextrins, but has no action on α-limit dextrins (Table IX). The yeast isoamylase also has no action on pullulan (*16*). In view of its ability to liberate both glucose and maltose from α,β-dextrins, the preparation of Kobayashi *et al.* (*45*) may have mixed direct and indirect debranching activities. The general concensus is that isoamylase like pullulanase will not liberate glucose. α-Maltose, or even α-maltotriose in the case of pseudomonad isoamylase, is the shortest $1 \rightarrow 6$-bonded stub (A chain) that is removed by direct debranching enzymes. By contrast, amylo-1,6-glucosidase will only remove α-glucose stubs. Because Kobayashi *et al.* (*45*) believed their isoamylase to have the capacity to remove both glucose and maltose, they wrote: ". . . there seems to exist no practical method for discriminating the actions of these two enzymes" (isoamylase and amylo-1,6-glucosidase). This appears to be an incorrect conclusion. The specificities of the two enzymes, as noted above, are sufficiently different in this respect to permit their distinction. In Section IV we survey the several types of difference.

43. E. Y. C. Lee, L. D. Nielsen, and E. H. Fischer, *ABB* **121,** 245 (1967).
44. E. Y. C. Lee, J. H. Carter, L. D. Nielsen, and E. H. Fischer, *Biochemistry* **9,** 2347 (1970).
45. Y. Sakano, T. Kobayashi, and Y. Kosugi, *Agr. Biol. Chem.* **33,** 1535 (1969).
46. Z. H. Gunja, D. J. Manners, and K. Maung, *BJ* **81,** 392 (1961).
47. O. Kjølberg and D. J. Manners, *BJ* **86,** 258 (1963).

Additional discrepancy exists between reports of Kjølberg and Manners (*47*), who stated that yeast isoamylase does not attack α-limit dextrins, an activity that they know to be in yeast but ascribe to a separate "limit dextrinase" (*39*), and of Kobayashi *et al.* (*45*) who claimed that the isoamylase does attack these dextrins.

E. Plant Isoamylase

As noted earlier, MacWilliam and Harris (*32*) and Manners *et al.* [e.g., (*34*)] applied the term "R-enzyme to a preparation from plant sources that attacks amylopectin and its β-dextrin but not oligosaccharide α-dextrins. Manners and Rowe (*48*) have obtained from sweet corn two such preparations, distinguished by the fact that one also attacks glycogen (termed isoamylase) and the other (termed R-enzyme) does not. Neither attacks pullulan.

Very likely these preparations should be classed as isoamylases, though the "R-enzyme" needs further documentation before its nature can be judged. The only positive enzymic property described for it is its ability to increase the intensity of iodine stain of amylopectin by relatively minute amounts. There is no proof that it is a debranching enzyme.

III. Indirect Debranching Enzymes

A. Introduction

In mammalian tissue and in yeast, glycogen is debranched by a two-component enzyme system, amylo-1,6-glucosidase/oligo-1,4 → 1,4-glucantransferase (*48a*). The action of the system on a segment of the phosphorylase limit dextrin (ϕ-dextrin) of glycogen is shown in Fig. 2. The first and necessary step is the action of the transferase which repositions the outer glucose units of the side chain, leading to the exposure of a single α-glucose unit linked 1 → 6 to the rest of the molecule. The 1 → 6 bond is then hydrolyzed by amylo-1 → 6-glucosidase, which is specific for the glucose unit. The overall action of the system is to redistribute the 1 → 4-linked glucose units and to release the glucose unit involved in the 1 → 6 bond as the sole low molecular weight product.

The two best-studied examples of this type of debranching system are those of rabbit muscle and of baker's yeast. Originally discovered by

48. D. J. Manners and K. L. Rowe, *Carbohydrate Res.* **9**, 107 (1969).
48a. Abbreviated to glucosidase-transferase.

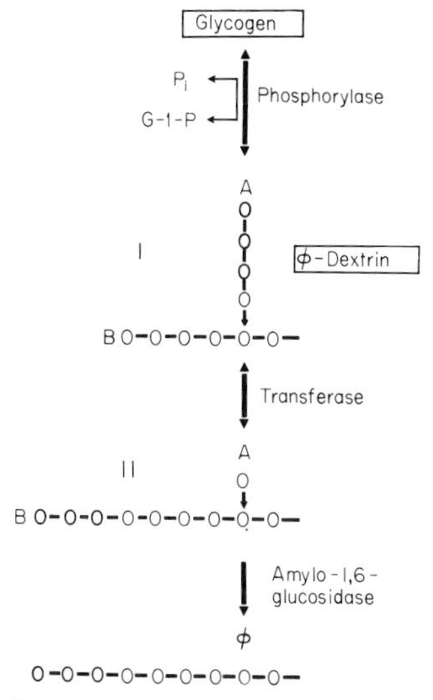

FIG. 2. The action of glucosidase-transferase on a segment of the phosphorylase limit dextrin structure [I, after Walker and Whelan (*50*)]. Structure II is that earlier proposed for the phosphorylase limit dextrin by Cori and Larner (*49*): (O) α-glucose unit; (**O**) glucose residues repositioned by transferase; (ø) reducing glucose unit; (—) 1 → 4 bond; and (→) 1 → 6 bond.

Cori and Larner in 1951 (*49*), the rabbit muscle debranching system was thought simply to be a glucosidase, "amylo-1,6-glucosidase," since glucose was the only low molecular weight product of its action on glycogen ϕ-dextrin. The implication of a transferase in the debranching process came about with the elucidation of the true structure of the ϕ-dextrin by Walker and Whelan (*50*). This structure is shown in Fig. 2, and it is seen that for glucose to be released by the action of the glucosidase, the "covering" glucose units must be removed. Transferase activity was shown to be present in preparations of rabbit muscle debranching enzyme by Brown and Illingworth (*51*), Abdullah and Whelan (*52*), and Brown *et al.* (*53*). When the amylo-1,6-glucosidase activity was purified to homogeneity, it was accompanied by the transferase, and the ratio of the two activities remained constant during purification (*54*, *55*).

49. G. T. Cori and J. Larner, *JBC* **188,** 17 (1951).
50. G. J. Walker and W. J. Whelan, *BJ* **76,** 264 (1960).
51. D. H. Brown and B. Illingworth, *Proc. Natl. Acad. Sci. U. S.* **48,** 1783 (1962).
52. M. Abdullah and W. J. Whelan, *Nature* **197,** 979 (1963).
53. D. H. Brown, B. Illingworth, and C. F. Cori, *Nature* **197,** 980 (1963).
54. D. H. Brown and B. Illingworth, *in* "Control of Glycogen Metabolism" (W. J. Whelan and M. P. Cameron, eds.), p. 139. Churchill, London, 1964.
55. D. H. Brown and B. I. Brown, "Methods in Enzymology," Vol. 8, 515, 1966.

The mechanism of debranching is such that only the glucose unit at the point of branching is released as a hydrolytic product; the transferred glucan segments remain on the polysaccharide where they are still available to the action of phosphorylase. The combined actions of phosphorylase and glucosidase-transferase in the presence of excess inorganic phosphate lead to the total degradation of glycogen to glucose 1-phosphate and glucose. At any intermediate stage, however, there is only one polymeric product, a partly degraded glycogen, and two monomeric products, α-glucose 1-phosphate and glucose. This is in contrast to direct debranching (see above), where a series of oligomers is formed. The glucosidase-transferase debranching system may therefore be regarded as an integral part of the overall phosphorolytic pathway for the degradation of glycogen since only with its participation is the *total* degradation of glycogen by phosphorylase possible.

Glucosidase-transferase has been studied in human tissues, mainly in connection with the hereditary disease where the enzyme system is deficient. This is glycogen storage disease type III (*56*; see Section III,H). Other than in rabbit muscle, the enzyme system has not been extensively investigated in lower animals.

Glucosidase-transferase has been identified in baker's yeast by Lee *et al.* and isolated to a state of homogeneity (*43, 44*). This enzyme system has similar properties to the rabbit muscle system. Yeast is the only unicellular organism which has been reported to contain this type of debranching enzyme system, but it does not seem to have been widely sought for. It may prove to be the dominant type of glycogen debranching mechanism, where glycogen is stored. This judgment is based on the relative inefficiency, or even incapability, of the direct debranching enzymes to attack glycogen (Section II), with the exception of the bacterial isoamylases (*7, 38*). Yeast also contains an isoamylase but its glycogen debranching capacity is low (*39*).

B. Purification and Physical Properties

The rabbit muscle enzyme system was first purified to a state of homogeneity by Brown and Illingworth (*54, 55*). The specific activity of their preparation (μmoles of glucose released per minute from glycogen ϕ-dextrin) was reported to be of the order of 4.0 IU/mg. Values of 8.3 and 7.6 IU/mg have been reported by other investigators for the

56. B. I. Brown and D. H. Brown, *in* "Carbohydrate Metabolism and Its Disorders" (F. Dickens, P. J. Randle, and W. J. Whelan, eds.), Vol. 2, p. 123. Academic Press, New York, 1968.

purified enzyme system (*57, 58*). The molecular weight determined by ultracentrifugation is 273,000 (*54*). Treatment with 3 *M* urea dissociated the system into subunits of $s_{20,w} = 6.6$ (*59*). Dissociation with sodium dodecyl sulfate and subsequent disc gel electrophoresis in sodium dodecyl sulfate revealed the presence of a single subunit of approximate molecular weight of 120,000, suggesting that the rabbit muscle system consists of two subunits of equal size (*60*).

Glucosidase-transferase from yeast has been purified to apparent homogenity by the criterion of disc gel electrophoresis (*44*). The specific activity of the yeast preparations was about 6 IU/mg, although further purification of an apparently homogeneous preparation led to an increase in specific activity to 8.4 IU/mg. The molecular weight of the system was about 280,000 by sedimentation equilibrium ultracentrifugation; the latter study also showed that the system dissociated at low protein concentrations. Determination of the molecular weight by sieving on Sephadex G-200 and by the disc gel electrophoresis method of Hedrick and Smith (*61*) gave values around 210,000. On sodium dodecyl sulfate disc gel electrophoresis the enzyme system dissociated into three subunits of approximate molecular weight 120,000, 85,000 and 70,000 (*60*).

The question has been posed whether glucosidase-transferase is a double-headed enzyme (*56*) or a multienzyme complex (*31*). This has not been answered despite the demonstration of subunits, which could also represent dissociation into separate enzymes. The reason is that the dissociated system is without activity (*59, 60*).

C. ASSAY OF GLUCOSIDASE-TRANSFERASE ACTIVITY

The generally used assay of glucosidase-transferase activity is the measurement of the glucose released from glycogen ϕ-dextrin (*44, 55, 62*). The assay is a direct measure of the number of branch linkages split by the combined actions of glucosidase-transferase and is a measure of the rate of the overall reaction. The unit of activity based on this assay is the amount of enzyme which releases 1 μmole of glucose per minute. The assay is not suitable for the determination of glucosidase-transferase in crude tissue homogenates owing to the presence of α-glucosidases

57. T. E. Nelson, E. Kolb, and J. Larner, *Biochemistry* **8,** 1419 (1969).
58. F. Huijing, E. Y. C. Lee, J. H. Carter, and W. J. Whelan, *FEBS Letters* **7,** 251 (1970).
59. D. H. Brown and B. I. Brown, *BJ* **100,** 8P (1966).
60. E. Y. C. Lee, J. H. Carter, and W. J. Whelan, unpublished work.
61. J. L. Hedrick and A. J. Smith, *ABB* **126,** 155 (1968).
62. J. Larner and L. H. Schliselfeld, *BBA* **20,** 53 (1956).

which also release glucose from glycogen. Such assays of crude homogenates are necessary in the diagnosis of glycogen storage disease type III, where this activity is absent, and a number of other methods have been developed. These have been reviewed by Hers *et al.* (*63*). Two of the substrates that have been used for specific assays are "fast B_5" and "B_7" (6^3-α-glucosylmaltotetraose and 6^3-α-maltotriosylmaltotetraose, respectively), which were used by Brown and Illingworth for the specific assay of the glucosidase and transferase activities, respectively (*6, 51*). The basis for the use of "B_7" for the specific assay of transferase activity, despite the fact that both activities are involved in the reaction, was the observation that the transferase action was the rate-limiting step (*51*).

The only truly specific substrates for the assay of amylo-1,6-glucosidase are the α-glucosyl-substituted Schardinger dextrins (*31, 64*). The glucosyl stub is not hydrolyzed by fungal amyloglucosidase and appears not to be acted upon by the other α-glucosidases present in tissue homogenates (*64*). Furthermore, the product of the reaction, Schardinger dextrin, is also immune to the action of α-glucosidases (*64*).

The incorporation of [^{14}C]glucose into glycogen by the reversion reaction of amylo-1,6-glucosidase is a widely used assay for glucosidase-transferase action (*62, 65*). The interpretation of whether this assay represents specifically the amylo-1,6-glucosidase activity alone or both activities is uncertain. This aspect is discussed in Section III,G. The original method of recovery of the labeled glycogen by precipitation procedures has been superseded by a more convenient method, that of counting the labeled glycogen on paper (*66, 66a*).

Several other methods have been described. These are (a) release of glucose labeled at the branching points (*63*), (b) measurement of the rate of phosphorolysis of glycogen ϕ-dextrin in the presence of phosphorylase and inorganic phosphate (*49*) or alternatively of arsenolysis in the presence of phosphorylase and arsenate (*63*), (c) measurement of the labeled formic acid released by periodate oxidation from glycogen labeled at the periphery (*63*), and (d) measurement of the change in iodine staining capacity of amylopectin or glycogen ϕ-dextrin (*66b*). The last two assays are of transferase activity.

63. H. G. Hers, W. Verhue, and F. Van Hoof, *European J. Biochem.* **2,** 257 (1967).
64. P. M. Taylor and W. J. Whelan, *ABB* **113,** 500 (1966).
65. H. G. Hers, *Rev. Intern. Hepatol.* **9,** 35 (1959).
66. T. E. Nelson and J. Larner, *Anal. Biochem.* **33,** 87 (1970).
66a. T. E. Nelson and J. Larner, *BBA* **198,** 538 (1970).
66b. T. E. Nelson, D. H. Palmer, and J. Larner, *BBA* **212,** 269 (1970).

D. Specificity of Glucosidase-Transferase

Amylo-1,6-glucosidase is specific for the hydrolysis of the 1→6 bond involving a single α-glucose unit attached to a chain of 1→4-bonded α-glucose units. Studies of the rabbit muscle system have shown that 6^3-α-glucosylmaltotetraose is hydrolyzed, whereas 6^3-α-glucosylmaltotriose is not, so that the glucose unit removed must be at a point of branching (*6*). The yeast system has not been studied with these substrates, but it would appear to have similar specificity requirements since panose is not a substrate (*60*).

The specificity of the rabbit muscle transferase has been studied by its action on labeled maltosaccharides (*51, 54*). Different specificities were observed for the donor and acceptor substrates. Thus glucose and the series maltose to maltopentaose were observed to be acceptors of glycosyl segments transferred from glycogen but were not acted upon when incubated alone with the enzyme system. Maltohexaose was the smallest oligosaccharide which was acted upon when incubated alone. In this way maltohexaose and higher oligosaccharides were found to be both donors and acceptors of maltosyl and maltotriosyl units. The rate of maltotriosyl transfer exceeded that of maltosyl transfer, and no evidence was found for glucosyl transfer. That the transferase acts via the transfer of glycosyl segments rather than sequential glucosyl transfer has been demonstrated by Hers *et al.* (*67*). The rabbit muscle glucosidase-transferase therefore appears to be adapted to the glycogen φ-dextrin structure (Fig. 2).

Using a system in which a single oligosaccharide was acted upon, the yeast transferase was also shown to catalyze the transfer of maltosyl and maltotriosyl units (*44*). The smallest oligosaccharide which was acted upon was maltotetraose, the mode of transfer with this substrate being of maltosyl units only. With maltopentaose and maltohexaose, both maltotriosyl and maltosyl transfer were observed, the latter being the predominant mode of transfer (*44*).

In terms of the specificity of the glucosidase-transferase systems toward polysaccharide substrates, glycogen φ-dextrin appears to be the best substrate for both the yeast and rabbit muscle systems (Table IV). The rabbit muscle system has a much slower action on native glycogen (*57*), and a very slow action on amylopectin. Brown *et al.* have demonstrated that the rate of action falls rapidly with increasing outer chain

67. H. G. Hers, W. Verhue, and M. Mathieu, *in* "Control of Glycogen Metabolism" (W. J. Whelan and M. P. Cameron, eds.), p. 151. Churchill, London, 1964.

TABLE IV
SPECIFICITY OF GLUCOSIDASE-TRANSFERASE TOWARD POLYSACCHARIDES[a]

Polysaccharide	Relative rate of release of glucose: Yeast	Relative rate of release of glucose: Rabbit muscle	Length of A chains
Shellfish glycogen			
φ-Dextrin	100	100	4
β-Dextrin	87	33	2, 3
"φ-β"-Dextrin	74	13	2
Amylopectin			
φ-Dextrin	29	33	4
β-Dextrin	43	13	2, 3
"φ-β"-Dextrin	12	1.4	2
Native polysaccharides			
Shellfish glycogen	82	—	(5.8)[b]
Rabbit liver glycogen	72	7	(8.6)[b]
Amylopectin	4.5	<0.1[c]	(12.6)[b]

[a] E.Y.C. Lee, J. H. Carter, and W. J. Whelan, unpublished results (1951).
[b] Average values, calculated as $[\overline{CL} \times (A/100)] + 2$, where A is the percentage degradation by β-amylase.
[c] Nelson *et al.* (*57*).

length of glycogen (*68*). In contrast the yeast enzyme acts on glycogen nearly as well as on the φ-dextrin, and it has a comparatively higher rate of action on amylopectin than the rabbit muscle system [(*60*), Table IV]. The fall in the rate of action of rabbit muscle glucosidase-transferase with increasing outer chain length has been ascribed to an inhibition of the enzyme system (*62, 68*). It should be noted, however, that the release of glucose from a polysaccharide substrate is the result of a sequential transferase-glucosidase action. If, as it seems, the muscle transferase prefers to transfer glycosyl segments containing three glucose units, then the rate at which the enzyme exposes glucose stubs for glucosidase action will decrease as the length of the outer chains of the substrate increases. It will be appreciated also, that since transferase action is reversible, the probability of a glucosyl stub being exposed will decrease sharply with increase in outer chain length. Furthermore, the exposure and release of glucosyl stubs means that the remaining substrate is itself undergoing an increase in outer chain length. Therefore, beginning with glycogen, from which glucose can be released, the macromolecule should undergo outer chain extension, and reaction will effectively cease when the probability of a glucose stub being exposed falls essentially to zero. This is likely to occur before debranching is complete

68. D. H. Brown, B. I. Brown, and C. F. Cori, *ABB* **116**, 479 (1966).

and is what is noted experimentally with the yeast enzyme (Section III,F). Therefore it seems incorrect to refer to inhibition by substrates of longer outer chain length. Transferase activity may occur at exactly the same rate regardless of chain length. It is the fact that the rate of reaction is measured by glucose release that prompts the idea of "inhibition" since glucose release must wait on exposure of a glucose stub. The greater rate of release of glucose from glycogen by the yeast system, relative to the muscle system, both being compared according to equal rates on glycogen ϕ-dextrin, suggests the ability of the yeast transferase to transfer longer chain segments.

Recently, Lee *et al.* (*60*) have studied the action of glucosidase-transferase on the limit dextrins of glycogen and amylopectin. In this study the phosphorylase limit dextrins, β-amylase limit dextrins, and "ϕ-β" limit dextrins (produced by the sequential actions of phosphorylase and β-amylase) of glycogen and amylopectin was used. These limit dextrins would be expected to have outer A chains of four, an equal mixture of two and three, and two units in length, respectively. The results are shown in Table IV, and they were interpreted as a reflection of the specificity of the transferase, on the assumption that the action of transferase is the rate-limiting step. The results confirm the specificity requirements of the transferase as deduced from studies carried out with maltosaccharide substrates (*54*) in that the rabbit muscle system is seen to be more specific for the ϕ-dextrins of both the glycogen and amylopectin series, where the optimal mode of transfer is of maltotriosyl segments. The yeast system, on the other hand, acted at a similar rate on the β-limit dextrin as it did on the ϕ-dextrin of glycogen, and in the case of the amylopectin limit dextrins, on the β-limit dextrin at a higher rate than on the ϕ-dextrin.

The "ϕ-β" dextrins of amylopectin and glycogen were much poorer substrates than the β- or ϕ-dextrins. These might not be expected to be substrates at all if the transferase could not catalyze glucosyl transfer. That hydrolysis did occur may stem from the experimental difficulty of preparing a limit dextrin in which every A chain has been shortened to two glucose units. An alternative explanation is that the transferase may be able to transfer branched segments and expose glucose units at inner branch points.

E. pH DEPENDENCE OF ACTIVITY

Yeast glucosidase-transferase displays a single pH optimum when either glycogen ϕ-dextrin or α-glucosyl Schardinger dextrin is used as

TABLE V
pH DEPENDENCE OF RABBIT MUSCLE GLUCOSIDASE-TRANSFERASE ACTIVITY

Assay method	pH optima	Buffer	Reference
Release of glucose from glycogen ϕ-dextrin	7.2–7.4	Glycylglycine	*62*
	5.0–6.4		*67*
	6.1–6.4	Citrate	*55*
	6.6	Citrate-phosphate	*31*
	6.5; 7.5	Citrate	*31*
	6.6	Anionic	*57*
	7.2	Cationic	*57*
Release of glucose from α-glucosyl Schardinger dextrin	5.5–5.9	Citrate-phosphate	*31*
	5.8; 7.5	Citrate	*31*
Release of glucose from 6^3-α-maltotriosylmaltotetraose (B_5)	6.1–6.4	Citrate	*55*
Incorporation of [^{14}C] glucose into glycogen	7.5–8.0		*67*
	6.4	Phosphate	*66*
	7.2	Tris	*66*
	5.6; 7.0		*69*
Incorporation of [^{3}H] glucose into Schardinger dextrin	5.5; 7.0	Citrate	*31*
Incorporation of [^{14}C] glucose into			
(a) glycogen and Schardinger dextrin	5.6; 7.0		*69*
(b) maltosaccharides	5.6		*69*
(c) glycogen β-dextrin	7.0		*69*
Change in iodine stain of amylopectin	6.0	Anionic and cationic	*66b*

a substrate, with a maximum in the pH range 6.0–6.8 (*44*). By contrast, a confusing state of affairs exists in regard to the rabbit muscle glucosidase-transferase (Table V) in that (a) a number of conflicting pH optima values have been reported and (b) different types of pH optima curve have been reported (i.e., symmetric, asymmetric, and biphasic). With regard to the activity determined with glycogen ϕ-dextrin, some resolution of the different reported values has been provided by Nelson *et al.* (*57*) who found that the pH optimum depended on the buffer used; a pH shift of 6.6 to 7.2 was observed when going from an anionic to cationic buffer. In addition, Nelson and Larner (*66*) have reported the same pH shifts in the determination of the pH activity curve using the reversion reaction (incorporation of [^{14}C]glucose into glycogen).

However, Taylor and Whelan (*31*) have reported biphasic pH optima when either ϕ-dextrin or α-glucosyl Schardinger dextrin was used as the substrate in citrate buffer, and a single optimum at pH 6.6 when citrate-phosphate buffer was used. This biphasic curve was also observed when the incorporation of [^{3}H]glucose onto Schardinger β-dextrin was measured (*31*).

Using the rabbit muscle system, Stark and Thambyrajah (*69*) confirmed that [^{14}C]glucose addition to Schardinger dextrin and glycogen shows two pH optima, at pH 7.0 and 5.6, but with glycogen β-dextrin and maltosaccharides containing 7 and 13 glucose units there are single optima, at pH 7.0 and 5.6, respectively. Nelson *et al.* (*66b*) have recently determined the pH optimum of the rabbit muscle transferase specifically by using the increase in the iodine stain of amylopectin. They reported a value of pH 6 for the transferase activity, which is not changed by the nature of the buffer (*66b*).

F. The Effect of Glucosidase-Transferase on Glycogen Structure

The major effect of the action of the glucosidase-transferase on glycogen or its phosphorylase limit dextrin is on the outer chains. The removal of branch point glucose is accompanied by a redistribution of the outer portions of the structure by the action of the transferase. As noted before, the only low molecular weight product is glucose, and the macromolecular nature of the substrate is preserved. Thus the net effect is an increase in the outer chain length. Historically, this effect provided important evidence in the implication of transferase activity in the debranching process; the lengthening of the outer chains led to a large increase in the iodine staining capacity of the glycogen (*52, 53*). The overall effects on glycogen structure are exemplified in the action of the yeast system on shellfish glycogen (*44*). The release of glucose reached a limiting value equivalent to 35% of the branch points, converting the glycogen from one of average chain length of 10 to one of 16. Further analysis (Table VI) showed that the effect is on the outer chains of the polysaccharide which were now similar in average length to those of amylopectin. This is seen also in the iodine stain spectrum of the product (Fig. 3). The increase in iodine stain brought about by glucosidase-transferase is much higher than that brought about by the action of a direct debranching enzyme such as pullulanase (see Section II) because of the chain-lengthening effect. This is shown in Table VII where the

69. J. R. Stark and V. Thambyrajah, *BJ* **120**, 17P (1970).

TABLE VI
COMPARISON OF GLYCOGEN AND AMYLOPECTIN WITH THE GLYCOGEN DEXTRIN FORMED BY THE ACTION OF YEAST GLUCOSIDE-TRANSFERASE[a]

Polysaccharide	Degree of β-amylolysis (%)	Average chain lengths		
		$\overline{CL}$	Outer $\overline{CL}$	Inner $\overline{CL}$
Glycogen	39	9.7	5.8	2.9
Glycogen dextrin[b]	64	15.8	12.1	2.7
Amylopectin	53	20	12.6	6.4

[a] From Lee *et al.* (*44*).
[b] Limit dextrin prepared by the exhaustive action of yeast glucosidase-transferase on shellfish glycogen.

actions of the two enzyme systems on amylopectin and its β-limit dextrin are listed. The increases in iodine stain as a result of glucosidase-transferase action are much higher despite lower extents of debranching as evidenced by the degrees of β-amylolysis (*43*). Nelson *et al.* (*66b*) have demonstrated that the rabbit muscle transferase can disproportionate the outer chains of amylopectin without any detectable action of the glucosidase.

That the glucose release reached a limiting value at 35% of the branch points is probably a reflection of the question discussed previously (Section III,D), *viz.*, that as the outer chain length increases, the rate at which glucosyl stubs are exposed by the action of transferase is

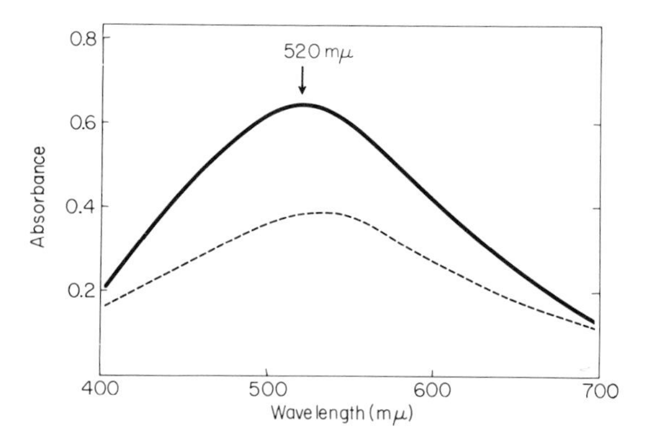

FIG. 3. Iodine stain spectrum of glycogen after prolonged action of yeast glucosidase-transferase [from Lee *et al.* (*44*)]: (——) debranched glycogen (0.2 mg/ml); (- - -) native amylopectin (0.1 mg/ml). Native shellfish glycogen has a negligible iodine staining capacity under the conditions used (0.05% KI, 0.005% I_2, 0.05 *N* HCl). Reprinted from *Biochemistry* **9**, 2347 (1970). Copyright (1970) by the American Chemical Society. Reprinted by permission of the copyright owner.

TABLE VII
COMPARISON OF THE ACTIONS OF YEAST GLUCOSIDASE-TRANSFERASE AND *Aerobacter* PULLULANASE ON AMYLOPECTIN AND ITS β-LIMIT DEXTRIN[a]

	Increase in iodine stain (%)		Degree of β-amylolysis (%)[b]	
Polysaccharide	Glucosidase-transferase	Pullulanase	Glucosidase-transferase	Pullulanase
Amylopectin	450	130	73	92
Amylopectin β-limit dextrin	340	170	52	100

[a] From Lee *et al.* (*43*).

[b] Before debranching, the degrees of β-amylolysis were 52% and 0, respectively.

reduced. However, amylopectin, with a similar average outer chain length to the glycogen end product, is a substrate (Table IV).

G. REVERSION REACTIONS OF GLUCOSIDASE-TRANSFERASE

It was originally demonstrated by Larner and Schliselfeld (*62*) that rabbit muscle "amylo-1,6-glucosidase" would catalyze the incorporation of [^{14}C]glucose into glycogen in what is known as the back incorporation reaction. This reversion reaction has been used as the basis of several assay procedures for the determination of amylo-1,6-glucosidase activity (*63*, *65*, *66*). It is clear, however, that the overall glucosidase-transferase system is involved in the incorporation of glucose into glyco-

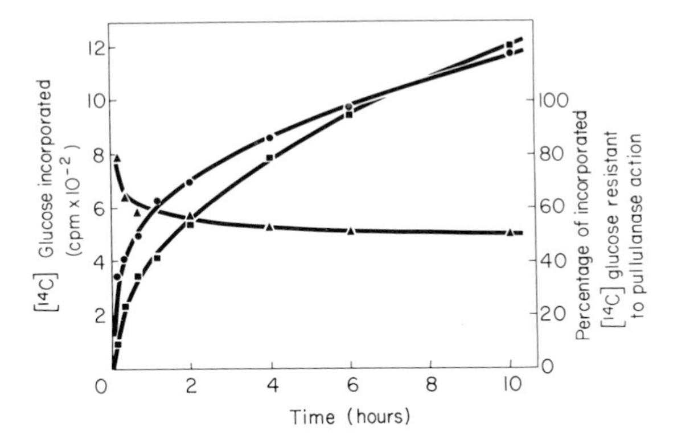

FIG. 4. Incorporation of [^{14}C]glucose into glycogen by yeast glucosidase-transferase (*60*). (●) Incorporated glucose resistant to the action of pullulanase, (■) incorporated glucose removed by pullulanase, and (▲) percentage of incorporated glucose resistant to pullulanase action.

gen. Studies by Hers *et al.* have shown that at least 35% of the glucose incorporated by the rabbit muscle system is "covered" by the action of transferase (*67*). This was shown by treatment of the labeled glycogen with pullulanase (see Section II,A), a debranching enzyme which will not remove side chains consisting of a single glucose unit; the amount of label which was removed by pullulanase action therefore represented the "covered" glucose units. In a similar study of the yeast system (*60*), the relative amounts of incorporated glucose which were present either as "covered" or single glucose stubs were determined as a function of time (Fig. 4). In the early stages of the reaction of the incorporation is mainly of single glucose stubs, and this eventually falls to about 50% of the total incorporation.

Glycogen is not the only polysaccharide which can act as an acceptor in the back reaction. The rates at which the rabbit muscle and yeast enzyme systems incorporate glucose into other polysaccharide substrates are shown in Table VIII. These rates are generally in the inverse order of the rate at which the polymers are debranched by the same enzyme system [cf. Table IV (*60*)]. Schardinger dextrins also act as acceptors in the reversion reaction (*5, 31*).

Nelson and Larner (*66a*) reported that rabbit muscle glucosidase-transferase synthesizes isomaltose from glucose, and found that the glucose released in the hydrolytic reaction on ϕ-dextrin is in the α form. Stark and Thambyrajah (*69*) noted maltotetraose to be the smallest oligosaccharide which will act as an acceptor of labeled glucose to form branched oligosaccharide (*69*).

TABLE VIII
RELATIVE RATES OF INCORPORATION OF [^{14}C] GLUCOSE INTO POLYSACCHARIDE SUBSTRATES BY GLUCOSIDASE-TRANSFERASE[a]

Substrate and concentration	Yeast	Rabbit muscle
(100 mg/ml)		
Rabbit liver glycogen	100	100
Shellfish glycogen	94	83
Glycogen ϕ-dextrin	50	58
Glycogen β-dextrin	22	24
(7 mg/ml)		
Amylopectin	100	100
Amylose	75	60
Amylopectin β-dextrin	55	27
Rabbit liver glycogen	17	34

[a] E. Y. C. Lee, J. H. Carter, and W. J. Whelan, unpublished results.

H. Glucosidase-Transferase in Glycogen Storage Disease

Several types of glycogen storage disease are known in which one or other of the enzymes involved in the metabolism of glycogen is absent (*56*). In type III glycogenosis a polysaccharide with a structure resembling that of a glycogen phosphorylase limit dextrin is accumulated, and it has been shown that this resulted from a deficiency in the debranching enzyme system.

The enzymic defect has been demonstrated in muscle, liver, leukocytes, and erythrocytes (*56*). While in most cases the debranching system is affected in all tissues, whatever assay method is used, there are known to be cases in which liver is affected whereas activity is present in muscle (*56*). On this basis Van Hoof and Hers (*70*) have proposed a subgrouping of type III disease, type IIIA designating the situation in which the defect is present in both liver and muscle, and type IIIB in which the defect is present in the liver only. On the assumption that two distinct enzymic activities are involved, Manners and Wright (*70a*) have suggested subgroups in which only one or both of the two activities are affected. Van Hoof and Hers (*70*) and Brown and Brown (*56*) have studied a number of cases and have used specific assays for the glucosidase and transferase activities. Illingworth and Brown (*71*) have reported one case in which there seems to be a lack of only the glucosidase activity in the liver with retention of transferase activity; Van Hoof and Hers (*70*) have reported several cases in which there seems to be a specific lack of the transferase.

The reversion reaction of glucosidase-transferase has been proposed by Huijing *et al.* (*58*) as the reason for the existence of the branch linkages in the liver "glycogen" isolated from patients with type IV glycogen storage disease. This disease is characterized by the total absence of branching enzyme (*56*), and the question arose as to the origin of the branch points. The reversal of rabbit muscle glucosidase-transferase action was shown to be capable of the *in vitro* synthesis of branch linkages (*58*) in the branching enzyme assay system of Brown and Brown [(*72*), Fig. 5]. The involvement of transferase in the reversion reaction plays an important role since the "covered" glucose side chain can serve as the locus of chain elongation, thus "trapping" the branch point

70. F. Van Hoof and H. G. Hers, *European J. Biochem.* **2**, 265 (1967).

70a. D. J. Manners and A. Wright, *BJ* **79**, 18P (1961).

71. B. Illingworth and D. H. Brown, *in* "Control of Glycogen Metabolism" (W. J. Whelan and M. P. Cameron, eds.), p. 336. Churchill, London, 1964.

72. B. I. Brown and D. H. Brown, "Methods in Enzymology," Vol. 8, p. 395, 1966.

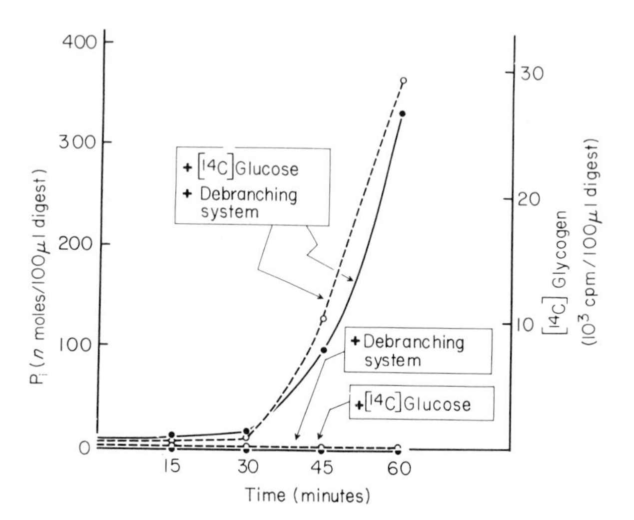

FIG. 5. Branching activity of rabbit muscle glucosidase-transferase in the assay system of Brown and Brown (*72*); taken from Huijing *et al.* (*58*). The system contained phosphorylase *a*, glucose 1-phosphate, and AMP. In the presence of both rabbit muscle glucosidase-transferase and [^{14}C]glucose, phosphate release (○) and incorporation of labeled glucose into polysaccharide (●) were observed. In the presence of either [^{14}C]glucose or glucosidase-transferase alone, no incorporation of labeled glucose or phosphate release was observed. From F. Huijing, E. Y. C. Lee, J. H. Carter, and W. J. Whelan, *FEBS Letters* **7**, 251 (1970). Reprinted by permission of North-Holland Publ., Amsterdam.

since glucosidase-transferase has a poor action on long chains (see Section III,D). The "glycogen" is therefore the product of glycogen synthetase, forming 1 → 4 bonds, and glucosidase-transferase, forming 1 → 6 bonds. The content of 1 → 6 bonds in the polysaccharide is about half that of normal glycogen, that is, about the same as in amylopectin (*56, 73*). In fact, the polysaccharide bears many resemblances to amylopectin, hence, the term *amylopectinosis* for this disease. The fine structure is, however, distinctly different from amylopectin (*73*) and accords with the explanation of the origin of the branch points.

I. OTHER INDIRECT DEBRANCHING ENZYMES

Walker and Builder (*74*) have described two separable enzymic activities in *Streptococcus mitis* which may represent a debranching system similar to glucosidase-transferase. The first activity is an α-1,6-glucosidase of somewhat different specificity to amylo-1,6-glucosidase; the second is a transferase activity. Thus both glucosidase and transferase activities are present as separate enzyme proteins in this organism, and

73. C. Mercier and W. J. Whelan, *European J. Biochem.* **16**, 579 (1970).

it has been suggested that the system may represent a more primitive form of the yeast and mammalian glucosidase-transferase systems (*74, 75*).

IV. Characterization of a Debranching Enzyme or Enzyme System

A glycogen or starch bearing tissue may be assumed to contain a mechanism for the specific splitting of the $1 \rightarrow 6$ bonds during catabolism. This assumption is reinforced by the failure to find any phosphorolytic enzyme capable of the total degradation of glycogen or amylopectin, i.e., able to split both $1 \rightarrow 4$- and $1 \rightarrow 6$-α-glucosidic bonds. There are certainly α-glucosidases that by themselves totally hydrolyze glycogen and amylopectin. These are known in fungi (*76*) and in mammalian lysosomes (*77*). A phosphorolytic pathway, however, seems to require the cooperation of a $1 \rightarrow 6$-bond hydrolase and a transferase (Section III) to accomplish total glycogenolysis or amylopectinolysis.

It is instructive to recall how the debranching enzymes were discovered. The first, amylosynthetase, was detected because of the increase in intensity of iodine stain of amylopectin, wrongly attributed to synthesis (see above). R-Enzyme was discovered in the same way but correctly categorized (*22*). The reason for the increased iodine stain has been attributed to a greater ease of complex formation once the unit chains of the dendritic molecule are set free (*22*).

Amylo-1,6-glucosidase was discovered as an impurity in preparations of rabbit muscle phosphorylase that caused the degree of phosphorolysis of glycogen to be greater than with pure phosphorylase preparations (*49*). It was not until the true structure of the phosphorylase limit dextrin was discovered (*50, 78*), and the fact that an oligotransferase had to act with the glucosidase in debranching (*51–53*), that it was realized that this debranching system also caused an increase in intensity of iodine stain (*52, 53*). The change in stain intensity is in fact much greater than for direct debranching. Pullulanase about doubles the intensity of the amylopectin-iodine stain; glucosidase-transferase causes a 450% in-

74. G. J. Walker and J. G. Builder, *BJ* **105,** 937 (1967).

75. E. Y. C. Lee, E. E. Smith, and W. J. Whelan, *in* "Miami Winter Symposia" (W. J. Whelan and J. Schultz, eds.), p. 139. North-Holland Publ., Amsterdam, 1970.

76. I. D. Fleming, *in* "Starch and Its Derivatives" (J. A. Radley, ed.), p. 498. Chapman & Hall, London, 1968.

77. P. L. Jeffrey, D. H. Brown, and B. I. Brown, *Biochemistry* **9,** 1403 and 1416 (1970).

78. M. Abdullah, P. M. Taylor, and W. J. Whelan, *in* "Control of Glycogen Metabolism" (W. J. Whelan and M. P. Cameron, eds.), p. 123. Churchill, London, 1964.

TABLE IX

COMPARISON OF THE SPECIFICITIES OF DEBRANCHING ENZYMES[a]

Enzyme	Amylopectin						Glycogen						Pullulan		α-Limit dextrins
	Native		φ-Dextrin		β-Dextrin		Native		φ-Dextrin		β-Dextrin				
Aerobacter pullulanase	C	—	C	M4	C	M2, M3	0	—	I	M4	I	M2, M3	C	M_3	+
Bean pullulanase (R-enzyme)	I	—	I	M4	I	M2, M3	0	—	0	—	0	—	C	M_3	+
Yeast isoamylase	I	—	?	—	I	M2, M3	I	—	?	—	?	—	0	–	*b*
Pseudomonad isoamylase	C	—	?	—	I	M3	C	—	?	—	I	M3	0	–	+
Glucosidase-transferase[c] (muscle or yeast)	I	G	I	G	I	G	I	G	I	G	I	G	0	–	+

[a] In the case of the polysaccharides, C, I, and 0 (left-hand column) denote, respectively, complete (or nearly so), incomplete, or no debranching. In the right-hand column, the characteristic product, if any, is given, G denoting glucose; M2 maltose; M3 maltotriose; M4 maltotetraose. In the case of the α-limit dextrins the "+" sign denotes that one or more of the mixed products of α-amylolysis of amylopectin and glycogen is hydrolyzed (see Table IIA and Section VI,D).

[b] Kjølberg and Manners (*47*) state that the enzyme has no action; Kobayashi *et al.* (*45*) disagree.

[c] Unique, among the enzymes listed here, in hydrolyzing α-glucosyl Schardinger dextrin (*31, 43, 44*).

crease (Table VII). The reason for the difference is that actual chain lengthening occurs in the latter case (see Section III,F). With either type of debranching, the degree of β-amylolysis increases because 1 → 6 bonds disappear.

Before the discovery of the transferase component associated with amylo-1,6-glucosidase, it was tacitly assumed that the latter had no direct action on glycogen or amylopectin. It could only act, it was thought, on the ϕ-dextrin. This idea was revised with the discovery of transferase. The two-component system does attack the native polysaccharide though, as explained earlier, only to a limited extent. Nevertheless, it could not be assumed any longer that an increased iodine stain and degree of β-amylolysis were evidence of a direct debranching enzyme. This was most clearly emphasized when Lee *et al.* (*43*) showed that yeast contains the glucosidase-transferase system. The earlier conclusions of the presence of an isoamylase, as judged from changes in iodine stain and degree of β-amylolysis, had to be revised (*39*).

The questions that emerge are how to detect a debranching enzyme and how to distinguish between the three types—pullulanase, isoamylase, and glucosidase-transferase. It seems that the first and third can be detected by the use of specific substrates and the second by an elimination process.

Pullulanase is detected by its ability to hydrolyze pullulan, with the production of maltotriose. The discovery of this polysaccharide has enormously assisted the enzyme's detection. The reason is that the earlier, and still-used detection method of increase in iodine stain fails to distinguish the type of debranching system and is useless if α, 1 → 4-bond hydrolases are present since these lower the iodine stain. Pullulan is resistant to endo-α-amylolysis, and to exo-β-amylolysis. It is susceptible to exo-α-glucosidases, but it is only slowly attacked because of its low content of nonreducing ends. We have recently used pullulan as the test substrate when purifying two enzymes from sweet corn, which, when purified were both shown also to attack amylopectin and α-limit dextrins [(*36*), Table IIB].

Amylo-1,6-glucosidase can be detected with α-glucosyl Schardinger dextrin (*31*). It is the only enzyme known that hydrolyzes the α-glucosyl residue, while the carrier dextrin itself is very resistant to amylolytic attack. Therefore one can purify an enzyme having this activity and, when pure, characterize it more fully, including examining it for transferase activity.

Isoamylase, at least that from *Pseudomonas*, is not yet known to have a specific substrate, but it can be distinguished from the others by its failure to attack their specific substrates (Table IX) and by tests for positive reaction on other substrates.

There is one substrate on which all three types of debranching enzyme act, and which can provide the beginning step in enzyme characterization. This is amylopectin ϕ-dextrin, the A chains of which are 4 units in length (Fig. 2). Pullulanase (*48*) and isoamylase (*78a*) form maltotetraose while glucosidase-transferase yields glucose.

Table IX shows the way in which the three types of enzyme act on amylopectin, glycogen, their ϕ- and β-dextrins, pullulan, α-limit dextrins, and α-glucosyl Schardinger dextrin. It will be seen that a screening of a suspected debranching enzyme preparation with these substrates will unequivocally characterize the enzyme type. Chromatography should be used to identify sugar products such as glucose, maltose, maltotriose, and maltotetraose, in addition to increase in reducing power, or iodine stain, or increase in β-amylolysis, though the last measurement is useful in indicating whether debranching occurs and is partial or complete. Confusion may result from use of trivial increases in iodine stain of amylopectin as the sole criterion for classifying an "enzyme" as "R-enzyme." An increase in iodine stain is not even an indication of debranching. Disproportionation of glycogen chains by phosphorylase (*79*), or by transferase component of glucosidase-transferase (*66b*), for example, increases the iodine stain.

Debranching enzymic activities should be expressed in meaningful and reproducible terms, certainly not by iodine stain change alone. All debranching involves hydrolysis, and this can be measured quantitatively. Such documentation will permit an evaluation of negative results, that is, a failure of a substrate to be hydrolyzed. As Drummond *et al.* (*24a*) have pointed out, amylopectin is apparently immune to potato R-enzyme and *Aerobacter* pullulanase if the enzyme is used in too low a concentration, even though its action on other substrates is still apparent.

V. The *in Vivo* Roles of Direct and Indirect Debranching Enzymes

It may have seemed from the foregoing that there is a great difference between the direct and indirect debranching enzyme systems. When the enzymes are considered in the context of the part they play in the catabolism of glycogen and amylopectin, similarities begin to appear. With mammalian liver or muscle or with plants only one debranching system or the other is found. Yeast and *S. mitis*, which contain both systems, or

78a. J. J. Marshall, unpublished result.

79. B. Illingworth, D. H. Brown, and C. F. Cori, *Federation Proc.* **20**, 86 (1961).

A. aerogenes, which produces extracellular pullulanase in the absence of pullulanase substrate, present different problems for explanation.

The indirect debranching system, consisting of glucosidase-transferase, is evidently very well designed to cooperate with phosphorylase in the breakdown of glycogen. At any incomplete stage of breakdown there will be present α-glucose 1-phosphate and glucose as the sole low molecular weight products, along with a macromolecular, partly degraded glycogen. The final result is a quantitative conversion into these two sugars.

The direct debranching enzyme (R-enzyme), however, substantially degrades amylopectin without requiring assistance from phosphorylase. This depolymerization virtually wipes out the ability of muscle phosphorylase to attack the chains (*75*). This enzyme is adapted to attack branched molecules. Potato phosphorylase, by contrast, acts equally well on the debranched amylopectin (*75*), but each linear chain leaves a residue of maltotetraose, not further attacked by phosphorylase, and not by R-enzyme. There is, however, in potatoes (*80*) and other plants (*81*) a transglycosylase, D-enzyme, with a specificity very similar to that of the transferase component of the indirect debranching enzyme complex. Indeed, it was the prior knowledge of the existence of potato D-enzyme that led to the prediction that muscle would contain a similar transferase (*50*). D-Enzyme action on maltotetraose, along with phosphorylase, can be predicted to yield 3 moles of α-glucose 1-phosphate and one of glucose (*75*).

Lee *et al.* (*75*) have pointed out that if R-enzyme and D-enzyme are both implicated in amylopectin catabolism, along with phosphorylase, the end products will be α-glucose 1-phosphate and glucose in the molar ratio $(n - 1):1$, where n is equal to $\overline{\text{CL}}$. This is exactly the same as for glycogen breakdown by phosphorylase and the indirect debranching system. Figure 6 depicts the two catabolic processes and their basic similarity.

It has been further speculated that the association of the glucosidase and transferase activities into a single structural unit or multienzyme complex may be an evolutionary development (*75*). An analogy is the fatty acid synthesizing system, which in yeast is multienzyme and in *E. coli* consists of separate enzymes (*82*). *Escherichia coli* stores glycogen, and an examination of its debranching system may be profitable. Walker

80. S. Peat, W. J. Whelan, and W. R. Rees, *JCS* p. 44 (1956); G. Jones and W. J. Whelan, *Carbohydrate Res.* **9**, 483 (1969).

81. D. J. Manners and K. L. Rowe, *Carbohydrate Res.* **9**, 441 (1969).

82. F. Lynen, *in* "Miami Winter Symposia" (W. J. Whelan and J. Schultz, eds.), p. 151. North-Holland Publ., Amsterdam, 1970.

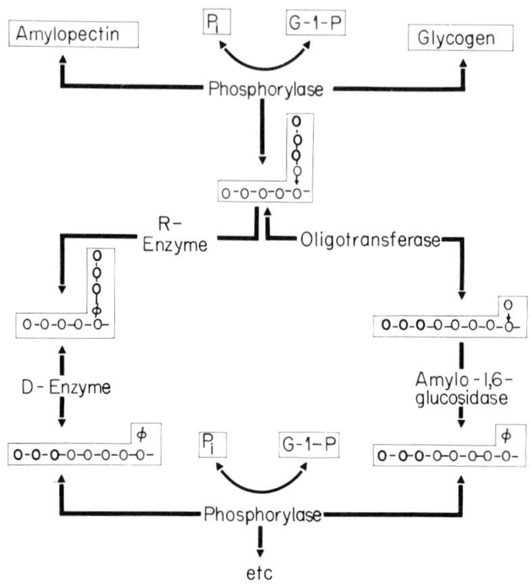

FIG. 6. The catabolism of amylopectin and glycogen as mediated by phosphorylase and debranching enzymes (*75*); direct for amylopectin (left-hand) and indirect glycogen (right-hand). Symbols as in Fig. 2. Reprinted by permission from "Miami Winter Symposia" (W. J. Whelan and J. Schultz, eds.), Vol. 1, p. 139. North-Holland Publ., Amsterdam, 1970.

and Builder (*74*) have already reported separable, though somewhat different, α-1,6-glucosidase and transferase activities in *S. mitis*.

VI. The Use of Debranching Enzymes in the Structural Determination of Glycogen and the Starch Components

The following is a summary of the wide variety of ways in which debranching enzymes have been used to explore the structures of glycogen, and the starch components, amylose and amylopectin [see also Lee *et al.* (*75*)]. It was the organic chemist who determined that these polymers are composed of α-D-glucopyranose units joined through $1 \rightarrow 4$ bonds and interlinked through $1 \rightarrow 6$ bonds. The ratios of the bond types can also be determined chemically. But nonenzymic methods have supplied little information on how the chains are arranged within the molecules or whether they are uniform or vary widely in length. It is the enzymologist who has provided such information as we now have, even though this still falls short of a definitive picture. The usual approach has been to dismember the molecules by using an exo enzyme to split the $1 \rightarrow 4$ bonds (β-amylase or phosphorylase) and/or a debranching enzyme (direct or indirect) for the $1 \rightarrow 6$ bonds, then to examine the nature of the fragments.

A. Determination of Average Chain Length

The enzymic determination of the average chain length ($\overline{CL}$) of glycogen and amylopectin requires an enzyme to split the $1 \rightarrow 6$ bonds and then a method of determining the proportion of reducing chain ends. Procedures used, presented in chronological order, are as follows:

(1) Muscle phosphorylase and glucosidase-transferase together convert glycogen and amylopectin quantitatively into α-glucose 1-phosphate and glucose in the molar proportion $(n - 1):1$, where n is equal to $\overline{CL}$. The glucose is that whose C-1 was involved in the $1 \rightarrow 6$ bond (Fig. 2) and the selective determination of the proportion of glucose gives the value of $\overline{CL}$ (*83*). The ability of this enzyme mixture to totally degrade glycogen has also been used for determination of glycogen concentration (*84*).

(2) R-Enzyme and β-amylase, the latter used in low concentration, convert amylopectin quantitatively into a mixture of maltose and maltotriose in the molar ratio $(n - 1.5):1$, where n is again equal to $\overline{CL}$ (*2*). In this case, chains with an even number of glucose units give only maltose, while those with an odd number also give 1 mole prop. of maltotriose. Separation and quantitiative determination of the two sugars permits a calculation of $\overline{CL}$.

(3) Method 2 is not applicable to glycogen because R-enzyme does not attack glycogen (*23*). Nor is the determination of the maltotriose easy. The method was modified (*85*) by substituting *Aerobacter* pullulanase for R-enzyme, permitting determination of glycogen $\overline{CL}$, and using a concentration of β-amylase sufficiently high to hydrolyze the maltotriose to maltose and glucose. The glucose is determined selectively with glucose oxidase and at the same time the sensitivity of the method is greatly increased.

(4) Glucosidase-transferase may also be used with β-amylase as the $1 \rightarrow 4$ bond hydrolase and with selective determination of glucose (*86*).

83. B. Illingworth, J. Larner, and G. T. Cori, *JBC* **199,** 631 (1952).
84. E. Bueding and J. T. Hawkins, *Anal. Biochem.* **7,** 26 (1964).
85. E. Y. C. Lee and W. J. Whelan, *ABB* **116,** 162 (1966).
86. J. H. Carter and E. Y. C. Lee, *Anal. Biochem.* **39,** 373 (1971).

This is the most practical of the published methods using glucosidase-transferase.

(5) The easiest of all methods for $\overline{CL}$ determination, because it uses a single enzyme, is the use of an isoamylase that totally debranches glycogen and amylopectin without the assistance of a 1 → 4-bond splitting enzyme. The enzyme is used to debranch glycogen and amylopectin and the copper-reducing power of the reducing chain ends set free is measured (*86a*).

B. Arrangement of the Unit Chains in Glycogen and Amylopectin

The three formulas that had been proposed by Haworth *et al.* (*87*), Meyer and Bernfeld (*88*), Staudinger and Husemann (*89*) to explain the structures of glycogen and amylopectin are depicted in Fig. 7. They may be distinguished by their molar content of A chains, *viz.*, Haworth *et al.*, ≃0; Meyer and Bernfeld, 50%; and Staudinger and Husemann, approximately 100%. The indirect (muscle) and direct (bean) debranching enzymes were used to study glycogen (*1, 2*) and amylopectin structures. In each case the strategy was to use a 1 → 4-bond splitting enzyme to convert the A chain into a "stub" that would be recognizable when split off by a debranching enzyme.

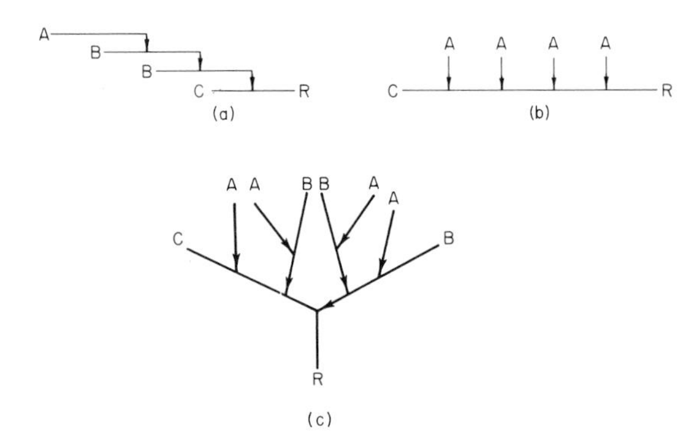

Fig. 7. Structures proposed for glycogen and amylopectin by (a) Haworth *et al.* (*87*), (b) Staudinger and Husemann (*89*), and (c) Meyer and Bernfeld (*88*). R indicates reducing end group. For definition of A, B, and C see text.

86a. Z. Gunja-Smith, J. J. Marshall, and E. E. Smith, *FEBS Letters* **13,** 309 (1971).
87. W. N. Haworth, E. L. Hirst, and F. A. Isherwood, *JCS* p. 577 (1937).
88. K. H. Meyer and P. Bernfeld, *Helv. Chim. Acta* **23,** 865 (1940).
89. H. Staudinger and G. Husemann, *Ann. Chem.* **527,** 195 (1937).

For glycogen and amylopectin, Larner *et al.* (*1*) used muscle phosphorylase to prepare the limit dextrin (ϕ-dextrin) which, at that time, before the discovery of the transferase component of the indirect debranching system, was supposed to contain A chains one glucose unit in length (Fig. 2). The ϕ-dextrin yielded glucose on treatment with muscle "amylo-1,6-glucosidase" and became susceptible to further phosphorolysis to produce a new ϕ-dextrin. This in turn yielded more glucose with "amylo-1,6-glucosidase," and so on. The results were consistent with a Meyer and Bernfeld molecule, and in terms of the amount of the glucose released at each stage and the fact that successive "tiers" of branch points were encountered the Haworth *et al.* and Staudinger and Husemann structures were discounted. This proof of the Meyer and Bernfeld structure, which is now firmly entrenched in the literature, we believe to be invalid. There are two interconnected reasons. First, reference to Section III shows that the assumed structure of the ϕ-dextrin was incorrect. Second, it was assumed that all the A chains, and only the A chains, would be removed on treatment with the debranching enzyme. This, we now know, is not so. Some A chains grow in length, because of transferase action, rather than disappear, while some B chains become A chains when the glucose stub (A chain) is removed, and themselves disappear. An inspection of the Haworth *et al.* and Staudinger and Husemann structures (Fig. 7), coupled with the newer knowledge of the nature of the indirect debranching system (Section III), will show that these structures could also give results similar to those given by a Meyer and Bernfeld structure.

Amylopectin structure was examined by making the β-amylase limit dextrin and treating it with R-enzyme (*2*). There were found maltose and maltotriose in roughly equimolar proportions, representing residues of even and odd length A chains, and in an amount near to that expected for a regularly rebranched (50% A chains) Meyer and Bernfeld structure. There was no proof that all A chains had been removed by R-enzyme, but the Haworth *et al.* and Staudinger and Husemann structures could nevertheless be excluded.

The release of maltose and maltotriose from β-dextrins by *Aerobacter* pullulanase, and their measurement, has been used to calculate the proportion of A chains (*90*). This method depends on the assumption, for which no proof was given, that all the A chains had been hydrolyzed by the pullulanase.

The structures of amylopectin (*91*) and glycogen (*73*) have subsequently been explored by debranching the polysaccharides with

90. G. N. Bathgate and D. J. Manners, *BJ* **101,** 3C (1966).
91. E. Y. C. Lee, C. Mercier, and W. J. Whelan, *ABB* **135,** 1028 (1968).

Aerobacter pullulanase and fractionating the unit chains on Sephadex G-50. Glycogen (human and rabbit liver) being immune to pullulanase, it is necessary to debranch the β-dextrin. This debranching is incomplete, and successive treatments with β-amylase and pullulanase are necessary for eventual complete degradation. The conclusion is emerging from these results that neither polysaccharide conforms in detail to the regularly rebranched Meyer and Bernfeld structure (*92*). Indeed, the evidence for the Meyer-Bernfeld model is only the near parity of A and B chains. This model simply happened to be the only one of the three proposed models (Fig. 7) that fitted the experimental findings. As French (*92a*) has subsequently pointed out, there are many other alternatives that also conform to an A:B ratio of unity.

A revised formula for both glycogen and amylopectin has been recently advanced by Gunja-Smith *et al.* (*39a*), which conforms to the 1:1 A:B ratio, but which differs from the Meyer-Bernfeld model in the following respects. In the regularly rebranched Meyer-Bernfeld model (Fig. 7c) each B and C chain carries one A chain. In the new model (Fig. 8) half the B chains carry two A chains, while the other half, plus the C chain, are substituted only by B chains. This model was arrived at through the use of *Cytophaga* isoamylase (*38*), taking advantage of the inability of the enzyme to cleave 1 → 6-links involving α-maltosyl residues (Section II,C). All the A chains of glycogen and amylopectin were converted into maltosyl residues by the successive actions of phosphorylase and β-amylase, to give the ϕ-β-dextrins. Debranching of this dextrin by isoamylase would cleave only B to B branch points. A Meyer-Bernfeld

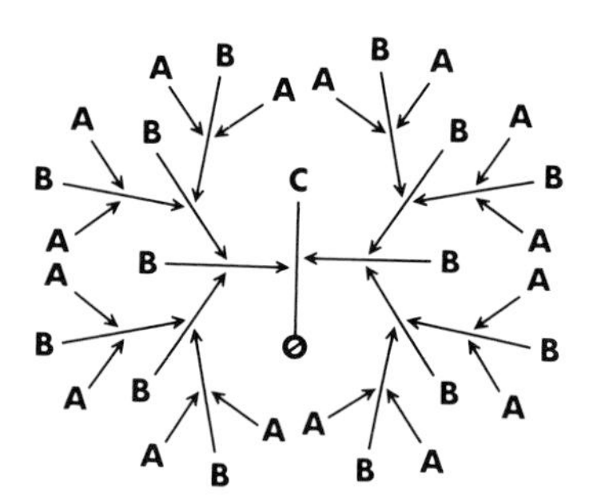

FIG. 8. Structure proposed for glycogen and amylopectin by Gunja-Smith *et al.* (*39a*). For definition of A, B and C chains see the text; ϕ = reducing chain end. Reprinted by permission of North-Holland Publ., Amsterdam.

92. C. Mercier and W. J. Whelan, unpublished results.

92a. D. French, *in* "Control of Glycogen Metabolism" (W. J. Whelan, ed.), p. 7. Academic Press, New York, 1968.

molecule would remain resistant to β-amylase even after such debranching. It was found, however, that with shellfish glycogen ϕ-β-dextrin the subsequent degree of β-amylolysis was 44%, and of waxy maize amylopectin ϕ-β-dextrin, 29%. It was on the basis of these results that the concept in Fig. 8 was proposed. Inspection of this model will show that approaching 50% of the molecule would be available to β-amylase after debranching of the ϕ-β-dextrin.

Yeast isoamylase (*47*) and *Aerobacter* pullulanase (*93*) have been used to prove that the anomalous (non-1 $\rightarrow$ 4) bonds known to be in amylose are α-1 $\rightarrow$ 6-glucosidic bonds.

C. Determination of Enzymic Action Pattern

R-Enzyme was used to prove that the A chains in amylopectin β-dextrin are 2 and 3 glucose units in length (*2*) and in the ϕ-dextrin are 4 units long (*50*). *Aerobacter* pullulanase was used to prove the same point for glycogen ϕ-dextrin (*78*).

The structures of the oligosaccharide limit dextrins formed from amylopectin and glycogen by various α-amylases were determined in part by hydrolyzing them with R-enzyme (*94*). This method has been refined by French and his co-workers (*5*), who have used a two-dimensional paper chromatographic method, the oligosaccharides being debranched on the chromatogram.

Verhue and Hers (*95*) and Brown and Brown (*96*), respectively, studied liver and muscle branching enzymes by using pullulanase to split off A chains formed by the branching enzyme. These, on examination, revealed that the specificity of both enzymes is directed toward creating A chains 7 glucose units in length.

Pullulanase was used to determine the unit-chain profiles of amylopectins synthesized by potato branching enzyme (Q-enzyme) from amylose or, with phosphorylase, from α-glucose 1-phosphate (*24, 25*). The unit chains were fractionated on Sephadex G-50 and the profiles compared with that from natural amylopectin. Though the synthetic polysaccharides seemed to be identical with natural amylopectin, having the same $\overline{CL}$, iodine stain, and degree of β-amylolysis, their profiles were very different. It was concluded that neither route of synthesis was equivalent to that occurring *in vitro*.

93. W. Banks and C. T. Greenwood, *ABB* **117,** 674 (1966).
94. W. J. Whelan, *Starke* **12,** 358 (1960).
95. W. Verhue and H. G. Hers, *BJ* **99,** 222 (1966).
96. D. H. Brown and B. I. Brown, *BBA* **130,** 263 (1966).

Another polysaccharide superficially identical to plant amylopectin is the "glycogen" found in the liver of a patient suffering from type IV glycogenosis (absence of branching enzyme). The unit-chain profile, however, is very different from that of the plant polysaccharide (*73*) and is in line with the explanation offered above (Section III,G) that in this glycogen the branch points are synthesized by the reversion reaction of the debranching enzyme system (*58*).

Acknowledgments

This review was written during the support period of grants to the authors from the National Institutes of Health (AM 12532) and the National Science Foundation (GB 8342). The authors wish to thank their colleague, Dr. J. J. Marshall, for helpful comments on the manuscript.

8

Bacterial and Mold Amylases

TOSHIO TAKAGI • HIROKO TODA • TOSHIZO ISEMURA

I. Introduction

Culture media of bacteria and molds are rich sources of amylases, which are used in large quantities for industrial purposes. For example, starch industries depend on the use of amylases. Therefore, the catalytic activities of amylases have been studied not only for their academic interest but also for their application to industry. Bacterial and mold amylases have been the objects of organic chemical and physicochemical studies because they are available in large quantities and can be easily purified and crystallized. Thus the properties of amylases have been extensively studied. Regrettably, none of the enzymes has been analyzed by X-ray diffraction, and therefore information on structure-activity relationships is limited.

II. Molecular Properties

A. Purification and State of Purity

A large number of bacterial and mold amylases have been isolated in crystalline form from various sources. Meyer *et al.* (*1*) first accomplished the crystallization of an amylase, *Bacillus subtilis* α-amylase. Since then the isolation and crystallization of the following α-amylases have been reported: *B. subtilis* (*2–6*), *Aspergillus oryzae* (*7–12*), *B. coagulans* (*13*), *A. candidus* (*14*), *Pseudomonas saccharophila* (*15*), *B. polymyxa* (*16*), *B. macerans* (*17*), *A. niger* (*18, 19*), *B. amyloliquefaciens* (*20*), and *B. stearothermophilus* (*21–23*). Because *B. subtilis* and *A. oryzae* α-amylases are the most common, purification procedures for them will

1. K. H. Meyer, M. Fuld, and P. Bernfeld, *Experientia* **3**, 411 (1947).
2. B. Hagihara, *Proc. Japan Acad.* **27**, 346 (1951).
3. T. Yamamoto, *Bull. Agr. Chem. Soc. Japan* **19**, 121 (1955).
4. J. Felling, E. A. Stein, and E. H. Fischer, *Helv. Chim. Acta* **40**, 529 (1957).
5. J. Fukumoto, Y. Yamamoto, and T. Ichikawa, *Proc. Japan Acad.* **27**, 352 (1951).
6. J. Fukumoto and S. Okada, *J. Ferment. Technol.* **41**, 427 (1963).
7. D. K. Roy, *Sci. Cult. (Calcutta)* **16**, 77 (1950).
8. L. A. Underkofler and D. K. Roy, *Cereal Chem.* **28**, 18 (1951).
9. E. H. Fischer and R. de Montomollin, *Nature* **160**, 606 (1951).
10. E. H. Fischer and R. de Montomollin, *Helv. Chim. Acta* **34**, 1987 (1951).
11. S. Akabori, B. Hagihara, and T. Ikenaka, *Proc. Japan Acad.* **27**, 350 (1951).
12. S. Akabori, B. Hagihara, and T. Ikenaka, *J. Biochem. (Tokyo)* **41**, 577 (1954).
13. L. L. Campbell, *JACS* **76**, 5256 (1954).
14. K. Takaoka, H. Fuwa, and Z. Nikuni, *Mem. Inst. Sci. Ind. Res., Osaka Univ.* **10**, 199 (1952).
15. A. Markovitz, H. P. Klein, and E. H. Fischer, *BBA* **19**, 267 (1956).
16. J. Robyt and D. French, *ABB* **104**, 338 (1964).
17. J. A. DePinto and L. L. Campbell, *Biochemistry* **7**, 114 (1968).
18. Y. Minoda and K. Yamada, *Agr. Biol. Chem. (Tokyo)* **27**, 806 (1963).
19. Y. Minoda, M. Arai, Y. Torigoe, and K. Yamada, *Agr. Biol. Chem. (Tokyo)* **32**, 110 (1968).
20. N. E. Welker and L. L. Campbell, *Biochemistry* **6**, 3681 (1967).
21. G. B. Manning and L. L. Campbell, *JBC* **236**, 2952 (1961).
22. S. L. Pfueller and W. H. Elliott, *JBC* **244**, 48 (1969).
23. K. Ogasahara, A. Imanishi, and T. Isemura, *J. Biochem. (Tokyo)* **67**, 65 (1970).

be described below. The *B. subtilis* α-amylase (*24*) was extracted from a commercial concentrate in powder form, and the *A. oryzae* α-amylase was obtained from commercial sources—Clarase from the Miles Chemical Co., Elkhart, Indiana, and Taka-diastase from the Sankyo Co., Tokyo, Japan. The amylase prepared from the latter is called Taka-amylase A.

Improved methods for purification and crystallization of several amylases have been published (*25*). The purification procedure of *B. subtilis* α-amylase involves starch adsorption, decoloration of dark pigment with an ion exchange resin, ammonium sulfate fractionation, and crystallization from an acetone precipitation. Purification should be carried out in the presense of diisopropylphosphofluoridate (DFP) in order to protect amylases from the action of proteases present in the bacterial extract (*26*). The procedures for crystallization of *B. subtilis* α-amylase are summarized in Table I.

TABLE I
PURIFICATION OF *B. subtilis* α-AMYLASE[a]

Culture filtrate
Adsorption to starch [0.3 saturated $(NH_4)_2SO_4$]
Enzyme adsorbed on starch
Eluted with *M*/30 Na_2HPO_4
Dialyzed against running water
Dialyzed solution
Decolored by Duolite A2 resin
Decolored solution
Salting out with $(NH_4)_2SO_4$ (0.7 saturated)
Dissolved in 0.01 *M* $Ca(OAc)_2$
Dialyzed against *M*/500 $Ca(OAc)_2$
Dialyzed solution
Addition of acetone (0°C, 60%)
Precipitate
Dissolved in 0.01 *M* $Ca(OAc)_2$, pH 10
pH adjusted to 6.0
Stored in a refrigerator
Crystal

[a] Data from Fukumoto and Okada (*6*).

Taka-amylase A has been crystallized from Taka-diastase (*11, 12, 14*). Akabori *et al.* (*11, 12*) reported an improved method involving ammonium sulfate fractionation, precipitation with Rivanol (2-ethoxy-6,9-diaminoacridinium lactate), and crystallization from aqueous acetone. In

24. *Bacillus subtilis* produces liquefying and saccharifying α-amylases. In this chapter, the former will be called *B. subtilis* α-amylase.
25. E. A. Stein and E. H. Fischer, *Biochem. Prep.* **8,** 27 (1961).
26. E. A. Stein and E. H. Fischer, *JBC* **232,** 867 (1958).

TABLE II
PURIFICATION OF TAKA-AMYLASE A FROM TAKA-DIASTASE[a]

Taka-Diastase (30 g)
Extracted with water (150 ml)
Addition of 0.25 *M* $Ca(OAc)_2$ (150 ml)
Filtered
Filtrate
Dilution with equal volume of water
Salting out with $(NH_4)_2SO_4$ (3/4 saturated)
Centrifuged
Precipitate
Dissolved in water
Dialyzed against running water
Dialyzed solution
Addition of 1% Rivanol solution (0.04 vol)
Filtered
Filtrate
Addition of 1% Rivanol (0.06 vol)
Precipitate
Dissolved in 0.5 *M* acetate buffer (pH 6.0)
Addition of acidic clay (10 g), filtered
Filtrate
Addition of acetone up to 60%
Centrifuged
Precipitate
Dissolved in 0.02 *M* $Ca(OAc)_2$
Addition of cold acetone to slight turbidity
Stored in refrigerator
Crystalline Taka-amylase A (ca. 0.8 g)

[a] Data from Akabori *et al.* (*11, 12*).

the process of purification, dialysis of the ammonium sulfate fraction must be carried out using a collodion bag instead of a cellulose tube, because of the presence of strong cellulase activity in the fraction. The procedures are summarized in Table II. Recently, Toda *et al.* established a more convenient and rapid method of crystallization, using DEAE-cellulose chromatography (*27*). Proteases contaminating the enzyme preparation can be removed by chromatography.

Crystallization of amylases require the presence of divalent cations, especially calcium ions. When crystallization is repeated, it becomes increasingly difficult because loss of the cations results in apparent high solubility of the enzymes (*28*).

27. H. Toda and S. Akabori, *J. Biochem.* (*Tokyo*) **53,** 102 (1963).
28. B. L. Vallee, E. A. Stein, W. N. Sumerwell, and E. H. Fischer, *JBC* **234,** 2901 (1959).

Most of the crystalline amylases have been shown to be homogeneous in sedimentation and electrophoretic analysis. It was found, however, that even after repeated crystallization, crystalline Taka-amylase A is still contaminated by traces of proteases which can only be removed by chromatography (*27*). Recently, polyacrylamide gel electrophoresis and analytical ion exchange chromatography have been used to examine the homogeneity of an amylase preparation. It has been shown by polyacrylamide gel electrophoresis that crystalline amylases from *B. macerans* (*17*), *B. amyloliquefaciens* (*20*), *B. stearothermophilus* (*22*), and *A. oryzae* (*29*) are homogeneous.

B. Amino Acid Composition

The amino acid composition of several α-amylases obtained from mold and bacterial origins are listed in Tables III and IV (*17, 22, 30–37*). These α-amylases differ from one another in their action patterns and physicochemical properties, although they catalyze the hydrolysis of $\alpha, 1 \rightarrow 4$-glucosidic linkage in polysaccharides. There are no striking similarities in their amino acid composition. *Aspergillus oryzae* α-amylase (Taka-amylase A) is poor in basic amino acid residues, which might contribute to its acidic character. But *B. subtilis* liquefying α-amylase contains neither cysteine nor cystine residue. *Bacillus subtilis* saccharifying α-amylase was found, however, to contain a sole cysteine residue whose sulfhydryl group was in a masked state (*36*). *Bacillus stearothermophilus* α-amylase was reported to contain a large number of proline residues, to which thermostability was attributed (*37*). An α-amylase obtained from an intimately related strain of *B. stearothermophilus*, however, showed no such high content of proline residues (*22*). *Aspergillus niger* produces two kinds of α-amylases, acid-stable and acid-un-

29. J. F. McKelvy and Y. C. Lee, *ABB* **132,** 99 (1969).
30. E. A. Stein, J. M. Junge, and E. H. Fischer, *JBC* **235,** 371 (1960).
31. K. Narita, H. Murakami, and T. Ikenaka, *J. Biochem.* (*Tokyo*) **59,** 170 (1966).
32. S. Akabori, T. Ikenaka, H. Hanafusa, and Y. Okada, *J. Biochem.* (*Tokyo*) **41,** 803 (1954).
33. Y. Minoda, M. Arai, and K. Yamada, *Agr. Biol. Chem.* (*Tokyo*) **33,** 572 (1969).
34. S. Akabori, Y. Okada, S. Fujiwara, and K. Sugae, *J. Biochem.* (*Tokyo*) **43,** 741 (1956).
35. J. M. Junge, E. A. Stein, H. Neurath, and E. H. Fischer, *JBC* **234,** 556 (1959).
36. H. Toda and K. Narita, *J. Biochem.* (*Tokyo*) **63,** 302 (1968).
37. L. L. Campbell and G. B. Manning, *JBC* **236,** 2962 (1961).

TABLE III

Amino acid	*A. oryzae* amylase [Ref. (*30*)]	Taka-amylase [Ref. (*31*)]	[Ref. (*32*)]	*A. niger* acid-stable amylase [Ref. (*33*)]
Aspartic acid	15.91	17.18	16.53	13.94
Threonine	8.38	8.90	10.86	6.84
Serine	6.04	7.20	6.48	8.59
Glutamic acid	8.09	8.60	6.95	7.62
Proline	4.22	4.71	4.18	3.21
Glycine	5.68	5.93	6.59	4.81
Alanine	5.92	6.28	6.80	4.66
Valine	6.03	6.46	4.69	5.60
Methionine	2.14	2.46	2.20	1.88
Isoleucine	6.34	7.17	5.20	5.52
Leucine	7.64	7.93	8.30	7.85
Tyrosine	10.88	11.55	9.55	9.71
Phenylalanine	3.99	4.36	4.25	3.48
Lysine	4.77	5.27	5.94	2.24
Histidine	1.82	2.22	2.02	1.65
Arginine	2.97	3.21	2.72	3.09
Tryptophan	3.78		3.97	3.88
Half-cystine	2.09		1.6	0.96
Cysteine				
Ammonia	1.68		1.50	1.42

stable enzymes (*18, 38, 39*). The amino acid composition of the acid-unstable enzyme is similar to that of *A. oryzae* α-amylase, and the enzyme has more basic amino acid residues than the acid-stable enzyme. Especially, the content of lysine residue of the acid-unstable enzyme is approximately twice that of the acid-stable enzyme. The acid-stable enzyme is rich in free carboxyl groups, and therefore shows a low isoelectric point of 3.44. The serine and proline contents of the acid-stable enzyme are also lower than those of the acid-unstable enzyme. The *B. macerans* amylase is characterized by its action of producing Schardinger dextrin from starch (*17*). The amino acid composition of this enzyme, however, is similar to those of other well-known amylases except it has a higher content of serine than other amylases.

38. Y. Minoda and I. Tsukamoto, *J. Agr. Chem. Soc. Japan* **35**, 482 (1961).
39. Y. Minoda, T. Koyano, M. Arai, and K. Yamada, *Agr. Biol. Chem.* (*Tokyo*) **32**, 104 (1968).

Amino Acid Composition of Bacterial and Mold α-Amylases
(grams of amino acid per 100 g of protein)

A. niger acid-unstable amylase [Ref. (*33*)]	*B. subtilis* N amylase [Ref. (*34*)]	*B. subtilis* amylase [Ref. (*35*)]	*B. subtilis* saccharifying amylase [Ref. (*36*)]	*B. stearothermophilus*, 503–4, amylase [Ref. (*37*)]	*B. macerans* amylase [Ref. (*17*)]
15.33	15.09	14.49	16.44	9.30	17.19
8.15	6.36	5.59	6.24	6.20	10.37
6.82	6.24	5.21	8.46	4.24	6.64
8.14	13.46	12.94	11.02	20.36	8.51
4.81	4.14	3.37	2.60	16.27	3.82
5.50	5.64	6.01	4.82	4.24	7.64
6.04	6.02	5.29	6.11	4.77	7.40
5.74	5.55	6.14	4.61	8.53	7.71
2.37	1.26	1.47	1.80	2.66	2.46
5.98	3.97	4.55	6.15	6.05	6.61
7.54	6.42	6.10	6.09	7.22	6.88
10.11	8.31	9.05	5.84	2.96	7.26
4.09	5.85	6.01	4.36	6.40	7.23
5.13	7.30	7.42	4.87	5.22	5.48
1.95	3.80	3.90	3.69	4.33	2.22
3.44	6.78	6.09	5.29	3.36	4.25
3.88	6.19	6.22		3.42	3.19
0.71	0	0	0.32		0
1.25	1.32	1.74	1.98	0.36	

C. Terminal Groups

By Sanger's DNFB (2,4-dinitrofluorobenzene) method, Taka-amylase A has been shown to have a single alanyl residue at the amino terminal of the molecule (*40*). An α-DNP-substituted peptide was isolated from the partial acid hydrolyzate of the DNP-amylase, and its amino acid sequence determined (*41, 42*). Three other DNP-peptides were also isolated, and the N-terminal sequence was determined as follows (*43*):

H_2N–Ala–Gly–Asp–Gln–Ser–Ala–Leu–Thr–

The C-terminal group of Taka-amylase A was determined by the

40. S. Akabori and T. Ikenaka, *J. Biochem.* (*Tokyo*) **42,** 603 (1955).
41. K. Narita and S. Akabori, *J. Biochem.* (*Tokyo*) **46,** 91 (1959).
42. A. Tsugita, *J. Biochem.* (*Tokyo*) **46,** 583 (1959).
43. A. Tsugita, T. Ikenaka, and S. Akabori, *J. Biochem.* (*Tokyo*) **46,** 475 (1959).

TABLE IV

Amino acid	*A. oryzae* amylase (51,750)[a] [Ref. (*30*)]		Taka-amylase A (54,000 ± 700)[a] [Ref. (*32*)]		Taka-amylase A (51,000)[a] [Ref. (*31*)]		*A. niger* acid-stable amylase (58,000)[a] [Ref. (*33*)]		*A. niger* acid-unstable amylase (61,000)[a] [Ref. (*33*)]	
Aspartic acid	61.86	(62)	67.4	(67)	65.7	(66)	60.7	(61)	70.3	(70)
Threonine	36.52	(37)	45.9	(46)	38.1	(38)	33.4	(33)	41.1	(41)
Serine	29.75	(30)	33.1	(33)	35.0	(35)	47.5	(48)	38.7	(39)
Glutamic acid	28.45	(28)	25.4	(25)	29.8	(30)	30.6	(31)	33.6	(34)
Proline	18.98	(19)	17.8	(18)	20.9	(21)	16.2	(16)	25.6	(26)
Glycine	39.18	(39)	47.4	(47)	40.4	(40)	37.4	(37)	44.7	(45)
Alanine	34.40	(34)	41.2	(41)	36.0	(36)	30.4	(30)	41.4	(41)
Valine	26.62	(27)	21.6	(22)	28.1	(28)	28.1	(28)	30.2	(30)
Methionine	7.43	(7)	8.0	(8)	8.4	(8)	7.3	(7)	9.7	(10)
Isoleucine	25.01	(25)	21.4	(21)	27.9	(28)	24.4	(24)	27.9	(28)
Leucine	30.14	(30)	34.2	(34)	30.8	(31)	34.7	(35)	35.3	(35)
Tyrosine	31.08	(31)	28.4	(28)	32.6	(33)	31.1	(31)	34.1	(34)
Phenylalanine	12.50	(13)	13.9	(14)	13.5	(14)	12.2	(12)	15.2	(15)
Lysine	16.89	(17)	22.0	(22)	18.4	(18)	8.8	(8)	21.6	(22)
Histidine	6.09	(6)	7.0	(7)	7.3	(7)	6.2	(6)	7.7	(8)
Arginine	8.83	(9)	8.5	(9)	9.4	(9)	10.4	(10)	12.1	(12)
Tryptophan	9.58	(10)	10.4	(10)	10.9	(11)	11.4	(11)	11.6	(12)
Half-cystine	8.99	(9)	3.6	(4)		(8)	5.2	(5)	3.6	(4)
Cysteine						(1)				
Total		(433)		(456)		(462)		(434)		(506)

[a] The numbers in these parentheses indicate molecular weights.

hydrazinolysis method (*44, 45*) and found to have three terminal groups, namely, serine, alanine, and glycine (*46*). However, recent reinvestigation by CPase digestion and hydrazinolysis techniques affirmed that serine was the sole C-terminal group (*47*). Therefore, Taka-amylase A has a single polypeptide chain with an alanyl residue at its N-terminal and a seryl residue at its C-terminal.

The N-terminal of *B. subtilis* α-amylase was determined to be valine by the DNFB method (*48*). An α-DNP-peptide was isolated from the

44. S. Akabori, K. Ohno, and K. Narita, *Bull. Chem. Soc. Japan* **25**, 214 (1952).
45. S. Akabori, K. Ohno, T. Ikenaka, Y. Okada, H. Hanafusa, I. Haruna, A. Tsugita, K. Sugae, and T. Matsushima, *Bull. Chem. Soc. Japan* **29**, 507 (1956).
46. T. Ikenaka, *J. Biochem.* (*Tokyo*) **43**, 255 (1956).
47. K. Narita, H. Murakami, and T. Ikenaka, *J. Biochem.* (*Tokyo*) **59**, 170 (1966).
48. K. Sugae, *J. Biochem.* (*Tokyo*) **47**, 170 (1960).

Amino Acid Composition of Bacterial and Mold α-Amylases
(moles of amino acid residues per mole protein)

B. subtilis N amylase (48,700)[a] [Ref. (*34*)]	*B. subtilis* amylase (48,675)[a] [Ref. (*35*)]	*B. subtilis* saccharifying amylase (41,000)[a] [Ref. (*36*)]	*B. stearothermophilus*, 503–4, amylase (15,600)[a] [Ref. (*37*)]	*B. stearothermophilus*, 1503–4, amylase (52,700)[a] [Ref. (*22*)]	*B. macerans* amylase (139,300)[a] [Ref. (*38*)]
55.13 (55)	53.0 (53)	50.68 (51)	10.90 (11)	61	180.0 (180)
26.01 (26)	22.82 (23)	21.46 (21)	8.11 (8)	37	121.2 (121)
28.98 (29)	24.12 (24)	32.98 (33)	6.28 (6)	30	88.0 (88)
44.61 (45)	42.81 (43)	30.71 (31)	21.60 (22)	35	80.6 (81)
17.53 (18)	14.26 (14)	9.25 (9)	22.03 (22)	16	46.2 (46)
36.62 (37)	38.92 (39)	26.31 (26)	8.80 (9)	47	141.7 (142)
32.92 (33)	28.85 (29)	28.10 (28)	8.34 (8)	35	115.7 (116)
23.08 (23)	25.48 (25)	16.14 (16)	11.35 (11)	29	91.6 (92)
4.14 (4)	4.81 (5)	4.94 (5)	2.77 (3)	8	22.9 (23)
14.76 (15)	16.89 (17)	19.21 (19)	7.20 (7)	15	65.4 (65)
23.86 (24)	22.66 (23)	19.03 (19)	8.58 (9)	31	73.0 (73)
22.30 (22)	24.31 (24)	13.20 (13)	2.54 (3)	25	55.8 (56)
17.24 (17)	17.69 (18)	10.82 (11)	6.04 (6)	21	61.0 (61)
24.30 (24)	24.68 (25)	13.67 (14)	5.56 (6)	28	52.2 (52)
11.93 (12)	12.24 (12)	9.75 (10)	4.34 (4)	9	19.9 (20)
18.94 (19)	17.01 (17)	12.47 (12)	3.00 (3)	16	34.0 (34)
14.80 (15)	14.84 (15)	11.40 (11)		22	21.8 (22)
			4.34 (4)		0
		0.74 (1)			0
(418)	(406)	(331)	(145)	(465)	

pronase digestion of DNP-amylase, and its amino acid sequence established as follows (*49*):

$$H_2N\text{–Val–Asx–Gly–Glx–Ser–}$$

The C-terminal group of *B. subtilis* α-amylase was determined to be lysine by the ^{3}H-label technique (*50*) [for this technique, see Matsuo *et al.* (*51*)]. *Bacillus stearothermophilus* α-amylase has been reported to contain two phenylalanines as N-terminals and two alanines as C-terminals by one group of investigators (*52*), and to contain one lysine and one leucine as N-terminals by another group (*22*). The N-terminal of *A. niger*

49. K. Sugae and Y. Honda, *J. Biochem.* (*Tokyo*) **47,** 307 (1960).
50. H. Kojima and K. Sugae, *J. Biochem.* (*Tokyo*) **64,** 713 (1968).
51. H. Matsuo, Y. Fujimoto, and T. Tatsuno, *Tetrahedron Letters* No. 39, 3465 (1965).
52. L. L. Campbell and P. D. Cleaveland, *JBC* **236,** 2966 (1961).

acid-stable α-amylase was reported to be leucine or isoleucine, and that of the acid-unstable α-amylase to be alanine (*53*).

D. Sulfhydryl and Disulfide Groups

Bacillus subtilis α-amylase has neither a sulfhydryl nor disulfide group (*34, 35*). Taka-amylase A, which has both, has therefore been used extensively to study sulfhydryl and disulfide groups. The half-cystine content of Taka-amylase A was first reported to be eight per molecule by amino acid analysis (*32*). This apparent even number of half cystines and the masked state of the sulfhydryl group delayed the detection of the sulfhydryl group in the enzyme. Nine half-cystines per molecule were later proposed for the enzyme (*30*). One sulfhydryl group per molecule before reduction and nine sulfhydryl groups per molecule after reduction were found to be titratable by *p*-mercuribenzoate in 8 *M* urea (*54*). Seon *et al.* carried out a thorough study on the reduction of Taka-amylase A by sodium borohydride or by mercaptoethanol (*55*). They also titrated nine sulfhydryl groups per the fully reduced Taka-amylase A molecule, and found the sole sulfhydryl group was unmasked by ethylenediaminetetraacetate (EDTA) as well as by 8 *M* urea. They pointed out that the reduction by sodium borohydride was unfavorable because it was accompanied by reductive cleavage of peptide bonds and the further reduction of the sulfhydryl group to hydrogen sulfide and alanine. When Taka-amylase A was reduced with sodium borohydride in an aqueous solution at 30°, Seon *et al.* found that the Taka-amylase A molecule first exposed two sulfhydryl groups and then another, stepwise (*56*). They concluded that the reduction of one particular disulfide group, probably located on the molecular surface, induced slow exposure of the masked sulfhydryl group. Such a stepwise appearance of the sulfhydryl groups could not be observed when reduction was carried out at a higher temperature or in the presence of denaturants; in such cases, nine sulfhydryl groups appeared simultaneously.

Kato *et al.* (*57*) and Toda *et al.* (*58*) showed that the sole sulfhydryl group in Taka-amylase A participates delicately in the enzymic activity.

53. Y. Minoda, M. Arai, and K. Yamada, *Agr. Biol. Chem.* (*Tokyo*) **33,** 5721 (1969).
54. T. Isemura, T. Takagi, Y. Maeda, and K. Yutani, *J. Biochem.* (*Tokyo*) **53,** 155 (1963).
55. B. K. Seon, H. Toda, and K. Narita, *J. Biochem.* (*Tokyo*) **58,** 348 (1965).
56. B. K. Seon, *J. Biochem.* (*Tokyo*) **61,** 606 (1967).
57. I. Kato, H. Toda, and K. Narita, *Proc. Japan Acad.* **43,** 38 (1967).
58. H. Toda, I. Kato, and K. Narita, *J. Biochem.* (*Tokyo*) **63,** 295 (1968).

When the amylase was incubated with iodoacetate or iodoacetamide at pH 8.0 at 50° in the presence of EDTA, the sulfhydryl group was alkylated, and the enzymic activity was completely lost. Interestingly, the activity could be recovered with *S*-carboxymethyl Taka-amylase A up to 15 %, but not with *S*-carboxyamidomethyl Taka-amylase A by the addition of calcium. They suggested that the sulfhydryl group was not the active site of the enzyme but likely to play a key role in maintaining the active configuration by chelating with the essential calcium atom (see Section II,F). They assumed that the carboxylate group in *S*-carboxymethylated Taka-amylase A can replace the sulfhydryl groups as a chelating site for calcium.

p-Mercuribenzoate bound to the enzyme could be removed by the addition of calcium, resulting in a regeneration of amylase activity and the sulfhydryl group. Therefore, it was suggested that *p*-mercuribenzoate and the calcium ion compete with each other for the sole sulfhydryl group in Taka-amylase A. Narita and Akao (*59*) determined the amino acid sequence around the sole cysteine residue of Taka-amylase A. A similar situation was presumed with the sole cysteine residue of *B. subtilis* saccharifying α-amylase (*36*), and a peptide containing the cysteinyl residue was isolated from the peptic digest of the amylase (*60*).

E. Carbohydrate Components

Taka-amylase A has been found to contain carbohydrates amounting to a small percent of its weight (*12, 61*). The carbohydrate component was found to consist of 8 moles of mannose, 1 mole of xylose, and 2 moles of hexosamine per mole of the enzyme (*62*). After proteolytic digestion of the enzyme, two closely related glycopeptides were isolated and their structure analyzed (*63*). The carbohydrate moiety was suggested to be linked to the enzyme through a hydroxyl group of serine residue. Hexosamine, however, was not found in the peptides isolated. In recent years, protein–carbohydrate linkages in several glycoproteins have been reported (*64–67*). Carbohydrate prothetic groups of almost all of the known glycoproteins contain hexosamine, and it has been

59. K. Narita and M. Akao, *J. Biochem.* (*Tokyo*) **58,** 348 (1965).
60. H. Toda and K. Narita, *J. Biochem.* (*Tokyo*) **63,** 302 (1968).
61. V. M. Hanrahan and M. L. Caldwell, *JACS* **75,** 4030 (1953).
62. H. Hanafusa, T. Ikenaka, and S. Akabori, *J. Biochem.* (*Tokyo*) **42,** 55 (1955).
63. A. Tsugita and S. Akabori, *J. Biochem.* (*Tokyo*) **46,** 695 (1959).
64. R. G. Johansen, R. D. Marshall, and A. Neuberger, *Biochem. J.* **78,** 518 (1961).
65. J. W. Rosevear and E. L. Smith, *JBC* **236,** 425 (1961).
66. S. Kamiyama and K. Schmid, *BBA* **63,** 266 (1962).
67. T. H. Plummer, Jr. and C. H. W. Hirs, *JBC* **239,** 2530 (1964).

proved that hexosamine links carbohydrate and protein moieties through an amide group of asparagine residue.

Reinvestigation was carried out, therefore, on the structure of the carbohydrate moiety of Taka-amylase A. The glycopeptide was isolated from Pronase digest of Taka-amylase A (*68*). The results of dinitrophenylation, hydrazinolysis, and periodate oxidation of the glycopeptide led to the conclusion that the carbohydrate moiety was linked at the reducing group of one of the glucosamine residues to the β-carboxyl group of an aspartic residue through amide linkage. The structure of the glycopeptide was determined to be as follows:

Ser–AspNH–carbohydrate

Most of the mannose linkages were presumed to be α-glycosidic because α-mannosidase liberated all mannose residues in the glycopeptide (*69*). Furthermore, it was shown by the action of *N*-acetyl-β-glucosaminidase on the glycopeptide that one *N*-acetylglucosamine residue at the nonreducing end had a β linkage (*69*). The sequential periodate oxidation proved the carbohydrate sequence to be as follows (*70*):

$$\begin{array}{l}(\mathrm{Man})_6\text{–GlcNAc–GlcNAc–NH} \\ \qquad\qquad\qquad\qquad\qquad\quad | \\ \qquad\qquad\qquad\qquad\quad \text{Ser–Asp}\end{array}$$

Recently, the carbohydrate moiety of *A. oryzae* α-amylase prepared from Clarase was reported to be heterogeneous (*29*). Several glycopeptides were isolated from Pronase digestion of the enzyme, and one was found to have a similar carbohydrate composition to that of Taka-amylase A (*70*).

Several amylases have been shown to contain carbohydrate moieties. *Rhizopus delmer* amylase contains D-mannose and 2-amino-2-deoxy-D-glucose (*71*). *A. phoenics* glucamylase contains D-mannose, D-glucose, and D-galactose (*72*). *Aspergillus niger* acid-stable and acid-unstable α-amylases contain 24 moles and 7 moles of mannose, and 4 moles and 1 mole of hexosamine per mole protein, respectively (*73*).

How these carbohydrate moieties are incorporated into the amylases

68. M. Anai, T. Ikenaka, and Y. Matsushima, *J. Biochem.* (*Tokyo*) **59,** 57 (1966).
69. H. Yamaguchi, T. Mega, T. Ikenaka, and Y. Matsushima, *J. Biochem.* (*Tokyo*) **66,** 441 (1969).
70. H. Yamaguchi, T. Ikenaka, and Y. Matsushima, *J. Biochem.* (*Tokyo*) **65,** 793 (1969).
71. J. H. Pazur and S. Okada, *Carbohydrate Res.* **4,** 371 (1967).
72. D. R. Linebock and W. E. Baumann, *Abstr. Paper, 158th ACS Meeting, New York* Carb. 74 (1969).
73. M. Arai, Y. Minoda, and K. Yamada, *Agr. Biol. Chem.* (*Tokyo*) **33,** 922 (1969).

and how they function, is still unknown. It has been suggested that the carbohydrate moiety of Taka-amylase A is not involved in its enzymic activity (*74*) and *A. oryzae* grown on a synthetic medium has been shown to produce Taka-amylase A free from carbohydrate moiety (*62*).

F. Calcium

α-Amylases have been classified as metalloenzymes having calcium as a cofactor (*75*). Stein *et al.* (*76*) completely removed calcium from *B. subtilis* α-amylase by chelation with EDTA or electrodialysis, the latter being much more effective than the former. In contrast to procedures reported previously, these techniques brought no irreversible denaturation, and thus yielded calcium-free amylase that could be fully reactivated upon restoration of calcium. Taka-amylase A was shown to be extremely resistant to the removal of calcium. The rate of calcium release from α-amylases varied markedly according to the origin of the enzymes (mammalian > bacterial > mold). Hsiu *et al.* (*77*) studied the catalytic property of *B. subtilis* α-amylase during the progressive removal by chelation or electrodialysis of the calcium bound to the enzyme. Removal of calcium was accompanied by a loss of activity that could be quantitatively reversed by restoration of the metal. The bacterial enzyme needed four or more gram-atoms of calcium per mole enzyme for full activity, whereas human salivary amylase needed one gram-atom. It was suggested that, by forming a tight metal-chelate structure, the metal produced intramolecular cross-links similar in function to disulfide linkages.

Calcium binding by *B. subtilis* α-amylase was carefully studied by Imanishi (*78*). The calcium-free α-amylase was prepared by electrodialysis according to the method of Stein *et al.* (*76*). Intactness of protein conformation of the calcium-free amylase was confirmed by measurements of optical rotatory dispersion and solvent perturbation on UV absorption and fluorescence spectra. It was also confirmed that the removal of calcium caused no appreciable aggregation of the enzyme molecule by sedimentation measurement. The conformation of the enzyme was found to become labile to changes of environment by the removal of cal-

74. S. Akabori, B. Maruo, H. Mitsui, and M. Nomura, *J. Gen. Appl. Microbiol.* **1**, 1 (1955).
75. E. H. Fischer and E. A. Stein, "The Enzymes," 2nd ed., Vol. 4, p. 313, 1960.
76. E. A. Stein, J. H. Hsiu, and E. H. Fischer, *Biochemistry* **3**, 56 (1964).
77. J. Hsiu, E. H. Fischer, and E. A. Stein, *Biochemistry* **3**, 61 (1964).
78. A. Imanishi, *J. Biochem.* (*Tokyo*) **60**, 381 (1966).

cium. By measurement of an optical rotatory parameter vs. temperature, it was shown that at least 6 g-atoms of calcium per mole enzyme was necessary for full protection of the molecular conformation. The amount of bound calcium was measured by the equilibrium dialysis technique; the data gave a linear Scatchard plot based on an equation in which equivalence and electostatic independence of binding sites were assumed. The plot indicated three binding sites for calcium per molecule of enzyme. The number roughly coincides with that reported by Hsiu *et al.* (*77*) as the minimum number of calcium bound per enzyme molecule for manifestation of the full enzymic activity. The first binding constant was the order of 10^6 at pH 6.7, much larger than that for calcium binding by bovine serum albumin and casein and comparable to that of calcium binding to EDTA homologs. Calcium binding stabilized the enzyme molecule by 7–8 kcal/mole enzyme. This stabilization is large compared to other factors generally considered to stabilize or destabilize globular conformation of protein molecules.

From consideration of enthalpy and entropy terms, the carboxylate group was suggested to be involved in the binding of calcium. These specific binding sites disappeared on denaturation of the enzyme molecule by 6 *M* urea; the denatured enzyme molecule was found to have binding sites with a much weaker affinity for calcium. Calcium atoms stabilizing the molecule but not essential for the maintenance of the enzymically active conformation may be bound to the molecule with much weaker affinity. The native enzyme molecule has been shown to bind more than three calcium atoms per molecule (*76*).

Taka-amylase A binds 10 g-atoms of calcium per mole enzyme, and 9 g-atoms among them are bound so loosely that they can be removed by dialysis against 0.02 *M* NaOAc with no effect on the enzymic activity (*79*). These loosely bound calcium atoms stabilize the enzyme molecule against denaturation (*79–81*). The undialyzable one is so firmly and specifically bound that it cannot be removed by dialysis against EDTA at pH 7–9 for 150 hr at low temperature (*76*). Incubation of Taka-amylase A with excess EDTA at 33° and pH 7–9.5 caused inactivation, but the activity could be recovered by the addition of calcium (*82*). Kato *et al.* (*57*) also showed that the firmly bound calcium could be removed with EDTA at pH 8.0 at 50°. The enzymic activity was lost but could be recovered

79. A. Oikawa and A. Maeda, *J. Biochem.* (*Tokyo*) **44,** 745 (1957).
80. B. Hagihara, T. Nakayama, H. Matsubara, and K. Okunuki, *J. Biochem.* (*Tokyo*) **47,** 537 (1956).
81. S. Fujita-Ikeda and T. Isemura, *J. Biochem.* (*Tokyo*) **47,** 537 (1960).
82. A. Tanaka, *Bull. Agr. Chem. Soc. Japan* **24,** 152 (1962).

by the addition of calcium. Essentially similar results were obtained by using polarography to detect changes in the molecular conformation (*83*).

Toda and Narita (*84*) showed that one atom of firmly bound calcium in Taka-amylase A could be replaced by other divalent cations such as magnesium, strontium, and barium without causing appreciable change in the amylase activity. Zinc and cadmium, however, could not reactivate Taka-amylase A inactivated with EDTA. Zinc was found to be firmly retained after prolonged dialysis. It is to be noted that magnesium and barium could not replace calcium in the refolding process of reduced Taka-amylase A (*85*). Seemingly these metals can replace calcium only when the intact conformation of the enzyme molecule is preserved, and consequently the binding site is intact. From measuring the effect of various chemical modifications on the sulfhydryl group, Toda *et al.* (*58*) suggested that the sole sulfhydryl group in the Taka-amylase A molecule is one of the groups forming the binding site of the most firmly bound calcium atom.

Both the acid-stable and acid-unstable α-amylases of *A. niger* were found to contain 1 g-atom of calcium, which could be removed with EDTA with a concomitant loss of enzymic activity (*86*). The enzymic activity of both enzymes was partially restored by the addition of calcium.

Bacillus subtilis saccharifying α-amylase was inactivated by incubation with EDTA and could be reactivated by the addition of calcium. The sole sulfhydryl group of the enzyme was also shown to participate in the calcium binding. The situation seems to be similar to the case of Taka-amylase A. The *B. subtilis* enzyme, however, was much more resistant to treatment with EDTA (*60*).

Fukumoto *et al.* (*87*) studied the effect of metal ions on the formation of amylase by *B. amyloliquefaciens*. The formation of amylase by the washed cell was increased to 40% by the addition of 10^{-3} *M* salts of calcium, strontium, or magnesium, but it was markedly inhibited by the addition of 1.5×10^{-5} *M* salts of mercury, cobalt, copper, or zinc. When the bacterial cell was washed with 10^{-2} *M* EDTA, the calcium content and amylase-forming activity decreased by up to 36 and 53–69%,

83. G. A. Molodova, I. D. Ivanov, and G. N. Nikolaev, *Izv. Akad. Nauk SSSR, Ser. Biol.* **30,** 359 (1965).
84. H. Toda and K. Narita, *J. Biochem. (Tokyo)* **62,** 767 (1967).
85. T. Takagi and T. Isemura, *J. Biochem. (Tokyo)* **57,** 89 (1965).
86. M. Arai, Y. Minoda, and K. Yamada, *Agr. Biol. Chem. (Tokyo)* **33,** 922 (1969).
87. J. Fukumoto, T. Yamamoto, D. Tsuru, and M. Kakumae, *J. Agr. Chem. Soc. Japan* **34,** 475 (1960).

respectively, although respiration of the cell remained almost unaffected. Using a culture medium containing strontium, amylase containing strontium instead of calcium was obtained in a crystalline state. It had the same enzymic activity as that of normal amylase but was found to be inferior to the normal amylase in pH and heat stability. They also found that EDTA (0.1–1 m*M*) inhibited the formation of α-amylase by *B. subtilis*. The inhibition was overcome by the addition of calcium or strontium. EDTA could remove calcium present in the cell wall fraction of the cell (*88*).

G. Molecular Weights

In the second edition of "The Enzymes," Fischer and Stein stated that all α-amylases investigated up to that time possessed very similar sedimentation constants, corresponding to molecular weights in the range of 50,000 (*75*). Several amylases of mold and bacterial origins studied since then also turned out to have molecular weights ranging from 40,000 to 60,000 as shown in Table V. Molecular weights of *B. subtilis* and *A. oryzae* α-amylases, also included in Table V, were measured by other groups and coincided with those cited in the earlier edition (*75*). The only exception is *B. macerans* amylase whose molecular weight has been reported to be 130,000 (*17*). Experimental data used to calculate that molecular weight, however, are not convincing; especially difficult to explain is the marked dependence of diffusion coefficient on concentration unless the conformation of the enzyme deviates appreciably from a globular one.

The monomer–dimer transformation of *B. subtilis* α-amylase through binding of zinc ion found by Fischer and Stein was reviewed in the second edition (*75*). No other example of such a dimerization has been reported with any other amylase. The phenomenon seems to be peculiar to *B. subtilis* α-amylase. It was further studied by Isemura and Kakiuchi (*89*). They found formation of higher aggregates than the dimer in the presence of higher concentration of zinc ion. It was also reported that the amylase became exclusively monomeric below pH 5. An imidazole group of histidine was suggested as the binding site of the zinc atom because the enzyme lost the dimerizing ability by photooxidation in the

88. N. Hamada, T. Yamamoto, and J. Fukumoto, *Agr. Biol. Chem.* (*Tokyo*) **31**, 1 (1967).
89. T. Isemura and K. Kakiuchi, *J. Biochem.* (*Tokyo*) **51**, 385 (1962).

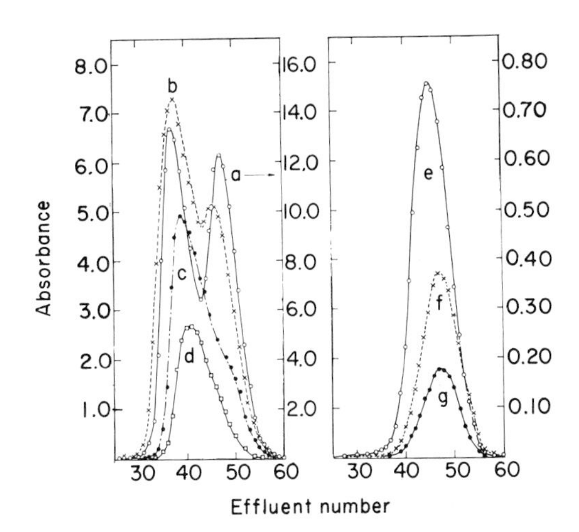

FIG. 1. Effect of protein concentration on the elution patterns of *B. subtilis* α-amylase from a Sephadex G-100 column equilibrated with 0.1 *M* NaCl–0.005 *M* $Ca(OAc)_2$, pH 7.5 (2.5 × 54 cm) (*90*). Two milliliters of the protein solution of following concentrations (in percent) were placed on the top of the column: (a) 11.4, (b) 5.68, (c) 2.85, (d) 1.05, (e) 0.40, (f) 0.25, and (g) 0.14. Elution rate, 45–65 ml/hr.

presence of methylene blue. Kakiuchi *et al.* (*90*) also studied the phenomenon by gel filtration. As shown in Fig. 1, the elution diagram changed markedly with protein concentration and did not reflect the composition of the equilibrium mixture. Since the pattern changed with the elution rate, they concluded that the rate of re-establishment of equilibrium perturbed by the separation was comparable to the flow rate through the gel column. These experiments also confirmed that the enzyme was in monomeric form under the condition generally used for the assay of enzymic activity. Kakiuchi (*91*) further studied the dimerization of the amylase by sedimentation equilibrium experiments. Theoretical analysis of the data obtained gave an equilibrium constant of 2.2×10^9 liter^2/mole^2 for the equilibrium, 2 monomers + $Zn^{2+} \rightleftharpoons$ dimer.

H. DENATURATION AND RENATURATION

Denaturation and renaturation of bacterial and mold amylases have been extensively studied. Most of the studies may have been motivated by the fact that the amylases are available in a large quantity and contribute to the understanding of protein behavior in general rather than to the behavior of this particular group of enzymes. However, a brief

90. K. Kakiuchi, S. Kato, A. Imanishi, and T. Isemura, *J. Biochem.* (*Tokyo*) **55,** 102 (1964).
91. K. Kakiuchi, *J. Phys. Chem.* **69,** 1829 (1965).

TABLE V
MOLECULAR WEIGHTS OF AMYLASES

Amylase	Molecular weight	Method[a]	Ref.
B. subtilis α-amylase	48,900 (monomer)	S and D	(*92*)
	96,900 (dimer)		
	47,000 (monomer)	SE	(*91*)
Taka-amylase A (*A. oryzae* α-amylase)	51,000	S and D	(*93*)
	52,600 ± 2,600	SE	(*94*)
	49,000[b]	GF	(*95*)
	50,000	S and D	(*96*)
A. niger acid-stable α-amylase	61,000	SE	(*97*)
A. niger acid-unstable α-amylase	61,000	SE	(*97*)
B. stearothermophilus thermostable α-amylase	48,000	SE	(*23*)
	52,700 ± 2,200	SE	(*22*)
B. subtilus saccharifying α-amylase	41,000	AE	(*98*)
B. macerans, cyclizing	139,000[c]	S and D	(*17*)

[a] Abbreviations: S and D, sedimentation and diffusion constants; SE, sedimentation equilibrium; GF, gel filtration; and AE, approach to equilibrium.

[b] Calculated by the reviewers from the curve in Fig. 5 in Fish *et al.* (*95*).

[c] See text.

review of the studies may be useful to those interested in amylases as enzymes.

Acid-denatured Taka-amylase A was found to retain an appreciable amount of ordered structure and to be able to regain the intact structure by neutralization (*99, 100*). But an irreversible change occurred when Taka-amylase A was exposed to an alkaline pH (*101*). Presumably, disulfide exchange reaction (*102*) is chiefly responsible for the irreversible alkaline denaturation. The presence of one sulfhydryl group besides four disulfide groups per molecule may favor the exchange reaction. The

92. E. H. Fischer, W. N. Sumerwell, J. M. Junge, and E. A. Stein, *Proc. 4th Intern. Congr. Biochem., Vienna, 1958*, Symp. 8, p. 124. Pergamon Press, Oxford, 1960.
93. T. Isemura and S. Fujita, *J. Biochem.* (*Tokyo*) **44,** 443 (1957).
94. J. F. McKelvy and Y. C. Lee, *ABB* **132,** 99 (1969).
95. W. W. Fish, K. G. Mann, and C. Tanford, *JBC* **244,** 4989 (1969).
96. S. Sirisinha and P. Z. Allen, *ABB* **112,** 137 (1965).
97. M. Arai, T. Koyano, H. Ozawa, Y. Minoda, and K. Yamada, *Agr. Biol. Chem.* (*Tokyo*) **32,** 507 (1968).
98. A. Yutani, K. Yutani, and T. Isemura, *J. Biochem.* (*Tokyo*) **65,** 201 (1969).
99. T. Takagi and H. Toda, *J. Biochem.* (*Tokyo*) **52,** 16 (1962).
100. T. Takagi and T. Isemura, *J. Biochem.* (*Tokyo*) **49,** 43 (1961).
101. T. Takagi and T. Isemura, *J. Biochem.* (*Tokyo*) **48,** 781 (1960).
102. E. V. Jensen, *Science* **130,** 1319 (1959).

Taka-amylase A molecule is stable between pH 5 and 10.5 judging from the pH-stability curve and pH dependence of optical rotatory properties (*100, 101*). Below pH 4.5 and above pH 10.5, exposure of masked carboxyl and phenolic hydroxyl groups, respectively, was observed by measuring the amphoteric properties (*100, 101*). Parallel studies have been made on amphoteric properties of *B. subtilis* α-amylase (*103, 104*). The enzyme molecule behaved similarly to the Taka-amylase A molecule against pH. However, *B. subtilis* α-amylase was found to be more resistant to alkaline pH (up to pH 12) than Taka-amylase. This alkali stability may be related to the lack of sulfhydryl and disulfide groups in the enzyme.

Taka-amylase A was found to be fairly stable against denaturation by a surface active agent, sodium dodecyl sulfate, by electrophoretic study (*105*). Cetyl quaternary ammonium compounds and sodium dodecyl sulfate were found to inactivate mold and bacterial α-amylases when used in excess amounts (*106*). Imanishi *et al.* (*107*) studied the effect of sodium dodecyl sulfate on *B. subtilis* α-amylase. Their results show that *B. subtilis* α-amylase is also fairly stable against the surface active agent. At neutral pH, the enzyme molecule seemed to retain some original structure even in the presence of 440 *M* excess of the detergent, and the structure was irreversibly destroyed by raising the temperature. The detergent seems to have a strong affinity to the unfolded polypeptide chain.

As reviewed in the previous edition of "The Enzymes" (*75*), α-amylases are stabilized by bound calcium. Fujita-Ikeda and Isemura (*81*) studied the effect of calcium on urea denaturation. Dialyzable calcium ions were found to stabilize the enzyme molecule markedly. At neutral pH region and 20°, the molecule was virtually stable against 6 *M* urea. In the absence of dialyzable calcium, urea of less than 4 *M* was sufficient to denature the enzyme (*108*). Size and shape of the urea-denatured Taka-amylase A molecule were also studied (*109*).

The effect of substrates on heat inactivation of Taka-amylase A was studied (*110*). The formation of enzyme–substrate and enzyme–product

103. T. Isemura and A. Imanishi, *J. Biochem.* (*Tokyo*) **51,** 172 (1962).
104. A. Imanishi, Y. Momotani, and T. Isemura, *J. Biochem.* (*Tokyo*) **55,** 562 (1964).
105. T. Isemura and T. Takagi, *J. Biochem.* (*Tokyo*) **46,** 1637 (1959).
106. Y. Pomeranz, *BBA* **73,** 105 (1963).
107. A. Imanishi, Y. Momotani, and T. Isemura, *J. Biochem.* (*Tokyo*) **57,** 417 (1965).
108. S. Fujita-Ikeda, *J. Biochem.* (*Tokyo*) **49,** 267 (1961).
109. T. Isemura and S. Fujita-Ikeda, *J. Biochem.* (*Tokyo*) **49,** 278 (1961).
110. G. Tomita and S. S. Kim, *Z. Naturforsch.* **22b,** 294 (1967).

complexes was found not only to stabilize the molecular conformation but also to prevent the formation of aggregates induced by heat denaturation.

The denaturation of *B. subtilis* α-amylase and Taka-amylase A under high pressure has been extensively studied (*111–115*). The enzymes can be inactivated under a pressure of about 6000 kg/cm^2. Effects of various factors such as protein concentration, pH, and ionic strength were studied. The product of pressure denaturation was studied by physicochemical methods. Compression under a moderate pressure (600–4000 kg/cm^2) was found to reactivate the heat-denatured and high pressure-denatured enzymes.

Taka-amylase A is one of the proteins that was found in the early 1960's to be renaturated from a randomly coiled state (*54, 116*). On the basis of present criteria of a randomly coiled state (*117*), the reduced Taka-amylase A molecule in 8 *M* urea or 6 *M* guanidine hydrochloride (*118*) is completely denatured to a randomly coiled state. Prior to the studies on the renaturation from the randomly coiled state, reversibilities of acid denaturation (*99*) and urea denaturation (*119*) were studied. It was pointed out that by use of Taka-amylase A the reoxidation of sulfhydryl groups in reduced proteins was catalyzed by heavy metal ions such as copper or iron; therefore, their amounts must be carefully controlled to get reproducible results in studies of renaturation, including the reoxidation step (*120, 121*). Presence of an optimum concentration for refolding of reduced Taka-amylase A was observed (*122*). It was suggested that Taka-amylase A can regain enzymic activity at intermediate stages of reoxidation (*123*).

Conformation of reduced Taka-amylase A was investigated in comparison with those of native and reduced carboxymethylated Taka-amylase A by measurements of sedimentation constant, UV difference spectrum, and optical rotatory dispersion (*124*). It was found that the

111. K. Suzuki and K. Kitamura, *J. Biochem.* (*Tokyo*) **54,** 214 (1963).
112. K. Miyagawa, K. Sannoe, and K. Suzuki, *ABB* **106,** 467 (1964).
113. K. Miyagawa, *ABB* **110,** 381 (1965).
114. K. Miyagawa, *ABB* **113,** 641 (1966).
115. K. Miyagawa, *Agr. Biol. Chem.* (*Tokyo*) **31,** 761 (1967).
116. T. Isemura, T. Takagi, Y. Maeda, and K. Imai, *BBRC* **5,** 373 (1961).
117. C. Tanford, *Advan. Protein Chem.* **23,** 121 (1968).
118. T. Takagi, *Kagaku No Ryoiki* **24,** 21 (1970) (in Japanese).
119. T. Takagi and T. Isemura, *J. Biochem.* (*Tokyo*) **52,** 314 (1962).
120. T. Takagi and T. Isemura, *BBRC* **13,** 353 (1963).
121. T. Takagi and T. Isemura, *J. Biochem.* (*Tokyo*) **56,** 344 (1964).
122. K. Yutani, A. Yutani, and T. Isemura, *J. Biochem.* (*Tokyo*) **66,** 823 (1969).
123. K. Yutani, T. Takagi, and T. Isemura, *J. Biochem.* (*Tokyo*) **57,** 590 (1965).
124. T. Takagi and T. Isemura, *BBA* **130,** 233 (1966).

reduced enzyme molecule has an effective molecular volume 3–6 times as large as that of the native enzyme molecule. Clearly, reduced Taka-amylase A is partially refolded, but further refolding requires the re-formation of disulfide bonds through reoxidation. Calcium was found to have a profound effect on the renaturation process of Taka-amylase A. Takagi and Isemura (*85*) reported that at least one mole of calcium ion per mole of Taka-amylase A was necessary for the recovery of intact structure through reoxidation in the presence of a catalytic amount of copper ion. They suggested that a part of the disulfide bonds was incorrectly formed in the absence of calcium. Later, Friedman and Epstein (*125*) reported that reoxidation in the absence of calcium ion led to an inactive protein, which contained about three sulfhydryl groups per molecule. Upon addition of calcium ion, this partially reoxidized product was fully reactivated. In this case, no metal ion was added as catalyst. Reoxidation of reduced Taka-amylase A by dehydroascorbic acid gave an inactive protein which had been fully reoxidized and could be reactivated by the addition of calcium. Presumably intact disulfide bonds were fully recovered. The difference in reformation of disulfide bonds shown in the two papers may originate in the reoxidation methods. Only strontium could replace calcium in the role of the renaturation process (*85*).

Introduction of 1-dimethylaminonaphthalene-5-sulfonyl (dansyl) group to the native Taka-amylase A molecule was found to interfere with the correct refolding (*126*). The hydrophobic nature of the residue was suggested as the major factor.

Effects of denaturing agents and proteolytic enzymes on the immunochemical reactivity of Taka-amylase A were investigated (*96*). Exposure of the enzyme to concentrated urea solutions under various conditions resulted in preparations giving precipitation reactions of partial identity with the native enzyme in immunodiffusion analysis. This result is interesting from the viewpoint of renaturation study. With increase of incubation time with urea (2–10 *M*, 37°, pH not described, but probably neutral), the number of precipitin lines on the side of antigen increased at first beside a line corresponding to the native (completely renatured) enzyme but decreased after a prolonged incubation. The higher the urea concentration, the sooner the trend changed. The authors interpreted the result by assuming that the lines corresponded to intermediates of protein denaturation. The lines, however, seem to correspond to intermolecular aggregates produced through an interchange of disulfide

125. T. Friedmann and C. J. Epstein, *JBC* **242**, 513 (1967).
126. T. Takagi, Y. Nakanishi, N. Okabe, and T. Isemura, *Biopolymers* **5**, 627 (1967).

bridges subsequent to the denaturation. The decrease in the number of lines may correspond to formation of aggregates too large in size to diffuse into the agar gel.

It was also found that *B. subtilis* α-amylase denatured by 8 *M* urea can refold to regain its enzymic activity (*127*). Fukushi *et al.* (*128*) studied the renaturation process of the enzyme and found that the recovered activity resulted from the completely renatured enzyme. They found that the presence of incompletely renatured enzyme molecule could be shown by electrophoresis in the presence of sodium dodecyl sulfate. Fukushi and Isemura further studied the renaturation of *B. subtilis* α-amylase with special reference to the folding process of the newly formed polypeptide of the enzyme in the bacterial cell (*129*). It was found that the cell produced a similar amount of the enzyme at 45° as at 37°. They assumed that the folding mechanism in the cell was somewhat different from that in the renaturation process *in vitro* since no renaturation was observed *in vitro* at 45°.

Yutani *et al.* (*130*) found that the rate and extent of reactivation of *B. subtilis* α-amylase was increased in the presence of various other proteins. Among them, bovine serum albumin was most effective. Seemingly weak interaction between the refolding protein and the coexisting protein helps prevent incorrect folding which leads to irreversible denaturation.

Reversibility of the heat inactivation and acid denaturation of *B. subtilis* α-amylase was also studied by another group (*131–133*). The enzyme was found to be inactivated with a number of cobalt complexes (*134*).

Denaturation and renaturation of *B. subtilis* saccharifying α-amylase were studied in comparison with those of *B. subtilis* liquefying α-amylase (*135*).

127. A. Imanishi, K. Kakiuchi, and T. Isemura, *J. Biochem.* (*Tokyo*) **54,** 89 (1963).

128. T. Fukushi, A. Imanishi, and T. Isemura, *J. Biochem.* (*Tokyo*) **63,** 409 (1968).

129. T. Fukushi and T. Isemura, *J. Biochem.* (*Tokyo*) **64,** 283 (1968).

130. K. Yutani, A. Yutani, and T. Isemura, *J. Biochem.* (*Tokyo*) **62,** 576 (1967).

131. T. Yamamoto, M. Nishida, and J. Fukumoto, *Agr. Biol. Chem.* (*Tokyo*) **30,** 994 (1966).

132. A. Nishida, J. Fukumoto, and Y. Yamamoto, *Agr. Biol. Chem.* (*Tokyo*) **31,** 682 (1967).

133. T. Maruyama, M. Niwa, T. Yamamoto, A. Nishida, and J. Fukumoto, *J. Biochem.* (*Tokyo*) **60,** 286 (1966).

134. Y. Pomeranz, *BBA* **77,** 451 (1963).

135. A. Yutani, K. Yutani, and T. Isemura, *J. Biochem.* (*Tokyo*) **65,** 201 (1969).

I. Fragmentation

For many years, enzyme chemists have been striving to degrade enzyme molecules to smaller fragments without loss of their enzymic activities. Substantial progress has been made in degradation of the Taka-amylase A molecule. As is usual with enzyme molecules, Taka-amylase A is resistant to proteolytic attack, but Toda and Akabori (*136*) found that phenylazobenzoylation made the enzyme molecule susceptible to proteolytic attack. *Streptomyces griseus* protease (Pronase) was most effective among various proteases. Phenylazobenzoyl Taka-amylase A (one mole dye per mole enzyme) partially digested by Pronase was isolated in a crystalline form, and it retained a sizable amount of the enzymic activity. The crystalline digested phenylazobenzoyl Taka-amylase A appeared as a distinct peak from that of the intact phenylazobenzoyl enzyme in DEAE-cellulose column chromatography. The molecular weight of the degraded product was estimated to be 36,000 by sedimentation experiment. A consistent value was obtained by determination of phenylazobenzoyl group bound to the enzyme molecule (*137*). Alanine was found as a sole N-terminal amino acid of the degraded enzyme. The N-terminal amino acid of the intact Taka-amylase A is also alanine. The action pattern of CPase suggested that the C-terminal region of the Taka-amylase A molecule had been attacked by Pronase (*138*).

Kato (*139*) found that the Taka-amylase A molecule could be degraded by sonic oscillation. Fifteen to twenty milliliters of 1% Taka-amylase A solution was subjected to sonic oscillation at 0–3° using a sonic oscillator of 10 kc, 100 W. Cleavage of peptide bonds (4 –7 per molecule) and an increase of reduced viscosity were observed. A crystalline active fragment was obtained by DEAE-cellulose column chromatography and gel filtration through Sephadex G-75. Its molecular weight was estimated to be 37,000. Also in this case, the N-terminal residue was alanine. In both cases mentioned above, Taka-amylase A was degraded to fragments of molecular weight 36,000–37,000. Seemingly Taka-amylase A has a labile part in the C-terminal region of the polypeptide chain. Further study of the sonication effect on Taka-amylase A brought an interesting phenomenon to light (*140*). Reduced carboxymethylated

136. H. Toda and S. Akabori, *J. Biochem.* (*Tokyo*) **53,** 95 (1963).
137. H. Toda, *J. Biochem.* (*Tokyo*) **53,** 425 (1963).
138. H. Toda, *J. Biochem.* (*Tokyo*) **54,** 1 (1963).
139. I. Kato, *J. Biochem.* (*Tokyo*) **63,** 472 (1968).
140. I. Kato, H. Toda, and K. Narita, *J. Biochem.* (*Tokyo*) **63,** 479 (1968).

Taka-amylase A, obtained after complete reduction of disulfide bonds, was placed in a sonic field in the presence of calcium ion. Surprisingly, there was a 15% revival of enzymic activity. Possibility of reformation of intact disulfide bonds was excluded. Almost the same number of peptide bonds were cleaved as in the case of sonication of the native enzyme. Results of optical rotatory dispersion and viscosity measurements suggested formation of a certain ordered structure. The investigators suggested that an intramolecular formation of β structure was responsible for the recovery of enzymic activity. Peptides obtained by tryptic digestion of reduced carboxymethylated Taka-amylase A were also found to regain enzymic activity by the application of sonic oscillation in the presence of calcium at pH 7.0 and 0–3° (10 kc, 100 W) (*141*). Two kinds of peptides were found to be essential to form the active aggregate, and they could be isolated in an almost homogeneous state. The pH optimum and pH-stability curve of the active aggregate formed from the two peptides were similar to those of the native amylase. It was suggested that the active site in the aggregate was identical or close to that of the native enzyme.

J. Thermostable and Acid-Stable Amylases

As stated in the beginning, bacterial and mold amylases are widely used for industrial purposes. Therefore, amylases that can survive extreme conditions attract attention. Especially, thermostable and acid-stable amylases have been extensively studied since high temperature and acidic pH are optimum conditions for the economic hydrolysis of starch. Actually, the consumption of such amylases is not large at present because normal amylases are low in price and are generally resistant to the reaction conditions, at least during the reaction period. The stable amylases, however, will be useful when they are chemically bound to some carrier and used repeatedly. Apart from industrial interests, the stable enzymes are attracting interest from a biological point of view in relation to the adaptation mechanism of living organisms to extreme conditions.

1. *Thermostable Amylases*

Enzymes of thermophilic organisms are now attracting interest because of the molecular mechanism of thermophily (*142*, *143*). The optimum temperature of *B. thermophilus*, for example, is 60–65°, while

141. I. Kato, H. Toda, and K. Narita, *J. Biochem.* (*Tokyo*) **63**, 487 (1968).
142. H. Hoffler, *Bacteriol. Rev.* **21**, 227 (1957).
143. T. D. Brock, *Science* **158**, 1012 (1967).

that of *B. subtilis* is 40°. *Bacillus stearothermophilus* α-amylase, one of the enzymes produced by thermophilic organisms, has been extensively studied.

Mannig and Campbell studied *B. stearothermophilus* α-amylase and described its isolation in crystalline form (*21*). The enzyme was reported to be totally different from all other α-amylases in that it had a small molecular weight of 15,600 (*144*) and consisted of two polypeptide chains linked by disulfide bonds (*52*). A high proline content of the enzyme was correlated with a low degree of α-helical structure (*37*). A large negative rotation characteristic of an unfolded protein was observed with the enzyme in its native state. The optical rotation was not significantly affected by 8 *M* urea, 4 *M* guanidine, or temperatures of 65° and 75°, nor was there any loss of enzymic activity under these conditions. This unusual unfolded structure, with the enzyme retaining activity, was postulated as the reason for the observed marked thermostability of the enzyme (*144*). It was thought that there was no reason that a completely unfolded (and hence denatured) protein could not function as an enzyme. The above results and interpretation attracted much attention because they were a challenge to the concept that an enzyme must have rigid and ordered structure to function.

Recently, Pfueller and Elliott restudied the enzyme (*22*). A strain of *B. stearothermophilus* called 1503-4, which seems to be a descendant of 503-4 used in the original work, was supplied by Campbell. The crystalline thermostable α-amylase obtained by Pfueller and Elliott was found to be different from that previously described by Mannig and Campbell. The purified enzyme obtained by a new procedure was found to have properties different from the previously reported unique ones; they resembled those of other α-amylases without thermostability. The molecular weight was approximately 53,000. The enzyme contained no cystine resembling other baterial extracellular enzymes. It was protected from denaturation above 50° by metal ions, particularly Ca^{2+}, and by protein such as bovine serum albumin and starch. The enzyme was capable of hydrolyzing starch at a temperature of 70° and above. In contrast to its thermostability, the enzyme was 50% inactivated in 24 hr at 6°.

A thermostable α-amylase from *B. stearothermophilus* was also crystallized and extensively studied by Ogasahara *et al.* (*23, 145, 146*). The molecular weight (48,000), the intrinsic viscosity (0.032 dl/g), the

144. G. B. Mannig, L. L. Campbell, and R. F. J. Foster, *JBC* **236**, 2958 (1961).

145. K. Ogasahara, A. Imanishi, and T. Isemura, *J. Biochem.* (*Tokyo*) **67**, 77 (1970).

146. K. Ogasahara, K. Yutani, A. Imanishi, and T. Isemura, *J. Biochem.* (*Tokyo*) **67**, 83 (1970).

optical rotatory dispersion, and the circular dichroism were measured (*23*). They all indicated that the enzyme was different from that described by Campbell's group and similar to those of *B. subtilis* α-amylase.

The enzyme reported by Mannig and Campbell remains a mystery. Pfeuller and Elliott carefully surveyed for the enzyme protein but could not get it (*22*). Sirisinha and Allen got an observation probably suggesting the conformational resemblance between *B. subtilis* α-amylase and *B. stearothermophilus* α-amylase (*147*). Namely, anti-*B. subtilis* α-amylase serum formed one or two precipitin lines with partially purified *B. stearothermophilus* α-amylase. Studies of other thermostable α-amylases have been reported (*148–151*).

2. *Acid-Stable α-Amylases*

Amylases from various sources are generally labile at acidic pH. Some molds, especially *A. niger*, are known to produce acid-stable α-amylases which are fairly stable down to pH 2. Minoda and Tsukamoto found that when a strain of *A. niger* was first cultured in a weak acidic medium (optimum pH 4.0), and then slightly neutralized (to pH 5.5), an acid-stable α-amylase was produced in addition to the ordinary acid-unstable α-amylase (*152*). Minoda *et al.* obtained the acid-stable α-amylase in a crystalline form (*18, 39*). The acid-stable enzyme retained 87% of its original activity when incubated at pH 2.2, 37° for 30 min, although the acid-unstable enzyme was completely inactivated by the same treatment. The former was also found to be superior to the latter in thermal stability. It should also be noted that the pH-stability curve of the acid-stable enzyme apparently shifted to the acidic side compared to that of the acid-unstable enzyme, as shown in Fig. 2. The acid stability seems to be gained in compensation for the stability in the weakly alkaline pH region (*19*). Both enzymes were found to have a molecular weight of ca. 60,000 (*97*). Takagi *et al.* (*153*) compared the two enzymes by optical rotatory dispersion (ORD) measurements. From the close simi-

147. S. Sirisinha and P. Z. Allen, *J. Bacteriol.* **90**, 1120 (1965).
148. K. I. Park and S. C. Hong, *Kungnip Kongop Yonguso Pogo* **15**, 50 (1965).
149. O. V. Kislukhina, *Prikl. Biokhim. i Mikrobiol.* **2**, 544 (1966).
150. E. P. Guzhova and L. G. Loginova, *Mikrobiologiya* **35**, 427 (1966).
151. L. G. Loginova, Y. Karpukhina, and E. P. Guzhova, *Izv. Akad. Nauk SSSR, Ser. Biol.* p. 595 (1967).
152. Y. Minoda and I. Tsukamoto, *J. Agr. Chem. Soc. Japan* **35**, 482 (1961).
153. T. Takagi, M. Arai, Y. Minoda, T. Isemura, and K. Yamada, *BBA* **175**, 438 (1969).

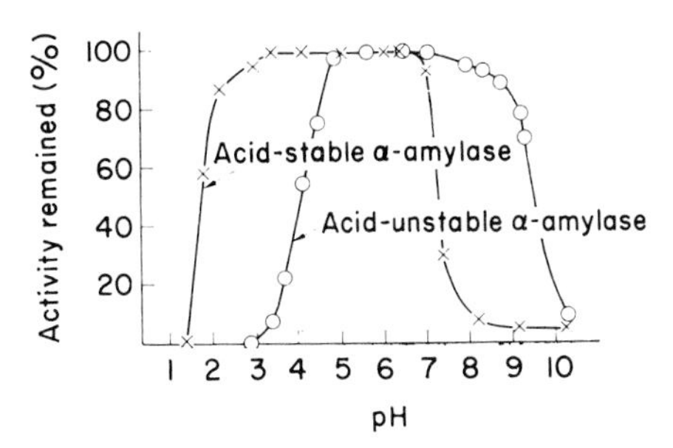

FIG. 2. Effect of pH on *A. niger* acid-stable α-amylase and acid-unstable α-amylases (*19*). Enzymes were incubated at each pH for 30 min at 37°C, and the remaining activities were assayed at pH 4.8.

larity of ORD curves of the two enzymes, it was inferred that there was no major difference in the mode of folding of the polypeptides composing them. Changes in amino acid composition, which enabled interactions between side chain groups in some limited regions of the acid-stable enzyme, are presumed to have given the molecular stability against pH, if the acid-stable enzyme were the product of evolution from the acid-unstable one.

K. CHEMICAL MODIFICATION

The chemical modification of α-amylases has been extensively studied. Earlier works on hog pancreatic amylase and other amylases were reviewed in the second edition of "The Enzymes" (*75*). Results of the last decade will be described here.

Ikenaka (*154*) has reported that Taka-amylase A was inactivated when 2 moles of tyrosine and 5 moles of lysine residues were dinitrophenylated by 2,4-fluorodinitrobenzene but retained 60% of its original activity when 11 out of 22 lysine residues were dinitrophenylated by 2,4-dinitrobenzene-1-sulfonate. These results suggest that ε-amino groups of lysine residues may not be essential for the amylase activity, but some specific phenolic groups of tyrosine residues may be closely related to the activity. It was also found that the acylation of Taka-amylase A with *p*-phenylazobenzoyl chloride at pH 6, which introduced one mole of phenylazobenzoyl group per mole of the enzyme, caused almost complete loss of the amylase activity, whereas the activity toward phenyl-α-maltoside increased to about 2.5-fold of the original activity

154. T. Ikenaka, *J. Biochem.* (*Tokyo*) **46**, 177 (1959).

(*155*). It was suggested that a reactive amino group in the vicinity of the active site might be involved in the modification.

In case of *B. subtilis* α-amylase, only 3 out of 25 lysine residues were dinitrophenylated with 2,4-dinitrobenzene-1-sulfonate without loss of the enzymic activity (*156*). Chemical modification of the enzyme with several reagents was also investigated. Among them, sulfanilic acid was most effective. Introduction of one azo group to a tyrosine residue of the enzyme caused an almost complete loss of the amylase activity. Onoue *et al.* (*157*) reported that one tryptophan residue in the amylase was oxidized by treatment with *N*-bromosuccinimide at pH 6.0 and 0°. The amylase activity was completely lost, but no appreciable conformational change was observed. One specific tryptophan residue in the amylase was suggested to be involved in the active site. When Taka-amylase A was exposed to *N*-bromosuccimide under the same conditions as above, the full activity remained even after longer treatment (*158*). The rate of inactivation was not significantly altered by the presence of 8 *M* urea or 0.005 *M* EDTA. Prolonged treatment with higher concentrations of these reagents, however, caused a decrease in the amylase activity and its affinity toward the antibody. Oxidation of the amylase with hydrogen peroxide in the presence of cupric ion caused a marked inactivation (*158*). Under mild conditions, *N*-bromosuccimide has been shown to oxidize tryptophan and tyrosine residues selectively (*159–161*), while hydrogen peroxide cleaves heterocyclic and aromatic amino acids in the presence of cupric ion, and oxidizes sulfur-containing amino acids (*162*). The specificities of the above-mentioned reagents are not established. No definite conclusion can be made, therefore, with regard to the involvement of a specific amino acid residue at the active site.

Tamaoki *et al.* (*163*) reported the reversible modification of amino groups in Taka-amylase A with 2-methoxy-5-nitrotropone. When Taka-amylase A was incubated with 50-fold reagent, 11 lysine out of 22 residues reacted with the reagent resulting in loss of amylase activity

155. T. Ikenaka, *J. Biochem.* (*Tokyo*) **46,** 297 (1959).
156. K. Sugae, *J. Biochem.* (*Tokyo*) **48,** 790 (1960).
157. K. Onoue, Y. Okada, and Y. Yamamura, *J. Biochem.* (*Tokyo*) **51,** 443 (1962).
158. S. Sirisinha and P. Z. Allen, *ABB* **112,** 149 (1965).
159. Y. Okada, K. Onoue, S. Nakashima, and Y. Yamamura, *J. Biochem.* (*Tokyo*) **54,** 477 (1963).
160. R. A. Phelps, K. E. Neet, L. T. Lynn, and F. W. Putnam, *JBC* **216,** 96 (1961).
161. T. Viswanatha and W. B. Lawson, *ABB* **93,** 128 (1961).
162. T. Viswanatha, W. B. Lawson, and B. Witkop, *BBA* **40,** 216 (1960).
163. H. Tamaoki, X. Murase, S. Minato, and K. Nakanishi, *J. Biochem.* (*Tokyo*) **62,** 7 (1967).

and increase of maltosidase activity. The modification could be reversed by treatment with 1 *N* hydrazine.

Isemura *et al.* (*164*) modified Taka-amylase A with poly-D,L-alanyl group by the interaction of *N*-carboxyanhydride of D,L-alanine with the enzyme in an appropriate buffer solution. Three samples of polyalanyl Taka-amylase A enriched with 25, 45, and 64 moles of alanine per mole of enzyme were prepared. These modified enzymes were purified by chromatography on DEAE-cellulose. In all three cases, it was found that 7–8 among the 23 amino groups in the enzyme had served as initiators for polyalanylation. The enzymic activity of the three polyalanylated enzymes was compared; the enzymic activity decreases progressively with the degree of polyalanylation when amylose was used as a substrate. But the enzymic activity was unaffected by the modification when a low molecular weight substrate such as phenyl-α-maltoside was used. These differences may reflect the effect of polyalanylation on the accessibility of the substrate to the active site of the enzyme.

Suzuki *et al.* (*165*) introduced sulfhydryl group (mercaptosuccinyl group) into Taka-amylase A using *S*-acetyl mercaptosuccinic anhydride. The enzyme was incubated with the reagent at pH 8.0 at 2° for 30 min, and the acetyl group was removed by treatment with hydroxylamine. When one mercaptosuccinyl group was introduced per mole of the enzyme, the amylase activity increased to 1.8-fold of the original activity.

Toda *et al.* (*58, 60*) reported some interesting findings concerning the involvement of the sole masked sulfhydryl group of Taka-amylase A and *B. subtilis* saccharifying α-amylase in their binding of calcium atoms. In their studies, the sulfhydryl group was variously modified. The details have been described in Section II,D of this chapter.

III. Catalytic Properties

A. Specificity of Amylases

The specificity of α-amylases is to catalyze the hydrolysis of $\alpha, 1 \rightarrow 4$-glucosidic linkage in polysaccharides or their degradation products. The $\alpha, 1 \rightarrow 6$-glucosidic linkage at the branching point in glycogen and amylopectin is not attacked by α-amylases. It is known that exoamylase degrades the polysaccharide chain by hydrolyzing every, or every alter-

164. T. Isemura, T. Fukushi, and A. Imanishi, *J. Biochem.* (*Tokyo*) **56,** 408 (1964).
165. S. Suzuki, Y. Hachimori, and R. Matoba, *BBA* **167,** 641 (1968).

TABLE VI
RELATIVE RATE OF HYDROLYSIS OF MALTOOLIGOSACCHARIDES BY *B. subtilis* SACCHARIFYING AMYLASE

Maltooligosaccharide	Conc.	Relative rate of hydrolysis
G_2	0.002 *M*	0
G_3	0.002 *M*	9
G_4	0.002 *M*	98
G_5	0.002 *M*	72
G_6	0.002 *M*	49
G_7	0.002 *M*	34
Maltodextrin (G_{23})	0.2%	85
Starch	0.25%	100

nate, glucosidic linkage from the nonreducing end of the chain. In contrast to exoamylase, α-amylase (endoamylase) acts on polysaccharides in several ways, although this action has been described as random action. It was demonstrated that the initial attack of *B. subtilis* saccharifying α-amylase on maltodextrin occurred much more readily at the third or more inner linkages from the nonreducing end (*166*). The relative rate of hydrolysis of maltooligosaccharides catalyzed by *B. subtilis* α-amylase is given in Tables VI and VII. *Bacillus subtilis* saccharifying α-amylase shows similar affinity for maltooligosaccharides having chain lengths from G_2 to G_{23}, whereas the affinity of liquefying α-amylase is markedly reduced for oligosaccharides with chain lengths less than G_{23}.

Matsubara *et al.* investigated the effect of aglycon on the rate of hy-

TABLE VII
RELATIVE RATE OF HYDROLYSIS OF MALTOOLIGOSACCHARIDES BY *B. subtilis* LIQUEFYING AMYLASE

Maltodextrin	Conc. (%)	Relative rate of hydrolysis
G_5	0.1	0
G_7	0.1	1
G_{10-12}	0.1	1
G_{15-20}	0.1	10
Maltodextrin (G_{23})	0.1	41
Starch	0.1	100

166. S. Okada, M. Higashihara, and J. Fukumoto, *J. Agr. Chem. Soc. Japan* **42**, 665 (1968).

drolysis of maltoside by Taka-amylase A (*167, 168*). Taka-amylase A catalyzes the maltosidic cleavage of various α-maltosides such as methyl-, ethyl-, phenyl-, and *p*-nitrophenyl-α-maltoside (*168*). Thus, Taka-amylase A has a broad aglycon specificity.

On the other hand, the glycon specificity of Taka-amylase A was examined by using *O*-methyl derivatives of phenyl-α-maltoside (*169*) and partially *O*-methylated amylose (*170*). It was found that phenyl 2′-*O*-methyl maltoside was not hydrolyzed by Taka-amylase A, whereas 4′-*O*-, 6′-*O*-, and 4′,6′-di-*O*-methyl maltosides were hydrolyzed to produce corresponding maltose derivatives. In any products identified, glucose residue at the reducing end was shown to be unsubstituted. Moreover, monoacetyl maltoside, in which the hydroxyl group at C-6 was acetylated, was found not to be hydrolyzed by Taka-amylase A nor to inhibit the hydrolysis of phenyl-α-maltoside (*171*). When partially *O*-methylated, amylose was digested by Taka-amylase A, it predominantly attacked glucosidic linkage of glucose residue, neither C-2 nor C-6 hydroxyl of which was substituted by a methoxy group, while 6-*O*-methyl glucosidic and 2-*O*-methyl glucosidic bonds were not hydrolyzed (*170*). Weil and Bratt (*172*) also reported that the 6-*O*-methyl glucosidic bond in 6-*O*-methyl amylose was resistant to hydrolysis catalyzed by several amylases. The affinity of the binding site of Taka-amylase A to a substrate was sensitively influenced by the stereostructural change at C-2 (*173*). The glycon specificity of α-amylases is seemingly more strict than their aglycon specificity.

B. Assay of Amylase Activity

Amylase action is characterized by simultaneous changes of the following properties of substrate: (a) decrease of viscosity, (b) increase of reducing groups, (c) change in iodine-staining property, and (d) change

167. S. Matsubara, T. Ikenaka, and S. Akabori, *J. Biochem.* (*Tokyo*) **46,** 425 (1959).
168. S. Matsubara, *J. Biochem.* (*Tokyo*) **49,** 232 (1961).
169. M. Isemura, T. Ikenaka, and Y. Matsushima, *J. Biochem.* (*Tokyo*) **65,** 77 (1969).
170. M. Fujinaga-Isemura, T. Ikenaka, and Y. Matsushima, *J. Biochem.* (*Tokyo*) **64,** 73 (1968).
171. T. Ikenaka, *J. Biochem.* (*Tokyo*) **54,** 328 (1963).
172. C. E. Weil and M. Bratt, *Carbohydrate Res.* **4,** 230 (1967).
173. S. Ono, K. Hiromi, Y. Nitta. N. Suetsugu, and M. Tagaki, *Abstr. 7th Intern. Congr. Biochem., Tokyo, 1967* Vol. IV, p. 768. Sci. Council Japan, Tokyo, 1968.

in optical rotation. Numerous methods for the assay of amylase activity have been described in the literature. They are based on one of the above phenomena, especially (a), (b), and (c). Amylase activity is expressed as (a) liquefying power, (b) saccharifying power, and (c) dextrinizing power, respectively, according to the method. Reducing groups are determined colorimetrically according to the new Somogyi–Nelson method (*174*) or the 3,5-dinitrosalicylic acid method (*175*). Dextrinizing power is assayed by determining the changes of blue value in the amylose–iodine complex (*174*). Soluble starch or amylose is commonly used as a substrate. Several oligosaccharides and synthetic maltosides such as phenyl-α-maltoside and *p*-nitrophenyl-α-maltoside are also used as substrates (*167, 176*). Recently, new, simple, and rapid assay procedures have been developed using new chromogenic substrates such as Cibachrom F3GA-amylose (*177*) or Remazol-brilliant blue–starch complex (*178*).

C. Mode of Amylase Action

Amylases have been classified in various groups according to their action mechanisms. But recently, marked differences were found in the mode of action of various α-amylases. For example, *B. subtilis* produces two types of α-amylase, saccharifying and liquefying (*5, 6*). The saccharifying amylase shows a larger reducing power than the liquefying amylase when acting on starch, as shown in Fig. 3.

A novel oligosaccharide mapping method has been developed to compare the action pattern of various amylases (*179*). The method involves chromatography of maltooligosaccharides containing ^{14}C in the glucose moiety of the reducing end, incubation of the oligosaccharides on the paper with an amylase for hydrolysis, separation of the hydrolytic products on the paper by chromatography in the second direction, and location of the radioactive hydrolytic products by autoradiography. This method is sensitive; the attacked glucosidic linkage in ^{14}C-labeled malto-oligosaccharides can be easily determined by detecting radioactive hydrolytic products. By this method the action patterns of the two *B. subtilis* α-amylases have been studied. The saccharifying amylase hydrolyzed the maltooligosaccharides from the nonreducing end by a mal-

174. H. Fuwa, *J. Biochem.* (*Tokyo*) **41,** 583 (1954).
175. P. Bernfeld, *Advanc. Enzymol.* **12,** 379 (1951).
176. A. P. Jansen and P. G. A. B. Wydeveld, *Nature* **132,** 525 (1958).
177. B. Klein, J. A. Foreman, and B. L. Searcy, *Anal. Biochem.* **31,** 412 (1969).
178. H. Rinderknecht, P. Wilding, and B. J. Haverback, *Experientia* **23,** 805 (1967).
179. J. H. Pazur and S. Okada, *JBC* **241,** 4146 (1966).

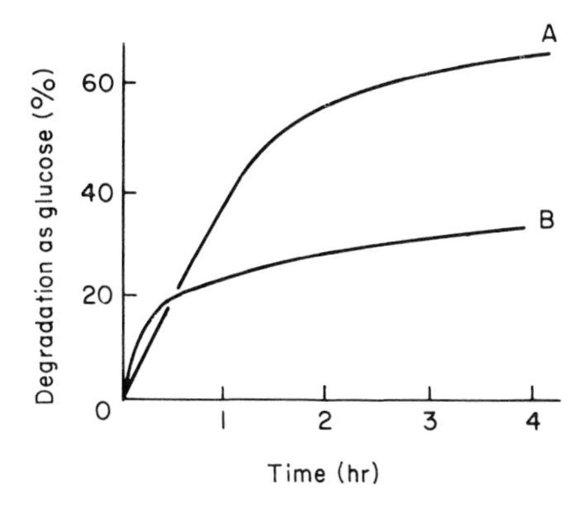

FIG. 3. Time-course of hydrolysis of starch by *B. subtilis* α-amylases (*5, 6*): 1% starch, 0.05 *M* acetate buffer, pH 6.0, 40°C. (A) Saccharifying type; and (B) liquefying type. Arrows indicate time at which iodine-starch color disappeared.

totriosyl unit; and the liquefying amylase hydrolyzed the maltooligosaccharides from the nonreducing end by a maltohexaosyl or maltopentaosyl unit (*180*). Okada *et al.* (*181*) further investigated action patterns of ten other kinds of amylases using the oligosaccharide mapping method. When maltooligosaccharides, labeled with ^{14}C at C-1 of glucose moiety of the reducing end, are used as substrates, the site of attack by an amylase is at a certain glucosyl linkage from the nonreducing end. α-Amylases were found to be divided into the following two groups according to the chain length of saccharide, with a new nonreducing end formed by the first attack: (1) *B. subtilis* liquefying, *B. stearothermophilus*, and malt α-amylases which hydrolyze by maltohexaosyl or maltopentaosyl unit; (2) *B. subtilis* saccharifying, *Endomycopsis*, and several fungal and animal α-amylases which attack by maltotriosyl units.

The presence of another type of amylase has been reported. Robyt and French (*182*) isolated an amylase from *B. polymyxa* and described its action pattern. The action catalyzes the degradation of the Schardinger dextrin to give maltose and maltotriose as products. This amylase shows characteristics of both α- and β-type amylases. When the amylase acts on amylose, maltoses are predominantly liberated. The upward change of optical rotation of the digest upon addition of alkali indicates that the products have β configuration. In this respect, the enzyme resembles β-amylase. However, when the enzyme acts on G_8 (where G is a glucose

180. S. Okada, M. Higashihara, and Fukumoto, *J. Agr. Chem. Soc. Japan* **42,** 665 (1968).

181. S. Okada, S. Kitahara, M. Higashihara, and J. Fukumoto, *Agr. Biol. Chem.* (*Tokyo*) **33,** 900 (1969).

182. J. Robyt and D. French, *ABB* **104,** 338 (1964).

unit), a predominant amount of maltose and small amounts of G_3, G_4, and G_6 are initially formed. Judging from the formation of a product having an odd number of glucose units (G_3) from a substrate having an even number of glucose units (G_8), the mechanism of action of *B. polymyxa* amylase is not like that of β-amylase which produces only maltose from substrates having an even number of glucose units. Since G_5 is not produced during the digestion, it was suggested that when G_3 is split from G_8, multiple attacks occur to give two molecules of G_3 and one molecule of G_2 (*183, 184*).

Additional information on the action pattern of amylases is given in Chapter 6 by Thoma, Spradlin, and Dygert, this volume.

D. Mechanism of the Reaction Catalyzed by Amylase

The kinetics and mechanism of reactions catalyzed by amylases have been reviewed in the second edition of "The Enzymes," and the effects of salt and pH on the amylase action described (*75*). Considerable progress has been made during recent years in the field of kinetics of amylase reaction owing to the development of several synthetic substrates and substrate analogs.

Ono *et al.* (*185–187*) established a general method for quantitative determination of anomer forms of sugar produced by amylases. The changes in optical rotation and reducing power during the enzymic reaction were measured simultaneously by using a simple substrate which had only one glucosidic linkage to be hydrolyzed. By this method, Ono *et al.* quantitatively determined that glucamylase (an exoamylase) produced α-glucose from phenyl-α-maltoside, maltose, or isomaltose (*185, 187a*) and Taka-amylase A (an endoamylase) produced α-maltose from phenyl-α-maltoside (*187*). But the action of *B. subtilis* saccharifying α-amylase, characterized by its high yield of glucose from starch compared with other amylases (*5, 132*), seems to be intermediate between the action of exo- and endoamylases. Yoshida *et al.* (*188*) demonstrated that the hydrolytic reaction of phenyl-α-maltoside catalyzed by this enzyme involves two parallel processes, one produces phenol and maltose

183. J. M. Balley and D. French, *JBC* **226,** 1 (1957).
184. M. Abdullah, D. French, and J. Robyt, *ABB* **114,** 595 (1966).
185. S. Ono, K. Hiromi, and Z. Hamauzu, *J. Biochem.* (*Tokyo*) **57,** 34 (1965).
186. Z. Hamauzu, K. Hiromi, and S. Ono, *J. Biochem.* (*Tokyo*) **57,** 34 (1965).
187. Z. Hamauzu, K. Hiromi, and S. Ono, *J. Biochem.* (*Tokyo*) **57,** 39 (1965).
187a. S. Ono, K. Hiromi, Y. Nitta, N. Suetsugu, and M. Takagi, *Symp. Enzyme Chem.* (*Tokyo*) **18,** 229 (1966).
188. H. Yoshida, K. Hiromi, and S. Ono, *J. Biochem.* (*Tokyo*) **62,** 439 (1967).

(process 1), and the other produces phenyl-α-glucoside and glucose (process 2). Either of these two processes may be dominant, depending on the substrate concentration. At lower substrate concentrations, process 1 is dominant, and at higher substrate concentrations, process 2 is dominant (*188, 189*). Therefore, under conditions where only process 1 occurs, quantitative determination of the anomeric form of maltose was carried out by the simultaneous measurement of phenol and optical rotation. It was demonstrated that maltose released from phenyl-α-maltoside by the action of *B. subtilis* liquefying α-amylase was exclusively of α- form, indicating that the configuration of the substrate linkage between glucose and phenyl residues was retained during the hydrolysis catalyzed by this enzyme (*190*).

Yoshida *et al.* (*189*) also demonstrated the transfer action of *B. subtilis* saccharifying amylase. At higher substrate concentrations, the intermedial production of maltotriose was shown by thin layer chromatography and paper chromatography (*190*). Further investigation of the mechanism of transfer gave evidence for glucosyl-transfer action of the enzyme (*191, 192*). Phenyl-α-glucoside was not hydrolyzed by the enzyme. Addition of glucose to the above system caused no change. Therefore, at least under the conditions studied, no evidence for synthetic action of the enzyme was obtained. However, when the mixture of phenyl-α-glucoside and maltose was incubated with the enzyme, phenol and glucose were formed, although phenyl-α-glucoside itself was not hydrolyzed, either in the presence or absence of glucose. From these findings, it was concluded that phenyl-α-maltoside was formed from maltose and phenyl-α-glucoside by glucosyl transfer action catalyzed by the enzyme, and then phenyl-α-maltoside was hydrolyzed into phenol and maltose. Maltose was also hydrolyzed into glucose by a side reaction of the enzyme. Methyl alcohol was also found to serve as an acceptor, judging from the formation of methyl-α-glucoside detected by thin layer chromatography. A similar transfer action was reported in the enzymic action catalyzed by Taka-amylase A (*193*).

Chemically modified amylose was used as a substrate to study amylase action (*172, 194, 195*). Ikenaka (*171*) prepared a monoacetyl-α-

189. H. Yoshida, K. Hiromi, and S. Ono, *J. Biochem.* (*Tokyo*) **65,** 741 (1969).
190. K. Hiromi, T. Shibata, H. Fukube, and S. Ono, *J. Biochem.* (*Tokyo*) **66,** 183 (1969).
191. H. Yoshida, K. Hiromi, and S. Ono, *J. Biochem.* (*Tokyo*) **66,** 183 (1969).
192. H. Yoshida, K. Hiromi, and S. Ono, *J. Biochem.* (*Tokyo*) **65,** 741 (1969).
193. S. Matsubara, *J. Biochem.* (*Tokyo*) **49,** 226 (1961).
194. B. J. Bines and W. J. Whelan, *Chem. & Ind.* (*London*) p. 997 (1960).
195. C. E. Weil and R. Rabbahn, *Carbohydrate Res.* **3,** 242 (1966).

maltoside and identified it as *O*-α-D-glucopyranosyl-(1,4)-phenyl-6-*O*-acetyl-α-D-glucopyranoside. He reported that monoacetyl maltoside was not hydrolyzed by Taka-amylase A and did not inhibit the enzymic hydrolysis of phenyl-α-maltoside. Therefore, it might be concluded that the hydroxyl group at C-6 participates in the formation of Michaelis complex of Taka-amylase A and a substrate. Fujinaga-Isemura *et al.* (*170*) obtained an interesting result in the study of enzymic hydrolysis of partially *O*-methylated amylose by Taka-amylase A. They prepared a partially *O*-methylated amylose where hydroxyl groups at C-2 and C-6 were substituted. All the oligosaccharides so far isolated from the Taka-amylase A digests of the *O*-methylated amylose had unsubstituted glucose at the reducing end. 6-*O*-Methyl glucosylglucose (i.e., 6′-*O*-methyl maltose) was isolated from the digest, but 2′-*O*-methyl maltose was not produced. These results gave evidence indicating that the glucosidic bonds in *O*-methylated amylose were hydrolyzed by Taka-amylase A, but the 2-*O*- and 6-*O*-methyl glucosidic bonds were not attacked by the enzyme. Previously, Ikenaka had shown that the maltoside linkage of phenyl 6-*O*-acetyl maltoside was not hydrolyzed by Taka-amylase A (*171*). Furthermore, according to Weil and Bratt (*172*), the 6-*O*-methyl glucoside bond in 6-*O*-methyl amylose was resistant to the hydrolysis catalyzed by all amylases tested, including hog pancreatic and *B. subtilis* α-amylases, sweet potato β-amylase, potato phosphorylase, and a gluc-amylase mixture from *A. niger*. It seems that a hydroxyl group at C-6 plays a very important universal role in the interaction between amylase and substrate. The isolation of 6′-*O*-methyl maltose also provides evidence that a glucosidic bond of glucose residue that is adjacent to a 6-*O*-methyl glucose residue can be hydrolyzed by Taka-amylase A.

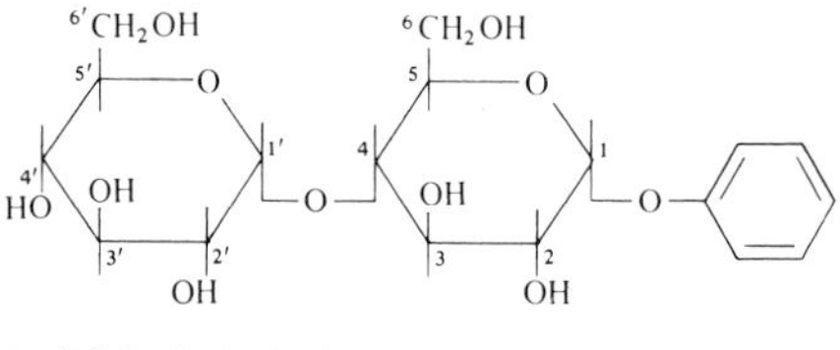

FIG. 4. *O*-Methyl derivatives of phenyl-α-maltoside:

Derivatives	Products
Phenyl-2′-*O*-methyl-α-maltoside	Not hydrolyzed
Phenyl-3′-*O*-methyl-α-maltoside	3′-*O*-Methyl maltose
Phenyl-4′-*O*-methyl-α-maltoside	4′-*O*-Methyl maltose
Phenyl-6′-*O*-methyl-α-maltoside	6′-*O*-Methyl maltose
Phenyl-4′,6′-di-*O*-methyl-α-maltoside	4′,6′-Di-*O*-methyl maltose

Contrarily 2′-*O*-methyl maltose cannot be isolated from the digest of an *O*-methylated amylose in which more 2-*O*-methyl glucose residues are contained. This finding shows that 2-*O*-methylation makes even the adjacent glucosidic bond resistant to enzymic action. Ono *et al.* (*173, 187a*) reported that the affinity at the binding site of Taka-amylase A to the substrate was sensitively influenced by the stereostructural changes at C-2. Thus a hydroxyl group at C-2 seems to be of as much or more importance in allowing amylose to be attacked by an amylase than a hydroxyl group at C-6.

The influence of hydroxyl groups of substrate on amylase action has been investigated using *O*-methylated derivatives of phenyl-α-maltoside (Fig. 4) (*169*). The derivatives were digested by Taka-amylase A and their products analyzed by gas chromatography and paper chromatography. Among the products, 3′-*O*-methyl maltose, 4′-*O*-methyl maltose, 6′-*O*-methyl maltose, and 4′,6′-di-*O*-methyl maltose were detected. No maltose derivatives with *O*-methyl substitution on the reducing end glucose residue were produced by enzyme digestion. Phenyl-2′-*O*-methyl-α-maltoside was found to be resistant to enzymic action. From these results, it has been concluded that in the maltose moiety of a substrate molecule, all hydroxyl groups on the reducing end glucose residue and the 2-hydroxyl group on the nonreducing end glucose residue participate in the amylase reaction (*169*). These results obtained from kinetics studies on *O*-methyl derivatives of phenyl-α-maltoside coincide with those obtained from studies on *O*-methylated amylose (*170*).

9

Cellulases

D. R. WHITAKER

I. Introduction

The Commission on Enzymes lists "cellulase" as the trivial name for enzymes which have the systematic name β-1,4-glucan 4-glucanohydrolase (EC 3.2.1.4). The systematic name has a precision which the original definition lacked but, if cellulases are to be discussed and not merely defined, one must still come to terms with their substrates. A strip of cellophane and a pine log can provide substrates for cellulases, but somewhere between these extremes the general problems associated with the degradation of a particular polymer become intertwined with other problems associated with the degradation of a particular type of cell or tissue. The emphasis in this chapter is on some of the more general aspects of cellulases. Some of the more specialized aspects are discussed in recent reviews and monographs (*1–4*).

II. Significance and Distribution

Enzymes which hydrolyze cellulose can be used by their parent organisms for at least three different purposes. First, they can be used as morphogenic agents which weaken cellulose-containing cell walls in preparation for growth or differentiation or for abscissions such as the dropping of seeds, flowers, or leaves. Plant cellulases and some fungal cellulases fulfill this role; some examples are mentioned later. Second, they can be used as invasive agents which, for example, enable a microbial plant pathogen to penetrate the tissues of its host. Third, they can be used as digestive agents which enable plant tissues to be penetrated by other enzymes and enable cellulose itself to be used as a carbon source. The cellulases of invertebrate animals and of many microorganisms have this function.

It is not always obvious whether a cellulase obtained from an invertebrate is of animal or of microbial origin. Some invertebrates, like the vertebrates, may rely on cellulolytic microorganisms in their digestive tracts to digest cellulose—the termite, for example, has a great variety of cellulolytic protozoa. However, some of the worst scourges of timber in the sea—the crustacea of the genus *Limnoria*—have digestive tracts which are virtually free of microorganisms (*5*). The same is true for certain saprophytic nematodes (*6*) and the silverfish, an insect notoriously fond of paper, has been raised aseptically without loss of its ability to digest cellulose (*7*). Other invertebrates may have a working population of cellulolytic microorganisms in their intestinal tract but produce cellulases in their digestive glands; the edible snail, *Helix*

1. E. T. Reese, ed., "Advances in Enzymic Hydrolysis of Cellulose and Related Materials." Pergamon Press, Oxford, 1963.
2. B. Norkrans, *Advan. Appl. Microbiol.* **9**, 91 (1967).
3. L. Jurášek, J. R. Colvin, and D. R. Whitaker, *Advan. Appl. Microbiol.* **9**, 131 (1967).
4. G. J. Hajny and E. T. Reese, eds., "Cellulases and their Applications." Amer. Chem. Soc., Washington, D. C., 1969.
5. D. L. Ray, ed., "Marine Boring and Fouling Organisms," p. 50. Univ. of Washington Press, Seattle, Washington, 1959.
6. M. Tracey, *Proc. 2nd Intern. Congr. Biochem., Paris, 1952* p. 242. Masson, Paris, 1953.
7. R. Lasker and A. C. Giese, *J. Exptl. Biol.* **33**, 152 (1956).

pomatia (*8*), several marine mollusks (*9–11*), a crayfish (*12*) and a sea urchin (*13*) are examples.

III. Assay and Detection

The progression from cellobiose to a cellulose fiber is a progression in chain length but there are other progressions as well. As chain length increases, there is an increasing tendency for chains to associate. The hydrogen bonds and van der Waals forces responsible for the association are weak individually but strong collectively. As a result, segments of adjacent chains can become so closely and regularly packed (i.e., "crystalline") that their hydroxyl groups are no longer accessible to the water molecules which could disrupt the hydrogen bonds between them. If the cellulose is a native cellulose (as opposed to a regenerated cellulose like cellophane), the cellulose will be in microfibrils, i.e., fine threads roughly 100 Å in width. If the cellulose is in a native fiber, there may be other progressions associated with its being a component of a cell and if it is a desiccated fiber, such as cotton, there may be still others associated with the drying process.

These and other considerations divide the insoluble substrates in common use into roughly three groups. Group I is headed by dried, native fibers with a high cellulose content such as de-waxed cotton fibers. Filter paper, various ground celluloses, as well as celluloses which, though partly degraded, are highly "crystalline" (for example, celluloses which have been given a mild acid hydrolysis to remove noncrystalline cellulose) could be added to this group. These are celluloses which, in comparison with those of group II, are slowly hydrolyzed by cellulases and resist hydrolysis by many cellulases. Assay procedures can be based on the loss of weight or on the formation of reducing sugars; detection procedures can be based, for example, on the formation of clear zones in cellulose-agar (*14, 15*). Group II includes celluloses which, though not extensively degrated, are extensively hydrated. Cellulose which has

8. G. A. Strasdine and D. R. Whitaker, *Can. J. Biochem. Physiol.* **41,** 1621 (1963).
9. S. Horiuchi and C. E. Lane, *Comp. Biochem. Physiol.* **17,** 1189 (1966).
10. G. Okada, T. Nisizawa, and K. Nisizawa, *BJ* **99,** 214 (1966).
11. L. A. Elyakova, V. V. Sova, and V. E. Vaskovsky, *BBA* **167,** 462 (1968).
12. Y. Yokoe and I. Yasumasu, *Comp. Biochem. Physiol.* **13,** 323 (1964).
13. E. Hultin and I. Wanntorp, *Acta Chem. Scand.* **20,** 2667 (1966).
14. G. S. Rautela and E. B. Cowling, *Appl. Microbiol.* **14,** 892 (1968).
15. J. G. Savory, C. C. Maitland, and K. Selby, *Chem. & Ind.* (*London*) p. 153. (1967).

been swollen in cold phosphoric acid (*16*) is one of the best substrates of this group. Regenerated celluloses such as cellophane also belong here. Group III includes substrates which are extensively degraded and hydrated. The cellodextrins obtainable by deacetylation of acetolyzed cellulose (*17*) are extreme examples of this type of substrate.

There are two categories of soluble substrates in common use. The first comprises the soluble cello-oligoglucosides in the series from cellobiose to cellohexaose. These are individual sugars and, as such, have a unique capability for characterizing certain catalytic properties of a cellulase in unequivocal, absolute terms.

The second category consists of soluble derivatives of cellulose such as carboxymethyl celluloses and hydroxyethyl celluloses. The former has been the more popular, but the latter is the substrate of choice if the effects of pH or ionic strength are to be interpreted in terms of their effects on the enzyme alone (*18, 19*). Films of carboxymethyl cellulose have been used to detect cellulase activity in tissue sections (*20*) and conventional assays can be based on the rate of formation of reducing sugar or, with greater sensitivity, on the initial rate of decrease in viscosity. If the substrate preparation has been adequately characterized, it is possible to relate the initial rate measured with a viscometer to the number of linkages cleaved per minute (*21, 22*). However, other preparations of the same substrate may give different rates. The resulting problem in data comparison could be overcome if the rates were expressible as a simple function of some readily determinable parameters of the substrate preparation. Almin and Eriksson (*22*) have made a careful study of the point, and it is evident from their findings that the two most easily determined parameters—the average degree of polymerization of the substrate, i.e., the average number of anhydroglucose residues per molecule, and the average degree of substitution (average number of substituting groups per residue)—do not provide sufficient information. Eriksson and Hollmark's data for *Stereum sanguinolentum* cellulase (*23*) suggest that the *initial* rates of degradation measured in a viscometer are determined mainly by cleavages at segments with three or more unsubstituted residues. Wirick (*24*) has also stressed the importance of extended sequences of unsubstituted residues. Such sequences are difficult

16. C. S. Walseth, *Tappi* **35,** 228 (1954).
17. D. R. Whitaker, *Can. J. Biochem. Physiol.* **34,** 488 (1956).
18. T. Iwazaki, K. Tokuyasu, and M. Funatsu, *J. Biochem. (Tokyo)* **55,** 30 (1964).
19. G. Pettersson, *ABB* **126,** 776 (1968).
20. A. T. Summer, *Histochemie* **13,** 160 (1968).
21. R. Werner, *J. Polymer Sci.* **C16,** 4429 (1969).
22. K. E. Almin and K. E. Eriksson, *ABB* **124,** 129 (1968).
23. K. E. Eriksson and B. H. Hollmark, *ABB* **133,** 233 (1969).
24. M. G. Wirick, *J. Polymer Sci.* **6,** 1705 (1965).

to characterize and, as discussed later, can differ by orders of magnitudes in their properties as binding sites for cellulases. The nature of the substituting group can also be important. Hydroxyethyl cellulose, for example, is reported to be a much better substrate for *Polyporus schweinitzii* cellulase than carboxymethyl cellulose (*25*). Possibly the factor involved is nonproductive binding by the carboxyl groups of the latter (*26*).

Assays of activity toward soluble derivatives of cellulose are often referred to as assays of "C_x activity" while assays of activity toward insoluble substrates in Group I are often referred to as assays of "C1 activity." The reasons are mentioned later.

IV. Induction and Repression

Several hormones are known to influence the production of cellulases with a morphogenic function. The application of indoleacetic acid to decapitated pea epicotyles leads to cellulase production (*27*). "Hormone A" stimulates cellulase production by male strains of the water mold, *Achlya*, the peak production of cellulase coinciding with the appearance of the hyphal branches which become male sexual organs (*28*). Abscisic acid induces the cellulase associated with abscission processes in cotton and bean plants (*29*).

The cellulase of pea epicotyles appears to be synthesized by membrane-bound ribosomes (*30, 31*); Davies and Maclachlan suggested that indoleacetic acid may be involved in the formation of polysomes at the membrane. Their findings strengthen some suggestive evidence that the cellulase of *Cellvibrio gilvus*, a cellulolytic bacterium, is also produced at ribosomes bound to the cell membrane (*32*).

A β-1,4-glucan is usually an essential component of the medium for cellulase production by a cellulolytic microorganism. Carboxymethyl cellulose (*33*) and hydroxyethyl cellulose (*34*) satisfy this condition for some organisms and, though the yield of cellulase may be reduced,

25. G. Keilich, P. J. Bailey, E. G. Afting, and W. Liese, *BBA* **185**, 392 (1969).
26. P. K. Datta, K. R. Hanson, and D. R. Whitaker, *BBA* **50**, 113 (1961).
27. D. F. Fan and G. A. Maclachlan, *Plant Physiol.* **42**, 1114 (1967).
28. D. des S. Thomas and J. T. Mullins, *Science* **156**, 84 (1967).
29. L. E. Cracker and F. B. Abeles, *Plant Physiol.* **44**, 1144 (1969).
30. E. H. Davies and G. A. Maclachlan, *ABB* **128**, 595 (1968).
31. E. H. Davies and G. A. Maclachlan, *ABB* **129**, 581 (1969).
32. S. A. Carpenter and L. B. Barnett, *ABB* **122**, 1 (1967).
33. G. L. Schneberger and W. W. Luchsinger, *Can. J. Microbiol.* **13**, 901 (1967).
34. M. A. Jermyn, *Australian J. Biol. Sci.* **20**, 221 (1967).

highly crystalline celluloses with various lattice structures can replace cotton cellulose in the growth medium for *Trichoderma viride* (*35*). If the induction and repression of cellulase synthesis is subject to the usual genetic control, and some preliminary data for *Neurospora crassa* cellulase (*36*) suggest that it is, it might be expected that a soluble product of cellulose digestion would be the effective inducing agent. Depending on the concentration, cellobiose can be both an inducer and a repressor for some organisms; the low concentration required for induction can sometimes be maintained by supplying the cellobiose as an ester which the organism's esterases can convert to the free sugar (*37*). Glucose is usually a repressor. Horton and Keen found that glucose concentrations as low as 5×10^{-4} *M* were sufficient to repress cellulase synthesis by the plant pathogen, *Pyrenochaeta terrestris* (*38*); as they pointed out, such ease of repression need not be a handicap for an organism which uses cellulases primarily for invasive purposes.

V. Production and Isolation

A. Cultural Conditions

Four culture methods are in common use for the production of microbial cellulases: (1) submerged culture in a liquid medium with forced aeration from spargers, (2) shake culture in a liquid medium, (3) stationary culture in a liquid medium, and (4) Koji-type processes in which the organism is grown on a moist, semisolid medium. The carbon sources used for industrial production are often exceedingly crude sources of cellulose such as wheat bran, straw, or cottonseed meal and most industrial cellulase preparations contain a variety of other enzymes.

B. Influence of Cultural Conditions on the Apparent Physical Properties of Crude Cellulases

The chromatographic and electrophoretic properties of the cellulase in a culture filtrate are often greatly influenced by the conditions of culture. For example, industrial preparations of *Trichoderma viride* cellulase give much more complex displacement patterns from DEAE-

35. G. S. Rautela and K. W. King, *ABB* **123,** 589 (1968).
36. M. G. Meyers and B. Eberhart, *BBRC* **24,** 782 (1966).
37. E. T. Reese, J. E. Lola, and F. W. Parrish, *J. Bacteriol.* **100,** 1151 (1969).
38. J. C. Horton and N. T. Keen, *Can. J. Microbiol.* **12,** 209 (1966).

Sephadex than the cellulase obtained by growing the same strain of the mold in shake culture with powdered cellulose as the carbon source (*39*). The cellulase of *Myrothecium verrucaria* has provided the most spectacular record of differences in cultural conditions leading to differences in electrophoretic properties (*40*); a more subtle example is the change in isoelectric point of *Chrysosporium lignorum* cellulases during culture of the organism (*41*).

One of the main factors is the formation of complexes with microbial polysaccharides. Some cellulases, that of *Myrothecium* cellulase, for example, can be produced free of bound polysaccharide (*42*). Others, such as those of *Stereum sanguinolentum* (*43*) and of *Penicillium notatum* (*44, 45*), are readily freed of bound carbohydrate. However, cellulases such as *Chrysosporum lignorum* cellulase (*46*) and some *Trichoderma viride* cellulases (*39, 47*) are not readily freed of carbohydrate and may be glycoproteins. The carbohydrate component of such enzymes cannot be assigned any function with certainty, but there are suggestions that it may stabilize the enzyme and may modify its enzymic properties (*47, 48*).

There are suggestions of other processes as well. Evidence of the formation of complexes with polypeptides has been reported for *Myrothecium* cellulase (*49*). Proteolysis may also be a factor since proteases are often present and have raised problems in the isolation of some cellulases (*47, 49*).

C. Isolation Methods

Cellulases present a few special problems in isolation work since several familiar materials are substrates for cellulases. Thus cellophane dialysis membranes must be protected by acetylation (*49*), for example, or replaced by collodion or animal membranes. Cellulose ion exchange resins have been used with buffers containing high concentrations of

39. Y. Tomita, H. Suzuki, and K. Nisizawa, *J. Ferment. Technol.* **46,** 701 (1968).
40. D. R. Whitaker, *Bull. Soc. Chim. Biol.* **42,** 1701 (1960).
41. K. E. Eriksson and W. I. Rzedowski, *ABB* **129,** 683 (1969).
42. R. Thomas and D. R. Whitaker, *Nature* **181,** 715 (1958).
43. K. E. Eriksson and B. Pettersson, *ABB* **124,** 142 (1968).
44. G. Pettersson, *ABB* **123,** 307 (1968).
45. G. Pettersson and D. L. Eaker, *ABB* **124,** 154 (1968).
46. K. E. Eriksson and W. Rzedowski, *ABB* **129,** 689 (1969).
47. G. Okada, K. Nisizawa, and H. Suzuki, *J. Biochem.* (*Tokyo*) **63,** 591 (1968).
48. T. Toda, H. Suzuki, and K. Nisizawa, *J. Ferment. Technol.* **46,** 711 (1968).
49. D. R. Whitaker, P. K. Datta, and K. R. Hanson, *Can. J. Biochem. Physiol.* **41,** 671 (1963).

cellobiose to inhibit the cellulase (*49*), but they are not resins of choice and fortunately the Sephadex ion exchangers have eliminated much of the need for them.

Pettersson's procedure (*44*) for the isolation of *Penicillium notatum* cellulase is a good example of an effective combination of purification procedures. Ion exchange chromatography on DEAE-Sephadex, gel filtration on Sephadex G-75, and preparative electrophoresis in Sephadex G-25 were applied in sequence with high recoveries of activity at each step. The enzyme at this point was homogeneous on disc electrophoresis but contained a small and variable amount of carbohydrate. Chromatography on hydroxylapatite reduced the carbohydrate content to less than 1% (*45*).

VI. "C1" Factors

As mentioned earlier, assays of activity toward dried fibers or preparations of crystalline cellulose are often designated in the literature as assays of "C1 activity," while assays with substrates such as carboxymethyl cellulose are designated as assay of "C_x activity." This notation originated with a hypothesis of Reese and his co-workers (*50, 51*) who postulated that a nonhydrolytic enzyme, C1, was responsible for the initial attack on native cellulose; a hydrolase, C_x, completed the degradation to soluble sugars. This hypothesis was subsequently modified to a "multiple C_x" theory which extended the specificity of some hydrolases (*52*) but, more recently, the C1 concept of a specific enzyme for the initial attack has again attracted attention.

Two papers of Selby and Maitland are particularly relevant to this revival. The first (*53*) reported that cellulase in culture filtrates of *Myrothecium verrucaria* could be fractionated by gel filtration on Sephadex G-75 into three components with molecular weights, estimated from the elution volume on Sephadex, 55,000, 30,000 and 5,300. (These estimates will be underestimates if elution was retarded by adsorption of enzyme on Sephadex.) The first and third of these components differed from the second in that they were more readily adsorbed by cotton, less active toward carboxymethyl cellulose, and more active toward

50. E. T. Reese, R. G. H. Siu, and H. S. Levinson, *J. Bacteriol.* **59**, 485 (1950).
51. E. T. Reese and H. S. Levinson, *Physiol. Plantarum* **5**, 345 (1952).
52. E. T. Reese, *in* "Marine Boring and Fouling Organisms" (D. L. Ray, ed.), p. 265. Univ. of Washington Press, Seattle, Washington, 1959.
53. K. Selby and C. C. Maitland, *BJ* **94**, 578 (1965).

cotton fibers in tests of loss of tensile strength. Combinations of the three components showed no evidence of synergism.

The second paper by Selby and Maitland (*54*) was on cellulase in culture filtrates of *Trichoderma viride*. This filtrate had a much higher activity toward cotton and their assay was based on determinations of the solubilization of cotton fibers over a period of 7 days. Gel filtration of the filtrate on Sephadex G-75, followed by ion exchange on DEAE-Sephadex and SE-Sephadex gave three components which influenced the degradation of cotton: (a) a C1 component with no activity toward carboxymethyl cellulose or cellobiose, (b) a component with high activity toward carboxymethyl cellulose and a low activity toward cellobiose, and (c) a component with a low activity toward carboxymethyl cellulose and a high activity toward cellobiose.

Individually none of these components gave appreciable solubilization of cotton but the combination (a) + (b) accounted for 35%, (a) + (c) for 20%, and (a) + (b) + (c) for approximately 100% of the filtrate's activity. The molecular weights of all three were estimated by gel filtration to be in the range 48,000–62,000; the C1 component was approximately 50% carbohydrate.

Wood extended these findings with essentially similar data for culture filtrates of *Trichoderma koningi* (*55*) and *Fusarium solani* (*56*). The C1 factors of each organism were capable of acting synergistically with the C_x component of the other. He also found that "swelling factor" activity, a measure of the effect of enzyme treatment on a fiber's capacity to swell in concentrated alkali (*57*), was associated with the C_x enzyme not with the C1 factor. This finding is of interest relative to Selby and Maitland's observations on *Myrothecium* cellulase since changes in swelling properties are usually evident before changes in tensile strength can be detected.

At present then, the C1 factor cannot be claimed to be an enzyme or to be operative on substrates other than dried fibers or to have a C1 function in the sense that it initiates a degradation. It may be that it has catalytic bond breaking or exchange properties which remain to be detected but nonenzymic effects are also possible. For example, as one possibility, when adsorbed on cellulose it may distort the local structure sufficiently to allow water molecules to hydrate previously unexposed segments of chains or it could be that the C1 factor simply protects the

54. K. Selby and C. C. Maitland, *BJ* **104,** 716 (1967).
55. T. M. Wood, *BJ* **109,** 217 (1968).
56. T. M. Wood, *BJ* **115,** 457 (1969).
57. P. Marsh, K. Bollenbacher, M. L. Butler, and L. R. Guthrie, *Textile Res. J.* **23,** 878 (1953).

cellulase from denaturation or irreversible adsorption. Protective effects are common and have been demonstrated with cellulases (*58*); however, bovine serum albumin cannot replace C1 factor (*54*).

VII. Physical and Chemical Properties of Cellulases

The cellulase of *Penicillium notatum* provides a good basis for comparison since Pettersson and his co-workers have characterized its physical and chemical properties very thoroughly. The enzyme has a molecular weight of 35,000, a negligible carbohydrate content, and the amino acid composition shown in Table I. It has no free sulfhydryl

TABLE I
AMINO ACID COMPOSITION OF *Penicillium notatum* CELLULASE[a]

Amino acid	Residues/mole	Amino acid	Residues/mole
Lysine	13	Glycine	28
Histidine	3	Alanine	33
Arginine	6	Valine	18
Aspartic acid + asparagine	40	Methionine	8
Threonine	29	Isoleucine	15
Serine	27	Leucine	21
Glutamic acid + glutamine	29	Tyrosine	15
Proline	11	Phenylalanine	13
Half-cystine	2	Tryptophan	13
		Amide groups	25

[a] Data taken from Pettersson and Eaker (*45*).

groups and an alanine residue as its single N-terminal amino acid (*45*). Thus the enzyme appears to be a single chain containing about 324 amino acid residues and one disulfide bridge. It is an acidic protein with an isoelectric point of 3.7 and, as shown by titration data, practically all its ionizing groups are exposed to the solvent (*59*). Its optical rotatory dispersion spectrum shows no evidence of pH-dependent changes in conformation between pH 4 and 8.5, and the spectrum is consistent with a conformation which would have 20–30% of the amino acid residues in α-helices (*59*). The native enzyme susceptible to degradation by endopeptidases (*60*).

Other fungal cellulases have been isolated which differ from this enzyme in molecular weight or in carbohydrate content or in the range

58. D. R. Whitaker, *Science* **116,** 90 (1952).
59. G. Pettersson and L. Andersson, *ABB* **124,** 497 (1968).
60. K. E. Eriksson and G. Pettersson, *ABB* **124,** 160 (1968).

of substrates which they degrade effectively. Table II compares a set of them (*39, 45, 47, 61–67*). All the molecular weights were estimated with the ultracentrifuge. The range of substrates listed for each enzyme is somewhat arbitrary and subject to some uncertainties, but briefly the properties reported for these enzymes are as follows. Enzymes (a), (b), and (c) have little or no activity toward cellobiose and they hydrolyze fibrous cellulose very slowly; they have moderate to high activity toward the higher oligoglucosides, toward soluble derivatives of cellulose, and toward swollen celluloses. Enzymes (c) and (d) may be similar. Enzyme (f) has a higher activity toward cellobiose and hydroxyethyl cellulose and little activity toward fibrous cellulose while (g) has a low activity toward the first two substrates and a high activity toward the third. Enzymes (h) to (j) do not hydrolyze cellobiose, but with some differences in their relative activities they hydrolyze all other substrates including fibrous cellulose. Table II shows no obvious correlations between activity and molecular weight, but there is possibly a suggestion that the carbohydrate components—mannose and hexosamine for (e), mainly, mannose and glucose for (h), (i), and (j)—are relevant to an ability to degrade fibrous cellulose.

A comparison of the amino acid composition in Table I with those reported for the other enzymes in Table II suggests that several features of *P. notatum* cellulase are typical features of fungal cellulases. These include the very high content of aspartic acid–asparagine residues, the high content of glutamic acid–glutamine residues, the low content of basic amino acids, and the comparatively high content of aromatic residues. It might be expected that fungal cellulases would tend to be fairly acidic proteins. Ahlgren and her co-workers have made the most precise observations on this point in the course of their applications of isoelectric focusing to the isolaton of cellulases from an *Aspergillus* (*68*) and from *Stereum sanguinolentum*, *Fomes annosus*, and *Chryosporium lignorum* (*69*). The range of isolectric points which they have observed

61. G. Pettersson, *ABB* **130,** 286 (1969).
62. P. K. Datta, K. R. Hanson, and D. R. Whitaker, *Can. J. Biochem. Physiol.* **41,** 697 (1963).
63. G. Pettersson, E. B. Cowling, and J. Porath, *BBA* **67,** 1 (1963).
64. G. Pettersson and J. Porath, *BBA* **67,** 9 (1963).
65. T. Iwazaki, K. Hayashi, and M. Funatsu, *J. Biochem.* (*Tokyo*) **57,** 467 (1965).
66. T. Iwazaki, R. Ikeda, K. Hayashi, and M. Funatsu, *J. Biochem.* (*Tokyo*) **57,** 478 (1965).
67. L. H. Li, R. M. Flora, and K. W. King, *ABB* **111,** 439 (1965).
68. E. Ahlgren, K. E. Eriksson, and O. Vesterberg, *Acta Chem. Scand.* **21,** 937 (1967).
69. E. Ahlgren and K. E. Eriksson, *Acta Chem. Scand.* **21,** 1193 (1967).

TABLE II
COMPARISON OF FUNGAL CELLULASES

Source	Ref.	Molecular wt	Sedimentation coefficient (S)	Carbohydrate content (%)	Range of substrates
(a) *Penicillium notatum*	(*45, 61*)	35,000	3.13	<1	Oligoglucosides to swollen celluloses
(b) *Myrothecium verrucaria*	(*62*)	49,000		<1	Oligoglucosides to swollen celluloses
Polyporus versicolor	(*63, 64*)				
(c) Cellulase B1		50,000	4.21	1	Range uncertain but both enzymes hydrolyze carboxymethyl cellulose and a degraded cellulose
(d) Cellulase D		11,400	1.53	33	
Trichoderma koningi	(*65, 66*)				
(e) Cellulase I		50,000	3.81	12	Oligoglucosides to fibrous cellulose
(f) Cellulase II		26,000	3.13	2	Cellobiose to hydroxyethyl cellulose
Trichoderma viride					
(g) Endoglucanase[a]	(*67*)	50,000	3.6	—	Oligoglucosides to swollen cellulose
(h) Cellulase II[b]	(*47*)		3.54	17	Oligoglucosides to fibrous cellulose
(i) Cellulase III[b]	(*47*)		3.75	16	Oligoglucosides to fibrous cellulose
(j) Cellulase IV[b]	(*47*)		3.62	10	Oligoglucosides to fibrous cellulose

[a] Li *et al.* used this term to distinguish the enzyme from a C1 component, an enzyme with a high activity toward a preparation of "crystalline" cellulose.

[b] Cellulases III and IV are probably derivatives of cellulase II; they are present in commercial preparations of *T. viride* cellulase but not in shake cultures with cellulose powder as carbon source; another component of the commercial products can be produced by incubating cellulase II in a glycerol medium (*39*).

is from pH 3.6 to 4.7. The isoelectric points of *P. notatum* cellulase (*59*) and of *Myrothecium cellulase* (*70*) are within this range.

VIII. Substrate Binding and Catalytic Properties

The soluble cello-oligoglucosides and their derivatives allow kinetic properties to be studied as a function of the degree of polymerization (DP) of the substrate from DP of two up to a DP of at least six. The following trends, first established with *Myrothecium verrucaria* cellulase (*71*), appear to be general trends within this range of DP: (a) as the DP is increased, K_m decreases and k_c the catalytic rate constant $(V_{max}/[E]_0)$, increases; and (b) interior linkages are cleaved faster than terminal linkages and the terminal linkage at the nonreducing end of a chain is cleaved faster than the terminal linkage at the reducing end. Table III illustrates various degrees of dependence of K_m on DP for enzymes from Table II. The binding sites of all these enzymes appear

TABLE III
K_m VALUES ($M \times 10^4$) FOR *P. notatum* AND *T. viride* CELLULASES

		T. viride Cellulases [Ref. (*48*)][b]			
Substrate	*P. notatum* cellulase [Ref. (*61*)][a]	Endo-glucanase [Ref. (*67*)][a]	Cellulase II	Cellulase III	Cellulase IV
Cellobiose		190			
D-Glucitol-β-cellobioside[c]					
Cellotriose	900	31			
D-Glucitol-β-cellotrioside[c]	910		2.4	3.2	2.2
Cellotetraose	250	28			
D-Glucitol-β-cellotetraoside[c]	13		2.0	2.5	0.5
Cellopentaose	10	7.0			
D-Glucitol-β-cellopentaoside[c]			1.1	0.5	0.2
Cellohexaose	6.7	1.0			

[a] $T = 40°$.
[b] $T = 30°$.
[c] Prepared by borohydride reduction of the oligoglucoside immediately below.

70. D. R. Whitaker, J. R. Colvin, and W. H. Cook, *ABB* **49**, 257 (1954).
71. D. R. Whitaker, *in* "Advances in Enzymic Hydrolysis of Cellulose and Related Materials" (E. T. Reese, ed.), p. 51. Pergamon Press, Oxford, 1963; "Marine Boring and Fouling Organisms" (D. L. Ray, ed.), p. 301. Univ. of Washington Press, Seattle, Washington, 1959.

to be capable of accommodating chains of five or six glucose units, but clearly their binding properties are not identical. The estimation of k_c can be complicated by inhibitions at higher substrate concentrations (*72*) or by transfer reactions (*48*), but the degree of dependence of k_c on the DP of the substrate seems also to vary within wide limits. For example, *T. viride* cellulase II (*48*) has a value of k_c for the hydrolysis of D-glucitol-β-cellopentaoside which is only about twice that for the corresponding cellotrioside, whereas *Myrothecium verrucaria* cellulase (*72*) has a value of k_c for the hydrolysis of methyl β-cellopentaoside which is more than a hundred times greater than that for the corresponding cellotrioside. There are few estimates of the absolute value of k_c. Hanstein and Whitaker (*72*) have reported the following values for hydrolyses by *M. verrucaria* cellulase at 28.6°: for methyl β-cellobioside, $k_c < 1.6$ min^{-1}; for methyl β-cellotrioside, $k_c = 16$ min^{-1}; for methyl β-cellotetraoside, $k_c = 570$ min^{-1}; and, for methyl β-cellopentaoside, $k_c >$ 1750 min^{-1}. The last two values bracket Werner's estimate of $k_c = 1125$ min^{-1} for the initial rate of degradation of a carboxymethyl cellulose by *Acetobacter xylinum* cellulase (*21*). The data in Table II should also be relevant to degradations of soluble derivatives of cellulose when, as discussed previously, the initial rate is determined by the distribution of segments with three or more unsubstituted glucose units. As shown by Hanson's theoretical analysis (*73*), the value of K_m for such degradations can be a complex weighted average of individual rate constants.

The effects of hydrogen ion, concentration, and certain inhibitors have given some clues regarding the side chains involved in enzymic activity, particularly for *P. notatum* cellulase. With hydroxyethyl cellulose as substrate, this enzyme shows a broad pH optimum from pH 4.5 to 7 and a sharp fall in activity on either side (*19*). This is a fairly typical pattern for fungal cellulases. [One remarkable exception, a cellulase of *Aspergillus niger*, is reported to have optimal activity toward hydroxyethyl cellulose at pH 2.3–2.5 (*74*).] An ionization with a pK_a of 7.5 could account for the fall in activity on the alkaline side, and since *P. notatum* cellulase is readily inhibited by diazonium-1*H*-tetrazole but can tolerate substantial nitration of its tyrosine groups and substantial succinylation of its amino groups, the reaction in question may be deprotonation of a histidine residue (*19*). There are also strong suggestions—from features of the perturbation of the enzyme's absorption spectrum by cellobiose and from features of the inhibition of

72. E. G. Hanstein and D. R. Whitaker, *Can. J. Biochem. Physiol.* **41**, 707 (1963).
73. K. R. Hanson, *Biochemistry* **1**, 723 (1962).
74. R. Ikeda, T. Yamamoto, and M. Funatsu, *Agr. Biol. Chem.* (*Tokyo*) **31**, 1201 (1967).

the enzyme by mercuric ions—that a trytophan residue is involved in substrate binding (*19, 60*).

Virtually nothing is known of the reaction mechanism for the catalytic process. It was shown that the cleavage process of *Myrothecium* cellulase releases reducing groups without a net inversion of configuration (*75*), and to this extent the process may be related to that of the α-amylases. There are little data on this point for other cellulases, but the enzymes which invert configuration all appear to be enzymes which degrade cello-oligoglucosides or cellodextrins by end-wise cleavage. One, the "exo-glucanase" of *T. viride* (*67*), cleaves glucose units from the nonreducing end of the chain. Another, in an enzyme preparation from *Cellvibrio gilvus* (*76*), appears to be an enzyme which cleaves cellobiose units from the nonreducing end but the specificity of the enzyme preparation was not absolute—cleavages occurred at more than one linkage, possibly as a result of contamination by an endoglucanase.

IX. Action of Cellulases on Cellulose and Related Substrates

As already mentioned, *Myrothecium* cellulase cleaves the linkages of cello-oligoglucosides at rates which depend on the location of the linkage in the chain, linkages at the ends being cleaved more slowly than linkages in the interior. The cellulases in Table III and a cellulase of *Aspergillus niger* (*77*) act similarly. This preference for interior linkages can approximate to a process of random cleavage when the chain is so long that only a small fraction of its linkages are near the ends, but unless the substrate is a highly degraded cellodextrin (as in the group III substrates discussed in the section on assay methods) or a swollen or regenerated cellulose (group II substrates) most of the chains will be so strongly associated that only a small fraction of their linkages will be accessible to enzymes. The production of oligoglucosides during the degradation of a cellodextrin by *Myrothecium* cellulase (*78*) and the changes in viscosity-average and number-average molecular weight during initial stages of the degradation of a swollen cellulose (*79*) by the same enzyme have been shown to be consistent with a random cleavage process. For most other cellulases, the evidence for random cleavage is

75. D. R. Whitaker, *Can. J. Biochem. Physiol.* **53**, 436 (1954).
76. W. O. Storvick, F. E. Cole, and K. W. King, *Biochemistry* **2**, 1106 (1963).
77. A. E. Clark and B. A. Stone, *BJ* **96**, 802 (1965).
78. D. R. Whitaker, *Can. J. Biochem. Physiol.* **34**, 488 (1956).
79. D. R. Whitaker, *Can. J. Biochem. Physiol.* **35**, 733 (1957).

limited to evidence obtained with soluble derivatives of cellulose. These substrates are discussed later.

The above substrates are readily degraded by almost all cellulases. Other enzymes are usually required if the lower oligoglucosides in the hydrolyzate are to be converted completely to glucose. These are often referred to as "cellobiases," but in some instances at least they have a much broader range than this name suggests. Thus a "β-glucosidase" of *Stachybotrys atra* degraded substrates with an average degree of polymerization of 12 (*80, 81*) while an exoglucanase of *Trichoderma viride* had appreciable activity toward cellodextrins (*67*). Both enzymes degrade their substrates by cleaving glucose units from the nonreducing end.

The degradation of group I substrates cannot be discussed in detail or with certainty. One aspect of the problem is illustrated by three sets of findings for *T. viride* cellulase. As indicated in Table I, Okada *et al.* (*47*) isolated three cellulases, possibly derivatives of one enzyme, which were each capable of degrading all the common substrates including fibrous cellulase. Li *et al.* (*67*) isolated an exoglucanase, an endoglucanase (Table I), and partly purified a C1 component to which they attributed activity toward crystalline cellulose. Selby and Maitland (*54*), as discussed earlier, attributed activity toward fibers to synergism with a C1 factor which, at present, can be assigned no enzymic activity.

The course of degradation itself is a complex function of the fine structure of cellulose. Halliwell has stressed the formation of "short fibers" as an intermediate stage (*82, 83*). Liu and King (*84*) have reported the formation of still smaller fragments, roughly 300–400 nm long and 50–60 nm wide, as a stage in the degradation of a crystalline cellulose although this distribution of particle size may have been influenced by the acid digestion used in the preparation of their substrate.

Soluble derivatives of cellulose might be regarded as synthetic heteropolysaccharides. Their degradation can never be strictly random since the probability of cleaving a linkage depends on its location relative to substituted glucose units. According to Klop and Kooiman's data for two fungal cellulases (*85*), a linkage –G′–O–G″– will resist cleavage

80. G. Youatt, *Australian J. Biol. Sci.* **11,** 209 (1958).
81. G. Youatt and M. A. Jermyn, *in* "Marine Boring and Fouling Organisms" (D. L. Ray, ed.), p. 397. Univ. of Washington Press, Seattle, Washington, 1959.
82. G. Halliwell, *in* "Advances in Enzymic Hydrolysis of Cellulose and Related Materials" (E. T. Reese, ed.), p. 71. Pergamon Press, Oxford, 1963.
83. G. Halliwell and M. Riaz, *BJ* **116,** 35 (1970).
84. T. H. Liu and K. W. King, *ABB* **120,** 462 (1967).
85. W. Klop and P. Kooiman, *BBA* **99,** 102 (1965).

unless two minimum conditions are met: G′ (linked at carbon 1) must be unsubstituted and G″ (linked at carbon 4) must either be unsubstituted or substituted at carbon 6. For reasons discussed earlier, the initial rate of change in visocisity relative to the initial rate of formation of reducing groups might be expected to depend, *inter alia*, on (a) the average degree of polymerization of the substrate; on (b) its distributions of unsubstituted anhydrocellobiose units, anhydrocellotriose units, etc. (these distributions can depend not only on the average degree of substitution but also on the reaction conditions employed in the preparation of the substrate); and on (c) the binding properties and catalytic properties of the enzyme. The ratio of the two initial rates can provide a useful empirical test for comparing cellulases and has been used, though clearly it has its limitations, as a measure of the "degree of randomness" of the degradation (*86*). This scale, for example, would rate cellulase IV of Table III as "more random" than cellulase II but "less random" than cellulase III (*48*).

Certain other soluble substrates of cellulases are natural heteropolysaccharides. The β-glucans of oats and barley, for example, contain both 1,3 and 1,4 linkages. The action patterns of *Streptomyces* cellulase (*87*), *A. niger* cellulase (*77, 88*), *T. viride* cellulase (*89*), and *P. notatum* cellulase (*61*) on these substrates have been reported. As far as one can tell, 1,4 linkages are the only linkages which are cleaved, but different cellulases give quite different patterns of oligoglucosides. (*61*).

X. Applications

Cellulases are produced commercially on quite a large scale, particularly in Japan. Most of the output is used for digestive tablets. Other current uses are related to the removal or softening of unwanted cellulose, for example, cellulose in certain food preparations [Toyama (*90*) has reviewed this application] and cellulose which interferes with an extraction process. The conversion of waste cellulose to sugar is a possible

86. K. Nisizawa, *in* "Advances in Enzymic Hydrolysis of Cellulose and Related Materials" (E. T. Reese, ed.), p. 171. Pergamon Press, Oxford, 1963.
87. A. S. Perlin, *in* "Advances in Enzymic Hydrolysis of Cellulose and Related Materials" (E. T. Reese, ed.), p. 185. Pergamon Press, Oxford, 1963.
88. A. E. Clarke and B. A. Stone, *BJ* **99,** 582 (1966).
89. O. Igarashi, M. Noguchi, and M. Fujimaki, *Agr. Biol. Chem.* (*Toyko*) **32,** 272 (1968).
90. N. Toyama, *in* "Advances in Enzymic Hydrolysis of Cellulose and Related Materials" (E. T. Reese, ed.), p. 235. Pergamon Press, Oxford, 1963.

application and high conversions of wood powder to glucose have been reported (*91*), but it is doubtful at present whether such a process could compete with other industrial processes for producing glucose.

Cocking and his co-workers have been the pioneers in the use of cellulases to prepare protoplasts from plants. Their first procedure used *Myrothecium* cellulase alone (*92*); their current procedures use a combination of cellulase and pectinase (*93*). They have now established the conditions for reproducible fusion of protoplasts from different species and have thereby opened a possible route for the production of hybrids from plants which cannot be crossed sexually (*94*). This apparently recondite application of cellulases may prove to be of great practical importance.

91. M. Katz and E. T. Reese, *Appl. Microbiol.* **16,** 419 (1968).
92. E. C. Cocking, *Nature* **187,** 962 (1960).
93. J. B. Power and E. C. Cocking, *BJ* **111,** 33P (1969).
94. J. B. Power and E. C. Cocking, *Nature* **225,** 1016 (1970).

10

Yeast and Neurospora Invertases

J. OLIVER LAMPEN

I. Introduction

Invertase (EC 3.2.1.26) or β-fructofuranoside fructohydrolase (β-*h*-fructosidase) acts typically on sucrose and related glycosides producing hydrolysis or, in varying degree, fructosyl transfer. The substrate must possess a terminal, unsubstituted β-D-fructofuranosyl residue, but the nature of the "afructon" moiety is of comparatively little importance for the enzymic action.

This chapter will deal with the invertases formed by yeast (*Saccharomyces* species) and by the fungus *Neurospora crassa* since these enzymes have been utilized for most recent biochemical and molecular studies. The long history of research on invertase, its function and distribution in nature, its relationship to other sucrose-cleaving enzymes (e.g., inulinases and α-glucosidases), and many facets of its catalytic action (substrate binding, pH dependence, action of inhibitors, etc.) have all been extensively examined in previous reviews (*1–3*). These features will be discussed only as necessary to the consideration of recent data. Major attention will be devoted to the relatively homogeneous preparations of both yeast and *Neurospora* enzymes which have become available within the last five years.

A. DETERMINATION

Invertase action has been detected by measuring the products of sucrose hydrolysis or of fructosyl transfer. Measurement has usually been carried out either polarimetrically or by determination of the reducing sugar formed (*4*). Recent workers have often used glucose oxidase to estimate the glucose formed (*5, 6*) since this procedure is relatively simple, minimizes any error from transfer reactions (which primarily involve the fructosyl moiety), and can be readily automated.

II. Yeast Invertase

Yeast invertase has been known the longest of any of the carbohydrases and figured importantly in much of the fundamental work on enzyme kinetics. Thus a great deal of information is available about its catalytic action under a wide variety of conditions and the effects of many different inhibitors. Yet until the last decade no reasonably homogeneous preparations of invertase were available, and very little of the descriptive data could be related to chemical or molecular features of the enzyme. It is no exaggeration to say that the great resistance of the catalytic activity of invertase to destruction during prolonged autolysis has been a major deterrent to elucidation of its molecular nature. Many purified preparations have been obtained from yeast allowed to autolyze for days, and even those subsequently handled very carefully [e.g., Berggren (*7*)] are clearly heterogeneous. In addition,

1. K. Myrbäck, "The Enzymes," 2nd ed., Vol. 4, p. 374, 1960.
2. C. Neuberg and I. Mandl, "The Enzymes," 1st ed., Vol. 1, Part 1, p. 527, 1950.
3. A. Gottschalk, "The Enzymes," 1st ed., Vol. 1, Part 1, p. 577. 1950; *Advan. Carbohydrate Chem.* **5,** 49 (1950).
4. S. Hestrin, D. S. Feingold, and M. Schramm, "Methods in Enzymology," Vol. 1, p. 231, 1955.
5. S. Gascon and J. O. Lampen, *JBC* **243,** 1567 (1968).
6. M. Messer and A. Dahlqvist, *Anal. Biochem.* **14,** 376 (1966). A. Dahlqvist, *Anal. Biochem.* **22,** 99 (1968).
7. B. Berggren, *Arkiv Kemi* **29,** 117 (1968).

interconversion of two invertase forms has been noted (*8*) during the extended incubation of homogenates initially prepared rapidly by the use of a French pressure cell.

Another potential source of heterogeneity is the existence in yeast of six different genes for invertase synthesis [SU_{1-6} (*9, 10*)]. The different genes should lead to the production of enzymes which would be similar but not identical. Finally, commercial baker's and brewer's yeasts, which are frequently chosen for isolation of invertase since they are readily available in quantity, are heteroploid and contain mixtures of the invertase genes. An autolysate of these yeasts will contain a complex of closely related enzymes which probably have also undergone enzymic alteration (*7*). It is important that studies on purification and molecular nature of invertase begin with an organism which has only one gene for invertase production (and thus the minimum number of enzyme forms) and that extraction be rapid and conditions selected that will minimize enzymic modification.

A. Localization and Multiple Forms

Almost all of the invertase produced by yeast is retained by the intact cell, although a few yeasts do release most of their enzyme (*11*). Early work, reviewed by Myrbäck (*1*), showed that the invertase of intact cells had the same pH-activity curve and sensitivity to inhibitors as enzyme in true solution. The experiments of Preiss (*12*), who determined invertase activity of yeast after irradiation with low voltage electrons, indicated that the bulk of the enzyme lies between 500 and 1000 Å from the outer edge of the cell, i.e., within the wall itself. More direct proof of an external location for invertase (i.e., outside of the permeability barrier of the plasma membrane) was provided by the demonstration that almost all of the activity was released during conversion of the cells to protoplasts with snail gut juice or with microbial enzyme preparations (*13–16*).

A separate form of invertase present only inside the cell membrane

8. J. Hoshino and A. Momose, *J. Biochem.* (*Tokyo*) **59,** 192 (1966).
9. O. Winge and C. Roberts, *Compt. Rend. Trav. Lab. Carlsberg, Ser. Physiol.* **25,** 419 (1957).
10. R. K. Mortimer and D. C. Hawthorne, *Genetics* **53,** 165 (1966).
11. L. J. Wickerham, *ABB* **76,** 439 (1958).
12. J. W. Preiss, *ABB* **75,** 186 (1958).
13. J. Friis and P. Ottolenghi, *Compt. Rend. Trav. Lab. Carlsberg* **31,** 259 (1959).
14. A. A. Eddy and D. H. Williamson, *Nature* **183,** 1101 (1959).
15. D. D. Sutton and J. O. Lampen, *BBA* **56,** 303 (1962).
16. M. Burger, E. E. Bacon, and J. S. D. Bacon, *BJ* **78,** 504 (1961).

was detected by Gascon and Ottolenghi (*17*) using gel filtration on Sephadex G-200 columns. The purified enzyme (*5, 18*) has a molecular weight of about 135,000 and is free of carbohydrate, in contrast to the larger external enzyme with a molecular weight of 270,000 about half of which is mannan. Beteta and Gascon (*19*) reported that the small form is present primarily in the yeast vacuole. Small amounts of the large form are also present in the vacuole fraction, but the great bulk of this material is outside the cell membrane.

Multiple forms of invertase had previously been demonstrated in cultures of *Candida utilis* (*20*) and in yeast autolyzates (*8, 21, 22*); however, the relation of these materials to the two well-characterized forms is not evident.

External invertase can be released from cells of *Saccharomyces fragilis* and *S. mellis* (but not *S. cerevisiae*) by treatment with thiols (*23, 24*), and this has led to the suggestion that the enzyme is retained within a mesh in which disulfide bonds play a critical role (*23*). Invertase is released from several *Saccharomyces* strains by phosphomannanase, which removes an outer wall layer of P-diester-linked mannan from the yeast cell, but very little protein or glucan (*25, 26*). Invertase might be bonded to the mannan through similar diester linkages since certain samples of the purified external enzyme contain phosphorous (W. Colonna, personal communication). Retention may also be effected by hydrogen or hydrophilic bonding between the mannan moiety of the invertase and the outer mannan layer, or by direct trapping of the molecule within a mannan mesh. One may conclude that external invertase is held within the wall or between the wall and cell membrane with the structures responsible for its retention varying from one species to another.

B. Biosynthesis

The formation of invertase by yeast and by *N. crassa* is primarily controlled by catabolite repression, particularly by hexoses. The effects

17. S. Gascon and P. Ottolenghi, *Comp. Rend. Trav. Lab. Carlsberg* **36,** 85 (1967).
18. S. Gascon, N. P. Neumann, and J. O. Lampen, *JBC* **243,** 1573 (1968).
19. P. Beteta and S. Gascon, *10th Meeting, Spanish Biochem. Soc., Madrid, 1970* p. 75.
20. R. Sentandreu, F. Lopez-Belmonte, and J. R. Villanueva, *FEBS Abstr.* p. 28 (1965).
21. E. Cabib, *BBA* **8,** 607 (1952).
22. T. Kaya, *J. Agr. Chem. Soc. Japan* **38,** 417 (1964).
23. D. K. Kidby and R. Davies, *BBA* **201,** 261 (1970).
24. R. Weimberg and W. L. Orton, *J. Bacteriol.* **91,** 1 (1966).
25. W. L. McLellan and J. O. Lampen, *J. Bacteriol.* **95,** 967 (1968).
26. W. L. McLellan, L. McDaniel, and J. O. Lampen, *J. Bacteriol.* **102,** 261 (1970).

of various levels and kinds of sugar have been examined in detail (*17*, *27–30*). A striking feature is that the level of internal invertase in yeast is relatively constant, varying only about threefold with changes in glucose concentration which cause a 1000-fold shift in the activity of the large external enzyme. Thus "high-glucose" yeast contains almost exclusively the internal form (*5*, *17*).

The biosynthetic relationship between the internal and external invertases has not been clarified. The two enzymes have almost identical kinetic parameters (K_m values, relative V_{max} on different substrates and pH-activity curves), low transferase activity, cross-react serologically; and were both absent in the three sucrose-negative mutants examined (*18*). The general occurrence of glycosylation during secretion of proteins by yeasts and other fungi (*31*, *32*) suggests that the internal enzyme may be the precursor of the external glycoprotein, but the reported presence of the internal enzyme in the yeast vacuole [probably the lysosome (*33*)] raises the possibility that it is produced by degradation of the glycoprotein form (*19*).

For purification and study of invertase it is convenient to have available an organism which forms the enzyme in high sugar concentrations. A mutant (FH4C) from a *Saccharomyces* strain (contains only the SU_2 gene) produces invertase as about 2% of its total protein even when growing in 5% glucose medium (*5*, *34*, *35*). Mutants of *N. crassa* that are relatively resistant to catabolite repression have been isolated by Metzenberg (*36*) and by Gratzner and Sheehan (*37*).

C. Preparation of Purified Invertase

The usual procedures for the extraction and purification of invertase have included a prolonged autolysis to release the enzyme and degrade much of the noninvertase protein, a heat treatment to precipitate the yeast gum (mostly mannan), and various precipitations or adsorptions

27. R. Davies, *BJ* **55,** 484 (1953).
28. A. Davies, *J. Gen. Microbiol.* **14,** 109 (1956).
29. H. Suomalainen and E. Oura, *ABB* **68,** 425 (1957).
30. F. Dodyk and A. Rothstein, *ABB* **104,** 478 (1964).
31. E. H. Eylar, *J. Theoret. Biol.* **10,** 89 (1965).
32. J. O. Lampen, *Antonie van Leeuwenhoek J. Microbiol. Serol.* **34,** 1 (1968).
33. P. Matile and A. Wiemken, *Arch. Mikrobiol.* **56,** 148 (1967).
34. J. O. Lampen, N. P. Neumann, S. Gascon, and B. S. Montenecourt, *in* "Organizational Biosynthesis" (H. J. Vogel, J. O. Lampen, and V. Bryson, eds.), p. 363. Academic Press, New York, 1967.
35. N. P. Neumann and J. O. Lampen, *Biochemistry* **6,** 468 (1967).
36. R. L. Metzenberg, *ABB* **96,** 468 (1962).
37. H. Gratzner and D. N. Sheehan, *J. Bacteriol.* **97,** 544 (1969).

TABLE I
RECENT PREPARATIONS OF YEAST INVERTASE

Reference	Carbohydrate (%)	IU (30°) per mg protein	Method of extraction[a]	Molecular weight
1951 Fischer and Kohtes (*38*)	70–80	800–1000	A	
1960 Anderson (*39*)	30	*ca.* 4000	A	112,000
1965 Myrbäck and Schilling (*40*)		3500	A	
1967–1968 Vesterberg and Berggren (*7, 41*)	30–40	*ca.* 2700–3000	A or M	270,000
1967 Neumann and Lampen (*35*)	50	2700–3000[b]	M	270,000
1970 Neumann and Lampen (*35*)[c]	47–52	4000–5000[b]	M	
1968 Gascon and Lampen (*5*)[d]	<3	2900[b]	M	135,000
1970 Gascon and Lampen (*5*)[c]		*ca.* 3500[b]	M	
1969 Andersen and Jorgensen (*42*)	13	3780	A	
1969 Greiling *et al.* (*43*)	77	1600	A	
1971 Waheed and Shall (*44*)	50	2770[b]	A	

[a] A, autolysis; M, mechanical breakage.
[b] Homogeneous by gel electrophoresis.
[c] Unpublished results; author's laboratory.
[d] Internal enzyme; others primarily external.

of the enzyme (*1, 2*). More recent attempts have utilized gel filtration, ion exchange chromatography, gel electrophoresis, and electrofocusing. Some of these preparations are listed in Table I (*5, 7, 35, 38–44*) with the reported specific activities converted to international units (IU) at 30° per mg of protein. The assignments of activity at 30° are somewhat arbitrary since the temperature coefficients reported vary in several instances. The best fractions tend to have a specific activity of about

38. E. H. Fischer and L. Kohtes, *Helv. Chim. Acta* **34,** 1123 (1951).
39. B. Andersen, *Acta Chem. Scand.* **14,** 1849 (1960).
40. K. Myrbäck and W. Schilling, *Enzymologia* **29,** 306 (1965).
41. O. Vesterberg and B. Berggren, *Arkiv Kemi* **27,** 119 (1967).
42. B. Andersen and O. S. Jorgensen, *Acta Chem. Scand.* **23,** 2270 (1969).
43. H. Greiling, P. Vögele, R. Kisters, and H.-D. Ohlenbusch, *Z. Physiol. Chem.* **350,** 517 (1969).
44. A. Waheed and S. Shall, *FEBS Abstr.* No. 481 (1967); *BBA* **242,** 172 (1971).

4000 IU/mg protein, although individual preparations have exceeded 5000 IU. It should be noted that Neumann and Lampen (*35*) used heating and ethanol precipitation and regularly obtained apparently homogeneous material with a specific activity of about 3000 IU. After elimination of these steps, potencies of 4000 to 5000 IU have been routine. This suggests that a significant amount of inactivation occurs during these treatments and that the partially inactive material subsequently fractionates with the native enzyme.

Many of the purified preparations (especially those in which the invertase was extracted rapidly rather than by prolonged autolysis) contain about 50% carbohydrate. Neumann and Lampen (*35, 45*) found that the mannan is covalently attached to the protein moiety of the enzyme, probably through glucosaminyl–asparagine bonds. The carbohydrate of Berggren's (*7*) purified invertase was not removed by dimerization with Hg^{2+} or by gel filtration in 4 *M* urea and 0.1 *M* mercaptoethanol; the mannan and protein were considered to be covalently linked. Greiling *et al.* (*43*) reported that their invertase preparation contained mannan linked to serine or threonine by O–glycosidic bonds. Few, if any, such alkali-labile linkages were detected by Neumann and Lampen (*45*) although this type of linkage is common in the bulk structural mannan of the yeast wall (*46*). Since the enzyme studied by Greiling *et al.* (*43*) contained almost 80% mannan and no evidence of homogeneity was presented, the O–glycosides may have been present in mannan peptides not linked to the invertase molecules.

Invertase samples purified by procedures which include prolonged autolysis frequently contain low amounts of carbohydrate, especially if the enzyme has been collected by precipitation with ammonium sulfate (*42, 47*). [The carbohydrate-free internal enzyme is readily precipitated in 70% saturated ammonium sulfate (*5*); the external glycoprotein enzyme is relatively soluble even in saturated ammonium sulfate (*35*).] In one instance (*48*) the mannan could be removed from the preparation by treatment with bentonite at pH 2.9 leaving an unstable enzyme; the carbohydrate-free internal invertase is also unstable at this pH (*18*). The best working hypothesis is that external invertase is a glycoprotein containing about 50% mannan and 2–3% glucosamine (probably the *N*-acetyl form). During prolonged autolysis much of the mannan can be removed, leaving a heterogeneous enzyme with a low or even negligible mannan content.

45. N. P. Neumann and J. O. Lampen, *Biochemistry* **8,** 3552 (1969).
46. R. Sentandreu and D. H. Northcote, *BJ* **109,** 419 (1968).
47. M. Adams and C. S. Hudson, *JACS* **65,** 1359 (1943).
48. E. H. Fischer, L. Kohtes, and J. Fellig, *Helv. Chim. Acta* **34,** 1132 (1951).

The carbohydrate moiety of external invertase is not essential for its catalytic activity since several highly active preparations have been almost devoid of carbohydrate (*42, 47, 48*). Also, the internal invertase has at least two-thirds the specific activity of the best external preparations [the value of this comparison is limited, however, since the amino acid compositions of the internal and external enzymes differ strikingly in the single strain examined (*18*)].

The carbohydrate moiety does increase the stability of the external enzyme to a variety of agents. Arnold (*49*) has shown that susceptibility to heat inactivation decreases with increasing carbohydrate level. The internal (carbohydrate-free) enzyme is more sensitive to acid pH than is the external enzyme (*18*), although it is more stable at alkaline pH. Also, the external enzyme is unaffected by condensed tannins, whereas the internal enzyme is sensitive (*50*).

D. Properties of Purified Enzyme

Most of the external invertase preparations of high specific activity listed in Table I were reasonably homogeneous in that they gave a single band upon polyacrylamide gel electrophoresis and have usually shown single though broad peaks upon gel filtration or ion exchange chromatography. The enzyme from *Saccharomyces* mutant FH4C (*35*) is free of substantial contamination with nucleic acid ($A_{280}:A_{260} = 1.8$) and has an $E_{280}^{1\%}$ (based on protein content) of 23.0. On ultracentrifugation it showed a major component with $s_{20,w}$ of 10.4 S (molecular weight 270,000 $\pm$ 11,000 daltons) and a minor peak, probably a dimer, which was also enzymically active. Material from the main peak did not aggregate when rerun under the same conditions; in contrast, the purified invertase from *Saccharomyces* strain LK2G12 aggregates readily and reversibly (*35*).

Although the specific activity per mg protein is reasonably uniform across the major peak from ion exchange chromatography (*35*), the carbohydrate content usually shifts, e.g., from 52 to 53% at the front of the peak to 47–48% at the back. This variation may result from partial enzymic removal of the mannan even during rapid purification, or the size of the carbohydrate side chains may vary naturally. In the one preparation examined (*45*) approximately 30 chains of mannan with a size range of 2,000–10,000 daltons were present per molecule of enzyme, linked to the protein through glucosaminyl–asparagine bonds. The

49. W. N. Arnold, *BBA* **178,** 347 (1969).
50. D. H. Strumeyer and M. J. Malin, *BJ* **118,** 899 (1970).

mannan associated with invertase has been reported (*51*) to be identical to the bulk mannan of the yeast wall, but the mannan of FH4C invertase (*45*) differs at least in that it contains few, if any, of the serine- or threonine-O–mannose linkages characteristic of the wall mannan (*43*, *46*).

The amino acid compositions of external invertase from brewer's yeast (*42*) and mutant FH4C (*18*) and of the internal enzyme from the mutant (*18*) are compared in Table II. The distribution of amino acids in the two external enzymes is similar and contrasts sharply with the internal enzyme, especially in the levels of glycine, tyrosine, histidine, and half-cystine. The external enzyme from mutant FH4C was not activated by cysteine (*18*) [although Hoshino and Momose (*52*) did report a fourfold

TABLE II
AMINO ACID COMPOSITION OF PURIFIED INTERNAL AND EXTERNAL INVERTASES

	Moles/135,000 g of protein		
	External		
Amino acid	[Ref. (*18*)]	[Ref. (*42*)]	Internal [Ref. (*18*)]
Glycine	71	74	115
Alanine	68	72	84
Serine	114	129	151
Threonine	84	106	80
Proline	65	64	63
Valine	69	68	73
Isoleucine	40	39	38
Leucine	83	76	77
Phenylalanine	80	85	77
Tyrosine	65	71	31
Tryptophan	33	[a]	30
Half-cystine	5	>1	0
Methionine	21	19	14
Aspartic acid	178	171	165
Glutamic acid	115	116	124
Arginine	27	27	32
Histidine	16	16	29
Lysine	60	54	85
Glucosamine	38	*a*	0
(Mannose)	(50%)	(13%)	(<3%)

[a] Not determined.

51. A. J. Cifonelli and F. Smith, *JACS* **77**, 5682 (1955).
52. J. Hoshino and A. Momose, *J. Gen. Appl. Microbiol.* **12**, 163 (1966).

activation of two forms from baker's yeast], nor was it sensitive to or reactive with sulfhydryl reagents (*35*). Three SH groups were accessible to iodoacetate in 8 *M* urea (*35*); the remaining two residues may exist as a disulfide bridge.

The pH-stability curves for the external and internal forms differ markedly (*18*). The external enzyme is inactivated at pH 8 and above, but it is stable at pH 3. The internal form is stable at pH 9, but it undergoes reversible inactivation at pH 5 or less. During this inactivation there was no change in the molecular weight of the internal enzyme.

E. Catalytic Properties

1. *Active Site*

To be cleaved by yeast invertase a substrate must contain an unsubstituted β-D-fructofuranosyl residue (*1, 3*). Gottschalk (*3*) has concluded that the enzyme combines primarily with this moiety through the glycosidic oxygen and the OH groups at C-6 and C-3 of fructose. There may be some specific interaction with the OH at C-2 of glucose since sucrose is the best substrate. Large "afructon" groups decrease the rate of cleavage, possibly through steric factors. Rupture of the glycosidic linkage occurs on the fructose side of the glycosidic oxygen (*53, 54*).

Binding of the substrate involves a group(s) with p$K \sim 3$ (*55–57*) which determines the acid side of the pH-activity curve. The most attractive possibility is a carboxylic group. The alkaline side of the pH-activity curve has the form of a dissociation curve of a weak acid with p$K \sim 7$, probably the imidazole group of a histidine residue (*1, 58*). This residue does not appear to be involved in substrate binding since the alkaline side of the pH curve is not dependent on substrate concentration (*59*) [such a dependence has clearly been demonstrated for the acid limb (*60*)]. Thus the active form of the enzyme requires at least an unprotonated acidic group (COO— ?) for substrate binding and a protonated imidazole residue which is apparently involved in substrate cleavage.

53. D. E. Koshland, Jr., and S. S. Stein, *JBC* **208**, 139 (1954).
54. D. E. Koshland, Jr., *Biol. Rev.* **28**, 416 (1953).
55. K. Myrbäck, *Z. Physiol. Chem.* **158**, 160 (1926).
56. K. Myrbäck, *Soc. Biol. Chemists, India, Silver Jubilee Souvenir* p. 204 (1955).
57. K. Myrbäck and E. Willstaedt, *Arkiv Kemi* **15**, 379 (1960).
58. S. Shall and A. Waheed. *BJ* **111**, 33P (1969).
59. R. Kuhn, *Z. Physiol. Chem.* **125**, 28 (1923).
60. K. Myrbäck and U. Bjorklund, *Arkiv Kemi* **4**, 567 (1952).

Sulfhydryl Groups. It has frequently been suggested that an SH group participates in the catalytic action of invertase. The activity of some preparations can be increased several fold by cysteine (*52*), and the enzyme can be inactivated by iodine and by Hg^{2+} or Ag^{2+} (1), although not by the more specific SH reagents (*35*). Also the iodine-invertase produced at pH 5.0 (which retains about 50% of the original activity) is much more resistant to Hg^{2+} or Ag^{2+} than is the native enzyme. This led to the suggestion that iodine reacts at pH 5.0 with an SH group that is partially masked in the native form of invertase (*1*). The concept has been questioned since cysteine, glutathione, and ascorbic acid do not produce reactivation (*61*), but it was supported by the observation of Shall and Waheed (*58*) that nonacidic thiols (mercaptoethanol and mercaptoethylamine) were effective. The latter workers (*58*) proposed that lack of reactivation by cysteine resulted from charge repulsion by a COO— group at or near the active center.

The mechanism of iodine inactivation and indeed the question of any catalytic role for the SH groups of external invertase must, however, be left undecided in light of the fact that internal invertase lacks cysteine residues yet shows catalytic properties very similar to those of the external enzyme (*18*). (Recent studies by A. Waheed in the author's laboratory have shown that internal invertase also reacts with iodine to yield a product with approximately half the original activity.) The most tenable assumption for the present is that iodine does not react with an SH of the external enzyme [nor with a tyrosine or histidine residue (*58*)] but possibly oxidizes a methionine residue to the sulfoxide [cf. Koshland *et al.* (*62*, *63*)].

2. *Inhibitors*

Inhibition of invertase by Zn^{2+} is reversible and noncompetitive, and it is strongly dependent on pH (*64*). The reactive group on the protein appears to be the unprotonated form of the imidazole group determining the alkaline side of the pH-activity curve. Myrbäck (*64*) proposed that Zn^{2+} (and Hg^{2+}) links two enzyme molecules to form a dimer. Dimerization by Hg^{2+} has been demonstrated by Berggren (*7*), who attempted unsuccessfully to separate the enzyme protein and the mannan in this manner.

61. K. Myrbäck and E. Willstaedt, *Arkiv Kemi* **13,** 179 (1957).

62. D. E. Koshland, Jr., D. H. Strumeyer, and W. J. Ray, Jr., *Brookhaven Symp. Biol.* **15,** 101 (1962).

63. M. E. Koshland, F. M. Englberger, M. J. Erwin, and S. M. Gaddone, *JBC* **238,** 1343 (1963).

64. K. Myrbäck, *Arkiv Kemi* **27,** 507 (1967).

Pyridoxal and pyridoxal phosphate are potent inhibitors of potato invertase (*65*), but they are much less effective against the yeast or *Neurospora* enzymes. In contrast, the fungal enzymes are very sensitive to inhibition by aniline whereas the potato enzyme is resistant (*65, 66*).

Invertase undergoes immediate reversible inactivation by concentrations of urea which do not produce major changes in physical properties (*67*). It appears to combine with *ca.* 1.3 molecules of urea per catalytic site (*68*). At high urea concentrations, e.g., 8 *M* at pH 5.0, or at pH values above 6.0, irreversible inactivation occurs along with changes in secondary and tertiary structure.

3. *Kinetics*

The values of K_s for soluble invertase and for the enzyme bound to cell walls were identical [determined from plots of the inhibition by aniline (*69*)]. The K_m for the wall preparations was lower (33 m*M*) than for the soluble enzyme (44 m*M*); this was considered to reflect some hindrance imposed by the site of the bound enzyme.

Invertase insolubilized by ionic binding to DEAE-cellulose was prepared by Suzuki *et al.* (*70*). The apparent pH optimum was displaced from pH 5.4 for the free enzyme to pH 3.4 for the bound; however, the latter value is probably not a true optimum pH since the DEAE-cellulose would probably provide a microenvironment of higher pH than that of the external buffer.

4. *Mechanism*

Following discovery of the activity of yeast invertase in transferring β-fructofuranosyl units to primary alcohols [for references, cf. Myrbäck (*1*)], it has frequently been proposed that a two-step reaction occurs with an active enzyme–fructose complex as the first step. This can be written as

$$\text{I EH} + \text{GOF} \rightleftharpoons \text{EF} + \text{GOH} \quad (1)$$

$$\text{II EF} + \text{HOH} \rightleftharpoons \text{EH} + \text{FOH} \quad (2)$$

$$\text{III EF} + \text{XOH} \rightleftharpoons \text{EH} + \text{XOF} \quad (3)$$

65. R. Pressey, *BBA* **159,** 414 (1968).
66. K. Myrbäck, *Z. Physiol. Chem.* **158,** 160 (1926).
67. A. M. Chase and M. S. Krotkov, *J. Cellular Comp. Physiol.* **47,** 305 (1956).
68. A. M. Chase, H. C. von Meier, and V. J. Menna, *J. Cellular Comp. Physiol.* **59,** 1 (1962).
69. M. V. Tracey, *BBA* **77,** 147 (1963).
70. H. Suzuki, Y. Ozawa, H. Maeda, and O. Tanabe, *Rept. Ferment. Res. Inst.* **31,** 11 (1967).

where E stands for enzyme, GOF for sucrose, GOH and FOH for glucose and fructose, and XOH any receptor sugar. Andersen (*71*) has recently shown that high concentrations of free fructose (but not glucose) can act as both donor and acceptor giving rise to fructose disaccharides. This provides strong support for the existence of an active enzyme–fructose complex. The qualitative and quantitative distribution of mono-, di-, tri-, and tetrasaccharides during sucrose cleavage is also consistent with a two-step reaction mechanism in which the first step is the formation of an enzyme–fructose complex (*72*).

III. *Neurospora* Invertase

The invertase of *N. crassa* has many features in common with that from *Saccharomyces* yeasts. Most of the enzyme is retained by the fungal cell and is external to the cell membrane; thus, it is freely accessible to exogenous substrates and has kinetic parameters and sensitivity to inhibitors identical with those of the free enzyme (*73*). By a variety of biochemical and histochemical procedures, the cell-bound enzyme has been shown to be partly in the cell wall and partly between the wall and the cell membrane (*73–76*). A portion of the enzyme is bound very tightly by the wall and can be extracted only with difficulty (*73, 75*).

Two sizes of active enzyme are present in *Neurospora* (*77, 78*), and there is a preferential release of the smaller form during growth (*79*). Since Manocha and Colvin (*80*) reported the occurrence of discrete pores in the cell wall of *Neurospora*, it has been suggested that release of the external enzyme occurs through these pores with molecular sieving favoring the passage of the smaller form.

The two sizes of *Neurospora* invertase are not differentially distributed as are those of yeast. Heavy (H) enzyme predominates both inside *Neurospora* protoplasts and in the walls of growing cells, but the light

71. B. Andersen, *Acta Chem. Scand.* **21,** 828 (1967).
72. B. Andersen, N. Thiesen, and P. E. Broe, *Acta Chem. Scand.* **23,** 2367 (1969).
73. R. L. Metzenberg, *BBA* **74,** 455 (1963).
74. E. P. Hill and A. S. Sussman, *J. Bacteriol.* **88,** 1556 (1964).
75. M. L. Sargent and D. O. Woodward, *J. Bacteriol.* **97,** 867 (1969).
76. P. L. Y. Chung and J. R. Trevithick, *J. Bacteriol.* **102,** 423 (1970).
77. F. I. Eilers, J. Allen, E. P. Hill, and A. S. Sussman, *J. Histochem. Cytochem.* **12,** 448 (1964).
78. R. L. Metzenberg, *ABB* **100,** 503 (1963).
79. J. R. Trevithick and R. L. Metzenberg, *J. Bacteriol.* **92,** 1010 and 1016 (1966).
80. M. Manocha and J. Colvin, *J. Bacteriol.* **94,** 202 (1967).

(L) enzyme is always present in small amounts (*81*, *82*). The L enzyme is favored at low pH and can be produced from the H form by heating or by high salt concentration either at acid or alkaline pH (*82*). It reaggregates in concentrated solutions in the absence of salt (e.g., during dialysis of an ammonium sulfate precipitate).

A. Purified Preparations

Neurospora crassa invertase was first obtained in a homogeneous state by Metzenberg (*78*) who purified the material extractable from mycelial powder by pH 5.0 buffer. The preparation was homogeneous by polyacrylamide gel electrophoresis and ultracentrifugation. It contained 2.4% hexosamine but no detectable neutral sugar. Shortly thereafter Eilers *et al.* (*77*) noted that crude extracts of *N. crassa* gave two bands of activity by gel electrophoresis. This observation was confirmed by Metzenberg (*82*) and the two components separated by gel filtration into heavy (H) and light (L) forms. Typically 65–85% of the activity was in the H form. The $s_{20,w}$ values for the H and L enzymes were 10.3 and 5.2, respectively. As already noted the H form can be converted to L enzyme by heat or high salt. During dissociation a transient form with $s_{20,w} = 9.8$ could be detected. This was considered to be the polymeric enzyme in a preclastic conformation (*82*).

From Metzenberg's data it can be suggested that *Neurospora* invertase occurs primarily as the polymeric H form which dissociates under appropriate conditions to yield four active subunits (if both molecules are assumed to be spherical). No evidence was obtained for more than one type of subunit (*82*). The L enzyme obtained by dissociation of the H enzyme seems to be identical to that present in cell extracts. Dissociation *in vitro* caused 31% loss of activity; but there was no change of activity during reaggregation, so that the true activity of the subunits may be the same in monomeric and in polymeric form.

Solutions of L invertase foamed copiously upon shaking, whereas H invertase showed little tendency to foam (*82*). This was considered to indicate that the hydrophobic side chains of H invertase are oriented inwardly and become exposed during dissociation. According to this concept hydrophobic bondings might serve to maintain the polymeric structure of the H form.

Pure invertase has recently been isolated from another high-producing strain of *N. crassa* (*37*) by Meachum *et al.* (*83*). The striking charac-

81. J. R. Trevithick and R. L. Metzenberg, *BBRC* **16**, 319 (1964).
82. R. L. Metzenberg, *BBA* **89**, 291 (1964).
83. Z. D. Meachum, Jr., H. J. Colvin, Jr., and H. D. Braymer, *Biochemistry* **10**, 326 (1971).

teristic of this material is that it contains approximately 11% mannose and 3% glucosamine covalently linked to the protein moiety. Proteinase digestion yielded peptides containing glucosamine and aspartic acid, suggesting that the glucosaminyl–asparagine bond predominates here as in yeast external invertase (*45*). If serine–O–glycoside linkages are present, their number must be small. The isolated enzyme had an $s^{\circ}_{20,w}$ of 10.5 and a molecular weight of 210,000. Four disulfide bonds were detected per mole. In the presence of a dissociating agent (6 M guanidine) and under reducing conditions, the $s_{20,w}$ was 5.2 and the molecular weight 51,500. The native enzyme thus appears to be a tetramer and, from tryptic peptide fingerprints, probably contains more than one type of subunit. As with Metzenberg's strain (*82*) both the 10.5 S and 5.2 S forms are found in crude cell extracts and both are enzymically active. The specific activity of Braymer's (*83*) purified enzyme is 1820 μmoles of sucrose split per minute per milligram of glycoprotein; the value for Metzenberg's material (*78*) is 1890 μmoles/minute.

B. Catalytic Properties

The substrate specificity of *Neurospora* invertase (*78*) is similar to that of the yeast enzyme. Sucrose is the most readily utilized substrate with a K_m of 6.1×10^{-3} M and the highest V_{max}. For raffinose, K_m is 6.5×10^{-3} M and the V_{max} 20% of that for sucrose. The values for β-methyl fructoside are K_m of 3.3×10^{-2} M and relative V_{max} of 30%. [These K_m values are notably lower than the 26×10^{-3} M for sucrose and 150×10^{-3} M for raffinose obtained with the purified yeast enzyme (*18*).] Trehalose and melezitose were not attacked. The activation energies (calculated from rates at 0° and 38°) are 10.8, 12.3, and 13.9 kcal for sucrose, raffinose, and β-methyl fructoside, respectively.

The pH-activity curve has a broad optimum at pH 4.5–6.0. The alkaline limb approximates that for yeast invertase, but the acid side falls off rapidly below pH 4.5 (*78*). *Neurospora* invertase is less sensitive than the yeast enzyme to inhibition by divalent cations. Zn^{2+} and Cu^{2+} caused only 33 and 79% inhibition, respectively, at 0.05 M. Both enzymes are strongly inhibited by aniline (*65, 78*). The sulfhydryl reagent *p*-hydroxymercuribenzoate inhibits the fungal enzyme and partial protection is afforded by the presence of substrate (sucrose). In contrast the purified yeast external enzyme was not inhibited by several sulfhydryl reagents, including *p*-mercuribenzoate (*35*).

11

Hyaluronidases

KARL MEYER

I. Introduction

The term hyaluronidase was introduced in 1940 (*1*) to denote enzymes which degraded hyaluronate. These enzymes were derived from micro-

1. K. Meyer, G. Hobby, E. Chaffee, and M. H. Dawson, *Proc. Soc. Exptl. Biol. Med.* **44**, 294 (1940).

organisms and animal tissues and measured by decrease in substrate viscosity and the production of reducing sugars, as reported from our laboratory in earlier papers (1936–1940). The importance of these enzymes became evident with the report of Chain and Duthie on the viscosity reducing effects on synovial fluid of extracts containing the "spreading factors" of Duran Reynals, foremost among them extracts of mammalian testes (*2, 3*). The term hyaluronidase has been used subsequently by many authors as a synonym of "spreading factors" (*4*). On the basis of these reports a very extensive literature has accumulated on hyaluronidases, their uses in histology for the demonstration of substrates in the tissues, for studies of diffusion of dissolved and particulate matter in the tissues, and in studies on the physiological and pathological breakdown of connective tissues and their therapeutic application (*5*).

In subsequent years it became obvious that enzymes degrading hyaluronate have varying substrate specificities and profoundly different reaction mechanisms. This makes it essential to deal with each type separately.

A. Testicular-Type Hyaluronidase, Synonym Hyaluronate Glycanohydrolase (EC 3.2.1.35)

Sources: (Ia)	Mammalian testis, specificially spermatozoa
(Ib)	Lysosomes
(Ic)	Submandibular gland, snake venom, bee venom
Substrates:	Hyaluronate, chondroitin, chondritin 4- and 6-S, in part dermatan sulfate (*6*)
Type of action:	endo-*N*-Acetylhexosaminidases
Main end product:	Tetrasaccharide (U·A·U·A) (*7*)

B. Leech Hyaluronidase, Hyaluronate Glycanohydrolase (EC 3.2.1.36)

Substrate:	Hyaluronate and oligosaccharides derived from hyaluronate only
Type of action:	β-Endoglucuronidase
Main end product:	Tetrasaccharide (A·U·A·U) (*7*)

2. The high viscosity of synovial fluid had been shown earlier to be mainly resulting from hyaluronate.

3. K. Meyer and M. M. Rapport, *Advan. Enzymol.* **13**, 199 (1952).

4. This is only partly correct. All hyaluronidases act as spreading factors, but not all spreading factors are hyaluronidases.

C. Bacterial Hyaluronidase (EC 4.2.99.1), Hyaluronate Lyase

Substrates:	Hyaluronate, chondroitin, chondroitin 4- and 6-sulfate, in part dermatan sulfate
Type of action:	endo-*N*-Acetylhexosaminidase by β elimination
Products:	3-(β-D-Gluco-4,5-en-uronido)2-acetamido-2-deoxy-D-glucose or D-galactose, respectively

II. The Substrates

A. Hyaluronate

Hyaluronic acid is a polymer of varying chain length composed of a disaccharide repeating unit, *N*-acetylhyalobiuronic acid, linked via the β-1,4 linkage of the *N*-acetylglucosaminyl groups. The deacetylated crystalline disaccharide, hyalobiuronic acid, is obtained by hydrolysis of hyaluronate with 2 *N* HCl in a yield of over 60% when the polymer is first digested to oligosaccharides by testicular hyaluronidase. Hyalobiuronic acid has the structure of 3-*O*-(β-D-glucopyranosyluronic acid)-2-amino-2-deoxy-D-glucose (*8*). The repeating unit containing the glucosaminidic bond has been shown to have the structure of 4-*O*-(2-acetamido-2-deoxy-β-D-glucopyranosyl)-D-glucuronic acid by two independent methods: first, by the production from hyaluronate of a Δ-4,5-unsaturated disaccharide in practically quantitative yield (see later) and, second, by direct chemical proof after reduction of the COOH group followed by methylation of the tetrasaccharide produced in about 90% yield from hyaluronate by testicular hyaluronidase (*9*). Hyaluronate is usually isolated from umbilical cord, from vitreous humor, from synovial fluid, from human mesothelioma, or from group A streptococci. When used as a substrate for hyaluronidases, it ought to be free of D-galactosamine, of ester sulfate, and of heavy metal with which many preparations are contaminated.

---

5. H. Gibian, *Ergeb. Enzymforsch.* **13**, 1 (1954).
6. Originally dermatan sulfate was designated as chondroitin sulfate B. This term is still widely used.
7. U designated uronic acid; A, *N*-acetylhexosamine.
8. B. Weissman, M. M. Rapport, A. Linker, and K. Meyer, *JBC* **205**, 205 (1953).
9. S. Hirano and P. Hoffman, *J. Org. Chem.* **27**, 395 (1962).

B. Chondroitin

Chondroitin is an isomer of hyaluronic acid in which the acetamidoglucosyl is replaced by the acetamidogalactosyl group. It usually is obtained by catalytic desulfation of chondroitin 4- or 6-sulfate (*10*). The natural polymer has been isolated from squid skin (*11*). A similar polymer which, however, contains about 2% sulfate ester groups has been isolated from bovine cornea (*12*).

C. Chondroitin 4- and 6-Sulfate (Chondroitin Sulfate A and C, Respectively)

These unbranched polymer chains contain repeating disaccharide units whose structures and bonds are isomeric with those of hyaluronate having D-galactosaminyl replacing the D-glucosaminyl groups (*13*). The sulfate group is present at the hydroxyl of the 4 or 6 position of the *N*-acetylgalactosamine. As a rule the polymers are obtained from mammalian cartilage after proteolytic digestion. The carbohydrate chains still contain a peptide chain to which the carbohydrate is covalently linked by an *O*-glycosidic bond of the serine hydroxyl group. The linkage region of the carbohydrate chains consists of an xylosyl-digalactosyl-glucuronyl tetrasaccharide which is not attacked by testicular hyaluronidase (*14*).

D. Dermatan Sulfate (Chondroitin Sulfate B)

Dermatan sulfate is isomeric with chondroitin 4-sulfate. It is a hybrid polymer containing in part, and to a variable extent, disaccharide units identical with chondroitin 4-sulfate, while the major portion is composed of disaccharide units of 3-*O*-(α-L-idopyranosyl uronic acid)-2-amino-2-deoxy-D-galactose, the 5-epimer of *N*-acetyl chondrosine. The disaccharide units as in the rest of the substrates of testicular hyaluronidase, are linked via β-1,4 linkages of the hexosaminyl groups to the uronidic moieties. Dermatan sulfate is linked to peptide chains by bonds and carbohydrate units similar or identical to those in the chondroitin sulfates.

10. T. G. Kantor and M. Schubert, *JACS* **79**, 152 (1956).
11. K. Anno, Y. Kawai, and N. Seno, *BBA* **83**, 348 (1964).
12. E. A. Davidson and K. Meyer, *JBC* **211**, 605 (1954).
13. P. Hoffman, A. Linker, and K. Meyer, *Federation Proc.* **17**, 1078 (1958).
14. L. Rodén, *in* "Biochemistry of Glycoproteins and Related Substances" (E. Rossi and E. Stoll, eds.), p. 185. Karger, Basel, 1968.

Testicular hyaluronidase splits dermatan sulfate only to a minor degree (see later). It usually is obtained from mammalian skin.

E. Additional Substrates of Hyaluronidases

Chondroitin sulfates of cartilage, of elasmobranch, and some other fishes contain, in addition to sulfate ester groups in the 6 position of the *N*-acetylgalactosamine, ester groups in the 2 or 3 position of the glucuronosyl groups. These chondroitin sulfates are sometimes referred to as chondroitin sulfate D (*15*).

Recently, a chondroitin sulfate has been isolated from squid cartilage in which some of the repeating disaccharide units are disulfated in the *N*-acetylgalactosaminyl moiety. These ester groups are located both in the 4 and 6 positions of a single galactosaminyl moiety (called chondroitin sulfate E) (*16*).

III. Prominent Hyaluronidases

A. Testicular-Type Hyaluronidases

1. *Testicular Hyaluronidase*

The major source of testicular-type hyaluronidase is the testis of mature bulls. The most significant contributions in the purification and physical properties of testicular hyaluronidase are described in two papers (*17, 18*). Brunish and Högberg prepared a highly active enzyme containing approximately 40,000 IU/mg. The purified enzyme was described as labile, as shown by a decreased yield without increased activity on repetition of the adsorption and elution method. The molecular weight by the Archibald method was 43,200. In the ultracentrifuge the material showed three components with sedimentation constants of 6.4, 4.1, and 1.5. The preparation had a carbohydrate content of 5.2% (with mannose as standard), 5.0% hexosamine and 0.39% sialic acid. (The spectrum of the color in the thiobarbiturate method was abnormal.) The isoelectric point of the protein(s) was not reported, although the adsorption and

15. M. B. Mathews, *BBA* **58,** 92 (1962).
16. S. Suzuki, S. Hidehiko, T. Yamagata, K. Anno, N. Seno, Y. Kawai, and T. Furuhashi, *JBC* **243,** 1543 (1968).
17. R. Brunish and B. Högberg, *Compt. Rend. Trav. Lab. Carlsberg* **32,** 35 (1960–1962).
18. C. L. Borders, Jr. and M. A. Raftery, *JBC* **243,** 3756 (1968).

elution procedures used appear to indicate a basic protein. Borders and Rafferty also applied ion exchange chromatography followed by repeated gel chromatography on Sephadex G-75. The enzyme isolated in a yield of 18% had an activity of 45,500 IU/mg and migrated as a single band on acrylamide gel electrophoresis at pH 4.3. The protein contained 5.0% mannose and 2.2% *N*-acetylglucosamine. The molecular weight (by gel filtration) was 61,000, a value independent of concentration over a 20-fold range. The isoelectric point and other physical properties of the enzyme were not reported. From its behavior on adsorption and from the amino acid analysis, it seems obvious that the protein contains an excess of basic amino acids over free carboxyl groups.

2. *Lysosomal Hyaluronidase*

Following earlier reports on the occurrence of hyaluronidase of the testicular type in a variety of mammalian tissues and fluids (*19, 20*), the lysosomal origin of a hyaluronidase from rat liver was demonstrated by Hutterer (*21*). Rat liver lysosomal hyaluronidase was purified and its properties studied by Aronson and Davidson (*22*). Lysosomes were prepared according to de Duve. After lysis, ammonium sulfate precipitation, and removal of foreign protein by hydroxyapatite, the active fraction was separated by gel filtration on Sephadex G-100. A peak representing the active fraction on ultracentrifugation was recovered. Disc electrophoresis of this fraction at pH 4.0 showed only a single band. The yield of enzyme was 10% with a 1300-fold purification and an activity of 4300 IU/mg. (i.e., less than one-tenth of the activity of the testicular enzyme described above). The molecular weight was given as 81,000 and 89,000.

Hyaluronidase of lysosomal origin was also obtained from the calvaria of infant rats (*23*). The hyaluronidase of synovial tissue (*19*) and of tadpole tissue (*24*) presumably is also of lysosomal origin.

3. *Submandibular Hyaluronidase*

A hyaluronidase which in its pH optimum (see later) resembles lysosomal hyaluronidase has been isolated from canine submandibular glands

19. H. Gibian, *in* "Einzeldarstellungen aus dem Gesamtgebiet der Biochemie" (O. Hoffmann-Ostenhof, ed.), Vol. IV, pp. 1–318. Vienna, 1959.
20. A. J. Bollet, W. M. Bonner, Jr., and J. L. Nance, *JBC* **238,** 3522 (1963).
21. F. Hutterer, *BBA* **115,** 312 (1966).
22. N. N. Aronson, Jr. and E. A. Davidson, *JBC* **242,** 437 and 441 (1967).
23. G. Vaes, *BJ* **103,** 802 (1967).
24. J. E. Silbert, Y. Nagai, and J. Gross, *JBC* **240,** 1509 (1965).

(*25*). Although the purification of the best fraction was only 91-fold, its specific activity was close to that of liver hyaluronidase (*22*). Enzymes of presumably similar properties have been reported in the venom of some snakes and in bee venom (*26*). The latter was claimed to possess no transglycosylation; however, the substrate used for transglycosylation was the hexasaccharide. Transglycosylation with this oligosaccharide is very slow (*27*). It should be noted that lysosomal enzyme of liver which exhibits transglycosylation did not transglycosylate with the hexasaccharide. Since the ratio of di- to tetra- to hexasaccharides appears to be similar to that of testicular and lysosomal enzymes, this lack of transglycosylation appears highly doubtful.

B. Hyaluronate-endo-β-glucuronidase (Leech Hyaluronidase)

Thus far hyaluronate-endo-β-glucuronidase has only been found in the leech, *Hirudo medicinalis*, and apparently is localized in the salivary glands. It is specific for hyaluronate and oligosaccharides from hyaluronate, of hexasaccharide size and larger, as produced by testicular-type hyaluronidase. It splits endo-β-glucuronidic bonds of these substrates producing mainly as end products a tetrasaccharide with glucuronate at the reducing end. Chondroitin sulfates or chondroitin are not split by the enzyme. Transglycosylation could not be demonstrated. The hexasaccharide isolated from hyaluronate digests with testicular hyaluronidase was split by this enzyme into two trisaccharides: U·A·U·A·U·A· → U·A·U plus A·U·A (*28*).

C. Hyaluronate Lyase (Microbial Hyaluronidase)

Like the testicular-type enzymes, the microbial hyaluronidases are endohexosaminidases. By β elimination they produce from their substrates Δ-4,5-unsaturated disaccharides [2-deoxy-2-acetamido-(4-deoxy-α-L-*threo*-hex-4-en-pyranosyluronic acid) D-glucose or D-galactose], respectively (*29, 30*) [(depending on whether the substrates correspond to

25. Y. H. Tan and J. M. Bowness, *BJ* **110,** 9 and 19 (1968).
26. S. A. Barker, S. J. Bayyuk, J. S. Brimacombe, and D. J. Palmer, *Nature* **199,** 693 (1963).
27. B. Weissmann, *JBC* **216,** 783 (1955).
28. A. Linker, P. Hoffman, and K. Meyer, *Nature* **180,** 810 (1957), *JBC* **235,** 924 (1960).
29. A. Linker, K. Meyer, and P. Hoffman, *JBC* **219,** 13 (1956).
30. A. Linker, P. Hoffman, K. Meyer, P. Sampson, and E. D. Korn, *JBC* **235,** 3061 (1960).

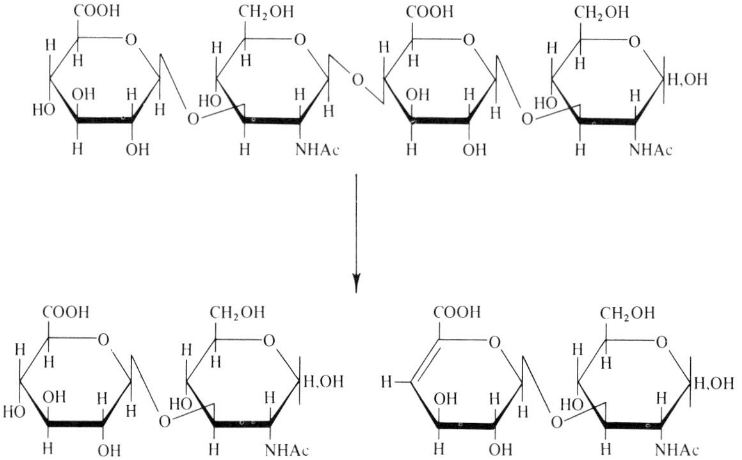

FIG. 1. Degradation of hyaluronate tetrasaccharide by hyaluronate lyase.

hyaluronate or chondroitin and its sulfate esters) (*31*)]. The desulfated unsaturated disaccharides of the three isomeric chondroitin sulfates are identical since the asymmetry at C_5 of the glucuronosyl and iduronosyl groups is abolished by the formation of the double bond (*32*). The mechanism of these enzymes and the structure of their products were first demonstrated with pneumococcal and streptococcal extracts and later with staphylococci, *Clostridium welchii, Proteus vulgaris,* and *Flavobacterium.* The mechanism of the reaction as β elimination was proven by the failure of the enzyme to act on *N*-acetylhyalobiuronic acid and the action on the normal tetrasaccharide. In the latter, the bacterial enzyme yields equal parts of *N*-acetylhyalobiuronic acid and the Δ-4,5-unsaturated disaccharide (*29*) (see Fig. 1). The mechanism of the reaction has been confirmed by carrying it out in $H_2{}^{18}O$ (*33*). The enzymic product, unsaturated uronides, is conveniently determined by measuring the increase of absorption at 230–232 nm (*34*).

From a species of *Streptomyces* an eliminase was obtained which yielded tetra- and hexasaccharides with Δ-4,5-unsaturated disaccharide at the nonreducing end linked to normal di- and tetrasaccharide, respectively. Unsaturated disaccharide per se was not formed (*35*).

31. The actions of pectinases and alginases provide other examples of similar elimination.

32. P. Hoffman, A. Linker, V. Lippman, and K. Meyer, *JBC* **235,** 3066 (1960).

33. J. Ludowieg, B. Vennesland, and A. Dorfman, *JBC* **236,** 333 (1961).

34. H. Greiling, H. W. Stuhlsatz, and T. Eberhard, *Z. Physiol. Chem.* **340,** 243 (1965).

35. T. Ohya and Y. Kaneko, *BBA* **198,** 607 (1970).

IV. Preparation and Assay of Hyaluronidases

The various hyaluronidases are assayed by physicochemical or chemical methods. With hyaluronate as substrate all hyaluronidases described above can be determined by viscometric or, more commonly, by turbidimetric methods (*19*). In the physicochemical methods, the initial reaction of the enzymes, namely, the breaking of a few glycosidic bonds in the center of the long chains, is determined. Thus no dialyzable products are formed after the viscosity of the hyaluronate has been reduced to approximately the control values. Comparing the turbidimetric with a microreductometric method, with hyaluronate as substrate, the extent of hydrolysis at 30 min corresponded to less than 2% of the total hexosaminidic groups (*3*). In general the physicochemical methods are far more sensitive than the chemical methods. The method of choice with hyaluronate as substrate is the increase in chromogen as measured by the Morgan–Elson reaction (*36*). This method measures the products of the enzymic reaction and not the remaining substrate as in the physicochemical methods.

The rate of the reaction is the same with hyaluronate and chondroitin. With chondroitin 4- and 6-sulfates as substrates, the rate of degradation is between one-sixth to one-fifteenth that of hyaluronate. This rate has been measured by turbidimetric or by reductometic methods. With chondroitin 6-sulfate the Morgan–Elson reaction can be employed. With chondroitin 4-sulfate the latter cannot be used since the substitution of the *N*-acetylgalactosamine in the 4 position prevents the chromogen formation (*37*).

With testicular, snake venom, and lysosomal hyaluronidases, the products of the exhaustive enzymic breakdown of hyaluronate are disaccharide (about 10%) and tetrasaccharide (about 90%). With nonexhaustive digestion higher even-numbered oligosaccharides are produced which are separated by ion exchange chromatography and gradient elution with acetic or formic acid (*38*).

With chondroitin 4- and 6-sulfate the main product on exhaustive digestion is disulfated tetrasaccharide. The degradation of the chondroitin sulfate is catalyzed by the same enzyme which degrades hyaluronate (*3*).

The production of the homolog series of oligosaccharides from hy-

36. K. Meyer, P. Hoffman, and A. Linker, "The Enzymes," 2nd ed., Vol. 4, p. 447, 1960.
37. M. M. Mathews and M. Inouye, *BBA* **53,** 509 (1961).
38. B. Weissmann, K. Meyer, P. Sampson, and A. Linker, *JBC* **208,** 417 (1954).

aluronate suggested that the products were formed by transglycosylation. Two methods have been utilized for the demonstration of transglycosylation:

(1) Hexasaccharides or preferably higher oligosaccharides gave both higher and lower oligosaccharides (*27*). At least 95% of the enzymic reaction was accounted for by transglycosylation. The rate increased with increasing chain length.

(2) Mixtures of hyaluronate and chondroitin sulfates on incubation with testicular hyaluronidases yielded, in addition to the homologous oligosaccharides, hybrid oligosaccharides composed of nonsulfated glucosamine containing disaccharide units and sulfated galactosamine-containing units (*39*). The reducing end group in these hybrids always appeared to be a nonsulfated disaccharide.

Neither the di- nor the tetrasaccharide is an acceptor in the transglycosylation. The production of disaccharide probably is an expression of the rate of transglycosylation vs. hydrolysis. Reduction of the reducing end group in higher oligosaccharides abolishes the transglycosylation. The production, in approximately 90% yield, of the tetrasaccharide is an indication of the specificity of the bonds formed by transglycosylation, in marked contrast to the relative nonspecificity of lysozyme, an endohexosaminidase otherwise quite similar to hyaluronidase.

Dermatan sulfate was originally reported not to be degraded by testicular hyaluronidase (*40*). Fransson and Rodén have shown that dermatan sulfate is degraded when tested viscometrically and chemically (*41*). Apparently in the hybrid molecule only repeating units containing the glucuronosyl moieties are split (*42*), while the hexosaminyl bond adjacent to the L-iduronosyl groups is refractory; i.e., the nonreducing end group is a glucuronosyl group. On the basis of the finding of carbohydrate chains unattached to peptide, it was postulated that the susceptible linkages were, in part, close to the linkage region. Whether or not the distribution of the other glucuronosyl moieties is random along the chain is unknown. In any case, the structural analysis of the dermatan sulfate fractions is complicated by heterogeneity of the fractions in which the ratio of L-iduronic acid to D-glucuronic acid is variable (*43*).

Greiling *et al.* (*34*) purified enzymes of *Staphylococcus aureus* and group A hemolytic streptococci by adsorption on DEAE-cellulose and

39. P. Hoffman, K. Meyer, and A. Linker, *JBC* **219,** 653 (1956).
40. K. Meyer and M. M. Rapport, *Science* **113,** 596 (1951).
41. L. A. Fransson and L. Rodén, *JBC* **242,** 4161 and 4170 (1967).
42. L. A. Fransson, *JBC* **243,** 1504 (1968).
43. P. Hoffman, A. Linker, and K. Meyer, *ABB* **69,** 435 (1957).

sequential elution with phosphate and phosphate-NaCl solutions. They concluded from their experiments the presence of three isoenzymes in staphylococci and of two in streptococci distinct by their Michaelis and inhibition constants with chondroitin sulfate and heparin. By electrofocusing Vesterberg *et al.* (*44, 45*) purified the enzyme from culture fluids of some strains of staphylococci. In one strain they found two main enzyme fractions with isoelectric points of pH 6.4 and 8.2. In a strain producing a higher concentration of the enzymes, they obtained by the same method two main fractions, one at pH 7.4 the other at 7.9. The relative amounts of the two enzymes varied in different preparations, and the two fractions behaved identically on gel filtration. The fraction with pH of 7.4 could be converted to that of 7.9 by prolonged incubation of the cultures.

Very significant contributions in the field of hyaluronate lyases and their products have been published from the laboratory of S. Suzuki (on the basis of their greater specificities for chondroitin sulfates, the enzymes were designated as chondroitinases). One enzyme was purified to apparent homogeneity from extracts of *Proteus vulgaris* adapted to chondroitin 6-S, called by the authors chondroitinase A,B,C [since it splits equally the three chondroitin sulfates A, B, and C (*46*)], the other from *Flavobacterium heparinum*, called chondroitinase A, C, which degrades chondroitin sulfates A and C, and is inactive with chondroitin sulfate B (dermatansulfate) (*47, 48*). Both types of enzymes degrade chondroitin and hyaluronate at a slower rate than the sulfated compounds. For the purification of chondroitinase A,B,C, the culture fluid (*Proteus vulgaris*) was fractionated with $(NH_4)_2SO_4$, the dialyzed solution adsorbed on DEAE-cellulose, which mainly retained inactive protein, followed by adsorption and elution on phosphocellulose. The active portion was concentrated by pressure dialysis. This fraction gave a single peak on acrylamide gel electrophoresis. The purification was 134-fold. By density centrifugation and gel chromatography, a molecular weight slightly lower than 150,000 was reported. From *Flavobacterium* culture fluid "chondroitinase A,C" was obtained by analogous procedures and proved homogeneous by acrylamide gel electrophoresis. The purification achieved with this fraction was 927-fold.

44. O. Vesterberg, T. Waldstrom, K. Vesterberg, H. Svensson, and B. Malmgren, *BBA* **133,** 424 (1967).

45. O. Vesterberg, *BBA* **168,** 218 (1968).

46. T. Yamagata, H. Saito, O. Habuchi, and S. Suzuki, *JBC* **243,** 1523 (1968).

47. H. Saito, T. Yamagata, and S. Suzuki, *JBC* **243,** 1536 (1968).

48. S. Suzuki, H. Saito, T. Yamagata, K. Anno, N. Seno, and Y. Kawai, *JBC* **243,** 1543 (1968).

V. Properties of Hyaluronidases

The main distinction between testicular and lysosomal hyaluronidases is their different pH optima which appear to be independent of the degree of purity of the enzyme. Thus, with hyaluronate as substrate, hepatic hyaluronidase had a pH optimum of 3.7 and no activity above pH 4.5 (*22*). Similar pH curves are reported for canine sublingual gland (*25*), for bone and kidney hyaluronidase (*23*), and for human serum hyaluronidase (*49*). Testicular hyaluronidase has a broad pH optimum, and at pH 5.0 has approximately 70% of the activity at pH 4.0 when assayed by the Morgan–Elson reaction. When measured by viscosity reduction or by turbidity methods, the enzyme is active at neutral pH.

Purification of leech hyaluronate endo-β-glucuronidase has been carried out by $(NH_4)_2SO_4$ fractionation and adsorption and elution on DEAE-cellulose resulting in a 91-fold purification with a yield of 8.3%. The enzyme had a pH optimum of 6.0 and a temperature optimum of 38° (*50*).

The hyaluronate lyases in general have pH optima higher than the hyaluronidases. Thus the chondroitinase A,B,C with chondroitin 4- and 6-sulfates and dermatan sulfate as substrates has a sharp pH optimum between pH 8 and 9, while with hyaluronate and chondroitin the pH-activity curve is broad with an optimum at ~pH 6.8 (*46*).

Testicular hyaluronidase has a remarkable heat stability (*36*) with an activity optimum at about 50°. The literature on the effects of neutral salts and of inhibitors is very extensive. As a rule the reaction is carried out in the presence of 0.15 *M* NaCl. However, NaCl is not essential for testicular or lysosomal hyaluronidase with an inhibitor-free hyaluronate preparation (*3, 22*). Impurities present in the substrates are either heavy metal, especially iron and copper salts, or sulfated polyanions like heparin, dermatan sulfate, and keratan sulfate.

VI. Kinetics of Hyaluronidases

A rational interpretation of kinetic data with testicular and lysosomal hyaluronidase is difficult since the breakdown products of the polysaccharide substrates continue to be acted upon by the enzymes. With low

49. M. de Salegui, H. Plonska, and W. Pigman, *ABB* **121,** 548 (1967).
50. H. Yuki and W. H. Fishman, *JBC* **238,** 1877 (1963).

enzymic concentrations the breakdown of the polymers as measured by turbidimetry follows a pseudomonomolecular reaction for the first 10–15 min. The end point of the complex formation of hyaluronate with protein, resulting in the turbidity, appears to be at a molecular weight of about 6000–8000 (*36*). However, the contributions to the turbidity of the higher fragments is not known. The reaction rate, in any case, differs with different preparations of hyaluronate. The interpretation is further complicated by the transglycosylation. The rate of this reaction increases with increasing chain length of the oligosaccharides. It seems possible that transglycosylation predominates over hydrolysis, even with substrates of high molecular weight. With the chemical methods, higher enzyme concentrations are used. Under specified experimental conditions the rate of chromogen formation with lysosomal enzyme was proportional to the enzyme concentration. The time curve gave a straight line for 15 min with a constant substrate and enzyme concentration. The hyaluronate had a K_m of 8×10^{-2} mg/ml (*22*). With hyaluronate lyase and related microbial enzymes kinetic studies are more meaningful since they are not complicated by transglycosylation and appear to be independent of the chain length of the polymers. The Michaelis constants for the different isoenzymes of *Staphylococcus* and group A hemolytic *Streptococcus* were reported as well as the inhibition constants with chondroitin sulfate and heparin (*34*).

VII. Biological Effects of Hyaluronidases

Hyaluronidases are utilized experimentally in the localization and identification of various mucopolysaccharides either in extracted fractions or in tissue sections. In the diagnostic use of the latter it is essential that the enzymes are free of proteases and of ribonucleases which often are contained in commercial preparations.

The different types of hyaluronate lyases reported from Suzuki's laboratory, and in addition, the specific 4- and 6-sulfatases are especially useful in the assay of isomeric chondroitin sulfates.

An isotope dilution method for the determination of hyaluronate in microgram quantities utilizing streptococcal hyaluronidase and reduction of the unsaturated disaccharides with Na-borotritiide has recently been reported (*51*).

The wide occurrence of hyaluronate and related mucopolysaccharides,

51. T. C. Laurent, E. Barany, B. Carlsson, and E. Tidare, *Anal. Biochem.* **31**, 133 (1969).

as well as the remarkable elaboration of three different types of hyaluronidases by structures as distinct as the poison glands of snakes and bees, the salivary glands of leech, and the submandibular glands of mammals, the apparently ubiquitous occurrence in lysosomes of mammalian cells and in mammalian testis points to the evolutionary advantage of such enzymes. The slow release of lysosomal enzyme seems indicated by its occurrence in mammalian serum where it, as well as its nonspecific inhibitors, have been studied extensively (*19*). The lowering of a diffusion barrier has been most elegantly demonstrated by the increase in filtration rate through a membrane of mouse fascia by Day (*52*). The large concentration of hyaluronidase in the head of mammalian spermatozoa apparently serves in the process of fertilization by dispersing the granulosa cell layer embedded in a jelly composed of protein and hyaluronate around the ovum. The enzyme has no effect on cervical mucus. An additional effect of spermatozoal hyaluronidase may be a growth-promoting action by oligosaccharides produced by the enzyme from its substrates (*53*). An increased local release of lysosomal hyaluronidase has been demonstrated in the metamorphosing tadpole (*24*) and probably plays a role in other remodeling processes.

A crude extract of tadpole tail fin has recently been shown to degrade hyaluronate and chondroitin 4-sulfate while it was inactive with chondroitin 6-sulfate. This finding of specificity of a vertebrate hyaluronidase requires confirmation with purified extracts and variations of the concentration and milieu conditions (*54*).

The once widely used therapeutic application of testicular hyaluronidase has been discontinued when it was found that these preparations, on repeated injection, became inactivated by antibodies against the enzyme and, more important, when the production of organ specific antibodies were observed which led to testicular degeneration.

While it appears probable that hyaluronidases influence the turgor and water content of tissues through depolymerization of hyaluronate, their role in the physiological water absorption in the kidney tubules (*36*) has not been substantiated. The irreversibility of the degradation of hyaluronate by hyaluronidases *a priori* makes such a hypothesis highly improbable.

52. T. D. Day, *J. Physiol.* (*London*) **117,** 1 (1952).
53. L. Ozzello, E. Y. Lasfargues, and M. R. Murray, *Cancer Res.* **20,** 600 (1960).
54. J. S. Silbert and S. DeLuca, *JBC* **245,** 1506 (1970).

12

Neuraminidases

ALFRED GOTTSCHALK • A. S. BHARGAVA

I. Historical Data on the Recognition of the Enzymes

There are but few examples in the history of enzymology where, as in the case of neuraminidase, the discovery of an enzyme has opened a wide field of biochemical research on an important group of natural compounds. Quite apart from the significance the discovery of the enzyme had for the advance in mucoprotein chemistry, the mere fact that it was found first in a virus was of great interest. Up to 1940 the view was generally held that viruses are devoid of enzymes and that this lack of enzymic equipment differentiated viruses basically from bacteria. This concept was challenged by G. K. Hirst. First, Hirst (*1*) and McClelland and Hare (*2*) observed that influenza virus agglutinates chicken red blood cells. In a second set of experiments, Hirst (*3*) showed that at 4° the influenza virus quickly adsorbed to the red cells and remained so for 18 hr; when the reaction was carried out at 37° the adsorption was equally rapid, but after 6 hr nearly complete elution of the previously adsorbed virus had occurred. Red cells from which the adsorbed virus had eluted spontaneously were not able to adsorb freshly added virus nor were they agglutinable by fresh virus suspensions (stabilized red cells). In contrast, the eluted virus was found to be functionally intact. Hirst's ingenious interpretation of the phenomena may be rationalized in the following manner:

Enzyme (E)	+	Substrate (S)	$\rightleftharpoons$	ES	$\rightarrow$	E + P
Virus particle with enzymes embedded in the virus coat		Receptors present at the red cell surface		Virus attached to the surface of two red cells (hemagglutination)		Red cells with altered receptors (stabilized cells)

Strong support for Hirst's assumption that an enzyme present at the virus surface was instrumental in "destroying" the cellular receptor was provided by Burnet's laboratory. Burnet *et al.* (*4*) observed that treatment of red cells by influenza virus rendered them agglutinable by any adult human serum. Burnet (*5*) correlated this observation with Friedenreich's (*6*) demonstration (1928) that erythrocytes on treatment with culture filtrates of diphtheroid bacilli and of vibrios became panagglutinable. Burnet and colleagues tested, therefore, culture filtrates of various bacteria for their ability to render red cells inagglutinable by influenza viruses. McCrea was the first to show that *Clostridium welchii* filtrates contained an enzyme capable of this performance; soon afterward it was found that much more active enzyme preparations could be obtained from *Vibrio cholerae* cultures (*4, 7*). Because of its virus

1. G. K. Hirst, *Science* **94**, 22 (1941); *J. Exptl. Med.* **75**, 49 (1942).
2. L. McClelland and R. Hare, *Can. J. Public Health* **32**, 530 (1941).
3. G. K. Hirst, *J. Exptl. Med.* **76**, 195 (1942).
4. F. M. Burnet, J. F. McCrea, and J. D. Stone, *Brit. J. Exptl. Pathol.* **27**, 228 (1946).
5. F. M. Burnet, *Physiol. Rev.* **31**, 131 (1951).
6. V. Friedenreich, *Acta Pathol. Microbiol. Scand.* **5**, 59 (1928).
7. J. F. McCrea, *Australian J. Exptl. Biol. Med. Sci.* **25**, 127 (1947).

receptor destroying activity the enzyme was termed "receptor-destroying enzyme" (RDE) (*8*).

Some information regarding the chemical nature of the substrate for the supposed viral enzyme and for the receptor-destroying enzyme was obtained from observations on the inhibition of influenza virus hemagglutination. Francis (*9*) discovered that influenza B virus, on heating at 55° for 30 min, though not losing its capacity to agglutinate red cells, was prevented from doing so by normal human serum. McCrea (*10*) isolated from human and rabbit sera a heat-stable mucoprotein fraction, identical with Rimington's serum mucoid fraction (*11*), inhibiting the hemagglutinating property of heat-inactivated influenza B virus (indicator virus). Burnet (*12*) found semipurified mucoid material prepared from human ovarian cyst content to be a potent virus hemagglutinin inhibitor, and Gottschalk and Lind (*13*) showed that of the various hen egg white fractions ovomucin strongly inhibited the virus hemagglutinin. Mucoid material isolated by DeBurgh *et al.* (*14*) from human red cells was also found to be an efficient inhibitor. The inhibitory capacity of all these mucoids was lost on treatment with active influenza virus or with the receptor-destroying enzyme.

These results taken together strongly suggested that the loss of biological activity of the red cell receptors and of the inhibitory mucoids on digestion with active influenza virus or with the receptor-destroying enzyme was due to the activity of two similar or isodynamic enzymes. Obviously, any further progress in this field had to come from a biochemical approach.

II. Occurrence

Neuraminidase is widely distributed in microorganisms and animal tissues. Viruses, bacteria, and avian and mammalian tissues are known to possess neuraminidase activity.

8. F. M. Burnet and J. D. Stone, *Australian J. Exptl. Biol. Med. Sci.* **25,** 227 (1947).

9. T. Francis, *J. Exptl. Med.* **85,** 1 (1947).

10. J. F. McCrea, *Australian J. Exptl. Biol. Med. Sci.* **26,** 355 (1948).

11. C. Rimington, *BJ* **34,** 931 (1940).

12. F. M. Burnet, *Australian J. Exptl. Biol. Med. Sci.* **26,** 371 (1948).

13. A. Gottschalk and P. E. Lind, *Brit. J. Exptl. Pathol.* **30,** 85 (1949).

14. P. M. DeBurgh, P. Yu, C. Howe, and M. Bovarnick, *J. Exptl. Med.* **87,** 1 (1948).

A. Viral Neuraminidase

The enzyme seems to be restricted in its occurrence to the group of myxoviruses. The name "myxovirus" was given by Andrews *et al.* (*15*) to a group of viruses because of the "special affinity of members of this group for certain mucins." These viruses are spherical RNA viruses, 800–1000 Å in diameter, enclosed by an envelope derived from the cytoplasmic membrane of the host cell. It has now been established that the viral neuraminidase is coded by the virus genome [see Fenner (*16*)]. The distribution of neuraminidase in myxoviruses is presented in Table I (*17–25*).

TABLE I
Occurrence of Neuraminidase in Myxoviruses

Myxoviruses	References
Myxo group	
Influenza A and B	Gottschalk and Perry (*17*), Kuhn and Brossmer (*18*), and Gottschalk (*19, 20*)
Fowl plague	Drzeniek *et al.* (*21*)
Paramyxo group	
Newcastle disease	Seto *et al.* (*22*) and Vasilera (*23*)
Measles	Howe *et al.* (*24*); see, however, Howe and Schluederberg (*24a*)
Para-influenza	Sokol *et al.* (*25*)

B. Bacterial Neuraminidase

The occurrence of neuraminidase in various bacteria has been described. Bacterial neuraminidases from *Vibrio cholerae*, *Diplococcus pneumoniae*, *Clostridium perfringens*, and *Streptococcus* are generally

15. C. H. Andrews, F. B. Bang, and M. F. Burnet, *Virology* **1**, 176 (1955).
16. F. Fenner, "Biology of Animal Viruses," Vols. 1 and 2. Academic Press, New York, 1968.
17. A. Gottschalk and B. J. Perry, *Brit. J. Exptl. Pathol.* **32**, 408 (1951).
18. R. Kuhn and R. Brossmer, *Angew. Chem.* **68**, 211 (1956).
19. A. Gottschalk, *BBA* **20**, 560 (1956).
20. A. Gottschalk, *BBA* **23**, 645 (1957).
21. R. Drzeniek, J. T. Seto, and R. Rott, *BBA* **128**, 547 (1966).
22. J. T. Seto, B. J. Hickey, and A. F. Rasmussen, Jr., *Proc. Soc. Exptl. Biol. Med.* **100**, 672 (1959).
23. L. Vasilera, *Vet. Med. Nauki* (*Sofia*) **5**, 39 (1968).
24. C. Howe, E. W. Newcomb, and T. Lee, *BBRC* **34**, 388 (1969).
24a. A. C. Howe and A. Schluederberg, *BBRC* **40**, 606 (1970).
25. F. Sokol, D. Blaskovic, and O. Krizanova, *Acta Virol.* (*Prague*) **5**, 153 (1961).

found in the culture medium. In other bacteria neuraminidase has been shown to be part of the cell, as, for instance, in *Corynebacterium diphtheriae*. Ada and French (*26*) have shown that neuraminidase from *V. cholerae* is an inducible enzyme. This induction could be achieved by free or bound sialic acid and by *N*-acetylmannosamine. Table II contains a list of bacteria possessing neuraminidase (*19, 27–34*).

The protozoan *Trichomonas foetus* was also found to contain neuraminidase (*35*).

TABLE II
OCCURRENCE OF NEURAMINIDASE IN BACTERIA

Bacteria	References
Vibrio cholerae	Gottschalk (*19*), Heimer and Meyer (*27*), and Faillard (*28*)
Clostridium perfringens	Popenoe and Drew (*29*)
Clostridium septicum	Gadalla *et al.* (*30*)
Diplococcus pneumoniae	Heimer and Meyer (*27*)
Corynebacterium diphtheriae	Blumberg and Warren (*31*) and Moriyama and Barksdale (*32*)
Lactobacillus bifidus	Shilo (*33*)
Streptococcus (group A, B, C, G, and L)	Hayano and Tanaka (*34*)
Streptococcus sanguis	Hayano and Tanaka (*34*)
Pseudomonas group	Shilo (*33*)

C. MAMMALIAN NEURAMINIDASE

Warren and Spearing (*36*) first showed the presence of neuraminidase in human and bovine plasma of commercial preparations of Cohn fraction VI. Later, neuraminidase was found in many tissues and organs of mammals [see Table III (*36–45*)] though the activity was low when compared with that of viral and bacterial neuraminidases.

26. G. L. Ada and E. L. French, *J. Gen. Microbiol.* **21**, 561 (1959).
27. R. Heimer and K. Meyer, *Proc. Natl. Acad. Sci. U. S.* **42**, 728 (1956).
28. H. Faillard, *Z. Physiol. Chem.* **305**, 145 (1956).
29. E. A. Popenoe and R. M. Drew, *JBC* **228**, 673 (1957).
30. M. S. A. Gadalla, T. G. Collee, and W. R. Barr, *J. Pathol. Bacteriol.* **96**, 169 (1968).
31. B. S. Blumberg and L. Warren, *BBA* **50**, 90 (1961).
32. T. Moriyama and L. Barksdale, *J. Bacteriol.* **94**, 1565 (1967).
33. M. Shilo, *BJ* **66**, 48 (1957).
34. S. Hayano and A. Tanaka, *J. Bacteriol.* **97**, 728 (1969).
35. E. Romanowska and W. M. Watkins, *BJ* **87**, 37P (1963).
36. L. Warren and C. W. Spearing, *BBRC* **3**, 489 (1960).
37. E. H. Morgan and C. B. Laurell, *Nature* **197**, 921 (1963).

TABLE III
OCCURRENCE OF NEURAMINIDASE IN MAMMALS

Species	Tissue	References
Human	Plasma	Warren and Spearing (*36*)
	Brain	Morgan and Laurell (*37*)
	Intestinal mucosa	Ghosh *et al.* (*38*)
Bovine	Plasma	Warren and Spearing (*36*)
	Brain	Morgan and Laurell (*37*)
	Platelets and erythrocytes	Gielen *et al.* (*39*)
Rat	Liver, kidney, spleen, brain, small intestine, mammary gland and testis	Carubelli *et al.* (*40*) and Mahadevan *et al.* (*41*)
Guinea pig	Brain	Morgan and Laurell (*37*)
Pig	Kidney	Tuppy and Palese (*42*)
	Brain	Leibowitz and Gatt (*43*)
Mice	Brain	Kelly and Greiff (*44*)
Rabbit	Kidney	Kuratowska and Kubicka (*45*)
Rabbit, bull, hamster, ram, human	Spermatozoa	Srivastava *et al.* (*45a*)

Subcellular distribution of neuraminidase in rat liver and kidney has been studied by Mahadevan *et al.* (*41*); the enzyme was shown to be localized in lysosomes. Taha and Carubelli (*46*) fractionated homogenates of rat liver in 0.25 *M* sucrose by differential centrifugation and investigated the intracellular distribution of neuraminidase. The highest relative specific activity was observed in the lysosomal fraction. The

38. M. K. Ghosh, L. Kotowitz, and W. H. Fishman, *BBA* **167,** 201 (1968).
39. W. Gielen, J. Etzrodt, and G. Uhlenbruck, *Thromb. Diath. Haemorrh.* **22,** 203 (1969).
40. R. Carubelli, R. E. Trucco, and R. Caputto, *BBA* **60,** 196 (1962).
41. S. Mahadevan, J. C. Nduaguba, and A. L. Tappel, *JBC* **242,** 4409 (1967).
42. H. Tuppy and P. Palese, *Z. Physiol. Chem.* **349,** 1169 (1968).
43. Z. Leibowitz and S. Gatt, *BBA* **152,** 136 (1968).
44. R. T. Kelly and D. Greiff, *BBA* **110,** 548 (1965).
45. Z. Kuratowska and T. Kubicka, *Acta Biochim. Polon.* **14,** 255 (1967).
45a. P. N. Srivastava, L. J. D. Zaneveld, and W. L. Williams, *BBRC* **39,** 575 (1970).
46. B. H. Taha and R. Carubelli, *ABB* **119,** 55 (1967).

supernatant fraction contained the largest percentage of the total emzymic activity present in the whole homogenate. However, when conditions that minimized the manipulation of the particles were applied, the supernatant fraction had a very low enzyme content. Changes of neuraminidase activity were detected in organs of female rats during pregnancy and lactation. Liver and mammary glands showed similar patterns with peaks of activity around 14–16 days after parturition. In brain, peaks were observed about the second week of pregnancy and the first week of lactation.

Changes in rat brain neuraminidase during development have also been investigated by Carubelli (*47*). Although increase in neuraminidase activity paralleled the increase in gangliosides and sialoglycoproteins, there is no direct proof that mammalian neuraminidase like the *Vibrio cholerae* neuraminidase is an inductive enzyme.

Besides the viral, bacterial, and mammalian neuraminidases, this enzyme has also been reported to be present in the chorioallantoic membrane of the chick embryo (*48*).

III. Purification of the Enzyme

The best sources of the enzyme are the influenza virus particle and *Vibrio cholerae*. In the virus the enzyme is a built-in constituent of the mosaic structure of the particle surface; in contrast, *V. cholerae*, during its growth under suitable conditions, produces and releases the enzyme into the culture medium.

A. Neuraminidase from Influenza Virus

1. *Purification of Virus Particles*

Purification of the influenza virus particle is best effected according to Knight (*49*). Briefly, the virus harvested from the allantoic fluid of infected chick embryos is adsorbed to chicken erythrocytes at 4° and eluted at 37° into phosphate buffer (pH 7.0). After removal of the red cells the virus is recovered from the eluate by centrifugation at 24,000 rpm for 15 min. The pellets containing the virus are resuspended and

47. R. Carubelli, *Nature* **219**, 955 (1968).
48. G. L. Ada, *BBA* **73**, 276 (1963).
49. C. A. Knight, *J. Exptl. Med.* **85**, 99 (1947).

spun at 5000 rpm for 5 min. This process of alternate high- and low-speed centrifugation is carried out three times.

2. *Isolation and Purification of Neuraminidase*

In recent years Rafelson *et al.* (*50*) and Drzeniek *et al.* (*21*) have described methods for the isolation and purification of viral neuraminidases. The preparation of Rafelson *et al.* was homogeneous in the ultracentrifuge ($s_{20,w} = 4.2$). Drzeniek *et al.* tested the purity of their neuraminidase by preparing antisera to the enzyme. Such antisera were found to be devoid of antibodies against other virus-specific surface antigens.

Laver and Valentine (*51*) succeeded in isolation and purification of the neuraminidase and hemagglutinin of BEL strain of influenza A_0 and the recombinant influenza virus X-7F1 possessing A_0 (NWS) type of hemagglutinin and A_2 type of neuraminidase. The complete separation of these two subunits of the influenza viruses was demonstrated by electron micrographs. The virus particles were disrupted at pH 9 with 1% sodium dodecyl sulfate (SDS), and the proteins were separated by electrophoresis on cellulose acetate strips in buffer containing 0.4% SDS [see Laver (*52*)]. The hemagglutinin of BEL virus and the neuraminidase of the recombinant virus were eluted separately from the strips with saline. The solutions containing the two virus components were kept in an ice bath overnight and SDS was removed by centrifugation. The material in the supernatant was further freed from SDS and at the same time concentrated by precipitation with cold acetone.

It may be mentioned that removal of SDS from the monomeric enzyme immediately gave rise to polymers, which had an appearance similar to that of the seeded heads of dandelions.

The neuraminidase of Newcastle disease virus (NDV) is less readily isolated than the neuraminidase of influenza virus (*21*).

B. Neuraminidase from *Vibrio cholerae*

Neuraminidase was prepared crystalline from *Vibrio cholerae* filtrates. The purification by Ada and French (*53*) involved methanol precipitation of the enzyme protein, adsorption of the enzyme to and elution from

50. M. E. Rafelson, Jr., S. Gold, and I. Priede, "Methods in Enzymology," Vol. 8, p. 677, 1966.
51. W. G. Laver and R. C. Valentine, *Virology* **38**, 105 (1969).
52. W. G. Laver, *JMB* **9**, 109 (1964).
53. G. L. Ada and E. L. French, *Nature* **183**, 1740 (1959).

human red cells, precipitation of the enzyme protein with ammonium sulfate, and adsorption onto hydroxylapatite, followed by elution with a gradient of phosphate-buffered saline at pH 6.8. The yield was increased by adding the dialysate of bovine colostrum to the nutrient medium.

Schramm and Mohr's technique (*54*) for purification of *V. cholerae* neuraminidase combined in one step specific binding of the enzyme to its substrate and column chromatography. Stromata of erythrocytes (sheep or calf) were mixed with Kieselgur to form a column. The enzyme protein strongly adsorbed to the column at pH 9, thus allowing removal of contaminating proteins by washing. Elution of the enzyme protein was effected at pH 5.5, and further purification and eventual crystallization was achieved by precipitation with ammonium sulfate.

The neuraminidase present in the culture filtrate of *Clostridium perfringens* was isolated and purified by Cassidy *et al.* (*55*). The three steps involved in the purification procedure were (1) ammonium sulfate precipitation between 50 and 85% saturation, (2) gel filtration on Sephadex G-75, and (3) DEAE-cellulose chromatography. The enzyme was purified 5500-fold and the recovery was about 22%. The purified enzyme preparation was free of *N*-acetylneuraminic acid aldolase, protease, RNase, DNase, DPNHoxidase, and β-galactosidase.

IV. Properties of the Enzyme

Some properties of the viral, bacterial, and animal neuraminidases are summarized in Table IV (*21, 48, 51, 56–63*). With regard to shape, Laver and Valentine (*51*) showed by electron microscopy that in SDS the neuraminidase was an elongated object. The elongated body had mean measurements of 85 × 50 Å. In shape it is closer to a cylinder (which with these dimensions has a volume of 167,000 Å^3) than to an ellipsoid (volume 110,000 Å^3). A volume of 116,000 Å^3 and the average protein

54. G. Schramm and E. Mohr, *Nature* **183,** 1677 (1959).
55. J. T. Cassidy, G. W. Jourdian, and S. Roseman, *JBC* **240,** 3501 (1965).
56. A. P. Kendal, F. Biddle, and G. Belyavin, *BBA* **165,** 419 (1968).
57. W. G. Laver, *Virology* **20,** 251 (1963).
58. H. Noll, T. Aolyagi, and J. Orlando, *Virology* **18,** 154 (1962).
59. E. Mohr and G. Schramm, *Z. Naturforsch.* **15b,** 568 (1960).
60. J. Pye and C. C. Curtain, *J. Gen. Microbiol.* **24,** 423 (1961).
61. W. G. Laver, J. Pye, and G. L. Ada, *BBA* **81,** 177 (1964).
62. J. T. Seto, R. Drzeniek, and R. Rott, *BBA* **113,** 402 (1966).
63. E. Balke and R. Drzeniek, *Z. Naturforsch.* **24b,** 599 (1969).

TABLE IV
SOME PROPERTIES OF VIRAL, BACTERIAL, AND ANIMAL NEURAMINIDASES

Source	Sedimentation coefficient (S)	Molecular weight	References
Influenza virus A_2	8.5	130,000 (From electron micrographs)	Laver and Valentine (*51*)
Influenza virus A_2	8.8	—	Drzeniek *et al.* (*21*)
Influenza virus A_2	8.0	220,000 (From sedimentation coefficient)	Kendal *et al.* (*56*)
Influenza B virus (LEE strain)	8.2	—	Laver (*57*)
Influenza B virus (LEE strain)	9.0	190,000 (From sedimentation coefficient)	Noll *et al.* (*58*)
Fowl plague virus	10.0	—	Drzeniek *et al.* (*21*)
Vibrio cholerae	1.3	10,000–20,000 (From sedimentation coefficient)	Mohr and Schramm (*59*)
Vibrio cholerae	5.3–5.5	90,000 (From sedimentation coefficient and diffusion coefficient)	Pye and Curtain (*60*) and Laver *et al.* (*61*)
Vibrio cholerae	5.4	—	Drzeniek *et al.* (*21*)
Vibrio cholerae	5.4	—	Seto *et al.* (*62*)
Clostridium perfringens	4.6	56,000 (From gel filtration)	Balke and Drzeniek (*63*)
Chick embryo chorioallantoic membrane	3.3	—	Ada (*48*)

partial specific volume of 0.73 cm^3/g indicate a molecular weight of about 130,000. The complex appearance of the neuraminidase suggests that it contains more than one kind of polypeptide chain [see also Drzeniek *et al.* (*64*)].

Neuraminidase produced by *Vibrio cholerae* 4Z strain which has a molecular weight of about 90,000 (*61*) is a protein molecule consisting most probably of a single polypeptide chain. It seems unlikely that this enzyme is simply a polymer of the enzyme with $10–20 \times 10^3$ molecular weight as reported by Mohr and Schramm (*59*).

Heat stability at 56° of the neuraminidase of myxoviruses has been tested (*21*); it was dependent on the physical state of the enzyme. Neuraminidase still bound to the NDV virus (strain Beaudette) was more stable than the isolated, soluble enzyme. On the contrary, the stability of the neuraminidases of fowl plague viruses was found to be the same in the bound and soluble form. As to the heat stability of A_2 influenza virus neuraminidase the bound A_2 Japan neuraminidase was highly stable, whereas the A_2 Singapore neuraminidase was labile.

Recently Tannenbaum *et al.* (*64a*) reported on the separation, purification, and properties of pneumococcal neuraminidase isoenzymes.

V. Substrate Specificity of Neuramidases

A. Configurational Specificity

Neuraminidases are able to split glycosides (ketosides) of α-D-configurated *N*-acylneuraminic acids. This was first suggested by Kuhn and Brossmer (*65*) and Gottschalk (*20*). On the basis of the specific optical rotation of *O*-acetyl-*N*-acetylneuraminosyl-(2 → 3)lactose, they concluded that according to Hudson's isorotation rules the ketosidic linkage of the trisaccharide probably has the α configuration. The statement that neuramidases are α-ketosidases became conclusive when crystalline anomeric α and β-ketosides of neuraminic acid were synthesized by Meindl and Tuppy (*66–69*), Faillard *et al.* (*70*), Kuhn *et al.* (*71*), and Yu and

64. R. Drzeniek, M. Frank, and R. Rott, *Virology* **36**, 703 (1968).
64a. S. W. Tannenbaum, J. Gulbinsky, M. Katz, and S. C. Sun, *BBA* **198**, 242 (1970).
65. R. Kuhn and R. Brossmer, *Chem. Ber.* **89**, 2013 (1956).
66. P. Meindl and H. Tuppy, *Monatsh. Chem.* **96**, 802 (1965).
67. P. Meindl and H. Tuppy, *Monatsh. Chem.* **96**, 816 (1965).
68. P. Meindl and H. Tuppy, *Monatsh. Chem.* **97**, 990 (1966).
69. P. Meindl and H. Tuppy, *Monatsh. Chem.* **97**, 1628 (1966).

Ledeen (*72*). The authors showed that α-ketosidically linked *N*-acylneuraminic acid was hydrolyzed by *Vibrio cholerae* and viral neuraminidases, whereas the β-anomer was not split at all.

The only naturally occurring linkage of ketosidically bound *N*-acetylneuraminic acid which is insusceptible to neuraminidase is cytidine 5′-monophospho-*N*-acetylneuraminic acid (CMP-NANA) (*73*). It is assumed from optical rotation studies that in this case the anomeric carbon atom of *N*-acetylneuraminic acid is in the β configuration (*74*).

B. Steric Hindrance in Natural Substrates for Neuraminidase Activity

All natural partners of α-ketosidically bound neuraminic acid are susceptible to neuraminidase. If, however, the vicinal carbon atom of the sugar partner, to which neuraminic acid is linked, is substituted, this substitution presents a steric hindrance for the action of neuraminidase. Thus, Kuhn and Wiegandt (*75*) have shown that the monosialoganglioside [G_{GNT}^{1} according to Wiegandt (*76*)] (see Fig. 1) is resistant to *Vibrio cholerae* neuraminidase. It is only after enzymic removal of galactose (linkage a) and *N*-acetylgalactosamine (linkage b) that neuraminidase will split the ketosidic linkage c. For the steric hindrance in other natural substrates see Sections V,E and V,F.

Earlier observations seemed to show that in some glycoproteins only 30–60% of sialic acid could be removed, though fragmentation of the macromolecule into small glycopeptides rendered all sialic acid susceptible to neuraminidase (*77, 78*). A complete removal of sialic acid from high molecular weight glycoproteins [e.g., ovine submaxillary glycoprotein (OSM)] was, however, achieved by prolonged or repeated treatment with *V. cholerae* neuraminidase (*55, 79*).

---

70. H. Faillard, G. Kirchner, and M. Blohm, *Z. Physiol. Chem.* **347,** 87 (1966).
71. R. Kuhn, P. Lutz, and D. L. McDonald, *Chem. Ber.* **99,** 611 (1966).
72. R. K. Yu and R. Ledeen, *JBC* **244,** 1306 (1969).
73. D. G. Comb, F. Shimizu, and S. Roseman, *JACS* **81,** 5513 (1959).
74. D. G. Comb, D. R. Watson, and S. Roseman, *JBC* **241,** 5637 (1966).
75. R. Kuhn and H. Wiegandt, *Chem. Ber.* **89,** 2013 (1963).
76. H. Wiegandt, *Ergeb. Physiol., Biol. Chem. Exptl. Pharmakol.* **57,** 190 (1966).
77. V. P. Bhavanandan, E. Buddecke, R. Carubelli, and A. Gottschalk, *BBRC* **16,** 353 (1964).
78. J. Labat and N. Berger, *Ann. Biol. Clin. (Paris)* **24,** 839 (1966).
79. A. S. Bhargava and A. Gottschalk, *BBA* **148,** 125 (1967).

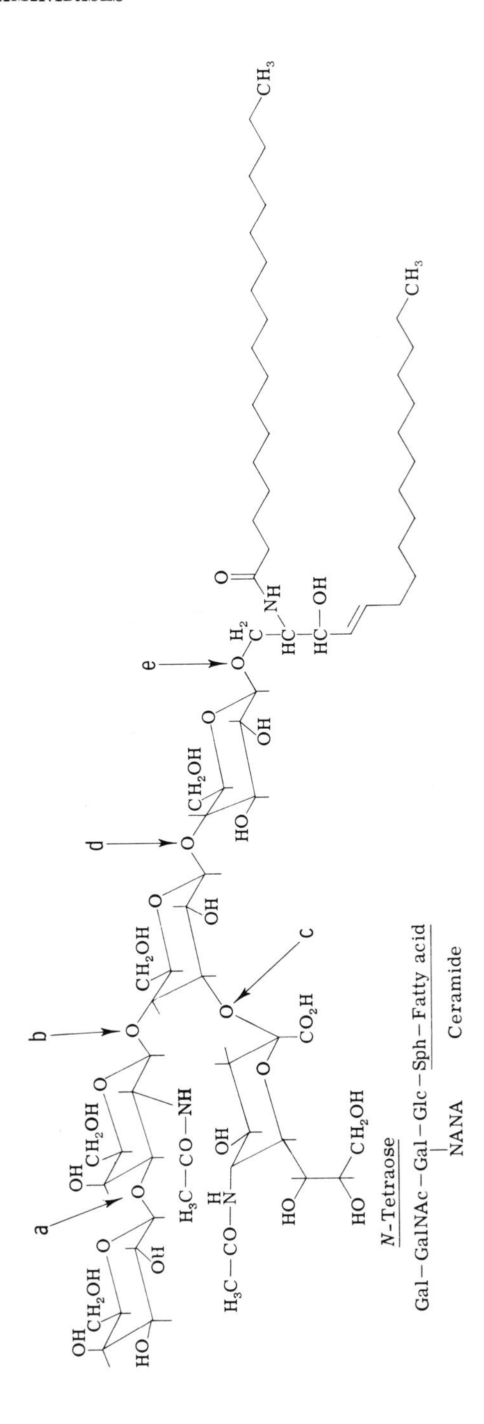

FIG. 1. Enzymic degradation of monosialoganglioside (G_{GNT}^{1}) from brain: (a) β-galactosidase, (b) β-*N*-acetylgalactosaminidase, (c) neuraminidase, (d) β-galactosidase, and (e) β-glucosidase.

C. Position of the Glycosidic Linkage

Four different positions of the α-ketosidic linkage of neuraminic acid to its sugar partners are known, namely, $(2 \rightarrow 3)$, $(2 \rightarrow 4)$, $(2 \rightarrow 6)$, and $(2 \rightarrow 8)$, as, for instance in *N*-acetylneuraminosyl-$(\alpha,2 \rightarrow 3)$lactose (see Fig. 2), *N*-acetylneuraminosyl$(\alpha,2 \rightarrow 4)$-D-galactose in α_1-acid glycoprotein, *N*-acetylneuraminosyl$(\alpha,2 \rightarrow 6)$-*N*-acetylgalactosamine in OSM (see Fig. 3) and disialyllactose [*N*-acetylneuraminosyl$(\alpha,2 \rightarrow 8)$-*N*-acetylneuraminosyl$(\alpha,2 \rightarrow 3)$lactose].

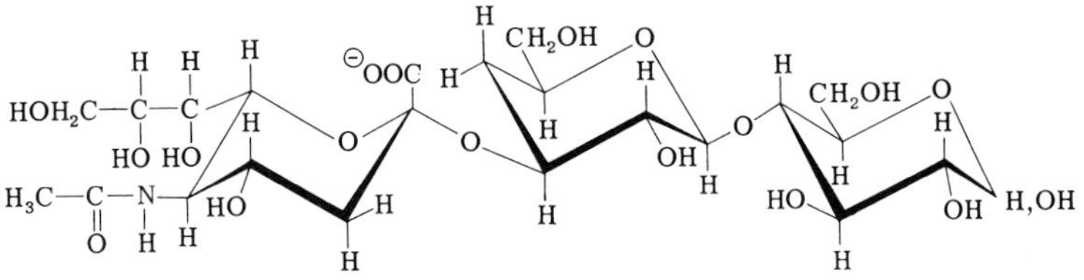

Fig. 2. *N*-Acetylneuraminosyllactose [α-D-*N*-acetylneuraminosyl(2-3)-β-D-galactopyranosyl(1-4)-D-glucopyranose].

It is interesting to note that all four types of linkage are cleaved by *Vibrio cholerae* neuraminidase (*80, 81*) and by neuraminidase from *Clostridium perfringens* (*55*). The viral neuraminidases, on the other hand, show differences in their ability to split the above types of α-ketosidic linkages. Under defined conditions of short time incubations the neuraminidase of NDV cleaves $(\alpha,2 \rightarrow 3)$ and $(\alpha,2 \rightarrow 8)$-ketosidic linkages, whereas fowl plague virus cleaves only the $(\alpha,2 \rightarrow 3)$-ketosidic linkage. Thus, NDV neuraminidase removes from disialyllactose both NANA residues, whereas fowl plague virus neuraminidase unable to cleave a NANA-$(\alpha,2 \rightarrow 8)$-NANA linkage fails to remove any NANA. These observations confirm the previous finding (*20*) that neuraminidase is an exoenzyme which liberates terminal *N*-acylneuraminic acid (*80, 81*).

The use of neuraminidases from *Vibrio cholerae*, NDV, and fowl plague viruses may become useful in the elucidation of the position of the ketosidic linkage of neuraminic acid to its partner. It should be stressed that the specificity of the viral neuraminidases, as outlined above, holds only for the standard conditions given below.

Substrates containing 1.6×10^{-7} mole total NANA were incubated for 15 min at 37°C either with 2–20 milliunits of *Vibro cholerae* neuraminidase at pH 5.5 in 0.05 *M* acetate buffer containing 0.15 *M* NaCl and

80. R. Drzeniek, *BBRC* **26,** 631 (1967).
81. R. Drzeniek and A. Gauhe, *BBRC* **38,** 651 (1970).

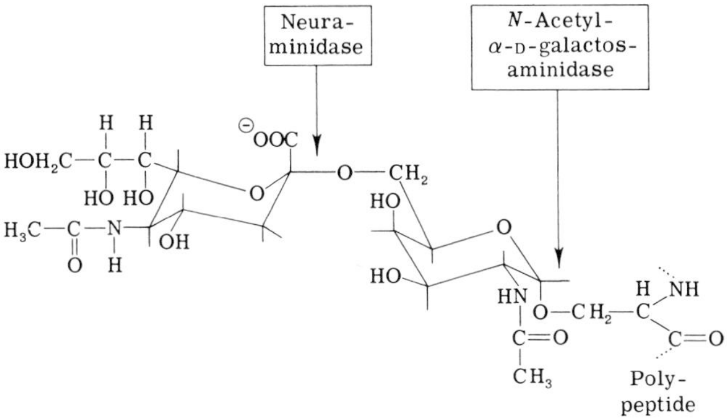

Fig. 3. The prosthetic group (disaccharide) of OSM linked *O*-glycosidically to a serine residue of the protein moiety.

7×10^{-3} *M* $CaCl_2$ or with 2–200 milliunits of viral neuraminidases at pH 7.0 in 0.1 *M* phosphate buffer in a total volume of 0.5 ml (*80, 81*).

With an excess of enzyme and prolonged treatment (24 hr) no difference between the bacterial and viral neuraminidases could be observed (*81*). However, more work is required to substantiate the claims that the various viral and bacterial neuraminidases can be used for structural analysis of the position of the ketosidic linkage of sialic acid to its partner (*80*).

D. *N*-Substitution of Neuraminic Acid

The *N*-substituent of neuraminic acid can either be, as in natural products, an acetyl or glycolyl residue, or as in synthetic substrates, a relatively small acyl group like the propionyl group or a voluminous substituent like the benzyloxycarbonyl residue.

The introduction of a larger substituent than acetyl like propionyl, butyryl, benzoyl, or benzyloxycarbonyl into synthetic neuraminic acid α-ketosides causes a decreased rate or a total block in the hydrolysis of these compounds. Thus, the benzyl-α-ketoside of *N*-propionylneuraminic acid was hydrolyzed more slowly than that of *N*-acetylneuraminic acid by *Vibrio cholerae* and influenza A virus neuraminidases. The α-ketosides of *N*-butyrylneuraminic acid and *N*-benzoylneuraminic acid (*69*) and of *N*-benzyloxycarbonylneuraminic acid (*82*) were resistant to neuraminidase action.

82. H. Faillard, C. F. Do Amaral, and M. Blohm, *Z. Physiol. Chem.* **350,** 798 (1969).

E. *O*-Substitution of *N*-Acylneuraminic Acid

In natural occurring substances neuraminic acid is often substituted at the free hydroxyl group of carbon atoms 4, 7, or 8 by an acetyl residue. Di- and tri-*O*-acetylated neuraminic acids are also known.

Early investigations by Faillard (*83*) indicated a very slow release of *O*-acetylated sialic acid from bovine submaxillary glycoprotein. Bovine submaxillary glycoprotein contains *N*-glycolylneuraminic acid, NANA, *N*-acetyl-7-*O*-acetylneuraminic acid, *N*-acetyl-8-*O*-acetylneuraminic acid, an *N*-acetyl-*O*-diacetylneuraminic acid which is probably *N*-acetyl-7,8-*O*-diacetylneuraminic acid. All these neuraminic acid derivatives are released by the neuraminidase of *Vibrio cholerae* and *Clostridium perfringens.*

N-Acetyl-4-*O*-acetylneuraminic acid, *N*-acetyl-4,7,8-*O*-triacetylneuraminic acid, present in horse submaxillary glycoprotein, and *N*-glycolyl-4-*O*-acetylneuraminic acid recently found in equine erythrocyte glycosphingolipids are resistant to neuraminidase. After removal by alkali of the 4-*O*-acetyl group the substrates become susceptible to the enzyme (*84, 85*).

From lactating mammary glands an *O*-sulfated *N*-acetylneuraminic acid containing trisaccharide was isolated (*86*) whose structure was established as *N*-acetylneuraminosyl(α,2 $\rightarrow$ 3)-D-galactopyranosyl-6-*O*-sulfate (β,1 $\rightarrow$ 4)-D-glucopyranose (*87*). This compound was hydrolyzed two times faster than *N*-acetylneuraminosyl(α,2 $\rightarrow$ 3)lactose by the lysosomal neuraminidase of rat liver (*88*).

F. Esterification of the Carboxyl Group of Neuraminic Acid

Colominic acid, isolated from *Escherichia coli* K235 by Barry and Goebel (*89*) is a low molecular weight homopolymer of *N*-acetylneuraminic acid joined by α-ketosidic-(2 $\rightarrow$ 8) linkages between the monomers (*90*). This compound was found resistant (*91, 92*), or partially

83. H. Faillard, *Z. Physiol. Chem.* **317,** 257 (1959).
84. R. Schauer and H. Faillard, *Z. Physiol. Chem.* **349,** 961 (1968).
85. S. I. Hakomori and T. Saito, *Biochemistry* **8,** 5082 (1969).
86. R. Carubelli, L. C. Ryan, R. E. Trucco, and R. Caputto, *JBC* **236,** 2381 (1961).
87. L. C. Ryan, R. Carubelli, R. Caputto, and R. E. Trucco, *BBA* **101,** 252 (1965).
88. D. R. P. Tulsiani and R. Carubelli, *JBC* **245,** 1821 (1970).
89. G. T. Barry and W. F. Goebel, *Nature* **179,** 206 (1957).
90. E. J. McGuire and S. E. Binkley, *Biochemistry* **3,** 247 (1964).
91. A. Gottschalk, "The Chemistry and Biology of Sialic Acids and Related Substances." Cambridge Univ. Press, London and New York, 1960.
92. D. Aminoff, *BJ* **81,** 384 (1961).

resistant to *Vibrio cholerae* and *Clostridium perfringens* neuraminidase; only 20% of *N*-acetylneuraminic acid was liberated from colominic acid within 1 hr according to McGuire and Binkley (*90*) and Cassidy *et al.* (*55*). After 24-hr incubation the compound was completely cleaved to free *N*-acetylneuraminic acid (*55*). These discrepancies were cleared up by the finding that colominic acid contains ester linkages between the carboxyl group of one monomer and the C9 (or C7) hydroxyl group of the adjacent NANA molecule. After saponification of this ester group by dilute alkali at room temperature 90% of *N*-acetylneuraminic acid was released in 1 hr by incubation with *C. perfringens* neuraminidase compared to 10–15% of *N*-acetylneuraminic acid liberated from colominic acid under identical conditions but not treated with alkali (*90*).

The methyl esters of bovine submaxillary gland glycoprotein (*93*), of *N*-acetylneuraminosyl(α,2 $\rightarrow$ 3)lactose (*71*), and *N*-acetylneuraminic acid α-methyl-ketoside (*68*) are not hydrolyzed or acted upon only at a very low rate by *Vibrio cholerae* neuraminidase—just as the specific inhibitory capacity of free NANA on the system neuraminidase *N*-acetylneuraminosyl(α,2 $\rightarrow$ 3)lactose is inactivated by esterification. The *N*-acetylneuraminic acid can be liberated after hydrolysis of the ester group by dilute alkali. Thus it seems that a negatively charged carboxyl group of the neuraminic acid molecule is necessary for the action of neuraminidases.

VI. Kinetic Data of Neuraminidases

Some kinetic data of neuraminidases of different origin are summarized in Table V. It may be mentioned that the pH optimum of the various neuraminidases often varies with the nature of the substrate. For instance, the pH optimum of influenza A virus (PR8 strain) neuraminidase with *N*-acetylneuraminosyl(α,2 $\rightarrow$ 3)lactose as substrate is 7.0, whereas the pH optimum of the same enzyme with α_1-acid glycoprotein as substrate is 5.4 [see Rafelson *et al.* (*94*)].

Neuraminidases from different sources are activated or inhibited or not influenced by calcium ions [see Table V (*36, 55, 94–99*)]. An inhibi-

93. A. Gottschalk, *Perspectives Biol. Med.* **5,** 327 (1962).
94. M. E. Rafelson, Jr., M. Schneir, and A. W. Wilson, Jr., *ABB* **103,** 424 (1963).
95. G. L. Ada, E. L. French, and P. E. Lind, *J. Gen. Microbiol.* **24,** 409 (1961).
96. R. C. Hughes and R. W. Jeanloz, *Biochemistry* **3,** 1535 (1964).
97. J. T. Cassidy, G. W. Jourdian, and S. Rosemann, "Methods in Enzymology," Vol. 8, p. 680, 1966.
98. A. Horvat and O. Touster, *JBC* **243,** 4380 (1968).
99. G. L. Ada and P. E. Lind, *Nature* **190,** 1169 (1961).

TABLE V

SOME KINETIC DATA OF NEURAMINIDASES OF VIRAL, BACTERIAL, AND ANIMAL ORIGIN WITH *N*-ACETYLNEURAMINOSYL(α, 2 $\rightarrow$ 3)LACTOSE AS SUBSTRATE

Source	pH optimum	K_m (mole/liter)	Effect of Ca^{2+}	Effect of EDTA[a]	References
Influenza A virus (Japan 305)	6.5	5×10^{-4}	No effect	Inhibition	Rafelson *et al.* (*94*)
Influenza A virus (PR8)	7.0	1.2×10^{-3}	No effect	Inhibition	Rafelson *et al.* (*94*)
Vibrio cholerae	5.6	1×10^{-3}	Stimulation	Inhibition	Ada *et al.* (*95*)
Diplococcus pneumoniae	6.5	1.8×10^{-3}	No effect	No effect	Hughes and Jeanloz (*96*)
Clostridium perfringens	4.5	2×10^{-3}	No effect	No effect	Cassidy *et al.* (*55, 97*)
Human plasma	5.5	—	Stimulation	Inhibition	Warren and Spearing (*36*)
Rat liver lysosomes	4.2	1.98×10^{-3}	Weak stimulation	Inhibition	Horvat and Touster (*98*)
Chick chorioallantoic membrane	4.5	1×10^{-3}	—	No effect	Ada and Lind (*99*)

[a] Ethylenediaminetetraacetate.

tion of 42 and 74% of rat liver and kidney neuraminidases, respectively, by Ca^{2+} at 50 m*M* with sialyllactose as substrate was reported by Mahadevan *et al.* (*41*).

VII. Inhibitors of Neuraminidase Activity

The first inhibitor of neuraminidase activity to be found was *N*-acetylneuraminic acid. Walop *et al.* (*100*), using *N*-acetylneuraminosyllactose as substrate and influenza A_2 Singapore virus particles as enzyme, showed that NANA competitively inhibited the enzyme reaction, the mean value of K_i being $5 \times 10^{-3}\,M$. Mohr (*101*) observed with the system *Vibrio cholerae* neuraminidase and calf erythrocyte stroma an inhibition of 52 and 62% with 900 and 1800 μg NANA/ml, respectively. *N*-Acetylneuraminic acid methyl ester did not effect any inhibition.

Recently, Meindl and Tuppy (*102*) demonstrated 2-deoxy-2,3-dehydro-*N*-acetylneuraminic acid to inhibit competitively the *Vibrio cholerae* neuraminidase when tested with phenyl-α-ketoside of NANA as substrate. The K_i value was $1 \times 10^{-5}\,M$. Viral neuraminidases were also found to be inhibited by this compound. A $K_i = 5.3 \times 10^{-6}\,M$ was observed for influenza A virus, Melbourne strain, neuraminidase (personal communication).

Another group of potent neuraminidase inhibitors are *N*-substituted oxamic acids. The most active inhibitors for both influenza A_2 Singapore neuraminidase and *Vibrio cholerae* neuraminidase were *N*-phenyloxamic acid, *N*-(2-pyridyl)oxamic acid, *N*-(2-thiazolyl)oxamic acid, and thiooxamilic acid (*103*).

Other inhibitors are tabulated (Table VI) (*50, 104, 105*).

VIII. Assay Method

The assay method for neuraminidase will depend on the substrate used. The most commonly used substrate in neuraminidase work is at

100. J. N. Walop, T. A. C. Boschman, and J. Jacobs, *BBA* **44,** 185 (1960).
101. E. Mohr, *Z. Naturforsch.* **15b,** 575 (1960).
102. P. Meindl and H. Tuppy, *Z. Physiol. Chem.* **350,** 1088 (1969).
103. J. D. Edmond, R. G. Johnston, D. Kidd, H. J. Rylance, and R. G. Sommerville, *Brit. J. Pharmacol.* **27,** 415 (1966).
104. K. W. Brammer, C. R. McDonald, and M. S. Tute, *Nature* **219,** 515 (1968).
105. H. Becht and R. Drzeniek, *J. Gen. Virol.* **2,** 261 (1968).

TABLE VI
INHIBITORS OF NEURAMINIDASE ACTIVITY

Inhibitors	Substrate	Source of enzyme	Conc. of inhibitor (*M*)	Inhibition (%)	References
1-Phenoxymethyl-3,4-dihydroisoquinoline	Purified *Collocalia* mucoid	Influenza A_2[a] Singapore	1.6×10^{-3}	33–38	Brammer *et al.* (*104*)
Merthiolate	*N*-Acetylneuraminosyllactose	Influenza A virus (Japan 305)	1×10^{-3}	100	Rafelson *et al.* (*50*)
Ascorbic acid	*N*-Acetylneuraminosyllactose	Influenza A virus (Japan 305)	5×10^{-2}	100	Rafelson *et al.* (*50*)
Cysteine	*N*-Acetylneuraminosyllactose	Influenza A virus (Japan 305)	5×10^{-2}	90	Rafelson *et al.* (*50*)
Glutathione (reduced)	*N*-Acetylneuraminosyllactose	Influenza A virus (Japan 305)	5×10^{-2}	90	Rafelson *et al.* (*50*)
Cu^{2+}	*N*-Acetylneuraminosyllactose	Influenza A virus (Japan 305)	1×10^{-2}	80	Rafelson *et al.* (*50*)
Hg^{2+}	*N*-Acetylneuraminosyllactose	Influenza A virus (Japan 305)	1×10^{-3}	100	Rafelson *et al.* (*50*)
Fe^{2+}	*N*-Acetylneuraminosyllactose	Influenza A virus (Japan 305)	1×10^{-2}	50	Rafelson *et al.* (*50*)
Fe^{3+}	*N*-Acetylneuraminosyllactose	Influenza A virus (Japan 305)	1×10^{-2}	50	Rafelson *et al.* (*50*)
Congo red	Bovine sialyllactose	Fowl plague virus	6×10^{-5}	64	Becht and Drzeniek (*105*)
Trypan red	Bovine sialyllactose	Fowl plague virus	2×10^{-4}	28	Becht and Drzeniek (*105*)
Congo red	Bovine sialyllactose	NDV	4×10^{-5}	100	Becht and Drzeniek (*105*)
Trypan red	Bovine sialyllactose	NDV	4×10^{-5}	100	Becht and Drzeniek (*105*)
1-Phenoxymethyl-3,4-dihydroisoquinoline	Purified *Collocalia* mucoid	*V. cholerae*	1.6×10^{-3}	33–38	Brammer *et al.* (*104*)

[a] Virus particle.

present still *N*-acetylneuraminosyl(α,2 $\rightarrow$ 3)lactose. It is commercially available, but the price is very high. If it is used, the liberated lactose can be determined according to Park and Johnson (*106*). The most desirable substrate would be a low molecular weight crystalline glycoside of α-*N*-acetylneuraminic acid. Meindl and Tuppy (*107*) have synthesized the crystalline α-phenylketoside of *N*-acetylneuraminic acid. The liberated phenol is determined by the Folin–Ciocalteu reagent (*108*, *109*). This method, however, can be applied only with purified neuraminidase preparations since proteins also react with this reagent.

The easily crystallizable 2-*m*-nitrobenzyl-*N*-acetyl-α-D-neuraminic acid was recommended by Meindl and Tuppy (*68*) as particularly suited for quantitative estimation of neuraminidase activity. The liberated NANA is determined by the Warren (*110*) or Aminoff (*111*) methods. The use of a crystalline low molecular weight synthetic substrate is superior to the use of natural substrates. It has been reported by several laboratories (*112–114*) that for reasons not quite understood *N*-acetylneuraminic acid bound to another sugar, as found in natural substrates, gives in some cases a positive Warren test.

IX. Biological Significance of Neuraminidase

As to the biological significance of the influenza virus neuraminidase it may be remembered that influenza virus infection of the respiratory tract spreads by droplets passing from individual to individual. The respiratory tract is lined by stratified columnar epithelial cells which are ciliated. The activity of mucus-producing cells embedded in the various layers of the mucosa keeps the surface of the lining cells covered by a thin layer of mucus. According to observations of Florey (*115*), cilia of the lining cells of the respiratory tract move a thin layer of mucus in a well-defined direction at a very appreciable rate. Such

106. T. J. Park and M. J. Johnson, *JBC* **181,** 149 (1949).
107. P. Meindl and H. Tuppy, *Monatsh. Chem.* **98,** 53 (1967).
108. O. Folin and V. Ciocalteu, *JBC* **73,** 627 (1967).
109. E. Layne, "Methods in Enzymology," Vol. 3, p. 447, 1957.
110. L. Warren, *JBL* **234,** 1971 (1959).
111. D. Aminoff, *Virology* **7,** 355 (1959).
112. R. Carubelli, V. P. Bhavanandan, and A. Gottschalk, *BBA* **101,** 67 (1965).
113. J. Eichberg and M. L. Karnovsky, *BBA* **124,** 118 (1966).
114. C. R. Brown, P. N. Srivastava, and E. F. Hartree, *BJ* **118,** 123 (1970).
115. H. Florey, "Lectures on General Pathology." University Press, Melbourne, 1954.

streams of mucus were found to be able to remove carbon particles dropped experimentally onto the mucosa. There seems to be little doubt that this mucus which contains about 4% sialic acid (*91*) functions as a protective coat, trapping foreign particles, among them bacteria and viruses. The possession of several hundreds of neuraminidase molecules at the surface of influenza virus particle will enable the virus to penetrate the protecting layer of the mucus; then the virus can adsorb at the surface of the host cells. For the adsorption onto and the penetration of the cell membrane the activity of viral neuraminidase does not seem to be necessary as evidenced by the behavior of indicator virus which does not exhibit any neuraminidase activity (*116*).

In addition to this function the newly formed influenza virus seems to require neuraminidase for its release from the receptor site at the outer surface of the host cell (*116, 117*).

116. A. Gottschalk, *Ergeb. Mikrobiol. Immunitätsforsch., Exptl. Therap.* **32,** 1 (1959).

117. J. T. Seto and R. Rott, *Virology* **30,** 731 (1966).

13

Phage Lysozyme and Other Lytic Enzymes

AKIRA TSUGITA

I. Introduction

Bacteriophages infect the host cell and lead to the replication of phage particles. After maturation, the phage particles are released from the host cell by lysis of the bacterial cell wall. The mechanism of the initial invasion of the phages into the host cell is not yet well elucidated, but the last step, release of the phages, is known to be carried out at least in part, by the action of phage-induced enzymes. These enzymes have been found in various phage lysates and are variously designated, as lysozyme, lytic enzyme, lysin, muralysin, endolysin, and virolysin.

A. Early History

There is no doubt that numerous bacteriologists in the early periods observed phage action in bacterial cultures prior to Twort's provocative paper. But in 1915, Twort (*1*) first described a filtrable "lytic agent" which causes lysis of staphylococci. In 1917, d'Herelle (*2*) reported his independent discovery of "bacteriophage" and shared credit with Twort for the discovery of phages, d'Herelle carried out much of the basic research on this agent including their lytic activity. In 1926, (*3*) he first suggested that "a lytic enzyme" was secreted by phage, and Bronfenbrenner and Muckenfuss, in 1928, (*4*) claimed that ". . . the ferment-like substance responsible for the lysis is different from the bacteriophage" using the same system. Steric, in 1929, (*5*) and Schuurman, in 1936, (*6*) both using *coli* phages brought forward evidence in favor of this view. Their evidence clearly indicates that an enzyme which diffuses out from the plaque into the surrounding agar, and which lyses heat-killed bacteria and causes a surface change in living bacteria, is produced by the phage-bacteria system. In 1933 and 1934, Wollman and Wollman (*7*) presented evidence that this enzyme is produced by the phage rather than the bacterium.

1. F. W. Twort, *Lancet* **II,** 1241 (1915).
2. F. d'Herelle, *Compt. Rend. Soc. Biol.* **165,** 373 (1917).
3. F. d'Herelle, "The Bacteriophage and its Behaviour." Williams & Wilkins, Baltimore, Maryland, 1926.
4. J. Bronfenbrenner and R. Muckenfuss, *J. Exptl. Med.* **45,** 887 (1927).
5. V. Steric, *Zentr. Bakteriol., Parasitenk., Abt. I. Orig.* **110,** 125 (1929); *Compt. Rend. Soc. Biol.* **126,** 1074 (1936).
6. C. J. Schuurman, *Zentr. Bakteriol., Parasitenk., Abt. I. Orig.* **137,** 438 (1936).
7. E. Wollman and E. Wollman, *Compt. Rend. Soc. Biol.* **112,** 164 (1933); E. Wollman, *ibid.* **115,** 1616 (1934).

Chemical changes remained ambiguous during this early period. First, Hetler and Bronfenbrenner (1928) (*8*) reported that lysed cultures of various bacteria contained more NH_2-nitrogen than control cultures and concluded that the increment was the result of hydrolyzed bacterial protein. Northrop (1938) (*9*) reported that during the lysis of *B. coli*, unhydrolyzed protein was liberated but that during the lysis of some other bacteria hydrolysis of protein and lysis occurred simultaneously. These observations had to be corrected after the importance of the discovery of egg white lysozyme by Meyer *et al.* in 1936 (*10*) became appreciated. In 1937, Dubos (*11*) separated a lysozyme-like enzyme from the bacteria, pneumococci, and the question naturally arose whether lysis of bacteria by phage is brought about by an enzyme of this type. Pirie (1939) (*12*), using a phage WLL on *B. coli* 14, observed an increase of reducing sugars in the medium during the lysis of bacteria. He also isolated a similar enzyme from uninfected coli cells which was clearly distinguished by its reactivity to a chemical inhibitor, iodoacetate. White (1937) (*13*) demonstrated that a weak cholera phage could be made much more active if egg white were added to the medium, and he concluded that this activation resulted from a more efficient lysozyme. Pirie (*12*) extended this idea by suggesting that an enzyme not of bacterial origin is released upon lysis of bacteria by phages and that it acts on the bacterial cell wall and thus increases the reducing sugar similar to the egg white lysozyme.

These early investigations did not reveal whether the lysis caused by phage induction was the result of a lytic enzyme possessed by the phage (*5, 14*) or of the activation of a host cell enzyme(s) during infection (*4, 12, 15, 16*).

B. Recent Developments

Because the discovery of phages was accomplished by finding their lytic activity to host cells, in early investigations the lytic enzymes were

8. D. H. Hetler and J. Bronfenbrenner, *J. Exptl. Med.* **48,** 269 (1928).
9. J. Northrop, *Proc. Soc. Exptl. Biol. Med.* **39,** 198 (1938).
10. K. Meyer, J. W. Palmer, R. Thompson, and D. Khorazo, *JBC* **113,** 479 (1939); K. Meyer, R. Thompson, J. W. Palmer, and D. Khorazo, *ibid.* p. 303.
11. R. Dubos, *J. Exptl. Med.* **65,** 873 (1937); K. Meyer, R. Dubos, and E. M. Smyth, *JBC* **118,** 119 (1937).
12. A. Pirie, *Brit. J. Exptl. Pathol.* **29,** 1939 (1939).
13. P. B. White, *J. Pathol. Bacteriol.* **43,** 591 (1936); **44,** 276 (1937).
14. T. F. Anderson, *J. Cell. Comp. Physiol.* **25,** 1 (1945).
15. J. Bordet and E. Renaux, *Ann. Inst. Pasteur* **42,** 1283 (1928).
16. E. Wollman and E. Wollman, *Ann. Inst. Pasteur* **56,** 137 (1936).

considered as one of the responsible biochemical agents in the lysis process. After 1940, investigations became more complex because of the varied mode of action of isolated lytic enzymes and their complicated roles in the life cycle of phage infection and also because of the great variety of experimental goals pursued in numerous laboratories.

The lysis of virulent bacteria or the inhibition of bacteria growth always has important practical aspects in medical science. Many biological agents besides phages or phage-induced lytic enzymes such as egg white lysozyme, bacterial autolysin, bacteriosine, and antibiotics have been investigated for their activity against infection in humans. Among these biological agents, phages or their lytic enzymes have generally not been found to be efficient tools compared to other agents such as antibiotics or autolysin (such as lysostaphin). Yet several bacterial phage systems such as *Klebsiella* phage (*17*) or streptococcal phages (*18*) are under investigation despite the brilliant results obtained with antibiotics. Interest in these systems increased after a polysaccharide depolymerase was found in *Klebsiella* phage system which can possibly be used in chemotherapy (*17*). Several phage depolymerases have been found in other gram-negative bacterial strains. None of the phage-induced lytic enzymes have yet been elucidated to the extent that egg white lysozyme has. Within a few years, new information can be expected for T2 and T4 phage lysozymes and possibly an endolysin of λ phage as well, because genetic information (*19, 20*), primary structures, and other results have been recently provided (*21–23*).

The autolytic activities of bacterial cells were found at about the same time (*11*) as the lytic activities of phage origin. Their specificities have been clarified in more detail than for the phage-induced lytic enzymes. They have become indispensable tools in investigating the structures of mucopolysaccharides and cell walls. Several comprehensive reviews cover the developments and current information (*24*). Whether the enzymes are synthesized under control of the bacterial genome or by the genetic information of lysogenized phages is still questionable in some

17. M. H. Adams and B. H. Park, *Virology* **2,** 719 (1956).
18. A. C. Evans, *Public Health Rept.* (*U. S.*) **49,** 1386 (1934).
19. G. Streisinger, F. Mukai, W. J. Dreyer, B. Miller, and S. Horiuchi, *Cold Spring Harbor Symp. Quant. Biol.* **26,** 25 (1961).
20. A. Del Campillo-Campbell and A. Campbell, *Biochem. Z.* **342,** 485 (1965).
21. M. Inouye and A. Tsugita, *JMB* **37,** 213 (1968).
22. A. Tsugita and M. Inouye, *JMB* **37,** 201 (1968).
23. D. S. Hogness, W. Doerfler, J. B. Egan, and L. W. Black, *Cold Spring Harbor Symp. Quant. Biol.* **31,** 129 (1966).
24. J. M. Ghuysen, *Bacteriol. Rev.* **32,** 425 (1968).

cases because some bacteria often carry lysogenic phages such as staphylococci or *Bacillus* bacteria. Weidel and his collaborators (*25*) successfully used T2 phage lysozyme as a tool to analyze the cell wall structure. Substrate specificities have been elucidated for several enzymes, such as endoacetylmuramidase of T2 (*26*) and T4 (*27*) phage lysozyme, endoacetylglucosamidase of streptococcal muralysin (*28*), peptidoglycan peptidases including staphylococcal virolysin (*29*), and *B. stearothermophilus* phage (*30*). Table I summarizes information for some lytic enzymes from various sources.

To clarify all the details in the life cycles of infected phages and their relationship to their host cells, it is important to understand the physiology of phage infection. This consists of the following steps: virus invasion, regulation of viral gene expression (cessation of host gene expression and trigger mechanism of phage metabolism), viral gene reproduction, and cessation of infected cell metabolism following lysis. Remarkable developments were recently accomplished on the regulation of gene expression and gene reproduction, but much of the compiled knowledge of the initial invasion and the lysis as the last step have yet to be obtained. Difficulties in the investigation of the penetrating mechanism and lysis mechanism resulted from the complex cell wall structure and the complex correlation between the cell metabolism and the lytic actions, including the initial penetration and the final breaking steps. Even with phages of *E. coli*, staphylococci, and streptococci, our knowledge is limited to a specified lytic enzyme or to limited steps of penetration and/or lytic processes. Various relationships of lytic activities are fragmental: for example, polysaccharide depolymerase promotes phage access to bacterial surface (*17*); muramysin may closely correlate with the host receptor (*30, 31*); and T2 or T4 phage lysozyme destroys the cell wall from inside (*32*). Still there is no system in which the total invasion and cell lysis processes have been completely clarified. In Section II,B,6, a summary of these processes in the T4 phage–*E. coli* system will be discussed.

25. W. Weidel, H. Frank, and H. H. Martin, *J. Gen. Microbiol.* **22,** 158 (1960).
26. D. Maass and W. Weidel, *BBA* **78,** 369 (1963).
27. A. Tsugita, M. Inouye, E. Terzaghi, and G. Streisinger, *JBC* **243,** 391 (1968).
28. S. S. Barkulis, C. Smith, J. J. Boltralik, and H. Heymann, *JBC* **239,** 4027 (1964).
29. D. J. Ralston, B. Baer, M. Lieberman, and A. P. Kruegur, *J. Gen. Microbiol.* **24,** 313 (1961).
30. R. M. Krause, *J. Exptl. Med.* **106,** 365 (1957).
31. W. R. Maxted, *J. Gen. Microbiol.* **16,** 584 (1957).
32. J. Emrich and G. Streisinger, *Virology* **36,** 387 (1968).

TABLE I
LYTIC ENZYMES INDUCED BY PHAGE INFECTION

Host	Phage	Name	Optimal pH	Molecular wt	Substrate specificity	Others
Gram-negative						
E. coli	T2	Lysozyme	6–7	18.6×10^3	*endo-N*-Acetylmuramidase	Sensitized cell
E. coli	T4	Lysozyme	7.2–7.4	18.7×10^3	*endo-N*-Acetylmuramidase	Sensitized cell
E. coli	N20F′		5.5 and 8.5	$11–13 \times 10^3$	*endo-N*-Acetylmuramidase	Sensitized cell
E. coli	λ	Endolysin	6–7	17.6×10^3	Similar to egg white and T2 lysozyme but details are different and unknown	(Lysogenic phage) sensitized cell
E. coli	F1, F5	Polysaccharide depolymerase	7.2		Big fragment No reducing material	
K. pneumoniae	Kp	Polysaccharide depolymerase	Neutral		No reducing sugar	Intact cell
P. aeruginosa	2	Polysaccharide depolymerase	7.5	180×10^3	Hexosamine, hexose Reducing sugar	Intact cell
P. putida	Pf15	Polysaccharide depolymerase	7.5		Reducing sugar	Intact cell
A. agilis	A-22	Polysaccharide depolymerase	7.5–8.5		Reducing sugar	Intact cell
A. cloacae	F12, 13, 14	Polysaccharide depolymerase	7.2		No reducing sugar	Intact cell

Gram-positive						
S. aureus	P1, P14	Virolysin	7.0–7.5		Polysaccharide depolymerase?	Sensitized cell
S. aureus	PAL					Intact cell
M. lysodeikticus	N1		6.5–7.0		Peptidoglycan peptidase no reducing sugar, no proteolytic	Intact cell
Group C streptococci	C1	Muralysin	6.1	100×10^3	*endo-N*-Acetylglucosaminidase	
Group C streptococci	B56		6		*endo-N*-Acetylglucosaminidase	
Group C streptococci	ϕY		6.2			Lysogenic phage, no hyaluronidase
Group A streptococci	ML3		6.6–6.9		Breaking glycosidic linkage between acetylamino sugar	
Group A streptococci	712′					
B. megaterium	G	Soluble enzyme	4–5		Nondialyzable reducing sugar	Sensitized cell, require intact phage
		Bound enzyme	9.5		Differ from "soluble enzyme"	Intact cell
B. stearothermophilus	TP-1		6.3 (6–7)		Peptidoglycan peptidase	Sensitized cell
B. subtilis	2C				No reducing sugar	

C. Application of Phage-Induced Lytic Enzyme System to Other Biologically Important Problems

T4 phage lysozyme was chosen for study of the genetic code because it is an easily isolated protein with low molecular weight. Comparison is made of the sequence of wild type lysozyme with that of pseudo-wild mutants carrying certain pairs of frame–shift mutations in the lysozyme gene. The genetic message is translated sequentially by triplets (frames), starting from a defined point. Deletions or insertions of one or more bases into the gene cause a shift of the reading frame (amino acid sequence) of the message (*33*). If the deletion of a base is followed by the addition of another base, which can be caused by proflavine mutagen, the reading frame and resulting amino acid sequence may be altered only in the region of the mutations.

Utilizing codons proposed on the basis of *in vitro* studies, a sequence of bases can be assigned that code both for the wild type sequence of amino acid and for the altered sequence of amino acids in the mutant strains. Thus some codons proposed on the basis of *in vitro* study were confirmed to be utilized *in vivo* and some of the codons have been newly identified (*34–43*). The same studies showed the direction of transcription (*44*) and translation (*34*) of the lysozyme gene, the mechanism of frame–shift mutations (*42*, *43*), and indicated essential and nonessential sequences for enzymic (*41*).

33. F. H. C. Crick, L. Barnett, S. Brenner, and R. J. Watts-Tobin, *Nature* **192,** 1227 (1961).

34. E. Terzaghi, Y. Okada, G. Streisinger, J. Emrich, M. Inouye, and A. Tsugita, *Proc. Natl. Acad. Sci. U. S.* **56,** 500 (1966).

35. Y. Okada, E. Terzaghi, G. Streisinger, J. Emrich, M. Inouye, and A. Tsugita, *Proc. Natl. Acad. Sci. U. S.* **56,** 1692 (1966).

36. M. Inouye, E. Akaboshi, A. Tsugita, Y. Okada, and G. Streisinger, *JMB* **30,** 39 (1967).

37. M. J. Lorena, M. Inouye, and A. Tsugita, *Mol. Gen. Gen.* **102,** 69 (1968).

38. Y. Okada, G. Streisinger, J. Emrich, J. Newton, A. Tsugita, and M. Inouye, *Science* **162,** 807 (1968).

39. A. Tsugita, M. Inouye, T. Imagawa, T. Nakanishi, Y. Okada, J. Emrich, and G. Streisinger, *JMB* **41,** 349 (1969).

40. Y. Okada, G. Streisinger, J. Emrich, A. Tsugita, and M. Inouye, *JMB* **40,** 299 (1969).

41. Y. Okada, S. Amagase, and A. Tsugita, *JMB* **54,** 219 (1970).

42. M. Imada, M. Eda, M. Inouye, and A. Tsugita, *JMB* **54,** 199 (1970).

43. G. Streisinger, Y. Okada, J. Emrich, J. Newton, A. Tsugita, E. Terzaghi, and M. Inouye, *Cold Spring Harbor Symp. Quant. Biol.* **31,** 77 (1966).

44. G. Streisinger, J. Emrich, Y. Okada, A. Tsugita, and M. Inouye, *JMB* **31,** 607 (1968).

With nonsense mutants of T4 lysozyme alterations of amino acids by a suppressive host have provided further knowledge on essential and nonessential amino acids in the sequence of T4 phage lysozyme. These results are discussed in Section II,B,5,b (*45, 46*).

The study of nonsense mutants in the presence of adequate host cells, which result in insertion of original or similar amino acids, have supplemented the information on lysozyme production. From this, a variation of suppression efficiencies that depend on the location of the nonsense code was observed—that is, the effect on the nonsense codon by an adjacent codon, which is suggested as the real ending mechanism of protein synthesis (*47, 48*).

The T4 phage lysozyme is a representative enzyme protein known to be synthesized as a "late" protein. After phage infection, "early" proteins are first synthesized, followed by late proteins. Thus, the lysozyme activity has been employed to study the switching mechanism from early to late protein groups in connection with synthesis of mRNA and the "trigger" role of RNA-polymerase (*49–51*). The mRNA specific for the lysozyme gene, as demonstrated by hybridization, was observed in a very early period and also during the late period after infection by the phage (*52, 53*). The very early mRNA of the lysozyme gene is not translated to the protein, but the late mRNA can be translated into the active lysozyme (*54, 55*).

Salser *et al.* (*56, 57*) reported successful *in vitro* protein synthesis of lysozyme protein under the control of mRNA, confirmed by using mRNA from nonsense mutants in the lysozyme gene and synthesis tools obtained from the suppressive hosts. This is one of a few cases in which *de novo* synthesis of a native protein, which has biological activity, has been carried out in an *in vitro* system. The above observation of utilization of mRNA for translation of lysozyme protein has been confirmed. DNA-

45. M. Inouye, E. Akaboshi, M. Kuroda, and A. Tsugita, *JMB* **50,** 71 (1970).
46. M. Inouye, H. Yahata, Y. Ocada, and A. Tsugita, *JMB* **33,** 957 (1968).
47. H. Yahata, Y. Ocada, and A. Tsugita, *Mol. Gen. Gen.* **106,** 208 (1970).
48. A. Tsugita, *12th Intern. Congr. Genet., Tokyo, 1968* Vol. II, p. 35 (1968).
49. R. R. Burgess, A. A. Travers, J. J. Dunn, and E. K. Bautz, *Nature* **221,** 43 (1969).
50. E. K. F. Bautz and J. J. Dunn, *BBRC* **34,** 230 (1969).
51. A. A. Travers, *Nature* **223,** 1107 (1969).
52. E. K. F. Bautz, F. A. Bautz, and J. J. Dunn, *Nature* **223,** 1022 (1969).
53. A. A. Travers, *Nature* **225,** 1009 (1970).
54. T. Kasai and E. K. F. Bautz, *JMB* **41,** 401 (1969).
55. T. Kasai, E. K. F. Bautz, and W. Szybalski, *JMB* **34,** 709 (1968).
56. W. Salser, R. F. Gesteland, and A. Bolle, *Nature* **215,** 588 (1967).
57. W. Salser, R. Gesteland, and B. Ricard, *Cold Spring Harbor Symp. Quant. Biol.* **34,** 771 (1969).

dependent *in vitro* protein synthesis was also recently reported by Schweiger and Gold (*58*).

D. Outline of Cell Wall Structure and Action of Lytic Enzymes

This section gives briefly some of the essential facts of the structure of cell walls and cytoplasmic membranes related to the mode of action of lytic enzymes. Detailed information can be found in the reports by Salton (*59*), Rogers and Perkins (*60*), Ghuysen *et al.* (*61*), and Ghuysen (*24*).

1. *Morphogenesis of Cell Wall and Membrane*

Salton (*59*) classified bacteria into four groups according to the macrostructure of cell walls.

Group 1 (most of the gram-positive bacteria including *Staphylococcus aureus* and *Micrococcus lysodeikticus* belong to this group): The walls, the width of which are 150–800 Å, consist of a thick amorphous cell wall and a cytoplasmic membrane. By the action of egg white lysozyme the cells can be converted into protoplasts that retain no residual cell wall component; that is, the cell wall and the membrane can be clearly separated.

Group 2 (most of the gram-negative bacteria, such as *E. coli,* belong to group 2): The multilayered wall and cytoplasmic membrane are composed of an 80–100-Å-thick double envelope. The layers of different electron densities can be observed in the electron microscope, and each is composed of chemically different envelopes as shown by differential extraction. The separation of the cell walls from the cytoplasmic membrane is not easily accomplished. Even extensive treatment with egg white lysozyme of the bacteria results only in the formation of spheroplasts, which contain residual peptidoglycans. Thus Salton used the words "multilayered wall" and "underlying membrane" instead of cell wall and cytoplasmic membrane.

Groups 3 and 4: The walls of group 3 bacteria consist of a simple

58. M. Schweiger and L. M. Gold, *Proc. Natl. Acad. Sci. U. S.* **63,** 1351 (1969).

59. M. R. J. Salton, "The Bacterial Cell Wall." Elsevier, Amsterdam, 1954.

60. H. J. Rogers and M. R. Perkins, "Cell Walls and Membranes." Spon, London, 1968.

61. J. M. Ghuysen, J. L. Strominger, and D. J. Tipper, *Comp. Biochem.* **26A,** 53 (1968).

single unit membrane and those of group 4 consist of even more complicated multilayered structures than those of group 2.

2. *Chemical Properties of Cell Walls*

Cell walls generally have a "basal layer" which renders rigidity, ductility, and elasticity to the bacteria and a "specific layer" which is specific for the bacterial species or strain.

a. Basal Layer. The chemical composition of the basal layer is common to most bacteria. It is essentially composed of a few amino acids and acetoamide sugars called "peptidoglycan" (mucopeptide, glycopeptide, or murein), *N*-acetylmuramic acid (M), *N*-acetylglucosamine (G), lysine [or diaminopimelic acid (DAP)], D-glutamic acid, D- or L-alanine in ratios of 1:1:1:1:2. Glycine, L-serine, L-threonine, D-aspartic acid, and additional L-alanine are also found in some species as constituents of the peptidoglycans. A peptidoglycan is constructed of "glycan strand peptide subunits" crosslinked by peptide "cross bridges."

Figure 1 illustrates a glycan strand where every *N*-acetylmuramic acid moiety is connected with a peptide subunit. As seen in Fig. 3, 50% of the *N*-acetylmuramic acid of the *M. lysodeikticus* peptidoglycan is not connected with peptide subunits (*62, 63*). In the peptidoglycan of *S. aureus*, 60% of the *N*-acetylmuramic acid residues are substituted with

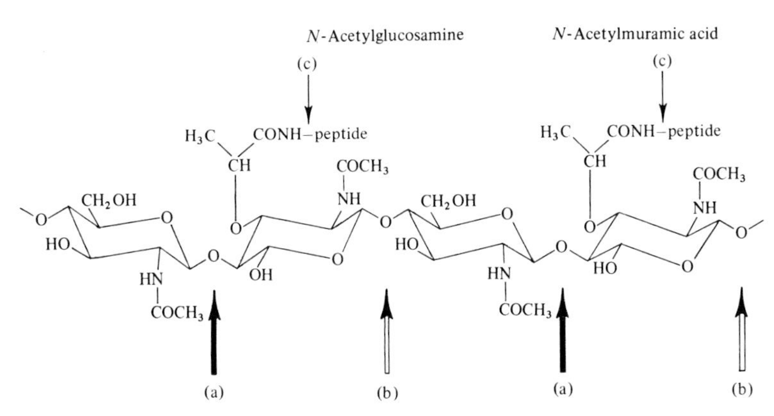

FIG. 1. A portion of peptidoglycan strand: (a) *endo-N*-acetylglucosaminidase, (b) *endo-N*-acetylmuramidase, and (c) *N*-acetylmuramyl-L-alanine amidase.

62. K. H. Schleifer and O. Kandler, *BBRC* **28**, 965 (1967).
63. J. M. Ghuysen, E. Bricas, M. Lache, and M. Leyh-Bouille, *Biochemistry* **7**, 1450 (1968).

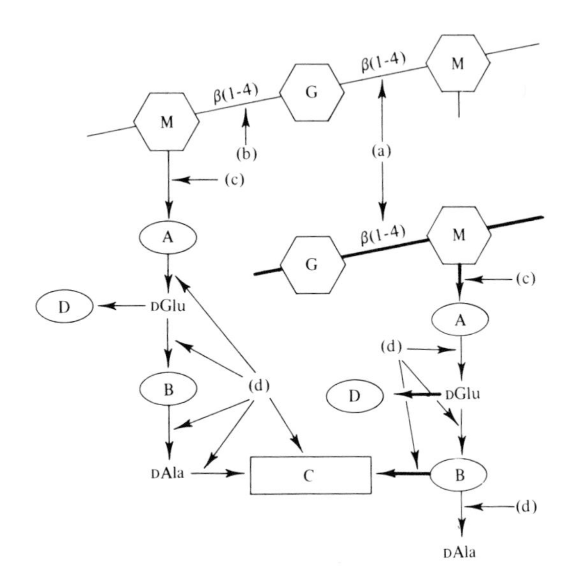

FIG. 2. Schematic presentation of the fundamental frame of peptidoglycan and of the sites of actions of lytic enzymes [cf. S. Kotani, *Protein, Nucleic Acid, Enzyme* (*Tokyo*) **13,** 1136 (1968) (in Japanese)].

(A) L-Ala (mostly found)
L-Ser *Butyribacterium rettgeri*
Gly *Microbacterium lacticum,* etc.

(B) L-Lys (Staphylococci, streptococci, pneumococci, micrococci, etc.)
meso-DAP
L-L-DAP; gram-negative bacteria including *E. coli,* some of gram-positive
D-D-DAP bacteria such *Bacillus*
Diaminobutyric acid; *Lactobacillus, Clostridium, Mycobacterium*
L-Orn: *Micrococcus radiodurans, Lactobacillus cellobiosus*
D-Orn: *Corynebacterium beta*
OH-L-Lys: *Streptococcus faecalis,* some strains of *S. aureus*

[C] Cross bridge

D-Ala–Gly–Gly–Gly–Gly–Gly$\overset{\epsilon}{-}$L–Lys; *S. aureus*
D-Ala-L-Ala-L-Ala-L-Ala$\overset{\epsilon}{-}$L-Lys; *S. thermophilus, S. faecalis*
D-Ala-L-Ala-L-Ala$\overset{\epsilon}{-}$L-Lys; group A streptococci
D-Ala-D-Asp$\overset{\epsilon}{-}$L-Lys; group N streptococci
D-Ala-*meso*-DAP; *E. coli, B. megaterium* KM1

(D) Amide (NH_2-); *S. aureus,* group A streptococci, etc.
Gly; *M. lysodeikticus*
(D-Glu)-δ-Orn-α-(D-Ala); bridge as *B. rettgeri, C. poinsettiae*

(a) site of action of *endo-N*-acetylglucosaminidase
(b) site of action of *endo-N*-acetylmuramidase
(c) site of action of *N*-acetylmuramyl-L-alanine amidase
(d) site of action of peptidoglycan peptidases

⟨M⟩ muramic acid and ⟨G⟩ glucosamine

an *O*-acetyl group on C6 (*24*). A variety of cross bridge peptides are shown in Fig. 2, and a few representative postulated structures are illustrated in Fig. 3 such as the peptidoglycans of *E. coli* (*64, 65*), *M. lysodeikticus* (*62, 63*), and *S. aureus* (*61*).

b. Specific Layers. The specific layers are species or strain specific and have various compositions: ribitol- and glycerol-type terchoic acid, teichuromic acid, polysaccharides, lipopolysaccharide protein complex, proteins such as M and T protein or R antigen of group A *S. pyogenes*, and lipids such as Wax B, C, D in *Mycobacterium*. The polysaccharides are also species specific and the content and structure differ. They are usually built from arabinose, mannose, muramic acid, etc.

3. *Mode of Action of Lytic Enzymes*

The mode of action of a variety of lytic enzymes is illustrated in Figs. 1 and 2. Endoacetylmuramidases such as the T4 or T2 lysozyme split the bond of *N*-acetylmuramyl-*N*-acetylglucosamine (Fig. 1) and endoacetylglucosaminidases such as streptococcal muralysin split *N*-acetylglucosaminyl-*N*-acetylmuramic acid (Fig. 1). Acetylmuramyl-L-alanine amidase splits the acetylmuramyl-L-alanine bond shown in Figs. 1 and 2, and peptidase splits peptide bonds, either intrapeptide bonds (endopeptidase), amino terminal (aminopeptidase), or carboxyl terminal amino acid (carboxypeptidase), from the peptidoglycan (Fig. 2) (*66*).

II. The Lytic Enzymes of *Escherichia coli* Phages

A. Introduction

Barring and Kozloff (*67, 68*) showed that incubation of various T series phages with *E. coli* B cell walls caused the release of as much as 15% of the total nitrogen from the host cell walls and suggested that the activity was located in the virus tail. There was some confusion concerning the invasion mechanism and lysis mechanism in phage–host interaction (*69*). Kozloff and Lute (*70*) found an ATP-dependent con-

64. W. Weidel and H. Pelger, *Advan. Enzymol.* **26,** 193 (1944).
65. I. Takebe, *BBA* **101,** 124 (1965).
66. M. R. J. Salton, *Bacteriol. Rev.* **21,** 82 (1957).
67. L. F. Barrington and L. M. Kozloff, *JBC* **223,** 615 (1956).
68. L. F. Barrington and L. M. Kozloff, *Science* **120,** 110 (1954).
69. D. D. Brown and L. M. Kozloff, *JBC* **225,** 1 (1957).
70. L. K. Kozloff and M. Lute, *JBC* **234,** 539 (1959).

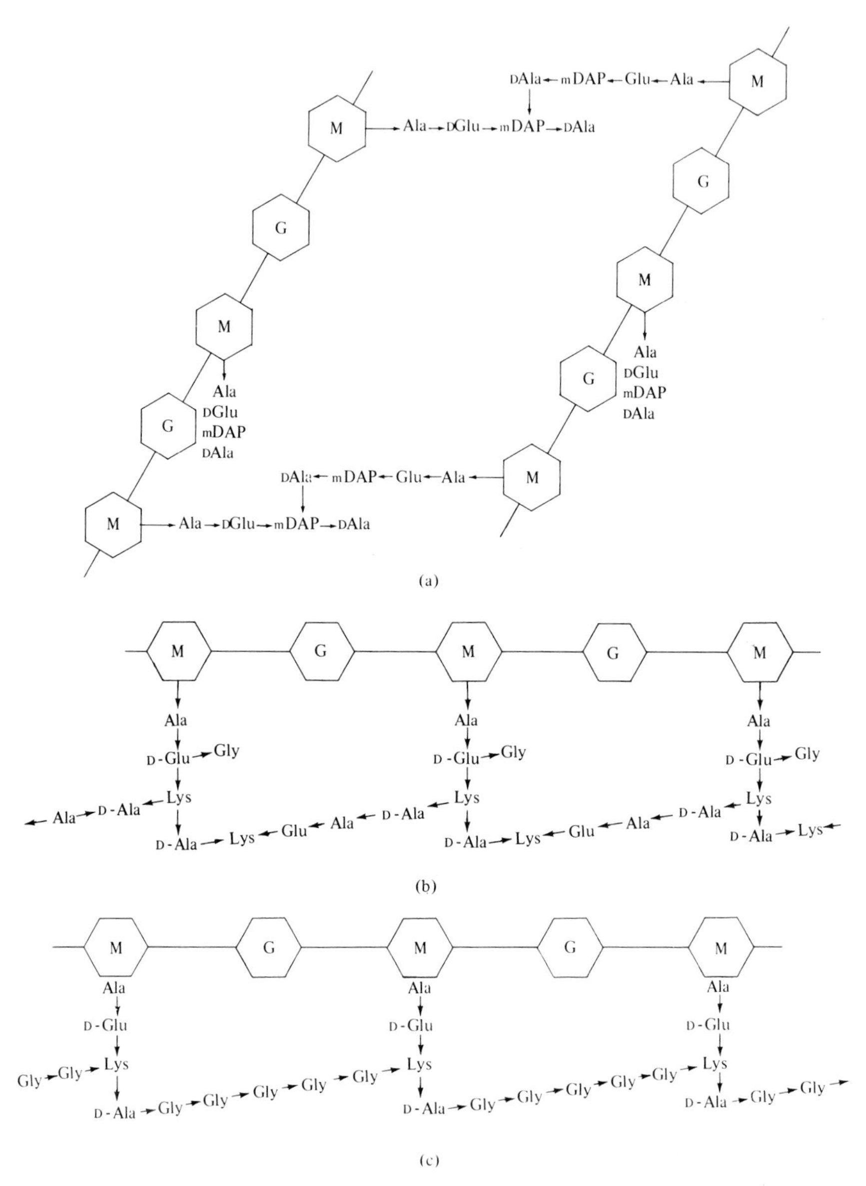

FIG. 3. Schematic presentations of peptidoglycans: (a) *E. coli,* (b) *M. lysodeikticus,* and (c) *S. aureus* Copenhagen. From Ghuysen (*24*).

tractile protein in the tail of the bacteriophage which was essential for successful viral invasion. This seemed to demonstrate that the invasion mechanism of the phage is different from the lysis mechanism of the host cell. (The ATPase was not observed clearly again by other workers.)

Brown *(71)* found that phage-free T_6 lysates are able to lyse chloroform-treated bacteria. Koch and Weidel *(72)* showed that the active part of the T2 phage tail is able to repeat its action on bacteria membranes, indicating that this action is enzymic in nature and confirming the results obtained by Barring and Kozloff. Further, Koch and Jordan *(73)* demonstrated no differences in the composition of the split product released by phage-free lysates as compared to that released by intact phages or phage ghosts. These authors also made the important discovery that the enzymically active protein is of low molecular weight and that host range specificity is associated with other parts of the phage rather than with the enzyme. Weidel and Primosigh *(74)* further confirmed that the enzymic action is entirely independent of the host range specificity proving that the phage adsorption and invasion mechanism is not connected to the enzymic action.

The enzyme was shown to be able to split off a component from isolated bacterial cell walls *(69)*, the chemical analysis of which was made by Koch *(75)*. The material contained glucosamine, muramic acid, glutamic acid, diaminopimelic acid, and alanine constituting a homogeneous compound with a molecular weight of 10,000 *(76)*. The increase of reducing sugar was observed during the enzymic reaction, repeating Pirie's observation *(12)*. A partial purification of this enzyme was carried out by Koch and Dreyer, and the enzyme preparation was shown to lyse *Micrococcus lysodeikticus*, commonly used as a substrate for egg white lysozyme, and to react competitively with egg lysozyme. This lysis enzyme was thus named "phage lyozyme" *(77)*.

Koch and Dreyer were also able to separate the lysozyme activity from the killing activity of the phage. Substantiating proof for the specificity of T2 phage lytic enzyme as lysozyme was obtained by Weidel's

71. A. Brown, *J. Bacteriol.* **71**, 482 (1956).
72. G. Koch and W. Weidel, *Z. Naturforsch.* **11b**, 345 (1956).
73. G. Koch and E. M. Jordan, *BBA* **25**, 437 (1957).
74. W. Weidel and J. Primosigh, *J. Gen. Microbiol.* **18**, 513 (1958).
75. G. Koch, *Fortschr. Botan.* **19**, 412 (1957).
76. W. Weidel and J. Primosigh, *Z. Naturforsch.* **12b**, 421 (1957).
77. G. Koch and W. J. Dreyer, *Virology* **6**, 291 (1958).
78. W. Weidel, H. Frank, and H. H. Martin, *J. Gen. Microbiol.* **22**, 158 (1960).
79. J. Primosigh, H. Pelzer, D. Maass, and W. Weidel, *BBA* **46**, 68 (1960).

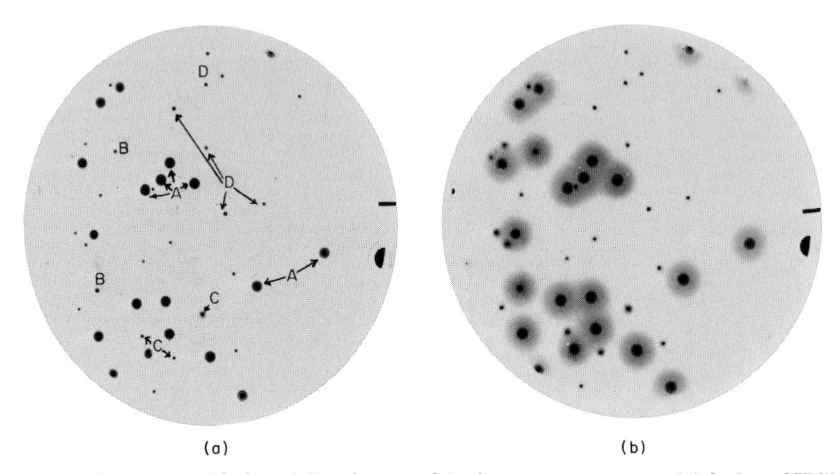

FIG. 4. Plaques and halos of T4 phage and its lysozyme mutants: (a) before $CHCl_3$ treatment and (b) after $CHCl_3$ treatment; A, wild type strain e^+rII; B, wild type strain e^+; C, pseudo-wild strain *e*JD13*e*JD5; and D, lysozyme negative strain *e*JD13 and *e*JD5.

group (*26, 76, 78, 79*), parallel to the purification and crystallization of the T2 enzyme (*80, 81*).

An early report stated that lysozyme purified from particles (tail) of phage T2 differed from that found free in the lysate, but later more exhaustive studies showed that the differences were not true (*81*). Sequential studies were carried out on the T2 enzyme and compared to similar studies on the T4 phage lysozyme. The two enzymes were found to be identical except for the alterations of three amino acids (*21*).

Streisinger *et al.* (*19*) reported a method of detecting mutants on the lysozyme gene and discovered mutations that affect both the lysozyme activity produced by phage T4 and the structure of the lysozyme molecule itself. The structural gene of lysozyme in T4 phage was designated by the letter *e*. A convenient recognition method for *e* gene mutations was introduced in the same paper (*19*), and genetic surveys not only of T4 phages but also of other bacterial phages became popular. Plaques of standard-type phage on plates incubated at 37° become surrounded by large halos after exposing the plate to chloroform vapor (Fig. 4). These halos result from the action of the phage lysozyme, which diffuses from the plaque and lyses chloroform-killed bacteria surrounding the plaque. Rare mutant plaques lack halos or are surrounded by small halos. Changing the conditions, such as temperature or pH or salt com-

80. W. Katz and W. Weidel, *Z. Naturforsch.* **16b**, 363 (1961).
81. W. Katz, *Z. Naturforsch.* **19b**, 129 (1964).

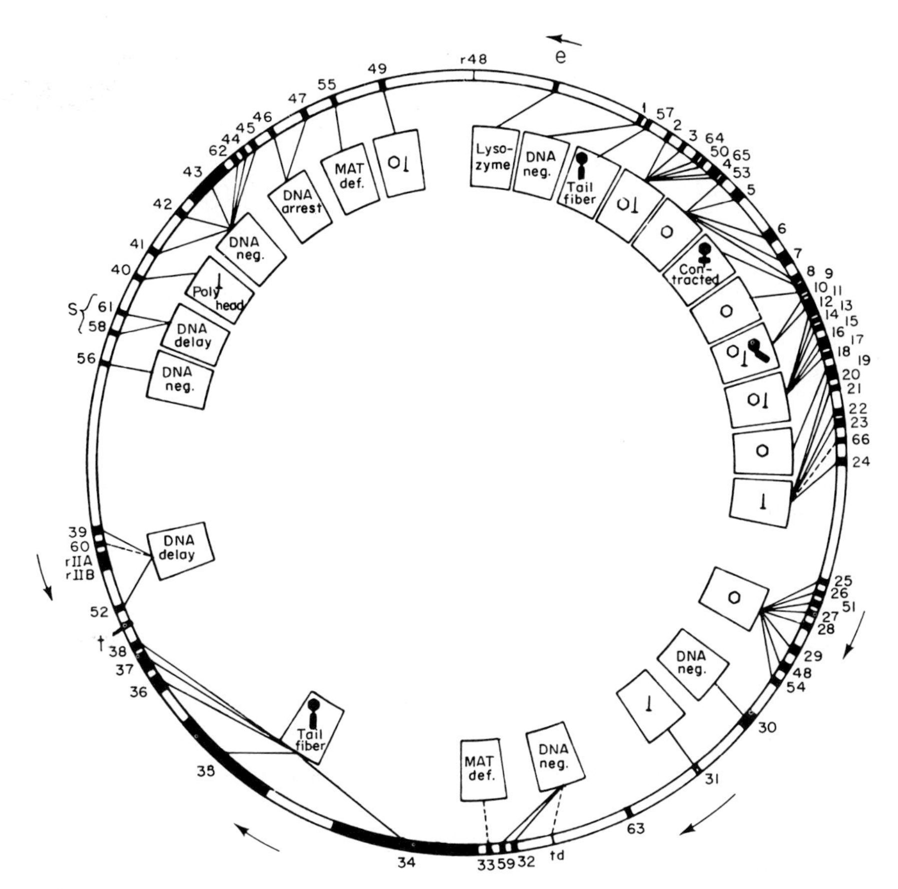

Fig. 5. Genetic map of bacteriophage T4. From R. S. Edger and W. B. Wood, *Proc. Natl. Acad. Sci. U. S.* **55**, 498 (1966).

position in the plates, is commonly employed in the above halo test. The relative positions of a large number of genes have been established for phage T4. Figure 5 shows the positions of the lysozyme gene with respect to certain other genes (*19*). Detailed genetic mapping within the lysozyme structure gene has also been accomplished (Fig. 6) (*44*). The T4 phage lysozyme was found mainly in the lysates, and only one lysozyme molecule for two phage particles was recovered in a probably nonspecific manner (*32*), which is contrary to the case of T2 phage lysozyme (*81*).

The lysozyme of the T4 phage has been purified (*27*). This protein is strongly basic and of a small size (MW = 18,000). The enzymic specificity was found to be similar to egg white lysozyme (as well as to

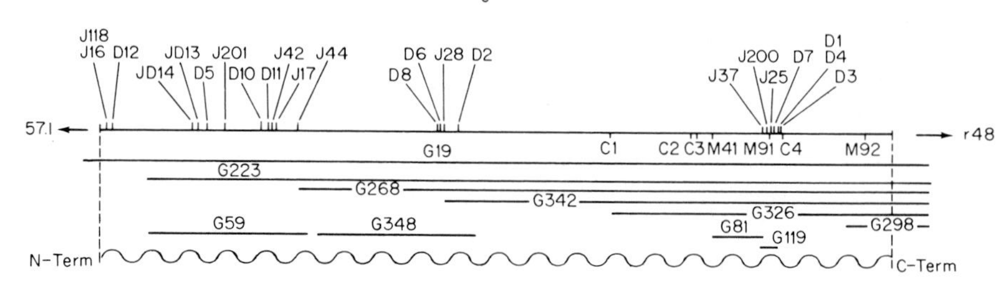

FIG. 6. Genetic map of lysozyme gene *e* of bacteriophage T4. From G. Streisinger, private communication, Streisinger *et al.* (*19, 43*), and this laboratory (*41, 42, 47*).

the T2 phage lysozyme), and its specific activity towards the host cell (wall) is higher than that of egg white lysozyme. However, this is not so apparent toward other substrates such as cell wall preparations of *M. lysodeikticus.* The amino acid sequence of T4 phage lysozyme (*22, 28, 82–86*) suggests no apparent structural similarity between T4 phage lysozyme and egg white lysozyme (*87*).

T4 phage lysozyme has been characterized as a lytic enzyme acting on the host cell wall from inside the cell (*88*). The other biological actions of this lysozyme have been clarified (*89, 90*) and are discussed in Section II,B,6.

The lysozyme of T7 phage was believed to be unstable until Pryme and Berentsen (unpublished) found that the assay methods used for the T4 lysozyme are unsuitable for the T7 lysozyme. They purified T7 lysozyme 100-fold from an infected lysate and showed that although the enzyme is virtually inactive under conditions of assay that are optimal for T4 lysozyme appreciable activity is detected by a tenfold reduction in salt concentration.

Besides the T series phages, lytic enzymes have been prepared from a virulent coli-phage N20F′ lysate. This phage was isolated from raw sewage obtained from the Nashville Sewage Plant by Moo-Penn and Wiesmeyer (*91*). The enzyme was well characterized and found to

82. M. Inouye and A. Tsugita, *JMB* **22,** 193 (1966).
83. M. Inouye, Y. Okada, and A. Tsugita, *JBC* **245,** 3439 (1970).
84. M. Inouye, M. Imada, E. Akaoshi, and A. Tsugita, *JBC* **245,** 3455 (1970).
85. M. Inouye, M. J. Lorena, and A. Tsugita, *JBC* **245,** 3467 (1970).
86. M. Inouye, M. Imada, and A. Tsugita, *JBC* **245,** 3479 (1970).
87. R. E. Canfield and A. K. Liu, *JBC* **240,** 1997 (1965).
88. F. Mukai, G. Streisinger, and B. Biller, *Virology* **33,** 398 (1967).
89. J. Emrich, *Virology* **35,** 158 (1968).
90. R. Josslin, *Virology* **40,** 719 (1970).
91. W. Moo-Penn and H. Wiesmeyer, *BBA* **178,** 318 (1969).

have two pH optima (*92*). Again, the specificity of this enzyme was defined as a muramidase (*93*).

Jacob *et al.* (*94*) observed lytic activity after induction of a λ phage, using *E. coli* K12 as host cell, and named this activity "endolysin." They partially purified endolysin (*95, 96*) and examined the specificity which still remains ambiguous. The structural gene, R, of this lytic enzyme is located near one end of the linear DNA; it is isolated from mature λ as indicated by its location as the most terminal gene on the right side of the genetic map of the vegetative λ (*20, 97*). The location of R has also been established by determining the gene content of fragments of λ-DNA, which vary in size but have in common the right end of the intact λ-DNA molecule (*23*). A map of the R cistron has been deduced from a set of temperature-sensitive and nonsense mutants (*98, 99*); and Black and Hogness (*100–102*) studied the purification, specificity, molecular weight, and preliminary protein structure on endolysin. Imada and Tsugita (*103*) purified the enzyme and subjected it to sequence studies. No homology in the primary sequence was found between λ-lytic enzyme and T4 phage lysozyme (*103*).

A polysaccharide depolymerase was found by Sutherland and Wilkinson (*104, 105*) in *E. coli* infected by phages.

B. T4 Phage Lysozyme

1. *Enzyme Assay*

The activity of phage lytic enzyme can be determined by various methods which are mainly based on either turbidimetry or the estimation of solubilized components of the cell wall. The choice of method

92. W. Moo-Penn and H. Wiesmeyer, *BBA* **178,** 330 (1969).
93. W. Moo-Penn and H. Wiesmeyer, *BBA* (1971) (in press).
94. F. Jacob, C. Fuerst, and E. Wollman, *Ann. Inst. Pasteur* **93,** 724 (1957).
95. F. Jacob and C. R. Fuerst, *J. Gen. Microbiol.* **18,** 518 (1958).
96. E. Work, *J. Gen. Microbiol.* **25,** 167 (1961).
97. A. Campbell, *Virology* **14,** 22 (1961).
98. P. M. Naha, *Virology* **29,** 676 (1966).
99. R. Thomas, C. Leurs, C. Dambly, D. Parmenter, L. Lambert, P. Brachet, N. Lefebvre, S. Mousset, J. Porcheret, J. Szpirer, and D. Wauters, *Mutation Res.* **4,** 735 (1967).
100. L. W. Black and D. S. Hogness, *JBC* **244,** 1968 (1969).
101. L. W. Black and D. S. Hogness, *JBC* **244,** 1972 (1969).
102. L. W. Black and D. S. Hogness, *JBC* **244,** 1982 (1969).
103. M. Imada and A. Tsugita, unpublished data.
104. I. W. Sutherland and J. F. Wilkinson, *J. Gen. Microbiol.* **34,** xiip (1964).
105. I. W. Sutherland and J. F. Wilkinson, *J. Gen. Microbiol.* **39,** 373 (1965).

depends on what kind of research is being done in the laboratory, and on how much enzyme is sufficient for measurement. While laboratories of bacteriology or microbial genetics are always ready to use growing bacteria or plates for assay, laboratories of biochemistry usually prefer dry cell preparations.

a. Turbidimetry of Growing Cells. The logarithmically growing cells of *E. coli* are harvested and suspended in a chloroform-saturated buffer with occasional shaking. The concentration of cells in the reaction mixture is adjusted to an absorbance at 450 nm of 0.6. The reaction proceeds in a linear fashion until the turbidity decreases to about 65% of the initial value. There is a good linear relationship between the changes in turbidity and the enzymic activity. One unit of enzymic activity is defined as the amount needed to catalyze a rate of decrease in optical density of 1.0/min (*106*; see also *71, 107*). The chloroform treatment removes the lipid component and possibly denatures the lytic enzyme in the host cell.

b. Turbidimetry of Dry Cell. Escherichia coli cells logarithmically growing are chilled, collected, and washed twice before treatment with chloroform-saturated buffer by gently mixing at room temperature for 1 hr. The cells are collected, lyophilized to dryness, and kept in a desiccator. A fresh suspension of lyophilized cells is mixed with the enzyme sample. The time required for a decrease in absorbance at 450 nm of 0.100 (from about 0.65 to 0.55) is measured at room temperature (*89*).

The reciprocal of the reaction time is proportional to the concentration of the enzyme (*27*). Activity in units (U) per milliliter is expressed by the formula $U = T_s/T \times 0.05$, where T is the time measured for the enzyme sample and T_s the time measured for a standard 0.05 mg/ml solution of egg white lysozyme. This method gives the egg white lysozyme equivalent units. Because the specific activity of the enzyme is very high, dilution of the enzyme sample is required. Since the T4 phage enzyme is basic in character which makes it adsorb to glass surfaces and is unstable in a very diluted solution, the protein is diluted in a buffer containing 0.2% gelatine and 0.5 M NaCl (*27*).

In a modified method, the harvested cells are frozen for about a week and then lyophilized, omitting the chloroform and heat treatment. If the cell powder is insufficiently sensitive to the enzyme, it may be transformed to a good substrate by storage at 37°. As an alternative method, the lyophilized cells are suspended the day before an assay is to be performed. The suspension is mixed with the sample and incubated, the

106. M. Sekiguchi and S. S. Cohen, *JMB* **8**, 638 (1964).
107. W. Weidel and W. Katz, *Z. Naturforsch.* **16b**, 156 (1961).

absorbance at 350 nm is followed. For each experiment, a standard curve for phage lysozyme or egg white lysozyme is constructed by plotting various concentrations of lysozymes against absorbance. The relative activity of a given sample is determined by referring the absorbance measured to the standard curve (*27, 89*).

The apparent absorbance results from light scattering. Several wavelengths have been used for measuring it, such as 350 (*27, 89*), 450 (*27, 106*), 600 (*20, 100*), and 650 nm (*91*). When checking the decrease of absorption at various wavelengths at every 50 nm from 350 to 650 nm, no practical reason of preference was found between 450 and 650 nm. The decrease at 350 nm, however, was somewhat smaller probably because of interference by absorption of DNA. The comparative value of absorbance will differ with optical characteristics of the instrument used. A wavelength showing the highest decrease upon enzymic action can be chosen for assay.

c. Spot Test on Plates. Advantages of this method are sensitivity and convenience, but only rough estimations can be obtained (*20*).

Top agar containing exponentially growing indicator bacteria was used for plating on plates containing bottom agar. If a more sensitive measurement is required, both agars are supplemented with sodium citrate (*35*). After incubation the plate is exposed to chloroform vapor.

Spot tests for measuring enzymic activity are carried out by applying the solution to be tested to the plate with reference to a standard solution of egg white lysozyme. The spot containing enzymic activity shows noticeable clearing after further incubation. The dilution end point, when compared with the standard enzyme, serves as an estimate of lysozyme activity (*19*). By this method about 0.1 μg of egg white equivalent enzymic activity can be determined.

d. Isotopic Assay. Barrington and Kozloff (*67*) labeled *E. coli* cells with ^{15}N and prepared a cell wall preparation according to Weidel (*108*) or Salton and Horn (*109*). The lytic activity of T2 phage was measured as nonprecipitable components formed during enzymic action.

Brown and Kozloff (*69*) labeled the cell with ^{14}C glucose and ^{14}C fructose of ^{14}C totally labeled amino acid mixture. The cell walls were prepared cautiously and subjected to measurement of lysozyme activity. The amount of liberated radioactivity was about 5% of the total counts in the cell wall preparation.

Recently, Schweiger and Gold (*58*) developed a more sensitive method

108. W. Weidel, *Z. Naturforsch.* **6b**, 251 (1951).
109. M. R. J. Salton and R. W. Horn, *BBA* **7**, 177 (1951).

for the detection of lysozyme activity based on liberation of radioactivity from *E. coli* cells grown to the stationary phase in the presence of ^{3}H-1,6-DAP. Another radioisotopic assay was reported by Pene (*110*) based on radioactivity rendered soluble in cold trichloroacetic acid solution cells with ^{3}H-DNA used as the enzyme substrate in the presence of excess deoxyribonuclease I and venom phosphodiesterase. Release of mononucleotides is proportional to the amount of lysozyme.

2. *Purification*

Because T4 phage lysozyme is a basic, small protein, several conventional methods designed for such proteins are used for its purification. The methods essentially utilize weak acidic resins for chromatography and Sephadex for molecular sieving. Besides these techniques the following two points are essential in purification: removal of DNA and protection of SH groups (*27, 34*).

a. Starting Lysate. Starting lysate must be made in a high titer. For this reason, and for easy inspection of contaminations, phages and host bacteria must be selected carefully. In the case of T4 phage lysozyme (*34*), a host bacteria strain B/1 was selected because of its resistance to T1 phage which is strongly infectious. The lysate is usually added with chloroform to complete the lysis. Most of the T4 lysozyme was found in the supernatant of the lysate. The lysozyme does not seem to be a structural component but rather is absorbed to the phage particles under certain culturing conditions (pH and ionic strength). Lysate containing phage particles, after removal of DNA, was poured directly onto an ion exchange column in an attempt to dissociate the enzyme from the phage particles.

This procedure has proved successful in the purification of T4 phage lysozyme (*27*), T2 phage lysozyme (*21*), and λ phage endolysin (*103*) as well as in the purification of lytic enzyme of N20F′ phage. Although the binding complex between the lysozyme and the phage particle may be more stable than the isolated lysozyme, pH below 5 to a great extent inactivates isolated enzymes from both T4 and λ.

b. Removal of DNA. Lysate containing DNA is a viscous solution, and the acidic character of DNA causes an association with basic proteins. An efficient method to remove DNA is by adding a basic dye, Rivanol (6,9-diamino-2-ethoxyacridine) to the lysate until no further precipitate is found. The precipitate, containing DNA, cell debris, intact cells, and acidic proteins is removed by sedimentation (*27*). This dye does not inhibit, to any noticeable extent, the lysozyme activity and has

110. J. J. Pene, *BBRC* **28**, 365 (1967).

been successfully used for T4 and T2 lysozymes and λ-endolysin in our laboratory (*21, 103*) and for N20F′ in another laboratory (*91*).

c. Chromatography and Concentration on Resin. Various weakly acidic resins with a carboxyl group as the functional group can be used for chromatographic separation of phage lysozyme just as they are for egg white lysozyme. Amberlite CG-50, was chosen for T4 phage lysozyme for ease of handling. The chromatographic conditions consist of changing the pH from a low to high value within the range of the enzyme's stability, and the ionic strength is increased by a gradient, or stepwise (*27*).

When a concentration process is required, the weakly acidic resin column can be used with stepwise change in the ionic strength of the eluent (*27*). A weakly basic resin can also be used for purification of the enzyme in the final step. Kretsinger used DEAE-Sephadex A-50 for additional purification and crystallization of T4 phage lysozyme (*111*).

d. Molecular Sieving. Since T4 phage lysozyme is a rather small molecule, molecular sieving is one of the most efficient purification methods. Sephadex G-75 or G-100 is used for T4 and T2 lysozymes (*21*) and λ-endolysin (*105*) and N20F′ lytic enzyme (*103*). The eluent must be maintained above 0.2 *M* in phosphate buffer to avoid irreversible adsorption of the enzymes to the Sephadex. The purification process, with purities and yields at each step is summarized in Table II.

TABLE II
RECOVERY AND SPECIFIC ACTIVITY AT EACH STEP OF PURIFICATION OF T4 PHAGE LYSOZYME[a]

Purification procedure	Lysozyme activity: Total (units)	Lysozyme activity: Recovery (%)	Specific activity (units/A_{280})
Lysate	7.5×10^4		0.17
1. Rivanol treatment	7.5×10^4	100	*b*
2. Initial IRC-50 column concentration	6.4×10^4	85.5	13.9
3. First column chromatography on IRC-50	6.0×10^4	80.0	120
4. Second column chromatography on IRC-50	4.4×10^4	58.0	200
5. First gel filtration on Sephadex G-75	3.8×10^4	50.2	240
6. Second gel filtration on Sephadex G-75	3.0×10^4	39.5	260

[a] Tsugita *et al.* (*27*).

[b] Absorbance was not measured because Rivanol remaining in the solution has a high absorbance at 280 nm.

111. R. Kretsinger, *JMB* (1971) (in press).

e. Crystallization. After three Amberlite IRC-50 and an additional DEAE-Sephadex column chromatography, the protein was precipitated by ammonium sulfate, the salt concentration lowered just enough to dissolve the protein and then raised by dialysis. The largest crystal obtained by Kretsinger was a needle about 0.4 mm long and 0.05 mm thick (*110*).

3. *Physicochemical Properties*

a. Homogeneity. The enzyme appeared homogeneous upon ultracentrifugation as evidenced by a single boundary observed with Schlieren optics. Homogeneity in molecular size was also shown by gel filtration on Sephadex G-75 (*27*). The enzyme was also homogeneous in solution electrophoresis and in acrylamide gel electrophoresis in the presence and absence of 5% sodium dodecyl sulfate (SDS) and 8 *M* urea (unpublished data from our laboratory).

b. Molecular Weight. The sedimentation coefficient of the enzyme ($s_{20,w}$) was 1.9 S. The molecular weight of the enzyme was estimated by the Archibald sedimentation equilibrium method, assuming that the partial specific volume of the enzyme was 0.741 arrived at from its amino acid composition. The value thus calculated was 19,000 ± 1,000 (*27*). From the amino acid sequence the molecular weight was calculated as 18,720 (*22*), which is compatible with the above value.

c. Absorption Spectrum. From the spectrum of the purified enzyme, the absorption coefficient at 280 nm (1 cm light path) was estimated as 1.28 per mg of enzyme per milliliter, and the ratio of the absorption of 280–260 nm was 1.92. This indicates the absence of nucleic acid components in this protein (*27*).

d. Stability of Enzyme. The enzyme proved to be stable when incubated at 37° for 20 hr in 0.1 *M* phosphate buffer, pH 6.0–6.5. The enzyme was inactivated by prolonged incubation at more acidic or alkaline pH values. Magnesium and sodium ions seem to exert a stabilizing effect. Enzymic inactivation was observed at temperatures above 40°, and 50% inactivation occurred at 53.5° (*27*).

4. *Chemical Properties*

a. Amino Acid Composition. T4 phage lysozyme consists of 164 amino acid residues [Asp_{11}, Asn_{11}, Thr_{11}, Ser_6, Glu_7, Gln_6, Pro_3, Gly_{11}, Ala_{15}, $Cys/2_2$, Val_9, Met_5, Ile_{10}, Leu_{16}, Tyr_6, Phe_5, Trp_3, Lys_{13}, His_1, Arg_{13}] (*22*). The total number of basic amino acids is 26, which exceeds the number of acidic amino acids by 8 and may account for the basic character of this protein. One of the three tryptophan residues plays an equally important role as it does in egg white lysozyme (*45, 46*). No common feature in chemical composition was seen between T4 phage lysozyme and egg white lysozyme (*22*) (Table III).

b. Primary Structure. The results of a preliminary amino acid sequence study of the enzyme was published by Inouye and Tsugita in 1966 (*82*) and the completed study was published in 1968 (*22*). The enzyme is a single polypeptide chain of 164 amino acids with a single NH_2-terminal methionine residue and a single COOH-terminal leucine residue (*34*). The sequence was deduced from the sequential analysis of four kinds of partial hydrolyzates: tryptic (*83*), chymotryptic (*84*) and peptic (*85*) digestions, and dilute acid hydrolysis (*86*). The protein was

TABLE III
AMINO ACID COMPOSITION OF VARIOUS LYTIC ENZYMES OF *E. coli* PHAGES AND EGG WHITE LYSOZYMES

	T4 lysozyme [Ref. (*22*)]	T2 lysozyme [Ref. (*21*)]	N20F′ lytic enzyme [Ref. (*92*)]	λ-*endo*-Lysin [Refs. (*101, 128*)]	Egg white lysozyme [Ref. (*87*)]
Asp	11 } 22	11 } 21	12	19	8 } 21
Asn	11	10			13
Thr	11	10	6	6	7
Ser	6	7	6	9 (10)[a]	10
Glu	7 } 13	7 } 13	11	16 (17)[a]	2 } 5
Gln	6	6			3
Pro	3	3	6	5	2
Gly	11	11	7	15	12
Ala	15	15	7	13	12
Cys/2	2	2	0	1	8
Val	9	10	8	7	6
Met	5	5	2	3	2
Ile	10	10	9	9	6
Leu	16	16	5	14	8
Tyr	6	6	2	5	3
Phe	5	5	2	5	3
Trp	3	3	1	3	6
Lys	13	13	7	12	6
His	1	1	1	3	1
Arg	13	13	6	12	11
NH_2	(17)[a]	(16)[a]	—	14	(16)[a]
Total	164	164	98	157 (159)[a]	129

[a] Numbers in parentheses indicate the integer values obtained by Black and Hogness (*101*).

5 10 15
H-Met-Asn-Ile-Phe-Glu-Met-Leu-Arg-Ile-Asp-Glu-Gly-Leu-Arg-Leu-Lys-Ile-Tyr-Lys-

20 25 30 35
Asp-Thr-Glu-Gly-Tyr-Tyr-Thr-Ile-Gly-Ile-Gly-His-Leu-Leu-Thr-Lys-Ser-Pro-Ser-

40 45 50 55
Leu-Asn-Ala-Ala-Lys-Ser-Glu-Leu-Asp-Lys-Ala-Ile-Gly-Arg-Asn-Cys-Asn-Gly-

60 65 70 75
Val-Ile-Thr-Lys-Asp-Glu-Ala-Glu-Lys-Leu-Phe-Asn-Gln-Asp-Val-Asp-Ala-Ala-Val-

80 85 90
Arg-Gly-Ile-Leu-Arg-Asn-Ala-Lys-Leu-Lys-Pro-Val-Tyr-Asp-Ser-Leu-Asp-Ala-Val-

95 100 105 110
Arg-Arg-Cys-Ala-Leu-Ile-Asn-Met-Val-Phe-Gln-Met-Gly-Glu-Thr-Gly-Val-Ala-Gly-

115 120 125 130
Phe-Thr-Asn-Ser-Leu-Arg-Met-Leu-Gln-Gln-Lys-Arg-Trp-Asp-Glu-Ala-Ala-Val-Asn-

135 140 145 150
Leu-Ala-Lys-Ser-Arg-Trp-Tyr-Asn-Gln-Thr-Pro-Asn-Arg-Ala-Lys-Arg-Val-Ile-Thr-

155 160
Thr-Phe-Arg-Thr-Gly-Thr-Trp-Asp-Ala-Tyr-Lys-Asn-Leu-OH

FIG. 7. The complete amino acid sequence of T4 phage lysozyme. Underlining indicates tryptic peptides, and the numbers correspond to those of the tryptic peptides from the amino terminal peptide. From Tsugita and Inouye (*22*).

found to have two cysteine residues as sulfhydryl groups, in contrast to the four S–S bridges in egg white lysozyme (*87*). This was shown by titration with *p*-mercuribenzoate (*22*). No common feature in the first-order structure was found between T4 phage lysozyme and egg white lysozyme (*87, 112*) although some structural similarity was noted in the nature of sequenced amino acids (*113*) (Fig. 7).

c. Tertiary Structure. While the tertiary structure of hen egg white lysozyme has been clarified by Phillips' group employing X-ray crys-

112. P. Jolles, *Proc. Roy. Soc.* **B167**, 350 (1967).
113. P. Dunnill, *Nature* **215**, 621 (1967).

tallography (*114, 115*) (see chapter by Phillips *et al.*, Volume VII), similar studies on phage lysozyme have hardly begun, although a trial was reported by Kretsinger (*111*). Preliminary study by X-ray crystallography showed a spacing along the needle axis, which is twofold screw axis of $a = 29.3$ Å and $b = 133$ Å. There is some uncertainty about the c axis because the macroscopic needles frequently were found to be finer needles grown together. The reflection was indexed with $c = 90$ Å and $\beta = 91.5°$. The volume of the unit cell was calculated as 351,000 Å. Using 2.37 Å/dalton as a mean value, and molecular weight, 8.0 molecules per unit cell or 4 per crystallographic asymmetric unit in space group P_{21}, were calculated. Since there is no evidence to suggest that the lysozyme functions as an oligomer, it is rather surprising to find 4 molecules per asymmetric unit.

5. *Catalytic Properties*

a. Substrate Specificity. The optimal pH for enzymic activity of the T4 phage lysozyme is 7.2–7.4. The activities at pH 5.5, 6.0, 6.5, 8.0, and 9.0 are less than 20, 60, 90, 75, and 45% of the maximum activity, respectively. The substrate specificity was examined by comparing the activities of egg white lysozyme and phage lysozyme, using lyophilized cells of *E. coli* and *Micrococcus lysodeikticus* as substrates. The T4 phage lysozyme is about six times more active than egg white lysozyme with *M. lysodeikticus* cells as substrate, and 250 times more active with *E. coli* cells.

The enzymic action of phage lysozyme was compared with egg white lysozyme, using *M. lysodeikticus* cell wall preparation as substrate, prepared by the method of Hara and Matsushima (*116*). After digestion of the cell wall preparation by the enzymes, the digest was reduced with sodium borohydride, followed by strong acid hydrolysis. The amounts of free muramic acid, muramicitol, glucosamine, and glucosaminitol in the hydrolyzates were analyzed on an amino acid analyzer (*27*). Table IV shows the results, and it can be seen that, as in the case of egg white lysozyme, phage lysozyme breaks the cell walls to give muramicitol, although the amount of muramicitol is less for phage lysozyme than for egg white lysozyme. This fact suggests that the phage enzyme is a muramidase.

114. C. C. F. Blake, D. F. Koenig, G. A. Mair, A. C. T. North, D. C. Phillips, and V. R. Sarma, *Nature* **206,** 757 (1965).
115. C. C. F. Blake, G. A. Mair, A. C. T. North, D. C. Phillips, and V. R. Sarma, *Proc. Roy. Soc.* **B167,** 365 (1967).
116. S. Hara and Y. Matsushima, *Bull. Chem. Soc. Japan* **39,** 1826 (1966).

TABLE IV
HYDROLYSIS OF *M. lysodeikticus* CELL WALLS BY T4 PHAGE LYSOZYME AND EGG WHITE LYSOZYME[a,b]

Compound	Without enzyme	Egg white lysozyme	Phage lysozyme
	(μmole/10 mg cell walls)		
Muramic acid	7.65	2.55	5.50
Muramicitol	0	4.90	1.45
Glucosamine	10.0	9.05	9.15
Glucosaminitol	0	0	0
Glutamic acid	10.3	10.0	10.9
Glycine	10.0	10.2	10.2
Alanine	19.9	20.9	20.0
Lysine	10.3	10.3	10.4

[a] Tsugita *et al.* (*27*).

[b] Cell walls of *M. lysodeikticus* (10 mg) were incubated with or without 0.2 mg of enzyme for 24 hr at 37° in 3 ml of 0.044 *M* phosphate buffer, pH 7.0. After incubation, the pH of the reaction mixture was adjusted to about 10 with sodium hydroxide, and 10 mg of sodium borohydride were added. The mixture was incubated for another 24 hr at room temperature. After reduction, the samples were dried and hydrolyzed with 6 *N* HCl for 24 hr at 105°. The hydrolyzate was analyzed at 45° with use of an amino acid analyzer.

b. Active Area in T4 Phage Lysozyme. Several groups reported (*117–119*) that Trp 62 in egg white lysozyme has an important role for its enzymic action. In T4 phage lysozyme three tryptophan residues, 126, 138, and 158 are present (*22*). An experimental approach to find if a tryptophan residue has an important role for catalytic function in the lysozyme made use of nonsense mutants and their revertants. Streisinger isolated several amber (genetic code = UAG) mutants using transitional mutagens. Genetic analysis showed three of them to be derived from tryptophan residues (code = UGG), and their location was determined by sequential studies of the suppressed proteins of the nonsense mutants (*46*) and of the proteins of the spontaneous reverted mutants of the *amber* mutants (*45, 46*).

Amber suppressive hosts, *su⁺1*, *su⁺2*, and *su⁺3* require an insertion of Ser, Gln, and Tyr, respectively, for *amber* codon. When individual mutants in the above three kinds of host bacteria are harvested, new lysozyme proteins are produced in which original tryptophan residues are replaced by Ser, Gln, or Tyr residues. Thus we can exchange the

117. K. Hayashi, T. Imoto, G. Funatsu, and M. Funatsu, *J. Biochem.* (*Tokyo*) **58,** 227 (1965).

118. L. N. Johnson and D. C. Phillips, *Nature* **206,** 761 (1965).

119. J. A. Rupley, *Proc. Roy. Soc.* **B167,** 41 (1967).

tryptophan residues by the other amino acid residues one by one and can measure the activity of each newly formed enzyme. Experimental results show that two of the tryptophan residues, 126 and 158, are replaced by all three of the amino acids above without marked change in enzymic activity, suggesting that these two trytophan residues are not essential for the catalytic action. However, when *amber* mutant for residue 138 was suppressed in $su^{+}1$ (Ser) or $su^{+}2$ (Gln), no lytic activity was detected; in $su^{+}3$ (Tyr), 50–60% of the specific activity of the wild type was retarded (*45*). The lysozyme protein of a revertant of this *amber* mutant which has a tyrosine residue (code = UAU or UAC) instead of a tryptophan residue at residue 138, restored about half of the wild type lysozyme activity (*47*). These observations were confirmed by isolating a triple recombinant of the three *amber* revertants, in which all three tryptophan residues are replaced by tyrosine residues. The triple recombinant also retained 50% of the activity. Canfield also reported that an essential Trp (residue 62) in egg white lysozyme was replaced by tyrosine in human lysozyme (*120*). The results are compatible with each other.

The lysozyme protein contains five glutamine residues at positions 69, 105, 122, 123, and 141, the code for which is CAA or CAG. Four *ochre* (UAA) mutants were isolated by Streisinger and Kneser (*121*) and also in our laboratory by using transitional mutagens (*47*). From the mode of mutation, these mutants were thought to be derived from only glutamine residues. The genetic mapping indicate a correspondence between the four *ochre* mutants and four glutamine locations in the peptide sequence, but with an ambiguity in positions 122 and 123 (*121*). This ambiguity was clarified recently by sequential analysis using the revertant derived from *ochre* mutant in positions 122 (*122*). The *ochre* mutants were suppressed by $su^{+}c$ (Tyr) without marked changes in their specific activities, indicating that replacement (Gln → Tyr) at these positions does not affect the enzymic activity. When the *ochre* mutant in position 141 was suppressed in $su^{+}c$, the same alteration (Gln → Tyr) caused about 50% decrease in its specific activity. The *ochre* mutant in position 105 could not be suppressed by $su^{+}c$ host, and no plaques were formed. This may indicate that this glutamine has a role in enzymic activity or in forming an active conformation (*47*). The lysozyme protein of a revertant of an *opal* mutant derived from the *ochre* mutant in position 122 was sequenced, the result of which indicated the amino acid replacement of Gln 122 → Trp. The specific activity of this revertant

120. R. E. Canfield, *Brookhaven Symp. Biol.* **21,** 137 (1968).
121. G. Streisinger and K. Kneser, private information (1969).
122. N. Ohta, Y. Okada, and A. Tsugita, in preparation.

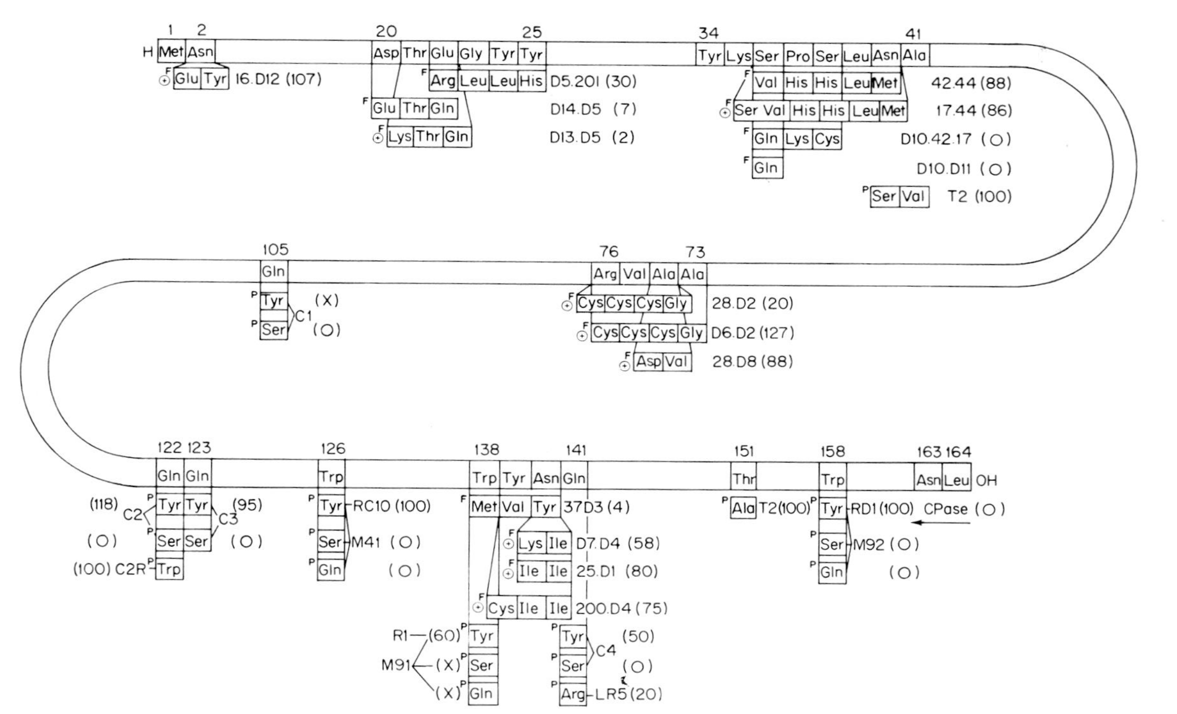

Fig. 8. Activities and alterations of amino acids in various mutant of T4 phage lysozyme (*41*). Here F stands for double frame shift mutation, D for revertant, P for point mutation (nonsense mutant or its revertant), C for *ochre* mutant, and M for *amber* mutant. Specific activity relative to the wild type lysozyme. (○) Similar activity as wild type one, and (×) almost zero activity. (⊕) Addition of one amino acid.

was almost identical with that of the wild type. Amino acid alteration in the revertant protein of an *opal* mutant derived from the *ochre* mutant in position 141 was found to be Gln → Arg, the specific activity of which was reduced 20% (*122*). (The code for the other residue, 69, Glu may be CAG.)

The amino acid alteration in T2 phage lysozyme also suggests that three amino acids, different from those for T4 phage lysozyme, are not essential for the enzymic activity: Asn 40 → Ser, Ala 41 → Val, and Thr 151 → Ala from T4 → T2 (*21*).

Pseudo-wild mutants carrying two (or three) frame-shifts of bases are altered in more than one amino acid sequence between these mutations. A comparison of the altered sequences in the mutant lysozymes and their specific activities with those of wild type lysozyme provides information about which amino acids are essential for enzymic activity. This information, together with that from nonsense mutants and their revertants, and T2 phage lysozyme, is summarized in Fig. 8. The specific activities of the individual frame shift mutants were obtained by direct measurement of the purified mutant proteins and/or indirectly from enzymic activity and serological activity of the crude lysate (*41*). In the figures, relative values of specific activities of the mutant lysozymes against those of wild type enzymes are listed.

Among 15 frame shift mutants, 8 were found to contain an additional amino acid (indicated by ⊕ in Fig. 8). In general, addition of an amino acid does not affect the catalytic activity, as can be seen in the mutant of *e*J16*e*J12, in the comparison of *e*J42*e*J44, and *e*J17*e*J44, and *e*JD7*e*JD4, *e*J25*e*JD1, and *e*J200*e*JD4. However, in one case the enzymic activity was found to be markedly decreased (*e*J28*e*JD2), but no decrease was observed in the other relevant mutant (*e*J28*e*JD8). One may think that certain kinds of altered amino acids in *e*J28*e*JD2 cause the decrease of enzymic activity, but the same alteration in the sequence without any addition of amino acid in the mutant, *e*JD6*e*JD2, does not cause any decrease of the activity. The addition of one amino acid does not affect enzymic activity, but when added with a particular kind of an amino acid, the catalytic activity is affected. One of the possible explanations for this is that a configurational change caused by the addition of an amino acid may expose the cysteine residue on the surface of the enzyme protein, which in turn affects the catalytic property or stability of the enzyme protein.

When the sequence Glu22–Gly23–Tyr24–Tyr25 is replaced by Arg–Leu–Leu–His (*e*J201*e*JD5), the specific activity drops to about 30%

of that of wild type enzyme. A further decrease in enzymic activity was observed when Asp20–Thr21–Glu22 are replaced by Glu–Thr–Glu (*e*JD14*e*JD5) (7%) and by Asp–Lys–Thr–Gln (*e*JD13*e*JD5) (2%). From these two replacements one may predict that the γ-carboxyl group of Glu 22 plays an important role in enzymic activity, not affected by alteration in *e*JD14*e*JD5 Asp 20 → Glu (*47*). Alternatively, Asp 20 may be rather important and cannot be replaced by any other amino acid including glutamic acid (*e*JD14*e*JD5) without some loss of the catalytic activity, also, that action of Asp 20 is interfered with by the adjacent lysine residue (*e*JD13*e*JD5). The latter possibility is more likely because it is compatible with the alteration found in *e*J201*e*JD5. Observations of a set of alterations in the frame shift mutants around Trp138–Tyr139–Asn140 again suggest an important role of tryptophan–residue 138. When the tripeptide is replaced by Met–Val–Tyr (*e*J37*e*JD3) the enzymic activity is lost to a marked extent, while the following replacements, not including the tryptophan, do not largely affect the enzymic activity: Trp–Tyr–Ile–Ile (*e*J25*e*JD1), Trp–Cys–Ile–Ile (*e*J200*e*JD4), and Trp–Tyr–Lys–Ile (*e*JD7*e*JD4). In the last mutant protein, the basic amino acid, lysine, when located adjacent to the tryptophan, may affect the function of the tryptophan residue. This kind of effect may be seen in the decrease of activity caused by a replacement at the 141 position (Gln 141 → Arg). Thus, the relationship between the amino acid sequence from position 138 to 141 and the enzymic activity can be expressed as follows: Trp–Tyr–Asn–Gln (100) = Trp–Tyr–Ile–Ile–Gln (100) = Trp–Cys–Ile–Ile–Gln (100) > Tyr–Tyr–Asn–Gln (50) ≃ Trp–Tyr–Lys–Ile–Gln (50) > Trp–Tyr–Asn–Arg (20) > Met–Val–Tyr–Gln (5) > Ser–Tyr–Asn–Gln (0) = Gln–Tyr–Asn–Gln (0), where numbers in parentheses indicate relative specific activities and underlining shows the sequences changed.

Endopeptidase digestion may furnish the information of essential amino acid. While aminopeptidase was unaccessible to the amino terminus of the native protein, carboxypeptidase released two amino acids from the carboxyl end without affecting the activity.

Summing up the above discussion the following amino acids or amino acid sequences are not essential for catalytic activity or for polypeptide conformation: Asn[2], Gly[23]–Tyr–Tyr[25], Thr[34]–Lys–Ser–Pro–Ser–Leu–Asn–Ala[41], Ala[73]–Ala–Val–Arg[76], Glu[122]–Gln[123], Trp[126], Tyr[139], Thr[151], Trp[158], and Asn[163]–Leu[164], while the following amino acids are suggested to be essential for catalytic activity: Asp[20], (Glu[22]), Gln[105], Trp[138], (Asn[140], Gln[141]) (amino acids in parentheses have rather indirect evidences).

6. *Roles of Lysozyme and the Other Lytic Actions in the Life Cycle of T4 Phage in the Host Cell*

The T4 phage infection causes lysis of the bacterial cell; at the beginning of phage research, it was thought that lysozyme was the only active agent for this lysis process. In the life cycle of the infecting phage, the lytic action is now known to be divided into two steps: (1) penetration of the phage into the host cell and (2) lysis of the host cell and multiplication of daughter phages. For the phage, the lysozyme plays a role only when it lyses the host cells during the reproduction of the phages (*32*, *90*, *123*).

In this section, the two mechanisms (penetration and lysis) are discussed in the light of current knowledge, including some hypothetical processes. The discussion, however, is limited to the T4 phage–*E. coli* system for which the most precise information is available.

a. Host Cell Recognition and Absorbance. The initial step of phage interaction with the host cell may be the recognition and absorbance prior to infection. A suggested recognition site is a lipopolysaccharide component of the bacterial cell wall (*124*, *125*). The host cell recognition was found to be dependent on tail fiber components of the phage particle (*123*). The phage particle has six tail fiber rods (20 × 800 Å) (*126*), and each fiber appears to consist of two morphologically almost identical pieces (20 × 400 Å) which, however, differ from each other in chemical composition (*127*, *128*).

The distal half of the fiber is controlled by genes 35, 36, 37, and 38 and the proximal part by gene 34 (*129*, *130*). Gene 57 is one of the tail fiber genes, but its role is not yet known. Furthermore, the distal part of the tail fiber has been isolated and was found to be responsible for host cell recognition (*128*, *131*). In the distal part of the role of gene 35 product and gene 36 product has become clear to some extent, and one of the remaining genes, 37 or 38, is now suspected to be responsible for the host recognition (unpublished work).

The components of the tail's short pins attached to the base plate

123. L. D. Simon and T. F. Anderson, *Virology* **32,** 279 (1967).
124. M. A. Jesaitis and W. F. Geobel, *J. Exptl. Med.* **96,** 409 (1952).
125. W. Weidel, *Ann. Rev. Microbiol.* **12,** 27 (1958).
126. S. Brenner, G. Streisinger, R. W. Horne, S. P. Champe, L. Barnett, S. Benzer, and M. W. Roes, *JMB* **1,** 281 (1958).
127. R. Takata and A. Tsugita, *JMB* **54,** 45 (1970).
128. S. Imada and A. Tsugita, *Mol. Gen. Gen.* **109,** 338 (1970).
129. R. S. Edgar and I. Lielausis, *Genetics* **52,** 1187 (1965).
130. J. King and W. B. Wood, *JMB* **39,** 583 (1969).
131. J. H. Wilson, R. B. Luftig, and W. B. Wood, *JMB* (1971) (in press).

were found to be responsible for the adsorption followed by fiber recognition (*123*). However, a fiberless particle has a decreased infectivity, suggesting that a fiber component might be involved in this adsorption step (*132*). Also, it was found that when treated with urea, which causes a degradation of the base plate (*133*), the phage particle is still able to infect the spheroplast of the host cell (*134*).

Weidel and his collaborator (*125, 135*) found that the receptors for T2 and T1 phages are lipoproteins, which are soluble in 90% phenol; but receptors for T3 and T7 as well as for T4 phages are lipopolysaccharide layers in the cell wall which can be solubilized by washing with water. The receptor particles of T5, which consist of lipopolysaccharides coated with lipoproteins, are located on the surface of the cell walls and are readily soluble in dilute alkali.

b. Penetration of Phage DNA into Host Cell. The next step of the infection is the penetration of phage DNA. The tail fibers bend and the "tail sheath" contracts giving the base plate access to the bacterial surface. By this time the tail pin has altered into a short tail fiber allowing the phage more access to the bacteria cell wall. The "tail core" penetrates the cell wall (about 120 Å thick), but not the cytoplasmic membrane, and injects phages DNA into the cell (*123*). A contracting enzyme was found to be associated with the tail of T2 phage (*136*). Mutants defective in genes 11 and 12 are capable of adsorbing the bacteria but cannot kill the host, probably because they lack the DNA injection mechanism (*137*).

Phage lysozyme has been reported to be associated with phage particles of the related phage T2, and it seems possible that the digestion of the bacteria cell wall with this phage-associated lysozyme is a crucial step in the penetration process. After comparing a lysozyme-deficient mutant (*e*) with the wild type strain, Emrich and Streisinger (*32*) questioned the idea that phage lysozyme is necessary for phage infection. They found that even this mutant phage causes an identical time course of the appearance of intracellular phages in bacteria, suggesting that the lysozyme does not play a role in the penetration step. Bacteria infected with a very high multiplicity of phage lyse immediately after

132. R. S. Edgar and W. B. Wood, *Proc. Natl. Acad. Sci. U. S.* **55,** 498 (1966).

133. C. To, Ph.D. Dissertation, Kansas State University, Manhattan, Kansas, 1968.

134. C. Veldhuisen and E. B. Goldberg, "Methods in Enzymology," Vol. 12, Part B, p. 858, 1968.

135. W. Weidel and E. Kellenberger, *BBA* **17,** 1 (1955).

136. P. P. Dukes and L. M. Kozloff, *JBC* **234,** 534 (1959).

137. J. King, *JMB* **32,** 231 (1968).

infection. This is called "lysis from without" (*138*). The *e* phage causes lysis from without equally well as the wild type phage does.

The amount of lysozyme associated with the purified wild type phage particle was calculated to 0.5 (*32*) or 0.01 molecule (*139*) per phage particle. No lysozyme activity is associated with the *e* phage particle. However, upon disruption, freezing and thawing of the purified phage, a new lytic activity appears that cannot be eliminated by treatment with antiphage lysozyme serum. The same lytic activity is found even in *e* phage after disruption. Although this new lytic activity is associated with phage particles, its level is extremely low compared to the level of the wild type phage lysate.

c. Changes in Host Metabolism and in Cell Wall or Cell Membrane. Puck and Lee (*140, 141*) found that after the phage has infected the cell, the host cell wall permeability increases rapidly. As for protein synthesis, most of the host-dependent protein synthesis is stopped immediately and the synthesis of "very early" and "early" proteins, directed by the phage genome, is initiated. By employing a trigger mechanism (phage specific σ factor of RNA polymerase) the late phage proteins appear at a certain time after infection. Oxygen uptake continues until the cell lysis (about 30 min after infection), independent of the presence of lysozyme, and the cessation of metabolism causes the lysis of the cell (*88*). A bacteria infected with phage, and later superinfected with the same phage, lyses later than a bacteria infected once. This phenomenon is called "lysis inhibition" (*142*).

The reason for the lysis inhibition is not known; some mutants such as *spackle* (*s*) (*89*) and rapid lysis (*r*) (*143*) do not exhibit lysis inhibition while others such as the *e* mutants do. During the delay of lysis caused by superinfection, the oxygen metabolism continues normally (*88*). A metabolic inhibitor such as KCN induces a premature lysis (*144*). The cessation of metabolism is followed by the disappearance of the cytoplasmic membrane and an increase of permeability.

Emrich found a mutant called spackle (*s*) (*89*) which does not induce lysis inhibition. Furthermore, bacteria infected with the *s* phage are more sensitive to lysis from without than bacteria infected with r_{I} or r_{II} mutants, which develop normal resistance to lysis from without. Domi-

138. M. Delbrück, *J. Gen. Physiol.* **23,** 643 (1940).
139. K. K. Mark, *New Asia Coll. Acad. Ann.* **11,** 31 (1969).
140. T. T. Puck and H. H. Lee, *J. Exptl. Med.* **99,** 481 (1954).
141. T. T. Puck and H. H. Lee, *J. Exptl. Med.* **101,** 151 (1955).
142. A. H. Doermann, *J. Bacteriol.* **55,** 257 (1948).
143. G. S. Stent and O. Maaløe, *BBA* **10,** 55 (1953).
144. A. H. Doermann, *J. Gen. Physiol.* **35,** 645 (1952).

nancy tests showed that s^+ is always dominant to s. Results indicate that the s mutant permits lysis of the phage-infected host even in the absence of lysozyme and effects a phage-directed synthesis of bacterial cell wall components (*89*). Although the role of the r gene is still ambiguous, the r gene product seems to connect with cell membrane or cell wall components to a certain extent, resulting in rapid cessation of infected cell metabolism and rapid lysis of the host cell (*145*). Another gene, the acridine sensitive gene, was also reported to affect the cell permeability (*146*). The genetic mapping for these genes, which differ from the e gene, is shown in Fig. 5.

Even the empty protein ghost of the T even phage can inhibit some host macromolecular synthesis and cause cell death (*147, 148*). The host infection completely inhibits synthesis of DNA, RNA, and protein within 2 min, and the sudden increase of permeability of host cell (*149*) and ghost-infected bacteria prevents the superinfected phage from multiplying (*147, 150*). The effect mediated by the ghost was shown not to result from phage lysozyme activity by using e mutant phage (*151, 152*). Thus, the mechanism of ghost infection is very similar to the phage infection except that the phage directs its own synthesis of cell wall which maintains the permeability characteristic of uninfected bacteria and a similar rate of oxygen uptake. The difference in permeability has not been studied in detail but oxygen uptake has been shown to result from the switching from bacterial metabolism to phage-directed metabolism. The extent of switching depends on the kind of phage. The immediate permeability change after ghost infection may be caused by an unknown mechanism which does not require protein synthesis (*152*) and does not cause any detectable soluble cell wall components to be released. A kind of conformational change in cell wall or cell membrane is postulated for this phenomena which successively causes the cessation of metabolism including DNA–cell membrane moiety in the bacteria.

d. The Lysis of Host Cell. Although we do not know yet what triggers the lysis at the latent lysis time, two lytic enzymic activities are

145. J. Séchaud, E. Kellenberger, and G. Streisinger, *Virology* **33**, 402 (1967).
146. S. Silver, *JMB* **29**, 191 (1967).
147. R. C. French and L. Siminovitch, *J. Microbiol.* **1**, 757 (1955).
148. R. M. Herriott and J. L. Barlow, *J. Gen. Physiol.* **40**, 809 (1957); **41**, 307 (1957).
149. I. R. Lehman and R. M. Herriott, *J. Gen. Physiol.* **41**, 1067 (1958).
150. V. Bonifas and E. Kellenberger, *BBA* **16**, 330 (1955).
151. D. H. Duckworth, *J. Virol.* **3**, 92 (1969).
152. D. H. Duckworth, *Virology* **40**, 673 (1970).

known to be involved in the process. One we call gene *t* product and the other is the conventional gene *e* product, lysozyme.

The "Tithonus" (*t*) mutant was isolated as a new class of lysis defective mutants by Josslin (*90*), and its product was found to be related to lipid metabolism in cell membrane.

The function of λ *s* gene is also required for cessation of phage-infected cell metabolism, just as is that of T4 *t* gene. This requirement also extends to the smaller coliphages, φX174 (*153*), S13 (*154*), and R17 (*155*). The former two DNA phages have a gene whose function seems analogous to that of the *t* gene, and the coat protein of RNA phage R17 also seems to resemble the *t* gene function. However, in φX174-infected cell lysate, no phage-induced lysozyme activity was found (*156*), and no lytic activity was proven to be in the R17 coat protein (*155*).

C. T2 Phage Lysozyme

1. *Purification*

T2 lysozyme was found to be associated with phage particles (*69*) and only a small fraction present in the supernatant of the lysate (*75*). The location of the lysozyme was thought to be at the "proximal tail" (*69*) or "tail fiber" (*80*). Katz (*81*) proved that the lysozyme from phage particles is identical with that of the supernatant. Katz and Weidel (*80*) prepared the lysozyme from phage disrupted by freezing and thawing followed by incubation with deoxyribonuclease to remove the DNA. After dialysis and treatment with a strongly basic resin (Dowex 1), the enzyme was purified by repeated chromatography on Amberlite CG-50. Katz (*81*) further purified the enzyme preparation by a Sephadex G-100 treatment and crystallized it in the presence of 0.15 *M* KCl and 0.02 *M* sodium bicarbonate buffer at pH 10.1, which is around the isoelectric point of the lysozyme.

Since it was found that the lytic activity of this enzyme was neutralized by antiserum against purified T4, the purification process for T4 phage lysozyme was used for the purification of T2 phage lysozyme. The lysozyme preparation thus obtained was pure enough to be subjected to the sequencing studies and amino acid analysis (*21*).

153. C. A. Hutchison, III and R. L. Sinsheimer, *Virology* **25,** 88 (1966).
154. R. Baker and I. Tessman, *Proc. Natl. Acad. Sci. U. S.* **58,** 1438 (1967).
155. N. D. Zinder and L. B. Lyons, *Science* **159,** 84 (1967).
156. R. Fujimura and P. Kaesberg, *Biophys. J.* **2,** 433 (1962).

2. *Physicochemical Properties*

a. Homogeneity. The purified enzyme from Katz's laboratory appeared homogeneous in sedimentation analysis both before (*80*), and after (*81*) further purification on Sephadex G-100 and crystallization. The enzyme purified in our laboratory (*21*) when chromatographed on IRC 50, gave single coincident peaks of UV absorption and enzymic activity with constant specific activity through the peak. Homogeneity was also shown by amino acid analysis, in which the ratios of each amino acid appeared to be near integer values (unpublished data).

b. Molecular Weight. From sedimentation analyses the molecular weight has been estimated to be between 21,000 and 15,200 (*80*) and about 14,000 (*81*). The molecular weight as calculated from the amino acid composition and the sequential analysis is 18,607 (*21*).

c. Absorption Spectrum. The purified T2 phage lysozyme shows almost the same UV spectrum and absorption coefficient as the T4 phage lysozyme (*21*).

d. Stability. As expected, the stability of this enzyme is quite similar to T4 phage lysozyme (*27*).

3. *Chemical Properties*

a. Amino Acid Composition. The T2 phage lysozyme consists of 164 amino acid residues [Asp_{11}, Asn_{10}, Thr_{10}, Ser_{7}, Glu_{7}, Gln_{6}, Pro_{3}, Gly_{11}, Ala_{15}, $Cys/2_{2}$, Val_{10}, Met_{5}, Ile_{10}, Leu_{16}, Tyr_{6}, Phe_{5}, Trp_{3}, Lys_{13}, His_{1}, Arg_{13}]. This is similar to that of the T4 phage lysozyme, except there is a decrease of one mole each of asparagine and threonine residues and an increase of one mole each of serine and valine residues (these amino acid residues are underlined in Table III). The number of acidic and basic amino acids is the same. This is compatible with the fact that the same purification method can be applied to both proteins and that chromatographic similarity is observed in both. The similarity of aromatic amino acid content explains why the absorption spectrum was much the same for both enzymes (*21*).

b. Primary Structure. From the amino acid composition of T2 phage lysozyme, its sequence was expected to differ from that of T4 phage lysozyme in only two places. By comparing the tryptic peptides of the two lysozymes, however, T2 was found to differ from T4 in three amino acid substitutions. These are Asn$\overset{40}{\rightarrow}Ser, Ala\overset{41}{\rightarrow}$Val and Thr$\overset{151}{\rightarrow}$Ala, from T4 lyso-

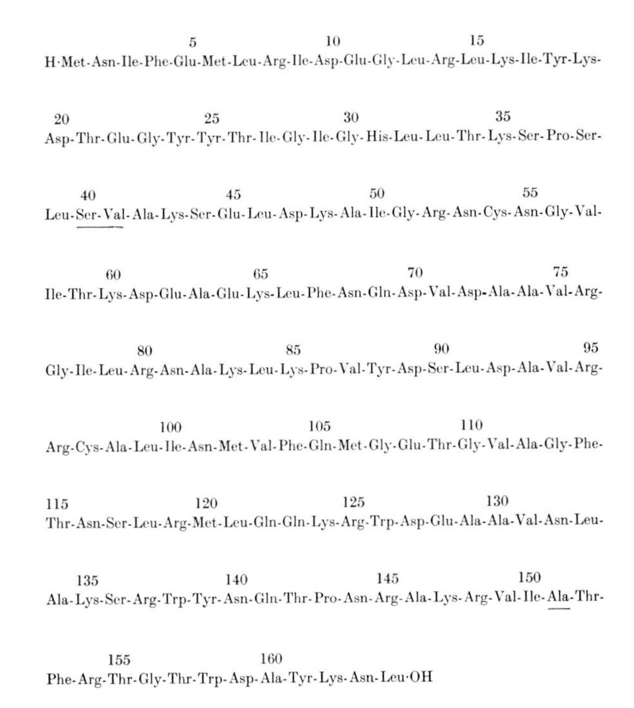

FIG. 9. The amino acid sequence of T2 phage lysozyme. Underlining indicates the positions which are different from the amino acid sequence of T4 phage lysozyme. From Inouye and Tsugita (*21*).

zyme to T2 lysozyme, all of which can be explained by transitional mutations consisting of single base alterations. The sequence is shown in Fig. 9 (*21*).

4. *Catalytic Properties*

The enzymic action of T2 phage lysozyme has been extensively studied; it has been considered to be representative of the general lysis action of bacteriophage.

Koch and Dreyer (*77*) studied the action of T2 phage lysozyme upon *E. coli* cell walls (*66*). Egg white lysozyme is inactive on cell walls incubated with the T2 enzyme. Similarly, no action of the T2 enzyme was found after incubation of cell walls with egg white lysozyme. Identical components were liberated from dinitrophenylated cell walls by either egg white lysozyme or T2 lysozyme. Also, T2 lysozyme lysed *Micrococcus lysodeikticus*, a substrate for egg white lysozyme (*77*).

In related studies, Weidel's group (*99, 108*) showed that identical fractions were released from *E. coli* cell walls by T2 enzyme or egg white lysozyme. Maass and Weidel (*26*) demonstrated that T2 phage lysozyme is *endo*-β-(1 → 4)-*N*-acetylglucosaminidase (*67*), while the other workers concluded its *endo*-β-(1 → 4)-*N*-acetylmulamidase action just as T4 lysozyme (*21*) and egg white lysozyme (*77*).

D. Lytic Enzyme of N20F′ Phage

1. *Purification*

Moo-Penn and Wiesmeyer infected *E. coli* W13a, a nonlysogenic strain, with N20F′ phage; the resulting lysate was stored overnight. Rivanol was then added to remove DNA, acidic protein, and cell debris, and the lysate was centrifuged, dialyzed, and immediately put on a CM-cellulose column. Chromatography was carried out using a discontinuous NaCl gradient at pH 7.0. The fractions containing lytic activity were separated on Sephadex G-75 column and concentrated by lyophilization (*91, 92*).

2. *Physicochemical Properties*

a. Homogeneity. The homogeneity of the purified enzyme was indicated by coincident UV absorption and enzyme activity peaks upon chromatography on SE-Sephadex G-25. The homogeneity was further shown by polyacrylamide disc gel electrophoresis, under both acid and alkaline conditions. Sedimentation velocity studies also indicated homogeneity, but suggested dimerization of the lytic protein (*92*).

b. Molecular Weight. Molecular weight determinations, performed by the Archibald method at pH 8.5, gave a molecular weight in the range of 10,000–12,000 at the meniscus for the protein, but approximately half this value at the cell bottom. This indicates that at pH 8.5 the samples are composed of more than one component. A similar determination at pH 5.5 indicated that the molecular weight at the meniscus and at the bottom are roughly the same (11,000–13,000). These data can be explained if the sample is regarded as being an equilibrium mixture of monomer and dimer (*92*).

The sedimentation coefficient, $s_{20,w}$, in 0.05 *M* tris-maleate at pH 5.5 and 8.5 was calculated as 1.99 and 1.90, respectively, while in 0.05 *M* tris at pH 8.5 and in low concentration of tris-maleate such as 0.01 *M* at both pH 5.5 and 8.5, the $s_{20,w}$ values increased to 2.47–2.48. These results were also obtained using two different lytic enzymes purified from two independently isolated spontaneous mutant strains, N20F′Ltu and N20F′Tu. These observations indicate that the concentration of maleate is important in determining the molecular form of the enzyme (*92*). The above studies, combined with observations on the effect of carboxylic acids on the enzymic activity at the two pH optima (*91*), lead to the conclusion that the pH 5.5 activity is associated with the monomer and the pH 8.5 activity is dependent on the formation of a dimer. This equilibration of monomer–dimer is a reversible function of the carboxylic ion concentration, where high concentrations of carboxylic ion favors the monomer formation. Sedimentation analysis showed that increased concentration of divalent cations such as Co^{2+} and Mg^{2+} produce the same effects as high concentrations of carboxylic acid. Moo-Penn *et al.* (*92*) further suggested that the splitting of the dimer might involve the binding of metal ions and carboxyl groups. This finding is somewhat similar to the observations on egg white lysozyme (*157, 158*). Those authors also have shown that by using the Archibald procedure at pH 6.6, the apparent molecular weight of egg white lysozyme at the bottom layers of the solution is consistently higher than at the meniscus.

c. Absorption Spectrum. The UV spectrum of the enzyme showed a maximum at 277 nm and a minimum at 249 nm at pH 8.5 (0.05 *M* tris buffer) with a ratio of maximum/minimum about 1.4; the absorption coefficient was not given (*92*).

d. Stability. The enzyme seems to be more stable than T4 or T2 phage lysozyme and λ phage lytic enzyme. This may result in part from the absence of cysteine residue(s) which the other three phage lysozymes contain. Thus the lytic enzyme of N20F′ can be purified and lyophilized in the absence of a reducing reagent. Heat inactivation was measured with crude enzyme extracts of N20F′ phage lysate. While activity at both pH optima was lost concomitantly with heating at pH 8.5, loss of activity at pH 5.5 occurred at a greater extent than with heating at pH 5.5 (*91*).

3. *Chemical Properties*

a. Amino Acid Composition. The amino acid compositions of the lytic enzymes of N20F′ phage and two spontaneous mutants have been analyzed. The enzyme of the wild type (N20F′ C) consists of 98 amino acids as follows: [Asp + Asn_{12}, Thr_6, Ser_6, Glu + Gln_{11}, Pro_6, Gly_7, Ala_7,

157. D. B. Smith, G. C. Wood, and P. A. Charlwood, *Can. J. Chem.* **34**, 364 (1965).
158. A. J. Sophianopoulos, C. K. Rhodes, D. N. Holcomb, and K. E. Van Holde, *JBC* **237**, 1107 (1962).

Val_8, Met_2, Ile_9, Leu_5, Tyr_2, Phe_2, Cys_0, Trp_1, Lys_7, His_1, Arg_6]. Differences from the amino acid composition of two spontaneous mutants were not sufficiently reliable for further discussion. The absence of half cystine residue and the presence of only one mole of tryptophan in all three proteins seem to be important (*92*).

b. Approaches in Sequence Study. The NH_2-end group is alanine and the COOH-terminus consist of (Asp–Thr–Ser)–Met–Ala–(Val–Ile). The fingerprint map of the tryptic peptides, as well as a comparative analysis of the terminal sequence and fingerprint on the two spontaneous mutant enzymes have been reported (*92*).

4. *Catalytic Properties*

a. Dual pH Optima in the Lytic Enzyme. The lytic enzyme has two pH optima, one at pH 5.5 and the other at pH 8.5. Relatively high concentrations of carboxylic acids (malate) and divalent cations (Mg^{2+} and Co^{2+}) enhance the pH 5.5 activity but inhibit the pH 8.5 activity. However, in the presence of low concentrations of carboxylic acid as well as divalent cations, the pH 5.5 activity is drastically reduced, but the pH 8.5 activity is not affected. As mentioned in the section on molecular weight, a monomer–dimer transition was observed for this protein. Table V summarizes the effects of various experimental conditions on the catalytic properties and the states of aggregation of the lytic enzyme at pH 5.5 and 8.5 (*92, 93*).

The experimental evidence indicated that the monomer is involved in the enzymic activity at pH 5.5, while the dimer is responsible for the activity at pH 8.5. In the phage lysate and in the purified enzyme, the ratios of enzymic activities at both pH were found almost identical. The same effects of carboxylic acids and cations were also observed on the enzymic activities of the phage lysates. These and other observa-

TABLE V

CATALYTIC PROPERTIES AND THE STATE OF AGGREGATION OF THE N20F′ LYTIC ENZYME[a]

Experimental conditions	pH	
	5.5	8.5
High carboxylic acid conc. 0.05 *M*	Monomer active	Monomer inactive
Low carboxylic acid conc. 0.01 *M*	Dimer inactive	Dimer active
High divalent cation	Dimer activate	Monomer inactivate

[a] From Moo-Penn and Wiesmeyer (*92*).

tions confirm that the single lytic protein exhibits two pH optima (*92*, *93*).

b. Substrate Specificity. The lytic enzyme of the N20F′ phage at both pH optima acts as a muramidase and thus has specificity as that of the T2 and T4 phage and egg white lysozymes (*92*). However, lysis of *E. coli* cell walls at pH 8.5 is seven times more rapid than with egg white lysozyme, as measured by the decrease in optical density; but the amount of reducing groups liberated by egg white lysozyme is approximately four times that of the lytic enzyme (*93*). These results suggest that although the same kinds of bonds are being broken, the bonds are at different locations. For example, the cell wall polymer may be subjected to sequential degradation or to point degradation, as suggested for the action of papaya lysozyme (*159*).

c. Active Area in the Lytic Protein. The phage lytic enzyme is inhibited at both pH optima by histamine, histidine, and indole derivatives, all of which also inhibit egg white lysozyme (*93*). Evidence indicates that trytophan might be involved in the substrate binding site of the enzyme. The inhibition of the phage enzyme by histamine causes a quenching of the fluorescence of tryptophan, which is also true for egg white lysozyme (*160*). Oxidation of the single tryptophan residue with *N*-bromosuccinimide resulted in almost complete loss of activity, with similar rates of loss of activity at both pH values.

In contrast, the extent of inhibition with histamine, histidine, and tryptophan derivatives is different at the two pH optima; more inhibition was observed at pH 5.5. Other inhibitors, *N*-acetylglucosamine and glucose, which also inhibit egg white lysozyme (*160*), mainly inhibit the activity at pH 8.5. This indicates that the enzymic sites, besides the tryptophan residue involved, can be different at the two pH optima. Also, KCl and NaCl inhibit the purified enzyme variably at the two pH optima; this kind of inhibition was not seen in T2, T4, or λ phage lysozyme (*93*).

E. λ PHAGE ENDOLYSIN

1. *Purification*

Black and Hogness (*100*) purified the enzyme from *E. coli* K12, doubly lysogenic for λ and λdg. Imada and Tsugita (*103*) used a bacterial strain,

159. J. B. Howard and A. N. Glazer, *JBC* **244,** 1399 (1969).
160. N. Shinitzky, V. Grioara, D. M. Chipman, and N. Sharon, *ABB* **115,** 232 (1966).

K12λCI 857, lysogenic for a heat inducible λ phage (Fig. 10). Extensive purification was obtained by conventional procedures. Purification steps of the endolysin by Black and Hogness are as follows: either pH 6.4 extraction from acetone dry cells or pH 3.2 precipitation, removal of DNA with use of streptomycin sulfate, Amberlite XE64 chromatography, CM-Sephadex C50 chromatography, ammonium sulfate precipitation (40–60% saturation), hydroxyapatite chromatography. The specific activity of the final preparation increased about 300 times.

Imada purified the endolysin by the process almost the same as for T4 and T2 lysozymes except using additional CM-Sephadex chromatography and acetone precipitation. The final preparation is 250-fold more purified than the original lysate.

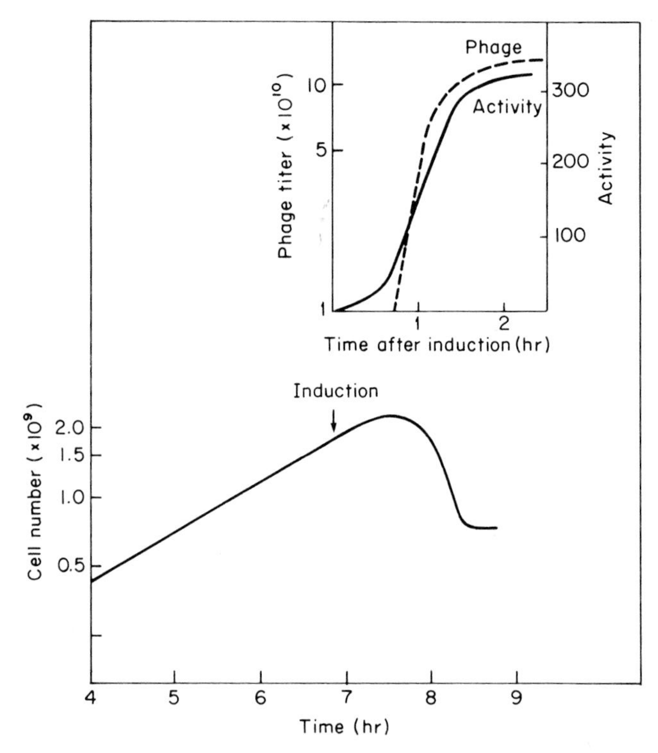

FIG. 10. Kinetics of appearance of λ-endolysin after heat induction. From Imada and Tsugita (*128*). A lysogenic bacteria of heat inducible λ phage, *E. coli* K12S (λCI 857) strain, was grown at 33° in Fraser medium. When the cell population reached 1.5×10^9/ml, the temperature was elevated to 43° for 20 min. The cultivation was further continued for additional 100 min at 37°. The induction profile is shown in inset in terms of phage titer on *E. coli* K12S and lytic activity for *E. coli* dried cell preparation.

2. *Physicochemical Properties*

a. Homogeneity. Black and Hogness' preparation of λ phage, when chromatographed on a hydroxylapatite column, gave a single peak of enzymic activity coinciding with the protein peak to $\pm 15\%$. Electrophoresis at pH 5 in starch gels subjected to polyacrylamide gel electrophoresis at pH 4.3 showed a single component. The enzyme showed a single symmetric peak in ultracentrifugation.

b. Molecular Weight. The sedimentation coefficient ($s_{20,w}$) of λ-endolysin is 2.06 S. The diffusion coefficient ($D_{20,w}$) was measured by analyzing the transient states during sedimentation to equilibrium; it was also calculated by measuring the spreading of the boundary formed in a synthetic boundary cell as a function of time. The values obtained for the $D_{20,w}$ were $10.2–10.5 \times 10^{-7}$ cm^2 sec^{-1}. Using the average value, and a partial specific volume, computed from the amino acid composition, of 0.732 cm^3/g a molecular weight for λ-endolysin of 1.78×10^4 was calculated. The molecular weight was also estimated from sedimentation equilibrium. Sedimentation at a relatively high speed and at lower speed gave values of 1.75×10^4 and 1.82×10^4, respectively (*100*).

Since egg white lysozyme undergoes a reversible dimerization between pH 5 and 9 (*158*), the sedimentation equilibrium of λ-endolysin was determined at various pH values. There was no significant changes as the pH was raised from 4.9 to 9.3.

Black and Hogness (*101*) calculated the minimum molecular weight of the λ phage endolysin to be about 17,873 from its amino acid composition, which is consistent with the value found independently by sedimentation analysis. From the amino acid composition derived from the present knowledge of sequential analysis (*103*), the minimum molecular weight is 17,558, which is almost the same as the value mentioned above.

c. Absorption Spectrum. From the spectrum of the purified λ-endolysin, the absorption coefficient at 280 nm was calculated to be 1.7–1.8 per milligram of enzyme per milliliter, and the ratio of the absorption at 280 nm to that at 255 nm is 2.0 (*103*).

d. Stability. The λ phage endolysin is more stable between pH 4.0 and 8.0 than T4 and T2 phage lysozymes. Below pH 4.0, a sudden decrease of enzymic activity was observed, while above pH 8.0 the decrease of enzymic activity occurs gradually until pH 9.5. The lysozyme is denatured to a large extent to the presence of Fe^{2+} ion. The addition of quinoline can prevent this denaturation and thus the addition of 10^{-3} *M* quinoline to the buffer employed through the purification process is recommended (*103*). Also, the presence of 10^{-3} *M* mercaptoethanol stabilizes the λ-endolysin since it protects the free sulfhydryl group of the cysteine residue in the protein (*100, 101, 103*).

3. *Chemical Properties*

a. Amino Acid Composition. λ Phage endolysin consists of 115 amino acid residues [Asp + Asn_{19}, Thr_{6}, Ser_{9}, Glu + Gln_{16}, Pro_{5}, Gly_{15}, Ala_{13}, Cys/2_{1}, Val_{6}, Met_{3}, Ile_{8}, Leu_{14}, Tyr_{5}, Phe_{5}, Trp_{3}, Lys_{12}, His_{3}, Arg_{12}]. This composition was obtained from a preliminary sequence study of the λ-endolysin (*103*). The results obtained from direct amino acid analysis carried out in two independent laboratories are as follows:

[Asp + Asn_{19}, Thr_{6}, Ser_{10}, Glu + Gln_{17}, Pro_{5}, Gly_{15}, Ala_{13}, Cys/2_{1}, Val_{7}, Met_{3} Ile_{8}, Leu_{13}, Tyr_{5}, Phe_{5}, Trp_{3}, Lys_{12}, His_{3}, Arg_{12}] total, 158 (*101*) [Asp + Asn_{17}, Thr_{7}, Ser_{9}, Glu + Gln_{16}, Pro_{5-6}, Gly_{14}, Ala_{13}, Cys/2_{1}, Val_{7-8}, Met_{3}, Ile_{8-9}, Leu_{13}, Tyr_{5}, Phe_{5}, Trp_{3}, Lys_{13}, His_{3}, Arg_{12}]; total, 153–157 (*103*).

Although the preparation methods of the enzyme were different, the amino acid compositions were in good agreement. The underlining indicates reported differences in the compositions from the preliminary sequential study. The amide content, roughly estimated from the amount of ammonia liberated after alkaline treatment, is less than 14 residues. From the sequential analysis, out of 19 moles of [Asp + Asn], 5 moles were found in the amidated form (Asn) and 14 moles were identified as aspartic acid. In 16 moles of [Glu + Gln], 9 moles were identified as glutamine and 7 moles as glutamic acid. All 14 moles of amide were confirmed. The number of strongly basic amino acid residues (Arg + Lys) is 24 moles, and the acidic amino acid residues about 21 moles, which indicates that the λ phage endolysin is a basic protein. The T4 and T2 phage lysozymes have 25 moles of basic amino acid residues and 18 moles of acidic amino acid residues, suggesting that the basic nature of these lysozymes is stronger than that of λ-endolysin. Some support for a weaker basic character of the λ-endolysin has been provided by its chromatographic behavior on IRC-50, and electrophoresis.

Black and Hogness compared the amino acid composition of λ phage endolysin with that of T4 phage lysozyme on the basis of five functional groups: aliphatic, hydroxyl, aromatic, acidic, and strongly basic groups. A similarity of amino acid composition of these two lysozymes in the groups compared was suggested, but little similarity was observed between the sequences of λ phage endolysin and egg white lysozyme. By a reaction with 5,5′-dithio-bis(2-nitrobenzoic acid), one mole of cysteine was found to be a free –SH group in the λ-endolysin (*101*).

b. Amino Acid Sequence. The NH_2-terminal sequence was determined by the cyanate method as methionyl-valine (*102*). The COOH-terminus was determined with hydrazinolysis as valine (*102*); carboxypeptidase did not act on this protein (*102, 103*). The protein containing one NH_2-terminal and two internal methionine residues was subjected to cyanogen bromide digestion. In addition to one homoserine residue (derived from the NH_2-terminal methionine), two homoserine residues containing peptide were separated and analyzed for their amino acid composition and NH_2- and COOH-terminal residues. From these results, together with the amino acid composition of λ phage endolysin, Black and Hogness (*102*) indicated the order of the three cyanogen bromide peptides as $\underset{1}{\text{Met}}\text{–}\underset{2}{\text{Val}}\text{–}\underset{3}{\text{Glx}} \cdots \underset{14}{\text{Met}}\text{–}\underset{15}{\text{Glx}} \cdots \underset{115}{\text{Met}}\text{–}\underset{116}{\text{Ile}} \cdots \underset{159}{\text{Val}}$.

Imada and Tsugita (*103*) have been sequencing the λ phage endolysin using the above method (cyanogen bromide) in addition to the conventional partial hydrolysis methods of tryptic, chymotryptic, peptic, and thermolytic (digestion with thermolysin) digestions. The sequence so far analyzed is shown in Fig. 11. The composition of the cyanogen bromide peptides are quite similar to those reported by Black and Hogness, but the fifteenth residue appears to be leucine instead of glutamic acid or glutamine as reported by Black and Hogness. The NH_2-terminal dipeptide sequence and the COOH-terminus as well as the amino acid next to the third methionine are identical in both studies. These amino acid residues or sequences are underlined in Fig. 11.

c. Homology between λ Phage Endolysin and T2 and T4 Phage Lysozymes. Black and Hogness (*102*) suggested the existence of structural similarities between λ phage endolysin and T2 and T4 phage lysozymes from comparison of their amino acid compositions and the identical amino acid at both NH_2-termini. Double diffusion of λ-endolysin and T4 lysozyme against anti-λ-endolysin serum and anti-T4-lysozyme-γ-globulin gave homologous antigen-antibody precipitin lines, but there was no evidence of heterologous interaction by this technique. The T4 lysozyme activity was indeed inhibited by the anti-λ-endolysin serum although the cross reaction was very weak; the amount of antiserum required per enzyme molecule was about 100-fold more than that required for the homologous reaction. No inhibition of the λ-lytic

5 10 15
Met-Val-Glu-Ile-Asn-Asn-Gln-Arg-Lys-Ala-Phe-Leu-Asp-Met-Leu-Ala-Trp(Asx,Asx,

20 25 30 35
Thr,Ser,Glx,Gly,Gly,Arg,Glx,Lys,Thr,Arg,Asx,Gly,His,Tyr,Asx,Val,Ile,Val,Gly,

40 45 50 55
Leu,Gly,Glu)Phe-Thr-Asp-Tyr-Ser-Asx-Pro-His-Arg-Lys-Leu-Val-Thr-Leu-Asn-Pro-

60 65 70 75
Lys-Leu-Lys-Ser-Thr-Gly-Ala-Gly-Arg-Tyr-Gln-Leu-Leu-Ser-Arg-Trp-Asp-Ala-Tyr-

80 85 90 95
Arg-Lys-Gln-Leu-Gly-Leu-Lys-Asp-Phe-Ser-Pro-Lys(Gln,Ser)Asp-Ala-Val-Ala-Leu-

100 105 110
Gln-Gln-Ile-Lys-Glu-Arg-Gly-Ala-Leu-Pro-Met-Ile-Asp-Arg-Gly-Asp-Ile-Arg-Gln-

115 120 125 130
Ala-Ile-Asp-Arg-Cys(Asx,Ser)Ile-Trp-Ala-Ser-Leu(Pro,Gly,Ala,Gly)Tyr-Gly-Gln-

135 140 145 150
Phe-Glu-His-Lys-Ala-Asp-Ser-Leu-Ile-Ala-Lys-Phe-Lys-Glu-Ala-Gly-Gly-Thr-Val-

155
Arg-Glu-Ile-Asp-Val

FIG. 11. The preliminary amino acid sequence of λ phage endolysin (*160a*). From Imada and Tsugita (*128*).

activity by anti-T4-lysozyme-γ-globulin was observed, even at levels about 20-fold greater than necessary for the complete inhibition of an equivalent weight of T4 phage lysozyme. Black and Hogness concluded that immunochemical cross-reactions occur between the λ and T4 enzymes and that the weakness of the cross-reaction between configurationally similar chains might result from different but functionally related side groups at various positions along the peptide chain.

In our laboratory, Inouye (unpublished data) has examined the effect of anti-T4 lysozyme on the enzymic activities of T4, T2, λ, and egg white lysozyme. Again, no inhibition of the λ-endolysin and egg white lysozyme was observed. Thus, structural similarities were not indicated. Comparison of the sequences of λ-endolysin and T4 lysozyme show no apparent similarity except for a few di- or tripeptide sequences such as Arg–Trp–Asp (70–72 in λ-endolysin and 125–127 in T2 and T4 lysozyme) which are known to be nonessential parts for the enzymic activity in T4 lysozyme, Arg–Cys (116–117 in λ-endolysin and 96–97

160a. The complete sequence has been established recently in our laboratory as follows: 18 → 41; Ser-Glu-Gly-Thr-Asp-Asp-Gly-Arg-Gln-Lys-Thr-Arg-Asn-His-Gly-Tyr-Asp-Val-Ile-Val-Gly-Gly-Glu, 47: Asx → Asp, 89 → 90: Ser-Gln, 120 → 121: Ser-Asn, and 127 → 130: Pro-Gly-Ala-Gly.

in T4 lysozyme), Phe–Leu–Asp–Met–Leu (11–15 in λ-endolysin) and Ile–Phe–Glu–Met–Leu (3–7 in T4 lysozyme).

4. *Catalytic Properties*

The optimal pH for enzyme action of λ phage endolysin is about 6–7 (*100*). The catalytic specificity of the λ-endolysin remains obscure. It was found 10 times more active than that of egg white lysozyme with *E. coli cell* walls as substrate. This value was obtained with impure preparations by Work (*161*) and Jacob and Fuerst (*95*) and was confirmed with purified enzyme by Black and Hogness (*101*). From our observations, purified λ-endolysin is 317 times more active than egg white lysozyme with *E. coli* cell walls (*103*). The latter value is almost equal to that of T4 lysozyme (*27*). Black and Hogness (*100*) tested tri- and penta-*N*-acetylglucosamine from chitin, but no catalytic activity was found with the λ-endolysin. In a previous study on an impure preparation of λ-endolysin with cell walls of various bacteria, Work and her collaborator (*96, 162*) isolated the same products as those produced by egg white lysozyme but observed some dissimilar products when using all cell walls of *E. coli*. She also observed (*161*) that the crude preparation of λ-endolysin is active on cell wall prep-

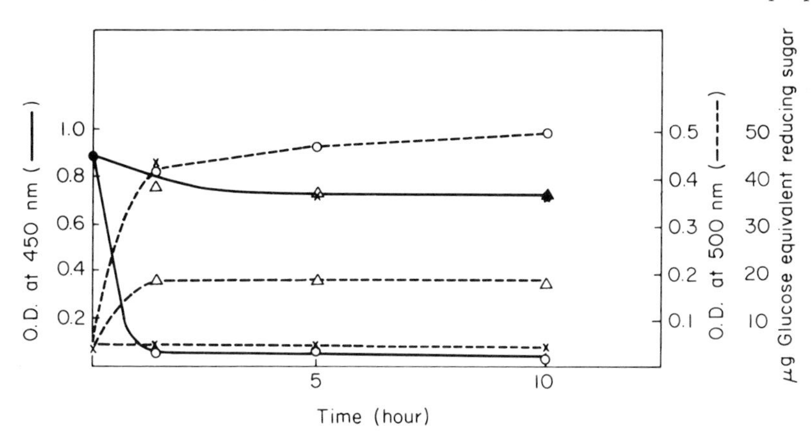

FIG. 12. Lytic activities of T4 and T2 phage lysozymes and λ phage endolysin. From Imada and Tsugita (*128*). The reaction mixture contained 10 mg *M. lysodeikticus* cell wall in 15 ml of 1:15 *M* Na phosphate buffer (pH 7.0). Lysozyme proteins of egg white, T4 phage, and λ phage endolysin were added at the final concentration of 33, 42, and 28 μg/ml, respectively. One milliliter aliquots were rapidly frozen in CO_2-acetone at indicated intervals in order to stop the reaction. The appearance of aldehyde groups resulting from enzyme cleavage was tested by Nalson–Somogyi reaction taking glucose as a reference. The activities are expressed by the decrease of optical density at 450 nm and the increased glucose equivalent reducing sugar.

161. E. Work, *BJ* **76,** 38p (1960).
162. E. Work and M. Lecadet, *BJ* **76,** 39p (1960).

arations of many strains of *E. coli*, *Salmonella typhimurium*, *Staphylococcus aureus*, and *B. megaterium* but not on *B. cereus* or *Micrococcus lysodeikticus*. Black and Hogness (*100*) found that the purified λ-endolysin has no effect on *M. lysodeikticus* cell wall; the catalytic rate of this enzyme is less than 0.001 the rate observed for egg white lysozyme. However, Imada (*103*) clearly observed a decrease in optical density of a suspension of *M. lysodeikticus* cell wall preparation incubated with purified enzyme, as shown in Fig. 12. The rate of the decrease in turbidity by λ-endolysin is almost the same as by T4 lysozyme. No increments in reducing sugar, muraminisitol after reduction, or free NH_2 group in the reaction mixture were detected. The enzymic action is markedly inhibited in citrate or barbital buffer.

F. Polysaccharide Depolymerase of F Series Phage

Phages F1, F5, F26, and F27 were selected for synthesizing depolymerase from sewage, using *E. coli* strains producing high mucoid, S53 and S23, derived from the K12 strain (*104*). The phages were selected because they formed plaques surrounded by a large halo when grown on mucoid bacterial strains (*163*). A correlation was found between the size of the halos and the amount of depolymerase formed. The host range of such phages was found to be specific and virulent for all K12 strains tested.

1. *Enzyme Assay for Polysaccharide Depolymerase*

A rough estimation of polysaccharide depolymerase activity was carried out by plate method on old cultures of *E. coli* strain S53 incubated at 30° for 48 hr on nitrogen-deficient medium and stored at 0° for at least 7 days before use. After addition of the enzyme solution, the plates were incubated and examined for dissolution of polysaccharide.

The enzymic activity was also assayed by measuring the decrease in viscosity of a polysaccharide solution, using a slime from *E. coli* strain S53, purified by the method of Wilkinson *et al.* (*164*). The prepared polysaccharide slime contained glucose, galatose, fucose, glucuronic acid, and glucuronolactone (see also Section I,D).

2. *Partial Purification of Depolymerase*

The phage-infected bacterial culture, obtained either by agar layer techniques (*165*) or by an aerated liquid culture, was dialyzed, con-

163. B. H. Park, *Virology* **2**, 711 (1956).
164. J. F. Wilkinson, W. F. Dudman, and G. D. Aspinall, *BJ* **59**, 446 (1955).

centrated, and centrifuged to remove the cell debris. The supernatant was then fractionated with ammonium sulfate at 40–50% saturation. The active material was applied to a column of DEAE-cellulose eluted with a changing salt concentration at neutral pH. The enzymically active material was obtained at a NaCl concentration of 0.1 *M*. Differential centrifugation showed the depolymerase thus partially purified to be associated with the viable phages (*105*).

3. *Stability of the Partially Purified Depolymerase*

The enzymic activity is entirely destroyed after 5 min at 70°. Exposure of the enzyme to pH values above 7.5 and below 6.0 also results in rapid loss of activity, and a decrease of phage titer was observed to be accompanied by a decrease in enzymic activity (*105*). The labile character of this enzyme is contrary to the phage-induced *Klebsiella aerogenes* depolymerase (*17*).

4. *Catalytic Properties and Biological Significance*

The enzyme acts on the substrate polysaccharide, resulting in a decrease in viscosity but no marked increase in reducing groups. Low molecular weight products are not produced by this reaction. Almost all the weight of the original polysaccharide can be precipitated by the addition of cold acetone after enzyme digestion. From an analysis of the composition of the precipitate and the soluble fraction of the digest, rather large fragments were demonstrated.

This enzyme is active on polysaccharides from various strains of *E. coli* including *E. coli* S17 (a strain in which none of the phages is capable of multiplication), as well as on polysaccharides from *Aerobacter cloacae*. However, the enzyme is not active on polysaccharides from *Klebsiella aerogenes*, and thus the depolymerase shows substrate specificity. The specificity is not necessarily correlated to phage specificity, indicating that the depolymerase does not play an essential role for phage infection to the bacteria. This possibility is confirmed by the fact that the substrate of the phage depolymerase is slime polysaccharide which does not occlude the bacterial surface. However, a particular phage (F6), which does not produce any depolymerase, is unable to infect the bacterial strain S53C, which is a capsulated variant of the S53 strain. The depolymerase has no effect on bacterial viability, suggesting that it is not essential for phage. It disperses the polysaccharide slime to expose the phage to the host surface.

165. M. H. Adams, *in* "Bacterio Phages," p. 504. Wiley (Interscience), New York, 1959.

III. Lytic Enzymes of Other Gram-Negative Bacteriophages

Lytic enzymes are known to be produced when infected by bacteriophages of other gram-negative bacterial strains. Investigations have concentrated on polysaccharide depolymerase partly because phage-induced polysaccharide depolymerase was first found in gram-negative bacteria, *Klebsiella pneumoniae* (*17*), which generally produce a marked amount of phage specific depolymerase. Furthermore, investigations were accelerated by the practical need for chemotherapeutic applications of the depolymerase (*17*). Apart from the *E. coli* phages, lysozyme activity has been reported only in very rare cases (*166*).

A. *Klebsiella pneumoniae* Phage Polysaccharide Depolymerase

In 1940, Humphries (*167*) described a preliminary study of the lytic enzyme production by a phage active on *Klebsiella pneumoniae* A, separation of the enzyme and the enzymic specificity. Adams and Park (*17*) extended this study and reported the properties of a similar enzyme, the *K. pneumoniae* B specific phage (*168*).

1. *Preparation of the Enzyme and Assay Methods*

The lysate was made by agar layer technique (*169*) using phage Kp and *K. pneumoniae* B (*168*) as host bacterium. From the lysate the phage was removed by high-speed centrifugation, by UV irradiation, and by filtration through Millipore filter membranes. The enzyme was scarcely affected by the above treatments, but it was completely precipitated by 50% saturated ammonium sulfate. The assays routinely involve visual estimation of the change in appearance of fully developed bacterial lawn on the plate by spotting a drop of the enzyme (*17*). The enzymic activity is also assayed conveniently by measurement of the decrease in viscosity (*17*).

2. *Enzyme Properties and Role*

Measurement of the activity is usually made at neutral pH. The enzyme is most active at 37° and less active at higher temperatures. The

166. L. W. Cohen, *Bacteriol. Proc.* p. 159 (1970).
167. J. C. Humphries, *J. Bacteriol.* **56,** 683 (1948).
168. B. H. Park, *Virology* **2,** 711 (1956).
169. M. Swantrom and M. H. Adams, *Proc. Soc. Exptl. Biol. Med.* **78,** 372 (1951).

enzyme does not affect the viability of the bacteria but markedly reduces the volume occupied by the bacterial cells (*161*), indicating that it reacts only on the surface of the bacteria, particularly on the strips or the capsule (*17*).

The enzyme reduces the viscosity of a capsular polysaccaride preparation and slightly increases the reducing sugar to about 1% of that expected from the complete hydrolysis of the polysaccharide. The product displays the precipitin reaction with the antibody against the original polysaccharide, indicating that the hydrolyzed products are still of a relatively large size.

3. *Others*

A part of the enzyme is attached to the phage particle as shown by the sedimentation analysis and by the fact that antiphage serum neutralizes the enzymic activity. Adams and Park (*17*) noted that the enzyme serves an essential function in the life cycle of the phage, permitting the phage to penetrate through the capsule to reach the host cell surface; the enzyme attached to the phage may facilitate efficient adsorption of phage Kp to the encapsulated host cells. Whether or not the enzyme molecules are built into the phage particle is unknown.

It is interesting to note that the enzyme can serve as a chemotherapeutic agent for mice infected with the strain of bacteria in which the enzyme was produced (*17*). The enzyme is produced under the control of phage genome and partly under the control of the host bacterial genotype.

B. Polysaccharide Depolymerase of Phage-Infected *Pseudomonas*

Phage-induced lytic enzymes were found in *Pseudomonas aeruginosa* (*170*) as well as in *Pseudomonas putida* (*171*). The enzymes are capable of digesting polysaccharides of the cell capsule or the extracellular slime, resulting in a decrease of viscosity. These enzymes are not detected in the uninfected host cell or in lysates of the same host cell when infected by other phages. Thus, these polysaccharide depolymerases were concluded to be synthesized in the bacterial cell under the direction of a specific phage genome.

1. *Polysaccharide Depolymerase of Pseudomonas aeruginosa Phage*

a. Enzyme Assay. Pseudomonas aeruginosa strain B1 is cultivated on sheets of cellophane overlaying the culture. Polysaccharide is ex-

170. P. F. Bartell, T. E. Orr, and P. Blakemor, *Bacteriol. Proc.* p. 142 (1964).
171. A. M. Chakrabarty, J. F. Niblack, and I. O. Gunsalus, *Virology* **32**, 532 (1967).

tracted from the slime layer precipitated by ethanol–acetone mixture (1:1). The precipitate is dialyzed, lyophilized, and serves for substrate (*172, 173*). The depolymerase activity is measured by the release of hexosamine from the polysaccharide substrate, and also by measurements of viscosity, using the polysaccharide substrate.

b. Purification. Large quantities of the crude enzyme were prepared from soft agar layer cultures of bacterial cells (*P. aeruginosa* strain BI) infected with phage 2. The cells were harvested, homogenized, and centrifuged. The supernatant, which contain both phages and the crude enzyme, was subjected to freezing and thawing and centrifuged repeatedly at high speed in order to remove the phages. Then the resulting supernatant fluid was fractionated with ammonium sulfate between 30 and 45% and passed twice through Sephadex G-200. The final step resulted in a purification of 1700-fold and a product free from bacteriophage. Concentration was carried out by dialysis against a suspension of polyethylene glycol and then by ultrafiltration with collodion sac (*172, 173*).

c. Homogeneity and Molecular Weight. The homogeneity of the purified preparation was proved by acrylamide gel electrophoresis (pH 8.0) and Sephadex G-200 gel filtration. Its molecular weight, estimated by gel filtration, was about 180,000 (*172, 173*).

d. Stability of the Depolymerase. About 50% of the activity is lost upon heating to 70° for 5 min, whereas the partially purified enzyme retained 90% of its initial activity following same treatment. From this observation, Bartell (*173*) suggested that depolymerase in *Pseudomonas* exists in two forms, one "freely diffusible protein," which is the less stable purified protein, and another protein "firmly attached to the phage particle," which is the partially purified enzyme. Similar distinctions have been made for other phage depolymerases (*17, 174*).

e. Catalytic Properties. The initial velocity of the enzymic reaction increases with elevation of temperature up to 45°. The optimal pH of the enzymic activity is 7.5. The enzyme reacts with polysaccharides of only certain strains of *P. aeruginosa* (10 out of 120 strains tested). This enzyme specificity is not always concomitant with susceptibility to phage.

The enzymic reaction with polysaccharide from *P. aeruginosa* BI results in a decreased viscosity as well as in a measurable increase in the level of hexosamine, hexoses, and reducing substances. The enzyme is

172. P. F. Bartell, T. E. Orr, and G. K. H. Lam, *J. Bacteriol.* **92,** 56 (1966).
173. P. F. Bartell, *JBC* **243,** 2077 (1968).
174. C. Eklund and O. Wyss, *J. Bacteriol.* **84,** 1209 (1962).

also active to mucocomplex obtained from *M. lysodeikticus* and appears with the phage at the latent period of phage infection.

2. *Polysaccharide Depolymerase of Pseudomonas putida*

Chakrabarty *et al.* found that a polysaccharide depolymerase is formed in *P. putida* after infection with a phage (*171*). The phage Pf15 forms plaques with large halos when plated to *P. putida* strain C1B cells. The enzyme can be separated from the phage by differential centrifugation and inactivation of residual phages by exposure to UV light. Further purification followed the method employed for *P. aeruginosa.*

The purified enzyme decreases the viscosity of polysaccharides of the host strain cells but not of the cells of the other strain, *P. putida* ClS. The enzyme releases a small quantity of reducing sugar, suggesting internal cleavages of the substrate. The production of the depolymerase was shown to be under control of the phage genome by the isolation of a depolymerase-less mutant phage by genetic recombination among host range and plaque mutants of Pf15.

C. *Azotobacter* PHAGE POLYSACCHARIDE DEPOLYMERASE

Capsule digesting enzymes were found in the lysate of *Azotobacter* bacteriophage (*174*) which forms large halos surrounding the plaques (*175*).

1. *Enzyme Assay and Purification of Enzyme*

The capsular polysaccharide is extracted from the cells and capsules with 10% NaCl and precipitated by adding acetone and ethanol. After three cycles of reprecipitation, the polysaccharide is lyophilized. The enzyme assay is carried out by measuring the viscosity in flow time.

Azotobacter agilis (*vinelandii*) and its phage A-22 were propagated. The resulting lysate was centrifuged repeatedly at various speeds to remove cell debris and the phages, and concentrated by dialyzing against polyethylene glycol.

2. *Enzyme Properties*

The enzyme digests the capsular polysaccharide with a concomitant increase of reducing sugar (*176*). The activity is maximum between pH 7.5 and 8.5 with a sharp drop below pH 7.

175. J. T. Duff and O. Wyss, *J. Gen. Microbiol.* **24**, 273 (1961).
176. T. S. Barker, C. Eklund, and O. Wyss, *Bacteriol. Proc.* p. 14 (1968).

The enzyme appears only after the phages lyse the host cells, but the enzyme formation in the bacterium begins soon after phage invasion. A sucrose density gradient centrifugation showed the enzyme to be associated with the phage particle and a purified phage preparation still contained a small fraction of enzyme. The amount of free enzyme was 1000-fold of that incorporated into the newly formed phage.

The enzyme produced in *Azotobacter agilis* A-22 phage digests the capsular polysaccharide from many but not all strains of *A. agilis* and *A. chroococcum.*

D. *Aerobacter cloacae* Polysaccharide Depolymerase

Sutherland and Wilkins (*105*) reported the presence of a polysaccharide depolymerase in the lysate of *Aerobacter cloacae* (NCTC5920) infected by the newly isolated specific phages F12, F13, and F14, together with the depolymerases of *E. coli* phages. The depolymerases are active against the coli polysaccharide but the enzyme induced by the phages of *A. cloacae* are not active to *E. coli* cells. No details were reported.

E. Lytic Enzyme of *Salmonella* Phage

Samonella typhimurium infected by phage 22 produces a marked amount of lytic enzyme. A preliminary survey was reported by Cohen (*166*) on the character of the enzyme which had a lysozyme-like action.

IV. Lytic Enzymes of Gram-Positive Bacteriophage

A. Staphylococci Phage Lytic Enzymes

Lytic enzymes from phage-infected bacteria for a staphylococcal system were discovered early in the history of phage research by d'Herelle in 1926 (*3*) and were confirmed by Bronfenbrenner and Muchenfuss (*4*). More recently, phage-induced cell wall and capsule-lysing enzymes were investigated in a *Staphylococcus aureus* system by Ralston and collaborators (*29, 177–181*). The lytic enzymes are produced with

177. D. J. Ralston, B. Baer, and A. P. Krueger, *Federation Proc.* **14**, 475 (1955).
178. D. J. Ralston, B. Baer, M. Lieberman, and A. P. Krueger, *Proc. Soc. Exptl. Biol. Med.* **89**, 502 (1955).
179. D. J. Ralston, B. Baer, M. Lieberman, and A. P. Krueger, *J. Gen. Physiol.* **41**, 343 (1957).
180. D. J. Ralston, M. Lieberman, B. Baer, and A. P. Krueger, *J. Gen. Physiol.* **40**, 791 (1957).
181. A. Gratia and B. Rhodes, *Compt. Rend. Soc. Biol.* **89**, 1171 (1923).

phages P14 and P1 and hosts *S. aureus* K1 and 145. The phage-induced lytic enzymes are designated "virolysins." Uninfected host bacteria produce another lytic enzyme, called "autolysin," when they are autolysed (*182*). These two lytic enzymes show the following similarities: Both enzymes lyse micrococci. Cells lyse only after they have been subjected to a damaging treatment. A constant proportion of cells can be lysed by both enzymes, independent of their concentration, and the residual cells remain resistant to the enzymes. The reaction velocity is proportional to the enzymic concentration and increase with temperature. Both enzymes are unaffected by phage antiserum. The behavior of both enzymes, when inhibited by various chemicals or by elevating temperature, is approximately similar. Both enzymes are precipitated by 40% saturated ammonium sulfate (*183*). However, the enzyme can be distinguished by their different pH optima, by antigenic cross-reaction, and by specificities for *Micrococcus lysodeikticus* (*185, 186*). "Lysis from without" occurs by phage infection in the presence of an extracellular lytic enzyme, either virolysin or autolysin (*179*).

Another lytic enzyme was found by infection of another kind of phage, PAL, on *S. aureus* 53ps. This enzyme is different from virolysin in catalytic character in that it can digest intact cells (*184*).

1. *Virolysin, Lytic Enzymes of Staphylococcus aureus Phage P1 and P14*

a. Enzyme Assay. Logarithmically grown cells of *S. aureus* K1 were harvested in saline solution and killed by heating to 56°–100° for 30 min. After this treatment, they could be stored in the cold. Lytic activity was measured by turbidimetry (*185*).

b. Partial Purification. The lytic enzyme virolysin is produced by infecting either *Staphylococcus aureus* K1 or 145 strain with the specific phage P1 or P14. The lysate was centrifuged to remove phage. The supernatant was then subjected to $(NH_4)_2SO_4$ fractionation (40% saturation) (*182*).

c. Stability of the Enzyme. Virolysin is unstable even at 37° and also labile against proteolytic action (*179*).

d. Catalytic Properties. Virolysin shows an optimal pH at 7.0–7.5, while the pH optimum of autolysin is 6.0–6.5. Sulfhydryl reagents inhibit

182. M. Lieberman, Ph.D. Dissertation, University of California, Berkeley, California, 1956.

183. D. J. Ralston, *Bacteriol. Proc.* p. 126 (1966).

184. S. A. Sonstein, J. M. Hammel, and A. Bondi, *Bacteriol. Proc.* p. 41 (1969).

185. J. S. Murphy, *Virology* **11**, 510 (1960).

the virolysin activity and the inhibition can be reversed by the addition of cysteine. EDTA also inhibits the activity but can be reversed by various divalent cations such as 1 m*M* Ca^{2+}, Mn^{2+}, Co^{2+}, Zn^{2+}, and Mg^{2+}. A higher Mg^{2+} concentration (5 m*M*) negatively affects this reactivation. Heat-killed logarithmic phase cells are more sensitive than heat-killed resting cells. Acetone treatment increases the sensitivity of the heated resting cells. Virolysin lyses living cells only if the homologous phage is present. Virolysin acts on all *S. aureus* strains but does not act on various heat-killed heterologous genera cells. The bacteria *Micrococcus lysodeikticus*, *Pseudomonas fluorescens*, *P. aeruginosa*, and *P. species* undergo slight but reproducible lysis by both virolysin and autolysin. *Micrococcus lysodeikticus* is found to be more susceptible to autolysin than to virolysin. The individual specificity of the remaining species was not studied.

Virolysin cannot act on *S. aureus* cells unless they are previously damaged by treatment with heat, acetone, UV light, or bacteriophages. Virolysin releases small peptide fragments from the peptidoglycan in cell walls, which causes the destruction of the phage receptor site (*182*).

e. The Role of the Lytic Enzyme in the Life Cycle of Phage Infection. The appearance of virolysin in the host cell represents *de novo* enzyme formation following infection (*179, 182*). Virolysin is detected first within 10–15 min (latent period 40–50 min) and increases linearly until lysis, while autolysin remains constant during the infection. The cell wall seems to be resistant to a high internal concentration of virolysin, as well as to external addition of virolysin, until the end of the latent period of phage life cycle.

An unknown "sensitizing reaction" must occur for the enzymes to be able to carry out their function, such as the removal of inhibitors, repressors, or stimulators. Although virolysin (as well as cell autolysin) appears to be one of the key lytic agents, external addition of large amounts of either phage or virolysin alone does not cause lysis from without but added together they accomplish the lysis. Absorption of phages to the host cell may sensitize the cell and render it susceptible to destruction by extracellular lytic enzymes. The detailed mechanisms of phage penetration, cell sensitization, and lysis at the latent period remain obscure.

2. *Lytic Enzyme of Staphylococcus aureus Phage PAL*

Sonstein *et al.* (*184*) found a lytic enzyme produced in *S. aureus* strain 53ps infected by a phage PAL. This enzyme, which is different from virolysin, can lyse intact cells. Sensitization of the cells was found to be unnecessary. Except for the above substrate specificity, the enzyme was found to be similar to virolysin. It was purified from phage-infected lysate by ammonium sulfate precipitation (50% saturation) and repeatedly passed through Sephadex G-200. The enzymic reaction liberates free amino terminal groups, part of which are dialyzable but does not result in the increment of reducing groups. A preliminary survey suggested that enzymic activity involves a specific peptidase activity.

B. *Micrococcus lysodeikticus* Phage-Induced Lytic Enzyme

A lytic phage, called N1, was originally isolated from sewage (*186*). It forms plaques with a clear central zone surrounded by a halo which increases its width with time of incubation on *Micrococcus lysodeikticus* strain 1, a strain sensitive to both egg white lysozyme and N1 bacteriophage. N1 phage also infects *M. lysodeikticus* strain BLR which, however, is insensitive to egg white lysozyme. The enzyme lyses cells and increases free amino groups in the substrate. The enzyme cannot be isolated directly from the phage particles but can be isolated from the phage-infected cell lysate (*187*).

1. *Purification*

The crude lysate was fractionated by ammonium sulfate precipitation (20–40% saturation), acid precipitation (pH 5.2–4.8), and selective adsorption of contaminating proteins on a calcium phosphate gel. The specific activity was increased 45- to 50-fold (*2*). This partially purified preparation is almost free from the initial phage titer (*187*).

2. *Catalytic Properties*

The N1 lytic enzyme is active against living cells as well as on lyophilized cell wall preparations of *M. lysodeikticus* (*186, 187*). Young cells are most sensitive and the sensitivity decreases to a minimum for stationary phase cells. Heated cells are more sensitive than nonheated ones. The enzyme is also active against *Sarcina flava* and *Sarcina subflova* but not to *Aerobacter aerogenes*, *Proteus vulgaris*, *E. coli*, *Pseudomonas aeruginosa*, *Serratia marcescens*, *Streptococcus lactis*, *S. faecalis*, *Lactobacillus arabinosus*, *Bacillus subtilis*, *B. polymyxa*, *B. cereus*, *B. megaterium*, *Staphylococcus aureus*, and *Sarcina lutea*.

186. J. M. Goepfert and H. B. Naylor, *Bacteriol. Proc.* p. 142 (1964).
187. J. M. Goepfert and H. B. Naylor, *Virology* **1**, 701 (1967).

Optimal conditions for the lytic activity are pH from 6.5 to 7.0, from 45° to 50°, and with 0.06 *M* NaCl. A rapid decrease of activity results below 0.05 *M* and a slight decrease above 0.06 *M* NaCl. The lysis does not release detectable amounts of amino sugars but releases free NH_2-groups at a linear rate. The amount of NH_2 groups released is not affected by prior treatment with egg white lysozyme. Formylation and acetylation of the cell walls results in a stimulation of the rate of lysis.

The strict specificity of the lytic enzyme seemed to be related to the amino acid composition of the peptidoglycan in the cell wall. The N1 enzyme lacks proteolytic and amidase activity, and a preliminary survey indicated a specific peptidase activity only.

Acting synergistically, egg white lysozyme and N1 enzyme caused lysis of cells (*M. lysodeikticus*, strain BLR) that are resistant to either enzyme acting separately. Studies of this synergistic action suggest that the role of the egg white lysozyme may be to affect the chemical group in the cell wall that is responsible for the wall's resistance to the N1 lytic activity.

While virolysin (*Staphylococcus* P1 phage) (*182*) and the lytic enzyme of *B. megaterium* lyse living cells only when the homologous phage is present, the N1 lytic enzyme does not require the presence of phage. However, the rate of lysis by N1 enzyme was observed to be stimulated when N1 phage was present at a phage-to-cell ratio of 10:1 or greater. The reason for this observation was not well understood, yet it was suggested that the phage infection may interfere with the normal cell metabolism, including certain effects on enzymes involved in cell wall synthesis; subsequently, this interference was correlated to the action of this particular enzyme (*187*).

3. *Stability*

The enzyme at 45° is most stable between pH 8 and 9. All lytic activity is destroyed by heating the enzyme to 55° for 5 min regardless of the pH. Chelating agents for divalent cations inhibit the lytic activity, but the inhibition is reversed by the addition of an excess of Ca^{2+}, Mg^{2+}, or Mn^{2+}. A sulfhydryl group is reversibly inhibited by SH reagent. The enzymic activity is not impaired by UV irradiation nor by the addition of the phage antiserum (*187*).

C. Streptococcal Phage-Induced Lytic Enzyme

Evans described a phenomenon of "nascent lysis" in a streptococcal phage system in 1934 (*18*). Krause (*30*) and Maxted (*31*) independently reported in 1957 that lysate from the infection of bacteriophage on group

C streptococci makes group A β-hemolytic streptococci lyse. Group A streptococci is the most toxic group for humans, at least in part because of a serologically distinguishable M protein. In the A group, the lytic activity is separated from the phage activity. Since 1957, various aspects have been studied by several research groups using a variety of streptococcal phage systems (*188–202*). One of the lytic enzymes was designated as mularysin, and the cell wall components solubilized by its action were characterized (*188, 189, 203, 204*). The results suggested an "endo-hexosamidase" for this enzyme.

1. *Enzyme Assay*

a. Assay with Living Cell. Group A streptococci, strain S23, are harvested at the logarithmic phase and washed and stored at —30°. Assay is performed by turbidimetry at 660 nm (*195*).

b. Assay using Cell Walls. The cells are shaken in a concentration of 10% (wet weight per volume) with an equal volume of glass beads. To remove lipid contaminants, the shaken cells are treated with laurylsulfate repeatedly, then washed and lyophilized. The cell wall preparation is further digested with trypsin to remove the type-specific antigens, and chloroform is added to suppress growth of adventitious bacteria.

For the assay of lytic activity, turbidimetry at 660 nm is performed with use of the sonicated suspension of the lyophilized cell wall preparation in the presence of EDTA and thioethanol (*28*).

c. Radiometric Assay. Group A streptococci are grown in the presence of ^{3}H-acetate and are harvested and washed. A 20% suspension of cell

188. F. S. Kantor and M. McCarty, *J. Exptl. Med.* **112,** 77 (1960).
189. R. M. Krause and M. McCarty, *J. Exptl. Med.* **114,** 127 (1961).
190. E. H. Freimer, B. M. Krause and M. McCarty, *J. Exptl. Med.* **110,** 853 (1959).
191. A. Markowitz and A. Darfman, *JBC* **237,** 273 (1962).
192. E. H. Freimer, *J. Exptl. Med.* **117,** 377 (1963).
193. L. D. Zeleznick, J. J. Boltralik, S. S. Barkulis, C. Smith, and H. Heymann, *Science* **140,** 400 (1963).
194. D. Kessler and R. M. Krause, *Proc. Soc. Exptl. Biol. Med.* **114,** 822 (1963).
195. C. C. Doughty and J. A. Hayashi, *J. Bacteriol.* **83,** 1058 (1961).
196. E. N. Fox and M. K. Wittner, *J. Bacteriol.* **89,** 496 (1965).
197. W. R. Maxted, *J. Gen. Microbiol.* **12,** 484 (1955).
198. B. Reiter and J. D. Oram, *J. Gen. Microbiol.* **32,** 29 (1963).
199. E. Kjems, *Acta Pathol. Microbiol. Scand.* **44,** 429 (1958).
200. E. N. Fox, *Bacteriol. Proc.* p. 142 (1964).
201. S. S. Barkulis, C. Smith, J. J. Boltralik, and H. Heymann, *Bacteriol. Proc.* p. 32 (1964).
202. J. Naylor and J. Czulak, *J. Dairy Res.* **23,** 126 (1956).
203. R. M. Krause, *J. Exptl. Med.* **108,** 803 (1958).
204. W. R. Maxted and H. Gooder, *J. Gen. Microbiol.* **18,** xiii (1958).

walls is ruptured by sonic treatment, followed by two freeze-thawing cycles. The cellular debris is collected, washed, and treated with trypsin followed by hyaluronidase. After appropriate washing, the cell wall preparation can be stored at —25° in a suspension of buffered saline. Before use, the cell walls are thawed and washed twice. The cell wall suspension at an adequate concentration is incubated with the enzyme at 37° for 1 hr. The supernatant liquid is assayed for radioactivity by a liquid scintillation spectrophotometer. This method is sensitive enough to study physiological aspects of the system (*179*).

2. *Lytic Enzyme Induced by Group C Streptococcal Phage*

a. C1 Phage Lytic Enzyme Muralysin. Purification of Muralysin (*28*). The lysate was prepared by infection with four times as much phage C1 as *Streptococcus pyogenes* strain 26RP66 at the exponential phase. Following lysis, the lysate was rapidly cooled in a crushed ice reservoir and EDTA and thioethanol were added. Streptomycin sulfate (final concentration of 0.4%) was added to remove DNA, which reduced the activity markedly (*194*). Ammonium sulfate fractionation—25% (*18*), 27.5% (*30*), or 50% (*203*)—was found to be valid to purify the enzyme. A high-speed centrifugation was used to remove phage particles. In the most purified preparation (*28*), the ammonium sulfate fractionations were repeated six times and the centrifugation procedure four times. In addition, Sephadex G-100 gel filtration and repeated calcium phosphate gel adsorptions were performed. The final preparation was 5000–6000-fold purified, compared to the starting lysate.

Purified muralysin preparation by analytical centrifugation gave a single, fast moving, homogeneous material ($s_{20,w} = 5.03$) (*28*). However, about 50% of the enzyme was inactivated during the run. From the sedimentation constant and from the behavior on Sephadex G-100, the molecular weight of muralysin is estimated to be more than 100,000. The purified preparation is relatively stable in a solution of 0.1 *M* sodium phosphate, pH 6.1. Freeze-thawing as well as lyophilization procedures cause loss of activity. The presence of a protecting sulfhydryl group in the enzyme and the avoidance of heavy metal ions help prevent the muralysin from denaturating.

The pH optimum for the lytic activity is 6.1. The enzyme is activated by monovalent cations (Na^+, K^+, or Li^+) (*195*). The optimal molarity for Na^+ is between 0.01 *M* and 0.1 *M* (*28*). Divalent cations (Mg^{2+}, Fe^{2+}, Ca^{2+}, Mn^{2+}, or Cu^{2+}) strongly inhibit the lytic activity. The inhibition by metal ions can be reversed by adding an excess of EDTA (*190*).

Reagents reactive to sulfhydryl groups inhibit the activity. The inhibition can be reversed by adding cysteine (*195*) or glutathione (*194*). The lytic activity is increased by the addition of several antibiotics such as penicillin, streptomycin, aureomycin, and bacitracin, when tested with growing streptococci (*194*). Some of these antibiotics are known to be effective chelating agents for metal ions. A possible explanation for the increment of lytic activity was thought to be this chelation. Antibiotic, Restocetin A, a carbohydrate containing polypeptidyl groups, inhibits the lytic activity, probably because of some structural similarity with the cell wall. However, this inhibition cannot be observed when using the extracted cell wall as substrate. The lytic action is not a result of the phage itself but of the lytic enzyme produced by the phage infection on the host cell (*197*, *203*). This was confirmed serologically with an antiserum preparation against the partially purified lytic enzyme and with a serum against the phage itself (*195*).

The physiological age of the cell is important in its susceptibility to lysis. Susceptibility decreases as the cell gets older and completely disappears when the cell stands at room temperature for 5 days (*195*), possibly because of changes in the surface structure of the host cell. The lytic enzyme induced by group C streptococcal phages shows high and complete activity against group A bacterial cells, lower activity against the host group C cells, and no activity against groups B and G cells. This finding is different from that of Maxted (*31*) who reported that the lytic factor from another phage (B563) of group C streptococci lyses groups A, B, C, and E streptococci and under certain conditions group H bacteria (*204*). This difference must result from the different kinds of phages employed. Thus the specificity of the lytic enzyme was concluded to result from the strain of phages, and was confirmed by the work of Reiter and Oram (*198*).

The purified group C carbohydrate inactivates C specific phages, suggesting that the primary absorption site of the group C phage is a group specific carbohydrate. However, this is not a general conclusion since Fox presented data contrary to this observation (*200*).

The degradation products of the group A streptococcal cell wall by muralysin do not contain low molecular weight compounds (*204*) but rather polysaccharide and several protein (*30*) constituents including M proteins (*188*). The enzyme specificity of muralysin was reported as an endoacetylglucosamidase from the results of the separation and characterization of mucopeptide products released from group A streptococcal cell walls. This specificity is different from that of egg white lysozyme which acts as an acetylmuramidase on mucopeptides isolated from the same cell walls (*199*). Kessler and Krause (*194*) presented evidence that the inactivation of the enzyme does not prevent the absorption of the phage to the cell.

b. B563 Phage Lytic Enzyme. One of Evans' original phages, B563, which is specific to group C *Streptococcus,* was used by Maxted to produce a lytic enzyme (*31, 188, 204*). The lytic enzyme is similar to muralysin. The enzymic activity remains unaltered on removal of the phage by high-speed centrifugation and by treatment with antiserum against the phage. It is precipitable by 40% ammonium sulfate. The activity diminishes in the absence of reducing agents during preparation and is susceptible to proteolytic activity.

The pH optimum for the enzymic activity and the stable pH are about the same as for muralysin. The group specificity of the lytic enzyme differs a little from C1 phage muralysin, even when produced in the same host. The enzyme readily lyses young living cultures but is much less active on heated suspensions, requiring the addition of a proteinase to complete the lysis. The mechanism of splitting is presumably the same as for the C1 phage muralysin, *N*-acetylhexosamidase, resulting in the liberation of polysaccharides and proteins including the M antigen (*31, 188*).

c. ϕY Phage Lytic Enzyme. A group C specific phage was isolated as a virulent mutant phage from a lysogenic culture. This phage produced clear plaques and induced a lytic enzyme in the streptococci strains C1 and 88. Upon phage lysis, the enzymic activity was observed almost entirely as a phage associated form. Equilibrium centrifugation of phage ϕY in CsCl was performed. Two distinct lytic activity peaks were demonstrated, one of which could be phage bound enzyme and the other presumably the tail fiber associated enzyme. However, no demonstrable enzymic activity was found after the phage was purified and the enzyme solubilized. No hyaluronidase activity was observed intracellularly or extracellulary during phage infection. The phage ϕY does not contain hyaluronidase which is found to be a constituent of another group A streptococcal bacteriophage (*199*). The lytic enzyme liberates oligo- and polysaccharides from streptococcal cell walls (*200*).

3. *Lytic Enzyme Induced by Other Group Streptococcal Phages*

a. Lytic Enzyme of Group A Streptococci Phages. Bacteriophages A1, A6, A12, and A25 cause lysis of group A streptococci. The designation A1, for instance, refers to the fact that the indicator strain used in propagation of the phage is a group A, type 1 *Streptococcus.* The group A phages do not lyse group C streptococci, nor do group C phages lyse group A bacterial cells. Investigation of the host range indicated that the host susceptibility for phages is primarily a group specific phenomenon but largely results from several factors such as the hyaluronic acid capsule and the surface protein (*30*).

The finding that group C cell wall carbohydrate specifically inactivates group C phage is not paralleled in the case of group A phages. Also, the fact that some of the heterologous phages can lyse the bacteria and that some of the homologous phages cannot indicates that the type specificity is not primarily related to the host range of phage. A lysogenized bacterium in group A type 12 *Streptococcus* was found to be resistant to all phages tested (*30*). The lytic enzyme of group A streptococci has not been studied extensively.

b. Lytic Enzyme of Group N Streptococcal Phages. Naylor and Czulak (*202*) observed halo formation with a lactic acid streptococcal phage. Reiter and Oram (*198*) purified the lytic enzyme from a phage (ML3) lysate of *Streptococcus lactis*. Their method consisted of acetone fractionation, salting out, and ion exchange chromatography on Amberlite CG-50. About 500-fold purification was achieved.

The characteristics of this lytic enzyme resemble these of the enzymes of group C streptococci; it lyses viable cells, is heat labile, is active in the presence of monovalent cations (0.15 *M*), and has an optimal pH of 6.6–6.9 (*198*). Lysis of viable cells is preceded by a lag period, the length of which varies with the initial enzyme concentration, but no delay is observed with cell wall preparations. The latter action results in nondialysable components, made up of sugars, amino sugars and amino acids, and the dialysable components consist of amino sugars and amino acids. Such dialysable components are not liberated by lytic enzymes of group C streptococcal phage (*204*). The effect of phage lytic enzyme on cell walls of group N streptococci appears to be the breaking of glucosidic linkages between *N*-acetylamino sugar peptide components, which is very similar to the specificity of egg white lysozyme on *M. lysodeikticus* wall. Strain specificity is not observed in group N streptococci, but the rate of reactions differs considerably. In the other groups, strains of group D are sensitive to ML3 phage lytic enzyme and strains of groups A, B, and C are insensitive.

Another lytic enzyme was prepared from the lysate of another phage infected to *Streptococcus lactis* C10, a strain unrelated to the above (*194*). This lytic enzyme has the same optimal pH and requires identical activation energy as the phage enzyme but is heat resistant. Also, the activity spectra are quite different. This and other experiments suggest that the specificity of phage lysin is determined by the phage.

The purified enzyme can lyse living bacteria; the crude enzyme preparation, which contains contaminating phage particles, readily lyses viable cells of heterologous phage-infected bacteria, but cannot lyse uninfected bacteria unless the residual phages are removed. The addition of a purified enzyme preparation at the time of the homologous phage infection at high multiplicity results in complete inhibition of activity, suggesting that the enzyme and phage compete for adsorption sites. This phenomenon contrasts with the observations that staphylococcal and *B. megaterium* phage-induced enzymes require their homologous phages (*29, 187*).

D. *Bacillus megaterium* Phage-Induced Lytic Enzyme

A virulent *Bacillus megaterium* phage G produces halos wide spreading around the plagues. The phage G and its lysate contain different lytic enzymes (designated as "bound enzyme" and "soluble enzyme," respectively) both of which dissolve *B. megaterium* KM strain. They were distinguished by serological tests as well as by other characteristics. Rabbit serum prepared against the soluble enzyme did not reduce the rate of cell wall dissolution by the bound enzyme. Also, the pH optima are clearly different: pH 5 for the soluble enzyme and pH 9.5 for bound enzyme. It was suggested that the bound enzyme acts in the initial phage penetration and the soluble enzyme acts in releasing the phage (*205*).

1. *Phage G Soluble Enzyme*

The soluble enzyme, which does not react with phage G antiserum, can be found in the cells in increasing amounts until the time of lysis, starting shortly before the first mature phage particles are produced (20–30 min after infection). Proflavin at a concentration which partially inhibits phage growth, induces maximum enzyme production. At a high titer (3×10^{13}/ml) the purified phage preparation possesses no lytic activity. The enzyme can be identified by ultracentrifugation and by the serological method. Thus, it is tentatively concluded that the enzyme is produced in the host cell by the genetic message of phage G and that it may play a role in an escape mechanism for phages by destroying the cell wall. Living cells are lysed only with an excess of phages together with the lytic enzyme [lysis from without (*138, 206*)], while dead cells are lysed by the enzyme only (*187*).

For assay, plates prepared with a heavy inoculation of *B. megaterium* KM are incubated 2–4 days at 33°, followed by an exposure to toluene vapor. Assays are performed by spotting 0.01 ml of the adequately diluted

205. J. S. Murphy, *Virology* **4**, 563 (1957).
206. A. P. Krueger and J. H. Northrop, *J. Gen. Physiol.* **14**, 223 (1930).

enzyme solution (*17*). Cell walls are prepared from the host bacteria by the method of Salton and Horn (*109*). The reduction in absorbancy of a cell wall suspension is measured in the assay for lytic activity.

The enzyme was purified from a supernatant of the lysate by ammonium sulfate precipitation. The supernatant fluid was then centrifuged at a speed sufficient to remove phage particles, and the resulting partially purified preparation (35-fold in specific activity) was subjected to starch gel electrophoresis which resulted in a single migrating band with 118-fold specific activity (*205*).

The enzyme rapidly lyses cell wall preparations with a concomitant increase in reducing sugars corresponding to a maximum of half of the total potential amount of reducing substances in the cell wall. The optimal condition for the enzymic activity is a pH between 4 and 5, at 30°. At 35° or below 30°, the enzymic activity is markedly reduced. The lytic activity is considerably specific since it is not active against *M. lysodeikticus*, *E. coli*, *Streptococcus hemolyticus*, and *B. cereus* (*187*, *205*).

The enzyme is denatured rapidly above 56° but is stable at 4° over a pH range from 4 to 9 in 0.5 *M* of various buffers. However, the enzyme is rapidly denatured in distilled water (*187*, *205*).

2. *Phage G Bound Enzyme*

This enzyme was found recently and details are not yet available. It is the first instance where a phage particle contains a lytic enzyme that differs from that in the lysate. The T2 phage particle also possesses a lysozyme, but this particle bound T2 lysozyme proved to be identical with the soluble enzyme found in its lysate.

For purification, phage G particles were purified from a lysate of phage-propagated *B. megaterium* KM strain by centrifugation and filtration through Supercel repeatedly, until no soluble activity could be detected by an assay method using heated bacteria as a substrate (*205*). The purified phage was suspended in 5 *M* urea resulting in a heavy viscous solution. This solution was treated by deoxyribonuclease and then dialyzed and centrifuged repeatedly. The resultant supernatant was crystal clear and was used as the purified bound enzyme.

The enzymic activity of the soluble enzyme is considerably reduced by the addition of an antiserum prepared against the soluble enzyme, as compared to the reduction resulting from the addition of normal rabbit serum. But the activity of the bound enzyme in the presence of the soluble enzyme antiserum did not differ from that in the presence of normal rabbit serum. Although both activities decrease considerably, this decrease may suggest the presence of phage structural components in the bound enzyme preparation which react with rabbit serum. The pH optimum of this enzyme was found between pH 9.5 and 10 (*187, 208*).

E. TP-1 Lytic Enzyme of *Bacillus stearothermophilus*

Welker found that a mitomycin-induced culture of *Bacillus stearothermophilus* (NCA 1503-4R), carrying bacteriophage TP-1, contains considerable quantities of a lytic enzyme (*207, 208*). He further concluded that the TP-1 lytic enzyme appears to be a peptidase, which hydrolyzes peptide bonds in the cell wall preparation (*209*).

For assay, acetone dried cells (*210*) of *B. stearothermophilus* were used. Twenty volumes of acetone were added to a washed log phase cell suspension and the mixture was kept at room temperature for 48 hr. Then the cells were centrifuged, suspended, and lyophilized.

For the assay cell walls of *B. stearothermophilus* prepared by the method of Sutow and Welker (*211*) were also used.

For purification, a crude lysate was treated with deoxyribonuclease and fractionated by ammonium sulfate precipitation (60% w/v). The precipitate was chromatographed on a CM-cellulose column. The enzyme fraction was then passed through a DEAE-cellulose column at pH 8 to remove yellowish contaminants. The pH of the eluate was adjusted to 6.5 and was further fractionated by ammonium sulfate precipitation (55–80%) repeated. The precipitate was extracted with 40% ammonium sulfate solution. The specific activity of the enzyme preparation at this step was about 2000 times higher than the original lysate and was free from phage TP-1 (*207*). Recently, the enzyme was further purified by Sephadex G-100, resulting in 8600-fold purification (*209, 210*).

Polyacrylamide gel electrophoresis of the purified enzyme revealed a single protein band accompanied by the lytic activity. The specific activity of this band is about the same as that of the material prior to the electrophoresis, which indicates a homogeneous preparation. A sucrose gradient centrifugation of the purified lytic enzyme also revealed that it was free from protein contaminants (*209, 210*).

The enzyme preparation was stable between pH 7.0 and 8.0 and also at pH 5.0 but unstable between pH 5.5 and 6.5 at 55°. A sodium phosphate concentration of at least 0.1 *M* was necessary to maintain activity

207. N. E. Welker and L. L. Campbell, *J. Bacteriol.* **89**, 175 (1965).
208. N. E. Welker and L. L. Campbell, *Bacteriol. Proc.* p. 126 (1966).
209. N. E. Welker, *Virology* **1**, 617 (1967).
210. E. Work, *Ann. Inst. Pasteur* **96**, 486 (1959).
211. A. B. Sutow and N. E. Welker, *J. Bacteriol.* **93**, 1452 (1967).

during a prolonged exposure of the enzyme above 55° at pH 6.3. The lytic enzyme in 0.1 *M* sodium phosphate was not inactivated after 1 hr exposure to 65.5° and only 1% inactivation occurred at 70.6° at pH 6.3.

The lytic enzyme is active over a pH range of 6.0–7.0 and is most active at pH 6.3. A sodium concentration of 0.005 *M* is required for minimal activity of the lytic enzyme. The activity is considerably stimulated by Mg^{2+}, Ca^{2+}, and Mn^{2+} ions (at 10^{-4} *M*) but inhibited by Hg^{2+}, Cu^{2+}, and Zn^{2+} ions (at 10^{-4} *M*) and EDTA. The EDTA inhibition can be reversed by the addition of excess Mg^{2+}, Ca^{2+}, or Mn^{2+}. Sulfhydryl reagents such as *p*-chloromercuribenzoate inhibit the lytic activity (*209*). Cell walls are rapidly solubilized by the lytic enzyme. The lytic action is completed after 15 min with a 70% reduction in optical density in a given system. The reduction in optical density was found to be parallel to the solubilization of NH_2-termini, with the release of small wall fragments. However, it was not accompanied by an increment of reducing substances. Electron micrographs suggest that the lytic enzyme partially degrades the cell wall with the release of small wall fragments. These observations indicate that the lytic enzyme induced is a peptidase which hydrolyzes peptide bonds in the cell wall peptidoglycan. This enzyme is specific to cell walls from a limited number of strains of *B. stearothermophilus*. The enzyme was shown not to be a proteolytic enzyme. Antiphage-serum does not interfere with the enzymic activity (*209*).

ACKNOWLEDGMENTS

The author wishes to thank Dr. S. Kotani, Dental School, Osaka University, and Drs. A. Imada and K. Nakahama, Takeda Pharmaceutical Co., Ltd., for their valuable discussions throughout the preparation of this review. He also appreciates the help of Dr. Bertil Grahn and Miss M. Kobayashi in preparing the manuscript.

14

Aconitase

JENNY PICKWORTH GLUSKER

I. Introduction

A. Function

Aconitase is a member of the Krebs cycle system of enzymes and catalyzes the interconversion of the anions of the three tricarboxylic acids: citric acid, (+)-isocitric acid, and *cis*-aconitic acid. The first two of these are isomers, differing only in the position of a hydroxyl group, and *cis*-aconitic acid is their dehydration product as shown in Eq. (1). Aconitase is of particular interest because of the stereospecificity of the reaction. This is shown below with Fischer formulas [Eq. (1)] and by Newman projections [Eq. (2)]. The hydrogen atom and the hydroxyl group, abstracted by the action of aconitase, are denoted by asterisks.

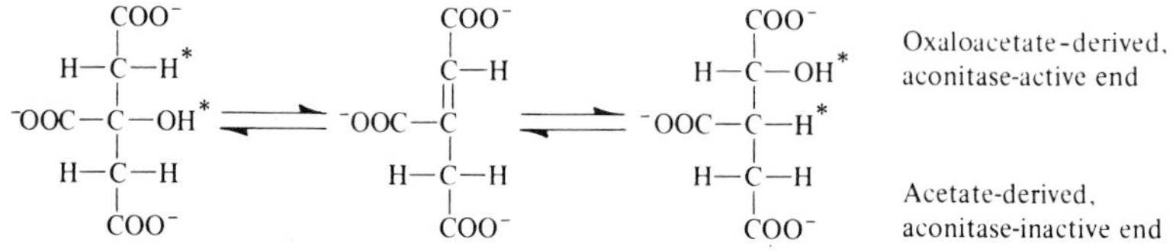

$$\begin{array}{ccccc} COO^- & & COO^- & & COO^- \\ H-C-H^* & & C-H & & H-C-OH^* \\ ^-OOC-C-OH^* & \rightleftharpoons & ^-OOC-C & \rightleftharpoons & ^-OOC-C-H^* \\ H-C-H & & H-C-H & & H-C-H \\ COO^- & & COO^- & & COO^- \\ \text{Citrate} & & cis\text{-Aconitate} & & (2R{:}3S)\ \text{Isocitrate} \end{array} \quad (1)$$

Oxaloacetate-derived, aconitase-active end

Acetate-derived, aconitase-inactive end

$$\text{Newman projections: } [COO^-,\ {}^*HO,\ COO^-,\ H,\ H^*,\ CH_2COO^-] \rightleftharpoons [COO^-,\ COO^-,\ H,\ CH_2COO^-] \rightleftharpoons [COO^-,\ H^*,\ COO^-,\ H,\ OH^*,\ CH_2COO^-] \quad (2)$$

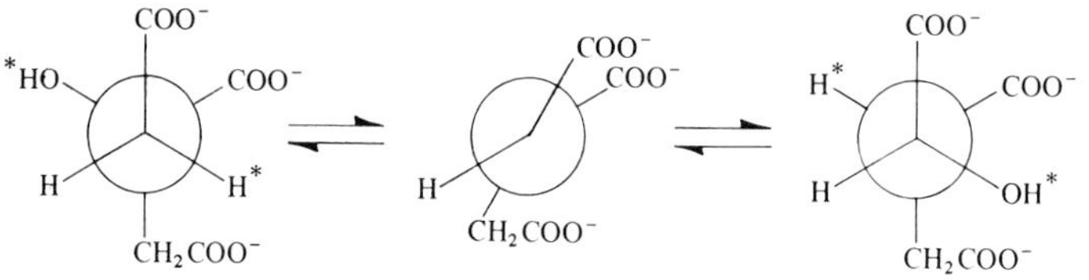

The enzyme was initially described by Martius (*1, 2*). The name *aconitase* was proposed by Breusch (*3*) but is now often replaced by the more formal one of *aconitate hydratase* or *citrate (isocitrate) hydro-lyase* (EC 4.2.1.3), although neither of these names gives a complete description of the enzyme function, which is both as an isomerase and as a hydratase (see Section III,F,4).

Flux through the mitochondrial enzyme is in the direction of citrate to isocitrate, which is degraded by the citric acid cycle. In the cytoplasm, flux is in the opposing direction, making citrate available for fatty acid biosynthesis and feedback regulation of glycolysis via phosphofructokinase.

B. Historical Background

Aconitase has posed numerous problems for the biochemist. Its purification and stabilization have been difficult (see Sections II,A and II,C). It has been suggested (*4*) that the two modes of hydration of *cis*-aconitate in the mechanism are effected by different enzymes or different catalytic

1. C. Martius, *Z. Physiol. Chem.* **247,** 104 (1937).
2. C. Martius, *Z. Physiol. Chem.* **257,** 29 (1938).
3. F. L. Breusch, *Z. Physiol. Chem.* **250,** 262 (1937).
4. K. P. Jacobsohn and J. Tapadinhas, *Compt. Rend. Soc. Biol.* **133,** 112 (1940).

sites on the same enzyme, one a citrate dehydratase and the other an isocitrate dehydratase. Ogston (*5*), however, pointed out that by analogy with fumarase it would be reasonable to believe that one enzyme site could perform both reactions, a proposal strongly supported by the detection of direct proton transfer in the conversion of citrate to isocitrate (*6*). However, the problem has been revived recently (*7*) (see Section III,H). For a while it was also questioned whether the citrate ion took direct part in the Krebs cycle since it was shown, by tracer experiments (*8*), that labeled carbon dioxide, in the enzymic carboxylation of pyruvate, goes *only* to the terminal carboxyl group adjacent to the carbonyl group in α-ketoglutarate. This asymmetric behavior was shown by Ogston (*9*) to be the expected result of the interaction of the symmetrical citrate with the enzyme which is itself asymmetric by, for example, attachment of the substrate to the enzyme at three points. Further tracer experiments (*10*) confirmed that citrate is, indeed, an intermediate in the cycle. This means that aconitase can distinguish between the two $-CH_2-COO^-$ groups of citrate and abstracts a proton from only one of them. There has been much discussion of whether *cis*-aconitate is an obligatory intermediate in the reaction catalyzed by aconitase. It has been shown (*6, 11*) that in the conversion of citrate to isocitrate a proton is transferred by the enzyme directly from one substrate to the other. Therefore, aconitase may be considered as isocitrate:citrate isomerase. In addition, the hydration of *cis*-aconitate in this reaction follows from the demonstration of the transfer of tritium from 2-tritio-2-methylcitrate to *cis*-aconitate to give tritiated isocitrate. This experiment shows that *cis*-aconitate has been hydrated by the enzyme (*6*) to form isocitrate at the same site as that at which 2-tritio-2-methylcitrate was dehydrated (see Section III,F,4). Thus, aconitase is also a hydratase.

C. Occurrence

Since aconitase is an enzyme in the tricarboxylic cycle it must appear wherever the cycle is operative. It is found in all respiring tissues of animals from protozoa to mammals particularly, in mammals, in the heart,

5. A. G. Ogston, *Nature* **167**, 693 (1951).
6. I. A. Rose and E. L. O'Connell, *JBC* **242**, 1870 (1967).
7. R. A. Peters and M. Shorthouse, *Nature* **221**, 774 (1969).
8. H. G. Wood, C. H. Werkman, A. Hemingway, and A. O. Nier, *JBC* **139**, 483 (1941).
9. A. G. Ogston, *Nature* **162**, 4129 (1948).
10. V. R. Potter and C. Heidelberger, *Nature* **164**, 180 (1949).
11. J. F. Speyer and S. R. Dickman, *JBC* **220**, 193 (1956).

kidney, and liver. It is also found in plants, yeast, bacteria, and molds.

There are also reports of a thermostable aconitase from *Bacillus stearothermophilus* (*12*) and of an aconitase-deficient mutant of *Bacillus subtilis* (*13, 14*) that was asporogenic.

Two enzymes were separated by Zweerink (*15*) from *Salmonella typhimurium* and also from *Escherichia coli* by density gradient centrifugation. One enzyme was identified as aconitase. The other, with a molecular weight approximately four times as large, could only catalyze the reaction of citrate to *cis*-aconitate (and so was designated citrate dehydratase) although, interestingly, there was evidence reported that this enzyme could dissociate into normal aconitase. Nielson (*16, 17*) has reported the presence of a high molecular weight protein in *Aspergillus niger* that like the one described above formed citrate exclusively from *cis*-aconitate. This enzyme, *aconitic hydrase*, is present in addition to aconitase and was separated from it by ammonium sulfate precipitation.

For further details and references on the distribution of aconitase in nature the reader is referred to (*18*).

D. Intracellular Distribution

Dickman and Speyer (*19*) described three aconitases in rat liver tissues: cytoplasmic aconitase occurring in the supernatant fraction after homogenization, "soluble" aconitase released from mitochondria by the freezing and thawing of the particles, and particulate-bound mitochondrial enzyme. The last two are collectively referred to as the mitochondrial enzyme. Different properties such as isoelectric points, pH optima, and sensitivity to fluorocitrate (*20*) and oxalomalate (*21*) (which inhibit the mitochondrial enzyme more powerfully than the cytoplasmic

12. L. Jannes and K. Kelopuro, *Suomen Kemistilehti* **34B,** 129 (1961); *Chem. Abstr.* **56,** 13330b (1962).
13. J. Stuyvaert, C. Liebecq, and Z. M. Bacq, *Intern. J. Radiation Biol.* **8,** 513 (1964); *Chem. Abstr.* **63,** 2098h (1965).
14. J. Szulmajster and R. S. Hanson, *Spores Symp., 3rd, Allerton Park, Ill., 1964*, p. 162. Burgess, Minneapolis, Minnesota, 1965; *Chem. Abstr.* **63,** 16824f (1965).
15. H. J. Zweerink, *Dissertation Abstr.* **28,** 2972B (1968).
16. N. E. Nielson, *BBA* **17,** 139 (1955).
17. N. E. Nielson, *J. Bacteriol.* **71,** 356 (1956).
18. J. M. Lowenstein, *in* "Metabolic Pathways" (D. M. Greenberg, ed.), Vol. 1, p. 146. Academic Press, New York, 1967.
19. S. R. Dickman and J. F. Speyer, *JBC* **206,** 67 (1954).
20. V. Guarriera-Bobyleva and P. Buffa, *BJ* **118,** 853 (1969).
21. A. Ruffo, personal communication (1970).

one) are reported and are used to distinguish the different fractions. The solubilized mitochondrial enzyme is reported to be less stable than the cytoplasmic enzyme and is less active than the enzyme in the intact mitochondrion, presumably because of inactivation during the solubilization procedure and also because of a more favorable environment within the mitochondrion. The experimentally observed differing sensitivities of tissues to aconitase inhibitors such as fluorocitrate (*20*, *22*) may be caused by the differing proportions of the various forms of aconitase in these tissues. There has, however, been no actual proof that these forms of aconitase differ in their primary structure.

The following values for the percentage activity of cytoplasmic aconitase in various tissues have been reported: 10–15% in rabbit cerebral cortex (*23*), 60–70% in human liver (*24*), 70–80% in rat liver (*20*), 49% in pig liver (*22*), 24% in pig kidney cortex (*22*), 2% in pig heart (*22*), about 50% in kidneys, predominately in the spleen and in small amounts in muscle of ICR mice and Slow lorises (*25*).

These observations on the intracellular distribution may be correlated with the function of the tissues in which they are found and with the function of the enzyme (see Section I,A). The liver is used for the synthesis of fatty acids and would be expected to contain more of the cytoplasmic enzyme than would the heart which is involved in the utilization of energy.

II. Molecular Properties

A. Purification of the Enzyme

The various purifications of aconitase which contain full details are listed in Table I (*22*, *26–31*). The purest sample to date is that of Villafranca and Mildvan (*28*).

22. R. Z. Eanes and E. Kun, *BBA* **227,** 204 (1971).
23. J. A. Shepherd and G. Kalnitsky, *JBC* **207,** 605 (1954).
24. J. A. Shepherd, Y. W. Li, E. E. Mason, and S. E. Ziffren, *JBC* **213,** 405 (1955).
25. A. L. Koen and M. Goodman, *BBA* **191,** 698 (1969).
26. J. M. Buchanan and C. B. Anfinsen, *JBC* **180,** 47 (1949).
27. J. F. Morrison, *BJ* **56,** 99 (1954).
28. J. J. Villafranca and A. S. Mildvan, *JBC* **246,** 772 (1971).
29. C. P. Henson and W. W. Cleland, *JBC* **242,** 3833 (1967).
30. M. J. Palmer, *BJ* **92,** 404 (1964).
31. E. E. Bruchmann, *Naturwissenschaften* **48,** 53 (1961).

TABLE I
PURIFICATION OF ACONITASE

Reference	Authors	Source	Factor of purification	Purity (%)
(*26*)	Buchanan and Anfinsen	Pig heart	23	30
(*27*)	Morrison	Pig heart	24	75–80
(*28*)	Villafranca and Mildvan	Pig heart	76	95
(*22*)	Eanes and Kun	Pig tissues		
(*29*)	Henson and Cleland	Beef liver	21	60–70
(*30*)	Palmer	*S. alba*	200	90
(*31*)	Bruchmann	*A. niger*	15–20	

B. PHYSICOCHEMICAL PROPERTIES

The following molecular data are listed by Villafranca and Mildvan (*28*) for the enzyme from pig heart. This aconitase was judged at least 95% pure with a value of $s_{20,w} = 6.16$. The molecular weight, determined by sedimentation equilibrium, is 89,000 (*28*). The diffusion coefficient, calculated with an assumed partial specific volume of 0.749, is $D_{20,w} = 6.68 \times 10^{-7}$ cm²/sec (*28*). The calculated frictional ratio, f/f_o, is 1.08 indicating that aconitase behaves in the ultracentrifuge as a globular protein with an axial ratio between 1.0 and 2.6 (*28*).

Aconitase and citrate dehydratase from *Salmonella typhimurium* (*15*) have approximate sedimentation values of 7.2 and 19.1, respectively, giving molecular weight estimates of 120,000 and 530,000.

The optical spectrum shows broad nonspecific absorption from 300 to 600 nm at 25°C, which is probably due to the bound ferric iron (see later). The enzyme has a spectrum more like that of transferrin or conalbumin than that of rubredoxin. There is no heme iron. The molar extinction coefficient at 490 nm is 3500 M^{-1} cm^{-1}; $E^{1\%}_{280\,nm} = 13.7$ (*28*).

The enzyme loses activity on standing, but this activity can be restored by incubation with ferrous ions and a reducing agent such as cysteine (*32, 33*). Since the enzyme is inactivated by mercurials there may be essential sulfhydryl groups present, but these have not been measured directly. It has been suggested that the enzyme is a monothiol enzyme (*34*). The activity is preserved at very low temperatures and the enzyme can be stored for months by freezing in liquid nitrogen (*28*). It is denatured below pH 5 (*34*).

32. S. R. Dickman and A. A. Cloutier, *AB* **25,** 229 (1950).
33. S. R. Dickman and A. A. Cloutier, *JBC* **188,** 379 (1951).
34. J. F. Morrison, *BJ* **58,** 685 (1954).

The isoelectric point of the pig heart enzyme is reported by Villafranca and Mildvan (*28*) to be 9.1. Eanes and Kun (*22*) reported values of 5.4 for the cytoplasmic enzyme and 7.5 for the mitochondrial enzyme from pig tissues. Morrison (*27*) found an isoelectric point between 5.7 and 8.6 for the pig heart enzyme. Smith found a value of 7.9 (*35*). The pig liver enzyme has a lower isoelectric point than the beef liver enzyme (*36*).

An electrophoretic study was made of the cytoplasmic isozyme variation in ICR mice and Slow lorises. No hybrid enzyme bands were indicated for heterozygous Slow lorises indicating (*25*) that aconitase is either a monomeric enzyme or a polymeric one in which a random association of subunits does not occur. In ICR mice there is suggestive evidence that aconitase isozymes might be polymeric in nature (*25*). No information on amino acid composition or direct evidence for the existence of subunits has been published to date.

Villafranca and Mildvan (*28*) reported that the enzyme is brown in color and contains 1.0 ± 0.1 moles iron per 89,000 g aconitase (by atomic absorption) or 0.7 ± 0.2 (by ESR spectroscopy). This is high spin ferric iron. It is thought to be structural rather than catalytic since it does not interact with the substrates and could not be removed by EDTA, which was present during the purification. It is not the same as the catalytic ferrous iron. There are 3 ± 1 water molecules in the inner coordination sphere of the aconitase-bound ferric iron as determined by nuclear magnetic resonance techniques.

C. Factors Affecting the Stability of Aconitase

The activity of aconitase decreases rapidly on storage or dialysis. This loss of activity may be decreased by covering the solution with mineral oil (*32*) or by keeping it in an inert atmosphere, since aeration reduces or removes the enzymic activity (*34*). However, the best way to stabilize or reactivate the enzyme is by the addition of ferrous ions and a reducing agent (*32, 33*).

The enzyme can apparently also be protected against loss of activity by the addition of substrate such as citrate (*26*) or *cis*-aconitate (*27*) or an inhibitor such as tricarballylate. This will protect the enzyme against the inactivating effects of certain buffers such as phosphate (*27, 33*). Furthermore, the pH dependence of the stability varies greatly with the nature of the buffer ions present.

The ratios of the rates of the various reactions catalyzed by aconitase

35. C. Smith, personal communication (1970).
36. J. McD. Blair, *European J. Biochem.* **8**, 287 (1969).

are the same in all buffers at the respective pH optima for the enzyme in the buffer. As the pH is changed from the optimum value the ratios may change greatly (*37*).

III. Catalytic Properties

A. Specificity

The substrates of aconitase, as shown in Eqs. (1) and (2), are citrate, *cis*-aconitate, and (2*R*:3*S*)isocitrate, which are all interconverted by the same enzyme or enzyme system. The stereospecificity of the enzyme is interesting. Of the two $-CH_2-COO^-$ groups in the citrate ion only the one that is derived enzymically from oxaloacetate is involved in the action of aconitase. The other $-CH_2-COO^-$, derived from acetate, is not affected by the reaction; that is, aconitase acts only on the *pro-R* $-CH_2-COO^-$ group of citrate. In fact, it acts on only one of those two hydrogen atoms of the $-CH_2-$ group. The stereochemistry of the action of aconitase on citrate was demonstrated by Englard (*38*) and by Hanson and Rose (*39*). Only one of the four possible isomers of isocitrate, the 2*R*:3*S* isomer, is a substrate of aconitase. The absolute configuration has been determined by chemical, X-ray crystallographic, and enzymic means (*40–45*). The active isomer is the anion of (+)-isocitric acid, also referred to as (*d*)-isocitric acid, *threo*-D_s-isocitric acid, or D_s,L_g-isocitric acid.

These experimental results on the absolute configurations of the substrates of pig heart or beef liver aconitase indicate that a *trans* addition of the elements of water to *cis*-aconitate has occurred (*46*). This is illustrated in Fig. 1. When addition occurs, C(2) of *cis*-aconitate is attacked *only* from one side of the plane of the double bond of *cis*-aconitate and C(3) is only attacked from the other side of this plane. It has been shown

37. J. F. Morrison, *Australian J. Exptl. Biol. Med. Sci.* **32,** 877 (1954).
38. S. Englard, *JBC* **235,** 1510 (1960).
39. K. R. Hanson and I. A. Rose, *Proc. Natl. Acad. Sci. U. S.* **50,** 981 (1963).
40. T. Kaneko and H. Katsura, *Chem. & Ind. (London)* p. 1188 (1960).
41. T. Kaneko, H. Katsura, H. Asano, and K. Wakabayashi, *Chem. & Ind. (London)* p. 1187 (1960).
42. O. Gawron and A. J. Glaid, III, *JACS* **77,** 6638 (1955).
43. O. Gawron, A. J. Glaid, III, A. LoMonte, and S. Gary, *JACS* **80,** 5856 (1958).
44. A. L. Patterson, C. K. Johnson, D. van der Helm, and J. A. Minkin, *JACS* **84,** 309 (1962).
45. A. Meister and M. Strassburger, *Nature* **200,** 259 (1963).
46. O. Gawron, A. J. Glaid, III, and T. P. Fondy, *JACS* **83,** 3634 (1961).

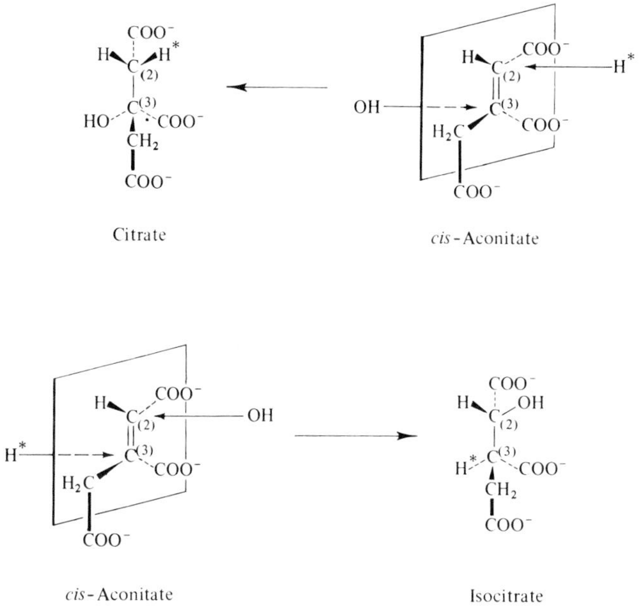

COO⁻
H C H*
(2)
C(3)
HO COO⁻
CH₂
COO⁻

Citrate

OH
H C COO⁻ H*
(2)
C(3)
H₂C COO⁻
COO⁻

cis-Aconitate

H*
H C COO⁻ OH
(2)
C(3)
H₂C COO⁻
COO⁻

cis-Aconitate

COO⁻
H C OH
(2)
H* C(3) COO⁻
CH₂
COO⁻

Isocitrate

FIG. 1. Stereochemistry of the conversion of *cis*-aconitate to citrate by aconitase. C(2) is attacked only from one side of the plane of the *cis*-aconitate ion and C(3) only from the other side. The hydrogen atom that is conserved in the conversion of citrate to isocitrate is marked with an asterisk.

(*6*) that the hydrogen atom which is abstracted from either citrate or isocitrate by the enzyme is retained by the enzyme and added again to *cis*-aconitate to give either hydroxyacid (see Section III,F,4). This suggests that the plane of the double bond in *cis*-aconitate can rotate during the reaction.

Gawron (*47*) showed that α-methyl-*cis*-aconitate is also a substrate of aconitase and can give methylcitrate and methylisocitrate as shown in Eq. (3). Thus there must be room in the active site for the more bulky methyl group replacing a hydrogen atom.

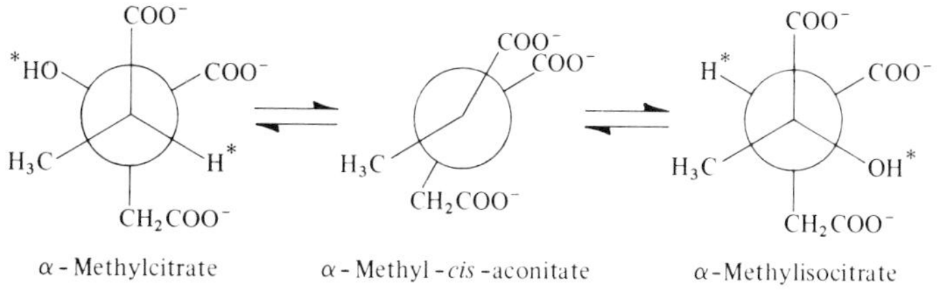

COO⁻
*HO COO⁻
H₃C H*
CH₂COO⁻
α-Methylcitrate

COO⁻
COO⁻
H₃C
CH₂COO⁻
α-Methyl-*cis*-aconitate

COO⁻
H* COO⁻
H₃C OH*
CH₂COO⁻
α-Methylisocitrate

(3)

47. O. Gawron and K. P. Mahajan, *Biochemistry* **5**, 2343 (1966).

None of the three isomeric monomethyl esters of *cis*-aconitic acid is a substrate of aconitase (*19*); that is, all three carboxyl groups are necessary for action. The competitive inhibitors (see Table IV) also contain three carboxyl groups with the same spatial relative arrangement as that found for the substrates.

B. Cofactors

1. *Ferrous Iron Requirement*

There appears to be a definite requirement of the enzyme from most sources for ferrous iron. Other metals such as Ca^{2+}, Mg^{2+}, Ba^{2+}, Mn^{2+}, Fe^{3+}, Co^{2+}, Ni^{2+}, and Cu^{+} (*32*) and Cu^{2+}, Cd^{2+}, Ru^{3+}, Rh^{3+}, Gd^{3+}, Zn^{2+}, and Ru^{+} have been tested over a wide range of concentrations (*48*), but none was found to cause any significant reactivation of the enzyme. It was suggested (*33*) that ferrous iron is involved in the interaction of the enzyme with substrate through a substrate–ferrous–aconitase complex and that this is the manner in which the three-point attachment of substrate, postulated by Ogston (*9*) to explain the asymmetric behavior of citrate with aconitase, is effected.

It was not possible to reactivate aconitase prepared from kidneys of iron-deficient rats by the addition of iron ascorbate although this treatment increased the activity of enzyme from kidneys of normal rats (*49*). Analogous results have been obtained in plants (*50*). This suggests that in iron-deficient tissue there is a lack of the enzyme system as a whole and that iron, in addition to activating, may also be an integral component of the enzyme system as has been found for the pig heart enzyme (*28*).

Villafranca and Mildvan (*28*) have shown, with purified pig heart aconitase, that the time dependence of enzymic activation by ferrous iron is sigmoidal in character which may be explained by a consecutive second-order process, shown in Eq. (4), with k_1 equal to $100\,M^{-1}$ sec^{-1}

$$\text{E(inactive)} \xrightarrow[k_1\ [\text{Fe}]]{} \text{E(Fe)(inactive)} \xrightarrow[k_2\ [\text{Fe}]]{} \text{E(Fe)}_2\text{(active)} \qquad (4)$$

and k_2 as $250\,M^{-1}$ sec^{-1}. Manganous ions are found not to activate aconitase but to decrease the rate and extent of activation by ferrous iron, suggesting competition at the active site of the enzyme, which was verified by binding studies (see Section III,H).

48. J. J. Villafranca and A. S. Mildvan, unpublished observations (1970).
49. E. Beutler, *J. Clin. Invest.* **38,** 1065 (1959).
50. H. I. Rahatekar and M. R. R. Rao, *Enzymologia* **25,** 292 (1963).

2. *The Role of the Reducing Agent*

Morrison (*34*) found that cysteine itself would not activate aconitase, that ferrous iron alone caused a slight increase in activity, but that ferrous iron and cysteine gave a much larger increase in activity. Cysteine can be replaced by ascorbic acid, the latter being only half as effective (*33*), or by hydroxylamine (*51*). Morrison (*34*) derived Michaelis constants for the enzyme–ferrous–activator and enzyme–activator complexes. For enzyme–ferrous complexes he found the following K_a values: 3.9×10^{-6} *M* with cysteine and 1.7×10^{-5} *M* with ascorbic acid. Villafranca and Mildvan (*48*) found 1.6×10^{-5} *M* with dithiothreitol. For enzyme complexes the following K_m values were found (*34*): $2.3–3.6 \times 10^{-3}$ *M* for cysteine, 1.2×10^{-3} *M* for ascorbic acid, and $4–6 \times 10^{-3}$ *M* for thioglycolate. It was suggested (*34*) that one ferrous ion and one molecule of reducing agent combine with aconitase.

The primary function of the reducing agents is to keep the iron in a reduced state (*33, 51*). The reducing agents may also act by maintaining protein reducing groups and possibly can make ferrous ions in the tissues available for activation of aconitase. They may also keep a monothiol group on the enzyme in a reduced form. In addition, they may serve as ligands for the ferrous iron to facilitate water exchange (see Section III,I).

C. Assay

Because of the multiple activities of aconitase, several assays have been described. Thus, the rate of formation of isocitrate from citrate can be followed by coupling with isocitric dehydrogenase and measuring the appearance of TPNH (*6*). The enzyme can also be assayed by monitoring the formation of *cis*-aconitate at 240 nm (*52*) and by following the formation of citrate from *cis*-aconitate or isocitrate in the presence of citrate lyase and malate dehydrogenase (*53*).

D. pH Optima

Morrison reported (*37*) values of the pH optima for the pig heart enzyme in various buffers as glycerophosphate, pH 7.5; phosphate, pH

51. E. B. Herr, Jr., J. B. Summer, and D. W. Yesair, *BBA* **20,** 310 (1956).
52. E. Racker, *BBA* **4,** 211 (1950).
53. J. F. Thomson, S. L. M. Nance, K. J. Bush, and P. A. Szczepanik, *ABB* **117,** 65 (1966).

7.7; *N*-ethylmorpholine, pH 8.1; and Veronal acetate, pH 8.6. Other reported values are pH 7.2 for aconitase from rhubarb leaves in phosphate buffer (*54*), pH 8–8.5 for aconitase from *Sinapis alba* (*30*), and pH 8.3 from *Aspergillus niger* (*55*). Dickman and Speyer (*19*) found the pH optimum of mitochondrially bound rat liver aconitase occurred at 5.8 and 7.3 and the soluble enzyme only at 7.3. This was also found for the enzyme from lupine mitochondria (*56*) and was considered to possibly result from an increased permeability of the mitochondrial membrane to substrate at lower pH values.

E. Equilibrium Concentration of the Tricarboxylic Acids

Citrate predominates in the equilibrium mixture although the free energies of the conversion, calculated by Burton and Krebs (*57*), are small. Values of $\Delta G°$ of $+2.04$ and -0.45 kcal are listed by them for the reactions of citrate to *cis*-aconitate and of *cis*-aconitate to isocitrate, respectively. The equilibrium concentrations of the tricarboxylic acids have been measured by several workers (*53, 54, 57–60*). The most accurate values (*53*) are $88.4 \pm 1.3\%$ for citrate, $4.1 \pm 0.2\%$ for *cis*-aconitate, and $7.5 \pm 1.0\%$ for isocitrate.

The equilibrium point of the aconitase reaction is affected by temperature and ionic strength (*36*) and also by metal ions (*36, 61*), the latter, probably owing to their differential affinity for the substrates. When α-methyl-*cis*-aconitate is the substrate, at equilibrium there is $27.1 \pm 0.2\%$ α-methyl-*cis*-aconitate and about equal amounts of the two hydroxyacids (*47*).

F. Kinetics

1. *Latent Period*

Krebs and Holzach (*62*) and Morrison (*63*) observed a lag period in the formation of isocitrate from citrate. The significance of this lag is not

54. J. F. Morrison and J. L. Still, *Australian J. Sci.* **9,** 150 (1947).
55. K. Taeufel and U. Behnke, *Nahrung* **10,** 443 (1966).
56. E. F. Estermann, E. E. Conn, and A. D. McLaren, *ABB* **85,** 103 (1959).
57. M. Saffran and J. Prado, *JBC* **180,** 1301 (1949).
58. H. Kacser, *BBA* **9,** 406 (1952).
59. H. A. Krebs, *BJ* **54,** 78 (1953).
60. L. V. Eggleston and H. A. Krebs, *BJ* **45,** 578 (1949).
61. P. J. England, R. M. Denton, and P. J. Randle, *BJ* **105,** 32C (1967).
62. H. A. Krebs and O. Holzach, *BJ* **52,** 527 (1952).
63. J. F. Morrison, *Australian J. Exptl. Biol. Med. Sci.* **32,** 867 (1954).

yet understood. It is decreased by high enzyme concentrations resulting in a virtually linear initial rate (*63, 64*). When citrate is added to aconitase the amount of *cis*-aconitate formed rises rapidly to a maximum and then falls off with time, while, as described above, the isocitrate concentration increases slowly at first, then more rapidly.

2. *Relative Reaction Rates*

Values for relative maximum velocities are listed in Table II. It seems that *cis*-aconitate can form citrate and isocitrate at equal rates, but that citrate and isocitrate each form *cis*-aconitate more rapidly than each other, i.e., that hydration or dehydration occurs more readily than isomerization. The V_{max} values with the three substrates fall in the order *cis*-aconitate $\geqslant$ isocitrate $>$ citrate. Activation energies of 12–13 kcal per mole are listed (*65*) for the reactions citrate to and from *cis*-aconitate and for isocitrate to either citrate or *cis*-aconitate.

TABLE II
RELATIVE MAXIMUM VELOCITIES OF THE REACTIONS CATALYZED BY ACONITASE

Source of enzyme	Pig heart [Ref. (*63*)]	Beef liver [Ref. (*29*)][a]	Beef heart [Ref. (*53*)]
Citrate → isocitrate	0.33	1.00	1.00
Citrate → *cis*-aconitate	1.00	1.04	0.89
Isocitrate → citrate	1.67	1.66	1.57
Isocitrate → *cis*-aconitate	3.00	1.71	1.46
cis-Aconitate → citrate	3.00	3.14	1.89
cis-Aconitate → isocitrate	3.00	3.10	1.67

[a] Values are listed ±7–10%.

The turnover number of aconitase is 13.5 per second per molecule of pig heart enzyme (*28*) for the reaction citrate to isocitrate.

3. *Michaelis Constants*

Values published are listed in Table III (*29, 30, 47, 48, 52, 53, 63, 65–69*) including values for α-methyl-*cis*-aconitate (*47*). The values vary

64. J. J. Villafranca, unpublished observations (1970).
65. J. Tomizawa, *J. Biochem.* (*Tokyo*) **40,** 339 and 351 (1953).
66. J. S. Britten, *BBA* **178,** 370 (1969).
67. J. F. Thomson and S. L. Nance, *ABB* **135,** 10 (1969).
68. P. Fortnagel and E. Freese, *JBC* **243,** 5289 (1968).
69. W. P. Hsu and G. W. Miller, *BBA* **151,** 711 (1968).

TABLE III
MICHAELIS CONSTANTS OF THE SUBSTRATES OF ACONITASE

Reference	K_m citrate (moles/liter)	K_m *cis*-aconitate (moles/liter)	K_m isocitrate (moles/liter)	Source of enzyme
	(all values to be multiplied by 10^{-4})			
(63)	36	1.2	4.8	Pig heart
(48)[a]	6.2	0.15	2.0	Pig heart
(66)		0.6		Pig heart
(52)	11		4.0	Pig heart, yeast
(65)	9	0.9	3.2	Rabbit liver
(29)	2.0	0.07[b]	0.34	Beef liver
(53, 67)	9.5	0.99	1.39	Beef heart
	8.5 (D_2O)	0.78 (D_2O)		
(68)	17			*Bacillus subtilis*
(69)			5.8	*Glycine max*. Merr
(30)	40–44	1	1.5	Mustard
(47)		1.9	Me *cis*-aconitate 6.0	Pig heart

[a] These are measured on highly purified enzyme.
[b] This value is probably too low *(67)*.

greatly, probably because of the different conditions of assay, modification of the enzyme during preparation, or buffer and ionic strength effects. A consistent finding is that the *cis*-aconitate ion has the lowest K_m value. The values of V_{max} are in the same order, that of *cis*-aconitate being the greatest.

4. *Kinetic Scheme*

Various kinetic models have been proposed *(11, 29, 58, 65, 70)*, but it has not been possible, from the data available, to distinguish among them. It seems, however, that a central complex is formed which can dissociate to yield free enzyme and any of the three acids.

Rose and O'Connell *(6)*, who studied the system extensively, deduced the scheme shown in Fig. 2. They found that at the beginning of the reaction with pig heart aconitase 3-*T*-isocitrate gave 2-*T*-citrate with no loss of tritium, showing that an intramolecular transfer of hydrogen had occurred. When a high concentration of *cis*-aconitate is present it is found that tritiated 2-methyl hydroxyacids give tritiated isocitrate; that is, tritium has been transferred from the 2-methyl hydroxyacids to

70. S. Aronoff and J. Z. Hearon, *ABB* **88**, 302 (1960).

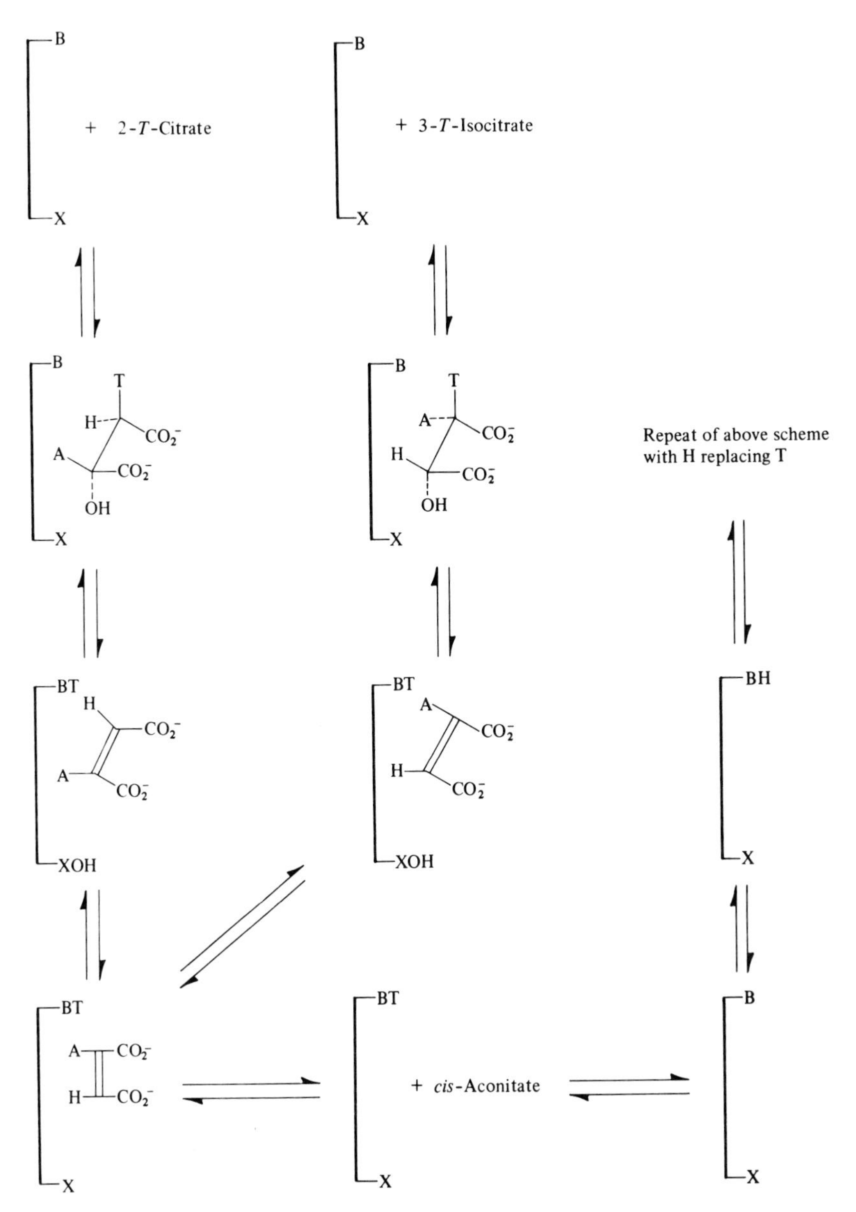

FIG. 2. The kinetic scheme of aconitase action of Rose and O'Connell (*6*). A, $-CH_2CO_2^-$; B, protein abstracting site; X, hydroxyl abstracting site.

free *cis*-aconitate. This indicates that the conjugate acid form of the enzyme, E-T, is slow to exchange with the medium, and can transfer its tritium to a different molecule of *cis*-aconitate before it dissociates. The dissociation of this proton is possibly a rate-determining step in the reaction The reaction of aconitase occurs with no transfer of ^{18}O-labeled hydroxyl; that is, in the conversion of citrate to isocitrate the latter receives a hydrogen atom from citrate but the hydroxyl is obtained from the medium and not from citrate.

Since the hydrogen atom that is abstracted is conserved by the enzyme but has been shown from experiments with tritium to be undiluted with hydrogen while on the active site, the hydrogen abstracting group cannot be $-NH_2$ (unless constrained) but is probably $-S^-$, $-COO^-$, or imidazole (*6*).

G. Inhibitors

1. *Iron-Binding Agents*

Compounds that can complex ferrous ions inhibit aconitase (*33*). This inhibition can be reversed by the subsequent addition of ferrous ions (*19, 33*). Various metal chelators have been tested as inhibitors of aconitase (see Table IV) (*68, 71*). The inhibition of aconitase by these chelators can, in *Bacillus subtilis*, prevent sporulation.

2. *Fluorocitrate*

Studies by Peters and his collaborators (*72, 73*) have shown that the toxic action of fluoroacetate, found as a natural product in the leaves of plants such as *Dichapetalum cymosum* and *Acacia georgina*, is a result of the "lethal synthesis" of fluorocitrate within the body from the fluoroacetate in the leaves. The product, fluorocitrate, is the inhibitor of aconitase (*74*), not the fluoroacetate itself. The amount of fluoroacetate in the leaves of these plants is sufficient to kill grazing cattle. Aconitase in these plants is apparently less strongly inhibited by fluorocitrate than is pig heart aconitase. For example, sycamore aconitase (*75*) is 2000-fold less sensitive to fluorocitrate than is the enzyme from animal sources. As a result of fluorocitrate poisoning there is an accumulation

71. E. B. Herr, Jr., J. B. Sumner, and D. W. Yesair, *ABB* **62**, 29 (1956).
72. F. L. M. Pattison and R. A. Peters, *in* "Handbuch der experimentellen Pharmakologie" (O. Eichler, ed.), Vol. **20**, p. 387. Springer, Berlin, 1966.
73. R. A. Peters, *Endeavour* **13**, 147 (1954).
74. W. D. Lotspeich, R. A. Peters, and T. W. Wilson, *BJ* **51**, 20 (1952).
75. D. H. Treble, D. T. A. Lamport, and R. A. Peters, *BJ* **85**, 113 (1962).

TABLE IV
SOME REPORTED INHIBITORS OF ACONITASE AND THEIR INHIBITOR CONSTANTS

Inhibitor	Enzyme source and conditions	K_i (M)	Type of inhibition[a]	Reference
		(a) Metal Chelators		
α-Picolinic acid	*Bacillus subtilis*	2.4×10^{-2}	nc	*(68)*
Quinaldic acid	*B. subtilis*	2.7×10^{-2}	nc	*(68)*
o-Phenanthroline	*B. subtilis*	5.0×10^{-4}	nc	*(68)*
α,α′-Bipyridyl	*B. subtilis*	1.2×10^{-3}	nc	*(68)*
		(b) Carboxylic Acids		
Fluorocitrate	Rat liver	10^{-6}	hc	*(77)*
Fluorocitrate	Pig heart	2.9×10^{-4}	lc	*(78)*
Oxalomalate	Potato mitochondria	2×10^{-4}	c	*(84)*
Oxalomalate	Yeast	1.4×10^{-4}	c	*(84)*
Oxalomalate	Pig heart	$1.3–2.3 \times 10^{-5}$	c	*(21)*
Oxalomalate	Rat liver mitochondria	1×10^{-6}	c	*(21)*
Oxalomalate	Rat liver cytoplasm	2.5×10^{-6}	c	*(21)*
γ-Hydroxy-α-keto-glutarate	Pig heart	$4.8–8.5 \times 10^{-4}$	c	*(21)*
trans-Aconitate	Pig heart	1.4×10^{-4}	lc	*(64)*
threo-L_s-Isocitrate	Beef heart	4×10^{-3}	c	*(53)*
Tricarballylate	Beef heart	2.3×10^{-3}	lc	*(64)*
Oxaloacetate	Pig heart	7×10^{-4} (slope) 16.5×10^{-4} (intercept)	pnci	*(64)*
1,2,3-Tricarboxy-cyclopentene-1	Pig heart	25×10^{-3}	c	*(86)*
1,2,3,4-Tetracarboxy-cyclopentane	Pig heart	15×10^{-3}	c	*(86)*
1,3,5-Tricarboxy-pentane	Pig heart	50×10^{-3}	c	*(86)*
Phthalic acid	Pig heart	100×10^{-3}	c	*(86)*
Trimesic acid	Pig heart	22×10^{-3}	c	*(86)*
Trimellitic acid	Pig heart	11×10^{-3}	c	*(86)*
Pyromellitic acid	Pig heart	5.3×10^{-3}	c	*(86)*

[a] Key: c, competitive; nc, noncompetitive; hc, hyperbolic competitive; lc, linear competitive; and pnci, parabolic noncompetitive inhibitor.

of citrate in the tissues. It is, however, not clear that the inhibition of aconitase is the sole cause of death. Other enzymes such as succinic dehydrogenase *(76)* are presumably also affected by fluorocitrate.

Fluorocitrate has been described as a hyperbolic competitive inhibitor

76. E. Kun, *in* "Citric Acid Cycle. Control and Compartmentation" (J. M. Lowenstein, ed.), Chapter 6, p. 318. Marcel Dekker, New York, 1969.

of the enzyme from rat liver ($K_i \sim 10^{-6}\,M$), which can then slowly inactivate the enzyme (*77*). With the pig heart enzyme fluorocitrate acts as a linear competitive inhibitor during the initial reaction ($K_i = 2.9 \times 10^{-4}\,M$) followed by an irreversible inactivation (*78*).

Of the four possible isomers of fluorocitric acid, the isomer that is derived from fluoroacetyl coenzyme A and oxaloacetate is a strong inhibitor of aconitase. The isomer of fluorocitric acid derived from acetyl coenzyme A and fluorooxaloacetate is not a strong inhibitor but has been found (*77, 79*) to be the mirror image (same diastereoisomer, opposite stereoisomer) of the active inhibitor of aconitase (see Section III,I). Methyl monofluorocitrate and α-difluorocitrate are neither inhibitors nor substrates (*77*).

When *cis*-aconitate is the substrate, fluorocitrate is a noncompetitive inhibitor [K_i (intercept) $= 1.5 \times 10^{-4}\,M$] (*64*). The conversion of citrate to *cis*-aconitate is inhibited more strongly by fluorocitrate than the reaction of isocitrate to *cis*-aconitase (*20*). These results are probably because *cis*-aconitate has a lower K_m value than does citrate or isocitrate and hence tends to displace the inhibitor more efficiently (*75*). Alternatively, it has been suggested (*20*) that two or more active sites may be involved.

3. *Other Carboxylic Acids*

a. Oxalomalate and γ-Hydroxy-α-ketoglutarate. Mixtures of glyoxylate and oxaloacetate inhibit citrate oxidation (*80*). This results from a condensation product of these two compounds (*81–83*) which competitively inhibits aconitase (see Table IV) (*21, 84*) and may thereby permit a regulation of the tricarboxylic acid cycle by glyoxylate. The condensation products are oxalomalate and γ-hydroxy-α-ketoglutarate. The ox-

$$\begin{array}{c} COO^- \\ | \\ C{-}H \\ \| \\ O \end{array} + \begin{array}{c} CO{-}COO^- \\ |\quad\quad | \\ CH_2{-}COO^- \end{array} \longrightarrow \begin{array}{c} {}^-OOC\quad CO{-}COO^- \\ |\quad\quad | \\ HO{-}C{-}CH{-}COO^- \\ | \\ H \end{array} \longrightarrow \begin{array}{c} {}^-OOC\quad CO{-}COO^- \\ |\quad\quad | \\ HO{-}C{-}CH_2 \\ | \\ H \end{array} + CO_2 \qquad (5)$$

Glyoxylate Oxaloacetate Oxalomalate = α-Hydroxy-β-oxalosuccinate γ-Hydroxy-α-ketoglutarate

77. D. W. Fanshier, L. K. Gottwald, and E. Kun, *JBC* **239,** 425 (1964).
78. H. L. Carrell, J. P. Glusker, J. J. Villafranca, A. S. Mildvan, R. J. Dummel, and E. Kun, *Science* **170,** 1412 (1970).
79. R. J. Dummel and E. Kun, *JBC* **244,** 2966 (1969).
80. F. D'Abramo, M. Romano, and A. Ruffo, *BJ* **68,** 270 (1958).
81. A. Ruffo, *Bull. Soc. Chim. Biol.* **43,** 705 (1961).
82. A. Ruffo, E. Testa, A. Adinolfi, and G. Pelizza, *BJ* **85,** 588 (1962).
83. A. Ruffo, E. Testa, A. Adinolfi, G. Pelizza, and R. Maratti, *BJ* **103,** 19 (1967).
84. B. Payes and G. G. Laties, *BBRC* **10,** 460 (1963).

alomalate readily loses carbon dioxide and is converted to γ-hydroxy-α-ketoglutarate. Rat liver mitochondrial aconitase is much more sensitive to oxalomalate than is the soluble enzyme (*85*). These inhibitors may fit the active site in a manner similar to *cis*-aconitate. The inhibition by oxalomalate becomes noncompetitive after preincubation with the enzyme (*21*). The absence of a third carboxyl group can explain the lower activity of the γ-hydroxy-α-hydroxyglutarate.

b. trans-Aconitate. The *trans*-aconitate ion is a linear competitive inhibitor of aconitase (*53, 57, 64*). Normal samples of *cis*-aconitate contain at least 6% of the *trans* isomer and this fact should be considered in kinetic studies with *cis*-aconitate. It is claimed (*57*) that some inactivation occurs at high inhibitor concentrations.

threo-L_s-Isocitrate, the mirror image form of the active isomer of isocitrate, is a feeble competitive inhibitor (*53*) and so is tricarballylate (*64*).

c. Carboxylic Acid Derivatives of Ring Compounds. Gawron (*86*) studied some aromatic acids and other compounds in which three or four carboxyl groups are held in a constrained manner. He found that they were weak inhibitors, the tetraacids being better inhibitors than the triacids, the diacids being noninhibitory. These compounds were assumed to be competitive inhibitors. 1,2,3-Tricarboxycyclopentene-1 was prepared as a compound which would contain a constrained grouping similar to the *cis*-aconitate ion, but it can be seen that it is only a very weak inhibitor, possibly because of its bulkiness.

4. *Other Inhibitors*

Aconitase is inhibited by compounds which react with sulfhydryl groups, such as *p*-hydroxymercuribenzoate (*33*) and, to a lesser extent, mercuric chloride, *o*-iodosobenzoate, ω-chloroacetophenone, ω-bromoacetophenone, and diphenylchloroarsine (*87*). Metals such as magnesium, calcium, and manganese displace the equilibrium toward citrate (*61*) (see Section III,E). Cyanide is a noncompetitive inhibitor (*88*). Nitrite acts as an inhibitor in certain bacteria (*89*) and so does hydrogen peroxide (*90*). There are also reports of inhibition by ferrous oxalo-

85. A. Adinolfi, V. Guarriera-Bobyleva, S. Olezza, and A. Ruffo, in preparation.
86. O. Gawron, personal communication (1970).
87. S. R. Dickman, "The Enzymes," 2nd ed., Vol. 5, p. 495, 1961.
88. J. Tomizawa and H. Fukini, *Symp. Enzyme Chem.* (*Tokyo*) **8,** 3 (1953).
89. J. W. T. Wimpenny and A. M. H. Warmsley, *BBA* **156,** 297 (1968).
90. E. E. Bruchmann, *Naturwissenschaften* **48,** 305 (1961).

acetate, although not oxaloacetate itself (*66*). Villafranca finds that oxaloacetate is a parabolic noncompetitive inhibitor, and that citramalate, D-malate, and L-malate are not inhibitors of pig heart aconitase when citrate or isocitrate are utilized as substrates (*64*). Citramalate is suggested (*91*) to be the inhibitor of aconitase in sour lemons, *Citrus limon* Linn., causing accumulation of citric acid, as opposed to sweet lemons, *Citrus limettiodies*, in which the aconitase is not, or less, inhibited.

H. Single vs. Dual Catalytic Site

Jacobsohn and Tapadinhas (*4*) suggested that two aconitases are present in equal amounts in most tissues—α-aconitase which catalyzes the interconversion of isocitrate and *cis*-aconitate, and β-aconitase which catalyzes the interconversion of citrate and *cis*-aconitate. At present it is believed that most of the evidence supports the hypothesis of a single active site that can catalyze the interconversions of all three substrates.

The evidence in favor of a single catalytic site may be summarized as follows. Several workers (*26, 27, 30*) find no differences in the relative rates of the reactions on purification. The pH optima are the same for all the reactions in a given buffer solution (*37*). If equal amounts of both substrates are used, the rate of formation of *cis*-aconitate should be the sum of the rates of dehydration of the single substrates if there are two active sites but only the average of these rates if there is one site. The latter was found experimentally (*63, 65, 92*). This work was continued with radioactive substrate (*92*). *cis*-Aconitate and labeled isocitrate were mixed with enzyme. The specific activity of *cis*-aconitate was always less than that of citrate showing that free *cis*-aconitate is not an obligatory intermediate in the reaction of isocitrate to citrate. Physical evidence for competition between citrate and isocitrate at the same manganese-binding site has been shown by NMR studies (*93*). It has been shown that manganous ions compete for the same metal binding site as ferrous ions and that isocitrate and citrate can each compete for this same manganese-binding site on the enzyme suggesting a single catalytic site and an E–M–S complex. It has also been shown by radioactive labeling experiments (*6*) that the hydrogen atom (tritium in this

91. E. Bogin and A. Wallace, *Proc. Am. Soc. Hort. Sci.* **89**, 182 (1966).
92. J. Tomizawa, *J. Biochem.* (*Tokyo*) **41**, 567 (1954).
93. J. J. Villafranca and A. S. Mildvan, *Abstr. Middle Atlantic Reg. Meeting, Am. Chem. Soc.*, No. BI14 (1970).

study) that is abstracted from citrate is added to the intermediate in the reaction to give isocitrate without any loss of tritium, initially, to the medium. This indicates the direct conversion of citrate to isocitrate without the obligatory intermediation of *cis*-aconitate, and it is difficult to see from this evidence how there could be two active sites unless they are adjacent or unless there are three sites.

The evidence cited for two active sites includes the fact that for many inhibitors the reaction of *cis*-aconitate is less inhibited than the reaction of isocitrate which is in turn less inhibited than the reaction of citrate. This has been much studied for fluorocitrate (*20*). An alternative explanation (*74*) is that these differences result from the different K_m values of the substrates. The K_i values are similar when citrate and *cis*-aconitate are substrates (*64, 78*) suggesting that this latter explanation is a good one. Peters (*7*) found that as pig heart aconitase ages the loss of activity is greater in the reaction of *cis*-aconitate formation from citrate than from isocitrate. Partial inactivation of a single active site of an enzyme is known (*94*). Alternatively, the partially purified enzyme preparation could have contained citrate dehydratase (*15, 17*). Racker (*52*) obtained a higher purification of an enzyme catalyzing the citrate to *cis*-aconitate step than the isocitrate to *cis*-aconitate step, a finding that may also have been a result of the presence of citrate dehydratase.

I. Mechanism of Action

It has been suggested (*95*) that C–H bond making is a rate-limiting step in the hydration of *cis*-aconitate but that C–H bond breaking is not rate-limiting in the dehydration. However, it was proposed (*53*) from kinetic isotope effects that the first (and rate-limiting) step in the dehydration of citrate or isocitrate is the rupture of the C–O rather than the C–H bond.

The scheme of Rose and O'Connell (*6*), shown in Fig. 2, stresses the importance of the two species E and EH. Citrate and isocitrate react only with the former and *cis*-aconitate with the latter which may, in part, explain the different behavior with, for example, fluorocitrate. The dissociation of the EH complex has been shown to be slow relative to the rate of exchange of *cis*-aconitate (*6*). Proton transfer to and from

94. R. F. Colman, *JBC* **243**, 2454 (1968).
95. S. Englard and S. P. Colowick, *JBC* **226**, 1047 (1957).

the medium may be rate-limiting steps in the dehydration and hydration reactions, respectively.

As seen in Fig. 1, the absolute stereochemistry of the aconitase reaction requires that the plane of the *cis*-aconitate ion rotate since the hydrogen atom that is abstracted is retained by the enzyme (*6*) and added again to the other side of the plane. This necessity for rotation was originally pointed out by Ogston (*5*) and further investigated by Gawron (*47*). A mechanism proposed by Gawron (*47*), involving two coupled histidines, cannot explain the retention of hydrogen by the enzyme. A more complete description of the mechanism, incorporating the ideas and results of Ogston (*5, 9*), Dickman and Speyer (*11*), Gawron (*46*), and

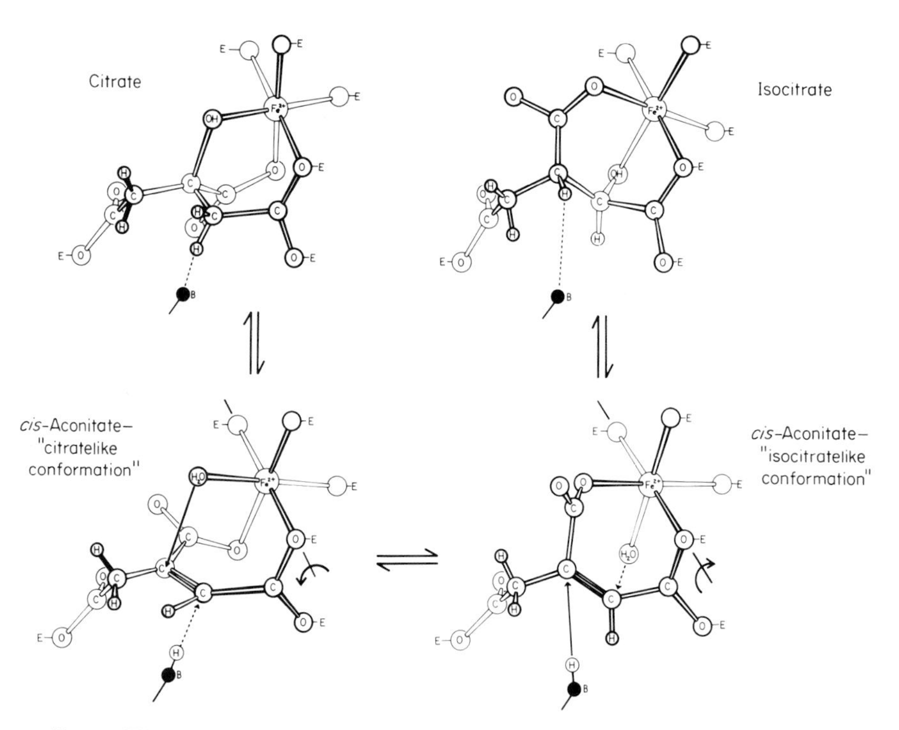

FIG. 3. The ferrous wheel mechanism proposed (*96*) for the action of aconitase. The various enzyme–ferrous substrates are depicted. These may dissociate to give free substrates. The probable points of attachment of the substrate to enzyme are denoted E and the proton abstracting group is denoted B. The addition of hydroxyl and hydrogen are *trans* and in a direction perpendicular to the plane of the double bond in *cis*-aconitate. The axis of rotation of the ferrous wheel which allows interconversion between the two possible conformations of the enzyme–ferrous–*cis*-aconitate complexes is indicated by a curved arrow.

Rose (*6*) has been proposed by Glusker (*96*) from X-ray crystallographic studies of the preferred conformations and modes of packing of the crystalline substrates. Results from such studies place stereochemical restrictions on the shape of the active site. The work has been extended (*78, 97*) to explain the manner of inhibition and inactivation by fluorocitrate. The three anions citrate, isocitrate, and fluorocitrate can crystallize in an extended shape and can all form tridentate chelates with cations which may be present. In the proposed "ferrous wheel" mechanism all three carboxyl groups are held on the enzyme, and the substrates (citrate or isocitrate) chelate to the ferrous iron via two carboxyl groups and a hydroxyl group as suggested by Dickman and Speyer (*11*). The hydrogen atom that is abstracted is added again from the same position on the active site of the enzyme to give either hydroxyacid again. This is illustrated in Fig. 3.

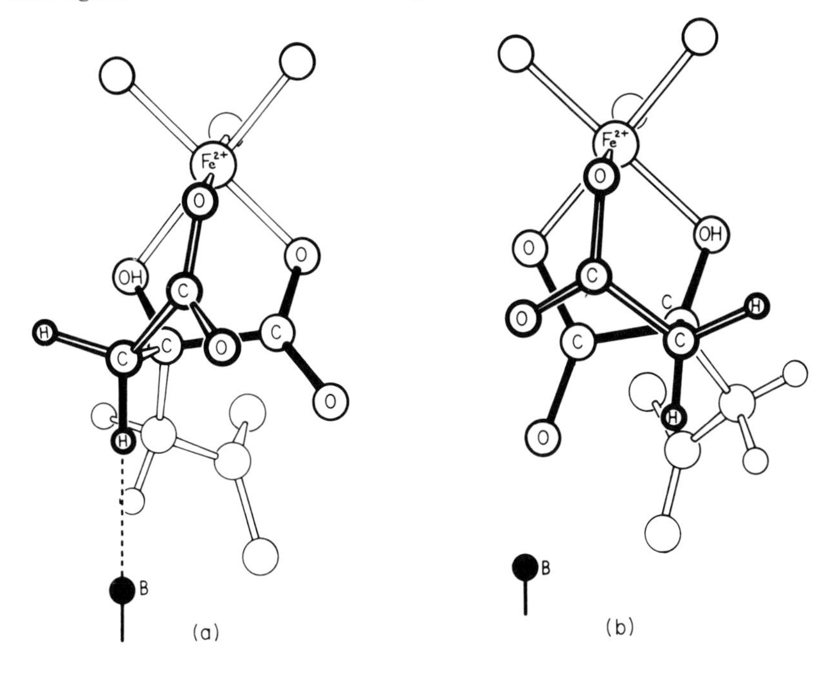

FIG. 4. Differentiation, by aconitase, of the two —CH_2COO^- groups of citrate. When citrate is attached in the "right way" (a) the hydrogen *trans* to the hydroxyl group is abstracted. When citrate is attached in the "wrong way" (b) the hydrogen atom *trans* to the hydroxyl group is too far from the proton abstracting group (solid circle) to be abstracted.

96. J. P. Glusker, *JMB* **38,** 149 (1968).
97. H. L. Carrell and J. P. Glusker, *Abstr. Am. Cryst. Assoc. Meeting, Tulane Univ., New Orleans, La.* No. F2 (1970).

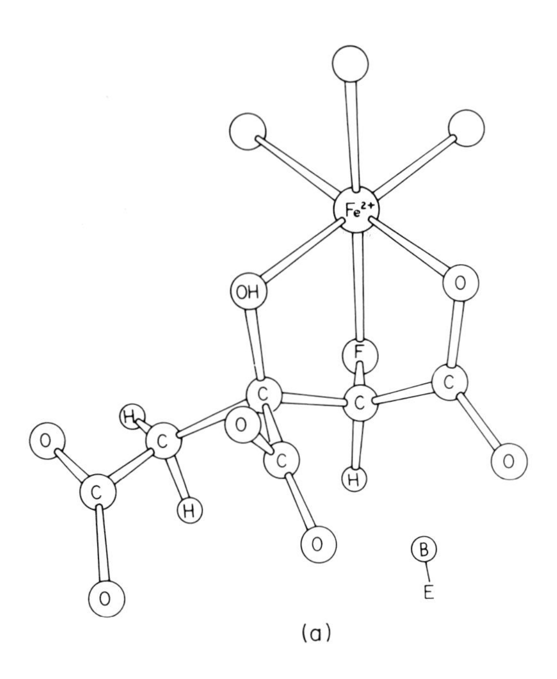

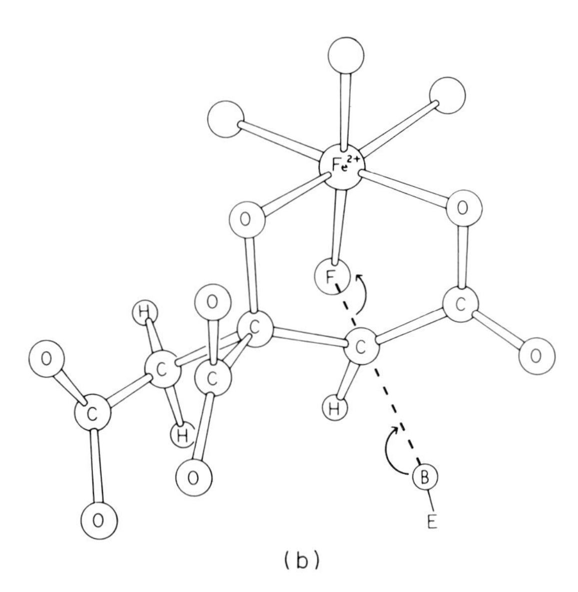

FIG. 5. Mechanism proposed *(78)* for inhibition (a) and inactivation (b) of aconitase by fluorocitrate. When fluorocitrate binds the wrong way it can inhibit,

In this mechanism the conformations of the citrate and isocitrate chelates are those which were found crystallographically. Removal of the elements of water gives enzyme-bound *cis*-aconitate in two conformations referred to as the "citrate-like" and the "isocitrate-like" conformations. The interchange between these conformers is effected by either a rotation in the ferrous coordination sphere or by a ligand–water substitution in the coordination sphere of iron. The latter is preferable because it does not suggest movements of groups on the active site of the enzyme. The addition of water occurs in a *trans* fashion perpendicular to the plane of the double bond. This plane rotates 90° as shown. The mechanism is consistent with the fact that α-methyl-*cis*-aconitate is also a substrate.

The function of the iron atom is probably to aid the attachment of the substrate to the surface of the active site of the enzyme and, by virtue of its positive charge, to facilitate the removal of the hydroxyl group from citrate or isocitrate by acting as an electron acceptor for a lone pair of electrons on the oxygen atom of the hydroxyl group, thus drawing electrons away from the C–O bond. Such iron–substrate–enzyme complexes were depicted by Morrison (*34*). It is not clear why ferrous iron alone can accomplish this catalysis.

This mechanism provides an explanation for the distribution made between the two ends of the citrate ion. As a result of chelation to the metal there is a "wrong way" and a "right way" for citrate to attach itself to the active site. As shown in Fig. 4, only when citrate is attached in the right way is the hydrogen atom *trans* to the hydroxyl group in a position where it can be abstracted by the hydrogen-abstracting group in the enzyme.

Consistent with this mechanism is the detection of an inactive aconitase–manganous–citrate bridge complex (*93*). The mean distance between the bound manganese and the protons of citrate (4.2 ± 0.6 Å), determined from the longitudinal relaxation rate of the protons of citrate, is consistent with the mean distance (4.47 Å) calculated from crystallographic data.

X-ray crystallographic studies (*78, 97*) of a racemic salt, rubidium ammonium hydrogen fluorocitrate, containing the inhibitory isomer (*79*), showed that the fluorine chelated to the metal and revealed the relative configuration of the two asymmetric carbon atoms is (1*R*:2*R* or 1*S*:2*S*, that is, *erythro*-fluorocitrate). If the same absolute stereochemistry is

---

but with time and under the influence of the metal the C–F bond may be broken and the proton abstracting group (solid circle) alkylated. An intermediate stage (b) in the alkylation is illustrated.

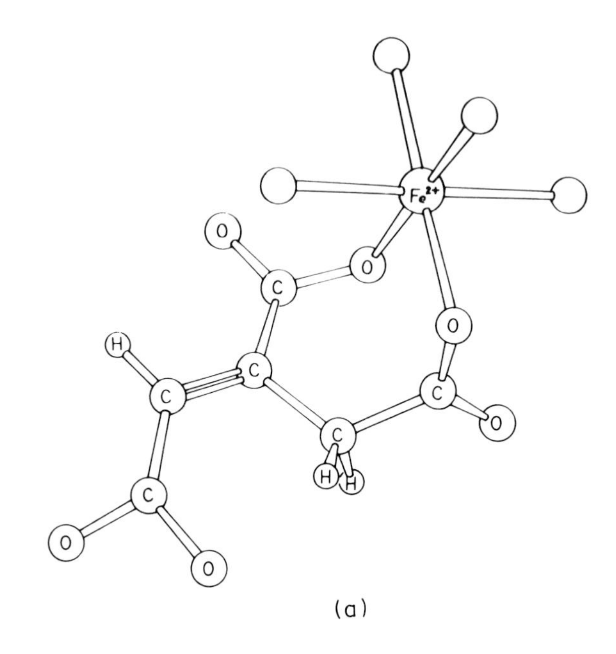

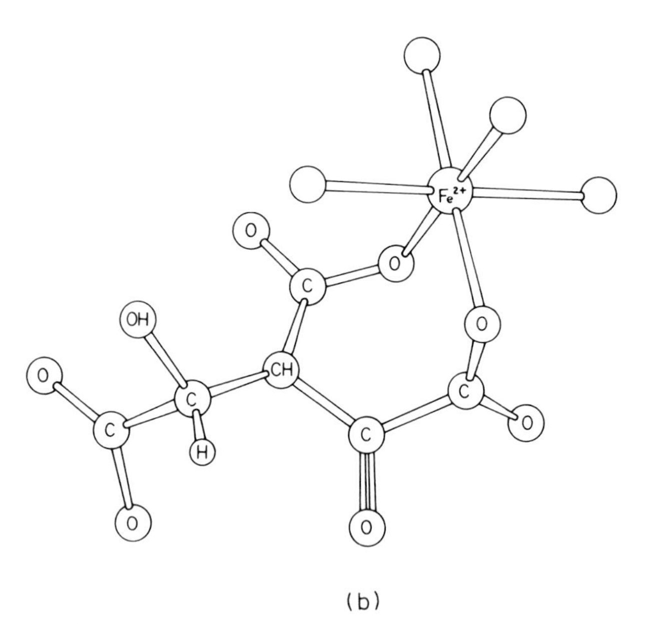

FIG. 6. Possible modes of attachment of the competitive inhibitors (a) *trans*-aconitate and (b) oxalomalate in the active site are shown.

assumed for the action of the citrate condensing enzyme (EC 4.1.3.7) with fluoroacetyl coenzyme A as with acetyl coenzyme A then the inhibitory isomer must be the $1R{:}2R$ form as shown in Fig. 5. This is being checked by absolute configuration studies on the resolved inhibitory isomer (*78*). A comparison of the conformations of the two substrates with that of fluorocitrate showed how it could act as a linearly competitive inhibitor by binding to the metal via the terminal carboxyl group the fluorine atom and the hydroxyl group, as shown in Fig. 5. To do this the fluorocitrate must fit in the active site in the wrong way (suggesting that citrate can do this part of the time). It was suggested that the fluorocitrate may then act as an inactivator by alkylating the active site near or at the hydrogen-abstracting group as a result of cleavage of the C–F bond. Further experiments, however, seem to indicate (*64, 98*) that the inactivation does not occur by defluorination.

These principles may be extended to other inhibitors, particularly the linearly competitive ones as shown in Fig. 6. Presumably *trans*-aconitate chelates with the double bond away from the active site as shown. An aconitase–manganous–*trans*-aconitate complex has been detected by NMR studies (*48*) with dimensions consistent with this structure. Thus the manganese to methyne hydrogen distance (~5.0 Å) is greater than the manganese to methylene hydrogen distance (~4.4 Å). In a similar way, in its interaction with aconitase, oxalomalate may be able to mimic the *cis*-aconitate ion.

Acknowledgments

The author wishes to thank Dr. J. J. Villafranca for many helpful discussions. She also wishes to thank Drs. Kun, Peters, Ruffo, Gawron, Mildvan, and Villafranca for making available copies of manuscripts prior to publication. The experimental results from the author's laboratory and the work on this chapter were supported by U. S. P. H. S. grants CA-10925 (formerly AM-02884), CA-06927, and RR-05539 from the National Institutes of Health, and by an appropriation from the Commonwealth of Pennsylvania.

98. R. Z. Eanes and E. Kun, unpublished observations (1970).

15

β-Hydroxydecanoyl Thioester Dehydrase

KONRAD BLOCH

I. Introduction

In certain bacterial systems the biosynthesis of long-chain unsaturated fatty acids is an anaerobic process (*1*). Such organisms, in contrast to

1. K. Bloch, P. E. Baronowsky, H. Goldfine, W. J. Lennarz, R. Light, A. T. Norris, and G. Scheuerbrandt, *Federation Proc.* **20,** 921 (1961).

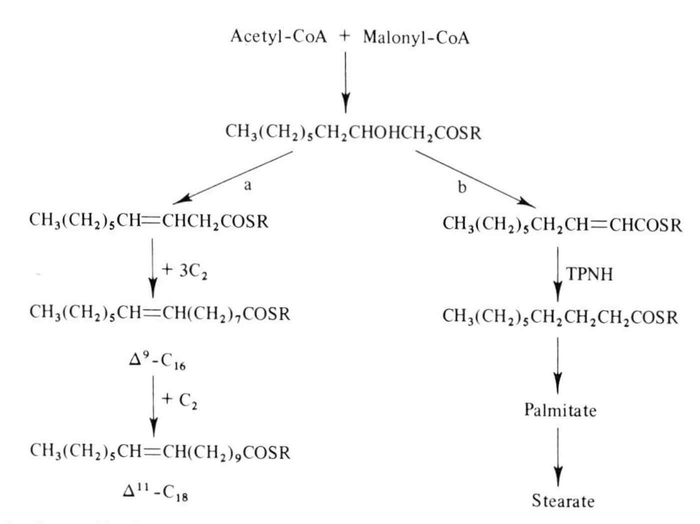

FIG. 1. Generalized scheme of fatty acid synthesis in *Escherichia coli*. Reaction a initiates the pathway to unsaturated acids and reaction b, the branch to saturated acids.

yeast, animal tissues, and plants, cannot convert stearate to oleate or palmitate to palmitoleate by oxidative hydrogen abstraction (desaturation). Instead, they dehydrate β-hydroxyacyl thioesters (principally C_{10}) to β,γ-enoate intermediates which they subsequently elongate to palmitoleate and vaccenate by the common reactions of fatty acid synthesis (*2*). The dehydrating enzyme that furnishes the initial olefinic products for this bacterial pathway to unsaturated fatty acids is β-hydroxydecanoyl thioester dehydrase (called dehydrase from here on). Owing to their association with dehydrase such bacterial fatty acid synthetases catalyze a branched pathway leading to long-chain unsaturated as well as saturated fatty acids from β-hydroxydecanoate as the last common intermediate (Fig. 1).

The suggestion, embodied in Fig. 1, that palmitoleate and vaccenate may arise in bacteria by direct chain elongation of shorter olefinic acids was first made by Hofmann *et al.* (*3*).

Certain mutants of *Escherichia coli* (*4*) and *E. coli* cultures inhibited by

2. G. Scheuerbrandt, H. Goldfine, P. E. Baronowsky, and K. Bloch, *JBC* **236,** PC70 (1961).
3. K. Hofmann, W. M. O'Leary, C. W. Yoho, and T. Y. Liu, *JBC* **234,** 1672 (1959).
4. D. F. Silbert and P. R. Vagelos, *Proc. Natl. Acad. Sci. U. S.* **58,** 1579 (1967).

3-decynoyl-*N*-acetylcysteamine (*5*) produce only saturated fatty acids. On addition of purified dehydrase to extracts of such cells, the capacity for synthesizing palmitoleate and vaccenate is fully restored. Dehydrase is, therefore, the essential enzymic component for olefin formation in these systems.

There is presumptive evidence for the occurrence of dehydrase or for the dehydrase mediated synthesis of unsaturated fatty acids in several species of bacteria. However, only dehydrase from *E. coli* has been thoroughly characterized to date.

Escherichia coli dehydrase catalyzes the reversible interconversion of three substrates, D(—)β-hydroxydecanoyl thioester, α,β-decenoyl thioester and β,γ-decenoyl thioester (*6–10*):

$$\mathrm{CH_3(CH_2)_5CH_2\overset{H}{\underset{OH}{C}}-CH_2COSR} \underset{a'}{\overset{a}{\rightleftharpoons}} \underset{trans}{\mathrm{CH_3(CH_2)_5CH_2CH{=}CH \cdot COSR}}$$

$$c \downarrow\uparrow c' \qquad\qquad b' \uparrow\downarrow b \tag{1}$$

$$\underset{cis}{\mathrm{CH_3(CH_2)_5CH{=}CH \cdot CH_2COSR}}$$

Reaction b′ [Eq. (1)] is the most rapid of the dehydrase-catalyzed reactions and convenient for assay purposes since it generates a product that has the α,β-unsaturated thioester chromophore ($\epsilon = 6700$ at 263 nm).

Dehydrase from *E. coli* B has been purified 1000-fold (8 steps) to a specific activity of 6000 [isomerization of 6000 nmoles of *cis*-β,γ-decenoyl-NAC (NAC stands for *N*-acetylcysteamine) to *trans*-α,β-decenoyl-NAC per minute per milligram of protein] in a 23% overall yield; 12 mg of pure enzyme are obtained per kilogram of frozen cells (*10, 11*). Varying the culture conditions (temperature, composition of the medium, and phase of growth) does not significantly alter the dehydrase content of *E. coli* B.

The present dehydrase preparations are homogeneous in the ultracentrifuge and migrate as a single entity on polyacrylamide gels (*10, 11*). Certain molecular properties of the enzyme will be discussed below.

5. L. R. Kass, *JBC* **243**, 3223 (1968).
6. W. Lennarz, R. Light, and K. Bloch. *Proc. Natl. Acad. Sci. U. S.* **48,** 840 (1962).
7. A. T. Norris and K. Bloch, *JBC* **238,** PC3133 (1963).
8. A. T. Norris, S. Matsumura, and K. Bloch, *JBC* **239,** 6353 (1964).
9. L. R. Kass, D. J. H. Brock, and K. Bloch, *JBC* **242,** 4418 (1967).
10. D. J. H. Brock, L. R. Kass, and K. Bloch, *JBC* **242,** 4432 (1967).
11. G. M. Helmkamp, Jr. and K. Bloch, *JBC* **244,** 6014 (1969).

II. Substrate Specificity

The thioester derivatives of the straight-chain 10-carbon acids are the most active substrates for dehydrase (*10, 12*). Relative activities for the dehydration reaction [a, Eq. (1)] and for isomerization [b′, Eq. (1)] calculated from initial velocities are given in Table I.

TABLE I
RELATIVE ACTIVITIES (INITIAL VELOCITIES) OF DEHYDRASE SUBSTRATES (NAC DERIVATIVES, 3×10^{-4} *M*)[a]

Chain	Dehydration[b] β-OH $\rightarrow$ α,β-enoate	Isomerization β,γ $\rightarrow$ α,β-enoate
Straight[c]		
C_8	0.05	0.00
C_9	0.56	0.28
C_{10}	1.00	1.00
C_{11}	0.77	0.75
C_{12}	0.03	0.00
Branched[d]		
iso-C_{10}	0.06	
ante iso-C_{10}	0.04	
iso-C_{11}	0.01	
ante iso-C_{11}	0.03	
iso-C_{12}	0.06	
ante iso-C_{12}	0.03	
D,L α-Methyl-C_{10}	0.003	

[a] Product formation was followed spectrophotometrically at 263 nm.
[b] All hydroxyacyl substrates were the racemic compounds.
[c] Data from Brock *et al.* (*10, 12*).
[d] Data from Endo and Bloch (*14*).

The data in Table I provide the documentation for the scheme shown in Fig. 1, which specifies that the dehydration step leading to the olefinic precursor of palmitoleate and vaccenate occurs only at the C_{10} stage. The negligible activity of the next lower (C_8) and the next higher (C_{12}) even-numbered homologues as dehydrase substrates rationalizes the absence of Δ^{11}-C_{16} and Δ^{13}-C_{18} acids in *E. coli* lipids (*13*). These isomers would be expected to arise by chain elongation of β,γ-octenoate and β,γ-dodecenoate, respectively. Although the activity of the C_9 and C_{11} substrates is substantial, dehydrase operates as a C_{10} specific enzyme in

12. G. M. Helmkamp, R. R. Rando, D. J. H. Brock, and K. Bloch, *JBC* **243**, 3229 (1968).
13. G. Scheuerbrandt and K. Bloch, *JBC* **237**, 2064 (1962).

the cell since odd-numbered fatty acids do not normally occur in *E. coli* lipids.

The introduction of branches into the carbon chain (C_9–C_{11} iso- and ante iso acids) also lowers activity drastically (*14*). Thus, the active site of dehydrase appears to be so designed as to accommodate only linear hydrocarbon chains and these cannot be longer than C_{11} or shorter than C_9.

A. Thioester Specificity

In bacterial fatty acid synthesis all acyl intermediates remain attached to acyl carrier protein (ACP) while undergoing their various transformations (*15, 16*). This is likely to be true also for the dehydrase-catalyzed reactions which do in fact proceed best with the ACP thioester derivatives (*10*). Simpler thiols, however, replace ACP quite effectively. The fatty acyl *N*-acetylcysteamine derivatives are about one-fifth as active dehydrase substrates as those of ACP and in this respect superior to the CoA, 4-phosphopantetheine, or pantetheine derivatives (Table II). Adequate reactivity and ease of preparation make the NAC derivatives convenient substrates for dehydrase studies.

The following broader aspects of substrate specificity are of interest. With the exception of dehydrase, the constituent fatty acid synthetase enzymes of *E. coli* repeat identical steps each time the carbon chain is lengthened. A relatively broad chain-length specificity for the elongating enzymes is therefore expected. On the other hand, dehydrase catalyzes a

TABLE II
Relative Activities (Initial Velocities) of β-Hydroxydecanoyl Thioesters (3×10^{-4} *M*) as Dehydrase Substrates[a]

Thioester moiety	Relative rate
ACP	1.00
N-Acetylcysteamine	0.18
Pantetheine	0.12
4-Phosphopantetheine	0.11
Coenzyme A	0.11
N-Acetyl-β-alanyl cysteamine	0.02

[a] Data from Brock *et al.* (*10*) and D. Jones, unpublished data.

14. K. Endo and K. Bloch, unpublished data.
15. P. W. Majerus, A. W. Alberts, and P. R. Vagelos, *Proc. Natl. Acad. Sci. U. S.* **51,** 1231 (1964).
16. S. J. Wakil, E. L. Pugh, and F. Sauer, *Proc. Natl. Acad. Sci. U. S.* **52,** 106 (1964).

nonrecurring step (at the C_{10} level) requiring it to be highly chain-length specific.

Thiolester specificities for dehydrase and for the elongating enzyme proper also differ strikingly. While the chain-length specificity of the elongating enzymes is broad (*17*), their requirement for ACP thioesters is nearly absolute. Fatty acyl CoA derivatives react either not at all or at rates two orders of magnitude less than the corresponding ACP substrates (*18–20*). For the elongating enzymes interactions between enzyme–protein and substrate–protein (acyl-ACP) are clearly more important than hydrophobic contacts with the acyl chains. By contrast, for the C_{10} specific dehydrase the detailed structure of the thioester moiety matters much less. In this instance the hydrophobic acyl chain appears to be the main determinant in the binding of substrate to enzyme and the protein moiety of ACP a dispensable feature.

B. Optical and Geometric Specificity

The thioesters of the D(—)β-hydroxy acid, of the *trans*-α,β- and of the *cis*-β,γ-enoate are the active dehydrase substrates (*8, 10*). The L(+), *cis*-α,β, and *trans*-β,γ isomers are inert both as substrates or as enzyme inhibitors. These steric requirements are in keeping with the assigned functions of dehydrase which are to convert the normal D-β-hydroxyacyl intermediates of fatty acid synthesis into precursors of *cis*-olefinic acids and to furnish *trans*-α,β-enoate for reduction to the saturated acid by the *trans* specific *E. coli* enoyl reductase (*21*).

III. Reaction Mechanism

A. Kinetic Studies

The kinetics of an enzyme catalyzing the interconversion of three substrates is likely to be too complex to furnish a clear picture of the reaction mechanism involved. Also, the kinetic behavior of an isolated multifunctional enzyme need not reflect occurrences in a multistep pathway of which the enzyme in question is a component. The utility of the kinetic informa-

17. R. E. Toomey and S. J. Wakil, *JBC* **241**, 1159 (1966).
18. A. W. Alberts, P. Majerus, B. Talamo, and P. R. Vagelos, *Biochemistry* **3**, 1563 (1964).
19. P. W. Majerus, A. W. Alberts, and P. R. Vagelos, *JBC* **240**, 618 (1965).
20. A. W. Alberts, P. W. Majerus, and P. R. Vagelos, *Biochemistry* **4**, 2265 (1965).
21. G. Weeks and S. J. Wakil, *JBC* **243**, 1180 (1968).

tion from dehydrase studies has, therefore, been limited to the narrowing of choices between various reaction mechanisms.

Various lines of evidence indicate that the functionally significant event catalyzed by dehydrase is the conversion of β-hydroxydecanoyl thioester to β,γ-decenoate. Thus, a crude *E. coli* fatty acid synthetase containing dehydrase will convert the ACP derivatives of the three dehydrase substrates [in the presence of malonyl-CoA and triphosphopyridine nucleotide (TPNH)] to products in the ratios shown in Table III (*10*).

TABLE III
CONVERSION OF ACYL-ACP DEHYDRASE SUBSTRATES TO LONG-CHAIN SATURATED AND UNSATURATED FATTY ACIDS BY *E. coli* FATTY ACID SYNTHETASE[a,b]

Substrate	Products, % of total	
	Δ^9-C_{16} + Δ^{11}-C_{18}	C_{16} + C_{18}
D,L-β-Hydroxydecanoyl-ACP	83	17
β,γ-*cis*-Decenoyl-ACP	51	49
α,β-*trans*-Decenoyl-ACP	11	89
Acetyl-CoA	82	18

[a] The conditions were those of a standard fatty acid synthetase assay, tubes containing, apart from the ACP-substrates, TPNH, malonyl-CoA, ACP, and an *E. coli* ammonium sulfate fraction.
[b] Data from Brock *et al.* (*10*).

In view of the uncertain purity of the chemically prepared ACP substrates, these data are only approximate. Nevertheless, the end product ratios (Table III) allow two conclusions: (1) Under the given conditions dehydrase catalysis heavily favors the pathway involving the β,γ-enoate intermediate; and (2) the two decenoic thioesters are interconvertible, isomerization of β,γ-enoate to α,β-enoate proceeding more readily than the reverse reaction. The ratios of end products (long-chain acids) synthesized are, of course, determined not only by the rates of the dehydrase-catalyzed reactions but also by the relative rates of removal of the two dehydrase products in subsequent steps. At any rate, in association with the fatty acid synthetase system, dehydrase clearly brings about the formation of long-chain unsaturated acids; it functions predominantly as a β-hydroxyacyl-β,γ-enoyl dehydrase.

Kinetic studies with isolated dehydrase reveal entirely different quantitative relationships [Fig. 2, Brock *et al.* (10)]. When substrate (NAC derivatives) conversions are allowed to go to completion, the same equilibrium concentrations of products obtain with all three substrates (25–28% α,β-enoate 2–3.5% β,γ-enoate; and 68–74% β-hydroxy acid). The thermodynamically favored α,β isomer is the predominant product at

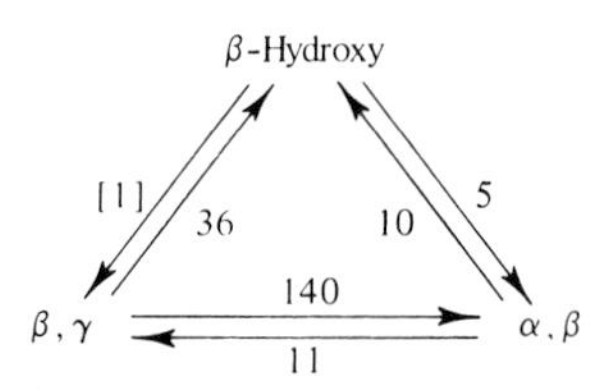

FIG. 2. Relative rates of interconversion of dehydrase substrates. The numbers represent relative initial velocities, normalized to the slowest of the rates (β-OH $\rightarrow$ β,γ). Substrate concentrations, 100 μM of NAC derivatives. Data from Brock *et al.* (*10*).

equilibrium (α,β-enoate:β,γ-enoate = 9:1) and also under initial velocity conditions (4–5:1). Yet in spite of this unfavorable equilibrium and the much more rapid α,β dehydration, dehydrase when associated with synthetase enzymes produces long-chain unsaturated fatty acids (derived from β,γ-decenoate) and long-chain saturated acids (derived from α,β-decenoate) in a ratio of 19:1. The explanation must be that β,γ-decenoate removal (by condensation) is much more rapid than α,β-decenoate removal (by reduction).

As for the chemical mechanism of the dehydrase-catalyzed reactions, the two issues of main interest are: (1) the sequence of events leading from hydroxy acid to the two enoic products, the choice being between formation of the two decenoates by two independent dehydration steps and a single dehydration to one enoate followed by isomerization to the other; and (2) the question whether dehydrase has true isomerase activity and if it is intrinsic, whether this activity is obligatory for dehydrase (dehydration–hydration) action. The initial reaction rates (composites of both rate constants and binding constants) allow some choice between various *a priori* schemes (Fig. 3). Among these, scheme (II) is ruled out by the rapid rate of the β,γ-decenoate–α,β-decenoate conversion which exceeds the rate of hydration of either α,β-decenoate or β,γ-decenoate by factors of 14 and 4, respectively. The further fact that isomerizations in either direction proceed at linear initial rates, without lag, also eliminates the intermediacy of β-hydroxy acid in the α,β–β,γ-decenoate conversion. The latter must be an isomerization in the mechanistic sense and direct rather than mediated by hydroxy acid. More evidence for intrinsic isomerase activity in dehydrase will be presented below.

Schemes (Ia) and (Ib) envision the alternative, single dehydrations to α,β- or β,γ-enoate followed by double bond isomerization. Chemically the more plausible dehydration is that to α,β-enoate. For this scheme (Ia) to be valid, the isomer (α,β–β,γ) ratio initially produced should, however, be at least as great as it is at equilibrium and not half (4–5:1, instead of 9:1). Furthermore, as shown in Fig. 2, going in the

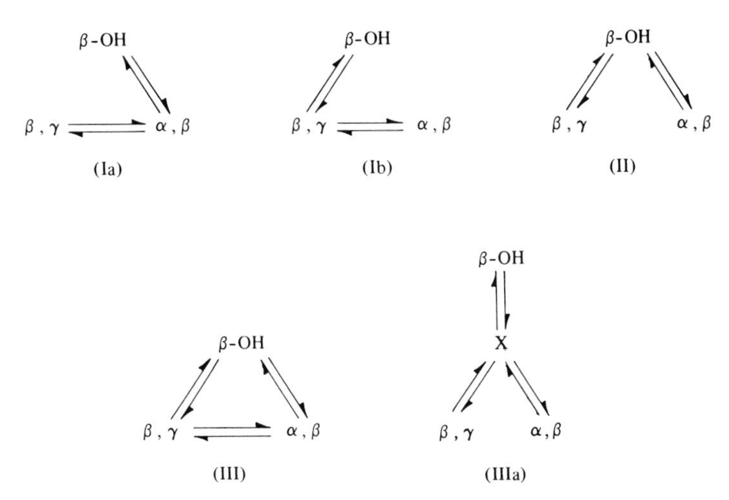

FIG. 3. Reaction schemes for dehydrase-catalyzed reaction.

reverse reaction β,γ-decenoate is hydrated 3–4 times more rapidly than α,β-decenoate. These data are not consistent with a role of α,β isomer as an obligatory intermediate in β,γ-decenoate formation. The argument, however, excludes only the intermediacy of the *free* α,β-enoate species (see below). As for the chemically less favorable scheme (Ib), this can also be discounted by kinetic arguments. If β,γ-enoate were a compulsory intermediate on the (reversible) path between β-hydroxy acid and α,β-enoate, then the rate of α,β-enoate formation should lag initially because of accumulation of β,γ isomer. This is not observed and moreover since dehydration to α,β-enoate is not slower than the rate of α,β–β,γ-enoate isomerization, β,γ-enoate cannot be on the path from β-hydroxy acid to α,β-enoate. The kinetic data also are at variance with scheme (III) which is merely a combination of schemes (I) and (II). Finally in (IIIa), a variant of scheme (III), an unspecified compound X is inserted centrally to connect the three dehydrase substrates. Evidence in favor of scheme (IIIa) and for the identity of intermediate X with enzyme bound α,β-decenoyl thioester is discussed below.

B. KINETIC ISOTOPE EFFECTS AND EXPERIMENTS WITH LABELED SUBSTRATES

Dehydrase catalysis requires as minimal events the abstraction and addition of hydrogen at C_α and C_γ (*22*). Examination of the rate of

22. R. R. Rando and K. Bloch, *JBC* **243**, 5627 (1968).

removal of these hydrogen atoms and of their stability during the reversible dehydration to the isomeric decenoates furnishes more explicit information on the mechanism of dehydrase action and resolves some of the problems raised by the kinetic analysis. In one such experiment the β-hydroxyacyl-α,β-enoate transformation was studied with α-D_2-1-^{14}C-β-hydroxydecanoyl-NAC as well as with the corresponding α-H_2 compound. A significant isotope effect, $k_H/k_D = 2.25$, was observed. The substitution of hydrogen by deuterium at C_α had no effect on K_m. Stretching of the C–H bond at C_α is, therefore, rate limiting in the β-hydroxy-α,β-enoate conversion. Either a carbanion or E_2 elimination mechanism is compatible with these results. No such deuterium isotope effect is observed in the dehydration of γ-D_2-1-^{14}C-β-hydroxydecanoyl-NAC, excluding hydrogen removal at C_γ as a control step in α,β-enoate formation. Similarly, when γ-^3H-1-^{14}C-β-hydroxydecanoyl-NAC is enzymically dehydrated, the $^3H:^{14}C$ ratio remains unchanged. This also shows that the C–H bonds at C_γ remain intact during the β-hydroxyacid-α,β-decenoate transformation. Dehydration to α,β-enoate is therefore straightforward and in line with chemical expectation.

Analogous labeling experiments designed to test the reaction path from β-hydroxy acid to β,γ-enoate led to a surprising result. When undergoing α,β-dehydration, α-^3H-^{14}C-β-hydroxydecanoyl-NAC lost the requisite atom of ^3H. Unexpectedly, the same loss of ^3H (from C_α) occurred also during conversion to β,γ-enoate. At various intervals after the start of the reaction the $^3H:^{14}C$ ratio in β,γ-decenoate was exactly the same as in the isomeric α,β-decenoate. Thus, the conversion of β-hydroxydecanoate to β,γ-decenoate cannot be a direct dehydration. It must include a deprotonation at C_α as an obligatory step. A decline of the $^3H:^{14}C$ ratio from 2 to 1 also attends the reverse process, the formal addition of water to β,γ-decenoate to form β-hydroxy acid. In view of the loss of one atom (and not of a fraction) of H at C_α both in the forward and in the back reaction, α,β-enoate becomes a mandatory intermediate between β-hydroxy acid and β,γ-enoate.

In conjunction with the kinetic analysis of the dehydrase system a two-step sequence for β,γ-enoate formation and transformation (via α,β-enoate) was considered [scheme (Ia) or (III), Fig. 3], but this was rejected as incompatible, *inter alia*, with the rate of hydration of β,γ-decenoate and the linearity of this process. To resolve the contradiction (of kinetic and isotopic results), we assume the putative C_α-deprotonated intermediate to be enzyme-bound rather than free α,β-decenoate [X in scheme (IIIa), Fig. 3]. Indirect evidence in support of this assumption is described below.

C. Trapping of Intermediates

Mercaptans add readily and irreversibly to α,β-unsaturated carbonyl compounds to form β-substituted thioesters (Michael addition).

$$RCH_2CH{=}CH \cdot \overset{\displaystyle O}{\overset{\|}{C}}{-}R \;+\; R'S^{\ominus} \longrightarrow R \cdot CH_2\underset{\displaystyle SR'}{\underset{|}{C}H}{-}CH_2{-}COR \qquad (2)$$

A transformation proceeding via a conjugated carbonyl intermediate and to more than one product might, therefore, be altered by mercaptans in such a way as to afford exclusively the Michael adduct of this intermediate. Indeed, in a system containing β,γ-decenoyl-NAC and enzyme, 10^{-2} *M* *N*-acetylcysteamine will completely quench the ordinarily rapid increase in absorption at 263 nm owing to α,β-enoate. Substrate is still consumed but the isomerized enoate is trapped by irreversible conversion to the nonabsorbing Michael addition product. More importantly, the dehydrase-catalyzed conversion of β,γ-decenoyl-NAC to β-hydroxy acid continues at nearly the normal rate in spite of the presence of *N*-acetylcysteamine. As will be recalled, the isotopic experiments described above required that this hydration proceed by way of α,β-decenoate as a compulsory intermediate. If this be the case, then the failure of the Michael reagent to interfere with the normal reaction path must mean that in the presence of enzyme the α,β-decenoate intermediate is shielded from reaction with mercaptan so that hydration or isomerization can proceed normally. Only after release or dissociation of α,β-decenoate from the enzyme will the mercaptan trap become effective. The above experiments are thus consistent with the postulate that enzyme-bound α,β-decenoate [X in scheme (IIIa), Fig. 3] is the central link between the three dehydrase substrates.

L(+)Keto-deoxyarabinose dehydrase studied by Portsmouth *et al.* (*23*) catalyzes a formally similar elimination of water, β,γ to a carbonyl function by way of an α,β-enoic conjugated system. In this instance isomerization is preceded and assisted by Schiff base formation between enzyme and substrate.

The double bond isomerization catalyzed by dehydrase proceeds with proton abstraction and addition and thereby differs from the intramolecular hydride transfer mechanisms demonstrated for Δ^5-3-ketosteroid

23. D. Portsmouth, A. C. Stoolmiller, and R. H. Abeles, *JBC* **242**, 2751 (1967).

isomerase (*24*) and dimethylallylpyrophosphate-isopentenylpyrophosphate isomerase (*25*).

IV. Protein Structure and Active Site

Molecular weight determinations of dehydrase by Sephadex gel filtration and sedimentation equilibrium analysis yield values of 34,500 ± 2,000 and 36,000 ± 1,500, respectively (*11*). Prosthetic groups have not been detected in the enzyme nor are metal ions required for catalysis. The homogeneous enzyme exhibits dehydrase and isomerase activity in a ratio of 1:15–17, a value that remains unchanged throughout purification. On dialysis against 3 *M* urea, the $s_{20,w}$ value of the enzyme changes from 3.35 to 2.09. The reduction in molecular size is associated with total and apparently irreversible loss of both dehydrase and isomerase activities. Exposure of dehydrase to 1% sodium dodecyl sulfate (SDS) at 60° for 20 min dissociates the enzyme to monomeric subunits having a molecular weight of 18,000–19,000 as estimated by SDS-polyacrylamide electrophoresis. Judging from the number of tryptic peptides, the associated dimeric enzyme appears to be composed of two identical subunits. This number agrees with that calculated from the amino acid composition for a polypeptide chain of molecular weight 18,000 (*11*).

A. Disulfide Bridges

Dehydrase contains 4 half-cystine residues per dimer, none titratable with *p*-mercuribenzoate or 5,5′dithio-bis(2-nitro)benzoate (DTNB) either in the presence or absence of urea (*11*). After treatment with mercaptoethanol in 5 *M* guanidine-HCl dehydrase reacts with iodoacetate at pH 8.5 to yield approximately 3 moles of carboxymethylcysteine after acid hydrolysis. Thus dehydrase contains two disulfide bonds, presumably as intrachain bridges, one each per monomer. Reduction of the native enzyme by mercaptoethanol does not alter its mobility during SDS-polyacrylamide electrophoresis.

Disulfide bridges are common in mammalian digestive hydrolases, such enzymes usually consisting of single polypeptide chains. They are found infrequently in intracellular enzymes. Some intracellular ex-

24. S. F. Wang, F. S. Kawahara, and P. Talalay, *JBC* **238**, 576 (1963).
25. B. W. Agranoff, H. Eggerer, U. Hennings, and F. Lynen, *JBC* **235**, 326 (1960).

amples, apart from dehydrase, are tryptophanase from *E. coli* (*26*) and histidinol dehydrogenase from *Salmonella typhimurium* (*27*). Though these three enzymes have subunit structure, they do not appear to have regulatory properties. The point may be raised whether the presence of intrachain disulfide bonds in oligomeric enzymes may not be incompatible with the conformational flexibility of the tertiary and quaternary structure that is thought to be characteristic of regulatory enzymes.

B. Active Site

Modifying reagents directed at cysteine, serine, or lysine residues in the polypeptide chain do not impair dehydrase activity. The enzyme is, however, sensitive to alkylating agents (bromoacetate at pH 7.0 and 1,3 dibromoacetone), to photooxidation, and to tetranitromethane. Histidine and tyrosine are, therefore, implicated as catalytically active amino acid residues. Inactivation rates given by the various reagents are shown in Table IV. The number of modified groups and of histidine and tyrosine residues remaining in the various inactive enzyme derivatives is listed in Table V.

TABLE IV
Rate of Inactivation of Dehydrase by Various Reagents at 25°

Reagent	$t_{1/2}$ (hr)
Photooxidation in presence of 0.05% methylene blue; pH 7.0	0.4
Bromoacetate, 0.05 *M*, pH 7.0	9.3
Bromoacetyl-NAC, 0.05 *M*, pH 7.0	31.5
1,3-Dibromoacetone, 0.05 *M*, pH 7.0	0.7
Tetranitromethane, 0.001 *M*, pH 8.0	1.25
3-Decynoyl-NAC, 0.008 m*M*, pH 6.0	0.05

Alkylation of dehydrase by bromoacetate reduces the content of histidine from 2 to 1.4–1.5 residues per monomer (molecular weight, 18,000) and affords 0.5 mole (uncorrected) of 3-carboxymethylhistidine. Calculated for the dimer (2.8–3.0 out of 4 histidines remaining and 1.0 carboxymethylhistidine appearing) these data would seem to suggest that alkylation by bromoacetate modifies only one histidine per mole of native dimeric enzyme. Moreover, if dehydrase is composed of two identical monomers as the number of tryptic peptides indicates, only one

26. Y. Morino and E. E. Snell, *JBC* **242**, 5602 (1967).
27. J. C. Loper, *JBC* **243**, 3264 (1968).

TABLE V

MODIFICATION OF HISTIDINE AND TYROSINE RESIDUES IN DEHYDRASE[a]

Amino acid[b]	Treatment							
	None	Photo-oxidation	Bromoacetate	1,3-Dibromo-acetone	Tetranitro-methane	3-Decynoyl-NAC	(1) 3-De-cynoyl-NAC (2) Bromo-acetate	(1) 3-De-cynoyl-NAC (2) Tetra-nitromethane
Histidine	2.1	0	1.4	1.5	—	1.2	1.4	—
3-Carboxymethylhistidine	—	—	0.5	—	—	—	0	—
Tyrosine	3.8	—	—	—	2.2	—	—	2.9
3-Nitrotyrosine	—	—	—	—	1.8	—	—	0.7

[a] Data from Helmkamp and Bloch (*11*).

[b] Moles of amino acid residues per 18,000 g of protein.

of the two subunits would appear to carry a catalytically active histidine. Further verification of this conclusion is needed.

The extremely rapid inactivation of dehydrase by 1,3-dibromoacetone and the faster rate of enzyme alkylation by bromoacetate at lower than higher pH suggest that the modifying agents are much more effective in the uncharged than in the anionic form. Studies with dehydrase inhibitors also indicate that anionic molecules have relatively little affinity for the enzyme. The normal substrates of the enzyme are, of course, neutral molecules also.

C. Nitration by Tetranitromethane

Tetranitromethane (10^{-3} M) rapidly inactivates dehydrase, reducing its tyrosine content from 4 to 2 residues per monomer (*11*). Concomitantly 2 moles of 3-nitrotyrosine are produced. Attempts to regenerate dehydrase activity by dithionite reduction of the nitrotyrosyl residues have not been successful. *N*-Acetylimidazole, another widely used tyrosine blocking reagent, does not inactivate dehydrase. Enzymic hydrolysis of acetylimidazole by dehydrase does not appear to be the reason why the phenolic hydroxyl group of the "active" tyrosine fails to react.

The participation of histidine and tyrosine in dehydrase catalysis is consistent with the pH dependence of dehydrase activity. The bell-shaped pH rate profile shows half-maximal values at 7.3 and at 10.2, respectively, indicative of functional imidazole and phenolic groups.

V. Dehydrase Inhibition by Substrate Analogs

Competitive inhibitors of dehydrase have not been found. Inert compounds in this category are the NAC derivatives of decanoic acid α,β-*cis*-decenoate and β,γ-*trans*-decenoate (*10*). α-Methyl-D,L-β-hydroxydecanoyl-NAC inhibits slightly with a K_I of 2×10^{-3} M (*14*).

A remarkably potent and specific inhibitor of all dehydrase-catalyzed reactions is the acetylenic substrate analog 3-decynocyl-NAC (*10*). At 10^{-6} M the acetylenic thioester completely abolishes both dehydrations and isomerization activities. Inactivation is extremely rapid and not reversible by raising the concentration of substrate. This noncompetitive mode of inhibition results from covalent binding of the acetylenic thioester to enzyme (*12*). Labeled inhibitor cannot be released from the enzyme by solvent extraction, dialysis, washing out with unlabeled inhibitor, or by Sephadex chromatography after protein denaturation. On titration of dehydrase with 3-decynoyl-NAC 1–2 moles of inhibitor are bound per mole of enzyme. 3-Decynoyl-NAC labeled in the *N*-acetylcysteamine portion also yields a labeled enzyme derivative. The thioester linkage of the inhibitor, therefore, remains intact during the reaction with dehydrase.

A. Reaction Site of Acetylenic Inhibitor

Acid hydrolysis of dehydrase fully inhibited by 3-decynoyl-NAC yields approximately 1 histidine residue less per monomeric subunit than the native enzyme (1.2 instead of 2, Table V) (*11*). Moreover, when enzyme so modified is subsequently treated with bromoacetate, no additional loss of histidine occurs showing that the "active" histidine is no longer alkylated. Carboxymethylhistidine cannot be detected in the hydrolysate. Reversing the order of these treatments leads to a similar result. Enzyme already alkylated fails to bind 3-decynoyl-NAC. Clearly, once the sensitive histidine in dehydrase is modified by one reagent, it can no longer respond to the other. The action of the acetylenic inhibitor therefore results from irreversible attachment to the histidine residue at the active (catalytic) site. Photooxidized dehydrase does not combine with 3-decynoyl-NAC.

After nitration with tetranitromethane, dehydrase also fails to bind 3-decynoyl-NAC and, conversely, prior exposure of dehydrase to 3-decynoyl-NAC protects, albeit partially, against modification by tetranitromethane. The latter sequence of treatments thus affords a derivatized enzyme containing only one 3-nitrotyrosine residue, whereas the number is two in dehydrase (monomer) modified by tetranitromethane alone. Attachment of inhibitor to enzyme appears to shield only one of the two exposed tetranitromethane-sensitive tyrosine residues against nitration.

The above results suggest that 3-decynoyl-NAC causes inhibition because of its structural resemblance to a dehydrase substrate. It must have a high affinity for both the catalytic and binding regions of the enzyme. Once bound, the inhibitor prevents further chemical modification of the already blocked histidine and interferes sterically with any access to tyrosine.

The various chemical modifications which abolish dehydrase activity (3-decynoyl-NAC, alkylating agents, or nitration) affect isomerization and dehydration equally whether inhibition is partial or complete (*11*). It is, therefore, not possible to specify which of the two amino acid residues functions in the proton transfers at C_α and C_γ, respectively, and

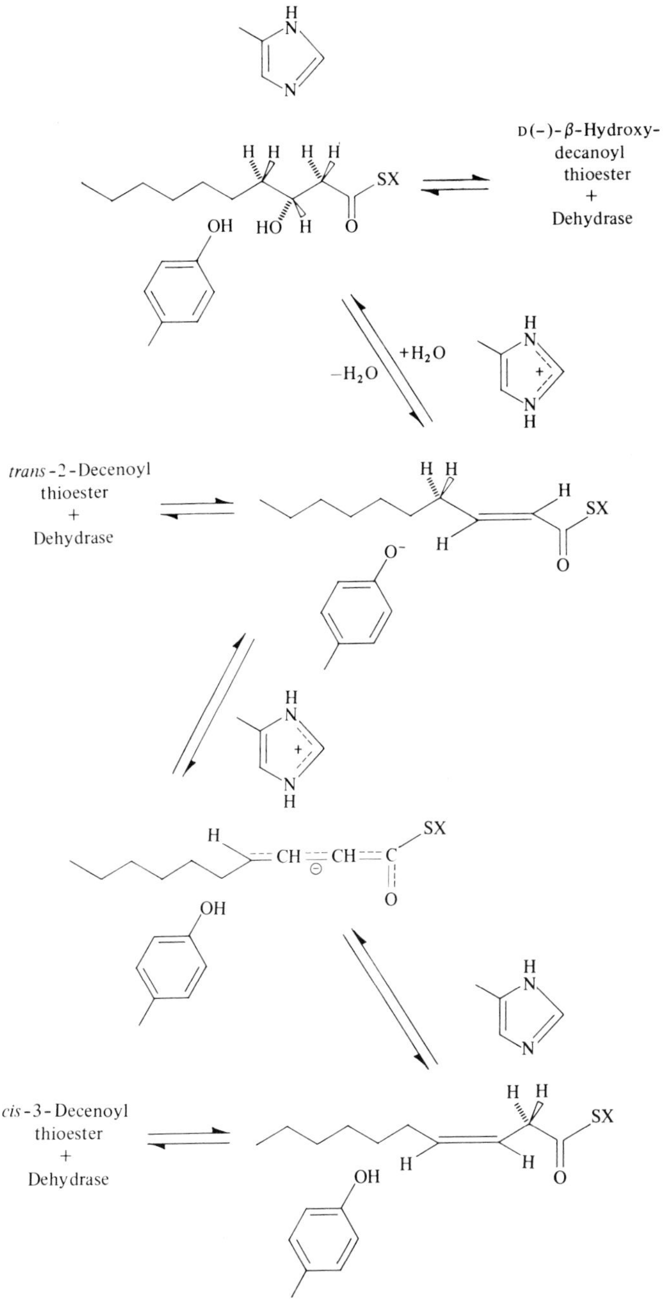

D(−)-β-Hydroxy-
decanoyl
thioester
+
Dehydrase

$+H_2O$

$-H_2O$

trans-2-Decenoyl
thioester
+
Dehydrase

cis-3-Decenoyl
thioester
+
Dehydrase

FIG. 4. A proposed mechanism of dehydrase-catalyzed reactions. From Helmkamp (28).

in the reversible removal of OH^- at C_β. A plausible mechanism of action of dehydrase proposed by Helmkamp (*28*) is shown in Fig. 4.

B. Inhibitor Specificity

Attachment of the acetylenic analog to the substrate site of dehydrase is also made likely by the manner in which the chain length of the inhibitor affects inhibitor potency (*12*). Figure 5 shows such results for homologous β,γ-acetylenic acyl-NAC derivatives. Plotting kinetic parameters for dehydrase substrates and for inhibitors as a function of

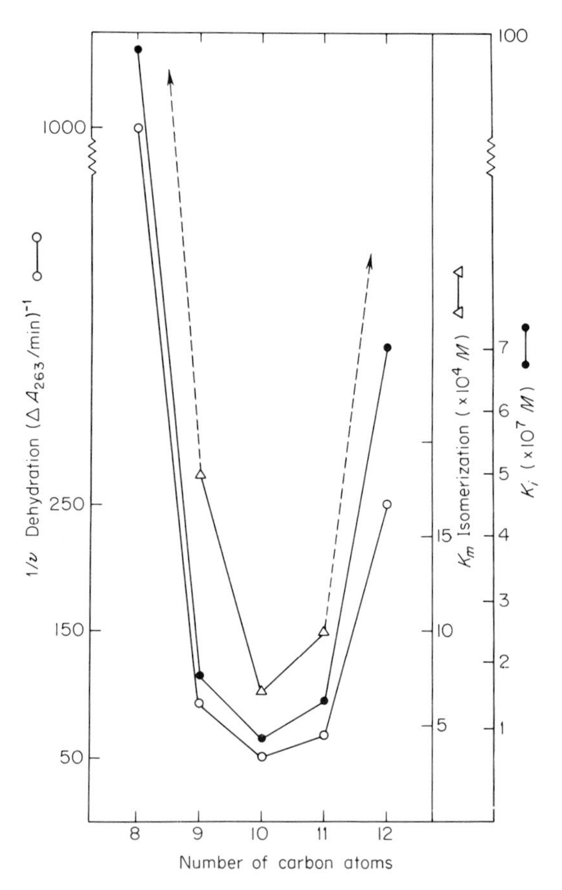

Fig. 5. Kinetic parameters of dehydrase substrates and acetylenic inhibitors as functions of carbon chain length. From Helmkamp *et al.* (*12*).

28. G. M. Helmkamp, Jr., Doctoral Dissertation, Harvard University (1970).

chain length affords essentially superimposable curves. For both inhibitors and substrates, C_{10} is the optimal chain length, the C_9 and C_{11} compounds are slightly less active, and C_8 and C_{12} thioesters are virtually inert. Free 3-decynoic or its methyl ester do not inhibit dehydrase even at very much higher concentrations. A thioester moiety and the same overall dimensions as for substrates are thus necessary features for inhibition. Equally critical is the location of the triple bond in the inhibitor molecule. The α,β- and γ,δ-acetylenic thioesters do not impair dehydrase activity even at relatively high concentrations (10^{-4} *M*).

C. Isomerization of 3-Decynoyl-NAC to 2,3-Decadienoyl-NAC

The unusually strict structural requirements for acetylenic dehydrase inhibitors and the fact that these substrate analogs inhibit noncompetitively have long been a puzzle. This has been solved and rationalized by the finding that 3-decynoyl-NAC is a dehydrase *substrate* and not an inhibitor per se (*29*). In a reaction strictly analogous to the β,γ-decenoyl → α,β-decenoyl conversion (A), dehydrase isomerizes the acetylenic thioester to the allenic 2,3-decadienoyl-NAC (B), the latter being the true inhibitor.

(A) $$CH_3(CH_2)_5C{=}CH\cdot\overset{H}{\underset{(H)}{C}}\cdot\overset{O}{\overset{\|}{C}}SR \quad (H^+) \rightleftharpoons CH_3(CH_2)_5CH_2CH{=}CH\cdot\overset{O}{\overset{\|}{C}}SR$$

(B) $$CH_3(CH_2)_5C{\equiv}C\cdot\overset{H}{\underset{(H)}{C}}\cdot\overset{O}{\overset{\|}{C}}SR \quad (H^+) \longrightarrow CH_3(CH_2)_5CH{=}C{=}CH\cdot\overset{O}{\overset{\|}{C}}SR$$

Both reactions involve a thioester facilitated abstraction of a proton from C_α and proton addition to C_γ. The first of the two processes, removal of hydrogen from C_α, is the rate-limiting step both in the inhibitor–enzyme reaction and in the reaction with normal substrate. Thus, the inhibition of dehydrase by the α-dideuterated analog $CH_3(CH_2)_5C{\equiv}C{-}CD_2CONAC$ proceeds with a kinetic isotope effect: $k_H:k_D = 2.60$ (*29*). This value is to be compared with the $k_H:k_D = 2.25$ observed for the enzymic dehydration of α-dideuterio-β-hydroxydecanoyl-NAC to *trans*-α,β-decenoyl-NAC (*22*). However, whereas the β,γ-enoate–α,β-enoate isomerization is an equilibrium reaction, the acetylene-allene transformation is irre-

29. K. Endo, G. Helmkamp, Jr., and K. Bloch, *JBC* **245**, 2493 (1970).

versible, at least operationally. The obvious reason is that the allenic product combines immediately with the enzyme in a nonenzymic reaction.

Synthetically prepared allenic thioester inactivates dehydrase 10–15 times more rapidly than the isomeric acetylene and, like the latter, in a noncompetitive manner (*29*). Deuterium substitution at C_α in 2,3-decadienyl-NAC does not slow the rate of dehydrase inhibition (*29*).

The realization that 3-decynoyl-NAC is not the inhibitor per se but a substrate for dehydrase at once explains why the structural requirements for the acetylenic analogs are so stringent. Like other substrate molecules they must have a specific chain-length and a thioester function. However, once the acetylene-allene isomerization has taken place, inhibition should not depend as critically on these structural features. Indeed, free 2,3-decadienoic acid, in contrast to free 3-decynoic acid inhibits dehydrase effectively, though at a slower rate than the NAC derivative. At pH 5 this inhibition by allenic acid is more rapid than at pH 7. Therefore, the protonated rather than the anionic allene is the inhibitory species.

Since allenes occur in optically active forms, an opportunity existed to test the steric specificity of these molecules. 2,3-Decadienoic acid has

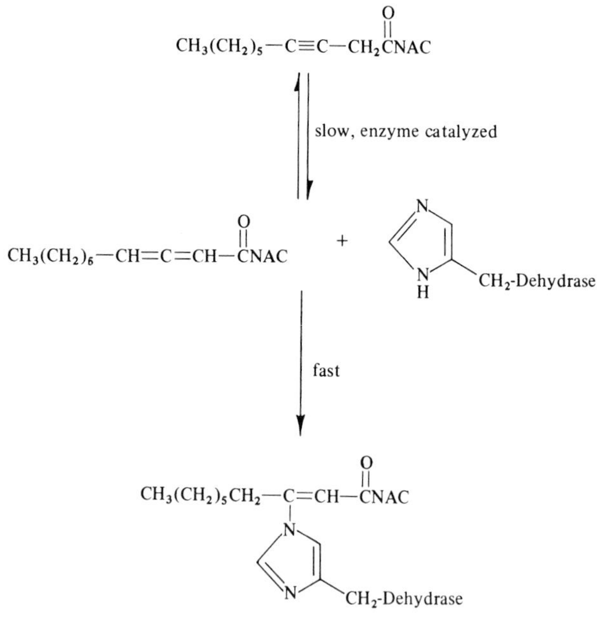

$$CH_3(CH_2)_5-C\equiv C-CH_2\overset{\overset{\displaystyle O}{\|}}{C}NAC$$

$$\Big\updownarrow \text{ slow, enzyme catalyzed}$$

$$CH_3(CH_2)_6-CH=C=CH-\overset{\overset{\displaystyle O}{\|}}{C}NAC \quad + \quad \text{(imidazole, N–H)}-CH_2\text{-Dehydrase}$$

$$\Big\downarrow \text{ fast}$$

$$CH_3(CH_2)_5CH_2-\underset{\underset{\displaystyle N\text{(imidazole)}-CH_2\text{-Dehydrase}}{|}}{C}=CH-\overset{\overset{\displaystyle O}{\|}}{C}NAC$$

FIG. 6. Proposed mechanism of dehydrase inactivation by 3-decynoyl-NAC. From Endo *et al.* (*29*). The assignment of the olefinic double bond to the α,β position in the adduct is arbitrary.

been resolved and only the dextrorotatory isomer found to inhibit dehydrase (M. Morisaki, unpublished). Optical specificity of this type of active site reagent has not been noted previously.

On exposure of dehydrase to 2,3-decadienoyl-NAC or to the free allenic acid, one histidine residue per subunit is modified, as on treatment with 3-decynoyl-NAC (*29*). Dehydrase inhibition or inactivation by 3-decynoyl-NAC is, therefore, the consequence of two consecutive events: (1) enzymic isomerization of 3-decynoyl-NAC to 2,3-decadienoyl-NAC; and (2) noncatalyzed, chemical modification of histidine residues by 2,3-decadienoyl-NAC. A possible mechanism for the interaction between 2,3-decadienoyl-NAC and enzyme-histidine based on the known susceptibility of allenes to nucleophiles is shown in Fig. 6. The unusual feature of the inactivation of dehydrase by 3-decynoyl-NAC is that the enzyme, by catalyzing the transformation of a substrate analog to an extremely reactive active site reagent, causes its own destruction.

D. 3-Decynoyl-NAC and *E. coli* Fatty Acid Synthetase

Escherichia coli fatty acid synthetase produces simultaneously long-chain unsaturated and saturated fatty acids, usually in a ratio of about 4:1 (*6*). According to Fig. 1, dehydrase is the key enzyme in the synthetase-catalyzed transformations leading to the olefinic products. Addition of the specific dehydrase inhibitor 3-decynoyl-NAC to an operating *E. coli* fatty acid synthetase should have a profound effect. Indeed, the inhibitor causes a marked shift of the fatty acid pattern in the predicted direction. The production of unsaturated fatty acids is almost totally abolished, but the synthesis of saturated acids continues at an undiminished rate and some β-hydroxydecanoate (the dehydrase substrate) accumulates (*30*). A synthetase selectively poisoned in this manner by the acetylenic inhibitor will, however, regain normal behavior (synthesis of unsaturated fatty acids) on addition of purified dehydrase. This restoration of the normal fatty acid spectrum proves unambiguously that dehydrase performs in the postulated manner. It also illustrates the highly selective action of 3-decynoyl-NAC. Except for dehydrase none of the synthetase enzymes are inhibited.

The above experiments, nevertheless, present a paradox. Uninhibited synthetase produces predominantly unsaturated fatty acids and, therefore, β,γ-decenoate must be the major product resulting from dehydrase catalysis in this system. On the other hand, isolated dehydrase forms

30. L. R. Kass and K. Bloch, *Proc. Natl. Acad. Sci. U. S.* **58**, 1168 (1967).

little β,γ-decenoate compared to α,β-decenoate which is the precursor of long-chain saturated acids. Two questions, therefore, arise: (1) Is the dehydrase product distribution (ratio of β,γ-decenoate to α,β-decenoate) a variable rather than a fixed quantity, modulated by the elongating enzymes or by other components of the elongating system? (2) Why does 3-decynoyl-NAC, the inhibitor of *all* dehydrase-catalyzed reactions, fail to block the formation of saturated fatty acids which must utilize α,β-decenoate as an intermediate?

As to point (1), the evidence suggests that dehydrase produces the isomeric decenoates in a fixed, thermodynamically controlled ratio (*10*). This being the case, the high proportion of unsaturated fatty acids elaborated by the complete synthetase system must result from differences in the rates of removal of the two dehydrase products. These removal processes involve respectively (Figs. 1 and 7) condensation of β,γ-decenoate with malonyl-ACP to 3-keto-5-dodecenoyl-ACP catalyzed by condensing enzyme and reduction of α,β-decenoate to decanoyl-ACP catalyzed by enoyl-ACP reductase. According to literature values, the turnover number of condensing enzyme is 5–20 times greater (*31*) than that of enoyl reductase (*21*). A kinetic explanation for the preferential channeling of dehydrase products into the unsaturated branch of the pathway is therefore possible.

The puzzle of the continued production of saturated fatty acids by 3-decynoyl-NAC-poisoned fatty acid synthetase has been resolved. Other enzymes capable of transforming β-hydroxydecanoylthioesters (but only the ACP not the NAC derivatives) have been discovered (*32, 33*). These more conventional, monofunctional dehydrases catalyze only the formation of α,β-enoate and thus generate specifically the intermediates for the saturated fatty acid pathway. At the same time, their chain-length specificity is much broader than that of the multifunctional C_{10} dehydrase. These "α,β" dehydrases must be insensitive to 3-decynoyl-NAC since this inhibitor does not suppress saturated fatty acid formation.

Figure 7 summarizes the available information on the competing enzymic events that operate at the level of β-hydroxydecanoyl-ACP. The branch to the left consumes the β-hydroxyacyl intermediate much more rapidly than the branch to the right, in line with the 10-fold difference in the *turnover numbers* of the respective dehydrases (*28,32*). Efficient further transformation of enzyme-bound α,β-enoate along the route to

31. M. D. Greenspan, A. W. Alberts, and P. R. Vagelos, *JBC* **244,** 6477 (1969).
32. C. H. Birge, D. F. Silbert, and P. R. Vagelos, *BBRC* **29,** 808 (1967).
33. M. Mizugaki and S. J. Wakil, *Federation Proc.* **27,** 647 (1968).

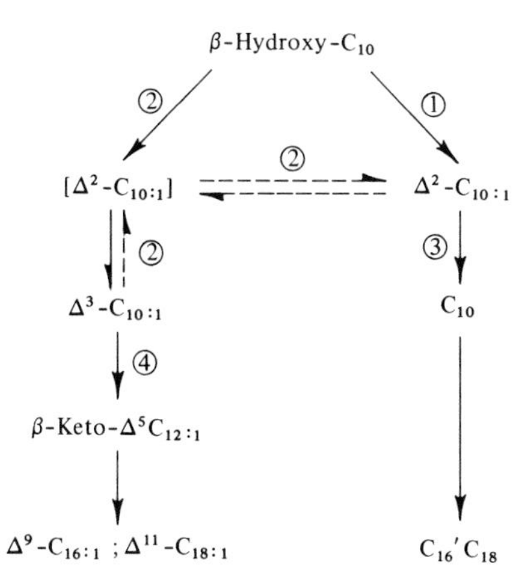

FIG. 7. Transformations of β-hydroxydecanoyl-ACP in the branched pathway to long-chain saturated and unsaturated fatty acids: (1) α,β-dehydrase [Birge *et al.* (*32*) and Mizugaki and Wakil (*33*)]; (2) multifunctional dehydrase; (3) enol reductase [Weeks and Wakil (*21*)]; and (4) condensing enzyme [Greenspan *et al.* (*31*)].

the left will ultimately result in a pattern of predominantly unsaturated fatty acids. The principal point is that generation of α,β-enoate by the multifunctional dehydrase is not essential for the saturated fatty acid branch. If the multifunctional dehydrase, nevertheless, produces α,β-enoate the reasons must be mechanistic; the route to β,γ-decenoate via α,β-enoate is chemically more favorable than direct dehydration.

VI. Dehydraseless Mutants

An *E. coli* K_{12} mutant requiring a long-chain unsaturated fatty acid for growth has been isolated by Silbert and Vagelos (*4*). Intact cells or extracts of the mutant organism synthesize only saturated fatty acids. The deficiency has been traced to the absence of β-hydroxydecanoyl thioester dehydrase. Addition of purified dehydrase to the mutant synthetase restores the normal capacity for synthesizing unsaturated and saturated fatty acids concurrently. In effect, the behavior of this mutant and its fatty acid synthetase is identical with that of wild-type *E. coli* synthetase poisoned by 3-decynoyl-NAC (*5*). The properties of the mutant confirm the essential role of dehydrase for the operation of the branch that leads to long-chain unsaturated acids.

VII. Distribution of Dehydrase

Dehydrase activity has been demonstrated by direct assay in extracts of *Pseudomonas fluorescens* (*34*). Anaerobic bacteria that synthesize olefinic acids [*Clostridia* and *Lactobacilli* (*13*)] and some facultative organisms (*Salmonella* and *Proteus vulgaris*) may be presumed to contain dehydrase. Inferential evidence for the presence or absence of dehydrase is provided by the sensitivity of microbial organisms to 3-decynoyl-NAC (*5*). By this test, dehydrase is present in all *E. coli* strains, *Salmonella typhimurium*, various *Pseudomonas* species, *Proteus vulgaris* and *Aerobacter aerogenes* (private communication, Dr. A. English, Ch. Pfizer & Co.), and absent in bacilli and mammalian yeast cells (*5*).

34. R. Essenberg, Doctoral Dissertation, Harvard University (1971).

16

Dehydration in Nucleotide-Linked Deoxysugar Synthesis

L. GLASER • H. ZARKOWSKY

I. Introduction

Dehydration reactions in deoxysugar synthesis are of two fundamentally different types. The first is exemplified by the synthesis of 2-keto-3-deoxyaldonic acids from the corresponding aldonic acids and are discussed in detail by W. A. Wood, Chapter 20, this volume. This chapter is concerned primarily with the synthesis of nucleotide-linked 6-deoxyhexoses and related reactions. These reactions are complex in that they involve not only a dehydration but also an intramolecular oxidation reduction step to yield a 4-keto-6-deoxyhexose, which after subsequent epimerization and reduction of the 4-keto group yields a nucleotide-linked 6-deoxyhexose. A typical example is the synthesis of

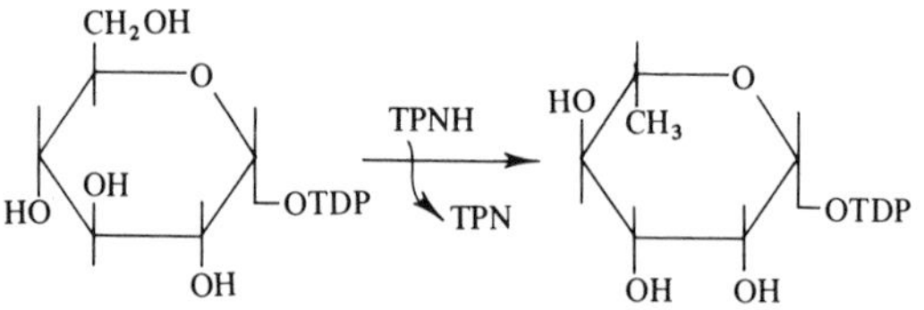

FIG. 1. Synthesis of dTDP-L-rhamnose. This multienzyme reaction has been demonstrated in a number of microorganisms. It involves not only reduction at C-6 but also inversion of the configuration at C-3, C-4, and C-5 of the hexose.

dTDP-L-rhamnose from dTDP-D-glucose as illustrated in Fig. 1. We will be primarily concerned with the mechanism of the first step in the reaction, the dTDP-D-glucose oxidoreductase reaction, which is the only one of the reactions of this type which has been studied in detail.

Listed in Table I are the known examples of this type of reaction and the biosynthetic pathways in which they occur (*1–16*). In most instances the properties of the enzymes catalyzing the reaction have not been studied in detail.

The proposed mechanism for the dTDP-D-glucose oxidoreductase is shown in Fig. 2. This mechanism was proposed, (*17*) when it was found that enzymes of this type either required catalytic quantities of diphosphopyridine nucleotide (DPN) (*1*) or, as shown later, appeared to contain enzyme-bound DPN (*18*). The intermediates shown in brackets (Fig. 2) were assumed to be enzyme bound. According to this mechanism a

1. L. Glaser and S. Kornfeld, *JBC* **236,** 1795 (1961).
2. J. M. Gilbert, M. Matsuhashi, and J. L. Strominger, *JBC* **240,** 1305 (1965).
3. H. Zarkowsky and L. Glaser, *JBC* **244,** 4750 (1969).
4. S. F. Wang and O. Gabriel, *JBC* **244,** 3430 (1969).
5. J. H. Pazur and E. W. Shuey, *JBC* **236,** 1750 (1961).
6. J. Baddiley, N. L. Blumson, A. Di Girolamo, and M. Di Girolamo, *BBA* **50,** 391 (1961).
7. W. A. Volk and G. Ashwell, *BBRC* **12,** 116 (1963).
8. M. Matsuhashi and J. L. Strominger, *JBC* **239,** 2454 (1964).
9. S. Matsuhashi, M. Matsuhashi, J. F. Brown, and J. L. Strominger, *JBC* **241,** 2483 (1966).
10. A. E. Hey and A. D. Elbein, *JBC* **241,** 5473 (1966).
11. G. Barber, *BBRC* **8,** 204 (1962); *ABB* **103,** 276 (1963).
12. V. Ginsburg, *JBC* **235,** 2196 (1960); **236,** 2389 (1961).
13. R. H. Kornfeld and V. Ginsburg, *BBA* **117,** 79 (1966).
14. D. W. Foster and V. Ginsburg, *BBA* **54,** 376 (1961).
15. E. C. Heath and A. D. Elbein, *Proc. Natl. Acad. Sci. U. S.* **48,** 1209 (1962).
16. A. Markovitz, *BBRC* **6,** 250 (1961); *JBC* **239,** 2091 (1969).
17. L. Glaser, *Physiol. Rev.* **43,** 215 (1963).
18. M. Matsuhashi, J. M. Gilbert, S. Matsuhashi, G. Brown, and J. L. Strominger, *BBRC* **15,** 55 (1964).

TABLE I
DISTRIBUTION OF NUCLEOSIDE DIPHOSPHATE HEXOSE OXIDOREDUCTASE[a]

Nucleoside diphosphate sugar	DPN requirement[b]	Final metabolic product	Source
TDP-D-Glu	+	L-Rhamnose	*Pseudomonas aeroginosa* (*1*)
	+B		*Escherichia coli* (*2–4*)
			Streptococcus faecalis (*5*)
			Streptomyces griseus (*6*)
		3-Amino-3,6-dideoxyhexose	*Xanthomonas campestris* (*7*)
		4-Amino-4,6-dideoxyhexoses	*E. coli* sp. (*8*)
CDP-D-Glu	+	3,6-Dideoxy-D-arabinohexose (tyvelose)	*Salmonella* sp. (*9, 10*) (see text)
		3,6-Dideoxy-D-xylohexose (abequose)	
UDP-D-Glu	?	L-Rhamnose	Plants (*11*)
GDP-D-man	+	L-Fucose	*Aerobacter aerogenes* (*12*)
	?		Other bacteria (*13*)
			Mammals (*14*)
	?	3,6-Dideoxyxylohexose (colitose)	*E. coli* (*15*)
		6-Deoxy-D-mannose + 6-deoxy-D-talose	*E. coli* (*16*)

[a] All enzymes listed in this table catalyze the reaction: nucleoside diphosphate-D-hexopyranoside → nucleotide diphosphate-4-keto-6-deoxy-D-hexopyranoside.

[b] (+) DPN must be added to the enzyme and (+B) DPN occurs tightly bound to the protein.

4-keto-hexose is formed in the first step, which is then converted to a 4-keto-5,6-glycoseen, and the double bond between C-5 and C-6 is then reduced by the DPNH formed in the oxidation of the hexose to a 4-keto-hexose.

The enzyme has been isolated in pure form from *Escherichia coli* B by two independent methods using standard fractionation procedures (*3*, *4*), and its properties are described in the next section.

II. The dTDP-D-Glucose Oxidoreductase

A. Molecular Properties

The molecular weight of the enzyme has been determined to be of the order of 80,000. The values obtained by various methods in two

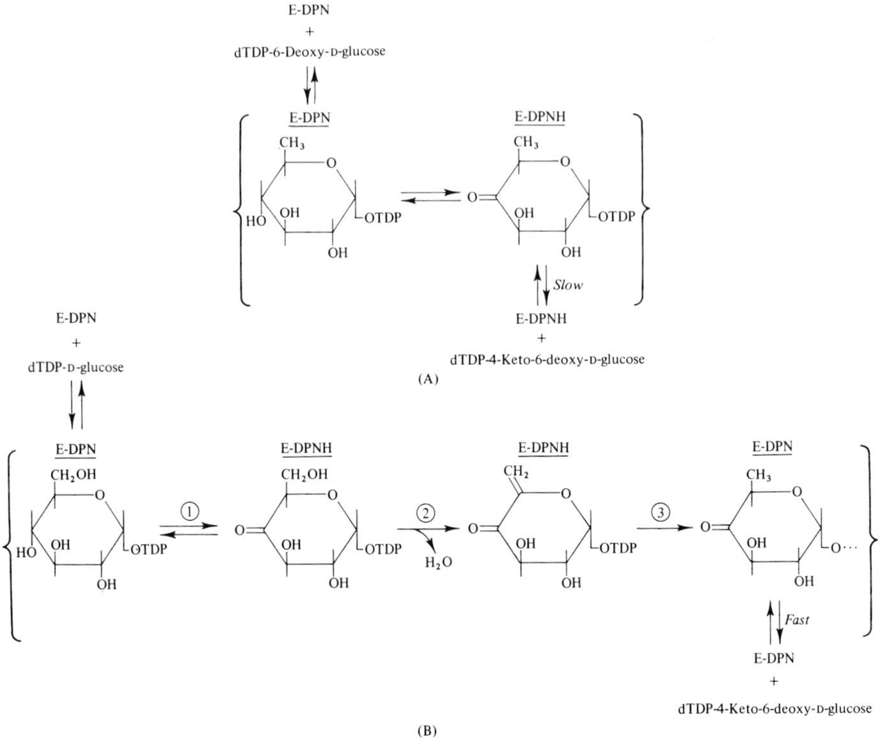

FIG. 2. Proposed mechanism for the dTDP-D-glucose oxidoreductase. Intermediates in brackets are presumed to occur as enzyme-bound intermediates. (A) The proposed reaction sequence with the substrate analog dTDP-6-deoxy-D-glucose. (B) The proposed reaction sequence with dTDP-D-glucose. For details see text.

TABLE II

PROPERTIES OF dTDP-D-GLUCOSE OXIDOREDUCTASE FROM *E. coli*

Property		Method	Ref.
(1) Native enzyme			
Molecular weight	78,000	Sedimentation equilibrium[a]	(*3*)
	78,000	Sephadex chromatography	(*19*)
	88,000	Sedimentation equilibrium[b]	(*4*)
DPN content	0.8 μmole/mole	Enzymic assay after acid denaturation	(*3*)
	0.94 μmole/mole	Alkaline denaturation and alkaline DPN fluorescence	(*4*)
(2) Apoenzyme			
Molecular weight	42,000	Gel exclusion chromatography	(*19*)

[a] Assuming $\bar{V}$ of 0.72.

[b] Assuming $\bar{V}$ of 0.745, sedimentation carried out in 1% sucrose.

different laboratories are detailed in Table II. The enzyme contains 1 mole of bound DPN per mole of enzyme. Removal of the DPN as discussed below dissociates the enzyme into subunits of molecular weight 40,000 as determined by gel filtration (*19*). It is not known whether these subunits are identical or different, but an examination of the kinetics of holoenzyme formation suggests that they may be functionally distinct. No amino acid analysis or protein structure work has been reported for this enzyme.

B. Kinetic Properties

The overall reaction follows deceptively simple kinetics. The apparent K_m and V_{max} are listed in Table III (*3, 4, 20, 21*). All thymidine diphosphate derivatives such as dTDP and dTDP sugars have measurable affinity for the enzyme. dUDP-D-glucose binds weakly to the enzyme and can be used as a substrate. Ribonucleotides are either not bound at all or so weakly that no binding has been detected. dTDP-6-deoxy-D-glucose,

TABLE III
Substrate Specificity of dTDP-D-Glucose Oxidoreductase

Substrate or inhibitor	K_m or K_i[a]	V_{max} (μmoles/μmole enzyme/min)	Ref.
Active			
dTDP-D-glucose	$K_m = 7 \times 10^{-5}$ M	545[b]	(*3, 4*)
dUDP-D-glucose	$K_m = 2.2 \times 10^{-3}$ M	545	(*3*)
dTDP-6-deoxy-D-glucose[c]	$K_i = 2.2 \times 10^{-4}$ M		(*20*)
Inactive			
dTDP and dTTP	$K_i = 2 \times 10^{-4}$ M		(*21*)
dTDP-6-deoxy-D-galactose	$K_i = 7 \times 10^{-4}$ M		(*21*)
Inactive, no binding detectable			
UDP-D-glucose, TMP			(*21*)

[a] All inhibitors tested are competitive with dTDP-D-glucose under usual assay conditions.

[b] The data of Wang and Gabriel (*4*) corrected to an extinction coefficient of 6.5 × 10^3 moles/liter at 320 mμ for the alkaline degradation production of dTDP-4-keto-6-deoxy-D-glucose yield a V_{max} of 740 μmoles/μmole enzyme/min.

[c] This substrate is active in a partial reaction; see text.

19. H. Zarkowsky, E. Lipkin, and L. Glaser, *JBC* **245**, 6599 (1970).
20. H. Zarkowsky, E. Lipkin, and L. Glaser, *BBRC* **38**, 787 (1970).
21. H. Zarkowsky and L. Glaser, unpublished observations (1970).

a substrate analog that can only participate in the first step of the mechanism in Fig. 2 will be discussed in detail in Section I,D.

The enzyme has a pH optimum at pH 8.0 and requires no additional cofactors. The overall reaction is not reversible. Partial reversal of the last step in the proposed mechanism has been looked for by hydrogen exchange, and no evidence was found that this step could be reversed (*22*).

C. Mechanism of the Reaction

1. *Intramolecular Hydrogen Transfer*

The mechanism shown in Fig. 2 predicts the transfer of hydrogen from C-4 of glucose to DPN, and then transfer of this hydrogen to C-6 of the sugar to yield the final product, dTDP-4-keto-6-deoxy-D-glucose. This hydrogen transfer has been demonstrated using dTDP-D-glucose-4-*d* (*22*), as well as dTDP-D-glucose-4-^{3}H (*23*). Hydrogen transfer from C-4 to C-6 of glucose has also been demonstrated with the CDP-glucose oxidoreductase (*24*). The mechanism shown in Fig. 2 makes no direct prediction about the hydrogen at C-5. It could remain bound to a group on the enzyme after removal at step 2 (Fig. 2), and be returned to C-5 of the sugar in step 3 (Fig. 2). Alternatively, it could go to a group on the protein which exchanges rapidly with the solvent, in which case if the reaction is carried out in D_2O one atom of deuterium should appear at C-5 of the product. This experiment has been carried out both with the dTDP-D-glucose oxidoreductase and the CDP-D-glucose oxidoreductase and in both cases there is incorporation of one deuterium atom from the solvent at C-5 of the 4-keto-6-deoxyglucose (*22*).

The same experiment is more ambiguous when carried out in HTO. Under these conditions there is isotope discrimination and only about 0.08 atom of ^{3}H (*25*) are incorporated per mole, which can all be shown to be located at C-5 of the sugar (*22, 26*).

It is important to consider whether the hydrogen transfer in the reaction is strictly intramolecular or not. If the hydrogen transfer is strictly intramolecular, then in the context of the proposed mechanism, the hydrogen that is transferred from the substrate to C-4 of the pyridine ring of DPN in step 1 (Fig. 2) is removed in step 3 (Fig. 2) of the proposed mechanism. If the hydrogen transfer is found to be inter-

22. A. Melo, W. H. Elliott, and L. Glaser, *JBC* **243,** 1967 (1968).
23. O. Gabriel and L. C. Lindquist, *JBC* **243,** 1479 (1968).
24. R. D. Bevill, *BBRC* **30,** 595 (1968).
25. L. Glaser, unpublished observations (1969).
26. A. Melo and L. Glaser, *JBC* **243,** 1475 (1968).

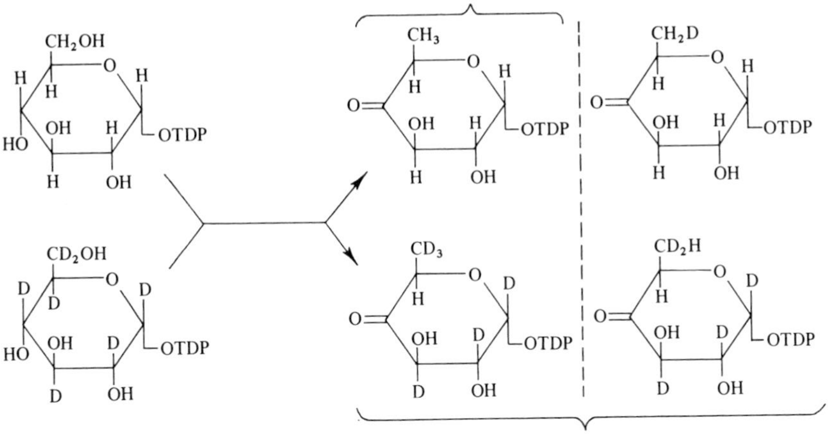

Products expected for intramolecular transfer

Products expected for intermolecular transfer

FIG. 3. Products expected from intramolecular or intermolecular hydrogen transfer in the dTDP-D-glucose oxidoreductase reaction.

molecular, it can mean either that the hydrogen transferred from DPNH in step 3 is not the same as the hydrogen transferred in step 2 or that additional enzyme-bound hydrogen carriers are involved.

That the hydrogen transfer from C-4 to C-6 was strictly intramolecular was shown by using a mixture of two deuterated substrates, the rationale of the experiment is shown in Fig. 3. The results indicated that the transfer was strictly intramolecular (*22*). In subsequent experiments when pure enzyme was available, this was confirmed by the fact that when the DPN on the enzyme was replaced by DPN-4-^{3}H, none of this label was transferred to the product of the reaction (*20*). Conversely when the substrate analog dTDP-6-deoxy-D-glucose-4-^{3}H was incubated with the enzyme, DPNH-4-^{3}H could be isolated from the enzyme and all of the tritium was shown to be on the B side of the nicotinamide ring (*27*).

Thus all of the above data are in agreement with the proposed mechanism as illustrated in Fig. 2. Two sets of observations in the literature are in apparent conflict with this mechanism, both were the result of experiments carried out with crude *E. coli* extracts. In the first, dTDP-D-glucose-3-^{3}H was converted to dTDP-4-keto-6-deoxy-D-glucose and a

27. S. F. Wang and O. Gabriel, *JBC* **245**, 8 (1970).

small fraction (7%) of the label was found at C-6 (*28*). In similar experiments, dTDP-D-glucose-5-^{3}H was converted to dTDP-4-keto-6-deoxy-D-glucose, and although most of the tritium appeared in the medium, a small fraction (approximately 10%) again appeared in C-6 (*29*). The most obvious explanation for these findings is that in each case, the labeled nucleotide was contaminated with small quantities of dTDP-D-glucose-4-^{3}H not detectable by the chemical degradation methods used, which upon incubation with the enzyme will give rise to dTDP-4-keto-6-deoxy-glucose-6-^{3}H. At the moment this appears to be the most likely explanation for these observations, although more complex explanations have been suggested (*28, 29*).

2. *Evidence for the Occurrence of Enzyme-Bound DPNH during the Reaction*

The proposed mechanism requires the formation of enzyme bound DPNH in the course of the reaction. The formation of enzyme-bound DPNH has been demonstrated in two ways.

(1) Spectral observations, using a dual wavelength spectrophotometer and very concentrated enzyme solutions, show that after substrate addition a new absorption band appears around 340 mμ and corresponds to the reduction of about 5% of the enzyme-bound DPN. This absorption band disappears when substrate is exhausted and reappears when fresh substrate is added (*3*).

(2) When the DPN on the enzyme is replaced with radioactive DPN (see Section I,E), then, after addition of substrate, about 5% of the radioactivity of the enzyme can be isolated as DPNH (*20*). Similarly, if the substrate analog dTDP-6-deoxy-D-glucose is added to the enzyme radioactive DPNH can be isolated (*20*). Conversely, if dTDP-6-deoxy-D-glucose-4-^{3}H is added to the enzyme containing nonradioactive DPN then DPNH-4-^{3}H can be isolated from the enzyme (*27*).

These data then are in agreement with the formation of DPNH in the course of the reaction. When substrate is added to the dTDP-D-glucose oxidoreductase, fluorescence changes can also be detected (*3, 4*). Using an activating wavelength in the 300–350 mμ range, there is an increased fluorescence emission with a peak in the 420–500 mμ region, suggesting the formation of DPNH. While it is quite possible that bound pyridine nucleotide is involved in this change, the following observations (*3*) suggest that it is not a result of the formation of DPNH which is an obligatory step in the catalytic reaction.

28. O. Gabriel and G. Ashwell, *JBC* **240,** 4125 (1965).
29. K. Hermann and J. Lehmann, *European J. Biochem.* **3,** 369 (1968).

(a) The fluorescence persists after substrate has been exhausted (contrary to the spectral data above) and decays very slowly for up to 20 min after substrate exhaustion.

(b) A second substrate addition does not yield an increase in fluorescence although the activity is unchanged.

(c) Preincubation of the enzyme with small quantities of product dTDP-4-keto-6-deoxy-D-glucose prevents fluorescence but not activity.

(d) Removal of the product by gel filtration restores the ability of the enzyme to respond to substrate addition with an increased fluorescence.

All these observations suggest strongly that upon substrate addition a conformational change occurs in the enzyme, which results in the described change in fluorescence, but that this change in fluorescence is not to be taken as evidence for DPNH formation in the course of the reaction.

An enzyme-bound dTDP-4-keto-D-glucose has never been isolated from this or similar enzymes. A recent claim (*30*) that addition of NaB^3H_4 to a reaction mixture containing dTDP-D-glucose and oxidoreductase yields dTDP-D-glucose 4-^{3}H cannot be used to demonstrate this since the source of the oxidoreductase was a crude extract of *E. coli*, and the quantity of product obtained far exceeded the quantity of oxidoreductase present.

3. *Isotope Effects*

The rate of dTDP-D-glucose oxidoreductase reaction with a variety of deuterated substrates is shown in Table IV (*21*). It is clear that substitution of the hydrogen at C-4 with deuterium leads to a large decrease in

TABLE IV

ISOTOPE EFFECTS ON THE dTDP-D-GLUCOSE OXIDOREDUCTASE[a]

Substrate	Relative velocity[b]	Fraction of enzyme-bound DPN reduced during steady state (%)
dTDP-D-glucose	1	4–5
dTDP-D-glucose-4-*d*	0.3	20–25
dTDP-D-glucose-5-*d*	1	4–5
dTDP-D-glucose-d_7	0.3	20–25

[a] Data from Zarkowsky and Glaser (*21*).

[b] The maximal velocity with dTDP-D-glucose is set at unity.

30. J. Lehmann and E. Pfeiffer, *FEBS Letters* **7**, 314 (1970).

the rate. This effect could be due to a deuterium effect on step 1 or step 3 (Fig. 2) of the proposed reaction sequence. The data in Table IV show that when dTDP-D-glucose-4-*d* is used as a substrate, the fraction of DPN presents as DPNH during the steady state is increased about 4-fold. This indicates that under these conditions the hydrogen transfer from DPNH to 4-keto-5,6-glycoseen is rate limiting.

As mentioned above, when the reaction is carried out in HTO, there is an 8–10-fold discrimination against tritium incorporation at C-5 of the product; however, when the reaction is carried out in D_2O, the velocity is unchanged. These results taken together suggest that the rate-limiting step is the transfer of hydrogen from DPNH to 4-keto-5,6-glycoseen and that the addition of hydrogen at C-5, from some hydrogen donor in equilibrium with the medium, follows the rate-limiting step.

Data obtained with a crude *E. coli* extract suggest that when dTDP-D-glucose-5-^{3}H is used as a substrate, there is a small discrimination against the tritium at C-5, which may indicate that with tritium, the rate removal of hydrogen at C-5 may become rate limiting (*29*).

D. Effect of Oxidation–Reduction State of Pyridine Nucleotide on Substrate Release

It is of interest to consider why nucleotide intermediates occur only firmly bound to the enzyme and are not released in the course of the reaction since they are structurally rather similar to both substrate and product. One possibility is, of course, that they occur as intermediates covalently bound to the enzyme. A more likely explanation, however, is that the protein conformation is different when the enzyme has bound to it DPNH as compared to DPN and that substrate release occurs from enzyme DPNH either not at all or very slowly.

This change of conformation on reduction of pyridine nucleotide is supported by the following observations (*19*, *20*).

When the substrate analog dTDP-6-deoxy-D-glucose is added to the enzyme at 16°, a rapid (less than 20 sec), reduction of about 20% of enzyme-bound DPN to DPNH takes place. If dTDP, a competitive analog of dTDP-6-deoxy-D-glucose is added to the enzyme, the DPNH is reoxidized to DPN. Essentially the same result is obtained if the enzyme is incubated with dTDP-6-deoxy-D-glucose for 20 sec or 5 min at 16°, before addition of dTDP.

During the incubation of the enzyme with dTDP-6-deoxy-D-glucose at 16° the enzyme does not lose activity. Since it is known that E-DPNH (E = enzyme) formed from the pure apoenzyme and DPNH (see below) is totally inactive under the usual assay conditions, it follows from these observations that product release from E-DPNH is either extremely slow or does not occur at all, as is illustrated in Fig. 2.

This scheme shows that E-DPN reacts with dTDP-6-deoxy-D-glucose to yield E-DPNH-dTDP-4-keto-6-deoxy-D-glucose. Product does not dissociate from E-DPNH under these conditions but can be displaced by reversal of these steps by dTDP which competes for binding to the enzyme. In agreement with this scheme it can be shown that in the usual enzymic assay, dTDP-6-deoxy-D-glucose behaves as a competitive inhibitor (*20*).

At the same time that the above experiments were carried out, a report appeared showing that when dTDP-6-deoxy-D-glucose was added to enzyme at 37° and high ionic strength, the inactive enzyme DPNH could be isolated after incubation at 37° for 15 min (*27*). The discrepancy between these two observations led to a reexamination of this problem. The results obtained, which are summarized below, show that dTDP-4-keto-6-deoxy-D-glucose can slowly dissociate from E-DPNH but that the rate of the reaction is 1000–4000-fold slower than the dissociation of the same compound from E-DPN in the normal reaction. These results confirm the conclusion that substrate dissociation from E-DPNH does not occur at a significant rate compared to the catalytic rate. The data on which this conclusion is based can be summarized as follows (*19*).

(a) Enzyme-DPNH prepared from apoenzyme and DPNH is inactive. It can be reactivated by preincubation with dTDP-4-keto-6-deoxy-D-glucose. This reactivation is rapid and occurs in less time than the fastest time measured (about 15 sec). Reactivation of E-DPNH present during the enzyme assay does not occur because dTDP-D-glucose also binds to E-DPNH and acts as a competitive inhibitor of the reactivation. A number of other thymidine derivatives including dTDP also prevent reactivation.

(b) The rate of free enzyme DPNH formation by dTDP-6-deoxy-D-glucose compared to the rate of product release is very slow at 37° as shown in Fig. 4; it reaches an apparent equilibrium which is dependent as expected on enzyme concentration, substrate concentration, and ionic strength appearing to represent not only the equilibrium defined by the equilibrium constant

$$K = \frac{(\text{E-DPN})(\text{dTDP-6-}d\text{-Glu})}{(\text{E-DPNH})(\text{dTDP-4-keto-6-}d\text{-Glu})}$$

but also the binding of all thymidine nucleotides present to both E-DPN and E-DPNH.

As can be seen in Fig. 4, the rate of E-DPNH formation is measured

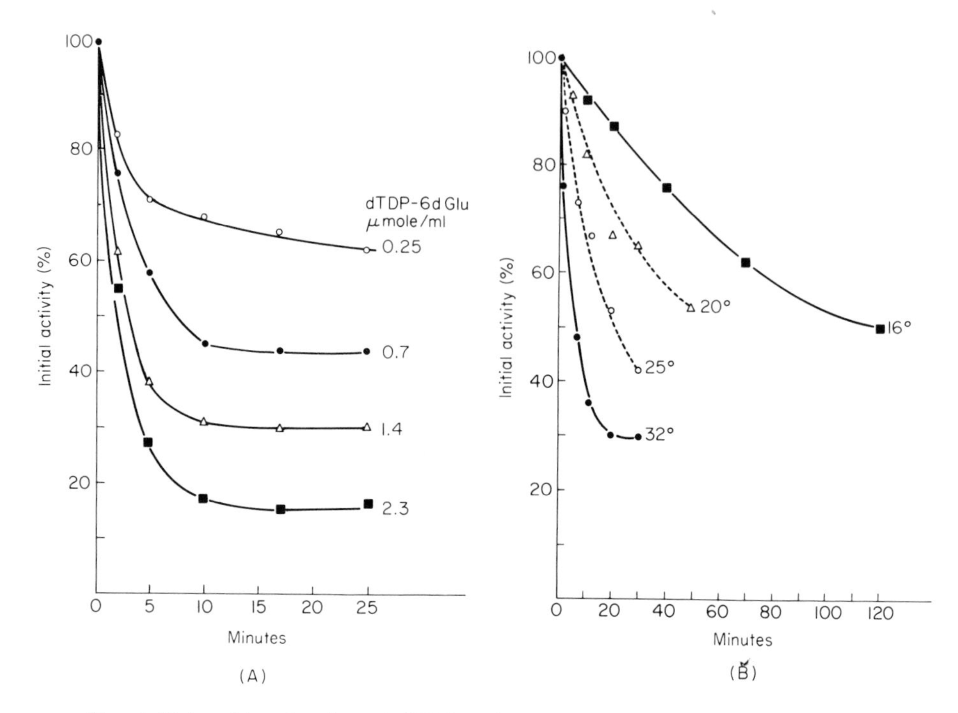

FIG. 4. Rate of inactivation of dTDP-D-glucose oxidoreductase by dTDP-6-deoxy-D-glucose. 14.4 units of dTDP-D-glucose oxidoreductase were incubated with dTDP-6-deoxy-D-glucose in a final volume of 0.3 ml, pH 8.0. At the time indicated, aliquots were assayed for residual activity. Addition of a large excess ($3 \times 10^{-5}\,M$) dTDP-4-keto-6-deoxy-D-glucose resulted in full recovery of activity in all cases. In (A) temperature was 37° with nucleotide additions as indicated. In (B) temperature was varied and the dTDP-6-deoxy-D-glucose concentration was $7 \times 10^{-4}\,M$.

in minutes at 37° and hours at 16°, compared to the normal enzyme turnover which is measured in fractions of a second.

E. Effect of Pyridine Nucleotide on Subunit Association

When dTDP-D-glucose oxidoreductase is incubated with high concentrations of *p*-hydroxymercuribenzoate, followed by the addition of 2-mercaptoethanol to bind the mercurial, a fraction of the enzyme becomes inactive but can be reactivated with the addition of DPN (*2, 3, 19*). The group to which the mercurial is attached is not known. While the extremely limited quantities of enzyme have not allowed adequate chemical characterization of the reactive groups, it is worthwhile to point out that formation of the apoenzyme in the presence of *p*-hydroxymercuribenzoate only occurs at higher temperatures (>15°) and is a slow reaction, suggesting that the reactive groups are not readily available. It is important to note that the apoenzyme once formed does not recapture all the DPN present in the original enzyme, and higher concentrations of added DPN are required to restore activity, although once the holoenzyme is reformed the bound DPN appears to be bound just as tightly as it was to the original holoenzyme. This will be discussed in detail below.

When the mixture of holoenzyme and apoenzyme is chromatographed on Sephadex G-100 the apoenzyme is readily separated from holoenzyme and appears to have a molecular weight of 40,000. Apoenzyme can only be reactivated by the addition of DPN. A variety of nucleotide analogs of DPN have been examined (Table V), none restores enzymic activity,

TABLE V

DPN SPECIFICITY FOR THE REACTIVATION OF dTDP-D-GLUCOSE OXIDOREDUCTASE[a]

Nucleotide	Binding	Activity
DPN	+	+
ADP-ribose	+	−
Acetyl pyridine DPN	+	−[b]
Thioniconimanide DPN	+	−[b]
Pyridinealdehyde DPN	+	−[b]
Deamino DPN	−	−
Other deamino DPN analogs	−	−

[a] Data from Zarkowsky and Glaser (*21*).

[b] These nucleotides give less than 1% of the activity given by DPN (*21*).

but those analogs which contain the adenosine diphosphate ribose portion of DPN, including ADP-ribose itself, react with the apoenzyme since they compete with DPN for enzyme activation. Adenosine diphosphate-ribose reaction with the apoenzyme has been shown to yield an inactive dimer of molecular weight 78,000 (*19*), containing tightly bound ADP-ribose.

The kinetics of reactivation of the apoenzyme by DPN are complex, and while they have not been investigated in detail a few important points are known. The rate of reactivation of the apoenzyme is slow, measured in a time scale of minutes and the apoenzyme appears to bind DPN with a dissociation constant of the order of 10^{-4} *M* at 37° (or 10^{-5} *M* at 16°). The simplest explanation of the kinetic data so far available is given in the following equation where M_1 and M_2 are the two monomers of molecular weight 40,000, and where M_1^* is a different conformation of M_1.

$$M_1 + \text{DPN} \rightleftharpoons M_1\text{-DPN} \xrightarrow{\text{slow}} M_1^*\text{-DPN} + M_2 \rightarrow \text{active enzyme}$$

The assumption that only one of the monomers binds DPN initially followed by a slow conformational change is based on the fact that the rate of activation of apoenzyme as a function of DPN concentration is directly proportional to the apoenzyme concentration while it would be proportional to the square of the apoenzyme concentration if the two monomers had to combine before the rate-limiting step to form holoenzyme.

It was originally assumed (*19*) that M_1 and M_2 are at least functionally distinct monomers. This assumption was made on the basis that if M_1 and M_2 really represent the same monomer unit, then inhibition of the rate of activation by excess DPN should be seen. For the mechanism shown this assumption is not necessarily correct. If the binding of DPN to M_1 is extremely rapid and readily reversible, and if the affinity of M_1^*-DPN for M_1 is very high, then the concentration of free M_1 (which is what is affected by excess DPN) would not significantly affect the rate of enzyme formation. Clearly, any decision of this point is somewhat speculative until the protein structure of the enzyme is determined.

The rate of holoenzyme formation is affected by dTDP and dTDP-D-glucose showing that the substrate binding site persists in the apoenzyme (*19*).

It is of interest to note that some oxidoreductases listed in Table I only show weak DPN binding, and DPN must be added to enzyme for activity (*25*). It will be very interesting to study such enzymes, particularly with reference to the role of the pyridine nucleotide in protein structure and the role of the pyridine nucleotide oxidation state on substrate binding.

III. Other Nucleoside Diphosphate Sugar Oxidoreductases

A. CDP-D-GLUCOSE OXIDOREDUCTASE

This enzyme has been partially purified from *Salmonella*. It has been shown that CDP-D-glucose-4-^{3}H is converted to CDP-4-keto-6-deoxy-D-glucose-6-^{3}H (*29*), and that when the reaction CDP-D-glucose to CDP-4-keto-6-deoxy-D-glucose is carried out in D_2O, one deuterium atom is incorporated at C-5 of the product (*22*). The reaction is unique in that the enzyme has a very high affinity for the product (CDP-4-keto-6-deoxy-D-glucose) which is a competitive inhibitor of the reaction. In spite of this high affinity for the product no reversibility of the reaction has been detected, including a lack of partial reversibility of the last step by measuring hydrogen exchange from the medium into CDP-4-keto-6-deoxy-D-glucose (*22*).

B. GDP-D-Mannose Oxidoreductase

No detailed work has been carried out specifically on the mechanism of this reaction, although it was the first reaction of this type described (*12*). In *Aerobacter aerogenes*, the reaction is subject to feedback control by GDP-L-fucose, as a part of a complex regulatory network shown below.

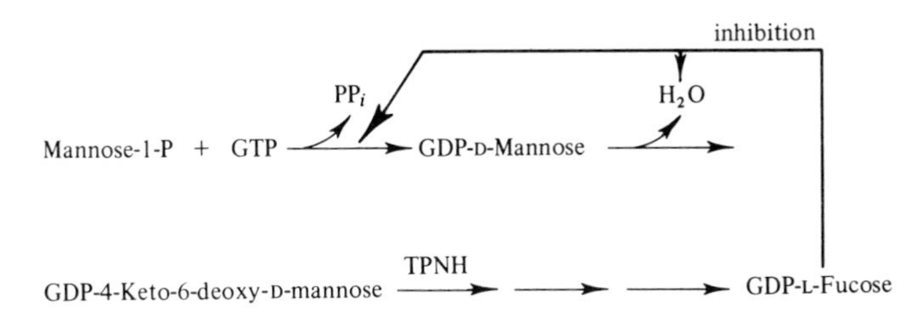

This is the only known sugar nucleotide oxidoreductase which is subject to allosteric regulation (*13*).

IV. Synthesis of 3,6-Dideoxyhexoses

The pathway of synthesis of 3,6-dideoxyhexoses will be mentioned briefly because it is quite possible that reduction at the 3 position of the hexose also proceeds by a dehydration mechanism (*31–33*) (see Table I).

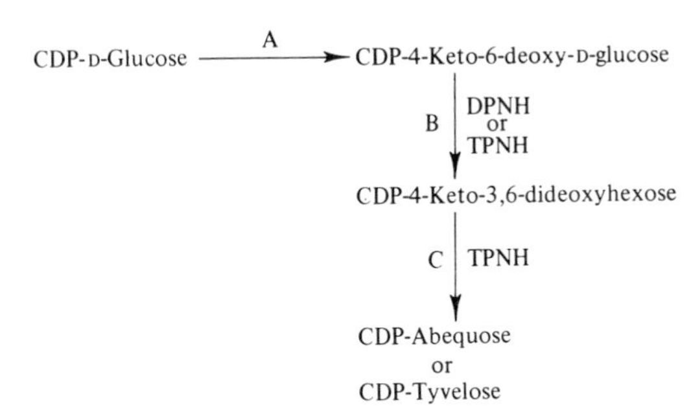

31. H. Nikaido and R. Nikaido, *JBC* **241,** 1376 (1966).
32. S. Matsuhashi and J. L. Strominger, *JBC* **242,** 3494 (1967).
33. H. Pape and J. L. Strominger, *JBC* **244,** 3598 (1969).

Reaction A is catalyzed by the CDP-glucose oxidoreductase as discussed above. Reaction B is catalyzed by two protein fractions labeled E1 and E3, and reaction C, the stereospecific reduction of 4-keto group, is catalyzed by a protein fraction labeled E2. Different *Salmonella* strains contain different E2 proteins depending on whether the 3,6-dideoxyhexose formed by these cells have the D-glucose or D-galactose configuration.

Reaction B is catalyzed by E1 and E3, appears not to involve thioredoxin or vitamin B_{12}, and proceeds without the incorporation of tritium from HTO into the product, although it has not been excluded that the lack of incorporation is either because of an unusually large isotope effect or because hydrogen incorporated at C-3 of CDP-4-keto-3,6-dideoxy-D-glucose can exchange with the medium and is therefore lost during isolation. The precise mechanism of this reaction remains to be determined.

Acknowledgment

Unpublished work in the author's laboratory has been supported by a grant from the National Science Foundation GB-6243X.

17

Dehydrations Requiring Vitamin B_{12} Coenzyme

ROBERT H. ABELES

I. Introduction

The requirement of a vitamin B_{12} coenzyme (*1a*) was first demonstrated in the conversion of glutamic acid to α-methylaspartate and succinyl-CoA to methylmalonyl-CoA (*1*). Subsequently, a number of reactions were discovered which appeared to be elimination reactions, and which also required vitamin B_{12} coenzyme. These reactions are: (1) CH_3–CH_2OH–$CH_2OH \rightarrow CH_3$–CH_2–CHO; (2) CH_2OH–$CH_2OH \rightarrow CH_3CHO$; (3) CH_2OH–CH_2OH–$CH_2OH \rightarrow CH_2OH$–$CH_2$–CHO; and (4) $CH_2NH_3^+CH_2OH \rightarrow CH_3CHO + NH_3^+$.

Reactions (1) and (2) are catalyzed by dioldehydrase (*2*, *3*), an enzyme isolated from *Aerobacter aerogenes*, which has been obtained in a highly purified state. The enzyme is induced when the organism is

1a. Vitamin B_{12} coenzyme = α-(5,6-benzimidazolyl)-5′-deoxyadenosyl cobamide.
1. See H. A. Barker, Chapter 14, Volume VI of this treatise.
2. R. H. Abeles and H. A. Lee, Jr., *JBC* **236,** 2347 (1961).
3. H. A. Lee, Jr. and R. H. Abeles, *JBC* **238,** 2367 (1963).

grown on glycerol under anaerobic conditions. Addition of propanediol in addition to glycerol further enhances the amount of enzyme produced, although the organism does not grow on propanediol alone. Reaction (3) is catalyzed by glycerol dehydrase. The first evidence of the presence of this enzyme and for the requirement of a B_{12} coenzyme was obtained with *Lactobacillus* (*4*). The enzyme from *Lactobacillus* has so far not been obtained in highly purified states. Pawelkiewicz (*5, 6*) and his associates have isolated a glycerol dehydrase from a strain of *Aerobacter*. 1,2-Propanediol and ethylene glycol are also substrates for this enzyme. This enzyme has been highly purified and extensively studied. The properties of this enzyme are very similar to dioldehydrase. Glycerol dehydrase activity is also present in *Aerobacter aerogenes* when grown in glycerol, but this enzyme has not been purified. An enzyme has been obtained from *Clostridium* sp. which catalyzes reaction (4) (*7*). This enzyme has been obtained in highly purified form, and its mechanism of action has been extensively studied by Babior (*8–10*).

In this chapter, the current state of knowledge regarding these enzymes will be summarized. This review will be divided into two main parts: (1) mechanistic studies and (2) properties of the apoenzyme. The major portion of the work carried out so far is concerned with the mechanism of action, and much less emphasis has been placed on the properties of the apoenzyme.

II. Chemistry of Substrate to Product Interconversion

The earliest mechanistic studies with dioldehydrase (*11, 12*) were done to establish whether the hydrogen, which replaces the OH group, when diol is converted to the corresponding aldehyde, is derived from the solvent or through intramolecular 1,2 shift. Experiments with deuterated substrates and nonisotopic substrates in D_2O established that no exchange with solvent protons occurs and therefore the OH group is displaced by

4. K. L. Smiley and M. Sobolov, *ABB* **97,** 538 (1962).
5. Z. Schneider and J. Pawelkiewicz, *Acta Biochim. Polon.* **13,** 311 (1966).
6. Z. Schneider, E. G. Larsen, G. Jacobson, B. C. Johnson, and J. Pawelkiewicz, *JBC* **245,** 3388 (1970).
7. B. H. Kaplan and E. R. Stadtman, *JBC* **243,** 1787 and 1794 (1968).
8. B. M. Babior, *JBC* **244,** 449 (1969).
9. B. M. Babior, *JBC* **245,** 1755 (1970).
10. B. M. Babior, *BBA* **167,** 456 (1968).
11. A. M. Brownstein and R. H. Abeles, *JBC* **236,** 1100 (1961).
12. R. H. Abeles and H. A. Lee, Jr., *Brookhaven Symp. Biol.* **15,** 310 (1962).

a hydrogen derived from C-1 of the substrate. Identical conclusions have been obtained with the other dehydrases (*8, 13*), and for that matter, in other B_{12} coenzyme-catalyzed rearrangements (*1*). These results suggest that the hydrogen is not transferred as a proton, although it must be kept in mind that proton transfers can occur in which no exchange with the solvent occurs. Nevertheless, since no exchange has ever been observed under many different conditions and with several different enzymes, it appears improbable that proton transfer is involved.

In the course of these studies, deuterium isotope effects were observed (*12*): With D,L-1,2-propanediol-1-d_2, $V^{H}_{max}/V^{D}_{max} = 10–12$, while with ethylene glycol-d_4 $V^{H}_{max}/V^{D}_{max} \sim 2.0–4.8$. The magnitude of the isotope effect observed with ethylene glycol-d_4 varied with different enzyme preparations. The reason for this variation is not clear. It is possible that the relative magnitude of certain steps in the reaction sequence changes, and under certain conditions the breaking of the C–H bond may no longer be predominantly rate limiting. The large isotope effect observed with deutero-1,2-propanediol argues against the occurrence of a 1,2-hydride shift, since relatively small isotope effects would be expected for reactions with symmetrical transition states (*14*).

The stereochemistry of the displacement of the C-2 hydroxyl group by hydrogen has been established for the conversion of 1,2-propanediol to propionaldehyde (*15, 16*). In separate experiments, (*R*) and (*S*)-1,2-propanediol-1-d_2 was converted to propionaldehyde in the presence of dioldehydrase and vitamin B_{12} coenzyme. The resulting propionaldehyde was converted to α-deuteropropionic acid, and its optical rotation determined and compared to that of α-deuteropropionic acid, which had been prepared by chemical synthesis. It was found that the replacement of the C-2 OH group of 1,2-propanediol proceeds with inversion of configuration with both isomers. The stereochemistry of other B_{12} coenzyme-dependent reactions has been examined (*1*). The reaction catalyzed by glutamate mutase proceeds with retention of configuration and that catalyzed by methylmalonyl-CoA isomerase with inversion.

The substrates of the dehydrases are all primary alcohols (*17*) and therefore either of two hydrogens could migrate in the course of the reaction. Since the two hydrogens are not stereochemically equivalent, it

13. B. Zagalak, *Bull. Acad. Polon. Sci., Ser. Sci. Biol.* **16,** 67 (1968).
14. F. H. Westheimer, *Chem. Rev.* **61,** 265 (1961).
15. B. Zagalak, P. A. Frey, G. L. Karabatsos, and R. H. Abeles, *JBC* **241,** 3028 (1966).
16. J. Retey, A. Umani-Ronchi, and D. Arigoni, *Experientia* **22,** 72 (1966).
17. In the case of ethanolamine ammonia lyase, it has been shown that carbon bonded to –OH becomes the aldehydic carbon of the product (*8*).

is likely that one hydrogen is stereospecifically selected. This point has been tested with dioldehydrase (*18*) and ethanolamine ammonia lyase (*8*), and in both cases stereospecific selection was observed. With dioldehydrase the stereochemistry at C-2 of the substrate determines which of the C-1 hydrogens is transferred. The pro (*R*)-hydrogens is transferred when the substrate is (*R*)-1,2-propanediol and the pro (*S*)-hydrogen with (*S*)-1,2-propanediol. A rationalization of steric control of a reaction at C-1 by C-2 has been presented based upon the assumption that the two OH groups and the methyl group of the substrate interact with defined areas of the enzyme (*18*). Three-dimensional models based on this assumption show that the position of any specific C-1 hydrogen relative to one of the points of interaction on the enzyme is determined by the configuration at C-2. Another example of control by C-2 of a stereospecific reaction at C-1 will be discussed below.

Through an elegant series of experiments, Arigoni and collaborators (*19*) have shown that in the reaction catalyzed by dioldehydrase, the C-2 OH group of the substrate is transferred to C-1 in the course of the reaction. This reaction therefore involves a group migration as do other reactions, i.e., methylmalonyl-CoA isomerase and glutamate mutase, which require vitamin B_{12} coenzyme. (*R*)-1,2-Propanediol-1-^{18}O, (*S*)-1,2-propanediol-1-^{18}O are converted to propionaldehyde in the presence of dioldehydrase and B_{12} coenzyme. The aldehyde obtained from the (*R*) isomer retained 8% of the ^{18}O whereas that obtained from the (*S*) isomer retained 88% of its ^{18}O. When (*RS*)-1,2-propanediol-2-^{18}O was used, 43% of the ^{18}O was retained. The following conclusions were reached:

(1) Stereospecific transfer occurs with both isomers from C-1 to C-2 leading to the formation of 1,1-gem-diol.

(2) The enzyme dehydrates the resulting gem-diol and selectively removes one of the two nonequivalent OH groups. Which of the two OH groups is removed is determined by the stereochemistry of the substrate.

This interpretation satisfactorily explains the experimental results, although more complex interpretations have been suggested (*20*). The stereochemical results regarding the transformation of (*R*)- or (*S*)-1,2-propanediol to propionaldehyde are summarized in Fig. 1.

18. P. A. Frey, G. L. Karabatsos, and R. H. Abeles, *BBRC* **18,** 551 (1965).
19. J. Retey, A. Umani-Ronchi, J. Seibl, and D. Arigoni, *Experientia* **22,** 502 (1966).
20. G. N. Schrauzer and J. W. Sibert, *JACS* **92,** 1022 (1970).

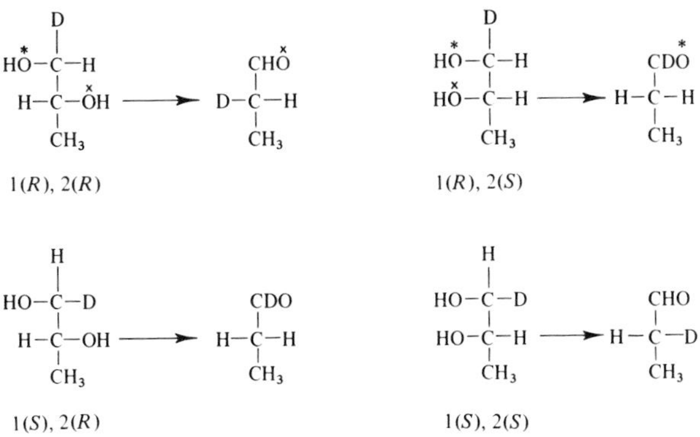

```
       D                                   D
 *     |                  x          *     |                  *
HO - C - H               CHO        HO - C - H              CDO
       |  x               |          x     |                  |
  H - C - OH  ---->  D - C - H      HO - C - H  ---->   H - C - H
       |                  |                |                  |
      CH3                CH3              CH3                CH3

1(R), 2(R)                          1(R), 2(S)

       H                                   H
       |                                   |
HO - C - D               CDO        HO - C - D              CHO
       |                  |                |                  |
  H - C - OH  ---->  H - C - H      HO - C - H  ---->   H - C - D
       |                  |                |                  |
      CH3                CH3              CH3                CH3

1(S), 2(R)                          1(S), 2(S)
```

FIG. 1. Reaction catalyzed by dioldehydrase.

III. The Nature of the Hydrogen Transfer

The mechanism of hydrogen transfer in these reactions has received considerable attention. Isotope experiments cited above indicate that in all cases a hydrogen from C-1 of substrate appears in the α position of the product aldehyde. Considerable evidence now exists which indicates that the hydrogen transfer involves a multistep process in which the coenzyme functions as an intermediate hydrogen carrier rather than a 1,2-hydrogen shift (*21–23*). Evidence for this mechanism was first obtained with dioldehydrase. When a mixture of nonlabeled and C-1 tritiated substrate molecules was added to dioldehydrase and vitamin B_{12} coenzyme, tritium was found in the product derived from the nonlabeled substrate. Therefore, intermolecular hydrogen transfer occurs. Subsequently, it was shown that when D,L-1,2-propanediol-1-^{3}H reacts with dioldehydrase, tritium is incorporated into the C-5′ position of the coenzyme. When the reisolated C-5′-tritiated coenzyme and unlabeled substrate is added to dioldehydrase, tritium is transferred to the reaction product. This also occurs with ethanolamine ammonia lyase (*10*), glycerol dehydrase (*13*), and other B_{32} coenzyme-dependent reactions (*1*). With dioldehydrase and ethanolamine ammonia lyase, tritium exchange between product and enzyme-bound coenzyme can be obtained. Furthermore, this exchange occurs under conditions where no conversion of

21. There is some disagreement concerning this point, as will be pointed out later in connection with a mechanism proposed by Schrauzer and Sibert (*20*).
22. R. H. Abeles and B. Zagalak, *JBC* **241,** 1246 (1966).
23. P. A. Frey, M. K. Essenberg, and R. H. Abeles, *JBC* **242,** 5369 (1967).

product to substrate is observed. This exchange, without reversal of the overall reaction, is good evidence for the existence of at least one enzyme-bound intermediate between substrate and product. With ethanolamine ammonia lyase exchange between acetaldehyde and enzyme-bound B_{12}-coenzyme-5′-^{3}H only occurs in the presence of ammonia. If the interpretation of the ^{18}O experiments carried out with dioldehydrase is extended to ethanolamine ammonia lyase, the following intermediate would be expected:

$$H_3C-\underset{\displaystyle H}{\overset{\displaystyle \overset{+}{N}H_3}{C}}-OH$$

The ammonia requirement supports the postulated intermediate and also suggests that in the reverse reaction, tritium transfer occurs after formation of the aldehyde-ammonia adduct.

When chemically synthesized vitamin B_{12}-coenzyme-5′-^{3}H was added to apoenzyme and unlabeled substrate, all of the tritium of the coenzyme was transferred to the product (*23*). This experiment not only confirmed the involvement of the C-5′ hydrogens of the coenzyme but also established that both C-5′ hydrogens, although they are not stereochemically equivalent (*24*), participate in the catalytic process. This suggests that an intermediate may occur in which the two hydrogens become stereochemically equivalent.

Based upon these results a sequential mechanism in which hydrogen is transferred from the substrate to the coenzyme and from the coenzyme to the reaction product was proposed. The kinetic feasibility of such a mechanism was tested (*25*). This was done with dioldehydrase by measuring the rate of tritium transfer from tritiated coenzyme to reaction product and the rate of tritium incorporation into the coenzyme when a tritiated substrate was used. These experiments showed that the coenzyme was labeled sufficiently rapidly so that all of the tritium found in the substrate could be derived from the coenzyme. A sequential mechanism in which the coenzyme functions as an intermediate hydrogen carrier is therefore kinetically permissible. In the course of these kinetic studies, some other properties of the system became apparent:

(1) When the rate of tritium transfer from ^{3}H-coenzyme–enzyme complex was compared with the rate of product formation it was found that the rate of product formation was 250 times faster than the rate

24. P. A. Frey, S. S. Kerwar, and R. H. Abeles, *BBRC* **29,** 873 (1967).
25. M. K. Essenberg, P. A. Frey, and R. H. Abeles, *JACS* **93,** 1242 (1971).

of tritium transfer from coenzyme to product, i.e., tritium is transferred in one out of 250 turnovers. This is a very large isotope discrimination even if allowance is made for statistical effects, since the postulated intermediate may contain three equivalent hydrogen atoms. The reason for the high isotope effect is not clear.

(2) When 1,2-propanediol-1-^{3}H reacts, the specific radioactivity of the coenzyme is 18–20 times that of the residual substrate and 200–400 times that of the reaction product. This accumulation of tritium in the coenzyme rules out the mechanism whereby tritium is introduced into the coenzyme by equilibration with either enzyme-bound substrate or product.

The following general mechanisms have been considered which could account for the isotope exchange experiments and are consistent with the experiment results.

$$SH_a + X^{H_c}_{H_b} \longrightarrow I \cdot X^{H_c}_{H_a H_b} \longrightarrow \begin{cases} PH_c + X^{H_a}_{H_b} \\ PH_a + X^{H_c}_{H_b} \\ PH_b + X^{H_c}_{H_a} \end{cases} \qquad (1)$$

SH and PH represent the substrate and product; I is an enzyme-bound intermediate derived from the substrate; and XHH represents the enzyme-bound coenzyme. Reaction sequence I represents a two-step sequence in which an intermediate exists containing three equivalent hydrogens, two derived from the C-5′ position of the coenzyme (H_c, H_b) and one from the substrate (H_a). In any one turnover any of the three hydrogens can appear in the product, although not necessarily with the same probability.

$$\begin{aligned} AH_a + X^{H_b}_{H_c} &\longrightarrow H_bA + X^{H_c}_{H_a} \quad \text{(first turnover)} \\ AH_d + X^{H_c}_{H_a} &\longrightarrow H_cA + X^{H_a}_{H_d} \quad \text{(second turnover)} \end{aligned} \qquad (2)$$

Mechanism II

This reaction sequence is a concerted process in which no intermediate containing three hydrogens exist. It is likely that such a process would be stereospecific for one of the two C-5′ hydrogens. Since the experimental data require that both C-5′ hydrogens participate, we have assumed that the displacement proceeds with inversion at the C-5′ carbon so that in two turnovers both C-5′ hydrogens of the coenzyme could be transferred to the reaction product.

Several lines of evidence exist now which favor mechanism I and are inconsistent with mechanism II:

(1) It has been established that a certain fraction of the reaction proceeds with intramolecular hydrogen transfer (*23*). Only mechanism I allows for both intermolecular and intramolecular hydrogen transfer (intramolecular hydrogen transfer would be the formation of PH_a from SH_a). Mechanism II leads to intermolecular transfer exclusively.

(2) When acetaldehyde is allowed to equilibrate with ^{3}H-coenzyme–enzyme complex, tritium is transferred to the aldehyde (*24*). Under these conditions, no transformation of aldehyde ethylene glycol can be detected. This result indicates the occurrence of an intermediate between glycol and aldehyde as proposed in mechanism I.

(3) When a substrate deuterated at C-1 and nonisotopic substrate is added simultaneously to ^{3}H-coenzyme–enzyme complex, the tritium specific activity of the product derived from the deuterated substrate is 4–7 times as high as that of the product derived from the nondeuterated substrate (*25*). Therefore, the presence of deuterium in the substrate affects the probability of tritium transfer from the coenzyme to the product. This result is predictable from mechanism I. According to this mechanism, the intermediate derived from the deuterated substrate would be $\mathrm{I}\cdot\underset{\mathrm{D}}{\overset{\mathrm{T}}{\mathrm{XH}}}$ and that from the nondeuterated substrate $\mathrm{I}\cdot\underset{\mathrm{H}}{\overset{\mathrm{T}}{\mathrm{XH}}}$ (H, D, and T represent hydrogen, deuterium, and tritium, respectively). It is apparent that the probability of tritium transfer to the reaction from the latter intermediate is lower than from the former.

The data thus far seem to require an intermediate form of the coenzyme containing three equivalent hydrogen atoms, two derived from the C-5′ position of the coenzyme and one from the substrate. Similar conclusions have been reached regarding methylmalonyl-CoA isomerase (*26*).

What is the structure of the intermediate? A suitable intermediate would be 5′-deoxyadenosine, which could be derived from the adenosyl moiety of the coenzyme through a process involving dissociation of the C–Co bond. Experimental evidence exists which indicates that dioldehydrase (*27*), ethanolamine ammonia lyase (*9*), and glycerol dehydrase (*28*) can catalyze the dissociation of the C–Co bond of the coenzyme. Furthermore, under certain experimental conditions, 5′-deoxyadenosine has been found as a result of enzymic action. Glycoaldehyde reacts with coenzyme–dioldehydrase complex (*27*) to form a new complex which

26. J. H. Richards and W. W. Miller, *JACS* **91,** 1498 (1969).

27. O. W. Wagner, H. A. Lee, Jr., P. A. Frey, and R. H. Abeles, *JBC* **241,** 1751 (1966).

28. A. Stroinski, Z. Schneider, and J. Pawelkiewicz, *Bull. Acad. Polon. Sci., Ser. Sci. Biol.* **15,** 726 (1967).

is enzymically inactive. The complex has the same spectrum as that observed in the presence of the substrate, and formation of the complex is K^+ dependent as is the overall catalytic process. When the complex is denatured under anaerobic conditions, an unidentified corrin can be detected spectroscopically, which is converted to hydroxocobalamin upon exposure to oxygen. In addition, 5′-deoxyadenosine derived from the adenosyl portion of the coenzyme and glyoxal derived from glycolaldehyde are isolated. When tritiated glycolaldehyde is used, tritium is found in 5′-deoxyadenosine. The reaction of the coenzyme-enzyme complex with glycolaldehyde appears to involve a reductive cleavage of the carbon–cobalt bond of the coenzyme in which glycolaldehyde is the reducing agent.

Formation of 5′-deoxyadenosine was also observed in the presence of ethanolamine ammonia lyase (*9*). When ethylene glycol is added to ethanolamine ammonia lyase, spectral changes similar to those observed with dioldehydrase and glycolaldehyde are seen. It was suggested that a new corrin observed upon addition of ethylene glycol may be an adduct of vitamin B_{12} and SH group on the enzyme. The adenosyl moiety of the coenzyme is converted to 5′-deoxyadenosine and ethylene glycol is converted to acetaldehyde. Tritiated 5′-deoxyadenosine is formed from tritiated ethylene glycol. The source of the α-hydrogen of acetaldehyde is not certain. The author proposes that it is derived from the enzyme. Other examples in which abortive reactions give rise to 5′-deoxyadenosine are known. There are several interpretations of these results: (1) These reactions are abortive because 5′-deoxyadenosine is formed, and the compound is not an intermediate in the normal catalytic process; (2) 5′-deoxyadenosine is derived from an intermediate, but is not actually an intermediate; and (3) 5′-deoxyadenosine is a normal intermediate, which accumulates when a poor substrate is used. Based on these experiments alone, no distinction can be made between these alternatives. It is safe to conclude that the enzyme has the capability of modifying the chemical reactivity of the coenzyme, and it appears probable that intermediate steps involving dissociation of the carbon–cobalt bond occur.

The following tentative reaction sequence involving the intermediate formation of 5′-deoxyadenosine and consistent with all requirements of mechanism I has been proposed for dioldehydrase (*25*), but it could be applicable with minor modifications to all reactions discussed here.

The first step involves an alkyl-metal exchange reaction; it is not necessarily a one-step reaction, and it is very likely that the enzyme plays an important part in activating the coenzyme and the substrate for this exchange. The activation may involve distortion of the corrin ring and reversible dissociation of one nitrogen ligand from cobalt, forming a coordinatively unsaturated complex to facilitate this exchange. The

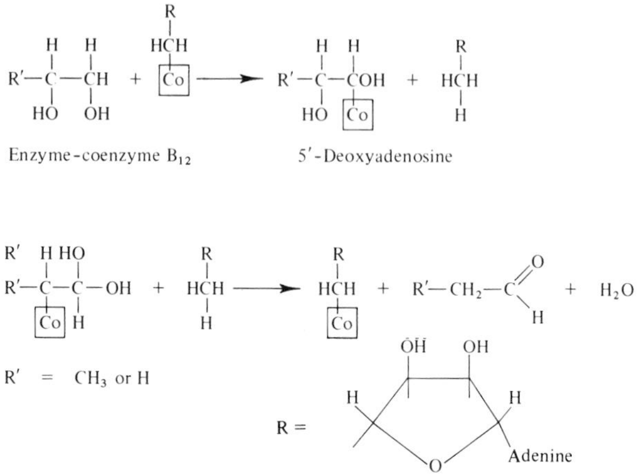

$$R'-\underset{HO}{\overset{H}{C}}-\underset{OH}{\overset{H}{C}}H + \underset{\boxed{Co}}{\overset{R}{HCH}} \longrightarrow R'-\underset{HO}{\overset{H}{C}}-\underset{\boxed{Co}}{\overset{H}{C}}OH + \underset{H}{\overset{R}{HCH}}$$

Enzyme-coenzyme B_{12} 5′-Deoxyadenosine

$$R'-\underset{\boxed{Co}}{\overset{H}{C}}-\underset{H}{\overset{HO}{C}}-OH + \underset{H}{\overset{R}{HCH}} \longrightarrow \underset{\boxed{Co}}{\overset{R}{HCH}} + R'-CH_2-C\begin{matrix}\!/\!\!/O\\ \backslash H\end{matrix} + H_2O \tag{3}$$

R′ = CH_3 or H

R = (ribose ring: OH, OH, H, H, O, Adenine)

second step shown above is a rearrangement, in which the cobalt moves from C-1 of the substrate to C-2, and the 2-hydroxyl group moves to C-1. The final product is formed by an alkyl-cobalt exchange similar to the initial reaction. This mechanism is consistent with all experimental facts and has the advantage that the carbon–cobalt bond is not lost during the course of the reaction. A serious drawback is that no satisfactory chemical model systems are known for the alkyl-metal exchange and the rearrangement of the alkyl-cobalamin, although a reaction described by Schrauzer may be a pertinent model for the rearrangement. α-Ethylcyanocobaloxime and β-ethylcyanocobaloximes are interconvertible through an intermediate π complex (*29*). This reaction provides an example of a

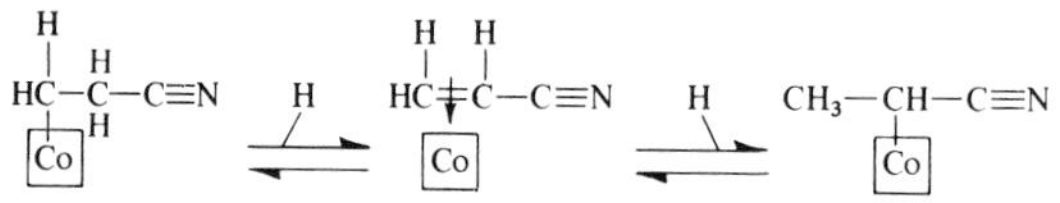

$$\underset{\boxed{Co}}{\overset{H}{HC}}-\overset{H}{\underset{H}{C}}-C\equiv N \;\underset{}{\overset{H}{\rightleftharpoons}}\; \overset{H}{HC}\doteqdot\overset{H}{C}-C\equiv N\,(\boxed{Co}) \;\overset{H}{\rightleftharpoons}\; CH_3-\underset{\boxed{Co}}{CH}-C\equiv N$$

rearrangement of an alkylcobaloxime; similar reactions may well occur with cobalamines.

An alternative mechanism has been proposed, based upon the following nonenzymic reaction of hydroxy alkyl-cobalamins (*20*):

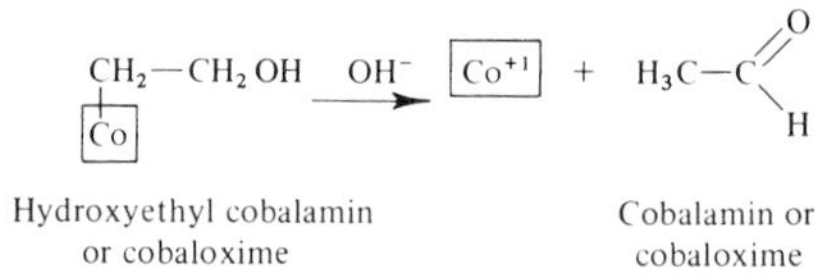

$$\underset{\boxed{Co}}{CH_2}-CH_2OH \xrightarrow{OH^-} \boxed{Co^{+1}} + H_3C-C\begin{matrix}\!/\!\!/O\\ \backslash H\end{matrix}$$

Hydroxyethyl cobalamin or cobaloxime — Cobalamin or cobaloxime

29. G. N. Schrauzer, J. H. Weber, and T. M. Beckham, *JACS* **92**, 7078 (1970).

It is proposed that this reaction involves a 1,2-hydride shift, although this point has not been demonstrated with the appropriate isotope experiments. Based upon this reaction, the following mechanism has been postulated:

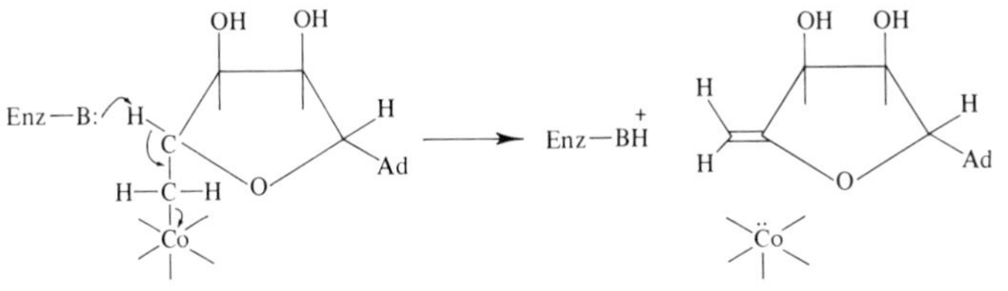

(a)

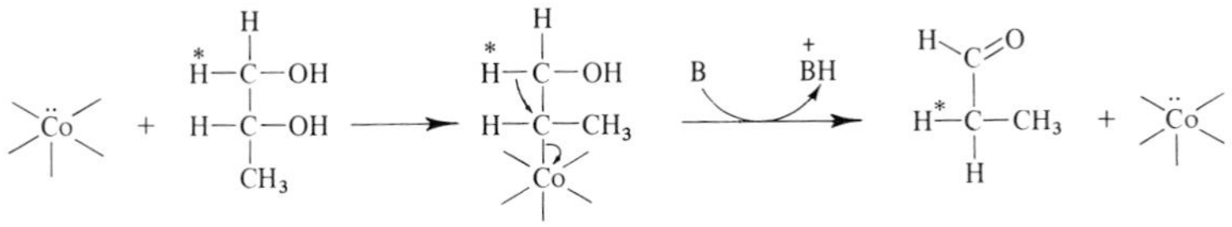

(b)

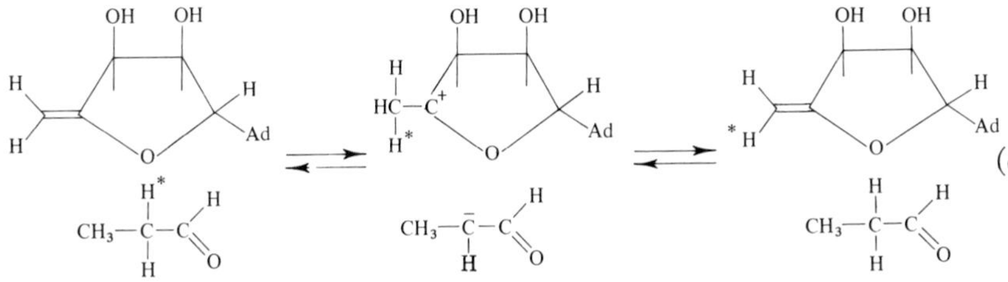

(c)

The mechanism involves an initial activation (a) of enzyme-bound coenzyme, which leads to the dissociation of the carbon–cobalt bond and the formation of highly nucleophilic $B_{12(s)}$ and 4′,5′-didehydro-5′-deoxyadenosine. The nucleophilic cobalt reacts with ethylene glycol to give hydroxyethyl cobalamin (b), which rearranges to give $B_{12(s)}$ and enzyme-bound acetaldehyde. The enzyme-bound acetaldehyde then exchanges tritium with the coenzyme (c) prior to its release from the enzyme. This exchange process accounts for the experimentally observed tritium incorporation into the coenzyme. The finding that the specific activity of the enzyme-bound coenzyme is 200–700 times that of the product makes this type of exchange highly unlikely (*30*). It should be pointed out that at this time no experimental evidence with the enzymic

30. P. A. Frey, M. K. Essenberg, S. S. Kerwar, and R. H. Abeles, *JACS* **92**, 4488 (1970).

system exists for the proposed activation of the coenzyme or for the intermediate occurrence of hydroxyethyl cobalamin. The two mechanisms discussed here differ fundamentally with respect to the mechanism of hydrogen transfer, but are in agreement on the following: (1) The catalytic process involves dissociation of the carbon–cobalt bond of the coenzyme, and (2) a new carbon–cobalt bond between substrate and coenzyme is formed.

IV. Enzyme–Coenzyme Interaction

Dioldehydrase combines with B_{12} essentially irreversibly. After coenzyme is added to dioldehydrase, the reaction mixture can be diluted or passed through Sephadex without loss of catalytic activity, suggesting that dioldehydrase combines initially irreversibly with the coenzyme. Further evidence for irreversible binding is the observation that addition of radioactively labeled coenzyme to enzyme–coenzyme complex does not result in the displacement of coenzyme from the complex by labeled coenzyme (*23*). The formation of catalytically active enzyme–coenzyme complex is a relatively slow process (*12*), dependent upon enzyme and coenzyme concentration. When coenzyme, at sufficiently low concentrations, is added to dioldehydrase and substrate, product formation is initially slow and gradually increases until it becomes linear. During the lag period changes take place in the properties of the enzyme and coenzyme: (1) The apoenzyme loses its sensitivity to *p*-mercuribenzoate; (2) the coenzyme loses its light sensitivity; and (3) at the end of the lag period, the enzyme is irreversibly bound to the enzyme.

The undissociable enzyme–coenzyme complex also forms in the absence of substrate. Under these conditions, the enzyme-bound coenzyme reacts with oxygen (*27*), although the free coenzyme is completely unreactive toward oxygen. The reaction with oxygen leads to loss of catalytic activity, dissociation of the carbon–cobalt bond with the formation of hydroxocobalamin. Several as-yet unidentified products derived from the adenosyl moiety of the coenzyme are formed. These results suggest that the interaction between coenzyme and apoenzyme leads to structural changes in the coenzyme, which affect its chemical reactivity.

Attempts have been made to obtain spectral evidence for the presence of a modified form of the coenzyme. In the absence of substrate, no positive results have been reported. In the presence of substrate, changes in the spectrum of enzyme-bound coenzyme have been observed with dioldehydrase (*27*) and ethanolamine ammonia lyase (*31–33*). The spectral changes are reversible, i.e., the spectrum reverts to that of B_{12} coenzyme when the substrate is consumed. In the presence of substrate, the spectrum of the enzyme-bound coenzyme is very similar to that of $B_{12r}(Co^{2+})$. An examination of the spectrum in the 590–800 nm region has led to the conclusion that the spectrum is that of $B_{12(r)}$ with the base on (*33*). In addition to the optical spectrum, the ESR spectrum of the enzyme–coenzyme complex of ethanolamine ammonia lyase and dioldehydrase has been investigated in the presence of substrate. A signal was obtained which resembled that obtained from $B_{12(r)}$ (*33, 34*). From the intensity of the signal, it was estimated that 0.1 molecule of coenzyme per molecule of ethanolamine ammonia lyase was in the radical form (*34*). No kinetic studies have been carried out to indicate whether the radical species could be involved in the reaction pathway nor have concomitant spectral measurements been made to determine how the extent of spectral change corresponds under the conditions of the ESR experiment to the intensity of the ESR signal. Babior (*9*) has proposed a mechanism similar to that shown in Eq. (3) except that the ligand exchange reaction proceeds through a radical mechanism.

V. Interaction with B_{12} Coenzyme Analogs

Coenzymes in which the dimethylbenzimidazole moiety of vitamin B_{12} coenzyme is replaced by adenine, benzimidazole (*2*), and chloroadenine have been used with dioldehydrase. All are active and have similar V_{max}, which is noteworthy, since the pK of chloroadenine is significantly different from adenine or benzimidazole (the chloroadenine-cobamide coenzyme is yellow at pH 7.0). This suggests that during the enzymic catalysis the cobalt may not interact with the nitrogen base. It would be expected, in view of the pronounced *trans* effects observed with cobamides, that the V_{max} of the reaction would be affected by changes in the pH of the nitrogen base if interaction between cobalt and nitrogen base occur. Cobinamide coenzyme is not enzymically active. The nitrogen base, therefore, must fulfill some role; possibly, it is involved in stabilizing a catalytically active protein conformation.

31. B. M. Babior and T. K. Li, *Biochemistry* **8,** 154 (1969).
32. B. M. Babior, *BBA* **178,** 406 (1969).
33. R. J. P. Williams, personal communication.
34. B. M. Babior and D. C. Gould, *BBRC* **34,** 441 (1969).

TABLE I
COENZYME ACTIVITY OF B_{12} COENZYME ANALOGS[a]

Analog	Relative activity		
	Ethanolamine ammonia lyase (*38*)	Dioldehydrase	Glycerol dehydrase (*37*)
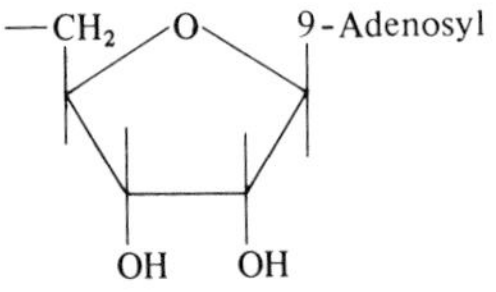 $—CH_2$ O 9-Adenosyl; OH OH	100	100	100
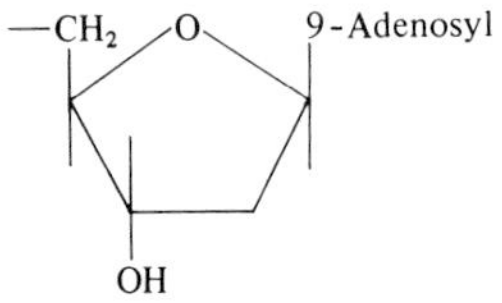 $—CH_2$ O 9-Adenosyl; OH	60		18
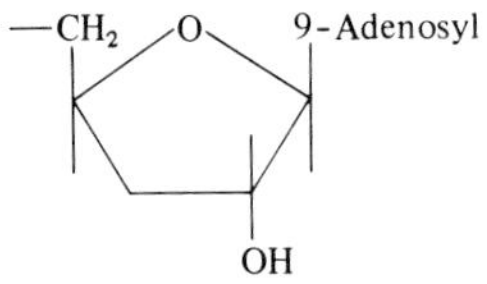 $—CH_2$ O 9-Adenosyl; OH			29
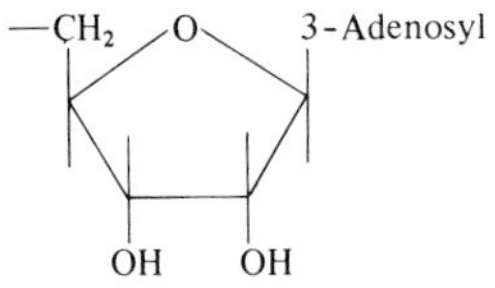 $—CH_2$ O 3-Adenosyl; OH OH			10
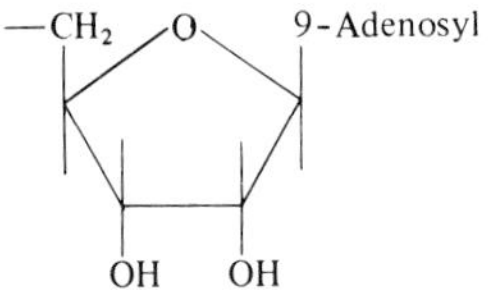 $—CH_2$ O 9-Adenosyl; OH OH		30	
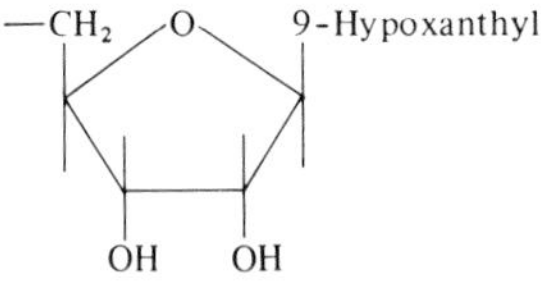 $—CH_2$ O 9-Hypoxanthyl; OH OH			

TABLE I (*Continued*)

Analog	Relative activity		
	Ethanolamine ammonia lyase (*38*)	Dioldenhydrase	Glycerol dehydrase (*37*)
$-CH_2$, O, 1-Uridyl, OH, OH (ring structure)	I		
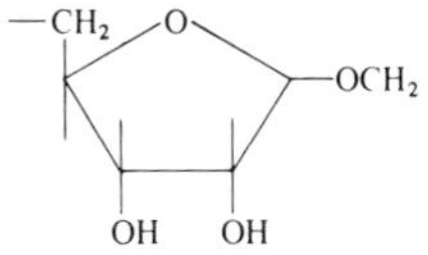 $-CH_2$, O, $-OCH_2$, OH, OH (ring structure)	I		
$-CH_2-(CH_2)_2-CH_2-Ad$	I	I	
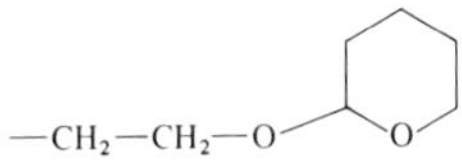 $-CH_2-CH_2-O-$ (tetrahydropyranyl, ring O)	I		
$-CH_2-CH_2OH$	I	I	
$-CH_3$	I	I	
OH or CN^-	I	I	I

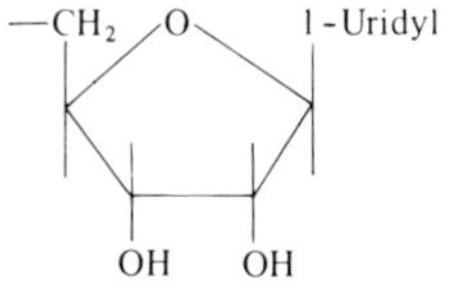

[a] Number indicates activity relative to vitamin B_{12} coenzyme. I indicates no coenzyme activity, but acts as inhibitor.

Coenzyme analogs in which the adenosyl portion of the coenzyme is replaced have been tested. The results obtained are summarized in Table I. Since the 2′- and 3′-deoxyadenosyl coenzyme can function as coenzymes, neither of these OH's can be essential for the catalytic process. The ribose ring oxygen is also not necessary. In a number of nonenzymic reactions, such as reaction with $NaBH_4$, CN, and H^+, where the carbon–cobalt bond is dissociated so that the electrons remain with the adenosyl moiety, the oxygen is required. These reactions do not occur when oxygen is replaced by $-CH_2-$. Since the coenzyme analog in which the

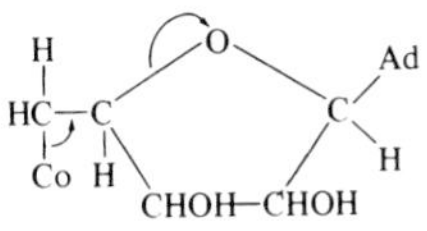

H, HC, C, Co, H, O, C, Ad, H, CHOH—CHOH

oxygen is replaced by $-CH_2-$ has catalytic activity (*35*), the type of electron displacement involved in these reactions appears not to be involved in the catalytic process.

When inhibitory analogs are added to dioldehydrase prior to addition of coenzyme, essentially no activity is observed when coenzyme is added (*36*). Catalytic activity cannot be restored by dialysis or Sephadex chromatography. The interaction of these compounds with the apoenzyme appears to be irreversible. Zagalak has developed a procedure by which hydroxocobalamin can be removed from the apoenzyme with the recovery of most of the catalytic activity (*37*). The procedure involves treatment with charcoal in the presence of Mg^{2+} and HSO_3^-.

The action of the inhibitory coenzyme analogs with ethanolamine ammonia lyase has been studied in detail, and results somewhat different from those with dioldehydrase were obtained (*38*). These compounds lead to inactivation of the enzyme whether they are added to the apoenzyme after vitamin B_{12} coenzyme or before. In no case is total inactivation obtained and a small fraction of catalytic activity remains. The author proposes that the inhibitor and apoenzyme combine to form a complex and that formation of the complex also leads to some irreversible change in the enzyme. This changed enzyme has lowered catalytic activity. The effect of corrin modifications have been investigated. A coenzyme analog has been proposed derived from 10-Cl-cobalamin (*39*). This coenzyme had the same apparent K_m as the unmodified coenzyme, but V_{max} was reduced by 40%. The authors concluded that since an electron withdrawing group reduces the rate of the reaction the catalytic process must involve a heterolytic cleavage of the carbon–cobalt bond, in which the electron remains with the adenosyl portion and Co^{3+} is formed. This interpretation must be viewed with caution since electronic effects in multistep processes, in which each step has different electronic requirements, are difficult to interpret.

VI. Properties of the Apoenzyme

The molecular weight of dioldehydrase was estimated to be 220,000–240,000 by Sephadex chromatography. Equivalent weights determined

35. S. S. Kerwar, T. A. Smith, and R. H. Abeles, *JBC* **245**, 1169 (1970).
36. R. H. Abeles and H. A. Lee, Jr., *Ann. N. Y. Acad. Sci.* **112**, 695 (1964).
37. B. Zagalak, personal communication.
38. B. M. Babior, *JBC* **244**, 2917 (1969).
39. Y. Tamao, Y. Morikawa, S. Shimizu, and S. Fukui, *BBRC* **28**, 692 (1967); *BBA* **151**, 260 (1968).

TABLE II
EFFECT OF CATIONS ON ASSOCIATION OF GLYCEROL DEHYDRASE SUBUNITS[a]

Buffer, 0.1 *M*	pH	Glycerol (conc. 10%)	Associated enzyme (%)
K^+ phosphate	8.6	−	95
K^+ phosphate	8.6	+	95
K^+ phosphate	6.5	−	5
K^+ phosphate	6.5	+	5
Na^+ phosphate	8.6	−	5
Na^+ phosphate	8.6	+	95
NH_4^+ phosphate	8.6	−	95
Cyclohexylamine, HCl	8.6	−	5

[a] From Schneider *et al.* (*40*).

from the amount of coenzyme and hydroxocobalamin bound are estimated to be 240,000, suggesting one coenzyme binding site per mole of enzyme. The molecular weight of ethanolamine ammonia lyase is estimated to be 520,000 from its sedimentation constant (*7*). Two moles of coenzyme are bound per mole of enzyme. The molecular weight of glycerol dehydrase is 240,000, and there is one coenzyme binding site per molecule. The enzyme can be dissociated into two protein fractions by chromatography on DEAE-cellulose. One of the protein fractions has a molecular weight of 200,000–240,000, and the other approximately 22,000. The individual fractions are catalytically inactive. Enzymic activity is restored on combination of the two fractions (*6, 40*). The association of the two fractions is dependent upon pH, presence of K^+ or substrate as illustrated in Table II. Unfortunately, no information is available at this time concerning the subunit composition of other dehydrases. Glutamate mutase has also been resolved into two protein fractions (*1*). The possibility is intriguing that one of the subunits may be primarily concerned with apoenzyme binding and possibly its chemical activation, while the other confers substrate specificity. It will therefore be of interest to examine coenzyme subunits obtained from different vitamin B_{12}-dependent enzymes.

40. Z. Schneider, K. Pech, and J. Pawelkiewicz, *Bull. Acad. Polon. Sci., Ser. Sci. Biol.* **14,** 7 (1966).

18

Enolase

FINN WOLD

I. Introduction

Enolase, 2-phospho-D-glycerate hydrolyase or phosphopyruvate hydratase (EC 4.2.1.11), catalyzes the dehydration of D-glycerate 2-phosphate to give enolpyruvate phosphate. The reaction, which is given in Eq. (1), is readily reversible. As one of the enzymes involved in glycolysis

$$\mathrm{H{-}\underset{\underset{H_2C-OH}{|}}{\overset{\overset{COO^-}{|}}{C}}{-}O{-}P(=O)(O^-)_2} \rightleftharpoons \mathrm{\underset{\underset{CH_2}{\|}}{\overset{\overset{COO^-}{|}}{C}}{-}O{-}P(=O)(O^-)_2} + H_2O \tag{1}$$

and fermentation, enolase is likely to be very nearly ubiquitous in the biological world.

The early studies on the enolase reaction, the discovery of the enzyme and the isolation of its substrates, the isolation of the enzyme from yeast, and the subsequent kinetic analyses establishing the requirement for a divalent metal ion, Mg^{2+}, the inhibition by fluoride in the presence of Mg^{2+} and phosphate, and the substrate specificity have been reviewed previously (*1–3*). This chapter will therefore primarily consider the new developments during the last decade. Enolase has now been purified to some degree from about 20 different sources and has been obtained pure from nine of these. It is thus possible to start to look for the features of its structure which appear to be required by all the different proteins which catalyze the enolase reaction, and these comparative aspects will be given some emphasis in the discussion of both structure and function. The role of the divalent metal ion has been studied extensively during the last 10 years and it appears that a dual function of the metal ion, both as a structural component of the active enzyme and as a part of the catalytic apparatus, must be considered. Many other features such as the substrate specificity, the subunit structure, the occurrence of isozymes in several species, and labeling with active site-specific reagents have also been explored. From this data the vague outline of a general structure-function model is emerging which appears to apply to all the enolases that have been studied to date. The major purpose of this chapter is to review this new information and to document this model.

The objects of this discussion, the various enolases that have been studied to date, are listed in Table I (*4–23*) with references to the respec-

1. O. Meyerhof, "The Enzymes," 1st ed., Vol. 1, Part 2, p. 1210, 1951.
2. T. Bücher, "Methods in Enzymology," Vol. 1, p. 427, 1955.
3. B. C. Malmström, "The Enzymes," 2nd ed., Vol. 5, p. 471, 1961.
4. E. W. Westhead, "Methods in Enzymology," Vol. 9, p. 670, 1966.
5. E. W. Westhead and G. McLain, *Biochem. Prep.* **11,** 37 (1966).
6. E. W. Westhead and G. McLain, *JBC* **239,** 2464 (1964).
7. H. Boser, *Z. Physiol. Chem.* **315,** 163 (1959).
8. T. G. Spring, Ph.D. Thesis, University of Minnesota (1970).
9. J. M. Chapman, C. Chin, and F. Wold, *J. Fish. Res. Board Can.*, **28,** 879 (1971).
10. R. P. Cory and F. Wold, *Biochemistry* **5,** 3131 (1966).
11. R. C. Ruth, D. M. Soja, and F. Wold, *ABB* **140,** 1 (1970).
12. J. A. Winstead and F. Wold, *Biochem. Prep.* **11,** 31 (1966).
13. B. G. Malmström, *ABB Suppl.* **1,** p. 247 (1962).
14. R. Czok and T. Bücher, *Advan. Protein Chem.* **15,** 315 (1960).
15. E. Y. Fedorchenko, *Ukr. Biokhim. Zh.* **30,** 552 (1958).
16. T. Baranowski, E. Wolna, and A. Morawiecki, *European J. Biochem.* **5,** 119 (1968).

TABLE I
BIOLOGICAL SOURCES FROM WHICH ENOLASE HAS BEEN PURIFIED[a]

Source	Reference
A. Highly Purified	
Yeast	*4–6*
Potato tubers	*7*
Escherichia coli	*8*
Lobster muscle	*9*
Rainbow trout muscle	*10*
Coho salmon muscle	*11*
Chum salmon muscle	*11*
Rabbit muscle	*4, 12–15*
Human muscle	*16*
Monkey muscle	*17*
B. Intermediate Purity	
Frog muscle	*18*
Turtle muscle	*18*
Chicken muscle	*18*
Beef muscle	*18*
Dog muscle	*18*
Mouse muscle	*18*
Beef brain	*19*
Human erythrocytes	*20*
C. Low Purity	
Brevibacterium fuscum	*21*
Leuconostoc mesenteroides	*18*
Marine planktonic algae	*22*
Peas	*23*
Soybeans	*18*
Snails	*18*

[a] Group A: highly purified (95–100% pure); group B: intermediate purity (25–90% pure); and group C: low purity (<5% pure). This purity classification is based on the assumption that all enolases have similar specific activity (about 100 μmoles of substrate converted per min per mg of protein).

tive isolation procedures. With the exception of potato enolase, which apparently has a very high content of sulfhydryl groups and is quite unstable, and also with the exception of ox brain enzyme, the enolases seem to be stable enzymes, quite easy to handle and store.

17. J. A. Winstead, *Abstr. 154th Natl. Meeting, Am. Chem. Soc., Chicago. Biol.* p. 208 (1967).
18. J. M. Cardenas and F. Wold, *ABB* **144,** 663 (1971).
19. T. Wood, *BJ* **91,** 453 (1964).
20. I. Witt and D. Witz, *Z. Physiol. Chem.* **351,** 1232 (1970).
21. N. Saito, *J. Biochem.* **61,** 59 (1967).
22. N. J. Anita, J. Kalmakoff, and A. Watt, *Can. J. Biochem.* **44,** 449 (1966).
23. G. W. Miller, *Plant Physiol.* **33,** 199 (1958).

II. Molecular Properties

A substantial body of information on the molecular properties of enolases has accumulated during the last decade. The original data on the yeast enzyme, reviewed 10 years ago (*3*), suggested that enolase was a single polypeptide chain with a molecular weight of 67,000 daltons. A reexamination of this data, together with the new data for other enolases, strongly suggest that all enolases are made up of two very similar or indeed, identical subunits, and have molecular weights in the range of 82,000–100,000 daltons. It also appears that Mg^{2+}, in addition to its role in the catalytic reaction, has an important structural function in stabilizing the active dimeric form of the enzyme. Superimposed on these general structural features are the individual molecular properties which in most instances give each enzyme its unique structural features that distinguish it from the other enolases. The experimental basis for this general picture of enolase structure will be examined in the following sections.

A. Criteria of Purity

The development of several high resolution analytical methods makes it possible to define enzyme purity much more rigidly than was the case in the past; especially the analytical gel electrophoresis systems and isoelectric focusing have provided powerful tools by which enzyme preparation can be subjected to very critical homogeneity tests. Specific activity has been and still is a convenient method for establishing the purity of an enzyme preparation, but it is only meaningful when the methods for both activity and protein determinations are internally consistent and when the reference specific activity of *pure* enzyme is known. Thus, constant specific activity cannot be an acceptable single criterion of purity. Yeast enolase illustrates this point well. In the original preparation the enzyme was crystallized as the Hg^{2+} salt (*24*) and could be recrystallized to constant specific activity (after removal of Hg^{2+}). When the enzyme was subsequently purified by chromatographic and electrophoretic methods, the resulting electrophoretically homogeneous yeast enolase had a specific activity some 20% higher than that obtained by crystallization. The purity of most of the pure enolases discussed in this chapter was established by electrophoresis on

24. O. Warburg and W. Christian, *Biochem. Z.* **310,** 384 (1941).

acrylamide or starch gels, and then cross-checked with specific activity under standardized conditions. It must be emphasized, however, that even these criteria may be misleading. In a recent study a yeast enolase preparation which was pure by the above criteria was found to resolve into at least three distinct components upon electrofocusing on acrylamide gel (*25*) in pH gradient between 6 and 8 (*26*). Other molecular properties did not appear to reflect this heterogeneity, and there is no documented explanation for the appearance of different forms upon electrofocusing (*26*). It is clear, however, that the problem of establishing the purity of enzymes is still a very real one and one that warrants continued and constant attention.

B. Chemical Properties

1. *Amino Acid Composition*

Complete amino acid analyses are available for enolase from yeast (*27–29*), rabbit (*30*), trout (*10*), coho salmon (*11*), chum salmon (*11*), *Escherichia coli* (*8*), and potatoes (*7*). The data are given in Table II. There does not appear to be any unique features in the amino acid make-up of these enzymes, and as in any such comparison, one may note similarities as well as differences. The most noticeable distinctions between enolases are found in the tryptophan and in the cystine-cysteine content.

2. *End Groups and Terminal Sequences*

The emphasis in the following section will be on the quantitative end-group analysis as documentation for the presence of two polypeptide chains in all the enolases studied to date.

a. The Amino Terminus. All the vertebrate enolases studied to date have been found to have "blocked" amino-terminal end groups. In the

25. G. Dale and L. Latner, *Lancet* **i**, 847 (1968); C. Wrigley, *Sci. Tools* **15**, 17 (1968).
26. K. G. Mann, F. J. Castellino, and P. A. Hargrave, *Biochemistry* **9**, 4002 (1970).
27. B. G. Malmström, J. R. Kimmel, and E. L. Smith, *JBC* **234**, 1108 (1959).
28. J. M. Brewer, T. Fairwell, J. Travis, and R. E. Lovins, *Biochemistry* **9**, 1011 (1970).
29. In both of the above papers (*27*, *28*) the molar quantities of amino acids are calculated on the basis of a molecular weight of 67,000. Since the molecular weight is now known to be 88,000 (*26*), the values must be corrected accordingly.
30. A. Holt and F. Wold, *JBC* **236**, 3227 (1961).

TABLE II

AMINO ACID COMPOSITION OF SOME ENOLASES[a]

Amino acid	*E. coli* 90,000 [Ref. (*8*)]		Yeast 88,000 [Refs. (*27, 28*)]		Trout muscle 91,000 [Ref. (*10*)]		Coho salmon muscle 100,000 [Ref. (*11*)]	Chum salmon muscle 100,000 [Ref. (*11*)]	Rabbit muscle 82,000 [Ref. (*30*)]		Potato [Ref. (*7*)]
	A[b]	B[b]	A	B	A	B	A(B)	A(B)	A	B	A
Lysine	83	75	80	71	88	80	86	89	82	67	63
Histidine	16	14	24	21	23	21	29	23	23	19	11
Arginine	23	21	30	26	26	24	32	31	37	30	32
Aspartate	94	85	98	86	109	99	114	114	94	77	121
Threonine	45	41	39	34	36	33	35	35	40	33	32
Serine	47	42	60	53	54	49	53	43	40	33	37
Glutamate	101	91	69	61	98	89	90	83	82	67	74
Proline	24	22	30	26	34	31	35	32	32	26	42
Glycine	100	90	78	69	89	81	91	82	87	72	53
Alanine	129	116	108	95	102	92	99	94	92	76	53
Cysteine	4	3–4	2.2	2	3.3	3	11	11	15	12	34
½-Cystine	4	3–4	0	0	6.6	6	3	4	0	0	—
Valine	53	48	75	66	62	56	57	61	69	57	89
Methionine	23	21	12	11	13	12	13	14	15	12	—
Isoleucine	71	64	45	40	62	56	58	62	52	43	84
Leucine	64	58	80	71	74	67	69	72	82	67	
Tyrosine	23	21	18	16	24	22	26	25	22	18	37
Phenylalanine	26	24	33	29	31	28	31	31	31	25	105
Tryptophan	3–4	3	9	8	3–4	3	3	6	12	10	9

[a] The analyses were obtained by standard amino acid analysis, except the data for potato enolase which are based on elution and quantification of DNP derivatives of the hydrolyzed protein and on quantification of the free amino acids as their copper complexes, after two-dimensional chromatography on paper.

[b] The results are reported as moles of amino acid per 100,000 g of enzyme in column A, and as moles of amino acid per mole of enzyme in column B.

case of rabbit muscle enolase the blocked amino terminus was identified as *N*-acetylalanine (*31*). Free *N*-acetylalanine was isolated by ion exchange chromatography of proteolytic (pronase) digests of the enzyme and was characterized by chromatography of the derivative itself as well as its hydrolysis product, alanine and its hydrazinolysis products, alanine and acetyl hydrazide. The yield of *N*-acetylalanine, based on alanine analyses after hydrolysis of the isolated derivative, was 1.2 moles/mole of enolase, more than can be accounted for by a single polypeptide chain. The suggestion that similarly blocked amino terminals may exist in three fish enolases is merely based on the negative results from exhaustive *N*-group analyses by both enzymic and chemical means (*10, 11*).

Yeast enolase has a free N-terminal amino group and recent quantitative studies have yielded 2.3 ± 0.25 moles of N-terminal alanine per mole of enolase by the methyl isothiocyanate method (*28*) and 1.85 ± 0.06 moles of N-terminal alanine per mole of enolase by the cyanate method (*32*). In connection with the end-group analysis, it was also shown that the N-terminal sequence of yeast enolase is Ala–Gly–Lys–Val–Gly–Asp–Thr–Gln– (*28*) confirming and extending earlier reports (*27*). There is no evidence that the two polypeptide chains of yeast enolase differ in the N-terminal sequence.

b. The Carboxy Terminus. The carboxy terminus of five enolases has been studied by enzymic digestion and hydrazinolysis. The results, some of which are summarized in Fig. 1, are consistent with two identical C-terminals per mole of enzyme in all five cases. The resistance to carboxypeptidase after the initial release of one or two carboxy-terminal amino acids is a constant feature of all the enolases studied, and makes the data easy to interpret. The assignment of Asn rather than Ile as the C-terminal amino acid in the salmon enolases is based on the absence of any free amino acid in the hydrazinolysis experiments and on the fact that the carboxypeptidase-catalyzed release of Asn is so much slower than that of Ile that the two amino acids should be released from a X–Ile–Asn sequence at approximately equal rates.

One feature of the quantitative end-group analyses of enolases with exopeptidases, the rather remarkable resistance of the enzymes to exopeptidase attack, requires consideration since these results (*32*) are in direct conflict with earlier data (*33*). In the earlier work as much as 20% of the total amino acid chain could be removed from either end of

31. J. A. Winstead and F. Wold, *Biochemistry* **3**, 791 (1964).
32. P. A. Hargrave and F. Wold, *JBC* **246**, 2904 (1971).
33. O. Nylander and B. G. Malmström, *BBA* **34**, 196 (1959).

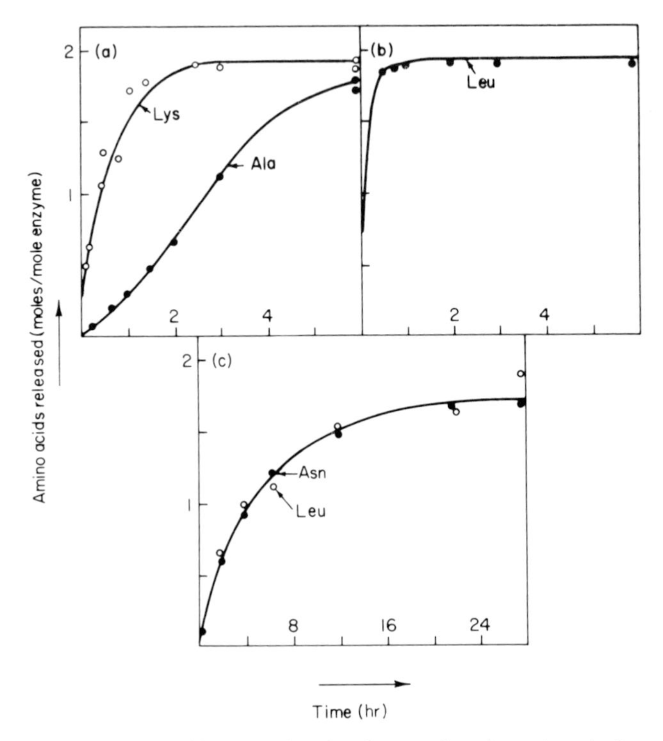

FIG. 1. The carboxypeptidase-catalyzed release of carboxy-terminal amino acids from enolases. (a) Rabbit muscle enolase digested with carboxypeptidase A and B [from Winstead and Wold (*31*)]. The data for monkey enolase are very similar to the curves in (a) (*17*). (b) Yeast enolase digested with carboxypeptidase A [from Hargrave and Wold (*32*)]. (c) Coho salmon enolase digested with carboxypeptidase A [from Ruth *et al.* (*11*)]. The data for chum salmon enolase are similar to those in (c) (*11*).

yeast enolase by exopeptidases without loss of activity. These findings have not been reproduced in the recent studies.

3. *The Identity of the Two Polypeptide Chains*

All the work to date on the end groups and the end-terminal sequences of enolases suggests that the two polypeptide chains that make up the active enzyme are identical. The failure to resolve the two chains under dissociating conditions is also consistent with this hypothesis. More convincing evidence that the total chains are indeed identical has been obtained for rabbit muscle enolase and yeast enolase. The evidence is based on the evaluation of the number of distinct peptides obtained by tryptic digestion of enolase. In the case of rabbit muscle enolase a careful analysis was made of the extent of the tryptic digestion as well as the number of tryptic cleavages which led to release of free Lys and free Arg, and the results were evaluated with these corrections imposed. It was calculated that for the actual tryptic digestion in question rabbit muscle enolase with 30 moles of Arg and 68 moles of Lys per mole of enzyme (Table II) should give 14–15 Arg-containing peptides and 35–45 total peptides if the two chains were identical or 28 Arg-containing peptides and 67–82 total ninhydrin peptides if the chains were totally different. The observed 14 ± 1 Arg-positive spots and 42 ± 2 ninhydrin spots in the peptide maps of rabbit muscle enolase are thus in excellent agreement with the identical chain count (*34*). The same conclusions can be drawn from similar studies on yeast enolase. This enzyme containing 26 moles of Arg and 70 moles of Lys per mole (Table II) should give 13 Arg peptides and 49 ninhydrin peptides if the chains are identical or 26 Arg peptides and 98 ninhydrin peptides if they are not. In this case it was assumed that the trypsin digestion was complete and that all cleavages gave unique peptides. The tryptic peptide maps of yeast enolase gave 8–12 Arg-positive spots and 46 ± 2 ninhydrin spots (*28*). Within the obvious limits in detecting subtle differences in amino acid sequences inherent in this technique, these two enolases thus appear to be dimers of two identical polypeptide chains. More conclusive evidence on this point must await the determination of the complete amino acid sequence of the two chains.

4. *Immunochemical Cross-Reactivity*

In an attempt to develop some parameter measuring structural similarities among various enolases, rabbit antibodies to yeast enolase and to trout muscle enolase and chicken antibodies to rabbit muscle enolase were prepared (*18*). The cross-reactivity of these antibodies with various enolases is summarized in Table III and serves as a start toward ordering the enolases into families of structurally similar enzymes. In the absence of any knowledge as to the nature of the antigenic structural determinants, no further conclusions can be drawn from these data at present.

C. Physical Properties

A large volume of physical data has been accumulated for the different enolases. Rather than discussing all of this in detail, it will be

34. J. M. Cardenas and F. Wold, *Biochemistry* 7, 2736 (1968).

TABLE III
IMMUNOCHEMICAL CROSS-REACTIVITY OF ENOLASES

Enzyme source	Immunochemical reaction[a]		
	C-Anti-RE	R-Anti-YE	R-Anti-TE
Rabbit	+	−	−
Dog	+		
Beef	+		
Mouse	+		
Turtle	+	−	−
Frog	+	−	−
Rainbow trout	+ (weak)	−	+
Coho salmon	−		+
Chum salmon			+
Ocean perch	−		+
Snail	−	−	−
Chicken	−	−	−
Yeast	−	+	−
L. mesenteroides	−	−	

[a] Precipitin band formation in micro-Ouchterlony plates is scored as +. The antisera used were anti-rabbit muscle enolase from chicken (C-Anti-RE), anti-yeast enolase from rabbits (R-Anti-YE), and anti-trout muscle enolase from rabbits (R-Anti-TE). Data taken from Cardenas and Wold (*18*).

tabulated at the end of this section. Three features of the physical properties will be discussed: the isozyme phenomenon as revealed by gel electrophoresis, the dissociation of the enolase dimer into the individual single chain subunits, and the metal binding and the role of divalent metal as a structural component in the active enzyme.

1. *Electrophoretic Mobility—Enolase Isozymes*

Electrophoretic mobility, just like immunochemical cross-reactivity, has been used as a quick means of comparing gross structural features of different enolases. Again, it is clear that any interpretation of electrophoretic mobility in terms of specific structural characteristics is impossible at this stage, and that these measurements thus are useful primarily for arbitrary classification (*18*). An important result of the analysis of electrophoretic mobility of enolases on starch and acrylamide gels has been the observation of multiple electrophoretically distinct active forms of the enzyme in certain species. The existence of enolase isozymes in yeast was first reported by Malmström (*35*), and has subse-

35. B. G. Malmström, *ABB* **70**, 58 (1957).

quently been studied in several laboratories (*6, 36*). Enolase isozymes have also been found in eight species of Salmonidae (*10, 11, 37*). Each species of fish contains three electrophoretically distinct enolases with relative electrophoretic mobility and in proportions so characteristic of each species that these criteria can be used to distinguish between the different species. The characteristic electrophoretic pattern of enolases does not seem to change with age or sex of the fish (*37*). Both the yeast and the fish isozymes have been scrutinized as to the possibility of simply being artifacts of isolation, and evidence for the appearance of new enolase forms during prolonged extraction of yeast (*6*) was obtained. However, these forms appear to be modifications of the three original forms, and the conclusion that the three enolase forms in both yeast and Salmonidae represent a real biological phenomenon seems to be well founded (*36, 37*). In attempts to assess the biological significance of the different forms, the isozymes have been separated and their individual molecular and catalytic properties determined. Some representative results for the chum salmon enolases are presented in Table IV. The thermal stability was studied to test one of the postulated roles of multiple enolase forms in *poikilotherms*, namely, that they require different enzyme forms with different temperature dependence to be able to adjust rapidly to the relatively large temperature fluctuations in their

TABLE IV
SOME PROPERTIES OF CHUM SALMON ENOLASE ISOZYMES[a]

		Isozyme No.		
		1	2	3
Forward reaction:	K_m (in μmoles/liter)	150	80	80
	Relative turnover No.	0.49	0.81	1.0
	K_I for lactate P (in μmoles/liter)	1800	700	1600
	pH optimum	6.9	6.9	6.9
Reverse reaction:	K_m (in μmoles/liter)	85	75	67
	Relative turnover No.	0.40	0.81	1.0
Thermal stability:	Temperature at which 50% activity is lost in 3 min	65°	64°	64°

[a] Data taken from Soja (*38*).

36. G. Pfleiderer, A. Neufahrt-Kreiling, R. W. Kaplan, and P. Fortnagel, *Ann. N. Y. Acad. Sci.* **151,** 78 (1968).
37. H. Tsuyuki and F. Wold, *Science* **146,** 535 (1964).

environments (*38*). The data in Table IV appear to eliminate that explanation. A similar lack of significant differences in the properties of the yeast isozymes has also been reported, with the suggestion that the enolase isozymes have no biochemical significance and may simply represent a biological accident (*36*). In spite of the lack of success in establishing a biological role for the enolase isozymes, there are good reasons to continue to investigate this phenomenon. One of these reasons is the fact that the occurrence of enolase isozymes may well be more general than the above discussion indicates. Using enolase specific staining methods, Dave *et al.* (*39*) investigated crude cell extracts by gel electrophoresis and found multiple enolase forms in both beef and rat muscle. Partially purified enolase from beef shows a single electrophoretic form (*18*), and it could thus well be that observation of a single enolase form after any purification step could be misleading.

2. *Subunit Structure*

The chemical evidence for two polypeptide chains in several enolases has developed along with physical studies on the dissociation of the enolase dimers, and good evidence for the two-subunit structure of several enolases has been established by different physical methods. The main approach has been careful molecular weight determinations in the absence and presence of dissociating agents by means of sedimentation equilibrium analyses. As an adjunct to this general approach, the recently developed methods of gel filtration in 6 *M* guanidinium chloride (*40*) and sodium dodecyl sulfate (SDS) electrophoresis (electrophoresis in SDS-containing buffers) on acrylamide gels (*41*, *42*) have been found very convenient and surprisingly accurate methods for determining the mass of the smallest covalent unit in these dissociating media. The studies on four enolases will be considered briefly below.

a. Rabbit Muscle Enolase. When rabbit muscle enolase (MW 82,000) was exposed to a solution of 20% aqueous dioxane (or acetone) containing 0.2 *M* ammonium sulfate and 5×10^{-3} *M* EDTA, rapid dissociation to subunits (MW 41,000 ± 2,000) was observed (*43*). Diluting the dissociated enzyme directly into the assay medium gave no activity. Based on this observation it was possible to assay for the rate of dissocia-

38. D. M. Soja, Ph.D. Thesis, University of Illinois (1967).
39. P. Dave, R. W. Kaplan, and G. Pfleiderer, *Biochem. Z.* **345**, 440 (1966).
40. W. W. Fish, K. G. Mann, and C. Tanford, *JBC* **244**, 4989 (1969).
41. A. L. Shapiro and J. V. Maizel, Jr., *Anal. Biochem.* **29**, 505 (1969).
42. K. Weber and M. Osborn, *JBC* **244**, 4406 (1969).
43. J. A. Winstead and F. Wold, *Biochemistry* **4**, 2145 (1965).

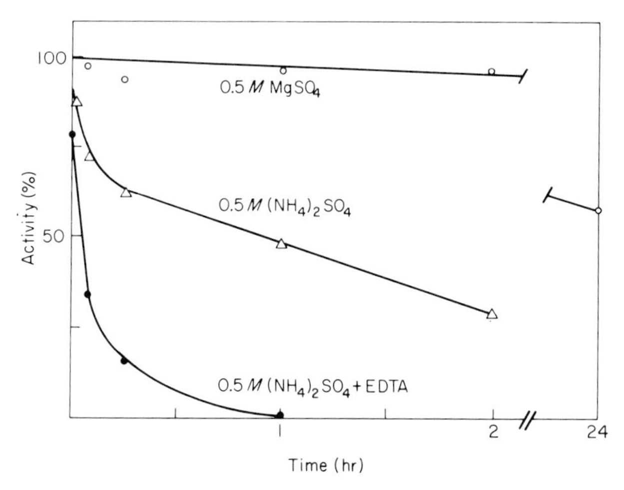

FIG. 2. The effect of Mg^{2+}, ammonium sulfate, and EDTA on the rate of dissociation of enolase in 20% aqueous dioxane. Rate of dissociation is measured as rate of formation of inactive enzyme. The data are taken from Winstead and Wold (*43*).

tion simply by following the rate of activity loss, and the effect of EDTA, Mg^{2+}, and ammonium sulfate on the rate of dissociation in 20% dioxane (or acetone) could be evaluated. Typical results are given in Fig. 2, and show that in the absence of ammonium sulfate or EDTA or with added Mg^{2+} the rate of dissociation is dramatically retarded. These results were confirmed by sedimentation velocity analyses, which showed well-resolved monomer ($s_{20,w} = 3.5$ S) and dimer ($s_{20,w} = 5.7$ S) peaks after 2 hr in the dioxane (or acetone) solvent with EDTA omitted, but the monomer peak only in parallel experiments in which EDTA was present. Thus, the presence of ammonium sulfate and the absence of Mg^{2+} are essential for rapid dissociation of rabbit muscle enolase in organic solvents. If the EDTA and dioxane (or acetone) were slowly dialyzed out, the dissociation was reversible, and 45% (or 87%) of active dimer could be obtained from the completely dissociated enzyme. In the same study it was also found that ammonium sulfate alone at high concentration could cause dissociation of rabbit muscle enolase. Direct measurements of molecular weight in the ultracentrifuge by the Archibald method showed that at infinite protein dilution the molecular weight of the enzyme in 2 *M* ammonium sulfate was 41,000. The molecular weight showed a dramatic dependence on protein concentration, however, and to allow for this as well as for the unusual viscosity and sedimentation velocity properties of the enzyme in 2 *M* ammonium

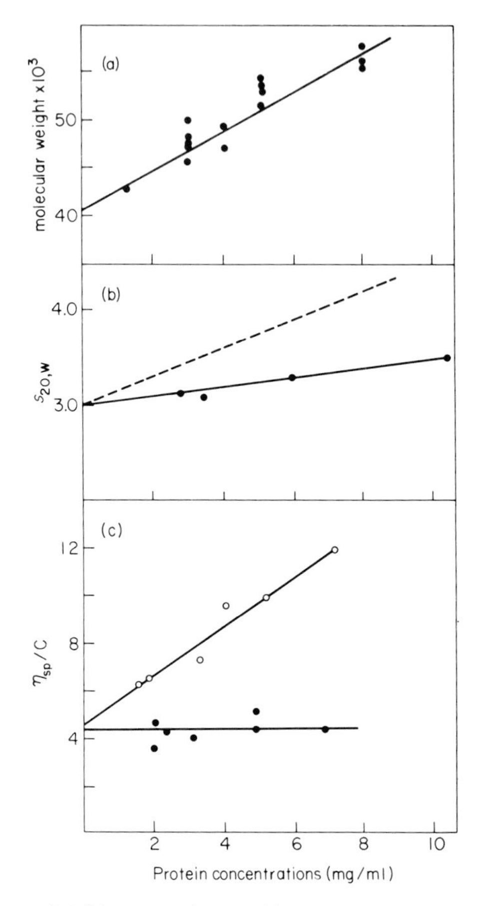

FIG. 3. The effect of 2 M ammonium sulfate on some molecular properties of rabbit muscle enolase. (a) The dependence of molecular weight on protein concentration in 2 M ammonium sulfate. (b) The dependence of the sedimentation coefficient on protein concentration in 2 M ammonium sulfate. The dotted line in (b) is the calculated concentration dependence of $s_{20,\mathrm{w}}$ for a simple monomer–dimer equilibrium using the data in (a), and assuming that the frictional coefficient is the same for both monomer ($s_{20,\mathrm{w}} = 3.0$) and dimer ($s_{20,\mathrm{w}} = 5.7$). The lower slope of the experimental curve strongly suggests that the dimer is an asymmetric form with a lower sedimentation coefficient than the native dimer. (c) The dependence of reduced viscosity on protein concentration in dilute buffer (●) and in 2 M ammonium sulfate (○). Again, the high viscosities at increasing protein concentrations suggest that asymmetric species are formed in 2 M ammonium sulfate. The data are taken from Winstead and Wold (*43*).

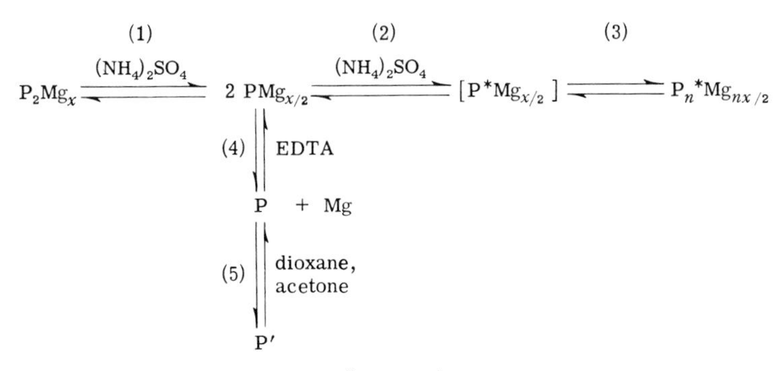

SCHEME 1

sulfate (Fig. 3), Scheme 1 was proposed as a model for the dissociation (*43*). According to this model ammonium sulfate alone [reactions (1), (2), and (3)] cause dissociation [reaction (1)] and a conformational change [reaction (2)]. The modified monomer, $P^*Mg_{x/2}$, is only present as an unstable intermediate, which at high protein concentration is stabilized by aggregation [reaction (3)]. If the conformational isomer P^* is highly asymmetrical in comparison to P, the scheme explains the high reduced viscosity and low sedimentation coefficient observed in $2\,M$ ammonium sulfate, in that the observed equilibrium would be between the compact $PMg_{x/2}$ (at low protein concentration) and the asymmetrical $P_n{}^*Mg_{nx/2}$ (at high protein concentrations). Scheme 1 also explains the dissociation in organic solvents as being affected through a conformational change of Mg-free monomer P to give P′ [reaction (5)]. This reaction cannot be readily reversible since direct addition of the dissociated enzyme (P′) into the assay medium gave zero activity. Since the rate of formation of P′ depends on the concentration of P, both ammonium sulfate [reaction (1)] and EDTA [reaction (2)] are required for rapid dissociation via this route. Scheme 1 adequately explains all the observations on the dissociation of rabbit muscle enolase, but more experimental detail is required to prove such a reaction scheme.

b. Yeast Enolase. According to the most recent work the molecular weight of yeast enolase is 88,000 and the subunit weight is 44,000, and it has been suggested that the lower molecular weight values observed in the past arises from the fact that the enzyme is partly dissociated even under conditions which presumably should favor dimer stability (*26*). (Throughout the discussion in this section as well as in other sections of this article the MW of 88,000 will be used for yeast enolase, and, whenever necessary, data in the literature calculated for a

MW of 67,000 have been recalculated here for the new molecular weight.) Enolase dissociates into subunits of one-half the native molecular weight when exposed to 1 *M* KCl in the presence of EDTA (*44*). Potassium bromide has also been shown to be an effective dissociating salt (*45*), while potassium acetate has no effect. In the presence of Mg^{2+}, the dissociation is very significantly suppressed; thus, again the stabilizing effect of Mg^{2+} on the dimer is observed. Yeast enolase also dissociates readily in other systems apparently without any requirement for Mg^{2+} removal, and good subunit molecular weights have been obtained by SDS gel electrophoresis, gel filtration in 6 *M* guanidinium chloride, and equilibrium sedimentation in 6 *M* guanidinium chloride (*26*).

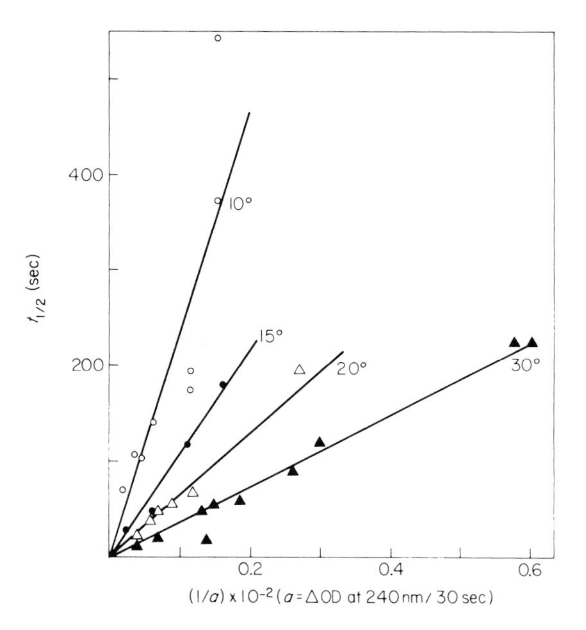

FIG. 4. The kinetics of reassociation of yeast enolase monomers produced in 1 *M* KBr. The integrated second-order rate equation for the process $2\ E \xrightarrow{k} E_2$ can be simplified to give $kt_{1/2} = 1/(E/2)$ predicting that $t_{1/2}$ is directly proportional to the reciprocal of the initial protein concentration (measured here as total recoverable activity) if the second-order combination of subunits is rate limiting. The data in the earlier portions of the activity recovery curves agree well with this second-order relationship. The data are taken from Gawronski and Westhead (*45*).

44. J. M. Brewer and G. Weber, *Proc. Natl. Acad. Sci. U. S.* **59**, 216 (1968).
45. T. H. Gawronski and E. W. Westhead, *Biochemistry* **8**, 4261 (1969).

The reassociation of the yeast enolase subunits sheds light on the structural and catalytic properties of the active enzyme. Using both recovery of activity and protein fluorescence changes as a measure of the rate of formation of the active dimer from inactive Mg^{2+}-free monomers in 1 M KCl, it was possible to assess the effect of Mg^{2+} in the reassociation process and to determine the stoichiometry of the Mg^{2+} binding in the process. The conclusion of these experiments is that it requires about 1.3 moles of Mg^{2+} for every 2 moles of subunits (or for every 1 mole of active dimer) to cause complete reassociation (*44*). The dissociation constant for the dimer was estimated to be about 10^{-7} M in the absence of both KCl and Mg^{2+}, and about 10^{-4} M in the presence of 1 M KCl, but without Mg^{2+}. The effect of Mg^{2+} is to lower the dissociation constant two orders of magnitude both with and without KCl (*44*). The reassociation at the protein concentrations (10^{-7} to 10^{-6} M) used in this work was reported to follow approximately first-order kinetics (*44*). Another kinetic analysis of the reassociation of yeast enolase at a concentration range of about 10^{-9} M gave good fit for second-order kinetics in the early portions of the recovery curves at all temperatures (*45*) (see Fig. 4). This becomes a crucial point in deciding whether there is any activity associated with the monomer (see Section III,C). Another dissociation–reassociation experiment is illustrated in Table V. In this experiment a mixture of an acetylated (^{14}C-acetate) derivative of yeast enolase and a molar excess of the native enzyme was first dissociated with KBr and was then allowed to reassociate. The results demonstrate that a hybrid enolase dimer was obtained with elution position, enzymic

TABLE V
THE HYBRIDIZATION OF YEAST ENOLASE[a]

Sample	Elution conductivity (mmho)	cpm/OD_{280}	Relative specific enzymic activity
Acetyl enolase	7.5–9.0	2870	81
Native enolase	0.5–1.0	0	100
Hybrid	3.0–5.2	1470	—
Rechromatographed hybrid	3.4–4.8	1260	*b*

[a] From Gawronski and Westhead (*45*). Partially acetylated enolase (^{14}C-acetate) and a 3-fold excess of native enolase were dissociated in 1 M KBr–0.01 M EDTA and allowed to reassociate by dialysis against 10^{-3} M Mg^{2+} in dilute buffer, pH 8. The hybrid enzyme was isolated by chromatography on DEAE-cellulose.

[b] Intermediate between the specific activity of the native and the acetylated enzyme.

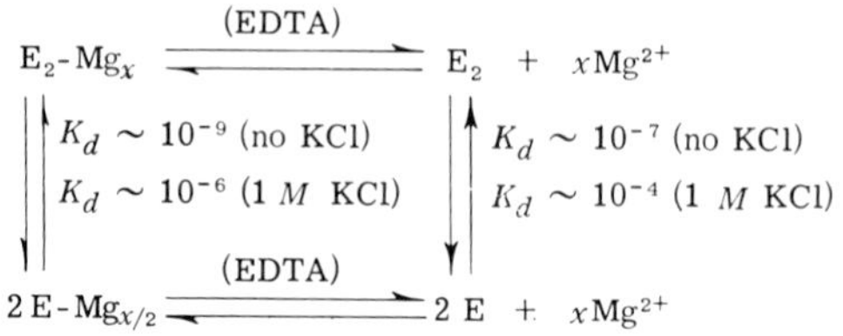

$$
\begin{array}{ccc}
E_2\text{-}Mg_x & \underset{}{\overset{(EDTA)}{\rightleftharpoons}} & E_2 + xMg^{2+} \\
\updownarrow \; K_d \sim 10^{-9} \text{ (no KCl)}, \; K_d \sim 10^{-6} \text{ (1 } M \text{ KCl)} & & \updownarrow \; K_d \sim 10^{-7} \text{ (no KCl)}, \; K_d \sim 10^{-4} \text{ (1 } M \text{ KCl)} \\
2\,E\text{-}Mg_{x/2} & \overset{(EDTA)}{\rightleftharpoons} & 2\,E + xMg^{2+}
\end{array}
$$

SCHEME 2

activity, and specific radioactivity exactly intermediate between the native enzyme and the acetylated starting material. Scheme 2 summarizes some of the features of the yeast enolase dissociation discussed above. Again the conclusions are tentative and will require more work to be confirmed.

c. *Coho Salmon Enolase.* Coho enolase (*46*) is the only fish enolase studied to date in which one of the three isozymes predominates to such an extent (90–95% of the total) that molecular properties can be studied without serious concern for ambiguities resulting from the multiple forms; its molecular weight is 100,000 ± 6,000 (*11*). The dissociation of coho enolase in salt solutions is not as clear cut as are those discussed above. In 2 *M* ammonium sulfate–0.005 *M* EDTA solutions containing 2-mercaptoethanol, sedimentation equilibrium analyses gave molecular weights as low as 58,000 ± 3,000 with a linear fringe displacement plot over the entire width of the cell. In 1 *M* KCl, on the other hand, heterogeneity with respect to the molecular weight was always observed in the ultracentrifuge, and only the molecular weights calculated from the meniscus region of the equilibrium concentration gradient approached the theoretical value for complete dissociation of identical subunits (50,000 ± 5,000). Sodium dodecyl sulfate electrophoresis gave a single band with a molecular weight of 46,000. The conclusion from all these experiments is that although complete dissociation is more difficult to achieve for coho enolase, there can be no serious doubt that this enzyme also contains two subunits of equal size.

d. *Escherichia coli Enolase.* The molecular weight of *E. coli* enolase has been determined from sedimentation–diffusion data (82,000 ± 8,200) and from sedimentation equilibrium (92,000) (*8*). The subunit molecular weight, determined by SDS electrophoresis is 46,000 ± 1,000, again in agreement with the general two-subunit model for enolases.

46. R. C. Ruth, Ph.D. Thesis, University of Illinois (1967).

3. *The Role of Magnesium Ion in Stabilizing the Enolase Dimer*

The requirement for divalent metal ion, notably Mg^{2+}, in the enolase reaction is well established and has been reviewed (*3*). The earlier studies, however, concentrated the attention on the catalytic properties of the enzyme, and only the role of Mg^{2+} as a metal activator (*47*) was considered. This role will be discussed in Section III,C. It seems clear, however, that in addition to being a metal-activated enzyme, enolase also qualifies as a metalloenzyme. The basis for this statement is first of all the observations discussed in a qualitative manner in Section II,C,2 above, demonstrating the role of Mg^{2+} in stabilizing the dimer of both yeast and rabbit muscle enolase. In addition to this evidence and considerably more convincing are the quantitative metal-binding studies carried out with yeast enolase.

Based on a significant difference in the fluorescence spectrum of metal-free enolase and the enzyme–Mg^{2+} complex it was possible to study the enzyme–Mg^{2+} interaction directly by the change in fluorescence yield, and the resulting titration curve gave a dissociation constant of 2×10^{-6} M and a stoichiometry of 1.3 moles of Mg^{2+} bound per 88,000 g of enzyme (*48*). In another study, using equilibrium dialysis and gel filtration, a titration curve like the one shown in Fig. 5 was obtained (*49*). This curve shows the binding of about 2 moles of Mg^{2+} per 88,000 g of enzyme with dissociation constants of about 5×10^{-6} M and 1.4×10^{-4} M. It appears that additional Mg^{2+} binding sites may be available at high Mg^{2+} concentrations. The difference in stoichiometry between this and the experiment discussed above is most likely because of the difference in techniques used and is readily explained on the basis that only the first, tightly bound Mg^{2+} can be observed by the fluorescence measurements. Thus it appears that yeast enolase binds at least 2 moles of Mg^{2+} in the absence of substrate. In the presence of substrate, additional Mg^{2+}-binding sites become apparent. These sites (Fig. 5) and the possible role of Mg^{2+} in the catalytic apparatus will be discussed in Section III,C. At this point let it only be suggested that according to Vallee's definitions (*47*), Mg^{2+} plays a dual role in this system and that enolase should be considered a metal-activated metalloenzyme.

47. B. L. Vallee [*Advan. Protein Chem.* **10,** 317 (1955)] has suggested that it may be useful to consider two distinct roles for metal ions in enzyme systems, one as an integral structural component of the enzyme protein, firmly bound in the so-called metalloenzymes; and a second as a required component of the catalytic apparatus, interacting less strongly with the enzyme at the catalytic site in the so-called metal-activated enzymes.

48. J. M. Brewer and G. Weber, *JBC* **241,** 2550 (1966).

49. D. P. Hanlon and E. W. Westhead, *Biochemistry* **8,** 4247 (1969).

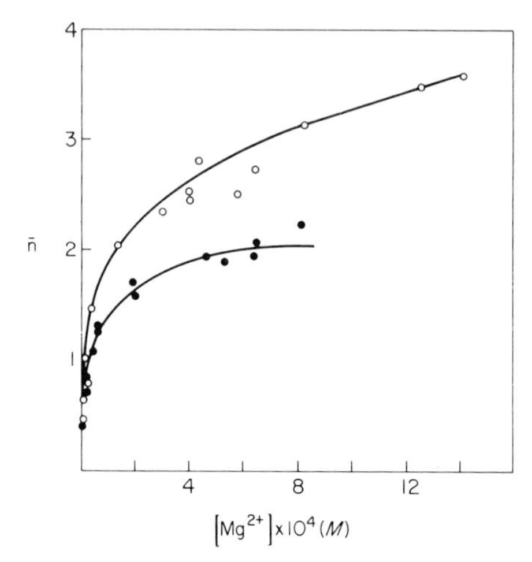

FIG. 5. A plot of $\bar{n}$, the number of Mg^{2+} ions bound per molecule of active yeast enolase, against free Mg^{2+} concentration in the absence (●) and in the presence (○) of an equilibrium mixture of glycerate 2-phosphate and enolpyruvate phosphate resulting from the addition of $1.2 \times 10^{-3}\,M$ glycerate 2-phosphate to the medium. The data were obtained by both gel filtration and equilibrium dialysis. The results in this figure are taken directly from Hanlon and Westhead (*49*), but have been replotted using 88,000 instead of 67,000 daltons for the molecular weights. From the data the following enzyme–Mg^{2+} dissociation constants can be estimated: Without substrate: $K_1 = 5 \times 10^{-6}\,M$, $K_2 = 1.4 \times 10^{-4}\,M$; with substrate: $K_1 = 5 \times 10^{-6}\,M$, $K_2 = 4 \times 10^{-5}\,M$, $K_3 = 3.2 \times 10^{-4}\,M$, and $K_4 = 1.28 \times 10^{-3}\,M$. [It must be emphasized that these values are simply estimated as the half-saturation of each step in the titration curve presented here, whereas in the original work (*49*) based on the lower molecular weight, the dissociation constants were evaluated properly by a least-squares curve-fitting technique.]

D. Summary of Molecular Properties

The available data on molecular properties of different enolases are compiled in Table VI. Only the data for pure enzymes are included in the tabulation, but some observations are also included in the footnotes for the partially purified enzymes. To briefly summarize the data, it seems fair to conclude that enolase is a dimer made up of two subunits and that each subunit in turn consists of a single polypeptide chain. The molecular weight range appears to be quite uniform for all the enolases studied; based on sucrose density gradient sedimentation, the partially purified enolases also fit this general statement. These gross structural properties may thus be uniform for all enolases, but based on other, more subtle, structural features (amino acid composition, immunochemical cross-reactivity, electrophoretic mobility, and extinction coefficient) there are enough differences between the different enzymes to make each a unique and distinct protein molecule.

III. Catalytic Properties

A. General Properties of the Enolase-Catalyzed Reaction

1. *The Substrates*

Water-soluble salts of both substrates for enolase can be prepared by unequivocal chemical synthesis (*50–52*) and are readily available commercially. Enolpyruvate phosphate has a broad ultraviolet absorption band with a maximum around 215 mm. The molar extinction coefficients for the monoprotonated acid, the magnesium salt, the potassium salt and the trianion form of enolpyruvate phosphate are significantly different, as the data in Fig. 6 show. Glycerate 2-phosphate is transparent above 220 nm, and this difference in spectral properties is the basis for the enolase activity assay. The phosphate ester of glycerate 2-phosphate is quite stable to hydrolysis in both acid and base, but at elevated temperatures an acid-catalyzed acyl migration takes place to give a mixture of the 2- and 3-phosphate ester of glycerate in a ratio of 4:1 in favor of the 3-phosphate (*50*). The phosphate ester of enolpyruvate phosphate, on the other hand, is readily hydrolyzed in acid and is also sensitive to iodine in weakly alkaline solution (*53*). The two substrates can be separated by paper chromatography (*54*) and by ion exchange chromatography on Dowex-1 Cl^- in gradients of LiCl (0–0.3 M) in 10^{-3} N HCl (*55*). A stepwise elution from Dowex-1 Cl^- has also been reported (*56*).

2. *The Activity Assay*

The most convenient assays for enzymic activity are based on the direct observation in the spectrophotometer of the appearance or disappearance of enolpyruvate phosphate on the basis of its ultraviolet absorption (Fig. 6). Changes in optical density must be corrected for

50. C. E. Ballou and H. O. L. Fischer, *JACS* **76,** 3188 (1954).
51. E. Baer and H. O. L. Fischer, *JBC* **180,** 145 (1949).
52. F. Wold and C. E. Ballou, *JBC* **227,** 301 (1957).
53. O. Meyerhof and P. Oesper, *JBC* **179,** 1371 (1949).
54. R. L. Bieleski and R. E. Young, *Anal. Biochem.* **6,** 54 (1963).
55. F. Wold, unpublished data (1955).
56. E. C. Dinovo and P. D. Boyer, *JBC* **246,** 4586 (1971).

TABLE VI
MOLECULAR PROPERTIES OF ENOLASES

Molecular property	Enolase source								
	Human muscle [Ref. (*16*)]	Rabbit muscle [Refs. (*30, 31, 43*)]	Monkey muscle [Ref. (*17*)]	Chum salmon muscle [Ref. (*11*)]	Coho salmon muscle [Ref. (*11*)]	Rainbow trout muscle [Ref. (*10*)]	Lobster muscle [Ref. (*9*)]	*E. coli* [Ref. (*8*)]	Yeast [Refs. (*3, 26, 28, 32*)]
Dimer molecular weight (in 1000 daltons):									
By sedimentation equilibrium		82		100	100	91		92	88
By sedimentation-diffusion	95[a]	78 ± 2	82					82 ± 8	
By gel filtration							87		
Monomer molecular weight (in 1000 daltons):									
By sedimentation equilibrium		41			50				44
By gel filtration									44
By SDS gel electrophoresis					46			46	48

Sedimentation coefficient (in Svedbergs):									
By direct measurement	5.9	6.0	5.7	5.9	5.9	5.5		5.4	6.0
In sucrose density gradients[b]		6.6	6.3	6.3					6.6
Diffusion coefficient (in $sec^{-1} \times 10^7$)		6.35	6.3					5.95	8.1
Partial specific volume (4°) (in ml per g):									
By direct measurement		0.728							
From amino acid composition				0.729	0.732				
Extinction coefficient at 280 nm		0.90	0.88	0.87	0.74	0.74		0.57	0.89
End-group analyses:									
Amino terminus		*N*-Ac–Ala							Ala
Moles a.a.[c] per mole		2							2
Carboxyl terminus		Lys	Lys	Asn	Asn				Leu
Moles a.a.[c] per mole		2	2	2	2				2

[a] From sedimentation-viscosity measurements.

[b] For comparison, partially purified enzymes gave the following values: turtle, 6.35; frog and beef, 6.8; dog, 6.2; and mouse, 6.7 (*18*).

[c] a. a., amino acid.

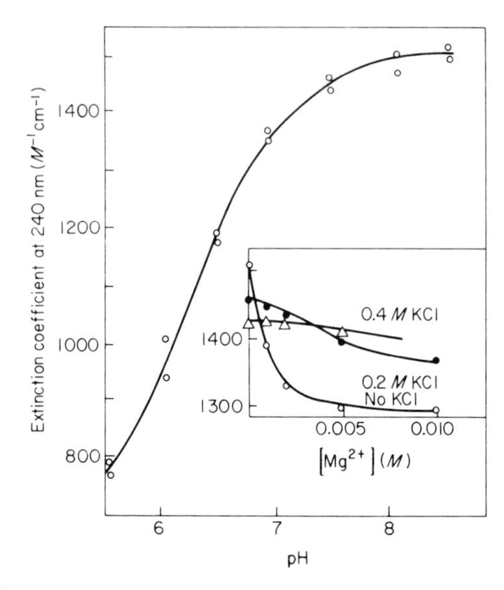

FIG. 6. The effect of pH, Mg^{2+}, and K^+ on the molar extinction coefficient of enolpyruvate phosphate (EP). From these data the molar extinction coefficient for the different forms of EP can be estimated to be: EP^{3-}, 1510 M^{-1} cm^{-1}; EPH^{2-}, 675 M^{-1} cm^{-1}; and $EP^{-3}Mg^{2+}$, 1290 M^{-1} cm^{-1}. The data are taken from Wold and Ballou (*52*).

the effect of pH and metal ions on the extinction coefficient of enolpyruvate phosphate to give a meaningful rate assay in terms of substrate concentration change per unit time. Several assay procedures have been used in different laboratories.

a. Direct Initial Rate Measurement following Optical Density at 240 nm. This is the original rate assay suggested by Warburg and Christian (*24*) and probably the one used most frequently (*12*).

b. Direct Initial Rate Measurement following Optical Density at 230 nm. This assay has greater sensitivity (the molar extinction coefficient of enolpyruvate phosphate is nearly doubled in going from 240 to 230 nm), but it may be impractical in certain cases since so many compounds absorb at the lower wavelengths and thus can interfere (*4, 5*).

c. Coupled Assay. In combination with an excess of pyruvate kinase, ADP, lactate dehydrogenase, and NADH, enolase rates can be determined directly by measuring rate of NADH disappearance either by its absorption (at 340 nm) or, with much greater sensitivity, by its fluorescence. This coupled assay is probably most useful in the determination of unknown quantities of enolase substrate. Since the equilibrium of the coupled system can be pulled essentially to completion in the direction of lactate, total NADH used is a direct measure of the total amount of D-glycerate 2-phosphate in the unknown sample. The coupled assay has also been used extensively for rate studies, however (*20, 57*).

3. *The Equilibrium*

Since D-glycerate 2-phosphate and enolpyruvate phosphate have different affinities for both protons and metal ions, the apparent equilibrium constant varies with both pH and metal ion concentration. The proton and cation dissociation constants have been determined for both substrates (*52*), and the pH and metal independent equilibrium constant has been found to be 6.3. This corresponds to a standard free energy change at 25° of 1080 cal/mole (*52*).

B. Kinetic Parameters—A Comparative Survey

The properties of "the active site" of enolase can be described by the normal kinetic parameters, K_m, V_{max}, and pH optimum. In addition, the inhibition by fluoride and the activation by Mg^{2+} are assumed to measure some property of the active site of the enolases. The fluoride inhibition was shown by Warburg and Christian (*24*) to result from the binding of a complex of Mg^{2+}, F^-, and HPO_4^{2-} (with the stoichiometry 1:2:1) to the enzyme. [This magnesium–fluoride–phosphate complex should not be confused with the magnesium salt of fluorophosphate, which has been shown to be completely inactive in this system (*58*).] The quantitative description of the fluoride inhibition is thus a function of the concentration of all three components in the complex, and is given as the inhibition index, $i =$ (fraction of inhibited enzyme/fraction of active enzyme) $\cdot$ $[Mg][HPO_4^{2-}][F]^2$. As should be expected, fluoride does not affect enolase in the absence of phosphate. The careful analysis of the magnesium activation has only been carried out for a few enolases; in most cases only the optimal Mg^{2+} concentration has been determined and found to be 10^{-3} M in all cases (at higher concentrations Mg^{2+} becomes inhibitory).

The kinetic properties of enolases, as measured by these parameters, are summarized in Table VII. There is some ambiguity in the data since the conditions under which the individual experiments were con-

57. R. Czok, *in* "Hoppe-Seyler/Thierfelder Handbuch der physiologisch- und pathologisch-chemischen Analyse," (K. Lang and E. Lehnartz, eds.), Vol. 6, Part A, p. 310. Springer-Verlag, Berlin and New York, 1964.

58. R. A. Peters, M. Shorthouse, and L. R. Murray, *Nature* **202**, 1331 (1964).

TABLE VII
KINETIC PROPERTIES OF ENOLASES

Source	$K_m \times 10^4\ M$	pH optimum	Mg^{2+} optimum $10^3\ M$	Fluoride inhibition ($i \times 10^{12}\ M^4$)	Ref.
Human muscle	0.3	6.8	—	—	*16*
Rabbit muscle	0.8	7.0	1	0.7	*18*
Monkey muscle	1.2	7.0	1	—	*17*
Mouse muscle	—	6.9	1	—	*18*
Dog muscle	0.7	7.0	—	0.8	*18*
Beef muscle	0.8	7.0	—	0.9	*18*
Human erythrocytes	0.6	7.5	1	—	*20*
Beef brain	—	—	1	—	*19*
Chicken muscle	0.7	7.1	1	1.0	*18*
Frog muscle	0.6	7.1	1	0.9	*18*
Turtle muscle	0.7	7.1	1	0.9	*18*
Snails	2.8	7.3	1	0.7	*18*
Lobster muscle	1.0	7.0	1	—	*9*
Trout muscle	0.4	6.9	1	1.4	*10*
Coho salmon muscle	0.4	7.1	1	1.2	*11*
Chum salmon muscle	0.4	6.9	1	1.5	*11*
Peas	2.5	8.0	1	7.0	*23*
Soybeans	1.9	7.6	—	1.5	*18*
Potatoes	8.3	—	—	1.0	*7*
Yeast	1.5	7.7	1	0.5	*18*
E. coli	1.0	8.1	1	0.5	*8*

ducted vary considerably. It is nevertheless felt that this is a useful first step toward a comparative study of the active site of different enolases and that interpretation in terms of comparative properties should be valid. The current conclusion from these measurements is that the catalytic properties of enolases have remained remarkably constant through evolution. Only the pH optimum shows significant variation and since these variations could reflect only minor microenvironmental effects on the pK of a single amino acid residue involved in binding or catalysis, the current hypothesis is that the structural components that make up the active site of enolases could be identical throughout the biosphere [see, for example, Cardenas and Wold (*18*)].

C. The Role of Magnesium Ion in Enolase Catalysis

The quantitative data pertaining to metal ion activation have been obtained primarily with the yeast enzyme. Several divalent metal ions such as Mg^{2+}, Mn^{2+}, Zn^{2+}, and Cd^{2+} can serve as activators (*3*). Manganese, which has a magnetic moment and thus affects the magnetic resonance of the water protons in its immediate hydration sphere, can be substituted for Mg^{2+}, and it now becomes possible to ask what happens to the proton magnetic resonance signal when the metal binds to the protein. In the case of yeast enolase, a large change in the relaxation rate of the nuclear magnetic resonance of the water protons was seen, indicating that the hydration sphere is drastically altered in the process of binding. Addition of substrate caused a substantial decrease in this relaxation rate change (*59*). Two possible interpretations were considered, that the metal functions as a bridge between the enzyme and the substrate or that the metal–enzyme complex forms independently of substrate, but that substrate binding causes a conformational change in the enzyme which in turn is reflected in the metal's hydration sphere (*59*). In view of the other available evidence, the second interpretation appears to be the most attractive one. Thus, three independent studies (*48, 49, 59*) demonstrate that 1.3–2 moles of metal ion are bound per mole of yeast enolase dimer in the absence of substrate. In addition there is evidence for a change in the enolase–Mn^{2+} complex when substrate is added (tentatively interpreted as a change in enzyme conformation) (*59*). Also, the data in Fig. 5 demonstrate that the presence of substrate creates two new Mg^{2+} binding sites (or increase the affinity for Mg^{2+} at two additional sites) in yeast enolase. The binding data appear to be best represented by the ordered addition of Mg^{2+} substrate and Mg^{2+} as illustrated in reaction (1) of Scheme 3. (As will be shown in Section III,E, enolases contain two substrate binding sites per dimer, and this feature has been incorporated in the representations in Scheme 3.) The only untested assumption in reacton (1) is that substrate does not bind to metal-free enzyme, E_2. In conjunction with the binding

Reaction (1) (from binding data):

$$^{*}E_2 + 2\,M \rightleftharpoons {}^{*}EM_2{}^{*}E + 2\,S \rightleftharpoons \mathrm{S{\diagdown}EM_2E{\diagup}S} + 2\,M \rightleftharpoons \mathrm{{}^{S}_{M}{>}EM_2E{<}^{S}_{M}}$$

Reaction (2) (from kinetic data):

$$^{*}E_2 + 2\,M \rightleftharpoons {}^{*}EM_2{}^{*}E \begin{cases} + 2\,S \rightleftharpoons \mathrm{S{\diagdown}EM_2E{\diagup}S} + 2\,M \\ + 2\,M \rightleftharpoons \mathrm{M{\diagup}EM_2E{\diagdown}M} + 2\,S \end{cases} \rightleftharpoons \mathrm{{}^{S}_{M}{>}EM_2E{<}^{S}_{M}}$$

SCHEME 3. $M = Mg^{2+}$; *E and E represent two different enzyme conformations.

59. M. Cohn, *Biochemistry* **2**, 623 (1963).

studies (*49*), Hanlon and Westhead also carried out a careful kinetic analysis of the yeast enolase system (*60*). In this work they showed that the K_m of substrate (K_m^S) was independent of Mg^{2+} concentration (in the range 10^{-5}–10^{-3} M) and that the K_m for Mg^{2+} (K_m^M) was independent of substrate concentration (in the concentration range from $[S] = K_m^S$ to $10 \times K_m^S$). These results are incompatible with any kind of ordered addition and can only be readily explained on the basis of the random addition illustrated in reaction (2), Scheme 3. There is no obvious explanation for the disagreement between these two models. It is possible that the vastly different enzyme concentrations used in the binding studies and in the kinetic studies could make the direct comparison of the data hazardous. It is also possible that the concentration ranges over which the constant K_m values were determined are too narrow, but according to the original report (*60*), these are not likely explanations.

The possibility that the third and fourth equivalent of Mg^{2+} which are bound in the presence of substrate have no function in the catalytic mechanism has also been considered. In view of the uniformly high Mg^{2+} concentration (10^{-3} M, see Table VII) required for optimal activity, this does not seem to be a very attractive possibility either. Qualitatively, the best present picture of the role of Mg^{2+} in the enolase reaction appears to be the one represented as the final form in both reactions in Scheme 3, the enolase dimer binding two moles of substrate and a total of 4 moles of Mg^{2+} per mole of catalytically active enzyme.

D. Substrate Specificity—Mapping the Active Site with Substrate Analogs

A number of substrate analogs which have been tested as inhibitors of the enolase reaction are listed in Table VIII (*61–66*). A comparison of the "fit" of these compounds as measured by the inhibitor constant, K_I (competitive inhibition), to the active site of enolases (most of the data have been obtained with the yeast and the rabbit muscle enzymes) makes it possible to formulate a description of the topography of the active site. First of all, the lack of inhibition by the compounds in

60. D. P. Hanlon and E. W. Westhead, *Biochemistry* **8**, 4255 (1969).
61. F. Wold and C. E. Ballou, *JBC* **227**, 313 (1957).
62. T. Nowak and A. S. Mildvan, *JBC* **245**, 6057 (1970).
63. F. Wold and C. E. Ballou, *JACS* **81**, 2368 (1959).
64. F. Wold, *J. Org. Chem.* **26**, 197 (1961).
65. R. Barker and F. Wold, *J. Org. Chem.* **28**, 1847 (1963).
66. F. Wold and R. Barker, *BBA* **85**, 475 (1964).

group B show that *both* the carboxylate and the phosphate are required for binding. Using the value 0.7×10^{-4} for the K_m of D-glycerate 2-phosphate as a reference, the data for the compounds in group C also suggest that the 3-hydroxymethyl group contributes to binding. Replacing the 2-hydrogen with a methyl group (L-lactate P) does not seem to affect the affinity, but the bulkier hydroxymethyl group in the 2-hydrogen position abolishes all binding (L-glycerate 2-P). The combination of a methyl group in place of the 2-hydrogen and the presence of the 3-hydroxymethyl group decreases the affinity as does bulky substituents on the 3-carbon. It is also interesting to note that D-*erythro*-2,3-dihydroxybutyric acid 2-phosphate is not a substrate for the enzyme; apparently the secondary alcohol is not dehydrated. The compounds in group D show the general decrease in affinity observed when one or more carbons are inserted between the carboxylate and the carbon to which phosphate is attached. This effect appears to be additive with the effects discussed above. Based on these observations the model in Scheme 4

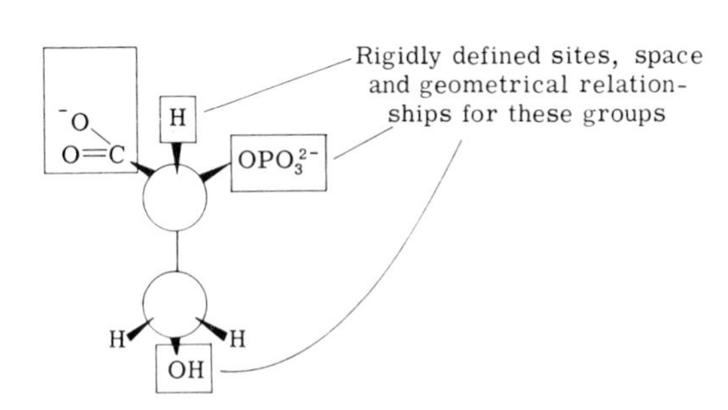

SCHEME 4

can be formulated, suggesting a fairly rigidly defined spatial relationship between the phosphate and the hydrogen and primary hydroxyl group involved in the dehydration. The presence of the carboxyl is essential for binding, but it appears that the geometrical restrictions on the carboxylate are less stringent than those imposed on the other groups. This model suggests that a compound such as D-erythronate 3-phosphate should be substrate for enolase with a K_m somewhat higher than the K_m for the natural substrate. The data in Table VIII (A) show that this prediction was correct (*66*). Although such a model gives no information about the active site components, it does provide the first outline of the size and shape of the binding pocket into which the substrate must be located for catalysis to take place.

TABLE VIII

THE RELATIVE AFFINITY OF SUBSTRATES AND SUBSTRATE ANALOGS FOR THE ACTIVE SITE OF ENOLASE[a]

A. Substrates

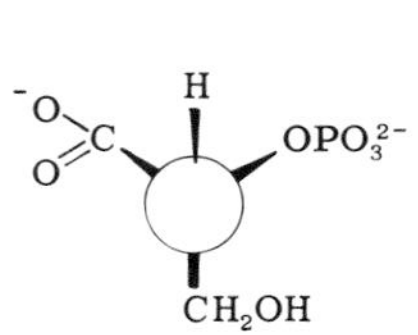

D-Glycerate 2-P	General structure	D-Erythronate 3-P
($K_m = 0.7 \times 10^{-4}\ M$)		($K_m = 3 \times 10^{-4}\ M$)

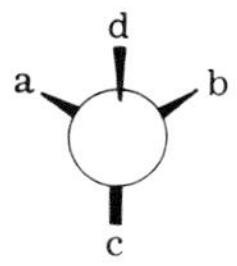

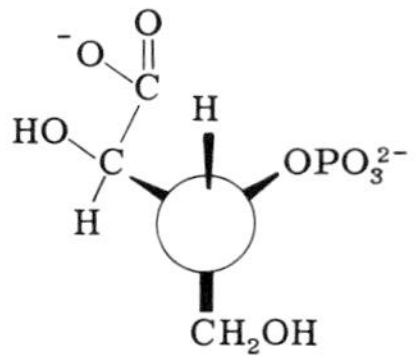

Substrate analogs	K_I	Ref.
B. Analogs with either COO^- or OPO_3^2 missing		
D-Lactate	No effect	(*61*)
D-Glyceraldehyde-P	No effect	(*61*)
β-Glycerol-P	No effect	(*61*)
Dihydroxyacetone-P	No effect	(*61*)
C. Analogs with a = COO^- and b = OPO_3^2		
D-Lactate-P (c = CH_3, d = H)	4×10^{-4}	(*61, 62*)
L-Lactate-P (c = H, d = CH_3)	4×10^{-4}	(*62*)

Glycolate-P (c = H, d = H)	4×10^{-4}	(*62*)
D-*erythro*-2,3-Dihydroxybutyrate 2-P (c = CH(OH)—CH_3, d = H)	6×10^{-4}	(*61*)
D-2,3-Dihydroxyisobutyrate 2-P (c = CH_2OH, d = CH_3)	10×10^{-4}	(*63*)
L-Glycerate 2-P (c = H, d = CH_2OH)	No effect	(*64*)
D-Erythronate 2-P (c = CH(OH)—CH_2OH, d = H)	No effect	(*65, 66*)
D. Analogs with a = C(R')(R)—COO^- and b = OPO_3^{2}		
β-Hydroxypropionate-P (R = R' = H, c = H, d = H)	5×10^{-4}	(*61*)
D-Glycerate 3-P (R' = OH, R = H, c = H, d = H)	5×10^{-4}	(*61*)
D-*erythro*-2,3-Dihydroxybutyrate 3-P (R' = OH, R = H, c = CH_3, d = H)	30×10^{-4}	(*61*)
D-2,3-Dihydroxyisobutyrate 3-P (R' = OH, R = CH_3, c = H, d = H)	No effect	(*63*)
L-Glycerate 3-P (R' = H, R = OH, c = H, d = H)	No effect	(*64*)
E. Analog with a = —CH(OH)—CH(OH)—COO^- and b = OPO_3^{2}		
D-Erythronate 4-P (c = H, d = H)	No effect	(*66*)

[a] For the sake of comparison only the carbon to which phosphate is attached (in position b) is viewed. The phosphate group is considered fixed on the enzyme surface, and the contribution of the other three constituents on the same carbon to binding is evaluated. The letters a, b, c, and d refer to the four positions indicated in the general structure under A.

E. The Number of Active Sites per Mole of Enzyme

The term *active site* implies that catalytic turnover takes place at that site, and since turnover can only be measured by kinetic analysis, a determination of the number of true active sites becomes an extremely difficult experimental task. Direct binding of substrate or substrate analogs, however, permits unequivocal determination of the number of substrate binding sites, and by this approach it has now been demonstrated that enolases from widely different sources have two sites, and by indirect arguments, that these probably also are catalytically active sites. Equilibrium dialysis of rabbit muscle enolase against ^{14}C-glycolate phosphate in the presence of optimal Mg^{2+} concentration gave the first clear evidence that 2 moles of substrate analog were bound per mole of enolase (*34*). Inactivation of rabbit muscle enolase by reaction with the active site-specific reagent glycidol phosphate also showed a clear stoichiometry of 2 moles of reagent being required for complete inactivation (*67*). This reaction will be discussed further in the next section. More recently a more extensive survey has been carried out with four different enolases using two substrate analogs, D-tartronate semialdehyde phosphate (TSP) and 3-amino enolpyruvate 2-phosphate (AEP). Originally, TSP was synthesized as a potential site-specific reagent for enolase (*68*) and was found instead to be a very potent competitive inhibitor. Subsequent studies showed that TSP can be converted to the corresponding enamine AEP by treatment with an excess of ammonia, and that AEP shows an even higher affinity for enolases than does TSP (*8*). D-Tartronate semialdehyde phosphate can be converted to the enolate form by treatment with Mg^{2+} at high pH, and both the TSP enolate ion and AEP show strong ultraviolet absorption with molar extinction coefficients around 10^4 liters/mole/cm (see Scheme 5). When either of these compounds are added to enolase in the presence of Mg^{2+}, binding takes place with characteristic spectral shifts, and the molar

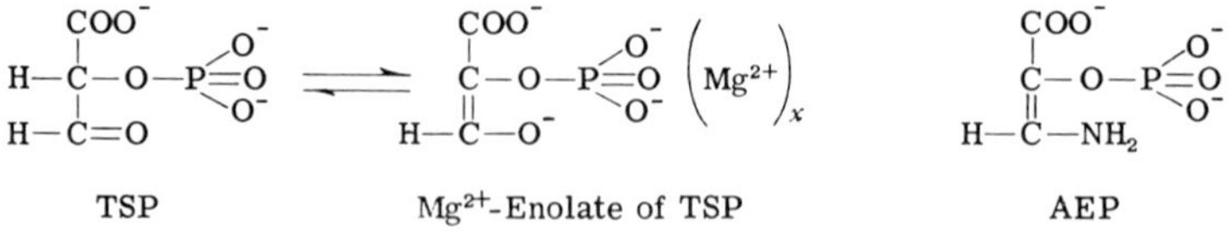

TSP ⇌ Mg^{2+}-Enolate of TSP (Mg^{2+})$_x$ AEP

Scheme 5

67. I. A. Rose and E. L. O'Connell, *JBC* **244,** 6548 (1969).
68. F. C. Hartman and F. Wold, *BBA* **141,** 445 (1967).

extinction coefficients for the enolase–analog complexes are again very large (10^4 liters/mole/cm). Using these difference spectra, direct titration of the enolases with the two analogs could be carried out (Fig. 7), and from the data both the dissociation constant (K_d) and the number

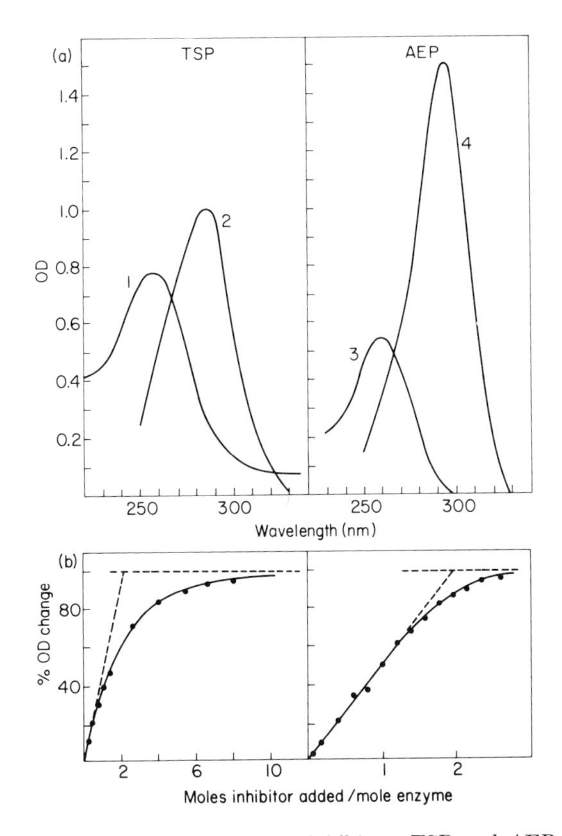

FIG. 7. Spectral properties of the enolase inhibitors TSP and AEP. (a) Curve 1: The ultraviolet spectrum of the enolate ion–Mg^{2+} complex of TSP produced in 0.01 M NaOH in the presence of 10^{-3} M Mg^{2+} (the free aldehyde at neutral pH has no absorption above 230 nm); $\lambda_{max} = 258$ nm, $\epsilon_{max} = 12{,}000$ M^{-1} cm^{-1}. Curve 2: The difference spectrum for the rabbit muscle enolase–TSP complex (enzyme contribution subtracted); $\lambda_{max} = 285$ nm, $\epsilon_{max} = 15{,}000$ M^{-1} cm^{-1}. Curve 3: The spectrum of AEP; $\lambda_{max} = 260$ nm, $\epsilon_{max} = 8{,}400$ M^{-1} cm^{-1}. Curve 4: The difference spectrum for the rabbit muscle enolase–AEP complex; $\lambda_{max} = 295$ nm, $\epsilon_{max} = 23{,}200$ M^{-1} cm^{-1}. The four spectra have been normalized to equimolar concentrations. (b) The spectrophotometric titration of rabbit muscle enolase with TSP, following the OD change at 285 nm (left) and with AEP, following OD change at 295 nm (right), showing saturation with 2 moles of inhibitor bound per mole of enzyme. The titration curves and the enzyme-inhibitor spectra were obtained in pH 7.0 buffer containing 10^{-3} M Mg^{2+}. The data are taken from Spring (*8*).

TABLE IX
DISSOCIATION CONSTANTS AND STOICHIOMETRY OF ENOLASE–INHIBITOR COMPLEXES[a,b]

Enzyme source	K_m	TSP K_I	TSP K_d	AEP K_I	AEP K_d	AEP n
Rabbit muscle	45	4.1	2.5	0.02	~0.1	1.95
Coho salmon muscle	40	—	1.6	—	~0.04	1.94
Yeast	100	14 (68)	15	—	~0.5	2.00
E. coli	100	19	15	0.09	~0.5	2.11

[a] Data from Spring (*8*).

[b] All K values are given in units of micromolar concentrations. The K_m values were determined as part of this experiment to allow direct internal comparison of all results. The K_I values were obtained by kinetic measurements and K_d values from direct spectrophotometric titrations. Because of the tight binding of AEP, the K_d values could not be determined precisely, but the end point n (moles of inhibitor bound per mole of enzyme) could be evaluated most precisely with AEP.

of sites could be accurately determined. In an attempt to correlate the direct binding data with actual catalytic site binding, the K_I for the two analogs was also determined with the idea that if their affinity as competitive inhibitors measured by the K_I matched the affinity measured by direct binding, the binding site for analog and for substrate are most likely identical. The results of these studies are summarized in Table IX and show that four enolases from widely different biological sources all have two analog binding sites per active dimer. It was also proposed that the general agreement between the K_I and K_d values support the proposition that the analog sites correspond to the catalytic sites (*8*).

F. THE ACTIVE SITE COMPONENTS

There is relative little information available on amino acids or peptide sequences involved in the active site of enolases. There is indirect evidence for an important histidine based on pK values derived by the effect of pH on Zn^{2+} binding (*69*) and on the kinetic parameters (*61*). More direct evidence for histidine in yeast enolase and for a carboxyl group in rabbit muscle enolase has been obtained from chemical modifications of these enzymes.

69. B. C. Malmström and L. E. Westlund, *ABB* **61,** 186 (1956).

1. *Carboxymethylation of Yeast Enolase*

Yeast enolase (*70*) was found to lose activity upon treatment with sodium bromoacetate in 0.5 *M* tris buffer at pH 7.0. When a parallel reaction was carried out in the presence of Mg^{2+} and substrate (or competitive inhibitor), the product retained essentially full activity. Analysis of the carboxymethylated substrate Mg^{2+} protected, active sample (3.5 His, 2.15 Met, and 1.3 Lys reacted) and the unprotected, inactive sample (3.5 His, 4.0 Met, and 1.3 Lys reacted) (*71*) showed that substrate-Mg^{2+} had protected 1.85 moles of methionine per mole of enzyme, and it was concluded that methionine was an active site component. It is of interest that neither Mg^{2+} alone or substrate alone, nor Mg^{2+} in combination with phosphate derivatives that are not competitive inhibitors of the enolase reaction, was effective in protecting the enzyme. The conclusion that there is an active site methionine has been justifiably questioned in view of the recent data on the dissociation of enolase subunits (*72*). It has been suggested that under the conditions of the reaction (0.5 *M* tris buffer) the enzyme is largely dissociated in the absence of magnesium and substrate, and that the addition of these compounds serve to stabilize the dimer form. Thus, the protected methionines may well be located, not in the active site, but in a region which is exposed upon dissociation (*72*).

2. *Photooxidation of Yeast Enolase*

Photooxidation of yeast enolase with methylene blue gives extensive modification of several residues and is not sufficiently specific to give interpretable results (*73*). If Rose Bengal is used as the activating dye, however, the oxidation is much more specific, and it was found that loss of activity was directly associated with the loss of a single residue of histidine. In this system also there is protection by the activating metal ion and substrate. Substrate by itself has no effect, Mg^{2+} by itself gives substantial protection, but only the combination of Mg^{2+} and substrate gives essentially complete protection (*74*). The protection by

70. J. M. Brake and F. Wold, *Biochemistry* **1,** 386 (1962).

71. The original report gave the results on the basis of a molecular weight of 67,000. The numbers given here are recalculated for a molecular weight of 88,000 (*26*).

72. E. W. Westhead and P. Boerner, *Abstr. 158th Natl. Meeting, Am. Chem. Soc., New York. Biol.* p. 252 (1969).

73. J. M. Brake and F. Wold, *BBA* **40,** 171 (1960).

74. E. W. Westhead, *Biochemistry* **4,** 2139 (1965).

Mg^{2+} suggests that the histidine residue may be the one proposed to be involved in metal binding (*69*), but a more direct involvement in the catalytic process cannot be excluded on the basis of the available data.

3. *The Specific Modification of Rabbit Muscle Enolase by Glycidol Phosphate*

As already stated above (Section III,E) glycidol phosphate (1,2-epoxipropanol 3-phosphate) reacts stoichiometrically with rabbit muscle enolase to give complete inactivation after incorporation of 2 moles of reagent per mole of enzyme. The inactive enzyme derivative, which is stable under normal conditions, can be cleaved by alkali or by treatment with hydroxylamine at pH 9, and it thus appears that the epoxide forms an ester with a carboxyl group in enolase. The rate of the reaction of glycidol phosphate with enolase is fairly slow (maximum rate at infinitely high reagent concentration is 6% inactivation per minute), but the presence of Mg^{2+} increases the rate. Enolpyruvate phosphate protects against inactivation apparently without any added Mg^{2+} [this fact may suggest that substrate does bind to the enzyme in the absence of the activating metal (see Section III,C)]. The high specificity of glycidol phosphate for enolase is rather surprising in view of the data in Table VIII, which show that only compounds with both a carboxyl group and a phosphate group have an appreciable affinity for the active site of enolases. However, in considering both steps of the reaction in which an active site reagent forms a covalent derivative with the enzyme (Scheme 6), it is clear that even if the enzyme–reagent complex

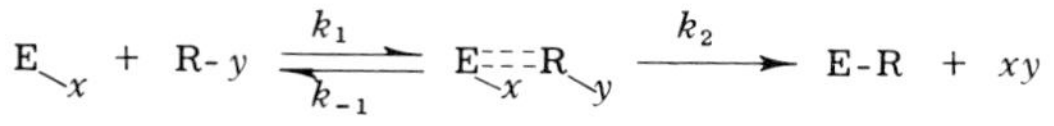

$$E_{\searrow x} + R\text{-}y \underset{k_{-1}}{\overset{k_1}{\rightleftharpoons}} E_{\searrow x}^{\cdots}R_{\searrow y} \xrightarrow{k_2} E\text{-}R + xy$$

SCHEME 6

is a relative improbable event, the reagent can still exhibit a high degree of specificity if the second (k_2) step is catalyzed in that particular complex only. All other nonspecific complexes would then be nonproductive in terms of covalent bond formation. For such a system, using a relatively stable reagent such as an epoxide, a slow, but specific reaction just like the one observed would be expected. Glycidol phosphate appears to be the first active-site-specific reagent for enolases, forming a sufficiently stable derivative to permit degradation and peptide isolation, and thus exploration of the active-site structure at the level of the amino acid sequence of the reacted peptide.

G. The Mechanism of the Dehydration Reaction

An understanding of the mechanism by which the removal of (or addition of) water is achieved in the enolase reaction is fundamental to the formulation of the enzymes role as the catalyst. Dehydration reactions can in the simplest terms be considered to proceed via any one of the pathways indicated in Scheme 7. Based on data from organic chemistry, it is generally found that the removal of the proton (1a and 2b) is faster or equal to the rate of OH^- removal and that other combinations of rate ratios probably need not be considered. Based on the data discussed in this section it appears that the enolase-catalyzed reaction proceeds via reaction (1), the carbanion mechanism in Scheme 7.

2. Carbonium intermediate

1. Carbanion intermediate

—C=C— $\underset{\text{fast}}{\overset{+H^{\oplus}\ \ b}{\rightleftharpoons}}$ —C(H)—$C^{\oplus}$— + $OH^{\ominus}$ $\underset{\text{slow}}{\overset{a}{\rightleftharpoons}}$ —C(H)—C(OH)— $\underset{\text{fast}}{\overset{a}{\rightleftharpoons}}$ —$C^{\ominus}$—C(OH)— $\underset{\text{slow}}{\overset{+H^{\oplus}\ \ b}{\rightleftharpoons}}$ —C=C— + $OH^{\ominus}$

—C(H)—C(OH)— ⇌ —C=C— + $H^{\oplus}$ + $OH^{\ominus}$

3. Concerted, base-catalyzed elimination (E_2)

Scheme 7

1. *Isotope Effects*

In a recent report the relative rates of the different steps in the enolase reaction have been studied by a detailed analysis of isotopic exchange reactions and interconversions (*56*). Scheme 8 illustrates the rationale of the experiments, and the results can be summarized as follows:

(1) The rate of C-2 hydrogen exchange with water (at equilibrium) is fast and shows a primary isotope effect. The rate of hydrogen exchange is considerably greater than the rate of ^{18}O exchange between H_2O and the C-3 hydroxyl.

(2) There is no primary isotope effect of the C-2 hydrogen in the conversion of G2P to EP, demonstrating that the cleavage of the C-2 hydrogen is not the rate-limiting step.

(3) and (4) The rate of ^{18}O exchange is equal to or slightly faster than the rate of ^{14}C exchange between substrate and product, showing that the release of OH^- is not the rate-limiting step.

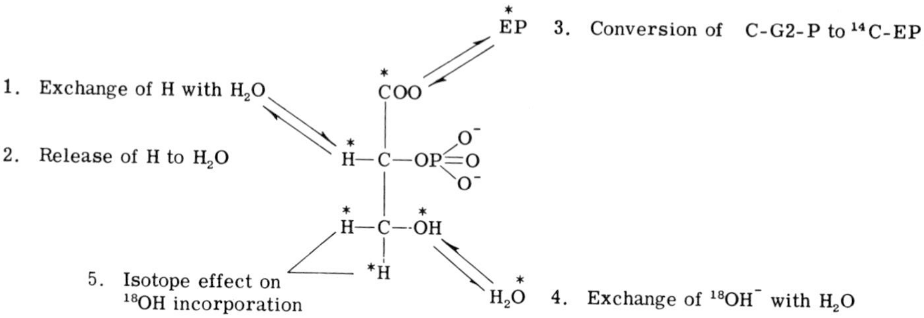

Scheme 8

(5) There is a pronounced reverse secondary isotope effect of the C-3 hydrogens on the rate of formation of G2P from EP (C-3 ^3H gives higher rates of G2P formation than does C-3 ^1H), showing that the rate-limiting step must be in the addition of OH^- to the enol double bond. The only reasonable precursor of the sp^3 hybridization of C-3 is the transition state involved in the OH^- addition, and this $sp^2 \rightarrow sp^3$ transition is thus proposed to be the rate-limiting step.

These and other considerations lead to the proposed mechanism given by Scheme 9. Simply on the basis of relative rates, the evidence for the carbanion intermediate and thus for reaction mechanism 1 in Scheme 7 seems very good.

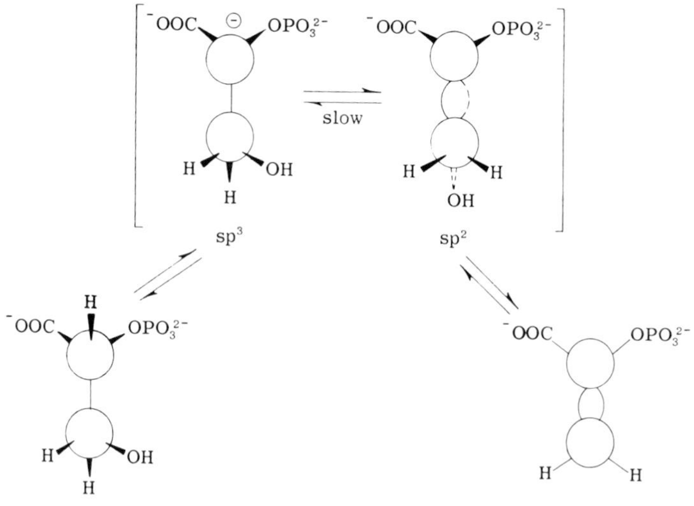

Scheme 9

2. *Indirect Evidence for the Carbanion Intermediate*

The results discussed in Section III,E for tartronic acid semialdehyde phosphate can be used as evidence for a facile proton extraction at C-2 as the first and fast step in enolase catalysis. Based on the similarity in the extinction coefficient of the Mg^{2+} and alkali produced enolate ion of TSP and the UV absorbing TSP–enolase complex, it seems inescapable to conclude that the enzyme can accomplish at pH 7 what can only be accomplished by strong base in the absence of enzyme namely, the extraction of the C-2 hydrogen, to give the enolate ion in both cases. It was therefore suggested that the active site of enolase contains a general base (such as a deprotonated imidazole) which is responsible for the fast catalytic step of the C-2 hydrogen removal (*8*). It is interesting that this mechanism also suggests a reason for the high affinity of TSP for the active site, namely, the contribution of the C-3 negative charge in the enolate ion to the total binding, the anion perhaps reacting with the enzyme-bound Mg^{2+} (*8*). This picture of the active site of enolase does not explain the very high affinity of AEP for the enolase site, however. Another inconsistency which is difficult to explain is the lack of hydrogen exchange with the solvent in the case of TSP (*8*). There does not seem to be any obvious reason why TSP should be different from the substrate in this reaction.

H. Monomer–Dimer Activity Relationships

Enolase isolated from animal, fish, yeast, and bacteria must be considered to have two subunits and two catalytic sites in the active enzyme. Whether the monomer is active or exists *in vivo* is uncertain. Until recently, in all situations in which monomers could be shown to predominate activity was missing. However, the conditions required to affect dissociation could well be strong enough to cause inactivation of any active monomer that might exist. In a recent report on gel filtration of dilute yeast enolase solutions in the presence of Mg^{2+} and substrate and at 43°, convincing evidence was presented that a catalytically active monomer indeed exists in very dilute solutions (*75*). However, the dissociation constant for the dimer–monomer equilibrium obtained in this work, approximately 10^{-9} *M* [this is in agreement with the earlier estimates (*44*)], suggests that no monomer ever exists *in vivo* since the enolase concentration in the yeast cell must be of the order of at least 10^{-6} *M*. Thus, it appears safe to conclude that although each monomer

75. S. Keresztes-Nagy and R. Orman, *Biochemistry* **10**, 2506 (1971).

has an intact active site the dimer is the only natural form of enolase.

The conclusion poses an interesting teleological problem: What makes the "two sites–two subunits" (dimer) structure of enolase so superior to the simple "one site–one subunit" (monomer) structure that only the former one is found in all forms investigated? There is no basis for an answer to this question at present, perhaps because no one has searched for an answer yet. Most of the "bidirectional" (enzymes apparently functioning in both the catabolic and anabolic direction of a pathway) glycolytic enzymes (and enolase among them) are considered rather "dull" enzymes void of sophisticated regulatory properties and just turning over substrates in either direction as the steady state concentrations fluctuate in response to other regulatory pressures. In the case of enolase, there is in fact good direct evidence that all *in vivo* rates are well taken care of by the quantities of enzyme present in the cell or tissue (*76*, *77*), and that these quantities apparently never reach levels low enough to make the enolase step the rate-limiting one in the total metabolic reaction chain. Even without any evidence for control, however, the most reasonable rationalization for the active dimer model is that it provides the basis for some regulatory mechanism. An interesting possibility is that the dimer represents a "desensitized" form of the enzyme and that a higher aggregate such as a tetramer with four active sites could show strong interaction between the sites and thus be under metabolite control.

76. O. H. Lowry and J. V. Passonneau, *JBC* **239**, 31 (1964).
77. T. Bücher and H. Seis, *European J. Biochem.* **8**, 273 (1969).

19

Fumarase and Crotonase

ROBERT L. HILL • JOHN W. TEIPEL

I. Fumarase

A. Introduction

1. *General Considerations*

Fumarase, or fumarate hydratase (EC 4.2.1.2), is a carbon–oxygen hydro-lyase which catalyzes the reversible hydration of fumarate to L-malate according to the following reaction:

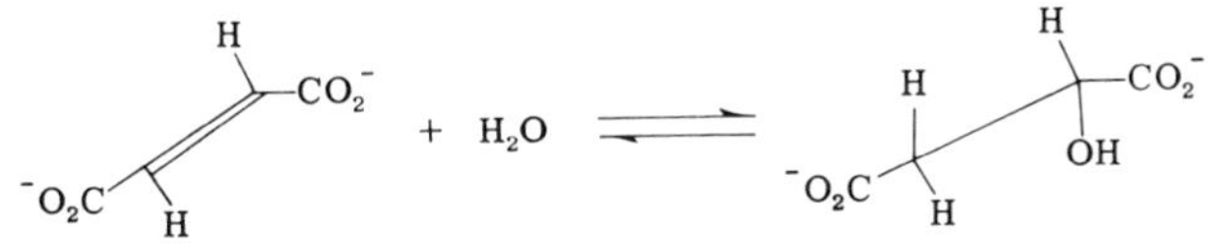

$$^{-}O_2C\text{–CH=CH–}CO_2^{-} + H_2O \rightleftharpoons {}^{-}O_2C\text{–}CH_2\text{–CH(OH)–}CO_2^{-}$$

Because of its fundamental role as an essential enzyme of the tricarboxylic acid cycle, fumarase is found in a wide variety of organisms including bacteria (*1–3*), yeast (*4, 5*), molds (*6*), plants (*7, 8*), inverte-

1. P. Mann and B. Woolf, *BJ* **24,** 427 (1930).
2. G. C. de Mello Ayres and F. J. S. Lara, *BBA* **62,** 435 (1962).
3. C. A. Lamartimore, H. D. Braymer, and A. D. Parson, *ABB* **141,** 293 (1970).
4. G. Favelukes and A. O. M. Stoppani, *BBA* **28,** 654 (1958).
5. S. Hayman and R. A. Alberty, *Ann. N. Y. Acad. Sci.* **94,** 812 (1961).

brates (*9*), and mammals (*10–13*). In contrast to other hydro-lyases, which require pyridoxal phosphate or metal ions as cofactors, fumarase displays no cofactor requirements.

Earlier reviews (*14–17*) of fumarase have documented the historical development of the enzyme and summarized our understanding of it through 1961. Thus, this review will only summarize many of the earlier studies in order to relate them to more current aspects of fumarase structure and function.

2. *Historical Development*

The development of our knowledge of fumarase and its action may be divided roughly into three parts: (1) its discovery and characterization of its reaction with substrates, (2) its crystallization and extensive investigation of its kinetics, and (3) structural analysis. Battelli and Stern in 1910 (*18, 19*) were the first to describe the action of fumarase when they observed that beef liver homogenates converted succinate to malate. Subsequently, Einbeck (*20*) demonstrated that this conversion consisted of two discrete steps: (1) the oxidation of succinate to fumarate and (2) the hydration of fumarate to malate. Battelli and Stern (*21*) confirmed these observations and gave the name fumarase to the enzyme responsible for catalysis of the hydration reaction. Although Einbeck believed that optically inactive malate was produced, Dakin (*22*) later showed that fumarase acted only upon L-malate. During the next thirty years these early observations were confirmed (*15*) but the exactness of most studies was limited severely by the impurity of the fumarase

6. K. P. Jacobson, *Biochem. Z.* **234,** 401 (1931).
7. W. S. Pierpoint, *BJ* **75,** 511 (1960).
8. D. S. Shih and L. B. Barnett, *ABB* **123,** 558 (1968).
9. P. W. Clutterbuck, *BJ* **22,** 1193 (1928).
10. E. L. Kuff, *JBC* **207,** 361 (1954).
11. J. A. Shepherd, Y. W. Li, E. E. Mason, and S. E. Zeffren, *JBC* **213,** 405 (1955).
12. V. Massey, *BJ* **51,** 490 (1952).
13. L. Kanarek and R. L. Hill, *JBC* **239,** 4202 (1964).
14. R. A. Alberty, "The Enzymes," 2nd ed., Vol. 5, p. 531, 1959.
15. S. Ochoa, "The Enzymes," 1st ed., Vol. 1, Part 2, p. 1217, 1951.
16. V. Massey, "Methods in Enzymology," Vol. 1, p. 729, 1955.
17. R. L. Hill and R. A. Bradshaw, "Methods in Enzymology," Vol. 13, p. 91, 1969.
18. F. Battelli and L. Stern, *Biochem. Z.* **30,** 172 (1910).
19. F. Battelli and L. Stern, *Biochem. Z.* **31,** 478 (1911).
20. H. Einbeck, *Biochem. Z.* **95,** 296 (1919).
21. F. Battelli and L. Stern, *Compt. Rend. Soc. Biol.* **84,** 305 (1921).
22. H. D. Dakin, *JBC* **52,** 183 (1922).

preparations available. In 1952, fumarase was first successfully crystallized from swine heart muscle by Massey (*12*) who demonstrated that it crystallized from ammonium sulfate solutions in an essentially homogeneous form. With the availability of pure enzyme, it was then possible for Massey, and Alberty and co-workers to characterize the catalytic properties of fumarase in considerable detail, including determination of the kinetic and thermodynamic parameters of the fumarase reaction and description of its stereospecificity, studies which have contributed significantly to current notions of the mechanism of action of fumarase. Subsequent methods for improving the yield of fumarase from heart muscle were developed by Kanarek and Hill (*13*) and provided an opportunity to examine many of its molecular properties such as its subunit structure and properties of its active site.

Clearly, investigations of fumarase have as their ultimate goal the complete understanding of its catalytic action as related to its structure. Studies performed to date have not realized this goal, but considerable insight into the enzyme is now possible, even though many suggestions concerning its structure–function relationships and mechanism of action must remain tentative until complete structural analysis is obtained.

3. *Preparation and Assay*

Crystalline fumarase was first prepared from swine heart muscle in 1951 by Massey (*12*), who, prior to crystallization from ammonium sulfate, employed calcium phosphate gels for adsorption of the enzyme from crude extracts. The crystalline enzyme was later obtained by Frieden *et al.* (*23*) solely by ammonium sulfate fractionation of muscle extracts. Kanarek and Hill (*13*) in 1964 described modifications of the procedure of Frieden *et al.* which improved substantially the yield of crystalline enzyme. In the modified procedure (*13, 17*), the heart muscle is not washed prior to extraction as in earlier methods, and over 90% of the fumarase was solubilized without difficulty. Fractionation of the extract with ammonium sulfate gives the crystalline enzyme in yields of about 70%. By these methods about 1 g of crystalline enzyme is obtained from 9 kg of muscle. The properties of the crystalline enzymes prepared by the different methods are indistinguishable except for slight differences in the ultraviolet absorption spectra, which have some bearing on the assay methods as discussed below.

Although crystalline fumarase appears to behave as a homogeneous protein under several conditions, it has been reported that multiple forms of the enzyme may be resolved by disc electrophoresis, electrofocusing,

23. C. Frieden, R. M. Bock, and R. A. Alberty, *JACS* **76**, 2482 (1954).

and chromatography on DEAE Sephadex (*24–26*). These results suggest that fumarase is composed of several isozymes; however, verification of such microheterogeneity awaits an exact determination of structural differences in the primary sequence of these forms.

Fumarase has been assayed by a variety of means, but it is generally agreed that the best method is that first proposed by Racker (*27*), which depends upon measurement of the absorption of fumarate in the ultraviolet. Fumarate absorption between 250 and 280 mμ is much greater than that of L-malate; thus, the velocity of the reaction can be measured conveniently and sensitively by spectrophotometric means.

The specific activity of fumarase has been measured by two methods which differ only by the means for measuring enzyme concentration. In the method of Frieden *et al.* (*23*), the concentration of fumarase is measured spectrophotometrically from its extinction coefficient at 250 mμ rather than at 280 mμ as described by Kanarek and Hill (*13*). The extinction coefficient for fumarase is 0.51 in 0.01 *M* phosphate buffer, pH 7.3 (1 mg/ml at 280 mμ); but the extinction coefficient at 250 mμ may vary considerably (*13*) as the result of turbidity in solutions of the enzyme, and therefore error may be introduced in absorption measurements due to light scattering. For these reasons, measurement of the specific activity based on determination of protein concentration at 280 mμ seems preferable. By this method, crystalline fumarase has a specific activity of 31,500–35,000 units/mg. Details of the assay procedure can be obtained from Hill and Bradshaw (*17*).

B. Molecular Properties

1. *Physical Properties*

Table I lists several of the physical properties of swine heart fumarase (*13, 23, 28–31*). The weight, number, and *z*-average molecular weights calculated from sedimentation equilibrium studies in the ultracentrifuge using a partial specific volume (0.738 ml/mg), which was calculated on the basis of the amino acid composition of the enzyme,

24. Y. C. Lui and L. A. Cohen, *Federation Proc.* **27**, 589 (1968).
25. P. E. Penner and L. H. Cohen, *Federation Proc.* **29**, 334 (1970).
26. C. E. Watson, C. Y. Kim, and W. T. Jenkins, *Federation Proc.* **29**, 939 (1970).
27. E. Racker, *BBA* **4**, 20 (1950).
28. L. Kanarek, E. Marler, R. A. Bradshaw, R. E. Fellows, and R. L. Hill, *JBC* **239**, 4207 (1964).
29. R. Cecil and A. G. Ogston, *BJ* **51**, 494 (1952).
30. N. Shavit, R. G. Wolfe, and R. A. Alberty, *JBC* **233**, 1383 (1958).
31. V. Massey, *BJ* **53**, 67 (1953).

TABLE I
PHYSICAL PROPERTIES OF SWINE HEART FUMARASE

1. Molecular weight:	
A. Sedimentation-equilibrium, 0.05 *M* phosphate buffer, pH 7.3	194,000 [Ref. (*28*)]
B. Sedimentation-equilibrium, 6 *M* guanidine hydrochloride, pH 4.25	48,500 [Ref. (*28*)]
C. Sedimentation-equilibrium, 6 *M* guanidine hydrochloride, 10^{-4} *M* 2-mercaptoethanol	48,500 [Ref. (*28*)]
D. Sedimentation-diffusion, 0.01 *M* phosphate buffer, pH 7.3	220,000 [Ref. (*23*)]
2. Sedimentation coefficient, 0.05 *M* phosphate buffer, pH 2.7, 0.1 *M* NaCl ($s^{\circ}_{20,w}$)	9.09 [Ref. (*23*)]
3. Diffusion coefficient	4.05×10^{-7} cm^2 sec^{-1} [Ref. (*29*)]
4. Isoelectric point, 0.1 *M* tris-acetate buffer	7.35 [Ref. (*30*)]
Isoelectric point, 0.1 *M* tris-acetate buffer, 0.09 *M* NaCl	6.6 [Ref. (*30*)]
5. Isoionic point	7.95 ± 0.05 [Ref. (*13*)]
6. Extinction coefficient (0.05 *M* phosphate buffer, pH 7.3, 1 mg/ml, 280 nm)	0.51 [Ref. (*13*)]
7. Optical rotatory dispersion parameters	
b_0 (0.01 *M* phosphate buffer, pH 7.4)	−311 [Ref. (*13*)]
mean residue rotation (0.01 *M* phosphate buffer, pH 7.4, 734 nm)	6980 [Ref. (*13*)]
8. f/f_0	1.24 [Refs. (*13*, *28*)]

are 194,000, 197,000, and 192,000, respectively. This value is somewhat lower than 220,000, which is the molecular weight reported earlier from sedimentation velocity-diffusion measurements. In all probability the higher molecular weight estimate is too large because of errors introduced into measurement of the diffusion constant as a consequence of the heterogeneity of the preparation examined. The sedimentation coefficient ($s^{\circ}_{20,w}$) is agreed to be about 9.1×10^{-13} sec^{-1}.

The molecular weight of fumarase in 6 *M* guanidine hydrochloride in the presence or absence of 2-mercaptoethanol is 48,500 as judged by sedimentation-equilibrium in the ultracentrifuge, or about one-fourth that of the native enzyme. The molecular weight of the enzyme in guanidine hydrochloride or dilute salt solution is unaffected by 2-mercaptoethanol.

The isoionic point of fumarase is 7.95 ± 0.05 at 25°. The isoelectric point of the enzyme differs considerably depending upon the buffers employed in estimating electrophoretic mobilities (*30*). In 0.1 *M* tris-acetate, its isoelectric point is 7.35, whereas in 0.09 *M* sodium chloride and in phosphate buffer, ionic strength, 0.21, the isoelectric points are 6.6 and 5.3, respectively. These data suggest that fumarase binds anions in the order, phosphate > chloride > acetate. Thus, fumarase can be expected to have different electrostatic properties in different salt solutions. The marked effect of different ions on the activity of fumarase has been noted earlier (*31*) and is considered in Section I,C below.

The optical rotatory dispersion spectrum of fumarase between 220 and 310 mμ is not unusual compared with other globular proteins and shows a maximum Cotton effect at 234 mμ. If the values of b_0 and the mean residue rotation reflect helical content, then fumarase can be judged to contain about 50% helix.

The ratio f/f_0, reflecting the shape of the molecule, is 1.24 for fumarase. This indicates that the enzyme is a compact, globular molecule in dilute salt solution.

2. *Amino Acid Composition, End Groups and Peptide Maps*

The amino acid composition of crystalline swine heart fumarase is given in Table II (*13*). The native enzyme (MW 194,000) contains a total of 1763 residues per molecule. Although fumarase contains all of the usual amino acids its content of tryptophan and half-cystine is particularly low. As discussed in Section I,B,5, the enzyme is devoid of disulfide bonds and the total half-cystine content can be accounted for as cysteine on the basis of its thiol content (*13*). The enzyme also contains about 351 residues of glutamic acid plus aspartic acid, about

TABLE II
THE AMINO ACID COMPOSITION OF SWINE HEART FUMARASE[a]

Amino acid	Residues per molecule	Amino acid	Residues per molecule
Lysine	129	Alanine	200
Histidine	54	Half-cystine	12
Arginine	54	Valine	131
Aspartic acid	177	Methionine	62
Threonine	100	Leucine	147
Serine	92	Isoleucine	97
Glutamic acid	124	Tyrosine	40
Proline	80	Phenylalanine	63
Glycine	143	Tryptophan	8
		Total No. of residues	1763

[a] Data from Kanarek and Hill (*13*).

half of which must represent glutamine and asparagine in view of the ammonium content of acid hydrolysates. This estimate of the amide content is also approximately in accord with the expected number of free carboxyl groups based on the isoionic point of fumarase (pH 7.95) and its total content of lysine plus arginine (183 residues).

NH_2-Terminal end group analysis of fumarase by the cyanate method reveals a total of 3.6 residues of alanine per molecule, assuming a molecular weight of 194,000 (*28*). Less than 0.3 residue each per molecule of lysine, aspartic acid, glutamic acid, and glycine was also noted in the analyses. It is unlikely that fumarase contains multiple end groups although this cannot be ruled out unequivocally by the presently available data. The COOH-terminal end groups have not been established.

Tryptic peptide maps of fumarase (*28*) reveal a total of 45–50 soluble tryptic peptides, or about one-fourth the expected number for a protein of the size of fumarase containing a single polypeptide chain with about 183 residues of lysine and arginine (see Table II). The number of peptides containing histidine, arginine, and tyrosine is also about one-fourth that expected for such a molecule.

3. *Subunit Structure*

Several kinds of observations have helped to establish the subunit structure of fumarase (*28*, *32*). First, the molecular weight of the native

32. J. W. Robinson, R. A. Bradshaw, L. Kanarek, and R. L. Hill, *JBC* **242**, 2709 (1967).

enzyme is 194,000 in solutions of low ionic strength, but it is about one-fourth (48,500) this value in 6 *M* guanidine hydrochloride. 2-Mercaptoethanol has no effect on the molecular weight. Second, NH_2-terminal end group analysis reveals 3.6 residues of alanine per molecule, assuming a molecular weight of 194,000. Third, tryptic peptide maps of fumarase reveal about one-fourth of the expected number of peptides based on the lysine and arginine content of the enzyme. These data indicate that fumarase is composed of four subunit polypeptide chains of identical sequence with a molecular weight of 48,500. Since mercaptoethanol has no effect on the molecular weight of the enzyme in guanidine hydrochloride or dilute salt solution, the enzyme is devoid of interchain disulfide bonds, and the four subunits must be combined in the native enzyme through noncovalent bonds. This is in accord with the observation that the total half-cystine content of the enzyme can be accounted for by the free sulfhydryl groups titratable with *p*-mercuribenzoate or other thiol reagents (*13*).

4. *Dissociation and Recombination of Subunits*

The enzymically active form of fumarase appears to be the tetrameric structure and, in general, any condition which results in dissociation of the enzyme, leads to loss of enzymic activity.

Precise determination of the molecular weight of fumarase at concentrations (5×10^{-4}–5×10^{-5} mg/ml) used to measure its specific activity have not been performed. Nevertheless, it is possible to show that at concentrations of about 10^{-4} mg/ml the enzyme chromatographs in phosphate buffers at pH 7.3 on columns of Sephadex G-200 in a manner indistinguishable from enzyme at concentrations of about 3 mg/ml (*33*). Furthermore, the enzyme retained its full enzymic activity at these low concentrations. For this reason and the fact that additional studies to be discussed below show that dissociation induced by several means invariably results in inactivation, it is likely that the active species of fumarase is tetrameric.

Although fumarase remains undissociated and fully active at very low concentrations of enzyme in phosphate buffers, at pH 7.3 it is rapidly inactivated at low enzyme concentrations in the absence of either phosphate, its substrates, or competitive inhibitors (*33*). In 0.01 *M* potassium acetate at pH 7.3 the rate of inactivation increases as the initial enzyme concentration decreases. In addition, the rate of inactivation is greater at pH 5 than at pH 7.3 in acetate solutions. At these low enzyme concentrations both malate and fumarate as well as inhibitors such as

33. J. W. Teipel and R. L. Hill, *JBC* **246**, 4859 (1971).

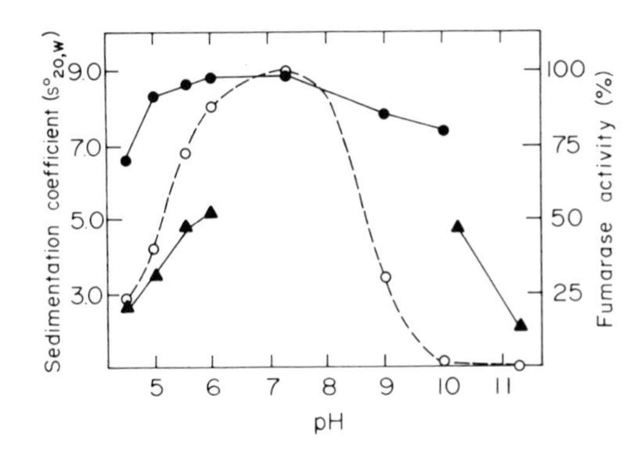

FIG. 1. The sedimentation coefficients and activity of fumarase in tris-acetate buffers as a function of pH. The activities were measured at pH 7.3 with L-malate after exposure to the pH indicated (*33*). (● and ▲) Sedimentation coefficients of fast and slow components, respectively, and (○) activity.

citrate prevent inactivation to a considerable degree. Since the rate of inactivation increases as the enzyme concentration decreases, it may be assumed that inactivation by dilution represents dissociation of the enzyme into its subunits.

The stability of fumarase as a function of pH in tris-acetate buffers is shown in Fig. 1. It is evident that the enzyme dissociates below pH 6 and above pH 10 under these conditions. From pH 4.5 to 6, two components are observed in the ultracentrifuge, one with a sedimentation constant close to that of the undissociated enzyme and a second component which presumably represents a lower molecular weight form. The activity of the enzyme under these conditions is roughly equal to the amount of undissociated enzyme. Above pH 10, fumarase dissociates extensively and is completely inactivated.

Enzyme dissociated at both acid and alkaline pH could be reassociated to the tetrameric structure with nearly full regain of enzymic activity by incubation of the enzyme at pH 7.3 (*33*).

In contrast to its instability in tris-acetate buffer, fumarase is quite stable between pH 5 and 9 in potassium phosphate and shows little tendency to dissociate (Fig. 2). Above pH 10, however, dissociation of the enzyme into an inactive forms is evident.

The effect of urea and guanidine hydrochloride on the dissociation and activity of fumarase is shown in Fig. 3. Based upon its sedimentation constant in urea solutions, fumarase appears to remain undissociated in solutions containing up to 2 *M* urea, but the enzyme is almost completely dissociated into its subunits in 2.4 *M* urea solutions. The magnitude of the sedimentation constant of the enzyme in 3 *M* urea (1.46 S)

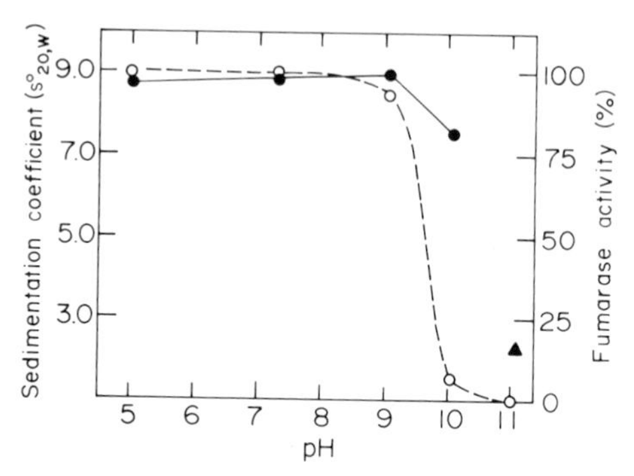

FIG. 2. The sedimentation coefficient and activity of fumarase in potassium phosphate buffers as a function of pH. All activities were measured at pH 7.3 with L-malate after exposure to the pH indicated (*33*). (● and ▲) Sedimentation coefficients and (○) activity.

suggests that the subunit chains are unfolded extensively. As the enzyme dissociates, it appears to be inactivated as judged by measurement of the activity in urea solutions; however, as described below, considerable reassociation of subunits can occur with regeneration of activity if the urea is removed.

The dissociation of fumarase by guanidine hydrochloride appears to proceed in two steps: (1) the formation of a species with a sedimentation constant of 6.5, and (2) dissociation of the 6.5 S component into a lower molecular weight species which is extensively unfolded (*33*). The enzyme appears to be progressively inactivated on formation of the 6.5 S species and completely inactivated when fully dissociated. Little is known about the 6.5 S species, but the optical rotatory dispersion spectrum

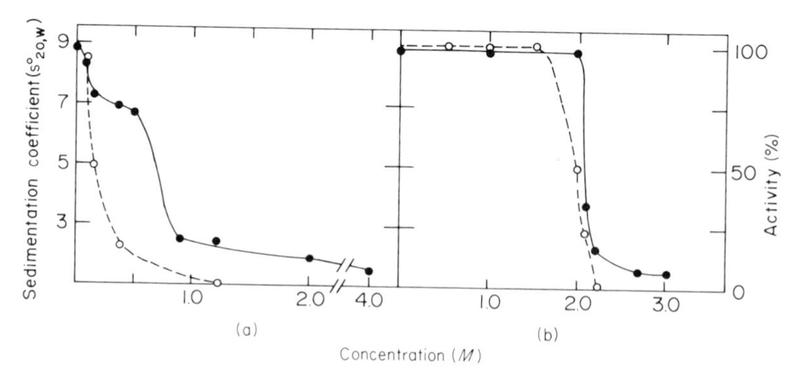

FIG. 3. Sedimentation coefficients and activity of fumarase as a function of (a) guanidine hydrochloride (*33*) and (b) urea concentration (*32*). (●) Sedimentation coefficients and (○) activity.

of fumarase in 0.5 *M* guanidine hydrochloride is very similar to that of the native enzyme.

The dissociation and inactivation of fumarase by urea or guanidine hydrochloride can be fully reversed by removal of these agents. Dialysis of fumarase solutions in 4 *M* guanidine hydrochloride or 6 *M* urea against water leads to little restoration of activity, but on dialysis against phosphate buffer or phosphate buffer containing malate over 90% of the fumarase activity is regained (Table III). The recombined enzyme can be crystallized and its properties are indistinguishable from those of the native, unaltered enzyme. These observations show that the extensively unfolded subunits are able to recombine completely to give the native enzyme.

Fumarase also dissociates on modification of its thiol groups with agents which are charged (*32*, *34*). Thus, carboxymethylation with iodoacetate or formation of the mixed disulfide of fumarase and cysteine or 2-thioethylamine leads to dissociation and formation of dimeric, enzymically inactive species. Reaction of the thiols with uncharged reagents such as *N*-ethylmaleimide or iodoacetamide does not cause dissociation, although the enzyme is inactivated by these reagents. The mixed disulfide derivatives of fumarase and cysteine or 2-mercaptoethylamine can be fully reactivated with formation of the native tetrameric structures by reaction with 2-mercaptoethanol.

5. *Thiol Groups*

The reactivity of fumarase with a variety of reagents has revealed something of the nature of the 12 thiol groups in the enzyme. In general, the conditions required to modify the thiol groups of fumarase by reagents such as *p*-mercuribenzoate, iodoacetate, or iodoacetamide suggest that the thiols are not in the active site but are buried in hydrophobic regions of the enzyme (*32*). Rather high concentrations of these reagents are required to modify the thiol groups, or the conformation of the enzyme must be perturbed somewhat if the thiol groups are to react more rapidly than observed in the native enzyme.

Although the thiol groups do not appear to be in the active site, modification of the thiol groups leads to inactivation of the enzyme with the loss in activity proportional to the number of thiol groups modified. For example, when 70% of the thiol groups in a fumarase preparation are modified by formation of a mixed disulfide with cysteine, 70% of the activity is lost. Furthermore, 70% of the fumarase has all of its thiol groups in mixed disulfide linkage, and the thiol groups in the

34. R. L. Hill and L. Kanarek, *Brookhaven Symp. Biol.* **17**, 80 (1964).

TABLE III

REACTIVATION OF FUMARASE DISSOCIATED WITH GUANIDINE HYDROCHLORIDE OR UREA[a,b]

Denaturant	Fumarase conc. (mg/ml)	Solvent	Recovery of activity (%)
Guanidine hydrochloride, 4 *M*	0.75	0.05 *M* 2-Mercaptoethanol, pH 7.3	16.5
	0.75	0.05 *M* Potassium phosphate, pH 7.3 0.05 *M* 2-Mercaptoethanol	84
	0.75	0.05 *M* L-Malate, 0.05 *M* potassium phosphate, pH 7.3, 0.05 *M* 2-mercaptoethanol	95
Urea, 6 *M*	0.09	0.01 *M* 2-Mercaptoethanol, pH 7.3	0
	0.09	0.01 *M* Potassium phosphate, pH 7.3	33
	0.09	0.05 *M* Potassium phosphate, pH 7.3	53
	0.09	0.05 *M* Potassium phosphate, pH 7.3 0.05 *M* L-Malate	73
Urea, 8 *M*	0.75	0.05 *M* Potassium phosphate, pH 7.3 0.01 *M* 2-Mercaptoethanol 0.05 *M* L-Malate	91

[a] Data from Teipel (*33*) and Hill and Kanarek (*34*).

[b] In these studies fumarase was dissolved in the denaturant at the final enzyme concentration shown and then dialyzed against at least 1000 volumes of the solvent indicated.

remaining 30% of the fumarase are unaltered. These studies suggest that the thiol groups in an individual molecule react in all-or-none fashion.

Several lines of evidence indicate that the thiol groups of fumarase are buried (*32*). First, the thiol groups react with iodoacetate or *p*-mercuribenzoate 200–300 times faster than in the native enzyme when the enzyme is dissolved in 8 *M* urea. Furthermore, modification of the thiol groups with reagents that yield a charged derivative results in dissociation of the subunits. The mixed disulfide derivative of fumarase with either cysteine or 2-thioethylamine appear to be dimeric under conditions where the unmodified enzyme is tetrameric. The *S*-carboxymethyl derivative also exists as a dimeric species; thus, it appears that a charge cannot be buried in the vicinity of the thiol groups. In contrast, uncharged thiol group derivatives, such as those formed with *N*-ethylmaleimide or iodoacetamide, exist in the normal, tetrameric form.

Indications that the thiol groups are in a hydrophobic environment are provided by studies on the reaction of fumarase with alkyl mercury nitrates, or with *p*-mercuribenzoate in aliphatic alcohols (*32*). The rate of reaction of the thiol groups with alkyl mercury nitrates is in the order methyl < ethyl < *n*-propyl < *n*-butyl. As the length of the aliphatic chain is increased, the reagent becomes more hydrophobic and a greater rate of reaction is observed. Examination of the kinetics of the reaction of these compounds as a function of temperature reveals that the difference in the free energy of activation ($\Delta F^{\ddagger}$) for the reaction of two alkyl mercurials differing in structure by one methylene group is about the order of magnitude estimated for the transfer of a methylene group from water to alcohol solution. The rate of reaction of *p*-mercuribenzoate with the thiols of fumarase is markedly different in dilute solutions of aliphatic alcohols (*32*). At equivalent concentrations of the alcohols, the rate of reaction with *p*-mercuribenzoate is greatest in *n*-pentanol and falls progressively in the order *n*-pentanol > *n*-butanol > *n*-propanol > ethanol > methanol. Thus, it appears that as the alcohols become more hydrophobic with increasing chain length, they are better able to disrupt those structures of the enzyme which normally hinder reaction of the thiols with *p*-mercuribenzoate.

Some evidence suggests that the thiol groups of fumarase may reside in or near the contact regions between subunits (*33*). Examination of the reaction with *p*-mercuribenzoate as a function of enzyme concentration shows that the thiol groups react more rapidly with decreasing enzyme concentration. Furthermore, at pH 5 in tris-acetate buffers, where the enzyme is considerably dissociated into dimeric species, the rate of modification by *p*-mercuribenzoate is markedly increased. Under these conditions where the thiol groups show an increased reactivity, the optical rotatory dispersion spectrum of fumarase is indistinguishable from that of the native enzyme. Thus, dissociation of the subunits by dilution of the enzyme or exposure to pH 5 does not appear to cause an appreciable alteration in the folding of the polypeptide chain subunits, but the reactivity of the thiol groups is enhanced markedly.

C. Catalytic Properties

1. *Kinetics of Reaction*

The kinetics of the reaction catalyzed by fumarase have been studied in great detail by measurement of initial rates as a function of substrate concentration, buffer composition, pH, temperature, and inhibitors. Because these studies have been reviewed extensively earlier (*14*), the following sections will not cover in depth all of the earlier kinetics studies but will only summarize some of the conclusions which give special insight into the mechanism of action of the enzyme as gained from more recent studies. The earlier reviews should be consulted for further details.

a. Substrate Concentration. At low concentrations of substrate the hydration of fumarate and the dehydration of malate follow Michaelis-Menten kinetics, but at concentrations more than about five times K_m, substrate activation is observed (*35–37*). At substrate concentrations greater than 0.1 M, inhibition occurs (*36, 38*). The substrate activation was initially explained by assuming that a second molecule of substrate can bind at a site(s) other than the active site and thereby increase the reactivity of the active site. This mechanism is consistent with Eq. (1), the simplest empirical rate expression found to fit the experimental data.

$$V_0 = \frac{V_{\max} + a/(\mathrm{S})}{1 + b/(\mathrm{S}) + c/(\mathrm{S})^2} \quad (1)$$

In this equation $V_{\max}$ is the maximum initial velocity, (S) is the substrate concentration, and a, b, and c are constants which depend upon pH, temperature, and buffer composition.

There are, however, several other mechanisms for explaining the substrate activation of fumarase which yield rate expressions identical in

35. R. A. Alberty and R. M. Bock, *Proc. Natl. Acad. Sci. U. S.* **39**, 895 (1953).
36. R. A. Alberty, V. Massey, C. Frieden, and A. R. Fuhlbrigge, *JACS* **76**, 2485 (1954).
37. M. Taraszka and R. A. Alberty, *J. Phys. Chem.* **68**, 3368 (1964).
38. S. Rajender and R. J. McColloch, *ABB* **118**, 278 (1967).

form to Eq. (1). One such mechanism is that of "negative cooperativity," originally proposed by Conway and Koshland to explain the kinetic behavior of glyceraldehyde-3-phosphate dehydrogenase (*39*). According to this interpretation fumarase possesses two or more identical binding sites for substrate which display cooperative interactions such that binding of substrate to the first site(s) *decreases* the affinity of substrate for the latter site(s).

The substrate activation of fumarase may also be explained if it is assumed that the enzyme possesses two or more independent, but non-identical, binding sites or, alternatively, if there are two or more forms of the enzyme present which have different affinities for substrate. The latter hypothesis appears attractive since, as noted above, there is some evidence that fumarase might exist in multiple isoenzymic forms.

Substrate inhibition at fumarate concentrations between 0.1 *M* and 1.0 *M* has been examined by infrared spectrometry (*38*). The inhibition can be explained by the nonproductive binding of two molecule of fumarate at the active site, in accord with a mechanism proposed earlier by Haldane (*40*).

b. Buffer Composition. The catalytic behavior of fumarase is altered considerably by the nature and concentration of anions (*31*). At low concentrations, phosphate, sulfate, borate, selenate, arsenate, and citrate activate fumarase and shift somewhat the pH optimum to a more alkaline pH. At higher concentrations the anions may be inhibitory. For example, phosphate activates fumarase at concentrations less than 5×10^{-3} *M* but is a competitive inhibitor at higher concentrations (*36*). Monovalent halides and thiocyanate are noncompetitive inhibitors at all concentrations tested. Acetate over a wide range of concentrations has little effect on the kinetic parameters of the enzyme, and for this reason tris-acetate buffers have been often chosen for kinetic studies of fumarase (*41*). The activation and inhibition of fumarase by di- and trivalent anions have been interpreted in the same manner as substrate activation by assuming that such ions can bind at noncatalytic sites and thereby alter the catalytic properties of the active site (*36*). Of course, a more precise mechanism for the effect of these anions remains to be established.

c. Hydrogen Ion Concentration. The effect of pH on K_m and $V_{\max}$ has been measured in tris-acetate (*42*) as well as in phosphate buffers (*43*,

39. A. Conway and D. E. Koshland, Jr., *Biochemistry* **7**, 4011 (1968).
40. J. B. S. Haldane, "Enzymes," p. 84. Longmans, Green, New York, 1930.
41. C. Frieden, R. G. Wolfe, Jr., and R. A. Alberty, *JACS* **79**, 1523 (1957).
42. C. Frieden and R. A. Alberty, *JBC* **212**, 859 (1955).
43. V. Massey and R. A. Alberty, *BBA* **13**, 354 (1954).

44); V_{max} was related to pH in both buffers by bell-shaped curves with maxima between pH 6 and 8, the exact values depending on substrate and buffer. These data suggest that two groups, one in the acidic form and one in the basic form, are intimately associated with the catalytically active form of the enzyme. Frieden and Alberty (*42*) have analyzed these data by the mechanism given in Eq. (2).

$$\begin{array}{ccccccc} EH_2 & & EH_2F & & EH_2M & & EH_2 \\ K_{bE}\Big\updownarrow & & K_{bEF}\Big\updownarrow & & K_{bEM}\Big\updownarrow & & K_{bE}\Big\updownarrow \\ F + EH & \underset{k_2}{\overset{k_1}{\rightleftharpoons}} & EHF & \underset{k_4}{\overset{k_3}{\rightleftharpoons}} & EHM & \underset{k_6}{\overset{k_5}{\rightleftharpoons}} & EH + M \\ K_{aE}\Big\updownarrow & & K_{aEF}\Big\updownarrow & & K_{aEM}\Big\updownarrow & & K_{aE}\Big\updownarrow \\ E & & EF & & EM & & E \end{array} \qquad (2)$$

F and M represent the doubly negatively charged substrates, fumarate and malate, respectively; EH represents the catalytically active form of the enzyme; HEF the enzyme–substrate complex with fumarate; and EHM the enzyme substrate complex with malate. The interactions between substrate and fumarase are defined by the six acidic dissociation constants (K) and the six rate constants (k) as shown. Six of the twelve constants could be assigned numeral values by analysis of the initial velocities in the forward and reverse reactions and the remaining six could be assigned limits (*45*). The values for these constants in 0.01 M tris-acetate buffers at 25° are listed in Table IV. The values differ somewhat at different ionic strengths. The second-order rate constants k_1 and k_6 approach values theoretically expected for diffusion-controlled reactions. In addition, the maximum and minimum values for k_2/k_1 are, within experimental error, equal to K_f, the Michaelis constant for fumarate.

TABLE IV
SPECIFIC RATE CONSTANTS AND pK VALUES FOR FUMARASE[a]

Rate constants	Minimum	Maximum	pK	Minimum	Maximum
$k_1 \times 10^{-9}$ (sec^{-1} M^{-1})	11	∞	pK_{aE}	6.2	6.2
$k_2 \times 10^{-3}$ (sec^{-1})	27	∞	pK_{aEF}	∞	5.3
$k_3 \times 10^{-3}$ (sec^{-1})	2.3	2.5	pK_{aEM}	6.6	6.6
$k_4 \times 10^{-3}$ (sec^{-1})	1.7	2.0	pK_{bE}	6.8	6.8
$k_5 \times 10^{-3}$ (sec^{-1})	46	∞	pK_{bEF}	7.3	7.3
$k_6 \times 10^{-9}$ (sec^{-1} M^{-1})	5	∞	pK_{bEM}	8.5	∞

[a] Data from Alberty and Pierce (*45*).

44. R. A. Alberty and V. Massey, *BBA* **13,** 347 (1954).
45. R. A. Alberty and W. H. Pierce, *JACS* **79,** 1526 (1957).

The values for pK_{aE} and pK_{bE} were found to be 6.2 and 6.8, respectively, which most closely approximate the pK of imidazole side chains of histidine. Thus, these studies suggest that two imidazole side chains may operate as catalytic groups in the active site, although other studies on the variation of initial velocities with temperature, as discussed below, indicate that pK_{aE} may represent the dissociation constant for a carboxyl group.

d. Temperature. The variation of K_m and V_{max} with pH has been examined as a function of temperature in tris-acetate buffers (*46*). The dissociation constants and rate constants indicated in Eq. (2) were estimated between 5° and 37°. The standard enthalpies of dissociation for K_{aE} and K_{bE} were -1.7 kcal mole^{-1} and 7.7 kcal mole^{-1}, respectively. The orders of magnitude for these enthalpies suggest that K_{aE} reflects the dissociation of a carboxyl group and K_{bE} that of an imidazolium group. If, indeed, K_{aE} reflects the dissociation of a carboxyl group, then it is clear that ionization of the group must be perturbed by its local environment in the catalytic site; otherwise, the pK would be expected to be about 3.5–4.5. In this respect, a carboxyl group in lysozyme which operates in the active site has been assigned a pK of about 6.7 (*47*).

Activation energies for the hydration and dehydration reactions have also been estimated as a function of pH and the heats of formation of enzyme–substrate complexes estimated (*48*). Interpretation of these studies is difficult, however, because of specific effects of phosphate ions on the enzyme.

e. Inhibitors. A number of structural analogs of fumarate and L-malate have been tested as competitive inhibitors (*49*). Most di- and tricarboxylic acids were found to be competitive inhibitors, but monocarboxylic acids or derivatives of fumarate or malate containing derivatized carboxyl groups were ineffective except at high concentrations. From these data, Massey (*49*) has suggested that good competitive inhibitors, just as the normal substrates, must be bound to positively charged regions of the active site through interactions with at least two negatively charged carboxyl groups in the inhibitor.

Wigler and Alberty (*50*) examined the effects of four inhibitors, succinate, D-tartrate, L-tartrate, and *meso*-tartrate, on the initial velocity of the fumarase reactions as a function of pH. From these studies, it

46. D. A. Brant, L. B. Barnett, and R. A. Alberty, *JACS* **85,** 2204 (1963).
47. J. A. Rupley, *Proc. Roy. Soc.* **B167,** 416 (1967).
48. V. Massey, *BJ* **53,** 72 (1953).
49. V. Massey, *BJ* **55,** 172 (1953).
50. P. W. Wigler and R. A. Alberty, *JACS* **82,** 5482 (1960).

TABLE V
SUBSTRATE SPECIFICITY OF FUMARASE[a]

Compound	V_{max} (mole/mole/min)	K_m $\times 10^{-5}$ M	K_I $\times 10^{-5}$ M	K_{eq}
Fluorofumarate	160,000	2.7		
Fumarate	48,000	0.5		4.4
Difluorofumarate	~41,000[b]	—		
Chlorofumarate	1,300	11	10	6.2
Bromofumarate	170	11	15	12.0
Acetylene dicarboxylate	125	14.5		
L-*trans*-2,3-Epoxysuccinate	3.6[c]	—	—	—
Iodofumarate	2.6	12	10	3.8
Mesaconate	1.4	51	49	0.18
Dimethylfumarate	<0.01		500	
trans-Aconitate	<0.01		0.73	
L-Malate	54,000	2.5		
L-*threo*-Chloromalate	340–680	2–14		
L-Tartrate	55.5	130	100	
L-*threo*-Hydroxyaspartate	0.9–1.8	750–1000	390–780	
L-Isocitrate	<0.01		0.13–0.26	
Citrate	<0.01		2.2	

[a] From Teipel *et al.* (*51*). Data are for a temperature of 25° and a pH of 7.3.

[b] Estimated from the K_m and V_{max} values reported for this compound (*61*). The absolute values for K_m and V_{max} for fumarate and difluorofumarate differ somewhat in different reports, but the relative values for kinetic constants appear to be about the same.

[c] From Albright and Schroepfer (*59*). V_{max} calculated relative to valúe for fumarate of 48,000 moles/mole/min.

was found that the dissociation constants, K_{aE} and K_{bE} were affected differently by D- and L-tartrate. This suggests that the two catalytic groups in the active site are nonequivalent and, thus, that the active site is unsymmetrical.

The free energy of formation between the hydroxyl groups of the tartrates and the active site were also determined by measuring $\Delta F°$ for displacement of succinate from the enzyme by the three types of tartrate. L-Tartrate and D-tartrate were weakly bound, but *meso*-tartrate was more strongly bound, in accord with the observation that the *meso* form is one of the best competitive inhibitors of the enzyme.

Teipel *et al.* (*51*) have reported the K_I values for a variety of competitive inhibitors which are also weak substrates for fumarase. These values, listed in Table V, are about equal to the K_m values for each

51. J. W. Teipel, G. M. Hass, and R. L. Hill, *JBC* **243**, 5684 (1968).

compound. The close correspondence between K_m and K_I for this series of compounds is in accord with the view that the apparent K_m for a substrate can be viewed as a true equilibrium constant for the dissociation of these substrates from the enzyme.

A recently discovered competitive inhibitor of fumarase, which does not fall into the class of dicarboxylic acids, is ATP (*52*). The nucleotides GTP, CTP, and UTP also competitively inhibit the enzyme, although not as effectively as ATP. The fact that all the nucleotide triphosphates are good inhibitors, plus the observation that tripolyphosphate is as effective an inhibitor as the nucleotides (*52a*), suggests that the negatively charged triphosphate moiety is primarily responsible for the tight binding to the enzyme.

2. *Substrate Specificity*

a. Stereospecificity. The hydrogen and hydroxyl groups of water are exchanged stereospecifically during the conversion of fumarate and L-malate. Only the L- isomer of malate may serve as a substrate (*22*), indicating the stereospecific removal of the hydroxyl group (*52b*). The stereospecificity for removal or addition of hydrogen was deduced from the observation that only one atom of nonexchangeable deuterium was introduced on hydration of fumarate with the enzyme in D_2O (*53–55*). In addition, essentially no deuterium was introduced into fumarate isolated from an equilibrium mixture of fumarate and malate.

It has also been established that the hydrogen removed from L-malate is *erythro* to the hydroxyl group (*56*). 3-Monodeutero-L-malate obtained by hydration of fumarate in D_2O was shown to have the *erythro* configuration by comparison with chemically synthesized *threo*-3-monodeutero-D,L-malate. The enzymically synthesized 3-monodeutero-L-malate on inversion possessed an NMR spectrum similar to *threo*-3-monodeutero-L-malate. Thus, if the carboxyl groups of L-malate are *trans* with respect to one another in the active site, these observations indicate that water is added with a *trans* stereochemistry across the double bond of fumarate, depicted as follows:

52. P. E. Penner and L. H. Cohen, *JBC* **244,** 1070 (1969).

52a. J. W. Teipel and R. L. Hill, unpublished observation (1969).

52b. The levorotatory isomer of malate has been assigned the configuration L-malate. By the *R/S* convention L-malate is (2*S*)-malate.

53. S. Englard and S. P. Colowick, *Science* **121,** 866 (1955).

54. S. Englard and S. P. Colowick, *JBC* **221,** 1019 (1956).

55. H. F. Fisher, C. Frieden, J. S. M. McKee, and R. A. Alberty, *JACS* **77,** 4436 (1955).

56. O. Gawron and T. P. Fondy, *JACS* **81,** 6333 (1959).

$$^{-}O_2C-CH=CH-CO_2^{-} + D_2O \rightleftharpoons {}^{-}O_2C-CHD-CH(OD)-CO_2^{-}$$

b. Unnatural Substrates. For many years it was thought that the substrate specificity of fumarase was quite rigid and that the enzyme acted only on its natural substrates, fumarate and L-malate. Eleven new substrates for fumarase, which are listed in Table V, have been recognized in the past three years. None of these substrates occur in substantial amounts if at all in animal tissues; thus, no biological significance can be attached to their discovery. The behavior of these substrates has provided, however, some insight into the catalytic properties of fumarase.

Consideration of the structures of the compounds listed in Table V reveals two general requirements for fumarase substrates:

(1) Two negatively charged carboxyl groups are required for a substrate, in accord with earlier suggestions that this was also a structural requirement for competitive inhibitors (*49*). Presumably, two regions in the active site bearing a positive charge would interact with the negatively charged carboxyl groups.

(2) The hydroxyl groups in all substrates must have the same configuration as found in L-malate. Furthermore, malate derivatives in which a substituent other than hydrogen is *erythro* with respect to the hydroxyl group, are not substrates, in accord with the stereospecificity discussed above.

Among the three diastereoisomeric forms of tartrate, only L-tartrate proves to be a substrate and is dehydrated to give oxalacetate as first shown by Nakamura and Ogata (*57, 58*). Because acetylenedicarboxylate is hydrated to give oxalacetate, the following interconversion of L-tartrate to acetylene dicarboxylate is also possible (*51*):

$$\text{L-Tartrate} \rightleftharpoons \text{oxalacetate} + H_2O \rightleftharpoons \text{acetylenedicarboxylate} + 2\ H_2O$$

The equilibrium for this reaction favors formation of oxalacetate, since oxalacetate serves as a very poor substrate for fumarase. The action of fumarase upon *threo*-β-hydroxyaspartate also gives rise to oxalacetate since on dehydration it yields aminofumarate which spontaneously deaminates to give ammonia and oxalacetate (*51*).

One of the most interesting new substrates found for fumarase is

57. S. Nakamura and H. Ogata, *JBC* **243**, 528 (1968).
58. S. Nakamura and H. Ogata, *JBC* **243**, 533 (1968).

L-*trans*-2,3-epoxysuccinate which is hydrated to *meso*-tartrate (*59*). This reaction is also stereospecific since only the L- isomer of the epoxide is hydrated.

The rate of hydration of the halofumarates is in the order fluorofumarate > fumarate > difluorofumarate > chlorofumarate > bromofumarate > iodofumarate. Fluorofumarate is hydrated fully three times faster than the natural substrate fumarate (*51, 60, 61*), and is the best substrate recognized at present. In contrast to the other halofumarates, however, fluorofumarate is hydrated to give α-fluoromalate which spontaneously decomposes to give oxalacetate as follows:

$$\begin{array}{ccc} H & & CO_2 \\ & C & \\ & \| & \\ & C & \\ {}^-O_2C & & F \end{array} + H_2O \rightleftharpoons \begin{array}{c} CO_2 \\ | \\ H-C-H \\ | \\ HO-C-F \\ | \\ CO_2^- \end{array} \rightleftharpoons \begin{array}{ccc} H & & CO_2^- \\ & C & \\ & \| & \\ & C & \\ {}^-O_2C & & OH \end{array} + HF$$

Difluorofumarate would appear to give fluorooxalacetate (*61*). Chloro-, bromo-, and iodofumarate are hydrated to give the corresponding β-halo-L-malates, in which the halo substituents are *threo* with respect to the hydroxyl group. This is in accord with the observation that *threo*-halo-L-malates are dehydrated by fumarase but that *erythro*-halo-L-malates are not. Teipel *et al.* (*51*) have suggested that the difference in the pattern of hydration of fluorofumarate and the other halofumarates is the consequence of the specific three-dimensional relationships among functional groups in the active site. As shown schematically in Fig. 4, oxalacetate may form two nonequivalent complexes with the active site, one as the consequence of its production by hydration of acetylenedicarboxylate and the other by dehydration of L-tartrate. This suggests that all substrates can bind potentially in either of two configurations, but if a substituent on the substrate is too large, only one type of productive binding is possible (van der Waals radii; hydrogen, 1.2 Å; fluorine, 1.35 Å; chlorine, 1.8 Å; bromine, 1.95 Å; and iodine, 2.15 Å). Thus, because of the small size of the fluorine atom, fluorofumarate and difluorofumarate can bind in the configuration which would give α-fluoromalate. The other halo derivatives could bind only in that configuration which would give the β-halo-L-malates, as also shown in Fig. 4. It would follow that that part of the active site catalyzing hydroxyl exchange may be sterically more restricted than the region catalyzing exchange of hydrogen ion. A comparison of the binding capacity of

59. F. Albright and G. J. Schroepfer, Jr., *BBRC* **40,** 661 (1970).
60. D. D. Clarke, W. J. Nicklas, and J. Palumbo, *ABB* **123,** 205 (1968).
61. W. G. Nigh and J. H. Richards, *JACS* **91,** 5847 (1969).

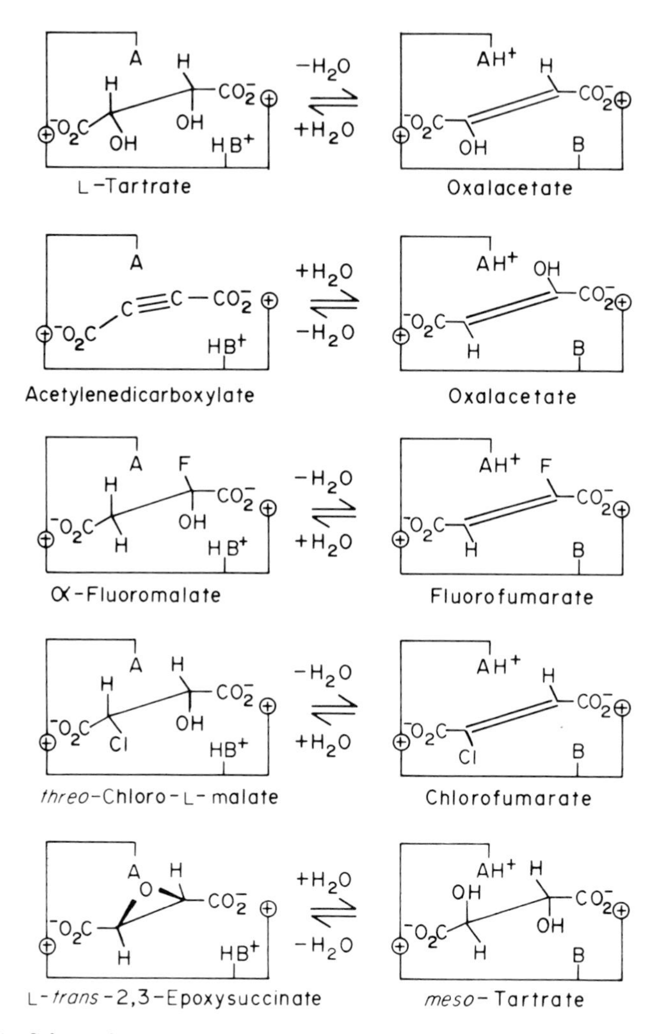

FIG. 4. Schematic representation of the active site complexes formed with fumarase and L-tartrate, acetylenedicarboxylate, oxalacetate, α-fluoromalate, fluorofumarte, *threo*-chloro-L-malate, chlorofumarate, L-*trans*-2,3-epoxysuccinate, and *meso*-tartrate. The active site region is schematically depicted as shown in Fig. 6.

(K_I) compounds given in Table V seems to bear this out. Thus, dimethylfumarate and citrate (α-carboxymethylmalate) are bound more poorly than mesaconate (monomethylfumarate) and L-isocitrate (*erythro*-carboxymethylmalate). The binding of dimethylfumarate and citrate would introduce a bulky group in the region catalyzing hydroxyl exchange, whereas binding of the latter two would not.

Figure 4 also shows the possible binding of L-*trans*-2,3-epoxysuccinate in the active site of fumarase. It can be assumed that enzymic hydration of this compound proceeds by the same mechanism as that in free solution, namely, a backside attack by water on the protonated intermediate. Furthermore, inversion occurs at the carbon atom where water attacks. From these assumptions, it follows that enzymic hydration of either the *trans*-D or *trans*-L derivative would yield *meso*-tartrate. The fact that only the L- isomer is hydrated enzymically, however, may result from the type of binding indicated in Fig. 4. The ring oxygen is depicted as projecting above a plane formed by the carbon backbone of the substrate. This would allow hydrogenation at the carbon near the B site, with inversion, and formation of *meso*-tartrate. If the A site is somewhat above the plane of the figure and the B site below, then the L-isomer fits such that the ring oxygen is nearer A than B, in accord with the view, as discussed above, that the B region is somewhat more sterically hindered than the A site. The D-isomer would be expected to fit such that the ring oxygen projects below the plane in Fig. 4. This could not be easily done if the B site is sterically hindered, thus suggesting why this isomer is not a substrate for the enzyme (*59*).

The observed differences in the rate of hydration of the fumarate derivatives listed in Table V are not in accord with those expected based upon present knowledge of the electrophilic addition of water to a double bond in free solution. In fact, the order observed, fluorofumarate > chlorofumarate > bromofumarate > iodofumarate, is almost exactly opposite that expected. These differences could be explained by assuming that fumarase hydrates double bonds by a different mechanism than that occurring in free solution, but other explanations seem more plausible. If it is assumed that the K_m values for fumarate and halofumarates reflect the binding constant for these compounds, which is a good assumption in view of kinetic and binding studies summarized in Tables V and VI, then chloro-, bromo-, and iodofumarate are all bound to about the same extent and only about five times less tightly than fluorofumarate and about 20 times less tightly than fumarate. On the other hand, the vast differences in the rate of hydration of these substrates is reflected by the marked differences in V_{max}. The rate differences may reflect the fact that a halogen atom on the carbon atom adjacent to the carbon atom, which carries the positive charge of the proposed carbonium ion intermediate, would tend to destabilize the carbonium ion and thereby reduce the rate. It has been suggested that this type of destabilization of an intermediate carbonium ion can explain why difluorofumarate is hydrated at a slower rate than fluorofumarate (*61*). Finally, to explain the fact that fluorofumarate is hydrated about three times faster than fumarate, it can be assumed that the carbonium ion intermediate could be stabilized by the fluorine atom which is on the charged carbon atom in the carbonium ion, thus facilitating the rate of hydration (*61*).

3. *Number of Catalytic Sites*

The reversible binding of fumarase with its substrates, fumarate and L-malate, and with two competitive inhibitors, citrate and *trans*-aconitate, has been examined by equilibrium dialysis (*62*). These studies reveal that there are four substrate or inhibitor binding sites per molecule of enzyme or an average of one site per polypeptide chain subunit. Spectrophotometric analysis of the binding of the enol tautomer of oxalacetate with fumarase leads to the same conclusion. The Michaelis constants and inhibition constants for these compounds as measured kinetically are about equal to the equilibrium constants determined by the binding studies (Table VI). The close correspondence of the kinetic and binding constants indicates that the affinity of fumarase for its substrates and inhibitors remains unchanged over a wide range of enzyme concentrations. Although it cannot be demonstrated unequivocally that the four binding sites are catalytic sites, the fact that both hydrated (citrate) and unhydrated (*trans*-aconitate) species bind the same number

TABLE VI
EQUILIBRIUM CONSTANTS AND KINETIC CONSTANTS FOR FUMARASE[a]

Compound	Conditions	$K_{eq}{}^{b} \times 10^{-5}\ M$	$K_I \times 10^{-5}\ M$	$K_m \times 10^{-5}\ M$
Fumarate[c]-L-malate	0.1 *M* Sodium acetate 0.01 *M* sodium phosphate, pH 7.5	50		62[d]
trans-Aconitate	0.1 *M* Sodium acetate 0.01 *M* sodium phosphate, pH 7.6	87	88	
Citrate	0.05 *M* Sodium acetate 0.01 *M* sodium phosphate, pH 7.6	210	350	
Oxalacetate	0.01 *M* tris-Acetate, pH 7.3	0.89	0.85	

[a] Data from Teipel and Hill (*62*).
[b] Measured from the slopes of Scatchard binding plots (*62*).
[c] Equilibrium mixture of 4.4 parts fumarate and 1 part L-malate.
[d] Calculated from the K_m values of fumarate and L-malate in 0.01 *M* sodium phosphate buffer at pH 7.6 as determined by Alberty *et al.* (*36*).

62. J. W. Teipel and R. L. Hill, *JBC* **243,** 5679 (1968).

of sites is in accord with the fact that catalytic sites must bind both types of substrates.

4. *Affinity Labeling of Active Site*

Some attempts have been made to identify structures in the active site of fumarase by affinity labeling. Bradshaw *et al.* (*63*) have examined the reaction of fumarase with several compounds whose structures are similar to either fumarate or L-malate and also contain a group capable of reacting with side chains of certain amino acids which may be in the active site. Eighteen substrate or inhibitor analogs were tested in these studies and of the seven compounds which inactivated fumarase, only two, iodoacetate and 4-bromocrotonate, inactivated the enzyme without extensively modifying the thiol groups. Among the compounds tested was D,L-bromosuccinate, which did not prove to inactivate fumarase significantly. This result is in contrast to that of Wagner *et al.* (*64*) who reported inactivation of fumarase by "α-bromosuccinate," perhaps with the formation of succinylhistidine. One of the most difficult problems associated with studies of this type is the high reactivity of most compounds used as affinity labels with thiol groups. Modification of the thiol groups inactivate the enzyme although available evidence suggests that the groups are not in the active site (see Section I,B,5). Thus, without careful consideration of the possibility of thiol group modification, attempts at affinity labeling of fumarase may be misleading.

Inactivation of fumarase to the extent of 80–90% occurs with both iodoacetate and 4-bromocrotonate with little or no alkylation of thiol groups. Under these conditions, three to six equivalents of each compound are introduced into the enzyme, as judged by incorporation with ^{14}C-labeled inhibitors. Inactivation by these compounds is markedly reduced by fumarate, malate, or competitive inhibitors. Furthermore, extensive alteration of the structure of fumarase such as occurs on modification of thiol groups was not found. On the basis of these observations, it is possible that structures in the active site of fumarase have been modified by these reagents. It was impossible to identify those amino acids modified by reaction with 4-bromocrotonate since the derivatives appeared to be unstable under the conditions of analysis; however, 3-carboxymethylhistidine and *S*-carboxymethylmethionine were identified in hydrolysates of iodoacetate-inactivated enzyme. The 3-carboxymethylhistidine was found in a single tryptic peptide of fumarase indicating that

63. R. A. Bradshaw, G. W. Robinson, G. M. Hass, and R. L. Hill, *JBC* **244,** 1755 (1969).
64. T. E. Wagner, H. G. Chai, and M. E. Charlson, *BBRC* **28,** 1019 (1967).

the alkylation is limited to only one of the 14 histidine residues in each subunit of enzyme. The absence of the 1- or 1,3-carboxymethyl derivatives of histidine also suggests a high degree of specificity for the reaction. In addition, 3-carboxymethylhistidine is formed on reaction with iodoacetate in the absence or presence of *trans*-aconitate, a competitive inhibitor of fumarase, although the rate of carboxymethylation is very much slower in the presence of inhibitor. The amount of 3-carboxymethylhistidine formed, however, never exceeds more than 0.5 residue per molecule or little more than 0.1 residue per subunit. Nevertheless, the total amount of histidine plus methionine carboxymethylated per molecule is 3.5–3.8 residues per molecule or about one residue per active site. Carboxymethylation of the methionine in fumarase also appears specific since only 2 of the 16 methionyl residues in each subunit were carboxymethylated as judged by analysis of the tryptic peptides from carboxymethylated fumarase. Thus, both histidine and methionine appear to be implicated in the active site of fumarase by these studies, but in view of the rather small amount of histidine modified and the lack of an obvious role for methionine in catalysis, the assignment of either of these residues to the active site as judged by affinity labeling must remain tentative.

Another excellent reagent which appears to be active site specific is bromomesaconate. Reaction of fumarase with bromomesaconate by Laursen *et al.* (*65*) resulted in rapid inactivation of the enzyme without thiol modification. Approximately 4.4 moles of the inhibitor were incorporated per 194,000 g of enzyme in good agreement with the number of active sites found per tetramer (see Section I,C,3). The labeled residues have not yet been identified.

D. Mechanism of Action

Most of the evidence concerning the mechanism of action of fumarase supports the view that a carbonium ion intermediate is formed during hydration of fumarate or the dehydration of L-malate. A mechanism of this type, which was first proposed by Alberty and co-workers (*66*), is presented schematically in Fig. 5. Several lines of evidence were first used to support this mechanism. The V_{max} and K_m for *erythro*-3-monodeutero-L-malate were found to be essentially identical to the corresponding values for L-malate (*55, 66*). This suggests that the breaking

65. R. A. Laursen, J. B. Bauman, K. B. Linsley, and W. C. Shen, *ABB* **130**, 688 (1968).

66. R. A. Alberty, W. G. Miller, and H. F. Fisher, *JACS* **79**, 3973 (1957).

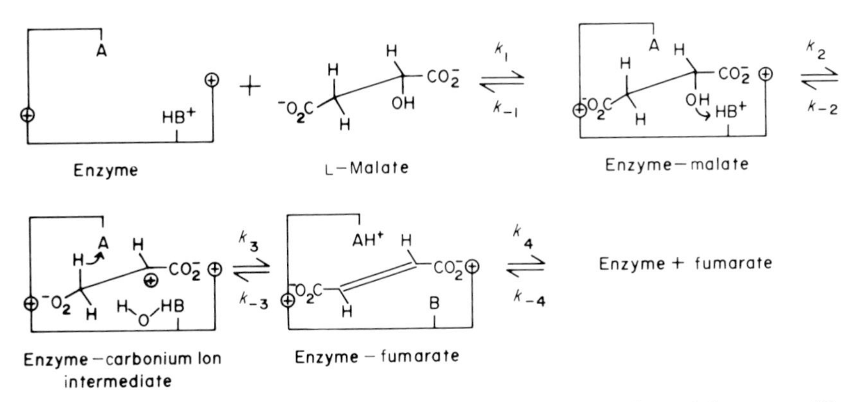

FIG. 5. Schematic representation of the mechanism of action of fumarase. The asymmetry of the active site is represented by four nonidentical groups indicated by ⊕, A and B. The ⊕ sites interact with the carboxylate groups of substrates, whereas the A and B sites correspond to the two acidic groups in the site.

of the hydrogen–methylene carbon bond was not rate limiting since, if it were, the rate of dehydration of *erythro*-3-monodeutero-L-malate should be decreased somewhat. In addition, the rate of exchange of the specific methylene hydrogen removed on dehydration does not exceed the rate of incorporation of hydrogen via addition to fumarate in the back reaction. These two observations suggested that the first step in the dehydration reaction was *not* the removal of hydrogen from malate to form a carbanion. It was thus proposed that a hydroxyl ion was first removed from malate (or a hydrogen ion added to fumarate in the hydration reaction) to form an enzyme-bound carbonium ion intermediate. Since no deuterium isotope effect was observed for the conversion of the carbonium ion intermediate (EX) to the enzyme–fumarate complex (EF) (rate k_3 in Fig. 5), and the dissociation rate of fumarate and association rate of malate were rapid (Table IV), the rate limiting step in the dehydration reaction was thought to be the conversion of the enzyme–malate complex (EM) to the carbonium ion complex (EX) (rate k_2 in Fig. 5).

Subsequent studies have provided further support that a carbonium ion is formed during catalysis. The secondary isotope effects observed for the dehydration of 2-^{3}H-1,4-^{14}C-L-malate and 2,3-^{3}H-1,4-^{14}C-L-malate are best explained by formation of a carbonium ion-like complex (*67*). Also, the action of fumarase on the different halofumarate substrates is as expected for a carbonium ion intermediate (*51, 61*). The observa-

67. D. E. Schmidt, Jr., W. C. Nigh, C. Tanzer, and J. H. Richards, *JACS* **91**, 5849 (1969).

TABLE VII
RELATIVE RATES OF EXCHANGE FOR ISOTOPICALLY LABELED SUBSTRATES OF FUMARASE[a]

Reaction	Relative rate
1. ^{18}O-Malate ⇌ ^{18}O-water (hydroxyl group of malate with solvent)	4.0
2. ^{14}C-Malate ⇌ ^{14}C-fumarate (carbon skeletons of both substrates)	2.5
3. ^{3}H-Malate ⇌ ^{3}H-water (exchangeable hydrogen on malate with solvent)	1.0
4. ^{2}H-Malate ⇌ ^{2}H-water (exchangeable hydrogen on malate with solvent)	1.0

[a] Data from Hansen *et al.* (*69*).

tion (*68*) that the rate of hydration of fumarate in H_2O is about twice that in D_2O appears inconsistent with this mechanism, but nonspecific effects of D_2O on the fumarase molecule may explain this apparent contradiction.

Recently, Hansen *et al.* (*69*) have measured the rates of exchange of the carbon skeleton of malate with fumarate, and the rates of exchange of the methylene hydrogen and hydroxyl groups with water using isotopically labeled substrates. The relative rates of exchange for these reactions are listed in Table VII. The rate of exchange of the hydroxyl group of malate is seen to be almost twice the rate of the fumarate–malate exchange and four times the rate of the exchange of the methylene hydrogen on malate with solvent. The rapid rate of exchange of the hydroxyl group relative to the methylene hydrogen provides further evidence for the existence of a carbonium ion intermediate, but it also indicates that the rate of conversion of the enzyme–malate complex to the carbonium ion complex is *not* rate limiting, as originally proposed by Alberty (*66*). To explain this inconsistency, Schmidt *et al.* (*67*) have suggested that the deuterium isotope effect is not observed for the conversion of the carbonium ion to fumarate because the transition state for this reaction is very similar to the carbonium ion itself; hence, this conversion may be slow and still not display a kinetic isotope effect. The finding by Hansen *et al.* (*69*) that the exchange of malate with fumarate is also faster than the methylene hydrogen of malate with water requires inclusion in the reaction scheme of an enzyme–proton complex without bound fumarate. The minimal reaction scheme for the fumarase catalysis consistent with these observations is shown in Fig. 6. In this scheme, the intermediate EX can probably be taken as the

68. P. A. Sere, G. W. Kosidic, and R. Lumry, *BBA* **50,** 184 (1963).
69. J. N. Hansen, E. C. Dinovo, and P. D. Boyer, *JBC* **244,** 6270 (1969).

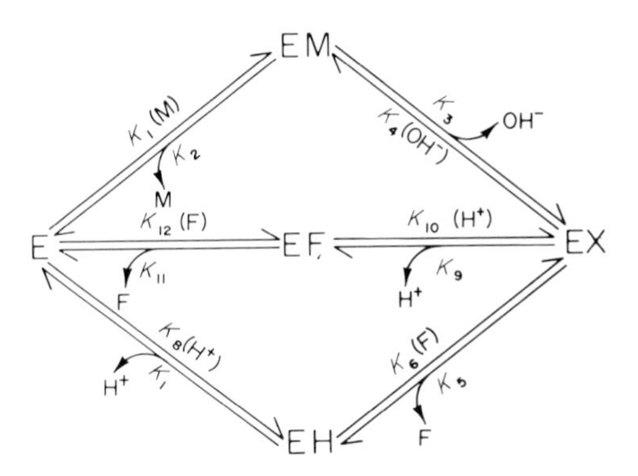

FIG. 6. The reaction pattern for fumarase based upon the exchange rates of isotopically labeled substrates (*69*).

carbonium ion complex discussed above. Additional discussion of isotope effects and isotopic exchanges in the fumarase reaction is given in Chapter 5 by Rose and Chapter 6 by Richards in Volume II of this treatise.

It has been calculated that the relative rates of formation of EX (Fig. 6) via the three pathways EM, EF, and EH are nearly the same; thus, no single step can be regarded as rate limiting in the catalysis (*69*). It is interesting that the retention of a proton on the enzyme which was abstrated from substrate has also been noted for aconitase (*70*), and it is possible that the proton is bound in the catalytic site to an imidazolium group which has been implicated from kinetic and chemical modification studies (see below).

The mechanism depicted in Fig. 5 indicates that two groups in the active site, shown as A and B, act in a concerted manner to effect formation of the carbonium ion intermediate and thus permit the interconversion of malate to fumarate. There is some experimental evidence that two groups, which are analogous to A and B and have the properties of weak acids, are essential for catalysis by fumarase. The kinetics of fumarase as a function of pH (see Section I,C,1) clearly indicate that two groups in the active site with pK values of about 6.3 and 6.8 are essential for catalysis. From these pK values it could be assumed that these groups are imidazolium side chains of histidyl residues, but study of the temperature dependence suggests that only one group may be an imidazole and the other a carboxyl group. Additional support for the possible involvement of a histidyl residue is provided by the fact that

70. I. A. Rose and E. L. O'Connell, *JBC* **242**, 1870 (1967).

a single histidyl residue is alkylated during inactivation of fumarase by iodoacetate (see Section I,C,4). Considerably more methionine than histidine is alkylated during reaction with iodoacetate, but it is unclear how the side chain of a methionyl residue could possess the properties of either group A or B. This does not exclude, however, the possibility that specific methionyl residues cannot be acting in another manner in the site. Finally, the studies of Hansen *et al.* (*69*) suggest that the enzyme extracts a proton from the substrate, and such an enzyme-bound proton could conceivably correspond to an imidazolium proton.

Although the reaction mechanism outlined in Fig. 5 gives a plausible indication of the nature of fumarase action and suggests certain structural features of the catalytic site, it is still impossible to correlate a great deal of our knowledge of fumarase structure with its function. It is likely that the functional, active enzyme is a molecule with a molecular weight of 194,000 and that there are four active sites per molecule, presumably one site per subunit. Nevertheless, it is unclear why fumarase must have four subunits and four active sites, although certainly a single subunit is unstable and when formed, has no catalytic activity. Of considerable interest, however, is the activation of fumarase produced by substrates. Substrates have been suggested to bind at noncatalytic sites and when bound at these sites alter the catalytic properties of the site. Phosphate ions also activate fumarase at low concentrations, perhaps by a mechanism not dissimilar from that for substrate activation. Such indirect effects of one site on the other may be interpreted in terms of subunit interactions although there is no evidence that subunit interactions alter the catalytic activity of the enzyme. Attempts to evaluate such possibilities may be of interest in the future, especially in view of the recent suggestion that ATP and magnesium levels *in vivo* may operate to regulate fumarase activity (*52*).

II. Crotonase

Little new information on crotonase has appeared since it was last reviewed in this series by Stern (*71*), and a full chapter devoted to it at this time seems unwarranted. Nevertheless, because it is similar to fumarase in that it requires neither pyridoxal phosphate nor a metal cofactor, a brief description of the few recent studies on crotonase structure and function will be given here.

71. J. R. Stern, "The Enzymes," 2nd ed., Vol. 5, p. 511, 1961.

Crotonase, or enoyl-CoA hydratase (L-3-hydroxyacyl-CoA hydrolyase; EC 4.2.1.19) from bovine liver catalyzes the reversible hydration of *cis*- and *trans*-$\Delta^{2,3}$-enoyl-CoA to the corresponding L(+)- or D(—)-β-hydroxyacyl-CoA derivatives as follows:

$$trans\text{-}\Delta^{2,3}\text{-enoyl-CoA} + H_2O \rightleftharpoons \text{L}(+)\text{-}\beta\text{-hydroxyacyl-CoA}$$

$$cis\text{-}\Delta^{2,3}\text{-enoyl-CoA} + H_2O \rightleftharpoons \text{D}(-)\text{-}\beta\text{-hydroxyacyl-CoA}$$

In higher organisms crotonase appears to be the only hydratase in fatty acid oxidation, in contrast to other enzymes acting during β-oxidation which show different substrate specificities with respect to the chain length of the acyl moiety. For example, there are four different fatty acyl-CoA dehydrogenases, each specific for a different range of chain lengths of fatty acyl-CoA (*72*). However, comparison of the V_{max} and K_m values for enoyl-CoA derivatives with chain lengths greater than C_6 has only recently been reported for the crystalline beef liver enzyme and is listed in Table VIII (*73*). Clearly, V_{max} decreases progressively with increasing chain length, whereas the K_m values increase with increasing chain length.

TABLE VIII
SUBSTRATE SPECIFICITY OF BOVINE LIVER CROTONASE

Substrate	V_m (mole/min/mole)	K_m (mole/liter)
Crotonyl-CoA (C_4)	340,000	2×10^{-5}
Hexenoyl-CoA (C_6)	151,000	2.4×10^{-4}
Octenoyl-CoA (C_8)	83,500	2.8×10^{-4}
Decenoyl-CoA (C_{10})	38,000	3×10^{-4}
Dodecenoyl-CoA (C_{12})	16,000	4×10^{-4}
Tetradecenoyl-CoA (C_{14})	5,000	4.2×10^{-4}
Hexadecenoyl-CoA (C_{16})	2,300	5×10^{-4}

Beef liver crotonase was reported earlier to possess a transferase activity termed "thiol-transcrotonylation" (*71*). Evidence in support of this activity was based on the observation that the rate of hydration in a mixture containing crotonase, crotonylpantetheine, and CoA was about 5–10 times greater than that observed in the absence of CoA. Crotonylpantetheine is a poor substrate for crotonase and is hydrated at a rate about 0.015% that of crotonyl-CoA. The increased rate was explained by assuming that the crotonyl group of crotonylpantetheine was transferred to CoA in order to form crotonyl-CoA, which was sub-

72. H. Beinert, "Methods in Enzymology," Vol. 5, p. 546, 1962.
73. R. M. Waterson and R. L. Hill, *Federation Proc.* **30,** 1114 (1971).

sequently hydrated to β-hydroxbutyryl-CoA. The reaction interpreted by others (*74*) as being of possible physiological significance, was proposed to proceed as follows:

$$\text{Crotonyl-pantetheine} + \text{CoA} \rightleftharpoons \text{crotonyl-CoA} + \text{pantetheine}$$
$$\text{Crotonyl-CoA} + H_2O \rightleftharpoons \beta\text{-hydroxybutyryl-CoA}$$

Recent studies have confirmed the observation that CoA stimulates the apparent rate of hydration when added to a mixture of crotonase and crotonylpantetheine; however, isolation of the products formed revealed neither crotonyl-CoA nor β-hydroxybutyryl-CoA but only free CoA, crotonylpantetheine, and β-hydroxybutyrylpantetheine (*73*). Thus, only crotonylpantetheine was hydrated and no evidence for transfer of the crotonyl group to CoA was obtained. Furthermore, acetyl-CoA was equally as effective as CoA in stimulating the hydration. Although these studies show that crotonase cannot catalyze transcrotonylation, the exact mechanism for stimulation of the hydration of crotonylpantetheine by CoA or acetyl-CoA is unknown.

The subunit structure of crystalline bovine crotonase has also been established (*75*). The molecular weight of the native enzyme is 164,000 as judged by sedimentation-equilibrium studies in the ultracentrifuge. In 6 *M* guanidine hydrochloride the molecular weight is 28,000, or about one-sixth that of the native enzyme. Mercaptoethanol is without effect on the molecular weight of the native or denatured enzyme, indicating that the six subunits are not combined through disulfide bonds. In addition, the enzyme is devoid of intrachain as well as interchain disulfide bonds, since its half-cystine content is equal to the total number of free thiol groups titratable with 5,5′-dithiobis(2-nitrobenzoate). The six subunits in crotonase appear to be identical as judged by tryptic peptide mapping, since about one-sixth of the expected number of tryptic peptides are observed based upon the lysine and arginine content of the native enzyme. In addition, only serine and aspartic acid were detected as the NH_2- and COOH-terminal end groups, respectively.

Acetoacetyl-CoA is the only CoA derivative which has been found to be an effective competitive inhibitor of crotonase at low concentrations (*73*). By kinetic analysis, the K_I for acetoacetyl-CoA is 1.6×10^{-5} *M*, or about equal in magnitude to the K_m for crotonyl-CoA. Spectrophotometric titration of the binding of acetoacetyl CoA suggests that

74. G. D. Greville and P. K. Tubbs, *Essays Biochem.* **4,** 155 (1968).
75. G. M. Hass and R. L. Hill, *JBC* **244,** 6080 (1969).

the enolate form binds specifically to crotonase with a binding constant equal to the K_I as determined by kinetic analysis. In addition, 6 moles of the inhibitor are bound per mole of native enzyme, or an average of one per subunit. This result suggests that each subunit contains a single active site.

20

6-Phosphogluconic and Related Dehydrases

W. A. WOOD

I. Introduction

A new type of hexose monophosphate pathway was discovered in *Pseudomonas saccharophila* by Entner and Doudoroff (*1*) which involves dehydration of 6-phosphogluconate and aldolytic cleavage of 2-keto-3-deoxy-6-phosphogluconate to pyruvate and D-glyceraldehyde 3-phosphate. This was originally considered to be a minor, if unique, variant of the hexose monophosphate system limited to a few organisms, mostly pseudomonads. In the nearly 20 intervening years, a much wider dis-

1. N. Entner and M. Doudoroff, *JBC* **196**, 853 (1952).

tribution and importance of this pathway in microorganisms has been established (*2*).

A large number of dehydrases for aldonic acids or their phosphate esters, as well as relevant kinases and aldolases for keto-deoxyaldonic and keto-deoxyaldaric acids and some of their phosphate esters have been found. At least three metabolic sources of dehydrase substrates have been reported, i.e., the aldonic acids largely derived by oxidation from monosaccharides, the uronic acids, and alginic and pectic acids via cleavage by α,β-elimination.

There are several variations of the Entner–Doudoroff pathway to the cleavage products: (1) phosphorylation of the aldonic acid, dehydration, and cleavage; (2) dehydration of the aldonic acid to the keto-deoxy acid followed by phosphorylation and cleavage; (3) dehydration and cleavage without phosphorylation; and (4) two successive dehydrations, in one case accompanied by decarboxylation, to yield α-ketoglutarate.

In two cases, alternate routes for the same keto-deoxy intermediate are found in different organisms. For instance, 2-keto-3-deoxy-L-arabonate may be cleaved to pyruvate and glyoxylate or it may be further dehydrated to α-ketoglutaric semialdehyde. Similarly, 5-keto-4-deoxy-D-glucarate is cleaved to pyruvate plus tartronic semialdehyde or it may undergo simultaneous dehydration and decarboxylation to α-ketoglutaric semialdehyde.

Dehydrases (*2a*) have been reported for hexonic acids (gluconic, 6-phosphogluconic, 6-phosphogalactonic, mannonic, galactonic, and altronic acids) and glucosaminic acid. The dehydrases for pentonic and methyl pentonic acids include those for D-fuconate, L-arabonate, D-xylonate, and, presumably, D-arabonate.

From a mechanistic point of view, the dehydrations are facilitated by carbanion-forming mechanisms which resemble those of aldolytic cleavage reactions. Especially interesting, therefore, is the fact that metal ion- and Schiff base-assisted catalysis has been reported for both aldolases and dehydrases. It may be that the fundamental difference between the aldolases and the dehydrases for keto-deoxy compounds

2. K. Kersters and J. DeLey, *Antonie von Leeuwenhoek; J. Microbiol. Serol.* **34**, 393 (1968).

2a. Although the term *dehydratase* has been more recently adopted to refer to enzymes carrying out dehydrations, the literature concerning many of the enzymes described in this chapter refer to dehydrases. Thus, dehydrase and dehydratase have equivalent meanings. An attempt will be made in this chapter to maintain the nomenclature used by the authors when citing their work.

lies in the type of attack on the hydroxyl group; that is, in the second phase of the mechanism. Whereas in aldolytic cleavage, nucleophilic attack on the hydroxyl hydrogen leads to carbon–carbon cleavage, in the dehydrases, an electrophilic attack would result in elimination of the hydroxyl group to complete the dehydration. At any rate, either a metal ion or a Schiff base structure plays an important role as an electron sink for both enzymes.

II. Metal Ion-Assisted Dehydrations

A. Dehydrases for Gluconic Acid

In many species of bacteria, gluconate is first phosphorylated and then dehydrated to 2-keto-3-deoxy-6-phosphogluconate. Notable exceptions involve *Clostridium aceticum* (*3*), a mutant strain of *Rhodopseudomonas spheroides* (*4*), and a subgroup of the *Achromobacter-Alkaligenes* group (*5*) in which dehydration to 2-keto-3-deoxygluconate precedes phosphorylation. These routes are utilized in the dissimilation of glucose, gluconate, 2-ketogluconate, fructose, and mannose.

1. *6-Phosphogluconate Dehydrase*

$$\begin{array}{c} COO^- \\ | \\ H-C-OH \\ | \\ HO-C-H \\ | \\ H-C-OH \\ | \\ H-C-OH \\ | \\ CH_2OPO_3^{2-} \end{array} \rightleftharpoons \left[\begin{array}{c} COO^- \\ | \\ C-OH \\ \| \\ H-C \\ | \\ H-C-OH \\ | \\ H-C-OH \\ | \\ CH_2OPO_3^{2-} \end{array} \right] \longrightarrow \begin{array}{c} COO^- \\ | \\ C=O \\ | \\ H-C-H \\ | \\ H-C-OH \\ | \\ H-C-OH \\ | \\ CH_2OPO_3^{2-} \end{array} + H_2O \qquad (1)$$

6-Phosphogluconate   2-Keto-3-deoxy-6-phosphogluconate

Entner and Doudoroff (*1*) showed that two steps were involved in the conversion of 6-phosphogluconate to pyruvate and glyceraldehyde 3-phosphate by *Pseudomonas saccharophila;* the product of the de-

3. J. R. Andressen and G. Gottschalk, *Arch. Mikrobiol.* **69**, 160 (1969).
4. M. Szymona and M. Doudoroff, *J. Gen. Microbiol.* **22**, 167 (1960).
5. K. Kersters, J. Khan-Matsubara, L. Nelen, and J. DeLey, *Antonie van Leeuwenhoek; J. Microbiol. Serol.* **37**, 233 (1971).

hydration reaction was isolated and characterized by MacGee and Doudoroff (*6*) as 2-keto-3-deoxy-6-phosphogluconate.

a. General Properties. Kovachevich and Wood (*7*) purified 6-PG (*7a*) dehydrase from *Pseudomonus putida* (*fluorescens*) about 27-fold and showed that the disappearance of 6-PG was faster than the rate of KDPG formation, presumably because of a slower rate of ketonization of the enol dehydration product. The reaction could not be reversed by incubating KDPG with the dehydrase in the presence of 6-PG dehydrogenase and NADP to detect and oxidize any 6-PG formed.

Purified 6-PG dehydrase is completely inactive in the absence of both a divalent metal ion and a thiol reducing agent. Fe^{2+}, Mn^{2+}, and Mg^{2+} ions reactivated, whereas Ca^{2+}, Zn^{2+}, Co^{2+}, and Fe^{3+} were unable to do so. The greatest activation was given by Fe^{2+} at 4×10^{-3} *M*, and Mn^{2+} and Mg^{2+} were 70 and 45% as active, respectively, at their optimum concentrations.

In the presence of Fe^{2+}, activation by thiols was greatest with GSH followed by cysteine and thioglycolate. Reactivation of the purified dehydrase requires a relatively critical program of additions. The dehydrase must be incubated with GSH 2–3 min before addition of Fe^{2+}. Increased incubation time with GSH in the absence of Fe^{2+} leads to inactivation. Also, the addition of Fe^{2+} before GSH results in only slight activation.

The K_m for 6-PG was 6×10^{-4} *M*. No evidence of sigmoidicity was found in the rate vs. concentration of ligand plots. The lag observed with Fe^{2+} resulted from precipitation of Fe^{2+} by phosphate ion in the assay. The curve of pH vs. velocity exhibited a broad pH maximum between pH 6.7 and 8.0. Above pH 8.0, dehydration velocity decreased rapidly.

With 1-^{14}C-6-PG as substrate, the isolated dehydration product was shown to have all of the radioactivity in the carboxyl group, releasable as CO_2 with ceric sulfate. The isolated product also served as a substrate for KDPG aldolase.

6-Phosphogluconate dehydrase is much more labile than KDPG aldolase, but it can be freed of KDPG aldolase by repeated acetone fractionation. Since the dehydration reaction is irreversible, large quantities of KDPG can be prepared by incubating 6-PG with the dehydrase (*7*).

6. J. MacGee and M. Doudoroff, *JBC* **210,** 617 (1954).

7. R. Kovachevich and W. A. Wood, *JBC* **213,** 745 (1955).

7a. Abbreviations used: 6-PG, 6-phosphogluconate; KDG, 2-keto-3-deoxygluconate; and KDPG, 2-keto-3-deoxy-6-phosphogluconate.

Hulcher (*8*) has reported that the 6-PG dehydrase of crude extracts of normally grown, or iron-deficient, *Pseudomonas mildenbergii* is stimulated by pyoverdine in the absence or presence of Fe^{2+} in the assay. Pyoverdine is one of a group of pigments secreted during growth in the absence of iron (*9*). It is a green, fluorescent peptide consisting, in part, of 4 threonine, 2 serine, 1 glutamic acid, 1 lysine, and an unidentified hydroxamic acid (*8*).

b. Mechanism. Meloche and Wood (*10*) visualized two possible mechanisms for formation of KDPG. One involved dehydration and ketonization of the enol as described by the above equation. The second alternative, attractive because of its participation in diol dehydration, involved hydride ion migration with no participation of protons from the solvent.

In tests to distinguish between these alternatives (*10*), it was shown that during the dehydration of 6-PG in T_2O by a 20-fold purified dehydrase, 1 proton from water was stably incorporated into KDPG. Further, there was a rapid back incorporation of tritium into 6-PG during the dehydration reaction. Finally, there was no exchange of tritium into KDPG indicating that the dehydrase could not catalyze the enolization of KDPG. The isolated ^3H-KDPG was cleaved by KDPG aldolase to yield pyruvate of the same specific activity; hence, tritium incorporation was at the C-3 position. It was also shown that the tritium in 2-^3H-6-PG was lost in the formation of KDPG. Thus, the mechanism involving enol formation and tautomeric rearrangement was clearly established, and the hydride ion migration mechanism was eliminated.

Although the back incorporation of tritium into 6-PG indicated that ketonization of the enol was rate limiting, it was necessary to establish that this process was enzyme-catalyzed or spontaneous. Accordingly, the dehydration was carried out in 100% D_2O, the monodeuterated KDPG isolated, and the steric disposition of the deuterium established by NMR analysis. If the reaction were spontaneous, the deuterium would be incorporated randomly at C-3. However, if the deuterium were incorporated stereospecifically, this would indicate an enzyme-catalyzed ketonization process, as has subsequently been established for 2-keto-3-deoxy-L-arabonate dehydratase (see Section III,A). This reasoning carries the assumption that enzymic tautomerization must lead to asymmetric incorporation. Interpretation of the NMR data for KDPG,

8. F. H. Hulcher, *BBRC* **31,** 247 (1968).
9. J. D. Newkirk and F. H. Hulcher, *ABB* **134,** 395 (1969).
10. H. P. Meloche and W. A. Wood, *JBC* **239,** 3505 (1964).

3-monodeutero-KDPG prepared enzymically, and 3,3-dideutero-KDPG prepared chemically all dissolved in D_2O by Dr. G. J. Karabatsos [see Meloche and Wood (*10*)] led to the conclusion that the deuterium was randomly incorporated at C-3.

2. *Gluconate Dehydratase*

$$\text{Gluconate} \rightarrow \text{2-keto-3-deoxygluconate} + H_2O \qquad (2)$$

Gluconate dehydratase has been reported in *Clostridium aceticum* (*3*) and has been purified to homogeneity by Kersters *et al.* (*5*) from a subgroup of the *Achromobacter-Alkaligenes* group defined by a 64–70% (G + C) content in the DNA. The dehydratase was induced by growth on gluconate and, to a smaller extent, by growth on glucose, but not on several other energy sources. The purification achieved was 99-fold with an 11% yield, and the preparation was homogeneous on disc gel electrophoresis. The pH optimum was 8.4 to 8.8, and the K_m for gluconate was 2×10^{-2} *M*. The molecular weight established by gel filtration was 270,000 ± 25,000.

Of 29 carbohydrates tested, the following aldonic acids were active: D-gluconate (100%), D-xylonate (88%), D-fuconate (13%), D-galactonate (13%), and L-arabonate (2%). Thus, the dehydratase acts on C-5 and C-6 aldonic acids with the C-2 and C-3 hydroxyl groups in the L-*threo* configuration. The C-2 carboxyl and the terminal primary hydroxyl or methyl group are also required.

D-Gluconate dehydratase of *Achromobacter* differs from similar dehydratases in the following: (1) a higher pH optimum and K_m for gluconate, (2) group specificity rather than single substrate or more restricted specificity, and (3) the lack of an absolute metal and thiol requirement.

B. Galactonate Dehydrase

$$\text{Galactonate} \rightarrow \text{2-keto-3-deoxygalactonate} + H_2O \qquad (3)$$

The conversion of galactose to galactonate, 2-keto-3-deoxygalactonate, and 2-keto-3-deoxy-6-phosphogalactonate has been demonstrated in *Pseudomonas saccharophila* (*11–13*) and *Gluconobacter liquefaciens*

11. J. DeLey and M. Doudoroff, *JBC* **227,** 745 (1957).
12. J. F. Wilkinson and M. Doudoroff, *Science* **144,** 569 (1964).
13. C. W. Shuster and M. Doudoroff, *Arch. Mikrobiol.* **59,** 279 (1967).

(*14*). Although KDPGal aldolase has been highly purified, the dehydrase activity has only been demonstrated in extracts.

C. D-Mannonate and D-Altronate Dehydrases

$$\begin{matrix}\text{D-Mannonate}\\ \text{D-Altronate}\end{matrix} \rightarrow \text{2-keto-3-deoxygluconate} + H_2O \qquad (4)$$

In *Escherichia coli*, the pathway of glucuronate utilization has been reported by Ashwell *et al.* (*15*) to involve isomerization to fructuronate, reduction to mannonate, dehydration to KDG, phosphorylation to KDPG, and cleavage to pyruvate and glyceraldehyde 3-phosphate. A similar pathway for galacturonate involves tagaturonate and D-altronate as intermediates (*15*).

The dehydrases are induced by the respective uronic acids but not by glucose, gluconate, or the alternate uronic acid. These dehydrases have also been found in *Aerobacter aerogenes* and *Serratia marcescens*.

D-Mannonate dehydrase has been purified 33-fold from D-glucuronate-grown *E. coli*. It is highly specific for D-mannonate; the pH optimum was 5.0–6.0. Thiol compounds did not activate, but Co^{2+} and Mn^{2+} did cause a 30–40% increase in activity. However, there was little loss in activity by dialysis against metal-chelating agents (*16*).

D-Altronate dehydrase was purified 15-fold from D-galacturonate-grown cells. It was more labile than D-mannonate dehydrase, but it could be stabilized partially by treatment with thiols. The pH optimum was more alkaline than that of D-mannonic dehydrase at 7.0–8.2. The dehydrase was highly specific for D-altronate and had an absolute requirement for ferrous ions; cysteine gave a 2- to 3-fold increase in activity. Attempts to reverse the reaction were unsuccessful (*16*).

D. Hexaric Acid Dehydrases

Glucaric and galactaric acids serve as energy sources for members of the Enterobacteriaceae and Pseudomonadaceae, and glucaric and galactaric dehydrases catalyze the first reaction in dissimilation of these hexaric acids in all bacteria (*17*, *18*). The further utilization of the corresponding keto-deoxyhexarates in the Enterobacteriaceae involves al-

14. A. H. Stouthamer, *BBA* **48,** 484 (1961).
15. G. Ashwell, A. J. Wahba, and J. Hickman, *JBC* **235,** 1559 (1960).
16. J. D. Smiley and G. Ashwell, *JBC* **235,** 1571 (1960).
17. H. J. Blumenthal and D. C. Fish, *BBRC* **11,** 239 (1963).
18. P. W. Trudgill and R. Widdus, *Nature* **211,** 1097 (1966).

dolytic cleavage to pyruvate and tartronic semialdehyde (*17*) or in the Pseudomonadaceae (*18–20*) involves a second dehydration and decarboxylation to α-ketoglutaric semialdehyde, followed by oxidation to yield α-ketoglutarate. Both dehydrases are induced by growth on either glucarate or galactarate (*17, 18*).

1. D-*Glucarate Dehydrase*

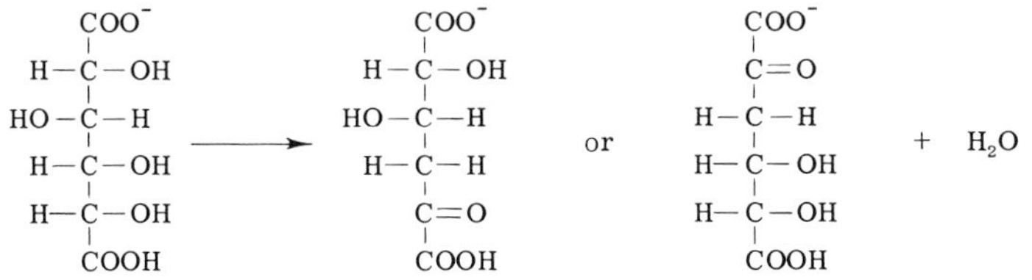

$$
\begin{array}{c}
COO^- \\ | \\ H-C-OH \\ | \\ HO-C-H \\ | \\ H-C-OH \\ | \\ H-C-OH \\ | \\ COOH
\end{array}
\longrightarrow
\begin{array}{c}
COO^- \\ | \\ H-C-OH \\ | \\ HO-C-H \\ | \\ H-C-H \\ | \\ C=O \\ | \\ COOH
\end{array}
\quad \text{or} \quad
\begin{array}{c}
COO^- \\ | \\ C=O \\ | \\ H-C-H \\ | \\ H-C-OH \\ | \\ H-C-OH \\ | \\ COOH
\end{array}
+ H_2O \qquad (5)
$$

D-Glucarate 5-Keto-4-deoxy-D-glucarate 2-Keto-3-deoxy-D-glucarate

D-Glucarate dehydrase has been purified 32-fold from glucarate-grown *E. coli* by Blumenthal (*21*), and similar dehydratases have been obtained from *Klebsiella aerogenes* and *Pseudomonas acidovorans* by Jeffcoat *et al.* (*20*). The *E. coli* enzyme has a pH optimum between 7.5 and 8.5. The dehydrase acts only on D-glucarate and D-idarate at 50% the glucarate rate. A divalent metal ion is required with Mg^{2+} serving best; Mn^{2+} and Co^{2+} are partially active. The equilibrium point is far in the direction of dehydration and reversibility could not be demonstrated. The K_m for D-glucarate was 8×10^{-4} *M*. Both 5-keto-4-deoxy and 2-keto-3-deoxyglucarate are produced in the ratio of 85:15. With L-idarate only one 5-keto-4-deoxyhexarate is formed since dehydration at either end of the molecule yields the same product (*21*).

2. D-*Galactarate Dehydrase*

$$\text{Galactarate} \rightarrow \text{5-keto-4-deoxygalactarate} + H_2O \qquad (6)$$

Galactaric dehydrase was purified 60-fold from galactarate-grown *E. coli* and completely separated from glucarate dehydrase by Blumenthal and Jepson (*22*). The activity in crude extracts is stable only for a few minutes. With both crude extracts and the purified dehydrase, D-galactarate gives partial stabilization.

19. S. Dagley and P. W. Trudgill, *BJ* **95,** 48 (1965).
20. R. Jeffcoat, H. Hassall, and S. Dagley, *BJ* **115,** 969 (1969).
21. H. J. Blumenthal, "Methods in Enzymology," Vol. 9, p. 660, 1966.
22. H. J. Blumenthal and T. Jepson, "Methods in Enzymology," Vol. 9, p. 665, 1966.

No requirement for or stimulation by metal ions has been demonstrated, and EDTA was not inhibitory. Cyanide and pyrophosphate inhibited strongly. The pH optimum was 8.0. Specificity studies were hampered by lability in the absence of galactarate. However, it was established that D-glucarate, mannonate, and D- and L-idarate were not dehydrated. The K_m for galactarate was 4×10^{-4} M. 5-Keto-4-deoxygalactarate was the only product formed. The dehydrase is also found in a number of other bacteria (*22*).

E. L-ARABONATE-D-FUCONATE DEHYDRATASES

$$\text{L-Arabonate} \rightarrow \text{2-keto-3-deoxy-L-arabonate} + H_2O \quad (7)$$

$$\text{D-Fuconate} \rightarrow \text{2-keto-3-deoxy-D-fuconate} + H_2O \quad (8)$$

In contrast to the majority of microorganisms where L-arabinose is either dissimilated via L-ribulose-5-phosphate to D-xylulose-5-phosphate or via L-arabitol, L-xylulose, xylitol, D-xylulose to D-xylulose-5-phosphates, the pseudomonads follow a pathway involving L-arabonate and 2-keto-3-deoxy-L-arabonate (*23–25*). 2-Keto-3-deoxy-L-arabonate is further utilized either by aldolytic cleavage of pyruvate + glyoxylate (*26*, *27*) or by dehydration to α-ketoglutaric semialdehyde and oxidation to α-ketoglutarate (*19*, *25*). Similarly, D-fucose is dissimilated via fuconate, 2-keto-3-deoxyfuconate to pyruvate and D-lactaldehyde (*27*).

Although L-arabonate dehydrase activity has been demonstrated in *P. fragi* (*19*, *28*) and *P. saccharophila* (*23–25*), and the purified gluconate dehydratase of *Achromobacter* (see above) also dehydrates L-arabonate and D-fuconate (*5*), a true dehydratase for these substrates has only recently been purified by Dahms from an unidentified pseudomonad (*26*, *27*).

Of a large number of altronic and uronic acids, only D-fuconate and L-arabonate were dehydrated. The K_m values were 4 and 4.3 mM, and V_{max} values were 0.47 and 1.0, respectively. The same dehydratase was induced by growth on either D-fucose, L-arabinose, or D-galactose. Since D-galactonate is not dehydrated, D-galactose (or D-galactonate) functions as a gratuitous inducer.

D-Fuconate dehydratase requires Mg^{2+} for activity, but Mn^{2+}, Fe^{2+},

23. R. Weimberg and M. Doudoroff, *JBC* **217**, 607 (1955).
24. R. Weimberg, *JBC* **234**, 727 (1959).
25. A. C. Stoolmiller and R. H. Abeles, *BBRC* **19**, 438 (1965).
26. A. S. Dahms and R. L. Anderson, *BBRC* **36**, 809 (1969).
27. A. S. Dahms, Ph.D. Thesis, Michigan State University (1969).
28. R. Weimberg, *JBC* **236**, 629 (1961).

Ca^{2+}, Co^{2+}, and Ni^{2+} were partially effective. Some activation by sulfhydryl compounds was observed.

F. D-ARABONATE AND D-XYLONATE DEHYDRASES

$$\begin{matrix}\text{D-Arabonate} \\ \text{D-Xylonate}\end{matrix} \rightarrow \text{2-keto-3-deoxy-D-pentonate} + H_2O \qquad (9)$$

The utilization of D-arabinose in pseudomonads is reported to involve D-arabonate, 2-keto-3-deoxy-D-pentonate (*28a*), and pyruvate and glycolate (*28, 29, 30*) as intermediates. In *Pseudomonas fragi*, D-arabinose and D-xylose are converted to α-ketoglutarate via the pentonic and 2-keto-3-deoxypentonic acids (*19*). In neither instance have the dehydrases been studied.

The ability of gluconate dehydrase to dehydrate D-xylonate has been described above.

III. Schiff Base-Assisted Dehydrations

In a manner related to the pyridoxal phosphate–catalyzed α,β-eliminations, it has been shown in two instances that Schiff base formation, presumably between a lysine residue on the enzyme and the carbonyl group of keto-deoxy acids, leads to dehydration. In the case of 2-keto-3-deoxy-L-arabonate, the second dehydration is accompanied by an internal dismutation to form α-ketoglutaric semialdehyde. In the second case with 2-keto-3-deoxyglucarate, dehydration by a Schiff base mechanism is accompanied by decarboxylation also to yield α-ketoglutaric semialdehyde (Fig. 1). Thus, these reactions join a long list of Schiff base-assisted enzymic catalyses.

The dehydration of glucosaminic acid which is only related to the dehydration of sugar acids by the similarities of structures, proceeds by pyridoxal phosphate-mediated catalysis as with the hydroxy amino acids, serine and threonine.

28a. For purposes of clarity in some cases, a 2-keto-3-deoxypentonic acid is named for the structure from which it was derived, i.e., 2-keto-3-deoxy-L-arabonate. Obviously, the only determinant of nomenclature of relevance to the carbohydrate remaining is the pentultimate secondary hydroxyl group which determines whether the D or L series is involved. Hence, 2-keto-3-deoxy-D- or L-pentonic acid is the only term warranted by the structure for any of the pentose-derived keto-deoxy intermediates.

29. N. J. Palleroni and M. Doudoroff, *JBC* **223**, 499 (1956).
30. N. J. Palleroni and M. Doudoroff, *J. Bacteriol.* **74**, 180 (1957).

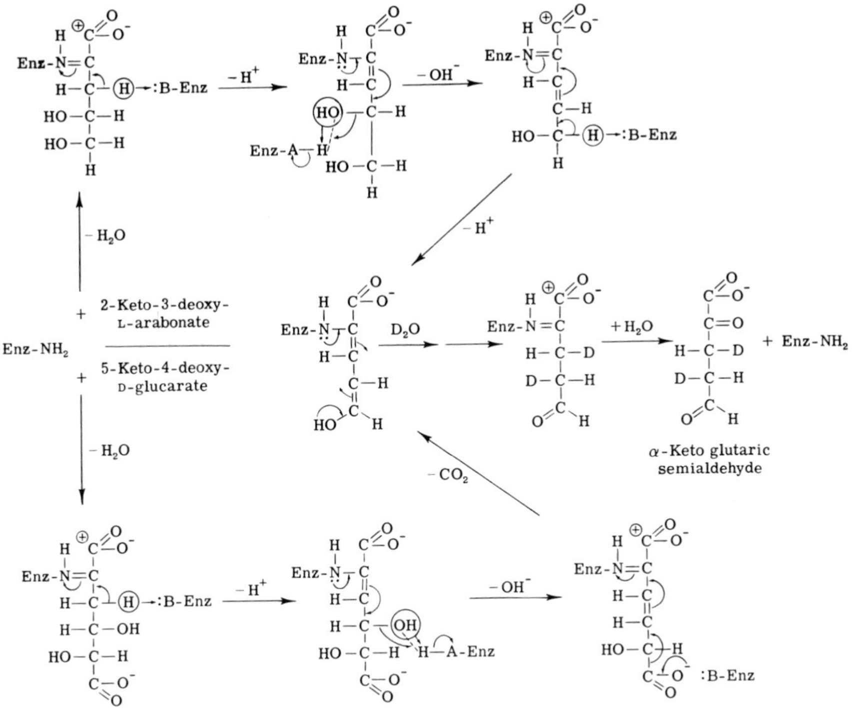

Fig. 1. Mechanism of ketodeoxy-L-arabonic and D-glucaric dehydrases.

A. 2-Keto-3-deoxy-L-arabonate Dehydratase

$$\text{2-Keto-3-deoxy-L-pentonate} \rightarrow \alpha\text{-ketoglutaric semialdehyde} + H_2O \qquad (10)$$

In the dissimilation of L-arabinose by pseudomonads via L-arabonate, 2-keto-3-deoxy-L-arabonate to α-ketoglutarate, NAD is required. This is in contrast to the pathway in an unidentified pseudomonad where pyruvate and glyoxylate are formed. In the absence of NAD, an intermediate derived from 2-keto-3-deoxy-L-arabonate accumulates which has been identified by Stoolmiller and Abeles (*25*) as α-ketoglutaric semialdehyde. Further oxidation in the NAD-requiring step yields α-ketoglutarate.

2-Keto-3-deoxy-L-arabonate dehydratase has been purified 415-fold and crystallized from extracts of L-arabinose-grown *Pseudomonas saccharophila* by Stoolmiller and Abeles (*31*). The preparation was homogeneous by disc gel electrophoresis. Based upon gel filtration, the molec-

31. A. C. Stoolmiller and R. H. Abeles, *JBC* **241,** 5764 (1966).

ular weight was estimated to be about 85,000 and a sedimentation velocity of 6.3 S was observed. Metal ions were not required, and several metal chelators and thiol reagents had no inhibitory effects. The K_m for 2-keto-3-deoxy-L-arabonate was $7 \times 10^{-5} M$. The pH optimum was 7.2.

Although the reaction resembles the dehydration of diols catalyzed by cobamide coenzyme, no B_{12} derivatives were found and diols did not serve as substrates.

Portsmouth *et al.* (*32*) have shown that treating the dehydratase with 2-keto-3-deoxy-L-arabonate and $NaBH_4$ causes inactivation and covalent binding of about 1 mole of substrate. In HTO, two tritium atoms are incorporated at the C-3 and C-4 positions per mole of α-ketoglutaric semialdehyde. Similarly, deuterated α-ketoglutaric semialdehyde with approximately two deuterium atoms was isolated and degraded to succinate. By optical rotatory dispersion measurements, it was shown that the deuterium at both C-3 and C-4 positions was incorporated asymmetrically.

It was shown that two analogs, α-ketobutyrate and α-keto-4-hydroxyvalerate inactivate in the presence of borohydride and, hence, form Schiff bases at the active site; these substrates also exchange tritium into position 3. Largely from these data, Portsmouth *et al.* (*32*) conceived the mechanism shown in Fig. 1. First, a Schiff base is formed. Eneamine formation, probably with base assistance, stereospecifically removes a C-3 hydrogen as a proton. Subsequent ketamine formation assisted by an acidic group on the enzyme shifts the double bond with removal of the OH group. At this step, the departure from an aldolytic cleavage mechanism occurs. If the group on the enzyme adjacent to C-4 were another base, cleavage between C-3 and C-4 would occur as it does in aldolases. The process continues with a second ketamine-eneamine tautomeric shift, presumably assisted by a base on the enzyme which labilizes a proton on C-5 to form the eneamine of α-ketoglutaric semialdehyde enol. Ketonization of the enol, regeneration of the ketamine form of Schiff base, and hydrolysis of the Schiff base thus yields α-ketoglutaric semialdehyde. The incorporation of tritium and deuterium stereospecifically at C-3 results from base-assisted ketamine generation from the eneamine form. The stereospecific incorporation of deuterium and, presumably, tritium at C-4 arises from the enzyme-catalyzed ketonization of the enol between C-4 and C-5. α-Ketobutyrate and 4-hydroxy-α-ketoglutarate incorporate tritium at C-3 because they are able to engage in Schiff base formation followed by the first ketamine-

32. D. Portsmouth, A. C. Stoolmiller, and R. H. Abeles, *JBC* **242**, 2751 (1967).

eneamine shift. However, without a hydroxyl group at C-4, the process can proceed no further (*32*).

A similar mechanism has been proposed for the concerted dehydration and decarboxylation which is catalyzed by a similar dehydratase for 2-keto-3-deoxyglucarate. The only difference is the loss of the carboxyl group rather than a proton from the corresponding carbon atom in the second eneamine tautomerization (see below).

Thus, both metal complexes and Schiff base adducts are capable of labilizing a hydrogen atom on the carbon α to the carbonyl group. The subsequent reaction path involving (a) hydroxyl elimination as in the simple dehydrases and (b) rearrangement of the C–C double bond followed by proton liberation and dismutation (keto-deoxy-L-arabonate dehydratase) or CO_2 elimination (keto-deoxyglucarate dehydratase) depend on the number and location of additional nucleophilic or electrophilic groups on the enzyme.

B. 5-Keto-4-deoxyglucarate Dehydratase

$$\text{5-Keto-4-deoxy-D-glucarate} \rightarrow \alpha\text{-ketoglutaric semialdehyde} + CO_2 + H_2O \quad (11)$$

A new kind of dehydratase which catalyzes simultaneous dehydration and decarboxylation of 5-keto-4-deoxy-D-glucarate has been discovered by Jeffcoat *et al.* (*33*) in glucarate-grown *Pseudomonas acidovorans*. The product is α-ketoglutaric semialdehyde which is the product derived from the dehydration of 2-keto-3-deoxy-L-arabonate (see above).

The dehydratase has been purified 28-fold to homogeneity by the criteria of disc gel electrophoresis and sedimentation analysis. It has a pH optimum of 7.0–7.3. Equimolar quantities of α-ketoglutaric semialdehyde and carbon dioxide were produced and 2-keto-3-deoxy-L-arabonate was shown not to be an intermediate. It was deduced that dehydration did not precede decarboxylation as a discrete step because the dehydrated intermediate 1,5-diketoadipate is symmetrical and the two carboxyl groups would not be distinguishable by the hypothetical decarboxylase. When glucarate-1-^{14}C was converted to α-ketoglutaric semialdehyde, the radioactivity appeared exclusively in the CO_2. Further, there was no divergence of dehydratase and decarboxylase activities during purification.

The purified dehydratase did not require metal ions for activity and was not inactivated by dialysis against EDTA. The dehydratase, like 2-keto-3-deoxy-L-arabonate dehydratase, was inactivated by treatment

33. R. Jeffcoat, H. Hassall, and S. Dagley, *BJ* **115,** 977 (1969).

with 5-keto-4-deoxy-D-glucarate plus $NaBH_4$. Covalent binding was demonstrated using ^{14}C-keto deoxyglucarate and tritiated dehydratase. No experiments have been performed on the mechanism. However, Jeffcoat *et al.* (*33*) proposed a sequence involving Schiff base formation and a series of tautomeric rearrangements similar to those established by Portsmouth *et al.* (*32*) and providing for the decarboxylation which is unique to this dehydratase (Fig. 1). 5-Keto-4-deoxy-D-glucarate dehydratase is the only example of dehydration and decarboxylation catalyzed by the same enzyme. In the case of 2-keto-3-deoxy-L-arabonate, a bifunctional dehydratase also exists in which dehydration is linked to enzyme-catalyzed ketonization of the enol. It might be that the dehydratase for 2-keto-3-deoxy-L-arabonate also catalyzes the simultaneous dehydration and decarboxylation of 5-keto-4-deoxy-D-glucarate because of the striking similarities of the mechanism.

C. Glucosaminic Dehydrase

$$\text{Glucosaminic acid} \rightarrow \text{2-keto-3-deoxygluconate} + NH_3 \qquad (12)$$

The dissimilation of glucosamine in some instances involves glucosaminic acid, 2-keto-3-deoxygluconate, and KDPG as intermediates (*34, 35*). Although superficially resembling the aldonic acid dehydrases, the mechanism involved is similar to that for the hydroxy amino acids (*35*). In those cases, only ammonia is produced; however, the portion of the overall α,β-elimination catalyzed by the enzyme is a dehydration. Merrick and Roseman (*36*) purified the dehydrase about 10-fold and showed it to be inactive on both isomers of serine and threonine. It had an absolute requirement for pyridoxal phosphate which was removed in fractionation. The dehydrase did not respond to allosteric effectors for the hydroxy amino acid dehydrases.

34. Y. Iminaga, *J. Biochem.* (*Tokyo*) **45,** 647 (1958).
35. W. A. Wood, *in* "Current Topics in Cellular Regulation" (B. L. Horecker and E. R. Stadtman, eds.), Vol. 1, p. 161. Academic Press, New York, 1969.
36. J. M. Merrick and S. Roseman, *JBC* **235,** 1274 (1960).

21

Carbonic Anhydrase

S. LINDSKOG • L. E. HENDERSON • K. K. KANNAN •
A. LILJAS • P. O. NYMAN • B. STRANDBERG

I. Introduction

The carbonic anhydrases are extremely efficient catalysts of the reversible hydration of carbon dioxide with maximum turnover numbers among the highest known for any enzyme. They have also been found to catalyze the hydrolysis of certain esters and related compounds and the hydration of aldehydes. The catalytic efficiency of the enzymes toward carbon dioxide is, however, several orders of magnitude greater than toward all other substrates so far investigated, and the physiological functions of the enzymes are ascribed to the carbon dioxide reaction. The carbonic anhydrases play an important role in respiration as well as in other physiological processes, where the rapid interconversion of carbon dioxide and bicarbonate ion is essential to the organism.

A. Historical Outline

In 1928, the important discovery was made (*1*) that the escape of carbon dioxide from hemolyzed blood is significantly faster than could be accounted for by the rate of the uncatalyzed conversion of bicarbonate to CO_2. The observation was confirmed in other laboratories [see Roughton (*2*, *3*) for historical details], and continued studies of the phenomenon led to the discovery of a nonhemoglobin erythrocyte catalyst for the reaction. The enzyme was first reported in 1932 by Meldrum and Roughton (*4*), who gave it the name *carbonic anhydrase*. Stadie and O'Brien (*5*), who appear to have worked independently of Meldrum and Roughton, confirmed the finding in 1933. A method for the purification of the enzyme and a detailed study of its properties, including the inhibition by low concentrations of cyanide, sulfide, and azide, was published in 1933 (*6*). In 1939, Keilin and Mann (*7*, *8*) demonstrated that carbonic anhydrase is a zinc-containing enzyme. This was the first clearly defined physiological function for this metal ion. These authors also discovered the inhibitory action of aromatic sulfonamides (*9*). Their proposal that the unsubstituted sulfonamide group, necessary for inhibition, should be bound to the zinc ion of the enzyme has, in fact, been verified by recent studies with X-ray diffraction and spectroscopy (see Section III,E,2). In 1940 to 1942, several papers were published dealing with the purification of the erythrocyte enzyme and the prop-

1. O. M. Henriques, *Biochem. Z.* **200,** 1 (1928).
2. F. J. W. Roughton, *Physiol. Rev.* **15,** 241 (1935).
3. F. J. W. Roughton, *Harvey Lectures* **39,** 96 (1943).
4. N. U. Meldrum and F. J. W. Roughton, *J. Physiol.* (*London*) **75,** 4P and 15P (1932); *Proc. Roy. Soc.* **B111,** 296 (1932).
5. W. C. Stadie and H. O'Brien, *JBC* **100,** lxxxviii (1933); **103,** 521 (1933).
6. N. U. Meldrum and F. J. W. Roughton, *J. Physiol.* (*London*) **80,** 113 (1933).
7. D. Keilin and T. Mann, *Nature* **144,** 442 (1939).
8. D. Keilin and T. Mann, *BJ* **34,** 1163 (1940).
9. T. Mann and D. Keilin, *Nature* **146,** 164 (1940).

erties of more or less purified preparations (*8, 10–12*). The molecular weight was determined to 30,000 (*11*) and the zinc content of 0.2% (*12*) should then correspond to one metal ion per enzyme molecule.

Lindskog (*13*) in 1960 presented a purification procedure for the bovine erythrocyte enzyme based on the high resolution offered by modern separation methods. Substantial evidence was presented showing that the enzyme preparation obtained was of high homogeneity. Shortly afterward purification methods for the human erythrocyte enzyme were reported from various laboratories (*14–18*). These studies showed that human red cells contain two forms of carbonic anhydrase differing considerably in catalytic and other properties. A similar polymorphism has since been encountered for a number of mammals (see Section II,B). The finding implied that many of the earlier studies on the catalytic properties of carbonic anhydrase had to be reevaluated since they have presumably been carried out on mixtures of the two types of enzyme.

The chemical studies of carbonic anhydrase have been surveyed by several authors. Reviews by Roughton and Clark (*19*), Gibian (*20*), Waygood (*21*), and Davis (*22*) cover the early literature. More recent reviews have been prepared by Maren (*23*) and Coleman (*24*). Our present knowledge of the molecular properties and mechanism of carbonic anhydrases is to a large extent based on studies on the mammalian erythrocyte enzymes, especially the forms from human and bovine blood. Reviews dealing exclusively with the mammalian red

10. D. A. Scott and J. R. Mendive, *JBC* **139,** 661 (1941); **140,** 445 (1941).
11. M. L. Petermann and N. V. Hakala, *JBC* **145,** 701 (1942).
12. D. A. Scott and A. M. Fischer, *JBC* **144,** 371 (1942).
13. S. Lindskog, *BBA* **39,** 218 (1960).
14. P. O. Nyman, *BBA* **52,** 1 (1961).
15. Y. Derrien, G. Laurent, M. Charrel, and M. Borgomano, *Haemoglobin-Colloq. Vienna, 1961* p. 28. Thieme, Stuttgart, 1962.
16. G. Laurent, C. Marriq, D. Nahon, M. Charrel, and Y. Derrien, *Compt. Rend. Soc. Biol.* **156,** 1456 (1962).
17. E. E. Rickli and J. T. Edsall, *JBC* **237,** PC258 (1962).
18. E. E. Rickli, S. A. S. Ghazanfar, B. H. Gibbons, and J. T. Edsall, *JBC* **239,** 2539 (1964).
19. F. J. W. Roughton and A. M. Clark, "The Enzymes," 1st ed., Vol. 1, Part 2, p. 1250, 1951.
20. H. Gibian, *Angew. Chem.* **66,** 249 (1954).
21. E. R. Waygood, "Methods in Enzymology," Vol. 2, p. 836, 1955.
22. R. P. Davis, "The Enzymes," 2nd ed., Vol. 5, p. 545, 1961.
23. T. H. Maren, *Physiol. Rev.* **47,** 595 (1967).
24. J. E. Coleman, *in* "Inorganic Biochemistry" (G. L. Eichhorn, ed.). Elsevier, Amsterdam, 1971 (in press).

cell carbonic anhydrase, particularly the human forms, have been published by Edsall (*25*) and Derrien and Laurent (*26*).

B. Distribution and Physiological Functions

Carbonic anhydrase is very widespread in nature and occurs in animals, plants, and certain bacteria. The distribution and physiological function in vertebrates, particularly mammals, have been discussed in detail by Maren (*23, 27*). In addition to their respiratory role of facilitating the transport of metabolic CO_2, the enzymes are involved in the transfer and accumulation of H^+ or HCO_3^- in a wide variety of organs. A comprehensive list of vertebrate tissues known to contain carbonic anhydrase is beyond the scope of this review, but it would include erythrocytes, kidney, gastric mucosa, and the eye lens, where high concentrations of the enzyme are found. The enzyme is present in gills and secretory organs of various types, such as the parotid and pancreatic glands, the rectal and alkaline glands, and the swim bladders of many species of fish.

The distribution and function in invertebrates has been treated by Polya and Wirtz (*28*) and by Addink (*29*). The enzyme has been found in species representing almost all the major groups such as *Coelenterata, Mollusca, Echinodermata, Arthropoda*, and *Annelida*. It occurs in the alimentary tract, gills, excretory and other organs. A detailed study of the carbonic anhydrase from a molluscan species (*29*) shows that it is a zinc-containing protein strongly resembling the mammalian enzymes with respect to catalytic properties and inhibition.

In animal cells, the carbonic anhydrases are believed to occur mainly dissolved in the intracellular water or only weakly bound to subcellular structures. In general, the enzymes are readily soluble after disruption of the cell, but in homogenates of certain vertebrate tissues such as kidney (*30, 31*), liver (*30, 32*), and brain (*30, 33*), as well as certain

25. J. T. Edsall, *Harvey Lectures* **62**, 191 (1968); *Ann. N. Y. Acad. Sci.* **151**, 41 (1968).
26. Y. Derrien and G. Laurent, *Exposes Ann. Biochim. Med.* **29**, 167 (1969).
27. T. H. Maren, *Federation Proc.* **26**, 1097 (1967).
28. J. B. Polya and A. J. Wirtz, *Enzymologia* **28**, 355 (1965); **29**, 27 (1965).
29. A. D. F. Addink, "Some Aspects of Carbonic Anhydrase of *Sepia officinalis* (L.)." Drukkerij Elinkwijk, Utrecht, 1968.
30. R. Karler and D. M. Woodbury, *BJ* **75**, 538 (1960).
31. T. H. Maren and A. C. Ellison, *Mol. Pharmacol.* **3**, 503 (1967).
32. T. H. Maren, A. C. Ellison, S. K. Fellner, and W. B. Graham, *Mol. Pharmacol.* **2**, 144 (1966).

invertebrate tissues (*29*), a significant fraction of the enzymic activity appears to be bound to particulate matter which has not yet been conclusively defined.

The physiological significance of the two distinct types of mammalian carbonic anhydrase is an intriguing problem. The two proteins have been shown to be governed by separate structural genes (*34, 35*). The two isoenzymes have been found in several tissues (*36, 37*). However, the quantitative relation between the two isoenzymes varies considerably for different organs suggesting that the two proteins are functionally independent enzymes. This finding would also indicate that the biosynthesis of the isoenzymes can be under separate control. This idea seems to be consistent with the discovery that only one type of the isoenzymes of the human red cell is quantitatively changed under certain pathological conditions (*38–40*) such as a disturbance in the level of thyroid hormone. The derepression of carbonic anhydrase synthesis taking place around the time of birth (*41*) occurs independently of the hemoglobin synthesis as indicated from studies with an ultramicromethod of the content of the various proteins in single cells (*42*). A recently described method (*43*), where enzyme synthesis can be followed in reticulocytes, seems to offer possibilities for the further investigation of carbonic anhydrase biosynthesis.

Several strains of the bacterial genus *Neisseria* have been shown to contain carbonic anhydrase (*44–46*). The enzyme has also been reported

---

33. W. D. Gray, C. E. Rauh, and R. W. Shanahan, *Biochem. Pharmacol.* **8,** 307 (1961).
34. R. E. Tashian, D. C. Shreffler, and T. B. Shows, *Ann. N. Y. Acad. Sci.* **151,** 64 (1968).
35. R. E. Tashian, *in* "Biochemical Methods in Red Cell Genetics" (G. J. Yunis, ed.), p. 307. Academic Press, New York, 1969.
36. M. J. Carter and D. S. Parsons, *BJ* **120,** 797 (1970); *BBA* **206,** 190 (1970).
37. J. E. A. McIntosh, *BJ* **114,** 463 (1969); **120,** 299 (1970).
38. D. J. Weatherall and P. A. McIntyre, *Brit. J. Haematol.* **13,** 106 (1967).
39. L. E. Lie-Injo, C. G. Lopez, and P. L. De V. Hart, *Clin. Chim. Acta* **29,** 541 (1970).
40. E. Magid, *Scand. J. Clin. & Lab. Invest.* **26,** 257 (1970).
41. R. Berfenstam, *Acta Paediat.* **41,** 32 (1952); E. Poblete, D. W. Thibeault, and P. A. M. Auld, *Pediatrics* **42,** 429 (1968).
42. D. Gitlin, T. Sasaki, and P. Vuopio, *Blood* **32,** 796 (1968).
43. N. L. Meyers, G. J. Brewer, and R. E. Tashian, *BBA* **195,** 176 (1969).
44. F. P. Veitch and L. C. Blankenship, *Nature* **197,** 76 (1963).
45. E. Sanders and T. H. Maren, *Mol. Pharmacol.* **3,** 204 (1967).
46. L. Adler, J. Brundell, S. O. Falkbring, and P. O. Nyman, (1971) (in preparation).

for some other bacteria (*44, 47*), but these reports still remain to be confirmed. The *Neisseria* enzyme is a zinc-containing protein resembling the mammalian enzymes in catalytic and molecular properties (*46*). Acetazolamide and other sulfonamides, well-known carbonic anhydrase inhibitors, strongly inhibit the bacterial enzyme (*45, 46*). Inhibitors of the sulfonamide type also act as bacteriostatic agents (*45, 46, 48–50*) suggesting that the enzyme might be of critical importance for the bacterial cell. The biosynthesis of carbonic anhydrase appears to be repressed at higher concentrations of CO_2 (*46*), and under similar conditions the bacteriostatic effect of sulfonamides diminishes (*45, 49, 50*). These findings strongly suggest that the enzyme is essential for the bacterial cell only at low CO_2 concentrations, a conclusion of possible value for a future elucidation of the physiological role of carbonic anhydrase in the bacterium.

The early literature on plant carbonic anhydrase has been summarized by Waygood (*21*). The enzyme occurs in higher plants such as spinach (*21, 51–55*), parsley (*56, 57*), and various grasses (*52, 54*). In lower plants, it has been reported for baker's yeast (*47*), mosses (*58*), and macroscopic (*59–61*) as well as microscopic (*62–67*) species of algae. The carbonic anhydrases, isolated in an apparently homogeneous state from parsley

47. W. T. Shoaf and M. E. Jones, *ABB* **139,** 130 (1970).
48. A. Forkman and A. B. Laurell, *Acta Pathol. Microbiol. Scand.* **65,** 450 (1965).
49. A. Forkman and A. B. Laurell, *Acta Pathol. Microbiol. Scand.* **67,** 542 (1966).
50. A. Forkman, *Acta Pathol. Microbiol. Scand.* **73,** 298 (1968).
51. K. Kondo, H. Chiba, and F. Kawai, *Bull. Res. Inst. Food Sci., Kyoto Univ.* **8,** 17 and 28 (1952); *Chem. Abstr.* **46,** 7183a and c (1952); L.-F. Yen and P. S. Tang, *Chinese J. Physiol.* **18,** 43 (1951); *Chem. Abstr.* **48,** 10849 (1954).
52. R. G. Everson and C. R. Slack, *Phytochemistry* **7,** 581 (1968).
53. C. Rossi, A. Chersi, and M. Cortivo, *NASA* (*Natl. Aeron. Space Admin.*), *Spec. Publ.* **NASA SP-188,** 131 (1969).
54. E. R. Waygood, R. Mache, and C. K. Tan, *Can. J. Botany* **47,** 1455 (1969).
55. R. G. Everson, *Phytochemistry* **9,** 25 (1970).
56. S. K. Fellner, *BBA* **77,** 155 (1963).
57. A. J. Tobin, *JBC* **245,** 2656 (1970).
58. E. Steemann Nielsen and J. Kristiansen, *Physiol. Plantarum* **2,** 325 (1949).
59. M. Ikemori and K. Nishida, *Ann. Rep. Noto Marine Lab.* **7,** 1 (1967).
60. M. Ikemori and K. Nishida, *Physiol. Plantarum* **21,** 292 (1968).
61. G. W. Bowes, *Plant Physiol.* **44,** 726 (1969).
62. S. Österlind, *Physiol. Plantarum* **3,** 353 (1950).
63. E. R. Waygood and K. A. Clendenning, *Can. J. Res.* **28,** 673 (1950).
64. C. D. Lichtfield and D. W. Hood, *Verhandl. Intern. Ver. Limnol.* **15,** 817 (1964).
65. E. Steemann Nielsen, *Physiol. Plantarum* **19,** 232 (1966).
66. M. L. Reed and D. Graham, *Plant Physiol.* **43,** S-29 (1968).
67. E. B. Nelson, A. Cenedella, and N. E. Tolbert, *Phytochemistry* **8,** 2305 (1969).

(*57*) and spinach (*53*), have comparatively high molecular weights, and their properties differ considerably from those of the mammalian and bacterial enzymes. Whether the plant carbonic anhydrases contain zinc or not is still a matter of controversy (*21, 54, 56, 57*). Several recent papers report about the inhibitory action of sulfonamides (*55, 60, 61, 68*). In general, the sulfonamides tested seem to be less strongly bound to the plant enzymes than to the mammalian forms, possibly explaining why several authors [(*56*); see also Waygood (*21*)] have reported a lack of inhibitory power.

Carbonic anhydrase has been stated to occur exclusively in the cytoplasm of the plant cell (*21*). However, several recent reports (*52–55, 67*) show that the enzyme can also occur in the chloroplasts. This seems to be the situation in plants carrying out CO_2 fixation according to the Calvin cycle, where ribulose diphosphate becomes carboxylated with CO_2 as the reactive species (*69*). Plants utilizing a C_4-dicarboxylic acid pathway, which appears to involve a carboxylation of phosphoenol pyruvate (*70*), have been shown to have a significantly lower carbonic anhydrase content with the enzyme absent or restricted to the cytoplasm (*52, 54*). In certain plants the CO_2 tension during precultivation determines the pattern of CO_2 fixation (*71*). This appears to parallel the carbonic anhydrase content of the cell (*66, 67*). These results would support an old idea (*72*) that carbonic anhydrase may play a role for the photosynthetic fixation of CO_2. The idea is further supported by the finding that in systems, where acetazolamide can act as an inhibitor of the enzyme *in vitro*, photosynthesis is inhibited at low inhibitor concentrations (*55, 59, 60, 68*).

The biosynthesis of the plant enzyme, as studied in unicellular algae (*66, 67*), appears to be repressed at increased concentrations of CO_2, a situation similar to that for the bacterial enzyme.

II. Molecular Properties

A. Methods of Isolation

Erythrocytes provide a readily available, abundant source of the carbonic anhydrases. There are about 1–2 g of carbonic anhydrase per

68. R. G. Everson, *Nature* **222,** 876 (1969).
69. T. G. Cooper, D. Filmer, M. Wishnick, and M. D. Lane, *JBC* **244,** 1081 (1969).
70. M. D. Hatch and C. R. Slack, *BJ* **101,** 103 (1966).
71. D. Graham and C. P. Whittingham, *Z. Pflanzenphysiol.* **58,** 418 (1968).
72. G. O. Burr, *Proc. Roy. Soc.* **B120,** 42 (1936).

liter of mammalian red cells (*8, 18, 23*). The isolation procedures first developed for the bovine and human isoenzymes have required only minor modifications when applied to erythrocyte carbonic anhydrases from other mammalian species.

The carbonic anhydrases can be liberated from the red cells together with hemoglobin by simple hemolysis in hypotonic solutions. It is generally necessary to remove the vast excess of hemoglobin before attempting the separation of the carbonic anhydrase isoenzymes. The hemoglobin can be removed by any of the following procedures: (a) denaturation and precipitation accomplished with a mixture of water, ethanol, and chloroform (*8, 13, 14*); (b) gel filtration on Sephadex G-75 (*18*); and (c) adsorption on DEAE-Sephadex or DEAE-cellulose (*73*). Method (a) is most convenient for large-scale preparations and does not seem to have any seriously harmful effects on the enzymes (see, however, Section II,G). Method (b) gives perhaps the most efficient recovery of the isoenzymes, while method (c) is effective for the complete removal of hemoglobin.

The carbonic anhydrase isoenzymes are the major proteins in the hemoglobin-free solutions obtained by any of the above procedures. Final purification of the isoenzymes has generally been accomplished with ion exchange chromatography or preparative electrophoresis. Chromatography on DEAE matrices has been the most widely used method, but the method of choice and the precise conditions vary somewhat with the species chosen to serve as the source of the enzyme. Isolation procedures similar to those outlined above have been used to obtain the erythrocyte carbonic anhydrases from man (*14, 16, 73–75*), rhesus monkey (*76*), various other primates (*34, 35, 75*), ox (*13, 75, 77*), horse (*75, 78*), pig (*75, 79*), dog (*75, 80*), rabbit (*37*), guinea pig (*36*), deer mouse (*34*), blue–white dolphin (*81*), and yellowfin tuna (*81*).

Methods have been described for the purification of carbonic anhydrase

73. J. McD. Armstrong, D. V. Myers, J. A. Verpoorte, and J. T. Edsall, *JBC* **241,** 5137 (1966).

74. G. Laurent, M. Charrel, F. Luccioni, M. F. Autran, and Y. Derrien, *Bull. Soc. Chim. Biol.* **47,** 1101 (1965).

75. H. Suyama, H. Sawada, H. Ueda, and I. Ohya, *Igaku To Seibutsugaku* **76,** 62 (1968); *Chem. Abstr.* **70,** 55424s (1969).

76. T. A. Duff and J. E. Coleman, *Biochemistry* **5,** 2009 (1966).

77. M. Liefländer, *Z. Physiol. Chem.* **335,** 125 (1964).

78. A. J. Furth, *JBC* **243,** 4832 (1968).

79. R. J. Tanis, R. E. Tashian, and Y. L. Yu, *JBC* **245,** 6003 (1970).

80. P. Byvoet and A. Gotti, *Mol. Pharmacol.* **3,** 142 (1967).

81. C. Shimizu and F. Matsuura, *Bull. Japan. Soc. Sci. Fisheries* **28,** 735 and 925 (1962).

from other organs of mammals (*8, 37*), from spinach (*53*) and parsley (*57*), and from the bacterium *Neisseria sicca* (*46*). In these procedures ion exchange methods and electrophoresis are utilized extensively, but the reader is referred to the original papers for detailed information.

Adsorption matrices designed for the specific adsorption of the carbonic anhydrases are currently under study (*82*). The principle involves the attachment of a sulfonamide to an insoluble matrix. These compounds are highly selective carbonic anhydrase inhibitors showing a high affinity for the enzyme, and a substantial degree of purification can be achieved in one step. Elution from the matrix is achieved by the addition of a dissolved inhibitor to the buffer, and in favorable cases a separation of the major isoenzymes can be accomplished at the same time. Once fully developed, these specific matrices will be an invaluable aid in the isolation of carbonic anhydrases from a wide variety of sources.

B. Polymorphism and Nomenclature

The polymorphism of mammalian carbonic anhydrases is in part resulting from the presence of two isoenzymes with distinctly different amino acid sequences and specific activities. However, additional polymorphism may be observed due to the presence of modified forms of each of the two isoenzymes. These modified forms may arise through mutational changes or result from secondary modifications occurring *in vivo* or *in vitro*.

In the earlier investigations of carbonic anhydrase isoenzymes, the various forms of the enzyme were designated in accordance with their chromatographic behavior or electrophoretic mobilities. On this basis a standard nomenclature was adopted for the human isoenzymes (*18, 83*), and it is shown in Table I together with older notations which occur in many of the references cited in this review (*14, 16–18, 35, 83, 84*). However, as the body of knowledge concerning the mammalian carbonic anhydrases has grown, it has become a generally accepted practice to classify the isoenzymes according to their specific activities in the CO_2 hydration reaction. In the prevalent nomenclature system, which we shall use in this chapter, the major high activity form is called C, and

82. J. Porath and L. Sundberg, *in* "The Chemistry of Biosurfaces" (M. Hair, ed.), Vol. II. Dekker, New York, 1971 (in press).

83. P. O. Nyman and S. Lindskog, *BBA* **85,** 141 (1964).

84. G. Laurent, M. Charrel, C. Marriq, and Y. Derrien, *Bull. Soc. Chim. Biol.* **44,** 419 (1962).

TABLE I
NOMENCLATURE FOR HUMAN AND BOVINE CARBONIC ANHYDRASE ISOENZYMES

Enzyme source	Activity type	Abundance[a] (%)	Standard nomenclature[b] [Refs. (*18*, *74*, *83*)]	Other designations [Refs. (*17*, *35*)]	[Ref. (*14*)]	[Refs. (*16*, *84*)]
Human	Low	5	A	I(+1)	II	X_2
	Low	83	B	I	III	X_1
	High	12	C	II	V	Y
Bovine	High	20	A			
	High	80	B			

[a] Approximate weight percent of total carbonic anhydrase in erythrocytes.

[b] As further elaborated on in the text, we shall use the letters B and C to denote the major low activity and high activity forms from a species. The minor low activity form of the human enzyme will be called *A*, and the bovine forms will be called *A* and *B*, with these symbols of the standard nomenclature italicized.

the major form with a relatively lower activity is called B (see Table I). This notation is devised to minimize the conflicts with the older systems, and it attempts to provide a rational framework for the comparison of carbonic anhydrases from different species.

The symbol A was earlier attributed to the most acidic form of the human and bovine carbonic anhydrases, respectively. In the few instances where these forms are specifically discussed in this chapter, we shall refer to them as human or bovine carbonic anhydrase *A*, where the italics indicate that this designation belongs to an earlier system, and that these forms do not represent a third activity type of carbonic anhydrase.

The human form *A*, representing about 5% of the total red cell carbonic anhydrase in man, is indistinguishable from the human B enzyme in terms of its specific activity, amino acid composition, and tryptic peptides (*85*). The human *A* form is always observed in blood taken from individuals (*35*), and mutational changes in the B enzyme are accompanied by the corresponding change in *A* (*86*). These data suggest that the human *A* is a chemically modified form of the B enzyme, possibly arising through the hydrolysis of amide groups of one or more glutamine or asparagine side chains (*87*, *88*). Recently, a number of

85. G. Laurent, D. Garçon, C. Marriq, M. Charrel, and Y. Derrien, *Compt. Rend.* **258,** 6557 (1964); G. Laurent, D. Garçon, M. Charrel, L. A. Ardoino, and Y. Derrien, *Bull. Soc. Chim. Biol.* **49,** 1021 (1967).

86. T. B. Shows, *Biochem. Genet.* **1,** 171 (1967).

87. S. Funakoshi and H. F. Deutsch, *JBC* **244,** 3438 (1969).

88. G. Laurent and Y. Derrien, *NASA* (*Natl. Aeron. Space Admin.*), *Spec. Publ.* **NASA SP-188,** 115 (1969).

additional electrophoretic variants of both the B and C isoenzymes have been isolated (*89*). These forms are present in trace amounts, and it has been proposed that they may also arise through amide hydrolysis or represent conformational isomers (*87*).

Bovine erythrocytes contain significant amounts of two electrophoretically separable forms (*13*) originally designated A and B in the order of mobility. To avoid confusion with the present notation, it should be strongly emphasized that both these forms have similar high specific activities and therefore belong to the C group of mammalian carbonic anhydrases. Both forms appear to have identical amino acid compositions (*83*). Presumably, one of the forms is a modified variant of the other form. Some evidence has been presented to indicate that two forms of the bovine enzyme are controlled by a pair of allelic genes (*90*). We shall allude to the original notation and call the major form of bovine carbonic anhydrase *B* only when direct comparisons with the form *A* are made. Since no low activity form has yet been found in the ox, we strive to avoid the letter notation in other contexts because it would imply an unproven form of polymorphism in this species.

The sequence homologies displayed by the human B and C forms (see Section II,F) strongly suggest that the two isoenzyme types arose through a process of gene duplication and subsequent independent evolution. Low activity forms B and high activity forms C have been found in many species, such as various primates (*34*), horse (*78*), pig (*79*), and guinea pig (*36*), representing several mammalian orders. Evidently, gene duplication occurred before the divergence of these orders. If these assumptions are correct, each group of the mammalian carbonic anhydrases should represent the evolutionary products of one of the duplicate genes. Generally, the members of one group should be structurally more closely related than the representatives of the two groups contained in one species. These predictions are borne out by the available comparative data as indicated in the subsequent sections.

In studies of primate carbonic anhydrases, Tashian *et al.* (*34, 91*) found that the B isoenzymes show a greater variability than the C isoenzymes. Furthermore, when only one type of carbonic anhydrase has been found in a mammalian or nonmammalian species, the enzyme usually has the activity characteristics of a C form. These observations allow the suggestion (*34, 91*), that the B enzyme has evolved more rapidy than

89. S. Funakoshi and H. F. Deutsch, *JBC* **243,** 6471 (1968).
90. G. Sartore, C. Stormont, B. G. Morris, and A. A. Grunder, *Genetics* **61,** 823 (1969).
91. R. E. Tashian, M. Goodman, V. E. Headings, J. DeSimone, and R. H. Ward, *Biochem. Genet.* **5,** 183 (1971).

the C enzyme and that the latter form may represent an older evolutionary type of carbonic anhydrase.

C. Immunological Properties

Rabbits, injected periodically with one of the purified human or bovine carbonic anhydrases, develop specific antibodies to the antigen over the course of about eight weeks (*92*). The human B enzyme yields antisera with a lower hemagglutination titer compared to the antisera induced by human C enzyme indicating that the B form is a more efficient antigen (*93, 94*).

The nature of the antigenic sites of the carbonic anhydrases is unknown, but they are evidently more or less independent of the active sites and different for the two isoenzymes. The enzyme–antibody complex retains some carbonic anhydrase activity, and the presence of a sulfonamide inhibitor does not affect the enzyme–antibody complex formation (*93*). Antisera specific for the human B enzyme will react equally well with the apoenzyme B and with the human *A* form but will not cross-react with the C enzyme, while antisera specific for the C enzyme show no cross-reactivity with the B enzyme or any of its modified forms (*93, 95–98*).

The remarkable specificity offered by the rabbit antisera provides a means for the detection and quantitative determination of the individual isoenzymes in a complex mixture. A radioimmunoassay has been used to determine the concentrations of the B and C enzymes in unfractionated hemolysates (*95, 99*). Qualitative immunological techniques have been employed to investigate the distribution of the B and C enzymes in various human tissues (*93*). Hansson (*96*) used fluorescent antibodies to demonstrate the simultaneous presence of both B and C enzymes in individual human red cells. Funakoshi and Deutsch (*89, 100*) have shown that the numerous electrophoretic variants of the human enzyme, present

92. J. M. Fine, G. A. Boffa, M. Charrel, G. Laurent, and Y. Derrien, *Nature* **200,** 371 (1963).
93. P. J. Wistrand and S. N. Rao, *BBA* **154,** 139 (1968).
94. L. Nonno, H. Herschman, and L. Levine, *ABB* **136,** 361 (1969).
95. V. E. Headings and R. E. Tashian, *Biochem. Genet.* **4,** 285 (1970).
96. H. P. J. Hansson, *Life Sci.* **4,** 965 (1965).
97. A. Micheli and C. Buzzi, *BBA* **89,** 324 (1964).
98. H. Suyama, H. Sawada, H. Ueda, and I. Ohya, *Igaku To Seibutsugaku* **77,** 15 and 21 (1968).
99. A. Micheli and C. Buzzi, *BBA* **96,** 533 (1965).
100. S. Funakoshi and H. F. Deutsch, *JBC* **245,** 2852 (1970).

in trace amounts in human erythrocytes, all cross-react either with the antiserum against the human B enzyme or with the antiserum against the human C enzyme.

The B- and C-type carbonic anhydrases from species relatively closely related to man are sufficiently similar to the human isoenzymes to cross-react with antisera prepared against the homologous human isoenzyme (*34, 93, 94, 98*). The relative order of serological similarity suggests an order among the primates that agrees with primate taxa based on morphological considerations (*94*). Tashian *et al.* (*34*) investigated the cross-reactions of purified isoenzymes or hemolysates from a variety of mammalian species with the specific antisera against human B and C forms. The results show that the immunological similarities extend over several nonprimate species but become less well defined or absent for species distantly related to man [cf. also Suyama *et al.* (*98*)].

D. Some Physical Properties

A representative sampling of physical constants for some of the carbonic anhydrases is presented in Table II (*18, 57, 73, 81, 83, 101–104*). In all cases investigated, the mammalian isoenzymes are monomeric proteins with molecular weights near 30,000. Similar molecular weights have been reported for the carbonic anhydrases from yellowfin tuna (*81*) and *Neisseria sicca* (*105*), whereas the enzymes isolated from parsley (*57*) and spinach (*53*) appear to be considerably larger molecules. However, in 6 *M* guanidine hydrochloride the parsley enzyme dissociates into subunits with molecular weights approximately equal to those of the mammalian enzymes (*57*).

The values of the sedimentation and diffusion constants for the human and bovine enzymes given in Table II are all taken from the work of Reynaud *et al.* (*101–103*), and they are in general agreement with the values reported earlier by other investigators (*13, 14, 18, 73, 81*). However, values approximately 15% lower than those given in Table II have been reported for the human isoenzymes (*73*). Theoretical considerations cited in the latter report suggest that the lower values may be more

101. J. Reynaud, P. Santini, M. Bouthier, J. Savary, and Y. Derrien, *BBA* **171**, 363 (1969).
102. J. Reynaud, G. Rametta, J. Savary, and Y. Derrien, *BBA* **77**, 521 (1963).
103. J. Reynaud, F. Luccioni, M. Bouthier, J. Savary, and Y. Derrien, *BBA* **221**, 367 (1970).
104. M. Jonsson and E. Pettersson, *Acta Chem. Scand.* **22**, 712 (1968).
105. J. Brundell, S. O. Falkbring, and P. O. Nyman, unpublished data (1971).

TABLE II
SELECTED PHYSICAL PROPERTIES OF SOME CARBONIC ANHYDRASES

Parameter	Human (*101, 102*)[a]			Bovine (*103*)[a]		Blue-white dolphin [Ref. (*81*)]	Yellowfin tuna [Ref. (*81*)]	Parsley [Ref. (*57*)]
	A	B	C	*A*	*B*			
Diffusion constant, $D_{20,w} \times 10^7$ (cm^2 sec^{-1})	9.8	10.7	10.0	9.4	9.4	8.5	9.4	
Sedimentation constant, $s^{\circ}_{20,w}$ (Svedberg units)	3.1	3.2	3.3	3.2	3.2	2.9	3.0	8.8
Intrinsic viscosity, $[\eta]$(cm^3 g^{-1})	3.65	2.74	*b*	2.7	2.7			
Molecular weight, MW $\times 10^{-3}$ (from s/D)	29.3	29.3	31.9	31.4	30.3	32.0	30.5	180
Molecular weight, MW $\times 10^{-3}$ (from composition)	*c*	28.8	29.8	*c*	29.5			(27.4)
Isoelectric point (pH)		6.6[a]	7.3[a]		5.9[a]			
$A^{1\%}_{280}$ (cm^{-1})	*c*	16.3[a]	18.7[a]	*c*	19.0[a]			11.3

[a] The hydrodynamic data on the human and bovine enzymes are taken from the work of Reynaud *et al.* (*101–103*). The isoelectric points for the human B and C enzymes are from Funakoshi and Deutsch (*100*) and for the bovine *B* form from Jonsson and Pettersson (*104*). The extinction coefficients, $A^{1\%}_{280}$, are from Nyman and Lindskog (*83*).

[b] Rickli *et al.* (*18*) and Armstrong *et al.* (*73*) reported that the intrinsic viscosity of the human C enzyme is very similar to that of the B form, but they did not give an exact value.

[c] The values for the human *A* form and for the bovine *A* form are presumably closely similar to those of the human B enzyme and the bovine *B* form, respectively.

correct, but the discrepancy between these values and those reported by other investigators remains unexplained. The molecular weights calculated from the sedimentation and diffusion constants given in Table II are in close agreement with the estimates reported by other workers and with estimates based on amino acid compositions.

In their native states the human B and C isoenzymes and the bovine forms behave as compact spheres in solution. They have frictional ratios near unity (*101–103*), and their intrinsic viscosities are unusually low for globular proteins. A significantly greater value of the intrinsic viscosity has been reported for the human *A* form, and an ellipsoidal axial ratio of 2.9 was calculated (*101*). However, this observation would require confirmation by independent methods since other structural properties such as the optical rotatory dispersion pattern (*106*) are indistinguishable from those of the B enzyme indicating extensive structural similarities between the *A* and B forms.

The isoelectric points of the human and bovine carbonic anhydrases given in Table II are recent determinations by the method of isoelectric focusing. The values given for the human C enzyme and the bovine enzyme agree well with values obtained by other procedures (*107*, *108*), whereas an isoelectric pH of 5.7 was previously reported for the human B enzyme (*109*).

Table II also lists the absorbancies at 280 nm for 1% solutions of some carbonic anhydrases. A difference of one residue per molecule in the tryptophan contents of the human B and C forms accounts for the observed difference in these extinction coefficients.

E. Composition

All of the mammalian carbonic anhydrases studied, as well as the *Neisseria* enzyme, are composed of one zinc ion and a single polypeptide chain containing approximately 260 amino acid residues. The parsley enzyme may be composed of six subunits, each containing approximately 260 amino acid residues and presumably also one zinc ion (*57*). However, homogeneous preparations of the spinach enzyme have been reported to contain no significant amounts of zinc (*53*). Seemingly, the old controversy (*21*) regarding the presence of zinc in plant carbonic anhydrase has not yet been completely settled.

106. D. V. Myers and J. T. Edsall, *Proc. Natl. Acad. Sci. U. S.* **53,** 169 (1965).
107. L. M. Riddiford, R. H. Stellwagen, S. Mehta, and J. T. Edsall, *JBC* **240,** 3305 (1965).
108. A. Nilsson and S. Lindskog, *European J. Biochem.* **2,** 309 (1969).
109. L. M. Riddiford, *JBC* **239,** 1079 (1964).

TABLE III

AMINO ACID COMPOSITIONS OF CARBONIC ANHYDRASES FROM VARIOUS SOURCES

Residue	Mammalian B type					Mammalian C type							
	Man[a] [Ref. (*83*)]	Rhesus monkey [Ref. (*76*)]	Horse [Ref. (*78*)]	Pig [Ref. (*79*)]	Guinea pig [Ref. (*36*)]	Man[a] [Ref. (*83*)]	Rhesus monkey [Ref. (*76*)]	Horse [Ref. (*78*)]	Ox [Ref. (*83*)]	Pig [Ref. (*79*)]	Guinea pig [Ref. (*36*)]	*N. sicca* [Ref. (*105*)]	Parsley[b] [Ref. (*57*)]
Lys	18	18	19	22	19	24	24	19	19	20	23	25	21
His	11	9	10	14	10	12	12	12	11	13	14	12	5
Amide NH_3[c]	26	—	22	—	—	21	—	32	25	—	—	35	—
Arg	7	7	5	6	5	7	8	9	9	9	5	5	5
Asp	31	36	32	34	25	29	30	27	32	27	25	28	24
Thr	14	13	12	10	11	13	11	12	15	13	12	20	10
Ser	30	30	28	28	33	19	18	18	16	17	22	15	18
Glu	22	22	25	20	23	25	26	26	24	28	25	29	21
Pro	18	17	18	17	16	17	16	16	20	17	17	13	22
Gly	16	15	23	18	18	22	22	23	20	22	20	18	22
Ala	19	16	15	19	17	13	12	17	17	13	19	25	17
Val	17	16	20	15	18	17	14	19	20	13	18	18	24
Met	2	1	2	0–1	1	1	2	1	3	3	3	2	4
Ile	10	10	9	15	11	9	10	7	5	10	8	7	8
Leu	20	19	21	23	20	25	24	22	26	23	25	17	21
Tyr	8	9	8	11	8	8	7	7	8	7	7	6	8
Phe	11	10	11	10	9	12	11	11	11	12	11	12	17
Trp	6	7	5	7	6	7	7	5	7	6	7	4	3
Cys	1	1	2	0	1	1	1	1	0	0	1	2	7
Total[c]	261	256	265	270	251	261	255	252	263	253	262	258	257

[a] The values for the human isoenzymes are taken from Nyman and Lindskog (*83*) with the exception that the values for glutamic acid have been reduced by 2 residues per molecule. The data shown in the table are in good agreement with results published from other laboratories (*18*, *73*, *110*).

[b] Estimated number of residues per subunit. The enzyme is believed to be composed of six equal subunits.

[c] The ammonia values are not included in the total.

Table III summarizes the current state of our knowledge concerning the amino acid compositions of some of the carbonic anhydrases. In addition, partial amino acid analyses have been reported for enzymes isolated from other primates (*34*), deer mouse (*34*), and spinach (*53*). In general, the enzymes are rich in the basic and acidic amino acids and have a noticeably high proline content. The sulfur-containing amino acids are rare in the mammalian and bacterial enzymes, and in most cases intramolecular disulfide bridges are impossible. The total absence of cysteine in some carbonic anhydrases should rule this amino acid out of consideration for involvement in metal binding or catalytic action (*83*). The plant enzymes, on the other hand, appear to contain several free sulfhydryl groups and can be inactivated with reagents such as *p*-mercuribenzoate (*53, 57*). In general, the addition of sulfhydryl-containing compounds such as cysteine has been quoted to be required for the stabilization of these enzymes (*21, 53, 57, 60*), but exceptions to this rule occur (*61*).

The mammalian B-type isoenzymes have consistently been found to have a considerably higher serine content than the C-type isoenzymes (*18, 73, 83, 110*). The bovine enzyme is a C form also by these criteria in addition to those of specific activity and immunological cross-reactivity. Furthermore, the serine content of the *Neisseria* enzyme, which is a high activity enzyme, is consistent with the trend established among the mammalian enzymes.

F. Primary Structure

The present knowledge of the primary structures of the carbonic anhydrases is summarized in Table IV. The complete amino acid sequence is not yet known for any carbonic anhydrase. The investigations have so far been concentrated on the human erythrocyte enzymes. The preferred isoenzyme has been the B form because of its abundance and accessibility. A limited amount of data is also available for the bovine enzyme.

The strategy applied in the primary structure studies of the carbonic anhydrases is straightforward. It is characterized by attempts to split the polypeptide chain into a small number of fragments using specific methods of cleavage (*111*). Methionyl peptide bonds have been split by

110. G. Laurent, M. Castay, C. Marriq, D. Garçon, M. Charrel, and Y. Derrien, *BBA* **77**, 518 (1963).

111. B. Andersson, P. O. Nyman, and L. Strid, *NASA* (*Natl. Aeron. Space Admin.*), *Spec. Publ.* **NASA SP-188,** 109 (1969).

TABLE IV

PRIMARY STRUCTURES OF CARBONIC ANHYDRASES[a]

Variants[b]	Glu *
Human B enzyme	Acetyl-Ala-Ser-Pro-Asp-Trp-Gly-Tyr-Asp-Asp-Lys-Asn-Gly-Gln-Pro-Glu-Trp-Ser-Lys-Leu-Tyr-Pro-Ile-Ala-
Human C enzyme	Acetyl-Ser-His-His-Trp-Gly-Tyr-------Gly-Lys(Asp,Gly,His,Pro,Gln)Trp-His-Lys-Asp-Phe-Pro-Ile-Ala-
Bovine enzyme	Acetyl-Ser-
Residue number	2 5 10 15 20
Variants[b]	Ala Val
Human B enzyme	-Asp-Gly-Asn-Asn-Gln-Ser -Pro-Val-Asp-Ile-Lys-Thr-Ser-Glu-Thr-Lys-His-Asp-Thr-Ser-Leu-Lys-Pro-Ile -
Human C enzyme	-Lys-Gly-Gln-Arg-Val-Asx(Pro,Ser,Glx)Ile-Asx-Thr-His-Thr-Ala-Lys-Tyr-Asp-Pro-Ser(Leu,Lys,Pro)Leu-
Residue number	25 30 35 40 45
	*
Human B enzyme	-Ser-Val-Ser-Tyr-Asn-Pro--Ala-Thr-Ala-------Lys-Glu-Ile--Ile -Asn-Val-Gly-His-Ser -Phe-His-Val-Asn-Phe -
Human C enzyme	(Ser,Val,Ser,Tyr,Asx,Glx)(Ser,Thr,Ala)Leu-Arg(Asn,Leu)(Ile,Asn,Ala,Gly,His,Asp,Phe)Glx(Val,Glx,Phe)-
Residue number	50 55 60 * 65 70
	*
Human B enzyme	-Glu-Asp-Asn-Asn-Asp-Arg-Ser-Val-Leu-Lys-Gly-Gly-Pro-Phe-Ser -Asp-Ser-Tyr -Arg(~81 residues)-
Human C enzyme	(Glx,Ser ,Asx,Asx,Asx)Lys-Ala-Val-Leu-Lys-Gly(Gly,Pro,Leu,Gly,Asx,Thr)Tyr-Arg-
Residue number	75 80 85 89

```
Human B enzyme    -Ile-Lys-Thr-Lys-Gly-Lys-Arg-Ala-Pro-Phe-Thr-Asn-Phe-Asp-Pro-Ser-Thr-Leu-Leu-Pro-Ser-Ser-Leu-Asp-
Residue number    171               175               180               185               190

                                                         *
Human B enzyme    -Phe-Trp-Thr-Tyr-Pro-Gly-Ser-Leu-Thr-His-Pro-Pro-Leu-Tyr-Glu-Ser-Val-Thr-Trp-Ile-Ile-Cys-Lys-Glu-
Residue number     195                200                205                210               215

Variants[b]                                                                                      Val      Pro
Human B enzyme    -Ser-Ile-Ser-Val-Ser-Ser-Glu-Gln-Leu-Ala-Gln-Phe-Arg-Ser -Leu-Leu-Ser -Asn-Val-Glu-Gly-Asp-Asn-Ala-
Human C enzyme                                               -Arg-Lys-Leu-Asn-Phe-Asp-Gly-Glu-Gly-Glu(Pro,Glu,
Bovine enzyme                                                -Arg(Thr,Leu,Leu,Phe,Asx,Ala,Glx,Gly,Asx,Pro,Glx,
Residue number       220               225                230               235               240

Variants[b]                Ile     Arg                                   Arg
Human B enzyme    -Val-Pro-Met-Glx-His-Asn-Asn-Arg-Pro-Thr-Gln-Pro-Leu-Lys-Gly-Arg-Thr-Val-Arg-Ala-Ser-PheOH
Human C enzyme     Glu)Leu-Met-Val-Asp-Asn-Trp-Arg-Pro-Ala-Gln-Pro-Leu-Lys-Asn-Arg-Gln-Ile -Lys-Ala-Ser-Phe-LysOH
Bovine enzyme      Glx,Leu)Met-Leu-Ala-Asn-Trp-Arg-Pro-Ala-Gln-Pro-Leu-Lys-Asn-Arg-Gln-Val-Arg-Gly-Phe-Pro-LysOH
Residue number              245                250               255               260              264
```

[a] The numbering refers to the human B enzyme and starts at its acetylated amino-terminal residue. The missing sequence, 90–170, may be a couple of residues shorter or longer; thus, numbering is tentative from residue number 171. Chemically modified residues are marked by an asterisk. See text for further information.

[b] Mutational interchanges in the human B enzyme and amino acid interchanges observed in the B isoenzymes from other primates.

treatment with cyanogen bromide (*112, 113*), and arginyl bonds have been hydrolyzed with trypsin (*111, 114, 115*). Tryptic hydrolysis of the lysyl peptide bonds has been prevented by the use of reversible chemical modifications of the lysyl side chains with methyl acetimidate (*115, 116*) or maleic anhydride (*114, 115*).

Several investigators have contributed to the primary structure determination of the human B enzyme (*26, 111, 112, 114, 115, 117–119*). The sequence given in Table IV is that of Andersson *et al.* (*117*) with the exception of the region **239–241** which has been written in accordance with Funakoshi and Deutsch (*118*) in their investigation of a mutant form of the enzyme. The amino-terminal portion of the polypeptide chain has also been investigated by Sciaky *et al.* (*114*). Their results are in complete agreement except for minor deviations at positions **41** and 69–70. The data for the human C enzyme refer to investigations by Henderson *et al.* (*115*) and Nyman *et al.* (*113*). They are to some extent tentative and may require future revision. For the bovine enzyme the carboxyl-terminal sequence was investigated by Nyman *et al.* (*113*) and Andersson *et al.* (*116*).

The results in Table IV show that the two main forms of the human erythrocyte enzyme are distinctly different proteins with unique amino acid sequences. However, similarities between the polypeptide chains exist, thus suggesting that the two proteins are related by common ancestry (see Section II,B). The known portions of the sequence of the bovine enzyme clearly show more extensive homologies with the human C form than with the B isoenzyme. The minor human form *A* has been shown to have the first 6 residues of the amino-terminal sequence identical with the B isoenzyme (*88*) in accordance with the opinion that these

112. C. Marriq, J.-M. Gulian, and G. Laurent, *BBA* **221,** 662 (1970).

113. P. O. Nyman, L. Strid, and G. Westermark, *European J. Biochem.* **6,** 172 (1968).

114. M. Sciaky, N. Limozin, N. Giraud, C. Marriq, M. Charrel, and G. Laurent, *BBA* **221,** 665 (1970).

115. L. E. Henderson, D. Henriksson, P. O. Nyman, and L. Strid, *in* "Oxygen Affinity of Hemoglobin and Red Cell Acid–Base Status. Alfred Benzon Symposium IV." Munksgaard, Copenhagen, and Academic Press, New York, 1971 (in press).

116. B. Andersson, P. O. Göthe, T. Nilsson, P. O. Nyman, and L. Strid, *European J. Biochem.* **6,** 190 (1968).

117. B. Andersson, P. O. Nyman, and L. Strid, unpublished results (1971); see "Atlas of Protein Sequence and Structure 1970–71" (M. O. Dayhoff, ed.), Vol. 5. Natl. Biomed. Res. Found., Silver Spring, Maryland, 1971 (in press).

118. S. Funakoshi and H. F. Deutsch, *JBC* **245,** 4913 (1970).

119. R. E. Tashian and S. R. Stroup, *BBRC* **41,** 1457 (1970).

two proteins are governed by the same structural gene (*35*). All the carbonic anhydrases so far investigated have their terminal α-amino groups acetylated (*120–122*).

Several mutant variants of human carbonic anhydrase B have been recognized by electrophoretic analysis (*35, 118*). Some of these have been characterized with respect to primary structure (*86, 118, 123*). Each of them appears to have a single amino acid residue in the sequence interchanged. The locations of these interchanges in the sequence of the human B enzyme are shown in Table IV at positions 240 and 257. A reported mutational interchange of a threonine for a lysine in the B enzyme (*86*) has not yet been located. Some of the amino acid replacements found in comparisons of the human B enzyme with the homologous B enzymes from other primate species (*119*) have also been included in Table IV at positions 4, 38, 47, 242, 245, and 247. Other such replacements have tentatively been assigned positions 154 and 170.

Table IV also shows the location of amino acid residues modifiable in the native enzymes by various chemical reagents (see Sections II,H and III,D for the structural and functional implications of these modifications). In the human B enzyme, His 204 can be modified with reagents such as bromo- or iodoacetate (*111, 124–126*), iodoacetamide (*124–126*), and bromopyruvate (*127*) with a concomitant loss of activity. His 67 also appears to belong to the active site region as evident from modification studies with sulfonamide inhibitors carrying a reactive chloroacetyl group (*128*). The tyrosyl residues at position 20, 88, (and tentatively at position 114) in the human B enzyme can be modified with tetranitromethane (*129*). In the human C enzyme, a histidyl residue modifiable with bromopyruvate has tentatively been assigned to the sequence region 60–66. From homology considerations it is believed to be in position 64 in the numbering of the B enzyme (*127*) or position 63 in the numbering of the C enzyme based on X-ray data (see Section II,G).

120. C. Marriq, F. Luccioni, and G. Laurent, *BBA* **105,** 606 (1965).
121. G. Laurent, C. Marriq, D. Garçon, F. Luccioni, and Y. Derrien, *Bull. Soc. Chim. Biol.* **49,** 1035 (1967).
122. C. Marriq, F. Luccioni, and G. Laurent, *Bull. Soc. Chim. Biol.* **52,** 253 (1970).
123. R. E. Tashian, S. K. Riggs, and Y. L. Yu, *ABB* **117,** 320 (1966).
124. P. L. Whitney, P. O. Nyman, and B. G. Malmström, *JBC* **242,** 4212 (1967).
125. P. L. Whitney, *European J. Biochem.* **16,** 126 (1970).
126. S. L. Bradbury, *JBC* **244,** 2002 and 2010 (1969).
127. G. O. Göthe and P. O. Nyman, unpublished results (1971).
128. P. L. Whitney, G. Fölsch, P. O. Nyman, and B. G. Malmström, *JBC* **242,** 4206 (1967).
129. F. Dorner, *JBC* **246,** 5896 (1971).

G. Crystal Structure Investigations

Well-defined crystals have been obtained of the human C enzyme (*130*), the human B enzyme (*131–133*), and both the B and C isoenzymes from rhesus monkey (*76*). So far, X-ray crystallographic investigations have been performed only on the human forms (*133–135*).

All the crystallizations have been performed in ammonium sulfate solutions near pH 8.5. In general, the crystal quality improves when the chloroform–ethanol treatment (see Section II,A) is avoided in the purification procedure (*134*). It has not been possible to obtain useful crystals from lyophilized preparations. In the case of the human B enzyme, enzyme–sulfonamide complexes yield the best crystals (*131, 133*). Crystals of the inhibitor-free enzyme can be obtained by seeding acetazolamide–enzyme B crystals into a solution of the native enzyme (*133*). The best crystals of the human C enzyme are obtained when the cysteinyl residue has been reacted with mercury compounds (*136*). Crystals of the mercury-free enzyme can be obtained by dialyzing out the mercurials against a buffer solution containing cysteine.

Compounds of various nature such as anionic or sulfonamide inhibitors can be dialyzed into crystals of both human isoenzymes (*132–134, 136–140*). The zinc ion can be removed from the crystallized enzymes by

130. B. Strandberg, B. Tilander, K. Fridborg, S. Lindskog, and P. O. Nyman, *JMB* **5,** 583 (1962).
131. J. E. Coleman, *JBC* **243,** 4574 (1968).
132. A. Liljas, K. K. Kannan, P. C. Bergstén, K. Fridborg, L. Järup, S. Lövgren, H. Paradies, B. Strandberg, and I. Waara, *NASA* (*Natl. Aeron. Space Admin.*), *Spec. Publ.* **NASA SP-188,** 89 (1969).
133. K. K. Kannan, K. Fridborg, P. C. Bergstén, A. Liljas, S. Lövgren, M. Petef, B. Strandberg, I. Waara, L. Alder, S. O. Falkbring, P. O. Göthe, and P. O. Nyman, *JMB* (1971) (in press).
134. A. Liljas, K. K. Kannan, P. C. Bergstén, I. Waara, K. Fridborg, B. Strandberg, U. Carlbom, L. Järup, S. Lövgren, and M. Petef, to be published.
135. K. K. Kannan, A. Liljas, I. Waara, P. C. Bergstén, S. Lövgren, B. Strandberg, U. Bengtsson, U. Carlbom, K. Fridborg, L. Järup, and M. Petef, *Cold Spring Harbor Symp.* **36** (1971) (in press).
136. B. Tilander, B. Strandberg, and K. Fridborg, *JMB* **12,** 740 (1965).
137. K. Fridborg, K. K. Kannan, A. Liljas, J. Lundin, B. Strandberg, R. Strandberg, B. Tilander, and G. Wirén, *JMB* **25,** 505 (1967).
138. I. Waara, A. Liljas, U. Bengtson, K. Fridborg, K. K. Kannan, B. Strandberg, and P. C. Bergstén, unpublished results (1971).
139. A. Liljas, "Crystal Structure Studies of Human Erythrocyte Carbonic Anhydrase C at High Resolution." Almqvist & Wiksell, Stockholm, 1971.

TABLE V
HEAVY ATOM DERIVATIVES OF HUMAN CARBONIC ANHYDRASE C USED IN THE HIGH RESOLUTION STRUCTURE DETERMINATION

Derivative	Occupancy (electrons)	Resolution (Å)	Position symbol and chemical group	Binding site in the protein[a]
E-MMTGA[b]	63	2.0	M_1(Hg)	Cys 210
E-AmSulf[b]	57	2.0	A_1(Hg)	Cys 210
	23		$A_2(SO_2NH_2)$	Close to Cys 210
	32		$A_3(SO_2NH_2)$	Zn^{2+} and Thr 197
	66		A_4(Hg)	His 63 and Glx 67
	23		A_5(Hg)	Between two molecules
E-I^-	28	2.5	I_3	Zn^{2+} and Thr 197
	20		I_6	Arg 243,Gln 11,Lys 22
	7		I_7	His 128
E-$Au(CN)_2^-$	70	2.5	G_8 [$Au(CN)_2^-$]	H_2O attached to Zn^{2+}, His 128

[a] The numbering of residues is based on the electron density map (cf. Table IV).
[b] Abbreviations: E, human carbonic anhydrase C; MMTGA, methylmercurithioglycolic acid; and AmSulf, 3-acetoxymercuri-4-aminobenzene-1-sulfonamide.

dialysis against 2,3-dimercaptopropanol-1, and other metal ions can subsequently be introduced [(*133, 136*); see also Section III,D].

The space group of the human B enzyme is $P2_12_12_1$, and the cell dimensions are $a = 81.5$ Å, $b = 73.6$ Å, and $c = 37.1$ Å. There is one molecule per asymmetric unit (*133*). The space group of the human C enzyme is $P2_1$ with one molecule per asymmetric unit (*130*). The cell dimensions are $a = 42.7$ Å, $b = 41.7$ Å, $c = 73.0$ Å, and $\beta = 104.6°$ (*134*). In general, the crystalline derivatives of the human carbonic anhydrases are quite isomorphous with the native enzymes.

The structure of the human C enzyme has been determined to a resolution of 2.0 Å (*134, 135*). Four heavy atom derivatives have been used, and the observed binding sites in the enzyme molecule are summarized in Table V. The detailed parameters of the heavy atom derivatives have been given by Liljas *et al.* (134). The numbering of residues employed in Table V and throughout this section refers to tentative positions in the sequence of the human C enzyme as deduced from the electron density maps. This numbering does not correspond exactly with the one

140. P. C. Bergstén, I. Waara, S. Lövgren, A. Liljas, K. K. Kannan, and U. Bengtsson, *in* "Oxygen Affinity of Hemoglobin and Red Cell Acid–Base Status. Alfred Benzon Symposium IV." Munksgaard, Copenhagen, and Academic Press, New York, 1971 (in press).

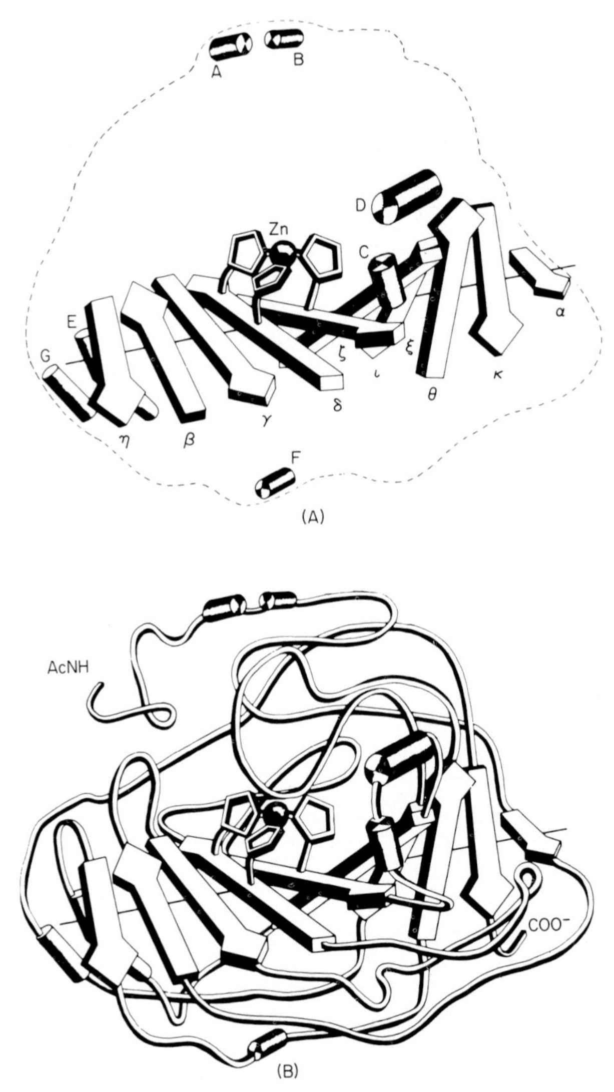

FIG. 1. (A) The main features of the secondary structure of human carbonic anhydrase C. Helices are represented by cylindrical shapes marked with Latin letters. The arrows correspond to the different strands of the large β structure denoted with Greek letters. The arrows are pointing from amino end toward carboxyl end. The β-structure strands are twisted around a common axis indicated with a solid line. The zinc ion is firmly connected to the β structure through its three ligands from the protein, all being histidyl residues. The dashed line represents the molecular boundary. (B) The main-chain folding of the human C enzyme from the acetylated amino end to the carboxyl end. The viewing angle is the same as in (A) looking into the active site cavity.

based on the chemically determined sequence of the human B enzyme (see Table IV).

1. *General Description of the Structure of Human Carbonic Anhydrase C*

The molecular boundaries in the electron density map are completely unambiguous, and the background density is very low throughout (*134*, *135*). The polypeptide chain can easily be followed in most places. However, some breaks in the continuity of the chain occur on the surface of the molecule. Most carbonyl groups are easily identified in the electron density map. Aromatic side chains are always flat, and in many cases the density is lower in the center of the ring. Side chains, which according to the chemically determined sequence (Section II,F) are forked, are in most cases seen as forked densities.

The molecule has an ellipsoid shape with the approximate dimensions 41 by 41 by 47 Å. The active site (Section II,G,4) is situated in a deep pocket leading into the center of the molecule, and the zinc ion is located almost at its bottom (*134*, *135*). These results are in good agreement with the low resolution structure (*137*). The tertiary structure obtained from the high resolution work (*134*, *135*) is shown in Fig. 1. The dominating feature of the molecule is a β structure that is passing through the whole molecule and divides it into two halves. The half of the molecule that in Fig. 1 is below the β structure consists of six pieces of polypeptide chain running almost perpendicularly to the chain direction of the β structure. This bottom layer contains an extended part of the large β structure. This "lower" half of the molecule also contains three helical segments (see Fig. 1), and between the six chain segments and the β structure there is a core of hydrophobic residues. The half of the molecule above the β structure contains another six pieces of the polypeptide chain. Also these chain segments are more or less perpendicular to the chain direction of the β structure and they include four helices. A few hydrogen bonds are formed between these six chains. The active site cavity is formed between the β structure and this upper layer (Section II,G,4). Figure 2 is a stereo drawing of all α-carbons.

2. *Secondary Structures*

Table VI gives the sequential order of the residues which are part of any detectable secondary structures. The major components are seven right-handed helical segments (A–G) and a β structure composed of 10 strands (α-κ). The schematic drawing in Fig. 3 shows the main-chain

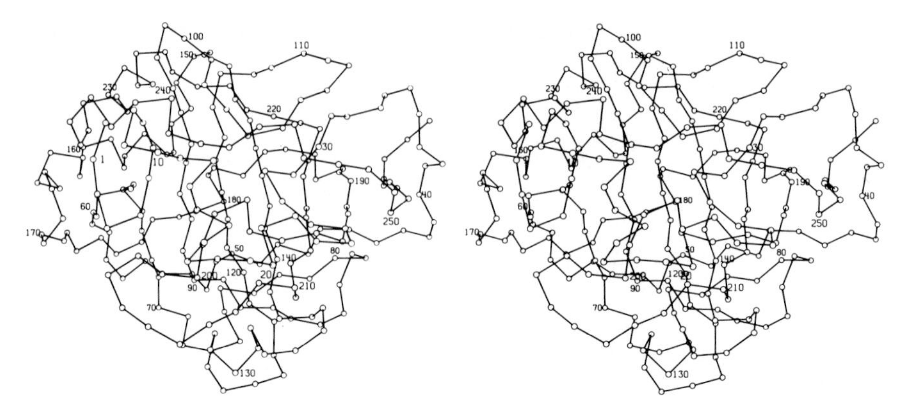

FIG. 2. A stereo drawing using the ORTEP program written by Dr. Carrol N. Johnson of all α-carbons in human carbonic anhydrase C.

hydrogen bonds and provides additional details of the secondary structure.

a. Helical Segments. As seen in Fig. 1, all the seven helical segments are found on the surface of the molecule. The helices comprise about 20% of the total number of residues (cf. Section II,H,1). The criteria

TABLE VI
THE SEQUENTIAL ORDER OF THE SECONDARY STRUCTURE COMPONENTS OF HUMAN CARBONIC ANHYDRASE C

Helix	β Structure	Residue Nos.
A		11–17
B		18–23
	α	38–40
	β	44–61
	γ	64–81
	δ	86–95
	ϵ	114–122
C		122–126
D		127–133
	ζ	138–149
E		154–163
	η	169–174
F		178–183
	θ	188–194
	ι	211–223
G		224–234
	κ	252–255

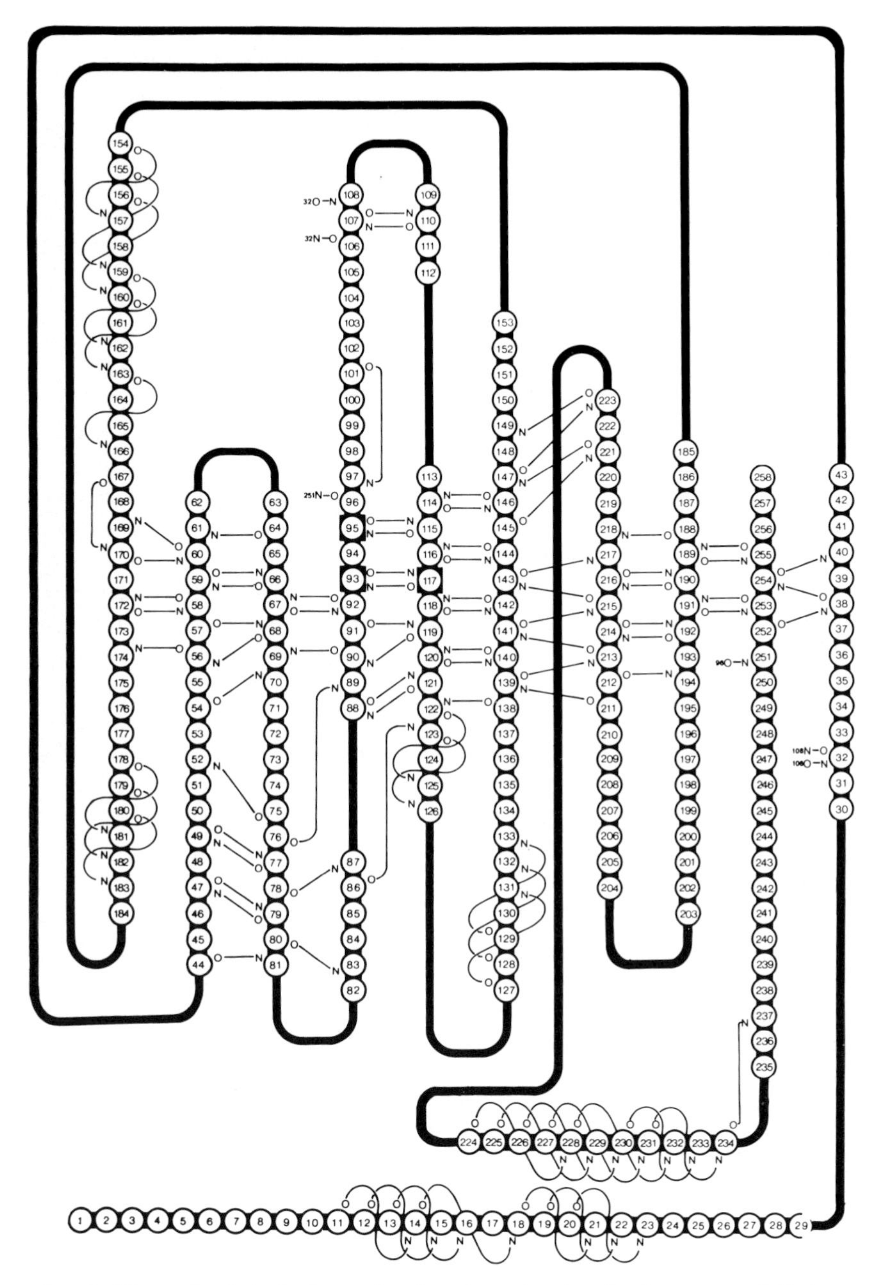

FIG. 3. The main-chain hydrogen-bond pattern showing the large area of β structure and the helices.

TABLE VII
HELICAL PARAMETERS IN HUMAN CARBONIC ANHYDRASE C

Helix	Residue Nos.	Unit rise (Å/residue)	No. of residues/turn
A	11–17	1.7	3.1
B	18–23	1.8	3.0
C	122–126	2.0	3.2
D	127–133	1.7	3.4
E	154–163	1.8	3.6
F	178–183	1.8	3.3
G	224–234	1.7	3.3
G′	224–232	1.6	3.7
α [Ref. (*141*)]		1.5	3.6
3_{10} [Ref. (*142*)]		2.0	3.0

employed for counting a residue as helical are that the residue has a helical conformation and at least one hydrogen bond, approximately parallel to the helical axis. The helical parameters are given in Table VII together with the parameters for ideal α and 3_{10} helices. Obviously, only a small fraction of the residues belong to the classic α-helix type (*141*). The helices D and G are close to true α helices if 3 residues at the carboxyl end of helix G are excluded. The helices A, B, and F are short and are close to 3_{10} helices (*142*). A stereo drawing of the E helix is shown in Fig. 4. In several of the helices, the peptide plane is inclined to the helical axis; thus, the carbonyl group is pointing out from the helix in a way similar to the one described by Nemethy *et al.* (*143*).

Venkatachalam (*144*) has discussed two types of single turns of the polypeptide chain with a hydrogen bond similar to the one in a 3_{10} helix. In type I, the side chains on either side of the central amide plane point away from the carbonyl group as in a 3_{10} helix, whereas, in type II, the carbonyl points almost to the same direction as the side chains. In carbonic anhydrase C (*135*) three turns of type I, and one turn of type II, have been observed (see Fig. 3).

b. β Structures. Totally about 37% of the amino acid residues in human carbonic anhydrase C are located in β structures (cf. Section II,H,1). The large β structure is built up of 10 backbone segments, all

141. L. Pauling, R. B. Corey, and H. R. Branson, *Proc. Natl. Acad. Sci. U. S.* **37,** 205 (1951).
142. J. Donohue, *Proc. Natl. Acad. Sci. U. S.* **39,** 470 (1953).
143. G. Nemethy, D. C. Phillips, S. J. Leach, and H. A. Scheraga, *Nature* **214,** 363 (1967).
144. C. M. Venkatachalam, *Biopolymers* **6,** 1425 (1968).

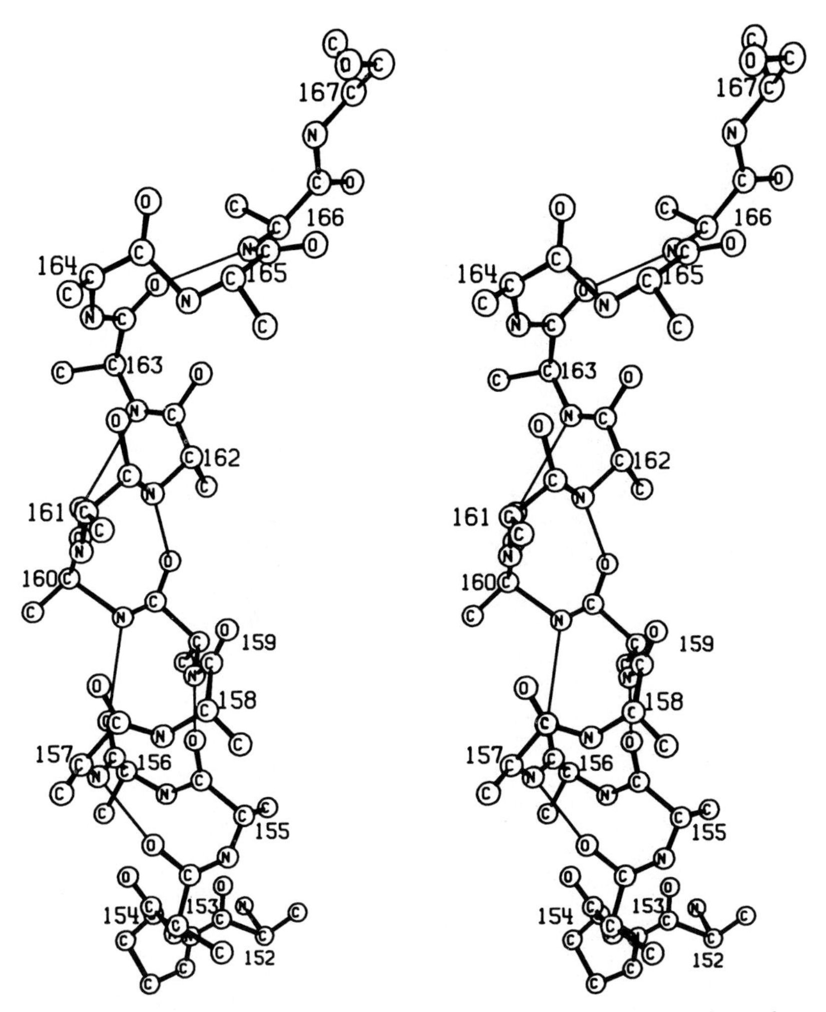

FIG. 4. An ORTEP stereo drawing of the E helix. Few hydrogen bonds are formed since the carbonyl groups are inclined to the helix axis.

running antiparallel to each other except for the chain pairs ζ–ι and κ–α which are parallel (Figs. 1 and 3). It is estimated that totally 91 residues belong to the large β structure, some of them not hydrogen bonded, and that 64 hydrogen bonds are formed within this structure element. One interesting feature is the fact that the β structure is twisted around a common axis, about 220° from the left edge η to the right edge α (see Fig. 1). This twist is related to a right-handed twist of about 20° between the peptide planes of successive residues along most chain segments of the β structure. The strands β, γ, and δ are extended on one side of the pleated sheet bending down below strands ε, ζ, ι, and θ; thus, a tunnel of β structure is formed closed by three residues joining strands α and β [cf. α-chymotrypsin (*145*)]. There are certain similarities, but also major differences, between the pleated sheets of carbonic anhydrase and carboxypeptidase A (*146*). In the latter protein, eight parts of the main chain form a twisted β structure in which neighboring chain segments run both parallel and antiparallel through the middle of the carboxypeptidase molecule.

The β structures are generally found in the interior of a protein molecule, and the residues attached to these structures are mainly hydrophobic. This is true also for the pleated sheet in human carbonic anhydrase C, except for the extension of strands β, γ, and δ and the active site region which is located on one part of the large β structure. These parts of the β structure are exposed to the solvent, and here a number of polar residues are found.

In addition to the large β structure, there is one smaller antiparallel β structure (Fig. 3) situated on the surface of the molecule (*135*). It consists of three short chain segments of 1–3 residues and forms some type of sandwich on the large pleated sheet with the β-structure planes almost parallel and about 10 Å apart. The chain direction in the large pleated sheet is perpendicular to the chain direction in the smaller one.

3. *Locations of Various Side Chains*

The known parts of the sequence and the clear shape of the side chains in the electron density map provide some information of general interest. The single cysteinyl residue in the human C enzyme is located on the surface of the molecule and is the binding site for mercurials (Table V and Section II,H,2). The single methionyl residue is situated in the interior of the molecule. Many of the phenylalanyl, tryptophanyl, and tyrosyl residues are wholly or partly buried (cf. Section II,H,2). Some of them are in the vicinity of charged or polar groups. At least two pronounced aromatic clusters are found. One is situated below the large β structure and contains probably 6 phenylalanyl residues in an asymmetric packing. The other (Fig. 5) is near the amino end, above the active site, and includes 1 tyrosyl, 1 phenylalanyl, 1 histidyl, and 2 tryptophanyl residues packed in an irregular manner (cf. Section II,H,1).

145. J. J. Birktoft, D. M. Blow, R. Henderson, and T. A. Steitz, *Phil. Trans. Roy. Soc. London* **B257,** 67 (1970).

146. W. N. Lipscomb, J. A. Hartsuck, G. N. Reeke, Jr., F. A. Quiocho, P. H. Bethge, M. L. Ludwig, T. A. Steitz, H. Muirhead, and J. C. Coppola, *Brookhaven Symp. Biol.* **21,** 24 (1968).

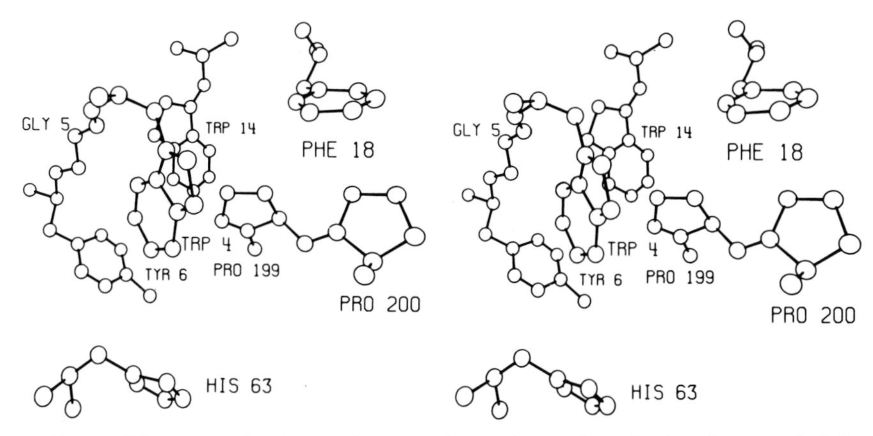

FIG. 5. The aromatic cluster close to the amino end with the rings stacked in an irregular way. His 63 is situated in the active site region.

Among the charged residues known from the chemical sequence, none has been found in the interior of the molecule. However, 3 of the arginyl residues seem to be hydrogen bonded on the surface of the protein rather than extending out in the solution. The lysyl residues seem to extend farther out in the solution and are seldom visible in the electron densities to their full length.

The groups identified as histidyl residues appear evenly distributed between the surface and the interior of the molecule (cf. Section II,H,2). Finally, it is of interest to note that 12 of the 17 prolyl residues are on the surface of the molecule pointing outward. Another 2 or 3 prolines are located on the surface but pointing inward from the main chain.

4. *The Active Site Region*

The active site region of human carbonic anhydrase C has been located by the natural label provided by the zinc ion that has been shown to be essential for activity [(*147*); see Section III,D]. The zinc ion has the highest electron density of the whole map. The positions of anionic and sulfonamide inhibitors also reveal the location of the active site. As in the case of some other enzymes (*146*, *148*, *149*) the active site is lined up on the β structure, but the zinc ligands in the human C en-

147. S. Lindskog and B. G. Malmström, *JBC* **237,** 1129 (1962).

148. M. J. Adams, G. C. Ford, R. Koekoek, P. J. Lentz, A. McPherson, M. G. Rossman, I. E. Smiley, R. W. Schevitz, and A. J. Wonnacott, *Nature* **227,** 1098 (1970).

149. H. W. Wykoff, D. Tsernoglou, A. W. Hanson, J. R. Knox, B. Lee, and F. M. Richards, *JBC* **245,** 305 (1970).

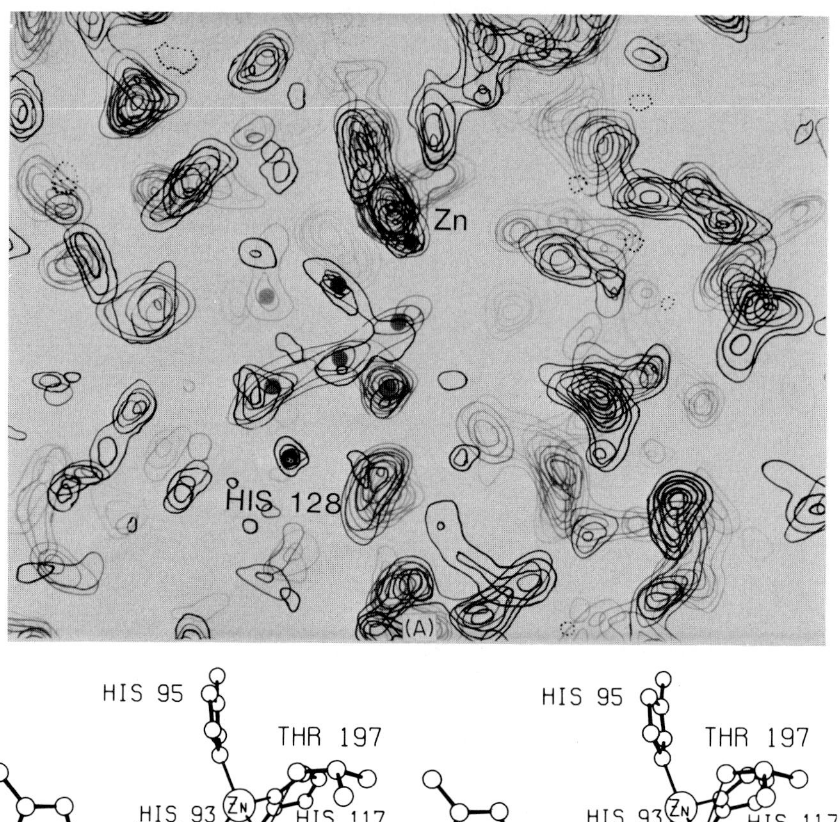

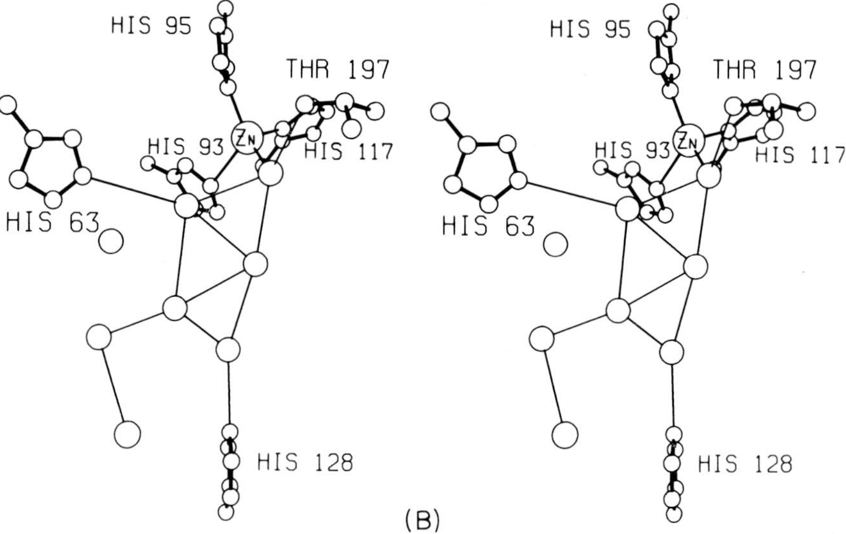

FIG. 6. (A) Electron density map in the region of the zinc ion and part of the active site. Between the zinc ion and His 128 there are eight peaks of density indicated by dots. These densities are poorly connected to the protein and are probably water molecules. One of them is intimately associated with the zinc ion. (B) An ORTEP stereo drawing of the same region as in (A) showing some of the important residues in the active site and the presumed hydrogen-bond arrangement.

zyme, in contrast to carboxypeptidase, come from the center of the β structure (Figs. 1 and 3).

a. The Metal Ion. The zinc ion is located almost in the center of the molecule. It has three ligands from the protein. All of them seem to be histidyl residues according to the electron density maps. The chemical sequence is not yet known for this region of the human C enzyme nor of the B form (Table IV). Two of the histidyl residues, 93 and 95, belong to chain δ in the β structure. Both these imidazoles bind to the zinc ion with their 3′-nitrogens. The third zinc ligand from the protein is His 117. This residue is situated in strand ϵ which is the chain neighboring strand δ. His 117 is hydrogen bonded to His 93. His 117 is binding to the metal with its 1′-nitrogen. The zinc ion has a fourth ligand which seems to be a water molecule or a hydroxyl ion in the active site cavity as illustrated in Fig. 6 (cf. Section III,G). The zinc coordination is closest to tetrahedral although rather distorted (cf. Section III,D). The greatest deviation from tetrahedral angles seems to be around 20°.

A number of different side chains are in the vicinity of the zinc ion and its ligands but are too far away to bind directly to the metal or are by nature unable to bind metals. Some of these residues are shown in Fig. 7.

The zinc coordination of human carbonic anhydrase C provides an

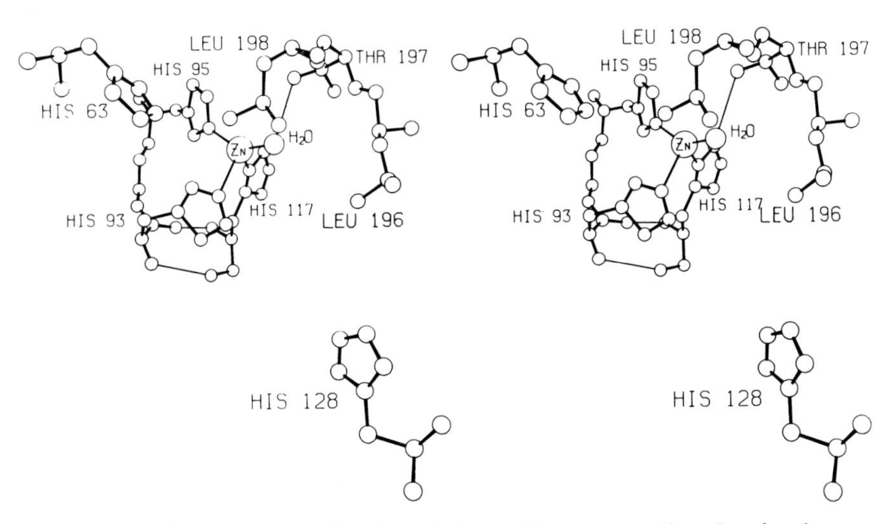

FIG. 7. An ORTEP stereo drawing of the residues surrounding the zinc ion as well as 2 histidyl residues farther out in the active site.

interesting comparison with the structures of insulin (*150*) and carboxypeptidase (*151*). The zinc ion in insulin has an octahedral environment. It binds to the 3′-nitrogens of 3 identical histidyl residues from three insulin molecules and to three water molecules. In carboxypeptidase, the zinc ion has a near-tetrahedral coordination from two 1′-nitrogen of two imidazoles, one glutamic acid, and one water molecule. In model complexes of histidines and metal ions, the 1′-nitrogen is usually binding to the metal ion (*152*).

b. Other Features of the Active Site. The active site cavity has a depth of about 15 Å. The zinc ion is located about 12 Å from the ellipsoidal surface of the molecule. The cavity is situated on β-structure strands γ, δ, ε, and ζ (Fig. 1). Helix D forms one of the walls to the active site and residues 195–208 form a loop above the active site in the orientation of this picture. The outer part of this loop is not visible in the electron density map probably because of poor stabilization from other parts of the protein.

Chemical modification studies indicate that possibly two reactive histidyl residues are located in the active site region [(*127*, *153*); see Section III,F]. This is in good agreement with the electron density map where, in addition to the zinc ligands, two possible histidyl residues are observed, His 63 and His 128, the former being tentatively localized by chemical methods (see Section II,F). These two histidyl residues are situated on opposite sides of the entrance to the inner portion of the active site cavity, thus defining the outer limits of the space where most inhibitors bind and where the catalytic events presumably take place (cf. Section III,G). The position of the complex anion, $Au(CN)_2^-$, used in the structure determination, is shown in Fig. 8 (cf. Section III,E,1).

In the inner part of the cavity only one aromatic residue, probably a phenylalanine, is observed. It is almost completely buried in the structure. Many other aromatic residues are found in the neighborhood of the active site. Both aromatic clusters are within 10 Å from the zinc ion and still less from the aromatic ring systems of sulfonamide molecules binding to the enzyme (see Section III,E,2). Most of the other side

150. M. J. Adams, T. L. Blundell, E. J. Dodson, G. G. Dodson, M. Vijayan, E. N. Baker, M. M. Harding, D. C. Hodgkin, B. Rimmer, and S. Sheat, *Nature* **224**, 491 (1969).

151. W. N. Lipscomb, J. A. Hartsuck, F. A. Quiocho, and G. N. Reeke, Jr., *Proc. Natl. Acad. Sci. U. S.* **64**, 28 (1969).

152. H. C. Freeman, *Advan. Protein Chem.* **22**, 257 (1967).

153. M. Kandel, A. G. Gornall, S. C. C. Wong, and S. I. Kandel, *JBC* **245**, 2444 (1970).

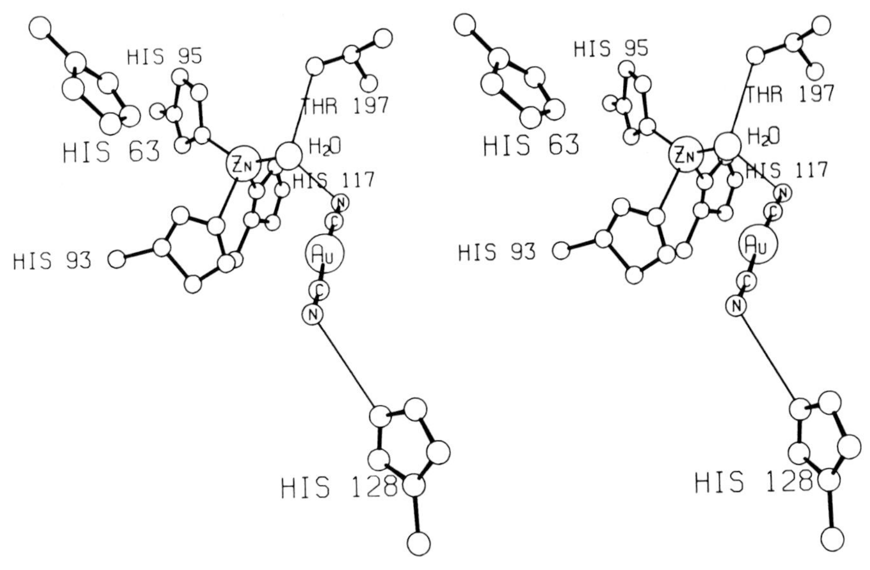

FIG. 8. An ORTEP stereo drawing of an $Au(CN)_2^-$ ion binding to the active site of the human C enzyme as determined by X-ray diffraction. One cyanide group is hydrogen bonded to the zinc-coordinated water molecule whereas the other cyanide group is within hydrogen-bond distance from His 128. Several water molecules have been displaced by the anion (cf. Figs. 6 and 7).

chains in the active site are forked such as valyl, threonyl, leucyl, glutamyl, or aspartyl residues. Since the corresponding part of the chemical sequence is very little known, it is too early to discuss the nature and function of these side chains in detail.

Eight peaks of electron density, some of them not well resolved from each other, are visible in the active site (Fig. 6). These densities are poorly connected to the protein except via the zinc ion, the histidyl residues and Thr 197. The sulfonamides (Section III,E,2) and anionic compounds (Fig. 8) bind in their place without any positive peaks occurring in difference electron density maps. Thus, these densities cannot be part of the protein being displaced by inhibitors. Therefore, it is likely that the densities represent molecules or ions from the solvent, i.e., water, ammonia, sulfate, tris, or chloride. The shape of the densities rules out sulfate and tris which, in addition, are very poor inhibitors. The shortest distances between the centra of these electron densities are in good agreement with the hydrogen bond distance in water, 2.8 Å. Although it seems probable that these densities are produced by eight water molecules arranged as indicated in Fig. 6. (cf. Section III,G), it cannot be excluded that some of them are ammonia or chloride. Such molecules would probably not drastically perturb a native ordered water structure.

H. Conformation in Solution

While the human C enzyme is as yet the only carbonic anhydrase that has been studied extensively by X-ray methods, physical and chemical investigations pertaining to the conformations in solution have been performed on several forms of the enzyme from mammalian as well as other sources. The results strongly suggest that the tertiary structures of various mammalian carbonic anhydrases are highly similar. However, the B and C isoenzyme types often show characteristic differences, and small but significant variations are usually observed within each of these groups.

1. *Spectroscopic Studies of the Native Enzymes*

In Volume II of this treatise, Timasheff (*154*) has discussed in some detail the optical rotatory dispersion (ORD), circular dichroism (CD), and infrared spectra of the carbonic anhydrases. Therefore, we shall only summarize the main features of the spectra in the light of what is now known about the crystal structure of the human C enzyme and add some new data concerning a bacterial and a plant carbonic anhydrase.

In the wavelength region 240–310 nm, the ORD and CD spectra of the mammalian carbonic anhydrases are unusually complex and rich in detail with multiple Cotton effects, undoubtedly resulting from aromatic chromophores. All these enzymes reveal common features as illustrated by the ORD curves in Fig. 9, but there are differences in detail. It has been proposed (*106, 155–157*) that the complex ORD and CD patterns arise from the presence of tightly packed aromatic residues, presumably with charged groups in their vicinity producing a specific asymmetric environment. The characteristic Cotton effect near 290 nm is present in all the mammalian carbonic anhydrases studied so far, and it can only result from tryptophan. Some features of the spectra are presumably owing to varying contributions of tyrosines (*155, 156*). It is of

154. S. N. Timasheff, "The Enzymes," 3rd ed., Vol. II, p. 371, 1970.
155. S. Beychok, J. McD. Armstrong, C. Lindblow, and J. T. Edsall, *JBC* **241,** 5150 (1966).
156. A. Rosenberg, *JBC* **241,** 5119 and 5126 (1966).
157. J. E. Coleman, *NASA* (*Natl. Aeron. Space Admin.*), *Spec. Publ.* **NASA SP-188,** 141 (1969).

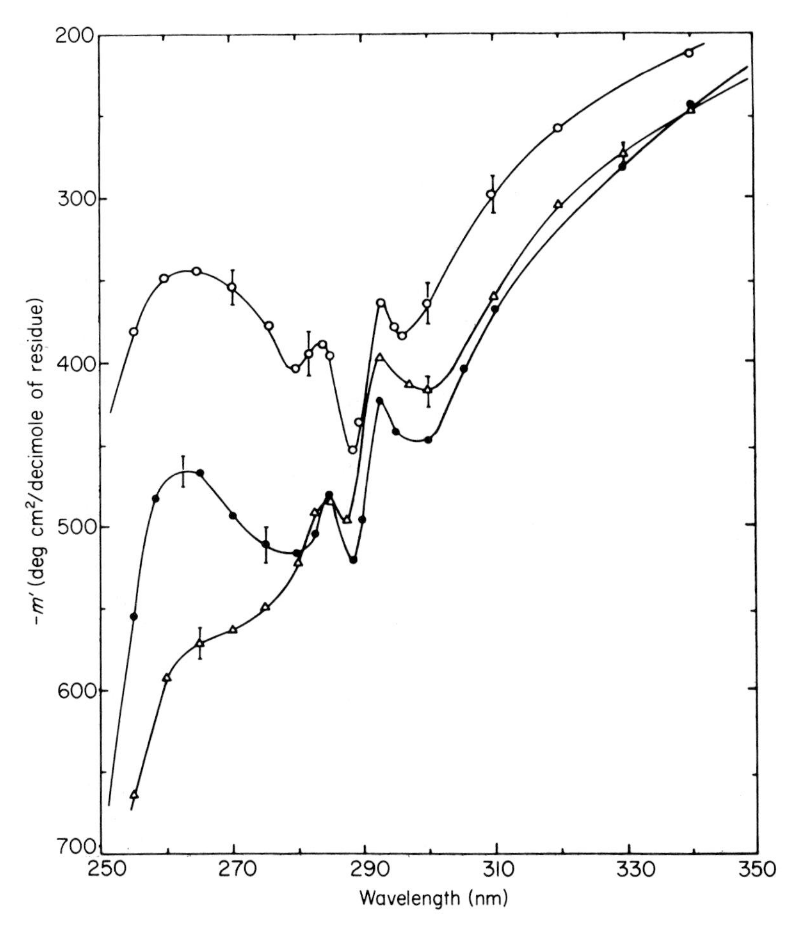

FIG. 9. Optical rotatory dispersion curves for (Δ) bovine carbonic anhydrase, (○) the human B enzyme, and (●) the human C enzyme. Conditions: 0.05 *M* phosphate buffer, pH 6.5, 25°. From Rosenberg (*156*).

interest to note that the aromatic residues in the amino-terminal sequence of B isoenzymes from several primate species have been shown to be invariant (*119*). Furthermore, the corresponding sequence of the human C enzyme seems to display a high degree of homology with respect to the B isoenzyme (Table IV). Many of the aromatic residues near the amino terminus of the human C form take part in the aromatic cluster shown in Fig. 5, and it can be tentatively concluded that this structural feature is common to several mammalian carbonic anhydrases and, at least in part, responsible for the complex aromatic Cotton effects. A more thorough investigation of the environment of aromatic side

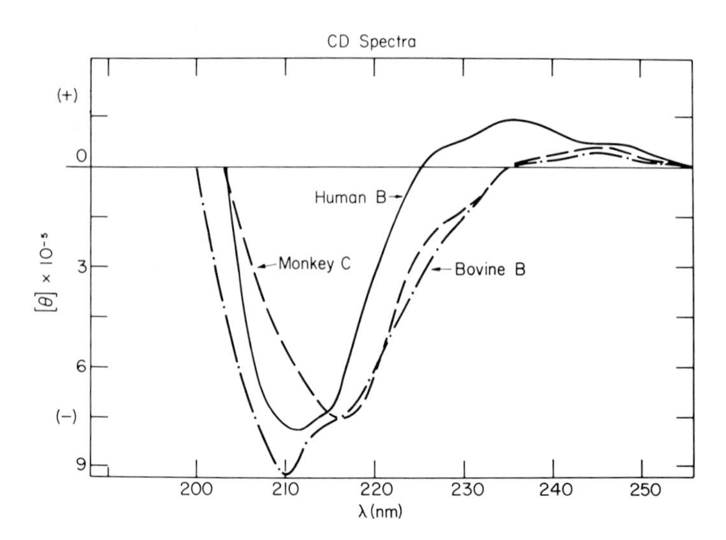

FIG. 10. Ultraviolet CD spectra of (——) the human B enzyme, (- - -) the rhesus monkey C enzyme, and (- · · -) the bovine enzyme. Conditions: 0.025 *M* tris buffer, pH 7.0, 23°. The ellipticities are calculated per decimole of protein rather than on the basis of mean residue weight. From Coleman (*157*).

chains, especially the distribution of charges, must await the completion of the sequence determination so that a detailed interpretation of the X-ray data can be performed.

In the far ultraviolet region, below 240 nm, the ORD patterns of the mammalian enzymes show several features in common indicating similar tertiary structures. A shallow but broad trough is found at 220 nm, a peak around 205 nm, and another trough at 196 nm. The absence of a trough at 233 nm and the numerically small b_0 values obtained from plots according to the Moffit–Yang equation have been taken to indicate a very low helix content (*73*, *106*, *155*, *156*). Similarly, the CD spectra lack the characteristic features of α-helical conformations. The spectra show subtle differences between the different enzyme forms as illustrated in Fig. 10. However, in most cases the spectra show extrema or shoulders near 217 nm and near 210 nm. A negative ellipticity peak at 217 nm is usually associated with antiparallel β structures (*158*), and the CD spectra of the carbonic anhydrases seem to be in accordance with the presence of such conformations. Quantitative estimates from the CD spectrum (*158*) of the contents of α helix and β structure of the human C enzyme do not correspond well to what is observed in the crystal structure (see Section II,G) where, however, the

158. N. Greenfield and G. D. Fasman, *Biochemistry* **8**, 4108 (1969).

helices are short with only a couple of turns each and mostly distorted. Presumably, the aromatic chromophores discussed above contribute also with CD bands below 240 nm (*155, 156*), thus obscuring the Cotton effects arising from the secondary structures.

Timasheff and co-workers (*159*) gave the first clear evidence for the presence of β structure in a carbonic anhydrase in studies of the infrared absorption spectrum of the bovine enzyme. The enzyme shows an absorption maximum at 1637 cm^{-1} and a weak shoulder at 1650 cm^{-1}. These frequencies correspond to the absorption maxima for polypeptides having an antiparallel β structure and an α-helical conformation, respectively.

The carbonic anhydrases from *Neisseria sicca* and parsley have considerably less intense aromatic Cotton effects than the mammalian enzymes. Both the bacterial and plant enzymes have a relatively low content of aromatic amino acids. The far ultraviolet CD spectrum of the bacterial enzyme shows a negative ellipticity band with an extremum at 208 nm and a broad shoulder between 215 and 223 nm (*105*). In fact, there is a striking similarity between the CD spectrum of the *Neisseria* enzyme and that of carboxypeptidase A (*158*). The latter enzyme has a helix content of 23–30% and 18% of a twisted pleated sheet consisting of both parallel and antiparallel strands (*146*). From the magnitude of the ellipticity at 208 nm (*158*), a helix content of approximately 10% can be estimated for the bacterial carbonic anhydrase, and it would thus seem possible that its tertiary structure is similar to that of a mammalian carbonic anhydrase.

The parsley enzyme shows a rather different picture. The ORD pattern displays a trough at 233 nm, which is characteristic of α-helical polypeptides. The α-helix content calculated from the magnitude of the mean residue rotation at 233 nm ($m'_{233} = -5500$ deg cm^2 $decimole^{-1}$) or from a Moffit–Yang plot of the rotation between 300 and 550 nm ($b_0 = -195$ deg cm^2 $decimole^{-1}$) is in both cases near 30% (*57*).

2. *Titration Studies and Chemical Modifications*

Hydrogen ion titrations and acid difference spectra of the human B and C isoenzymes (*107, 109, 160*) and the bovine enzyme (*108*) reveal the presence of masked histidyl, tyrosyl, and tryptophanyl residues. No evidence has been found for buried carboxyl groups or lysyl residues. All the lysyl groups of the human B and the bovine enzymes can be amidinated without loss of activity (*108, 124*). The titration curve of the human B enzyme is reversible between pH 4 and 11.5. Only 4 out

159. S. N. Timasheff, H. Sussi, and L. Stevens, *JBC* **242**, 5467 (1967).
160. L. M. Riddiford, *JBC* **240**, 168 (1965).

of 11 histidines titrate freely, while the remaining 7 residues become unmasked and take up protons near pH 4.2 where the protein unfolds (*109*). Closely similar results have been obtained for the human C form (*107*) and the bovine enzyme (*108*). It seems reasonable to assume that the histidyl residues liganded to the zinc ion (see Fig. 6) are among the nontitratable ones, while those situated in the active site region, chemically modifiable (see Fig. 6 and Section III,F), presumably are also titratable. The individual pK values of the titratable histidines are presently being studied by NMR spectroscopy (*161*).

The single sulfhydryl group of the human C enzyme reacts with mercurials [(*136*); see Section II,G], while the sulfhydryl group of the native human B enzyme appears to be masked (*17*).

Spectrophotometric titrations of tyrosyl ionizations have shown that 3 out of 9 residues in the human C enzyme, 4 out of 8 in the human B enzyme, and only 1 out of 8 in the bovine enzyme titrate freely (*107–109*). Nitration of tyrosyl residues with tetranitromethane gave similar estimates regarding the availability of these side chains (*162*). No activity losses were observed during nitration of the native enzymes near neutral pH. The modifiable tyrosyl residues have been localized in the amino acid sequence of the human B enzyme [(*129*); see Section II,F]. Iodination of tyrosyl residues in the human B enzyme leads to loss of activity and major conformational changes (*162*).

Acid denaturation of the human isoenzymes gives rise to unusually large spectral changes, and it has been suggested that most or all of the tryptophans are buried in the interior of the native molecules (*107, 160*). On the other hand, solvent-perturbation difference spectra indicate that about 40% of the tryptophans are exposed to the solvent in the native proteins (*163*). Presumably, the arrangement of several aromatic residues in clusters as shown in Fig. 5 has substantial effects not only on the ORD and CD patterns but also on the absorption spectrum, and a sizable fraction of the acid difference spectrum may result from the exposure of these clustered residues to the solvent.

3. *Stability and Denaturation*

In addition to the discontinuity in the titration curves and the blue shift of the absorption spectrum, referred to in Section III,H,2, the acid denaturation of the human and bovine carbonic anhydrases are associated with an expansion of the molecule, a marked decrease of the

161. R. W. King and G. C. K. Roberts, *Biochemistry* **10,** 558 (1971).
162. J. A. Verpoorte and C. Lindblow, *JBC* **243,** 5993 (1968).
163. S. Lindskog, A. Nilsson, and K. Utterberg, unpublished results (1971).

solubility, and large changes of the ORD including the disappearance of the aromatic Cotton effects (*155, 156*). While these phenomena indicate a substantial unfolding of the polypeptide chain, it is evident that the acid-denatured molecules are not in a completely random state. Thus, solvent-perturbation difference spectra indicate that a fraction of the aromatic residues are shielded from the solvent (*163*), and the ORD patterns develop characteristics of α-helical structures. The α-helix content would be 10–20% as estimated from b_0 values and the magnitude of the trough at 233 nm (*106, 155, 156*).

The mammalian enzymes are unusually stable at alkaline pH. However, the human B and C isoenzymes differ markedly in their stabilities toward alkaline denaturation. Laurent *et al.* (*164*) showed that the C enzyme is irreversibly denatured after exposure to pH 12.7 for 2 min, whereas the B enzyme retains nearly all its zinc and enzymic activity after a similar treatment. Riddiford *et al.* (*107*) found that the human C enzyme loses about 50% of its activity in 3 hr at pH 10.5 and 25° and that the increase in denaturation rate between pH 10.5 and 12 is roughly similar to that for the B enzyme between pH 12 and 13. The bovine enzyme can be exposed to pH 13.0 for 30 min at 0° without any irreversible activity loss (*108*). Alkaline denaturation is associated with complex changes in the ORD and CD patterns. A detailed discussion of these phenomena is given by Beychok *et al.* (*155*).

The human B and C isoenzymes also show differences in their stabilities toward denaturation by urea and guanidine hydrochloride. The differences begin to show up at 1 *M* guanidine hydrochloride where the C enzyme shows rapid changes in the absorption spectrum, whereas the B enzyme is unchanged (*165*). The C enzyme undergoes complete unfolding in 6 *M* urea almost instantly in contrast to the B form which requires more than an hour to undergo the transformation. At lower urea concentrations (2 *M*) a smaller, reversible, conformational change takes place leading to an exposure of tyrosyl residues; thus, they can all react with tetranitromethane (*162*). This reaction leads to irreversible denaturation, however.

Attempts to achieve a refolding of the urea-denatured human isoenzymes have so far given poor yields of native material (*165, 166*), possibly because of the presence of a sulfhydryl group in each of these forms. The bovine enzyme contains no sulfhydryl group, however, and Wong

164. G. Laurent, D. Garçon, M. Charrel, and Y. Derrien, *Compt. Rend. Soc. Biol.* **157**, 2028 (1963).

165. J. T. Edsall, S. Mehta, D. V. Myers, and J. McD. Armstrong, *Biochem. Z.* **345**, 9 (1966).

166. U. Carlsson, L. E. Henderson, and S. Lindskog, unpublished results (1971).

and Tanford (*167*) have reported the complete renaturation of this enzyme after unfolding in 6 *M* guanidine hydrochloride. This observation has been confirmed in our laboratory (*166*). The presence of zinc ions appears to accelerate the renaturation process (*167*).

The zinc ion undoubtedly contributes to the stability of the protein, but it is not required under most conditions for maintaining the native conformation. The ORD curves of the apoenzymes at neutral pH are not significantly different from those of the native enzymes (*156, 168, 169*). However, small differences in the ultraviolet absorption spectrum (*170*), and in fluorescence (*171*), between apoenzyme and zinc-containing enzyme have been reported. At pH 5.5, the human B apoenzyme loses the aromatic Cotton effects and there are also additional changes of the ORD (*168*). These changes can be reversed by the addition of zinc ions (*168*). However, prolonged exposure of the apoenzyme to pH 5.5 produces an irreversible conformational change. The binding of the metal ion is further discussed in Section III,D.

4. *Hydrogen Exchange Studies*

Hydrogen–tritium exchange patterns are in accordance with the human and bovine carbonic anhydrases having compact structures. Rosenberg and Chakravarti (*172*) measured the temperature dependence of the slowest exchanging hydrogens and reached the conclusion that two exchange mechanisms are operating, one dominating at high temperatures and proceeding through cooperative thermal unfolding and another involving local rearrangements of various parts of the molecule. They also studied the pH dependence of the exchange kinetics of the bovine enzyme at 25° and concluded that there is a loosening of the structure at lower pH. Tashian (*173*) found no evidence for such a pH-dependent expansion of the human B and C enzymes, but his studies were performed at a lower temperature (4°). The exchange kinetics of both human isoenzymes are similar, but the C enzyme shows a somewhat more rapid exchange than the B enzyme.

Emery (*174*) has reported that the bovine apoenzyme has an increased

167. K. P. Wong and C. Tanford, *Federation Proc.* **29,** 335 (1970).
168. J. E. Coleman, *Biochemistry* **4,** 2644 (1965).
169. S. Lindskog, *BBA* **122,** 534 (1966).
170. R. W. Henkens and J. M. Sturtevant, *JACS* **90,** 2669 (1968).
171. J. M. Brewer, T. E. Spencer, and R. B. Ashworth, *BBA* **168,** 359 (1968).
172. A. Rosenberg and K. Chakravarti, *JBC* **243,** 5193 (1968).
173. R. E. Tashian, *Compt. Rend. Trav. Lab. Carlsberg* **37,** 359 (1970).
174. T. F. Emery, *Biochemistry* **8,** 877 (1969).

number of slowly exchanging protons compared to the zinc-containing enzyme.

III. Catalytic Properties

A. The Catalyzed Reactions

The major reaction catalyzed by the carbonic anhydrases can be formulated

$$CO_2 + H_2O \rightleftharpoons H^+ + HCO_3^- \tag{1}$$

The reaction is in principle well suited for kinetic analysis since it is reversible and can be studied in both directions over a reasonably wide range of pH. In the following, the terms hydration and dehydration will be used to indicate in which direction the reaction is studied. Another factor contributing to make carbonic anhydrase mechanistically attractive is the very simplicity of the catalyzed reaction. Furthermore, the reaction proceeds readily in the absence of enzyme, and a good deal of information has been collected about the nonenzymic reaction and its catalysis by various low molecular weight substances. Kinetic studies at alkaline pH are cumbersome, however, because of the direct reaction between CO_2 and OH^-. The thermodynamic and kinetic aspects of these nonenzymic interconversions have been thoroughly reviewed by Edsall and Wyman (*175*) and Edsall (*176*).

Carbonic acid, H_2CO_3, is neglected in Eq. (1) because it is a rather strong acid, $pK_{H_2CO_3} = 3.8$ at 25° (*175*). The interconversion between H_2CO_3 and HCO_3^- is very fast (*177*); thus, these chemical species can be considered at equilibrium during the course of reaction (1). Hence, in almost the whole pH range of interest for enzymic studies, the concentration of free H_2CO_3 is small compared to that of HCO_3^-. The value of the equilibrium constant K_{eq}, for the reaction of Eq. (1) is practically identical with that of the apparent first dissociation constant of carbonic acid, K_1 (*175*). At 25° and zero ionic strength, $pK_1 = 6.352$ (*175*). Thus, the molar ratio of the concentrations of dissolved CO_2 and H_2CO_3 is approximately 400 at equilibrium. The value of pK_2 for carbonic acid is 10.3; thus, CO_3^{2-} can be neglected at pH values below 8.

175. J. T. Edsall and J. Wyman, "Biophysical Chemistry," Vol. 1, p. 550. Academic Press, New York, 1958.

176. J. T. Edsall, *NASA (Natl. Aeron. Space Admin.), Spec. Publ.* **NASA SP-188,** 15 (1969).

177. M. Eigen and G. Hammes, *Advan. Enzymol.* **25,** 1 (1963).

The ability of carbonic anhydrase to catalyze the hydrolysis of certain esters such as *p*-nitrophenyl acetate and β-naphthyl acetate was independently discovered in a number of laboratories (*178–181*). A large number of esters and similar substances have subsequently been tested as carbonic anhydrase substrates. With many of these substrates, the hydrolytic activity is thought to be an intrinsic property of the enzymic active site. One of the major criteria for this is a parallel behavior of the esterase and the CO_2 activities with respect to pH and inhibitors (see Section III,C,2).

The carbonic anhydrase-catalyzed hydration of aldehydes (Section III,C,3) was first described by Pocker and Meany (*182*). The hydration reactions

$$RCHO + H_2O \rightleftharpoons RCH(OH)_2 \tag{2}$$

proceed at relatively slow rates in the absence of a catalyst, and they can be conveniently measured spectrophotometrically at wavelengths where the unhydrated carbonyl group is absorbing.

B. Assay Methods

Roughton and Clark (*19*) and Davis (*183*) have reviewed the earlier carbonic anhydrase assay methods and given critical evaluations of these methods. Here, we shall call attention to the procedures that have now become standard operations in research on these enzymes and point out recently developed improvements on the older methods.

Manometric methods (*19, 184*) are seldom used nowadays, and most of the presently employed procedures utilize the production or consumption of H^+ in the reversible reaction (1). Where semiquantitative data will suffice, most research groups use a simple assay provided by mixing a weakly buffered enzyme solution containing a pH indicator with CO_2-saturated water, and recording the time required for a visible color change (*184, 185*). Generally accepted conditions for the assay at

178. R. E. Tashian, C. C. Plato, and T. B. Shows, *Science* **140,** 53 (1963).
179. F. Schneider and M. Liefländer, *Z. Physiol. Chem.* **334,** 279 (1963).
180. B. G. Malmström, P. O. Nyman, B. Strandberg, and B. Tilander, *in* "Structure and Activity of Enzymes" (T. W. Goodwin, J. T. Harris, and B. S. Hartley, eds.), p. 121. Academic Press, New York, 1964.
181. Y. Pocker and J. T. Stone, *JACS* **87,** 5497 (1965).
182. Y. Pocker and J. E. Meany, *JACS* **87,** 1809 (1965).
183. R. P. Davis, *Methods Biochem. Anal.* **11,** 307 (1963).
184. F. J. W. Roughton and V. H. Booth, *BJ* **40,** 309 and 319 (1946).
185. F. J. Philpot and J. St. L. Philpot, *BJ* **30,** 2191 (1936).

2°, using bromothymol blue as the indicator, have been defined (*18*). The assay is suitable for use during the isolation of the enzyme, where the number of activity units can provide a rough estimate of the recovery of the enzyme. The inherent difficulties in controlling the CO_2 concentration and determining the precise end point together with other factors, such as the instability of the enzyme at very low concentrations, contribute heavily to the errors in the procedure, and there is considerable lack of reproducibility when the results from various laboratories are compared. Improved precision may be obtained at the cost of simplicity. The pH changes occurring during the reaction can be followed with greater precision in a thermostated spectrophotometer (*13, 186*), and the Veronal buffer commonly used in this assay can serve as the pH indicator (*187*). However, the pH interval covered during the course of these assays is broad, and other methods have to be used for the accurate estimation of initial enzymic rates.

The pH interval can be minimized and the assay conducted at higher temperatures with the aid of a stopped-flow apparatus (*188–191*). With a sufficiently sensitive detection system, the initial rates can be determined with minimal pH changes. The observed absorbancy changes can be converted to units of substrate concentration by means of a calibration curve, which is related to the ionization constants of the buffer and the indicator, respectively. A linear relationship is obtained when the buffer and the indicator have identical pK values, and the conversion to concentration changes is accomplished by multiplying the absorbancy change with a "buffer factor" (*188, 191*). Khalifah (*191*), in his studies on the human carbonic anhydrases, utilized a set of non-inhibitory buffer indicator pairs with approximately equal ionization constants covering a pH range from 6 to 9. The stopped-flow pH-indicator methods of assay provide the greatest accuracy and flexibility in studies of the kinetics of the enzyme-catalyzed reversible hydration of CO_2. However, the cost of the instrumentation restricts the general application of the method.

McIntosh (*192*) has improved a useful alternative method suitable for limited kinetic studies. The hydration reaction is carried out at a

186. A. M. Clark and D. D. Perrin, *BJ* **48,** 495 (1951).
187. P. O. Nyman, *Acta Chem. Scand.* **17,** 429 (1963).
188. C. Ho and J. M. Sturtevant, *JBC* **238,** 3499 (1963); B. H. Gibbons and J. T. Edsall, *ibid.* p. 3502.
189. J. C. Kernohan, W. W. Forrest, and F. J. W. Roughton, *BBA* **67,** 31 (1963).
190. J. C. Kernohan, *BBA* **81,** 346 (1964).
191. R. G. Khalifah, *JBC* **246,** 2561 (1971).
192. J. E. A. McIntosh, *BJ* **109,** 203 (1968).

constant pH with the aid of a modified automatic titrimeter. The rate of CO_2 hydration can be determined directly from the rate of base consumption. In the neutral pH range and at low temperatures, it is possible to measure rates up to 15 times faster than the rate of the nonenzymic reaction. At higher pH values and temperatures, the rate of the uncatalyzed reaction increases, and the response time of the apparatus becomes limiting. Hansen and Magid (*193*) have employed a similar apparatus in kinetic studies of the dehydration reaction utilizing HCO_3^- as substrate. The reaction is followed at low temperatures in the neutral pH region, while the evolved CO_2 is swept away from the reaction mixture with an inert carrier gas. The method allows the determination of initial rates as a function of substrate concentration during the course of one experiment because the product is removed and no back reaction can occur.

The inherent technical difficulties involved in the use of CO_2 or HCO_3^- as substrates have led to an increased application of the esterase activity for assay purposes. Convenient spectrophotometric procedures suitable for use with pure enzymes have been described employing *p*-nitrophenyl acetate or β-naphthyl acetate as substrates (*35, 76, 123, 181, 194*). The latter compound has been extensively used for the detection of carbonic anhydrase isoenzymes after electrophoretic separations on gel media (*35, 178*).

Some sulfonamide inhibitors (Section III,E,2) have high affinities for carbonic anhydrase ($K_i < 10^{-8}\ M$); thus, they can be utilized in combination with one of the previously mentioned assays to titrate the total concentration of enzymic active sites (*195, 196*).

C. Kinetic Properties

1. *The Interconversion of CO_2 and HCO_3^-*

Many of the important features of the kinetics and inhibition of carbonic anhydrase were elucidated in the classic studies by Roughton and co-workers (*6, 19, 184*), who mainly employed manometric techniques to follow the reaction. In recent times, the assay methods have been improved and, perhaps more importantly, the presence of different forms of the enzyme has been discovered in several animal species (see

193. P. Hansen and E. Magid, *Scand. J. Clin. & Lab. Invest.* **18**, 21 (1966).
194. J. A. Verpoorte, S. Mehta, and J. T. Edsall, *JBC* **242**, 4221 (1967).
195. T. H. Maren, A. L. Parcell, and M. N. Malik, *J. Pharmacol. Exptl. Therap.* **130**, 389 (1960).
196. J. C. Kernohan, *BBA* **96**, 304 (1965).

Section II,B). However, the work of Roughton and colleagues has been followed up by a few investigators only, presumably because of the limitations still inherent in the presently available experimental methods (see Section III,B), and much important kinetic information remains to be collected.

The carbonic anhydrases follow a simple Michaelis-Menten behavior with respect to both CO_2 and HCO_3^-:

$$v_0/[E_0] = k_{cat}[S]/(K_m + [S]) \tag{3}$$

DeVoe and Kistiakowsky (*197*) observed a deviation from Eq. (3) for the hydration reaction catalyzed by a sample of human enzyme. Presumably, however, their preparation represented a mixture of high and low activity forms since this kind of isoenzyme composition had not yet been discovered at the time of their studies.

a. Bovine Carbonic Anhydrase. Kernohan (*190, 196*) has presented self-consistent data of the pH dependence of both the hydration and dehydration reactions at 25° catalyzed by the bovine enzyme. However, as pointed out by Kernohan (*196*), the results are to a degree influenced by the use of 80 m*M* Cl^- in the assay medium, which leads to a non-negligible inhibition (see Section III,E,1). Kernohan confirmed the finding of Roughton and Booth (*184*) that $K_m^{CO_2}$ (see Table VIII) is approximately independent of pH and buffer composition over a wide range of conditions. Similar results were obtained by DeVoe and Kistiakowsky (*197*) at 0.5° and in a more limited pH region. Kernohan

TABLE VIII

SELECTED KINETIC PARAMETERS OF BOVINE AND HUMAN CARBONIC ANHYDRASES[a]

Enzyme	Hydration reaction		Dehydration reaction	
	k_{cat} ($sec^{-1} \times 10^{-6}$)	K_m (m*M*)	k_{cat} ($sec^{-1} \times 10^{-4}$)	K_m (m*M*)
Bovine[b]	1	12	40	26
Human C[c]	1.4	9	8.1	22
Human B[c]	0.2	4	0.3	16

[a] See text and Figs. 11 and 12 for the pH dependence of these parameters.

[b] Hydration reaction: pyrophosphate buffer, pH 9, 25° (*196*). Dehydration reaction: Imidazole buffer, pH 6.7, 25°, and 14 m*M* Cl^- (*190*). See text concerning the effect of Cl^-.

[c] Hydration reaction: 1,2-dimethylimidazole buffer, pH 8.75, 25° (*191*). Dehydration reaction: *N*-2-hydroxyethylpiperazine-*N'*-2-ethanesulfonate (HEPES) buffer, pH 7.3, 2° (*198*). Note the different temperatures.

197. H. DeVoe and G. B. Kistiakowsky, *JACS* **83**, 274 (1961).

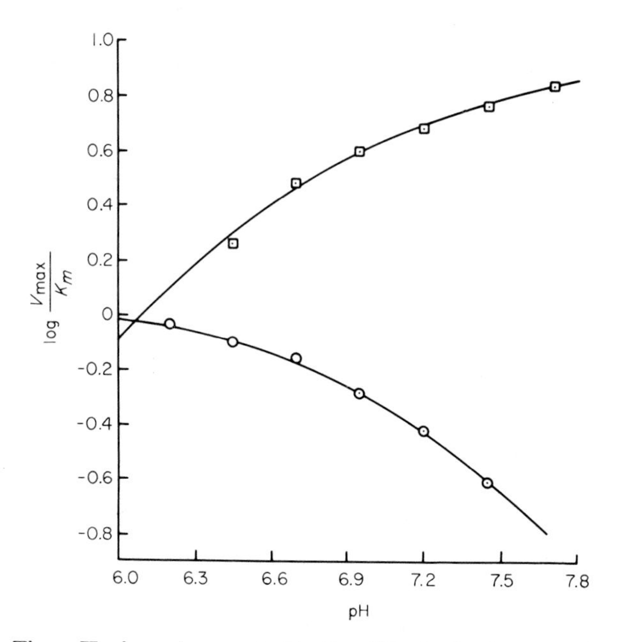

FIG. 11. The pH dependence of □ the CO_2 hydration reaction and (⊙) HCO_3^- dehydration reaction catalyzed by bovine carbonic anhydrase. The activities are expressed as $\log(V_{max}/K_m)$, but since $K_m^{CO_2}$ is pH independent (see text) $k_{cat}^{CO_2}$ is proportional to $V_{max}^{CO_2}/K_m^{CO_2}$. Conditions: 25°, 80 m$M$ Cl^-, 0.128 μM enzyme. Note that the curves should intersect at pH $= pK_1'$ according to Eq. (4). At the ionic strength and temperature of these experiments $pK_1' = 6.12$ (*175*). From Kernohan (*190*).

further found that the pH dependence of $k_{cat}^{CO_2}$ is in accordance with the basic form of a group with a pK of 6.9 being required for activity (Fig. 11). With HCO_3^- as substrate, $K_m^{HCO_3^-}$ was unmeasurably large under the conditions employed by Kernohan (80 mM Cl^-), but $k_{cat}^{HCO_3^-}/K_m^{HCO_3^-}$ behaved as if the acidic form of a group having pK = 6.9 were required for activity (Fig. 11). Thus, the Haldane relation is obeyed (*190*)

$$\frac{k_{cat}^{CO_2}}{K_m^{CO_2}} \cdot \frac{K_m^{HCO_3^-}}{k_{cat}^{HCO_3^-}} = \frac{K_1'}{[H^+]} \tag{4}$$

where K_1' is the apparent first dissociation constant of carbonic acid. DeVoe and Kistiakowski (*197*), working with dilute phosphate buffers at 0.5°, found that $k_{cat}^{HCO_3^-}$ was independent of pH, and $K_m^{HCO_3^-}$ was inversely proportional to the hydrogen ion concentration between pH 7.0 and 7.3. Because of the different conditions employed, the results of Kernohan and DeVoe and Kistiakowsky cannot be combined, but either set of data is consistent with the following simple kinetic scheme or minor variations of it:

$$\begin{array}{ccccccc} E^{n} + CO_2 & \underset{k_{-1}}{\overset{k_{+1}}{\rightleftharpoons}} & (ECO_2)^{n} & \underset{k_{-2}}{\overset{k_{+2}}{\rightleftharpoons}} & EH^{n+1} + HCO_3^- \\ K_h \updownarrow & & \updownarrow K_h & & \\ EH^{n+1} + CO_2 & \underset{k_{-1}}{\overset{k_{+1}}{\rightleftharpoons}} & (EHCO_2)^{n+1} & & \end{array} \qquad (5)$$

where K_h is the acid dissociation constant of the activity-linked group and E^n and EH^{n+1} have the basic and acidic form of this group, respectively. Alternative schemes, where H_2CO_3 is the direct reactant in the dehydration reaction, could be excluded because the calculated second-order rate constant for the combination of enzyme and H_2CO_3 was larger than for a diffusion-controlled reaction (*190, 197*). In Eq. (5) it is assumed that the CO_2 binding site and the activity-linked titratable group are independent, while HCO_3^- only combines with EH^{n+1} having the acidic form of this group. Kernohan's results are in concordance with a rapid CO_2 binding step (i.e., k_{-1} being much greater than k_{+2} and than $k_{-2} \cdot [HCO_3^-]$). He concluded that $K_m^{CO_2}$ should represent the dissociation constant for this substrate. The molecular events behind the schematic mechanism of Eq. (5) are discussed in Section III,G.

b. Human Carbonic Anhydrase. The kinetic properties of the human carbonic anhydrases B and C have been less intensely investigated. A brief study was reported by Gibbons and Edsall (*199*). Recently, the pH dependence of the hydration reaction has been reported by Khalifah (*191*) and that of the dehydration by Magid (*198, 200*). These workers have employed different conditions for their measurements, Khalifah working at 25° and Magid at 1°; thus, direct combinations of their results are not possible. However, it appears that the human C enzyme is kinetically very similar to the bovine enzyme, which is also a high activity isoenzyme (Fig. 12 and Table VIII). Magid's results on the pH dependence of $k_{cat}^{HCO_3^-}$ and $K_m^{HCO_3^-}$ for the human C enzyme correspond with the findings of DeVoe and Kistiakowsky on the bovine enzyme. Presumably a scheme similar to Eq. (5) is applicable also to this enzyme form. The human B form is less active than the C form (Table VIII), but its kinetic properties show certain similarities. The values of $K_m^{CO_2}$ are of

198. E. Christiansen and E. Magid, *BBA* **220,** 630 (1970).
199. B. H. Gibbons and J. T. Edsall, *JBC* **239,** 2539 (1964).
200. E. Magid, *BBA* **151,** 236 (1968).

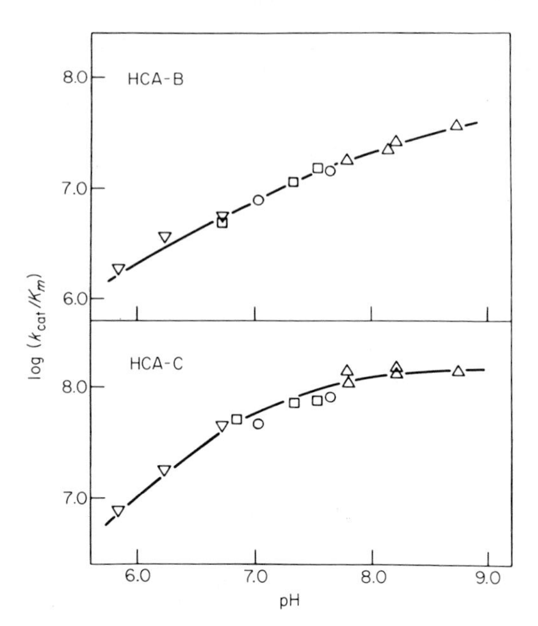

FIG. 12. The pH dependence of the CO_2 hydration reaction at 25° catalyzed by the human B enzyme (HCA-B) and the human C enzyme (HCA-C). The different symbols denote various noninteracting buffer indicator pairs. For both enzymes $K_m^{CO_2}$ is pH independent (see text); thus, the pH-rate profiles of $k_{cat}^{CO_2}$ are similar to the curves shown here. From Khalifah (*191*).

the same order of magnitude for both isoenzymes (Table VIII), and pH independent in both cases (*191*). As seen in Fig. 12, $k_{cat}^{CO_2}$ for the B enzyme also increases with pH, but the pH-rate profile cannot be described as a simple titration curve. Presumably, the catalytic mechanisms of the high and low activity forms are similar, but the ionization states of additional groups appear to influence the catalytic rate in the case of B enzyme. Magid's earlier results (*200*) indicated that the kinetic parameters for the dehydration reaction catalyzed by the human B enzyme behaved quite differently with respect to pH than in the case of the C enzyme. Later studies showed, however, that the phosphate buffer employed had a pronounced effect on the results (see Section III,E,1), and data obtained in a noninteracting buffer give a picture that seems to comply with a moderately modified Eq. (5) (*198*).

2. *Hydrolytic Reactions*

a. Specificity. The most intensely studied category of ester substrates for carbonic anhydrase is a group of phenolic esters of aliphatic carboxylic acids (Table IX). Nitrophenyl acetates are efficiently hydrolyzed,

TABLE IX

APPARENT SECOND-ORDER RATE CONSTANT k_{enz} FOR THE HYDROLYSIS OF VARIOUS ESTERS CATALYZED BY BOVINE CARBONIC ANHYDRASE[a]

Substrate	k_{enz} (M^{-1} sec^{-1})
Phenyl acetate	3
o-Nitrophenyl acetate	100
m-Nitrophenyl acetate	240
p-Nitrophenyl acetate	960
p-Nitrophenyl propionate	68
p-Nitrophenyl butyrate	6
p-Nitrophenyl caproate	2
p-Nitrophenyl trimethylacetate	0.2

[a] In tris sulfate buffer, pH 7.6, 25°. From Thorslund and Lindskog (*202*). See the original paper for further experimental details.

while unsubstituted phenyl acetate or α- and β-naphthyl acetates are hydrolyzed much more slowly. This relation resembles that of water or hydroxide ion hydrolysis of these esters, but the electronic effects of ring substitution seem to be even more pronounced for the enzymic reaction (*201*). The relative catalytic rates for a series of carboxylic acid esters of *p*-nitrophenol (see Table IX) suggest that the introduction of a bulky acyl moiety presents a steric situation interfering with rapid enzymic hydrolysis (*202, 203*). The substrate specificity seems to vary somewhat for different forms of the enzyme (*34, 194*). Tashian *et al.* (*34*) have shown that the C-type isoenzymes from some primates are similar in this respect, whereas the B-type isoenzymes show relatively large variations in the catalytic rates observed for different substrates.

Kaiser and Lo (*204*) found that the hydrolysis of a cyclic sulfonate ester, 2-hydroxy-5-nitro-α-toluenesulfonic acid sultone (I), is greatly

O_2N CH_2 SO_2 O

(I)

accelerated by carbonic anhydrase. The apparent catalytic second-order rate constant k_{enz} is 2.8×10^5 M^{-1} sec^{-1} for the bovine enzyme at pH 7.4 and 25°, which is the largest value found for any ester substrate of

201. Y. Pocker and J. T. Stone, *Biochemistry* **7**, 3021 (1968).
202. A. Thorslund and S. Lindskog, *European J. Biochem.* **3**, 117 (1967).
203. Y. Pocker and D. R. Storm, *Biochemistry* **7**, 1202 (1968).
204. E. T. Kaiser and K.-W. Lo, *JACS* **91**, 4912 (1969).

carbonic anhydrase. In this case also, the nitro group has a large effect on the catalytic rate (*205*).

In addition to the typical ester substrates, carbonic anhydrases have been found to catalyze the hydrolysis of some compounds often used for the chemical modification of proteins, such as carbobenzoxy chloride (*180*), 1-fluoro-2,4-dinitrobenzene (*206*), and sulfonyl chlorides (*128*).

b. Effects of Substrate Concentration. The enzyme-catalyzed initial velocities appear to follow Michaelis-Menten kinetics, but the solubilities of the esters usually limit investigations to substrate concentrations smaller than the estimated K_m. Therefore, K_m values reported in the literature must be regarded as approximate since they have generally been obtained by long extrapolations. Published K_m values vary considerably from one laboratory to another. Thorslund and Lindskog (*202*) found that the rate of *p*-nitrophenyl acetate hydrolysis catalyzed by the bovine enzyme is proportional to the substrate concentration over the whole measurable range. On the other hand, Pocker and co-workers have published extensive reports concerning effects of pH and substrate structure (*203, 207*), D_2O, and temperature (*208*) on K_m and k_{cat}, but it is doubtful whether these kinetic parameters could be estimated with a satisfactory accuracy. Verpoorte *et al.* (*194*) found that the human carbonic anhydrases yield pH-independent values of K_m with *p*- and *o*-nitrophenyl acetates as substrates, but the magnitudes of some of these K_m values could not be reproduced by Whitney (*125*).

Efforts have been made to find evidence for the formation of an acyl intermediate in carbonic anhydrase–catalyzed ester hydrolysis. No "burst" of *p*-nitrophenol could be detected in the initial phase of the reaction (*194, 207*), and no transesterification to added methanol could be demonstrated (*194*). Carbonic anhydrase is not inactivated by diisopropyl fluorophosphate (*180, 209*).

c. Effects of pH on the Enzymic Rates. The pH-rate profiles of *p*-nitrophenyl acetate hydrolysis are sigmoidal for the bovine enzyme as well as for the human B and C enzymes (*194, 202, 207, 210*), and there seems to be a close correspondence with the pH dependence of the CO_2 hydration reaction. However, the complex behavior of the human B form in the CO_2 reaction does not appear to be paralleled in the esterase

205. K.-W. Lo and E. T. Kaiser, *Chem. Commun.* p. 834 (1966).
206. P. Henkart, G. Guidotti, and J. T. Edsall, *JBC* **243**, 2447 (1968).
207. Y. Pocker and J. T. Stone, *Biochemistry* **6**, 668 (1967).
208. Y. Pocker and J. T. Stone, *Biochemistry* **7**, 4139 (1968).
209. M. Liefländer and R. Zech, *Z. Physiol. Chem.* **349**, 1466 (1968).
210. J. E. Coleman, *JBC* **242**, 5212 (1967).

reaction, thus suggesting some subtle mechanistic differences for the two reactions.

Pocker and Storm (*203*) studied a series of homologous carboxylic acid esters of *p*-nitrophenol, and they reported that the pH-rate profiles deviate increasingly from the simple sigmoidal pattern as the acyl moiety becomes longer or branched. Furthermore, they observed steep rate increases at alkaline pH values. This phenomenon was observed for the acetate ester only at temperatures above 25° (*208*). Pocker and Storm (*203*) further reported that the sulfonamide inhibitor, acetazolamide (see Section III,E), does not inhibit the hydrolysis of the long-chain or the branched esters until the inhibitor concentrations are raised to levels much exceeding the K_i values obtained with CO_2 or *p*-nitrophenyl acetate as substrates. Although a noncompetitive inhibition pattern was reported for all the esters, Pocker and co-workers interpreted their results in terms of substrate-induced conformational changes affecting the affinity for the inhibitor. It seems more likely that there are alternative binding sites and/or "catalytic" groups in the enzyme, and that such "secondary" catalysis dominates in the case of the less good ester substrates (cf. also Section III,C,3). Some data pertaining to a possible secondary binding site for acetazolamide in the bovine enzyme have been reported (*211*), which may be related to the inhibition of the postulated "secondary" catalytic ester hydrolysis.

3. *The Hydration of Aldehydes*

The carbonic anhydrase-catalyzed hydration of acetaldehyde (*182, 212*) shows several features in common with CO_2 hydration and ester hydrolysis catalyzed by the same enzyme. It seems probable that this reaction is also connected with the active site of the enzyme. Thus, the pH-rate profile resembles that of the CO_2 hydration reaction, having an inflection at pH 7.0 for the bovine enzyme and at pH 7.3 for the human C enzyme (*182, 212*). The pH dependence is essentially expressed in k_{cat}, while K_m appears to be approximately constant between pH 6 and 8.

Pocker and co-workers have also studied a series of higher aliphatic aldehydes (*213*) as well as the isomers of pyridine aldehyde (*214*). Representative kinetic parameters for these substrates are given in

211. S. I. Kandel, S. C. C. Wong, M. Kandel, and A. G. Gornall, *JBC* **243,** 2437 (1968).
212. Y. Pocker and J. E. Meany, *Biochemistry* **4,** 2535 (1965).
213. Y. Pocker and D. G. Dickerson, *Biochemistry* **7,** 1995 (1968).
214. Y. Pocker and J. E. Meany, *Biochemistry* **6,** 239 (1967).

TABLE X
SELECTED KINETIC PARAMETERS FOR THE HYDRATION OF ALDEHYDES CATALYZED BY BOVINE CARBONIC ANHYDRASE[a]

Substrate	pH	k_{cat} (sec^{-1})	K_m (M)
Acetaldehyde	7.64	880	0.61
Propionaldehyde	7.45	2.8	0.20
Isobutyraldehyde	7.46	0.5	0.14
2-Pyridine aldehyde	7.53	72	0.013
3-Pyridine aldehyde	7.2	33	0.007
4-Pyridine aldehyde	7.54	205	0.012

[a] From the data of Pocker and co-workers (*212–214*). 0.01 *M* dimethylmalonate buffers, 0°.

Table X. The published data suggest, perhaps even more clearly than in the case of the "bulky" ester substrates, that more than one enzymic mechanism might be operating.

First, the pH-rate profiles change appearance from the almost "normal" pattern of acetaldehyde through propionaldehyde to the other aldehydes, which show a weak pH dependence between pH 5 and 6, but give constant activities in the region where the CO_2 hydration-linked group is titrating. Since K_m is reported to be pH independent in all cases, these differences are not likely to be reflections of substrate-dependent changes of the effective pK of the catalytic group but would rather suggest that different groups in the enzyme might be involved with different substrates.

Second, acetazolamide does not inhibit with a "normal" K_i for any of the studied aldehydes. The values of K_i are reported to vary from 0.6 μM for acetaldehyde to 30 μM for isobutyraldehyde (*213*) as compared to approximately 0.01 μM in the CO_2 reaction at comparable pH and temperature (*23, 215*). Furthermore, Pocker and Dickerson (*213*) reported a variation in the formal type of inhibition, and for several aldehydes the hydration reaction is only partially inhibited by acetazolamide. Clearly, further experiments are required to clarify the carbonic anhydrase-catalyzed hydration of these aldehydes.

D. THE METAL ION COFACTOR

The metal ion is required for the catalytic function of carbonic anhydrase. Apoenzymes have been prepared by dialysis against 1,10-phenanthroline at pH 5.0–5.5 (*17, 76, 147, 168, 216*). The metal-free enzymes are

215. S. Lindskog, *JBC* **238,** 945 (1963).
216. S. Lindskog and P. O. Nyman, *BBA* **85,** 462 (1964).

inactive with respect to the hydration of CO_2 and the hydrolysis of *p*-nitrophenyl acetate, but the addition of stoichiometric quantities of Zn^{2+} rapidly restore these activities.

1. *Thermodynamics and Kinetics of Zn^{2+} Binding*

The apparent stability constant K_{app} for Zn^{2+} increases with pH. Between pH 5 and 10, $\log K_{app} = (5.0 + 1.0\,\text{pH})$ for the human B enzyme, and similar values have been obtained for the human C and bovine enzymes (*216*). Protons are produced in the binding reactions (*215*), and as a first approximation this reaction at neutral pH proceeds as

$$Zn^{2+} + E^{n} \rightleftharpoons (Zn - E)^{n+1} + H^{+} \tag{6}$$

where E^n denotes the apoenzyme having the net charge n.

Henkens *et al.* (*217*) performed a calorimetric study of the binding and found that, in fact, heat is absorbed. The stability is thus linked to a substantial entropy increase ($\Delta S' = +88$ cal deg^{-1} mole^{-1} at pH 7 and 25°).

Henkens and Sturtevant (*170*) measured the rate of Zn^{2+} binding in a stopped-flow apparatus. The observed rates vary to some extent with ionic strength and pH. At pH 8 and 25°, the apparent second-order rate constant for the recombination reaction is approximately $3 \times 10^4\ M^{-1}$ sec^{-1} and independent of ionic strength. This value is almost 1000-fold smaller than the rate constants for reactions with Zn^{2+} and low molecular weight ligands in aqueous solution (*177*). The results show that most of the pH dependence of the Zn^{2+}-apoenzyme stability constant must be contained in the dissociation rate constant in approximate correspondence with Eq. (6).

Calculations of half times for Zn^{2+} dissociation from a combination of rate and equilibrium data would give values of about seven years at pH 8, one and one-half years at pH 7, and three weeks at pH 5, giving strong reasons why exchange of enzyme-bound Zn^{2+} with extraneous $^{65}Zn^{2+}$ could not be observed at neutral pH (*218*). However, dialysis against 1 mM 1,10-phenanthroline at pH 5 leads to the formation of bovine apoenzyme *B* with a half time of about two days (*147*), suggesting that the chelating agent not only affects the position of the equilibrium by decreasing the free metal ion concentration but also changes the dissociation mechanism, presumably through the formation of a transient ternary complex before the metal ion is released from the enzyme. Phenanthroline does not inhibit the enzyme after short incubation times

217. R. W. Henkens, G. D. Watt, and J. M. Sturtevant, *Biochemistry* **8,** 1874 (1969).
218. R. Tupper, R. W. E. Watts, and A. Wormall, *BJ* **50,** 429 (1952).

(*147, 219*); thus, the steady state concentration of the ternary complex is probably small. Certain other chelating agents containing sulfhydryl groups have been reported to facilitate zinc ion removal via an initial, inactive complex (*220*).

2. *Metal Ion Specificity*

A large number of metal ions have been tested for their ability to activate apocarbonic anhydrase (*76, 147, 168, 202, 216*). Of these, only Co^{2+} yields a specific activity of similar magnitude to that of the native enzyme. Low activities have occasionally been found with Mn^{2+}, Fe^{2+}, and Ni^{2+} (*147, 202, 216*). The metal ion specificity appears similar with CO_2 or with an ester substrate and for different forms of the enzyme as illustrated in Table XI. While the interchangeability of Zn^{2+} and Co^{2+} is a property common to several zinc metalloenzymes such as carboxypeptidase and alkaline phosphatase (*221–223*), the metal ion specificity of carbonic anhydrase is unusually narrow.

The activities of the zinc and cobalt carbonic anhydrases must reflect special properties of these complexes because most other metal ions tested can bind to the same chelating site in the enzyme (*147, 168, 210*), but they yield practically inactive products. The apparent stability constants for the binding of Mn^{2+}, Co^{2+}, Ni^{2+}, Cu^{2+}, Zn^{2+}, Cd^{2+}, and Hg^{2+} to human apoenzyme B at pH 5.5 follow a similar relative order as is commonly observed for low molecular weight complexes in solution (*216*),

TABLE XI
RELATIVE SPECIFIC ACTIVITIES OF COBALT(II) CARBONIC ANHYDRASES[a]

Enzyme	CO_2 hydration	*p*-Nitrophenyl acetate hydrolysis
Rhesus monkey B [Ref. (*76*)]	35	230
Human B [Refs. (*83, 125*)]	50	111
Human C [Ref. (*83*)]	55	—
Bovine [Refs. (*83, 202*)]	50	97

[a] Expressed as percent of the activities of the corresponding Zn(II) enzymes when measured by standard assay methods.

219. R. P. Davis, *JACS* **81,** 5674 (1959).
220. S. Carpy, *BBA* **151,** 245 (1968).
221. B. L. Vallee and R. J. P. Williams, *Proc. Natl. Acad. Sci. U. S.* **59,** 498 (1968).
222. S. Lindskog, *Struct. Bonding* (*Berlin*) **8,** 153 (1970).
223. B. L. Vallee and W. E. C. Wacker, *in* "The Proteins" (H. Neurath, ed.), 2nd ed., Vol. 5, p. 103, Academic Press, New York, 1970.

thus indicating that no great strain is imposed on the protein when cations of different ionic radii and stereochemical preferences are bound. This does not mean that the coordination geometries attained must be independent of the nature of the metal ion. The ligand groups may show some flexibility as to their exact positions, and the number of metal ion-coordinated water molecules may vary. The large Hg^{2+} ion binds in essentially the same position as Zn^{2+} as shown by X-ray methods on the human C enzyme (*137*), but the center of Hg^{2+} is shifted about 1 Å into the active site cavity (cf. Fig. 7). Electron paramagnetic resonance (EPR) spectra of the inactive Cu(II) enzymes are similar to those of common model complexes (*216, 224*), indicating that the chelating site can adjust itself toward the particular geometry preferred by this ion (*225*). A property common to Zn^{2+} and Co^{2+} is their tendency to form more or less regular tetrahedral complexes [cf. Lindskog (*222*) and Dennard and Williams (*226*)], and carbonic anhydrase like carboxypeptidase A appears to be specially equipped to form a somewhat distorted tetrahedral metal coordination.

Thus, the stereochemical specificities of the metal ions might be one factor determining why Zn^{2+} and Co^{2+} are functional in carbonic anhydrase while other metal ions are not. It does not suffice to explain the reactivity of the metal ion in the zinc and cobalt enzymes, however, because in that respect there are no counterparts among low molecular weight complexes. Evidently, the proteins provide "unique" metal binding conditions when compared to the simpler systems [(*221, 227*); see also Section III,G].

3. *Cobalt(II) as a Spectroscopic Probe of the Active Site*

While the coordination of Zn^{2+} in the human C enzyme is now known from the X-ray studies (see Fig. 6), it had previously been possible to make inferences about the bonding of Co^{2+} in this and other forms of the enzyme from the spectroscopic and magnetic properties of the cobalt-substituted enzymes (*228*). These derivatives have a reddish blue color with maximal molar absorptivities of 300 to 400 M^{-1} cm^{-1} (*215*), features indicative of a tetrahedral-like geometry (*222, 226*). The spectral forms of the Co(II) enzyme itself, shown in Fig. 13, are "unique" in the sense

224. J. S. Taylor, P. Mushak, and J. E. Coleman, *Proc. Natl. Acad. Sci. U. S.* **67**, 1410 (1970).
225. R. Malkin and B. G. Malmström, *Advan. Enzymol.* **33**, 177 (1970).
226. A. E. Dennard and R. J. P. Williams, *Transition Metal Chem.* **2**, 116 (1966); R. J. P. Williams, *Protides Biol. Fluids, Proc. Colloq.* **14**, 25 (1967).
227. B. G. Malmström, *Pure Appl. Chem.* **24**, 393 (1970).
228. S. Lindskog and A. Ehrenberg, *JMB* **24**, 133 (1967).

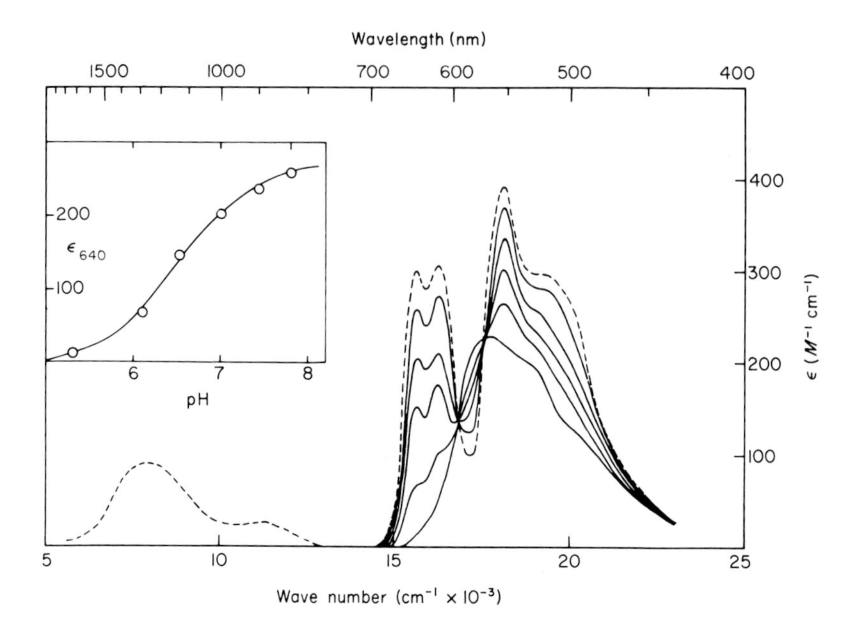

FIG. 13. The pH dependence of the metal-linked absorption spectrum of bovine Co(II) carbonic anhydrase. The broken curve represents the basic spectral form. The solid curves were obtained in imidazole-sulfate buffers, at pH 7.80, 7.00, 6.55, and 6.10 in order of decreasing ϵ_{640}. The spectrum of lowest intensity was obtained by extrapolation and represents the acidic spectral form. The near-infrared band is shown only for the basic spectral form. Insert: Spectrophotometric titration at 640 nm. The curve was calculated for the titration of a single group with $pK = 6.6$. From S. Lindskog, Cobalt(II) in metalloenzymes. A reporter of structure–function relations. *Struct. Bonding* **8,** 153–196 (1970). New York-Heidelberg-Berlin: Springer.

that they bear no close resemblance to any known model system [cf., however, Holt *et al.* (*229*) and Dobry-Duclaux and May (*230*)], but strikingly similar spectra have been recorded for other cobalt-substituted zinc metalloenzymes (*222, 223*). However, in combination with some inhibitors, particularly CN^-, Co(II) carbonic anhydrase develops spectra having the characteristic features of tetrahedral coordination as illustrated in Fig. 14 (*228, 231*). In view of these results, and since Zn^{2+} and Co^{2+} are functionally interchangeable, it seems likely that both ions are bound to the chelating site in essentially the same manner. Furthermore, the absorption spectra of B- and C-type enzymes from different

229. E. M. Holt, S. L. Holt, and K. J. Watson, *JACS* **92,** 2721 (1970).

230. A. Dobry-Duclaux and A. May, *Bull. Soc. Chim. Biol.* **50,** 2053 (1968); **52,** 1447 (1970).

231. F. A. Cotton, D. M. L. Goodgame, M. Goodgame, and A. Sacco, *JACS* **83,** 4157 (1961).

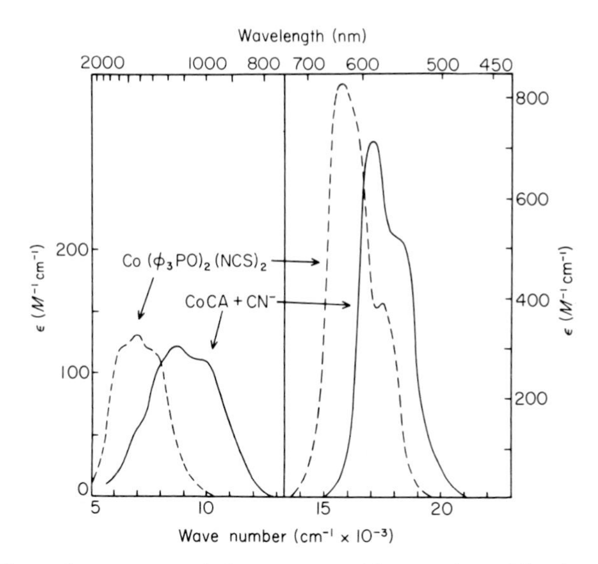

FIG. 14. Absorption spectra of the monocyanide complex of bovine Co(II) carbonic anhydrase (solid curve, CoCA + CN^-), and the pseudotetrahedral complex, $Co(II)[(C_6H_5)_3PO]_2(SCN)_2$ [broken curve, $Co(\phi_3PO)_2(NCS)_2$]. From Lindskog and Ehrenberg (*228*) and Cotton *et al.* (*231*), respectively.

species are closely similar suggesting essentially identical structures of the metal binding sites in all cases. However, Coleman (*131*, *232*) observed that the cobalt-linked absorption bands of the human and monkey B enzymes are optically inactive, whereas distinct circular dichroic bands are present for the high activity forms. These results suggest that there are some differences in the metal environments of the B- and C-type isoenzymes.

As indicated in the previous paragraph and shown in Figs. 13 and 14, the absorption spectra of Co(II) carbonic anhydrases are very sensitive to pH and to the interaction with inhibitors. These phenomena link events occurring in the immediate vicinity of the metal ion with the catalytic process. The pH–rate profiles displayed by the Co(II) enzyme in the CO_2 hydration reaction (*233*) and in the hydrolysis of *p*-nitrophenyl acetate (Fig. 15) are very similar to those of the zinc-containing enzyme (cf. Figs. 10 and 11) suggesting analogous catalytic mechanisms. The pH dependence of the cobalt-linked absorption spectrum (Fig. 13) closely parallels the corresponding pH dependencies of these catalytic reactions (Fig. 15) giving strong evidence that the titrating group in

232. J. E. Coleman, *Proc. Natl. Acad. Sci. U. S.* **59**, 123 (1968).
233. S. Lindskog, *Biochemistry* **5**, 2641 (1966).

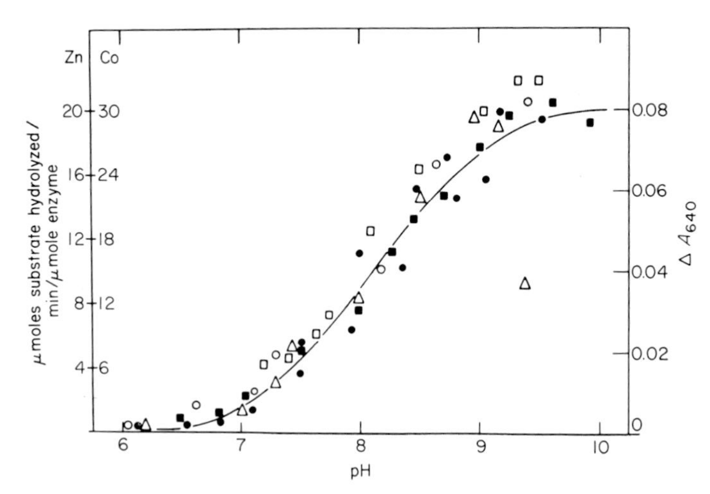

FIG. 15. Correlation between the pH dependence of *p*-nitrophenyl acetate hydrolysis catalyzed by the Zn(II) and Co(II) forms of human carbonic anhydrase B and the pH dependence of the metal-linked absorption spectrum of the Co(II) enzyme. Symbols: (●,■) esterase activity of the zinc enzyme, (△,□) esterase activity of the cobalt enzyme. (○) A_{640} at the pH of measurement minus A_{640} at pH 6.0, Δ A_{640}. From Coleman (*210*).

the active site of the enzyme is intimately associated with the metal ion cofactor. As will be further discussed in Section III,G, a simple, but not unique, molecular interpretation of these findings is that the "catalytic" group in carbonic anhydrase is represented by a metal-coordinated OH^- ion.

Further reference to the probe properties of cobalt will be given in Sections III,E and F, and a more detailed discussion can be found in a recent review (*222*).

E. INHIBITORS

1. *Anions*

The carbonic anhydrases are inhibited by most monovalent anions. This inhibition was discovered by Meldrum and Roughton (*6*) and was first systematically studied by Roughton and Booth (*184*), who concluded that it is caused by the specific binding of the anions to the enzyme. Some of the most potent anionic inhibitors such as CN^- are typical "metal poisons," and it has been postulated already by Keilin and Mann (*8*) that their action involves a binding to the metal ion.

a. Kinetic Studies. In a study of the anion inhibition of a preparation of human carbonic anhydrase, Davis (*22, 234*) reported that the results can be explained in terms of a general ionic strength effect on the catalytic transformation. A reinvestigation of the pH dependence and the specificity of anion inhibition was undertaken by Kernohan (*196*) working with the bovine enzyme. He found that the inhibitor constants for Cl^- and NO_3^- are considerably different and that the inhibition can be best described as resulting from the binding of one anion per active site. The mass of presently available data give ample proof of the specific nature of the enzyme–anion interaction.

Kernohan (*196*) also showed that the inhibition by anions depends on the ionization state of the activity-linked titratable group (cf. Fig. 11). The inhibitors bind strongly only when this group is in the acidic form (cf. Fig. 16A). These results can be summarized by the following equilibria:

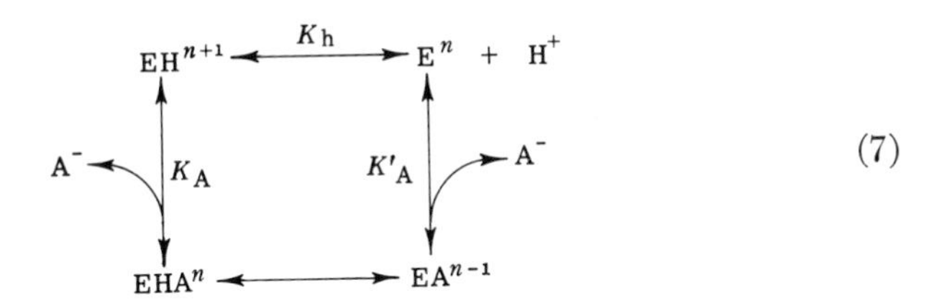

(7)

In Eq. (7) A^- represents the anionic inhibitor, and K_h is the acid dissociation constant of the activity-linked group [cf. Eq. (5)]. The inhibitor dissociation constant K_A is much smaller than K_A'; the latter generally has a magnitude of approximately 1 *M*. As a first approximation, the formation of EA^{n-1} can be neglected. The anion inhibition is then formally equivalent to a competition between A^- and OH^- for the protonated activity-linked group (see Section III,G).

Kernohan (*196*) found that $K_m^{CO_2}$ is independent of Cl^- and NO_3^-, thus suggesting independent binding sites for substrate and inhibitors. The effects of anions on the dehydration reaction are complex and have not yet been fully analyzed. On the basis of Eqs. (5) and (7), competitive inhibition might be expected. Kernohan's results indicate, however, that $K_m^{HCO_3^-}$ as well as $k_{cat}^{HCO_3^-}$ increase with the concentration of Cl^- (*190*), but that $k_{cat}^{HCO_3^-}/K_m^{HCO_3^-}$ decreases as should be expected from the results on the hydration reaction combined with the Haldane relation of Eq. (4). Christiansen and Magid (*198*) found similar effects on $K_m^{HCO_3^-}$ and $k_{cat}^{HCO_3^-}$ for the human B enzyme in the presence of phosphate buffers, but under

234. R. P. Davis, *JACS* **80**, 5209 (1958).

TABLE XII

APPARENT ANION INHIBITION CONSTANT K_i FOR BOVINE CARBONIC ANHYDRASE, pH 7.55, 25°[a]

Inhibitor	K_i (M)
HS^- ($+H_2S$)[b]	1.9×10^{-6}
CN^- ($+HCN$)[b]	3.2×10^{-6}
NCO^-	3.9×10^{-5}
NCS^-	5.9×10^{-4}
N_3^-	5.9×10^{-4}
ClO_4^-	1.6×10^{-2}
HCO_3^-	2.6×10^{-2}
HSO_3^-	3.0×10^{-2}
NO_3^-	4.8×10^{-2}
CH_3COO^-	8.5×10^{-2}
I^-	8.7×10^{-3}
Br^-	6.6×10^{-2}
Cl^-	0.19
F^-	1.2

[a] From the inhibition of *p*-nitrophenyl acetate hydrolysis. Data for HS^-, CN^-, and NCO^- from Thorslund and Lindskog (*202*). Other data from Pocker and Stone (*235*). The pH variation of K_i is illustrated in Fig. 16.

[b] Calculated on the basis of the total cyanide and sulfide concentrations, respectively.

their conditions of measurement, phosphate had a net activating effect. The dehydration reaction of the human C enzyme was much less influenced by phosphate.

In recent studies of anion inhibition, the esterase assay with *p*-nitrophenyl acetate as substrate has often been preferred to the CO_2 reaction because the K_A values can be determined with greater simplicity and accuracy by means of the esterase reaction. Thus, the data of Verpoorte *et al.* (*194*), Bradbury (*126*), and Whitney (*125*) show that the pH dependence of anion inhibition of the human carbonic anhydrases resembles that of the bovine enzyme. Data for the bovine enzyme obtained by the esterase assay and illustrating the large span of enzyme–anion binding strengths are shown in Table XII (*202, 235*). The pH-inhibition profiles for the three most powerful inhibitors of Table XII (NCO^-, CN^-, and HS^-) are illustrated in Fig. 16. While the pK_a of HOCN is about 4, so that cyanate is present as an anion in the whole pH range studied, HCN and H_2S titrate in this pH region. As seen in Fig. 16, the apparent pK_i values for cyanide and sulfide (calculated on the basis of total inhibitor concentrations) decrease at both acidic and alkaline pH.

235. Y. Pocker and J. T. Stone, *Biochemistry* **7**, 2936 (1968).

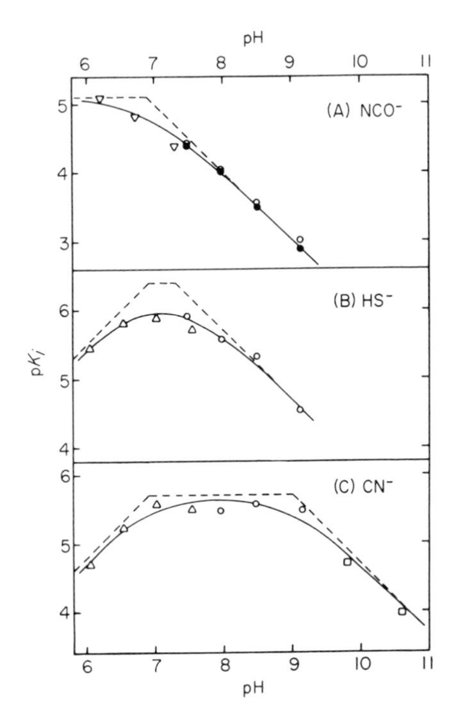

FIG. 16. The pH dependence of the inhibition of bovine carbonic anhydrase by cyanate, sulfide, and cyanide. Apparent inhibitor dissociation constant K_i is based on the sum of the concentrations of the anion and its conjugate acid. The K_i values were estimated from the inhibition of the esterase reaction with *p*-nitrophenyl acetate as substrate. The open symbols represent Zn(II) enzyme in various buffer systems. Filled circles represent the Co(II) enzyme. The curves have been calculated assuming that Eq. (7) is valid and that the only enzyme–inhibitor complex formed has the composition EHA^n [see Eq. (7)]. The pK values indicated by the intersections of the broken lines have been used in these calculations. From A. Thorslund and S. Lindskog, Studies of the esterase activity and the anion inhibition of bovine zinc and cobalt carbonic anhydrases. *European J. Biochem.* 3, 117–123 (1967). Berlin-Heidelberg-New York: Springer.

This is in accordance with Eq. (7) if it is assumed that HCN and H_2S do not bind to EH^{n+1}. Of course, these equilibrium data do not distinguish between the possible pathways by which an enzyme–inhibitor complex of the composition EHA^n can be formed from a mixture of E^n, EH^{n+1}, HA, and A^- (cf. Section III,E,2). Coleman (*210*) studied the interaction of cyanide and sulfide with the human B enzyme by titrating the amount of H^+ or OH^- produced in the binding reaction at various pH values. His results are in complete agreement with the formation of an EHA^n complex.

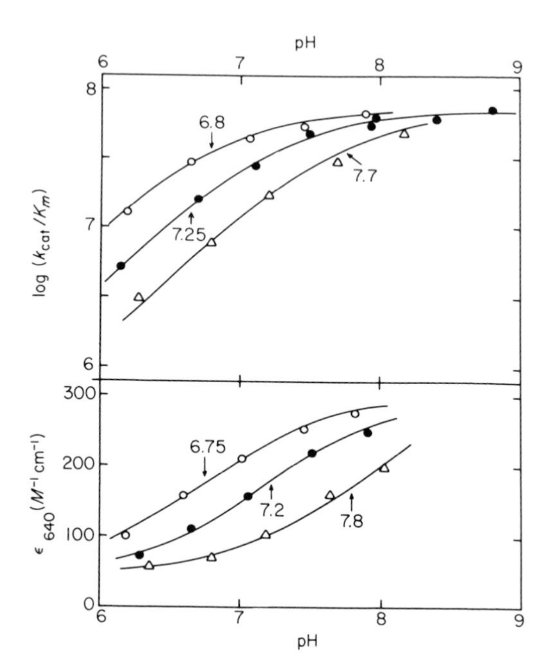

FIG. 17. The pH-rate profiles (top) and pH dependence of ϵ_{640} (bottom) of bovine Co(II) carbonic anhydrase at three Cl^- concentrations, from the left, 0.025 *M*, 0.1 *M*, and 0.5 *M*. The curves are calculated and represent titrations of single groups having the indicated p*K* values. From S. Lindskog, Cobalt(II) in metalloenzymes. A reporter of structure–function relations. *Struct. Bonding* **8**, 153–196 (1970). New York-Heidelberg-Berlin: Springer.

b. Spectroscopic Studies. As shown in Section III,D,3 the binding of an anion inhibitor to Co(II) carbonic anhydrase is associated with drastic spectral changes (*157*, *210*, *215*, *233*). These changes essentially parallel the inhibition of CO_2 hydration as illustrated in Fig. 17 and of the *p*-nitrophenyl acetate hydrolysis (*125*, *201*), thus suggesting a close interaction between the anion, the metal ion, and the "catalytic" group. Furthermore, the anions induce changes in the optical activities (ORD and CD) of the cobalt-linked absorption bands (*157*, *168*, *169*, *232*). The EPR spectra of the Co(II) enzymes are also influenced by inhibitors (*236*). While the ordinary violet enzyme–cyanide complex (see Fig. 14) is of the high-spin type like the free enzyme and the other inhibitor complexes (*228*, *236*), millimolar concentrations of CN^- cause the formation of a yellowish, low-spin type of complex, presumably containing two cyanide ions (*224*, *236*). The EPR spectra indicate that this complex is five-coordinated with an almost axial symmetry, and

236. E. Grell and R. C. Bray, *BBA* **236**, 503 (1971).

the observed nitrogen superhyperfine structure would suggest that an axial ligand is one of the histidine residues in the chelating site of the enzyme.

Coleman has shown that the inactive Cu(II) enzymes can bind certain anions and that concomitant changes of the ORD (*76, 168, 232*) and EPR (*224*) patterns take place.

Ward (*237*) studied the nuclear magnetic resonance of $^{35}Cl^-$ in the presence of Zn(II) and Co(II) carbonic anhydrases. He observed that the specific binding is associated with a considerable broadening of the signal, presumably resulting from an enhancing effect of the metal ion on the magnetic relaxation rate of the chlorine nucleus. The relaxivity varied with pH in accordance with Eq. (7) suggesting that the anion, at least formally, is displaced by OH^- from the immediate environment of the metal ion. While the high activity enzymes from man, horse, and ox gave pK_h values for this process in agreement with the kinetic studies on the inhibition of the esterase reaction, it is interesting to note that the B forms from man and horse gave pK_h values near 9, about two units larger than observed from inhibition. This finding constitutes another example of the complex behavior displayed by the low activity forms as compared to the simple patterns shown by the high activity forms (cf. Fig. 12).

Riepe and Wang (*238*) found that when N_3^- binds to the bovine enzyme, the infrared absorption of this anion is shifted to a position approximately corresponding to what is observed for some model complexes of Zn^{2+} and Co^{2+}.

The simplest molecular interpretation of these spectroscopic studies is that the metal ion–anion interaction involves a direct coordination of the inhibitor. However, it is clear that the free energy of anion binding depends on special properties of the active site where the positive charge of the metal ion constitutes an essential factor in the total electrostatic environment. Thus, all the anion inhibitors bind much more strongly to the enzyme than expected for simple complexation to a metal ion in aqueous solution (*222, 226*). This is particularly striking for NO_3^- and ClO_4^- (Table XII) which do not normally form stable metal ion complexes. While the most potent anion inhibitors are typical "metal poisons" and appear to be strongly influenced by the metal ion, the relative binding strengths of those inhibitors characterized by $K_i > 10^{-3}$ (cf. Table XII) seem rather to conform to the lyotropic or Hofmeister series of anion–protein interactions. This series is related to the size and charge of

237. R. L. Ward, *Biochemistry* **8,** 1879 (1969); **9,** 2447 (1970); R. L. Ward and K. J. Fritz, *BBRC* **39,** 707 (1970).
238. M. E. Riepe and J. H. Wang, *JBC* **243,** 2779 (1968).

the anions and reflects their effects on the water structure around nonpolar groups (*239*). For example, while the stabilities of Zn^{2+} coordination increase in the order $I^- < Br^- < Cl^-$ (*226*), the inhibitory powers increase as $Cl^- < Br^- < I^-$ in agreement with the Hofmeister series. It has been pointed out (*233, 235*) that there are striking similarities between the patterns of anion inhibition of carbonic anhydrase and the nonmetalloenzyme acetoacetate decarboxylase from *Clostridium acetobutyricum* (*240*).

Only preliminary X-ray data are available concerning the binding of anions of the type discussed in this section (*132, 138*). The anions bind in the vicinity of the metal ion, but at the pH of crystallization (pH 8.5) they do not seem to be within a direct binding distance. Possibly, some of these anions are similarly located as the complex ion, $Au(CN)_2^-$, which was used in the structure determination and shown (Fig. 8) to be linked to the metal ion via a water bridge.

2. *Sulfonamides*

Certain aromatic sulfonamides are very powerful and selective inhibitors of the animal and bacterial carbonic anhydrases. The unsubstituted $R–SO_2NH_2$ group is required for inhibitory efficiency, and any substitution such as methylation of acetylation of the sulfonamide function leads to a greatly reduced or abolished inhibitory power (*23*). This class of inhibitors is of great importance in studies of the physiological functions of carbonic anhydrases (*23*). One of the most commonly used sulfonamides, acetazolamide or Diamox (II), has found some therapeutical application in the treatment of glaucoma. A vast number of sulfonamide inhibitors have been synthesized, and a comprehensive survey of the empirical relations between sulfonamide structure and inhibitory

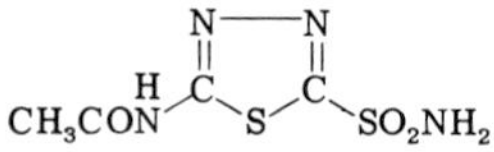

N—N
‖ ‖
CH₃CONH–C   C–SO₂NH₂
S

(II)

power has been performed by Bar (*241*). We shall limit our discussion in this chapter to studies concerning the molecular nature of the enzyme–inhibitor interaction and its implication for the catalytic mechanism of the enzyme.

239. P. H. von Hippel and T. Schleich, *in* "Structure and Stability of Biological Macromolecules" (S. N. Timasheff and G. D. Fasman, eds.), p. 417. Dekker, New York, 1969.
240. I. Fridovich, *JBC* **238,** 592 (1963).
241. D. Bar, *Actualites Pharmacol.* **15,** 1 (1963).

a. Effects of pH on Inhibitor Equilibria. The sulfonamide group is weakly acidic, and the apparent pK_i decreases both at acidic and alkaline pH in analogy to the pK_i values of CN^- and HS^- shown in Fig. 16. The pK_a values governing binding correspond to those of the activity-linked group and the $R\text{–}SO_2NH_2$ group, respectively (*242–244*), showing that the enzyme–inhibitor complex has the composition EHA^n in the symbolism of Eq. (7). Difference spectra obtained by King and Burgen (*245*) for the complexes between the human B and C enzymes and *p*-nitrobenzene sulfonamide vs. the enzyme plus unbound inhibitor are strongly suggestive of the presence of an ionized $R\text{–}SO_2NH^-$ in the complex. Chen and Kernohan (*246*) reached the same conclusion from observations of the shift of the fluorescence emission spectrum of dansylamide, 5-dimethylaminonaphthalene-1-sulfonamide (III), binding to the bovine enzyme.

H_3C CH_3
N

SO_2NH_2

(III)

Acetazolamide (II) has a second titratable group (CH_3CONH–). It has been shown that the ionization of this group has but a small effect on the binding (*247*).

Clearly, one factor determining the stability of a carbonic anhydrase–sulfonamide complex is the acidity of the sulfonamide group. However, a simple correlation between pK_a and pK_i, as originally proposed by Miller *et al.* (*248*), seems only to be valid in restricted series of closely related inhibitors (*247*).

b. The Sulfonamide–Metal Interaction. The presence of Zn^{2+} or Co^{2+} in the enzyme is required for strong binding of sulfonamides (*215*). The apoenzymes, or other metallocarbonic anhydrases, show much weaker binding. (*249*). Sulfonamides perturb the metal-linked absorption spec-

242. J. C. Kernohan, *BBA* **118,** 405 (1966).
243. S. Lindskog and A. Thorslund, *European J. Biochem.* **3,** 453 (1968).
244. P. W. Taylor, R. W. King, and A. S. V. Burgen, *Biochemistry* **9,** 3894 (1970).
245. R. W. King and A. S. V. Burgen, *BBA* **207,** 278 (1970).
246. R. F. Chen and J. C. Kernohan, *JBC* **242,** 5813 (1967).
247. S. Lindskog, *NASA (Natl. Aeron. Space Admin.), Spec. Publ.* **NASA SP-188,** 157 (1969).
248. W. H. Miller, A. M. Dessert, and R. O. Roblin, Jr., *JACS* **72,** 4893 (1950).
249. J. E. Coleman, *Nature* **214,** 193 (1967).

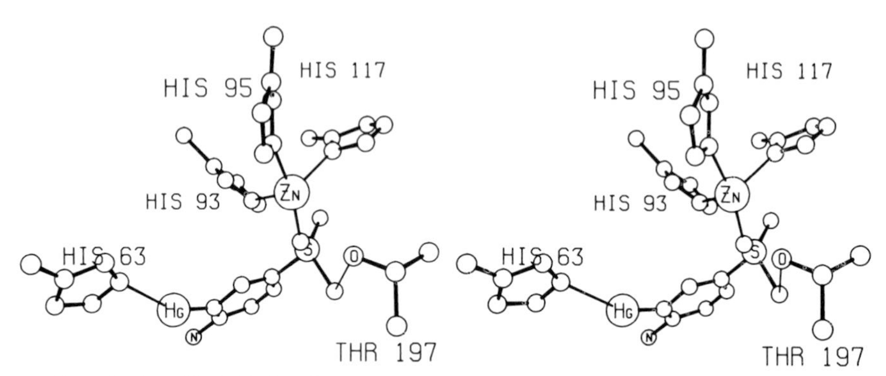

FIG. 18. An ORTEP drawing of a molecule of 3-acetoxymercuri-4-aminobenzene-1-sulfonamide binding to the active site of human carbonic anhydrase C as determined by X-ray diffraction. The sulfur atom of the sulfonamide group is located approximately 2.9 Å from the zinc ion, and one of the oxygen atoms or the nitrogen atom is binding directly to the metal ion. Another atom of the sulfonamide group is hydrogen bonded to Thr 197. The mercury atom of the inhibitor interacts with His 63. All water molecules in the inner part of the active site cavity have been displaced by the inhibitor (cf. Fig. 6).

trum of the Co(II) enzyme in a strikingly similar manner as some strongly binding anionic inhibitors (*76, 168, 215, 216*). Concurrently, the optical activity of the visible absorption bands undergo considerable changes (*131, 157, 169, 232*). These observations imply a close interaction with the metal ion, thus involving subtle changes in the symmetry of the coordination sphere [cf. Lindskog (*222*)]. Competition experiments with sulfonamides and anion inhibitors show that the active site cannot simultaneously bind both types of inhibitors (*125, 202, 210*).

The results of the X-ray investigation of the human C enzyme clearly demonstrate that the sulfonamide group is directly bound to the metal ion (*132, 137, 140*). The position of 3-acetoxymercuri-4-aminobenzene-1-sulfonamide (cf. Table V) is shown in Fig. 18.

c. Other Sulfonamide–Enzyme Interactions. The X-ray data indicate that in addition to the sulfonamide–metal ion linkage other interactions between the sulfonamide group and the active site wall are possible. Furthermore, the mercury atom of the inhibitor shown in Fig. 18 is seen within binding distance of His 63. The position of another sulfonamide, 1-amino-2,4-disulfonamide-5-chlorobenzene, has also been determined (*140*). In addition to the bond to the metal ion, this compound interacts with His 63 via the other sulfonamide group and with His 128, presumably via the amino group.

Hydrophobic contacts with other parts of the active site region may also be present. This is further indicated by the shifts of the absorption spectra and fluorescence emission spectra observed in several cases (*131, 246, 250, 251*). Chen and Kernohan (*246*) found that the fluorescent inhibitor (III) efficiently quenches the tryptophan fluorescence of the bovine enzyme. They estimated that the effective average distance between tryptophan and the inhibitor is 16 Å. Galley and Stryer (*252*) studied the phosphorescence spectrum of the complex between the bovine enzyme and *m*-acetylbenzene sulfonamide and concluded that the triplet exitation energy of the inhibitor is transferred to a tryptophan in its immediate environment. The X-ray studies of the human C enzyme have not yet revealed any direct tryptophan-sulfonamide contacts, however. Two or three tryptophanyl residues are seen at a distance of about 8 Å from the sulfonamide ring (*140*).

Circular dichroism studies of some azosulfonamides show that their fixation results in an induced optical activity of these chromophores (*253*). Significant differences were found for the CD spectra of the azosulfonamide complexes with B- and C-type carbonic anhydrases from various species showing that the isoenzymes must display structural variations in their active sites (*157*). These studies are discussed in greater detail by Timasheff (*154*) in Volume II of "The Enzymes."

d. Rates of Sulfonamide Binding. Maren *et al.* (*254*) and Leibman *et al.* (*255*) observed that in some cases it was necessary to preincubate enzyme and inhibitor to obtain a full expression of the inhibition in their assay. Kernohan (*242*) estimated association rates for the combination of benzene sulfonamide with the bovine enzyme following the rate of inhibition of the CO_2 reaction. The esterase reaction has been employed in similar experiments on the binding rates for several sulfonamides to the bovine zinc and cobalt enzymes (*243, 247*). Taylor *et al.* (*244, 256*) combined the fluorescence quenching technique, previously applied by Chen and Kernohan (*246*), with a stopped-flow apparatus for direct measurements of the combination rates for a large number of sulfonamides to the human B and C enzymes. Their extensive results support the earlier observation (*242, 247*) that the pH dependence of K_i is almost

250. J. Olander and E. T. Kaiser, *JACS* **92,** 5758 (1970).
251. R. Einarsson and M. Zeppezauer, *Acta Chem. Scand.* **24,** 1098 (1970).
252. W. C. Galley and L. Stryer, *Proc. Natl. Acad. Sci. U. S.* **60,** 108 (1968).
253. J. E. Coleman, *JACS* **89,** 6757 (1967).
254. T. H. Maren, A. L. Parcell, and M. N. Malik, *J. Pharmacol. Exptl. Therap.* **130,** 389 (1960).
255. K. C. Leibman, D. Alford, and R. A. Boudet, *J. Pharmacol. Exptl. Therap.* **131,** 271 (1961).
256. P. W. Taylor, R. W. King, and A. S. V. Burgen, *Biochemistry* **9,** 2638 (1970).

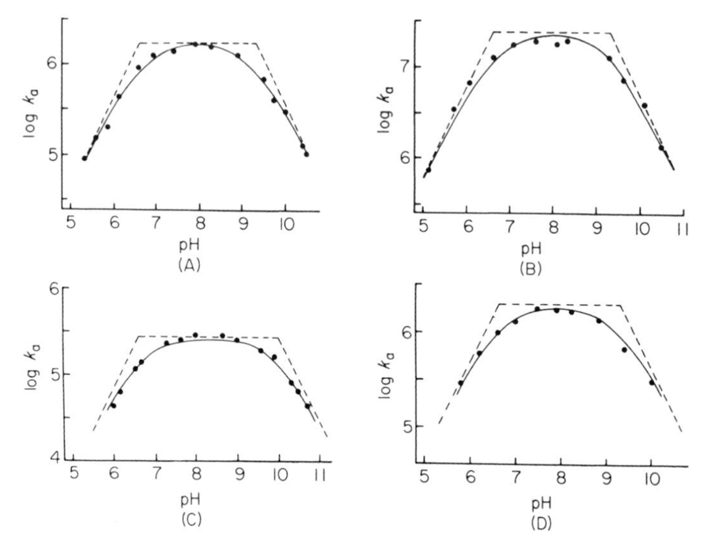

FIG. 19. The pH dependence of the apparent association rate constants between various sulfonamides and human carbonic anhydrase C. (A) *p*-Nitrobenzenesulfonamide and Zn(II) enzyme. (B) *p*-(Salicyl-5-azo)benzenesulfonamide and Zn(II) enzyme. (C) Dansylamide (III) and Zn(II) enzyme. (D) *p*-Nitrobenzenesulfonamide and Co(II) enzyme. The curves have been calculated in a manner similar to those in Fig. 16. The p*K* of the activity-linked group was assumed to be 6.60, and the p*K* values of the sulfonamides were taken as 9.30 for *p*-nitrobenzenesulfonamide and salicylazobenzenesulfonamide and 10.00 for dansylamide. From Taylor *et al.* (*244*). Copyright (1970) by the American Chemical Society. Reprinted by permission of the copyright owner.

completely expressed in the apparent second-order association rate constants. Some of the data of Taylor *et al.* are shown in Fig. 19. Furthermore, a substantial part of the affinity differences displayed by various sulfonamides is reflected in different magnitudes of the association rate constants; thus, the (pH-independent) dissociation rate constants show a rather limited variation.

As discussed by Kernohan (*242*), the kinetic data are in accordance with the direct formation of a complex EHA^n [cf. Eq. (7)] by either of two pathways

$$E^n + AH \underset{k_d}{\overset{k_{a1}}{\rightleftharpoons}} EHA^n \tag{8}$$

or

$$EH^{n+1} + A^- \underset{k_d}{\overset{k_{a2}}{\rightleftharpoons}} EHA^n \tag{9}$$

It has been proposed that one rationale for the relatively slow equilibration reaction might be that the pathway of Eq. (9) is predominant since the concentration of the ionized form of the sulfonamide A^- is usually a minor fraction of the total concentration at neutral pH. Lindskog (*247*) calculated the values of k_{a2} for a number of inhibitors and found that they approached but did not exceed what can be expected for a diffusion-controlled reaction between EH^{n+1} and A^-, about $10^9 M^{-1} sec^{-1}$. Taylor *et al.* (*244*), however, found one inhibitor, *p*-(salicyl-5-azo)benzenesulfonamide, where k_{a2} would have to be $1.12 \times 10^{10} M^{-1} sec^{-1}$. Consequently, Eq. (9) cannot dominate in this case, and they proposed that the reaction essentially proceeds according to Eq. (8) but that a rapid proton transfer step precedes the formation of the final complex to yield an ionized sulfonamide group. However, as long as the rate-limiting step in the binding process is not associated with or followed by the dissociation of a proton, there are several equally possible multistep mechanisms. It seems reasonable to assume that there is an initial formation of a labile complex more or less independently of the ionization states of the sulfonamide and the metal-linked group, and that the rate-limiting step comprises the final formation of a stable $R{-}SO_2NH^-$ metal ion bond after a rapid redistribution of protons.

e. Interdependence of Sulfonamide and Substrate Binding. Acetazolamide and sulfanilamide were found by Leibman and Greene (*257*) to be competitive inhibitors of the HCO_3^- reaction catalyzed by the bovine enzyme. In most earlier studies (*22, 23*), sulfonamides were reported to be noncompetitive with respect to CO_2. However, in a kinetic study of the dog enzyme, Maren and Wiley (*258*) found that the magnitude of the apparent K_i depends on whether the inhibitors are incubated with the enzyme or the substrate before mixing. Lindskog and Thorslund (*243*) observed biphasic reaction curves in stopped-flow experiments when bovine Co(II) enzyme and sulfanilamide had been preincubated. The initial phase of the CO_2 reaction gave a noncompetitive inhibition pattern resulting from the pseudo-irreversible behavior of the enzyme–inhibitor reaction, whereas the rates of the latter phase were compatible with a competitive inhibition. It has further been shown that CO_2 slows down the combination of sulfanilamide and benzene sulfonamide with the bovine enzyme in a competitive fashion (*243, 259*). However, Lindskog (*247*) reported preliminary observations that CO_2 had no

257. K. C. Leibman and F. E. Greene, *Proc. Soc. Exptl. Biol. Med.* **125,** 106 (1967).
258. T. H. Maren and C. E. Wiley, *J. Med. Chem.* **11,** 228 (1968).
259. J. C. Kernohan, *BJ* **98,** 31P (1966).

effect on the rate of combination between acetazolamide and the human B enzyme.

The types of inhibition observed in the ester and aldehyde reactions (*194, 203, 213, 214*) are, in part, discussed in Sections III,C,2 and 3.

3. *Other Inhibitors*

Carbonic anhydrases are inhibited by low concentrations of some metal ions (*22*). The human C enzyme appears to be much more sensitive to Cu^{2+} than the B enzyme (*260*). While chelating agents such as EDTA or 1,10-phenantroline do not appear to inhibit at "normal" concentrations (see, however, Section III,D), imidazole has recently been found to act as a competitive inhibitor of the human B enzyme, while no effect was found on the C enzyme (*191*). Inhibition by mercaptans has also been reported (*220, 261, 262*), and Roussin's salt, $K[Fe_4S_3(NO)_7]$, inhibits at low concentrations (*263*). Phenols, alcohols, acetone, acetonitrile, and various other organic substances have been found to be mildly inhibitory (*194, 202, 235*). A marked product inhibition by *o*-nitrophenol was observed during the hydrolysis of *o*-nitrophenyl acetate catalyzed by the bovine enzyme (*202*). See also earlier reviews for further information (*19, 22*).

F. Active-Site-Directed Chemical Modifications

1. *Anionic Reagents*

Bromoacetate (*124, 125, 153*) and iodoacetate (*126*), respectively, react specifically with the 3′-N of 1 histidine residue (His 204) in human carbonic anhydrase B. Concomitantly, most of the enzymic activity is lost. The modified enzyme retains a low but significant CO_2 activity as well as some esterase activity, however (*124–126*).

The modification reactions proceed through the initial formation of a reversible enzyme–anion complex. The chemical modification reaction is sufficiently slow so that the reversible inhibition can be studied separately. The binding of these compounds is controlled by the activity-linked group in complete analogy with other anions. The haloacetates bring about similar spectral changes of the Co(II) enzyme as other

260. E. Magid, *Scand. J. Haematol.* **4,** 257 (1967).
261. S. Schwimmer, *Enzymologia* **37,** 163 (1969).
262. H. Keller and U. H. Peters, *Z. Physiol. Chem.* **317,** 228 (1959).
263. A. Dobry-Duclaux, *Bull. Soc. Chim. Biol.* **48,** 887 (1966).

anions and compete with them for binding to the enzyme. The rate of the irreversible step is proportional to the concentration of the reversible enzyme–reagent complex. From the pH dependence of the modification rate at saturating levels of iodoacetate, Bradbury (*126*) calculated a pK_a of 5.8 for the modifiable histidine. In a similar experiment with the bromo derivative shown in Fig. 20, Whitney (*125*) obtained $pK_a = 5.6$. These values, of course, refer to the acidity of the group in the initial enzyme–anion complex. Since it is unlikely that the negative charge introduced by the reagent will cause a decrease of the apparent pK_a of a group in the enzyme, it can be concluded that His 204 is not identical with the activity-linked group. As is evident from Eq. (7), when $K_A \ll K'_A$ this group should obtain a substantially increased pK_a in the anion–enzyme complex.

The neutral reagent, iodoacetamide, also reacts with His 204, but the reaction is less specific than in the case of the haloacetates (*124*, *125*). Iodoacetamide seems to shift the pK_a of the modifiable histidine to a value below 5 (*125*).

While the high activity forms (C type) from ox and man are reversibly inhibited by bromoacetate, specific alkylation is not achieved (*153*).

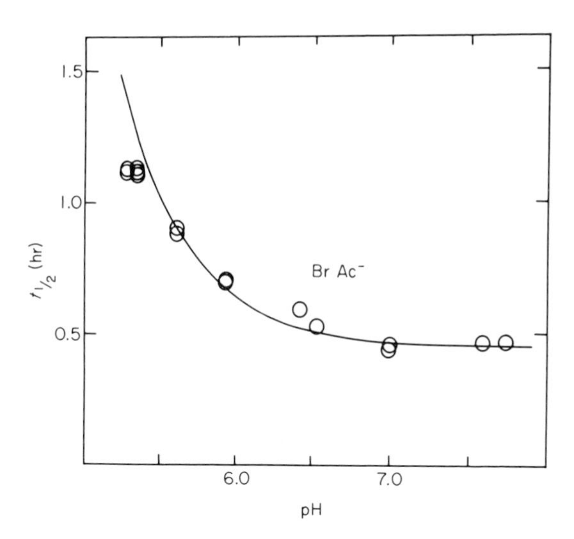

FIG. 20. The pH dependence of the irreversible reaction of bromoacetate with the human B enzyme at 25°. The reaction was followed by the loss of esterase activity. The half-time $t_{1/2}$ represents calculated values for enzyme saturated with inhibitor in a reversible enzyme–anion complex. From P. L. Whitney, Inhibition and modification of human carbonic anhydrase B with bromoacetate and iodoacetamide. *European J. Biochem.* **16**, 126–135 (1970). Berlin-Heidelberg-New York: Springer.

Activity losses have been observed, but the half-times are several days as compared to a few hours for the B-type enzymes from man and horse. The results indicated that in addition to a partial alkylation of a presumed active site histidine, unspecific carboxymethylation occurs in the C-type enzymes. It should be noted that the human C enzyme probably has a leucyl residue in the position corresponding to His 204 in the human B enzyme (see Section II,F).

Specific alkylation of a histidyl residue in the human C enzyme has, however, been achieved by use of bromopyruvate (*127*). This compound, which modifies His 204 in the human B enzyme very rapidly ($t_{1/2} = 3$–4 min at saturation, pH 7.6 and 25°), appears to alkylate the 3′— N of a single residue in the C form with a half-time at saturation of 13 hr at pH 6.8 and 25°. The modified residue has not yet been localized with certainty, but preliminary sequence work (Section II,F) and model building suggest that His 63 (C numbering) is a likely candidate (see Figs. 6 and 7). The product of this modification reaction has a substantial residual activity, about 30% in the *p*-nitrophenyl acetate assay but only 4% in a standard CO_2 assay (*127*).

Whitney *et al.* (*124*) showed that the few percent residual activity of monoalkylated human B enzyme has a pH-rate profile of similar sigmoid shape but shifted to higher pH compared to the native enzyme. The optimal activity, reached at alkaline pH, is lower for the modified enzyme than for the native enzyme, however. Obviously, this activity is an intrinsic property of the alkylated enzyme, suggesting that His 204 is not critically involved in catalysis. The pH-dependent change of the metal-linked absorption spectrum of the correspondingly modified Co(II) enzyme undergoes a similar shift toward alkaline pH, but all the major features of the spectra remain, showing that the metal environment is not substantially affected by the modification (*125*). The spectrum of the modified Co(II) enzyme resembles that of the unmodified Co(II) enzyme rather than that of the reversible Co(II) enzyme–bromoacetate complex. This observation indicates that the alkylation reaction involves a displacement of the anionic function away from the near vicinity of the metal ion. While unreacted bromoacetate competes with other anions and sulfonamides for a binding site on the native enzyme, the modified enzyme can bind another inhibitor molecule, for example, bromoacetate, which, in the case of the modified Co(II) enzyme, gives rise to a spectrum quite similar to that of the reversible complex of bromoacetate with unmodified Co(II) enzyme (*125*). Taylor *et al.* (*244*) further showed that sulfonamides can bind to the modified enzyme although with a reduced binding strength which is reflected in an increased dissociation rate (cf. Section III,E,2,d).

2. *Sulfonamide Reagents*

Whitney *et al.* (*128*) tested a number of sulfonamides, to which chloroacetyl groups had been attached, for their ability to label human carbonic anhydrase B. In two instances, *N*-chloroacetyl chloro-

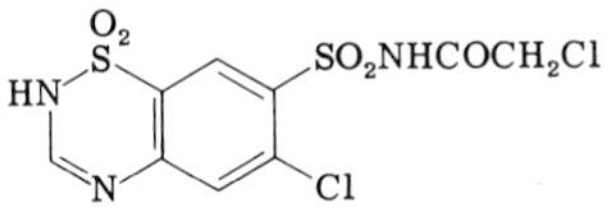

O_2 S HN N $SO_2NHCOCH_2Cl$ Cl

(IV)

thiazide (IV) and the related substance *N*-chloroacetyl cyclothiazide, specific modification and complete inactivation were observed. One equivalent of 3′-carboxymethyl histidine was recovered after acid hydrolysis during which the sulfonamide *N*-acyl bond is split. Sequence studies (Section II,F) showed that the alkylated residue is His 67 (*111*). In the human C enzyme, there is no histidine in the corresponding position which is probably located a considerable distance from the zinc ion in the tertiary structure (*135*). If the B enzyme is conformationally similar to the C enzyme, this would imply that compound (IV) initially binds to the metal ion through the $-SO_2NH-$ function in the heterocyclic ring [compound (IV) does not modify the human C enzyme (*128*)].

The high activity forms from ox and man were successfully labeled by Kandel *et al.* (*153*), who used bromoacetazolamide—a reactive derivative of compound (II) having a $BrCH_2CONH-$ side chain. Complete inactivation was obtained concomitantly with the formation of one 3′-alkylated histidyl residue. The location of this residue in the primary structure has not yet been ascertained, but model building with the human C enzyme suggests His 128. Human and equine B enzymes could be partially modified and inactivated by bromoacetazolamide, but the reaction was slow.

G. The Catalytic Mechanism

The minimal requirements for the function of a carbonic anhydrase are the presence of binding loci for the substrates, CO_2 and HCO_3^-, and a mechanism whereby, in effect, a water molecule is rapidly split into H^+ and OH^-; thus, a reactive OH^- can act on CO_2 in the hydration reaction, and a reactive H^+ can act on HCO_3^- in the dehydration reaction forming CO_2 and H_2O. The enzymes have extremely high turnover numbers, and the bond making and bond breaking processes as well as the proton transfers involved must occur with great rapidity.

1. *Substrate Binding*

The independence of $K_m^{CO_2}$ on pH and anionic inhibitors has been taken to imply that this parameter represents a dissociation constant (see Section III,C) and that CO_2 does not bind to the metal ion. Furthermore, CO_2 does not perturb the absorption spectrum of the Co(II) enzyme (*233*).

Assuming that sulfonamides are competitive with respect to CO_2, Riepe and Wang (*238*) measured the infrared difference spectrum between CO_2-containing enzyme solutions and sulfonamide-inhibited CO_2-containing enzyme solutions at pH values where the equilibrium of the catalyzed reaction is displaced toward CO_2. They observed a difference peak at 2341 cm^{-1}, which they ascribed to the active-site bound CO_2. This CO_2 binding was competitive with respect to NO_3^-, N_3^-, and N_2O. Since dissolved aqueous CO_2 absorbs at 2343.5 cm^{-1} and its absorption is shifted to lower frequencies in less polar media, they proposed that the CO_2-binding site is hydrophobic. Khalifah (*191*) and Christiansen and Magid (*198*) have not been able to detect any inhibition of the hydration or dehydration reactions by high concentrations of N_2O, however, raising doubts as to the relevancy of the infrared observations for the catalytic mechanism of carbonic anhydrase.

The binding of HCO_3^- appears to be linked to the metal ion, and this substrate has been reported to compete with other anions (and, formally, with OH^-), but the kinetic data on anion inhibition of the dehydration reaction are somewhat confusing (see Section III,E,1). This substrate may be directly bound to the metal ion as was originally proposed by Smith (*264*). However, the spectrum of the HCO_3^- complex of the bovine Co(II) enzyme is not distinctly different from that of the free enzyme at low pH, as opposed to the spectra induced by most anionic inhibitors (*233*). This might mean that HCO_3^- instead is linked to the metal ion via one or more water molecules (cf. Fig. 8).

2. *The Catalytic Reaction*

The mass of available data, summarized in the previous sections, strongly imply that the basic form of a group in the enzyme, having a pK near 7 and being closely linked to the metal ion, is critically in-

264. E. L. Smith, *Proc. Natl. Acad. Sci. U. S.* **35**, 80 (1949).

volved in the catalysis of CO_2 hydration (as well as the rapid esterase and aldehyde hydration reactions). This seems to apply to all the mammalian carbonic anhydrases so far studied, although the kinetic results on the human B enzyme (see Section III,C,1) indicate a somewhat more complex situation in this case. The chemical nature of this "catalytic" group has not been unambiguously ascertained, but a large number of schemes for the molecular mechanism of carbonic anhydrase-catalyzed reactions have been publicized (*191, 203, 204, 207, 210, 226, 238, 244, 257, 265, 266*). Most investigators seem to agree that the simplest, and perhaps most likely, model involves a zinc-bound water molecule dissociating above pH 7 to form a zinc-bound hydroxide ion, $E{-}Zn^{2+}{-}OH^-$.

It has been argued (*203, 208*) that a pK of 7 is unlikely for such a process since it implies at least a 100-fold stronger binding of OH^- to the enzyme than observed in simpler zinc complexes. However, as pointed out in Section III,D, metal ion binding in carbonic anhydrase shows several "unique" properties, and one of them is the strong binding of monovalent anions (see Section III,E,1).

Although there is no direct evidence for a zinc–hydroxide complex in the enzyme, the crystal structure studies described in Section II,G suggest the existence of a cluster of structured water in the active site including one molecule coordinated directly to the metal ion (see Fig. 6). Furthermore, Fabry *et al.* (*267*), using the paramagnetic Co(II) enzyme, demonstrated by proton magnetic relaxation measurements that at alkaline pH at least one rapidly exchanging proton is positioned within a short distance of the metal ion. The observed relaxivity varied with pH in conformity with the titration of the activity-linked group.

Assuming that the "zinc–hydroxide" mechanism is correct, the catalytic role of the metal ion would primarily be to promote the rapid dissociation of H_2O into H^+ and OH^- and to stabilize an OH^- at pH values where free OH^- ions are not abundant. The hydration of CO_2 may take place by a direct transfer of OH^- to CO_2 (*210, 238*), but it is also possible that the metal-linked OH^- acts as a general base to accept a proton from a water molecule that is attacking the CO_2 molecule either directly or via an ordered water structure (*191*). This mechanism, where the electrophilic character of the zinc ion is directed toward a water molecule, is somewhat different from that of carboxypeptidase A,

265. J. H. Wang, *NASA (Natl. Aeron. Space Admin.), Spec. Publ.* **NASA SP-188,** 101 (1969).
266. J. H. Wang, *Proc. Natl. Acad. Sci. U. S.* **66,** 874 (1970).
267. M. E. Fabry, S. H. Koenig, and W. E. Schillinger, *JBC* **245,** 4256 (1970).

where the metal ion presumably acts to polarize the carbonyl group of the susceptible peptide bond in the substrate molecule (*146*).

However, it is doubtful whether the zinc-promoted polarization of water is sufficient to explain the catalytic efficiency of carbonic anhydrase, since the metal-linked OH^- is a much weaker base than free OH^- ions, and presumably also a much less reactive nucleophile. Khalifah (*191*) has suggested that in addition to its above-mentioned role the metal ion exerts a polarizing effect on the substrates, mediated perhaps by water molecules. This mechanism has some features in common with those proposed by Breslow (*268*) and by Dennard and Williams (*269*) for the catalysis of CO_2 hydration by a Cu(II)-glycylglycine-OH^- complex and certain oxyanions, respectively. Wang (*265, 266*) suggested that a histidyl residue in the vicinity of a zinc-bound OH^- facilitates the transfer of a proton from the OH^- to an oxygen atom of CO_2 so that the product becomes coordinated via a charged O atom. Wang's formulation, containing a transiently coordinated O^{2-}, constitutes another attempt to compensate for a lower nucleophilicity of a metal-linked OH^- as compared to that of free OH^- ions.

Pocker *et al.* (*203, 207*) have proposed another mechanism including a metal-bound water molecule, but they ascribed the activity-linked pK of 7 to a histidyl residue connected to it via hydrogen bonds. In their original formulations the proposals of Wang and Pocker *et al.* seem unlikely (*222*), and they are not supported by the modification experiments and the X-ray evidence.

The active site pocket of the human C enzyme does contain a number of histidyl residues (Figs. 6 and 7), but definite functional roles cannot yet be ascribed to them. The observed water cluster appears to connect the metal ion with these histidines through a system of hydrogen bonds, however, and this arrangement may have some functional importance. For arguments given in Section III,F, His 204 in the human B enzyme is probably not critically involved, and the properties of other labeled carbonic anhydrases, although less well characterized, in no case suggest crucial roles for the modifiable histidyl residues.

Clearly, many questions remain concerning the molecular mechanism of carbonic anhydrase action. Therefore, we refrain from discussing the relative merits of the various proposed reaction schemes in greater detail. More structural information as well as an increased knowledge of enzymic function, bearing on the substrate loci and on intermediate steps in the catalytic process, shall undoubtedly be extracted in the near

268. E. Breslow, *in* "The Biochemistry of Copper" (J. Peisach, P. Aisen, and W. E. Blumberg, eds.), p. 149. Academic Press, New York, 1966.

269. A. E. Dennard and R. J. P. Williams, *JCS, A* p. 812 (1966).

future. In this search, the naturally occurring variants of carbonic anhydrase and artificially induced modifications of enzyme structure are likely to be of continuing importance.

ACKNOWLEDGMENT

The authors are very grateful to all those who provided them with their manuscripts on carbonic anhydrase prior to publication.

Author Index

Numbers in parentheses are reference numbers and indicate that an author's work is referred to, although his name is not cited in the text.

G

H

K

L

N

O

P

Q

R

U

V

W

Wollman, E., 344, 345, 361

Y

Z

Subject Index

B

C

D

E

G

H

K

L

M

N

Q

R

U

V

W

X

Y

Z